Compressed Sensing

Compressed sensing is an exciting, rapidly growing field which has attracted considerable attention in electrical engineering, applied mathematics, statistics, and computer science. Since its initial introduction several years ago an avalanche of results have been obtained both of a theoretical as well as practical nature, and various conferences, workshops, and special sessions have been dedicated to this growing research field. This book provides the first detailed introduction to the subject, highlighting recent theoretical advances and a range of applications, as well as outlining numerous remaining research challenges. After a thorough review of the basic theory, many cutting-edge advances in the field are presented, including advanced signal modeling, sub-Nyquist sampling of analog signals, hardware prototypes, non-asymptotic analysis of random matrices, adaptive sensing, greedy algorithms, the use of graphical models, and the separation of morphologically distinct data components. Each chapter is written by leading researchers in the field, and consistent style and notation are utilized throughout. An extended introductory chapter summarizes the basics of the field so that no prior knowledge is required. Key background information and clear definitions make this book an ideal resource for researchers, graduate students, and practitioners wanting to join this exciting research area. It can also serve as a supplementary textbook for courses on computer vision, coding theory, signal processing, image processing, and algorithms for efficient data processing.

Yonina C. Eldar is a Professor in the Department of Electrical Engineering at the Technion, Israel Institute of Technology, a Research Affiliate with the Research Laboratory of Electronics at the Massachusetts Institute of Technology, and a Visiting Professor at Stanford University. She has received numerous awards for excellence in research and teaching, including the Wolf Foundation Krill Prize for Excellence in Scientific Research, the Hershel Rich Innovation Award, the Weizmann Prize for Exact Sciences, the Michael Bruno Memorial Award from the Rothschild Foundation, and the Muriel & David Jacknow Award for Excellence in Teaching. She is an Associate Editor for several journals in the areas of signal processing and mathematics and a Signal Processing Society Distinguished Lecturer.

Gitta Kutyniok is an Einstein Professor in the Department of Mathematics at the Technische Universität Berlin, Germany. She has been a Postdoctoral Fellow at Princeton, Stanford, and Yale Universities, and a Full Professor at the Universität Osnabrück, Germany. Her research and teaching have been recognized by various awards, including a Heisenberg Fellowship and the von Kaven Prize by the German Research Foundation, an Einstein Chair by the Einstein Foundation Berlin, awards by the Universität Paderborn and the Justus–Liebig Universität Gießen for Excellence in Research, as well as the Weierstraß Prize for Outstanding Teaching. She is an Associate Editor and also Corresponding Editor for several journals in the area of applied mathematics.

Compressed Sensing
Theory and Applications

Edited by

YONINA C. ELDAR
Technion-Israel Institute of Technology, Haifa, Israel

GITTA KUTYNIOK
Technische Universität Berlin, Germany

CAMBRIDGE
UNIVERSITY PRESS

University Printing House, Cambridge CB2 8BS, United Kingdom

Cambridge University Press is part of the University of Cambridge.

It furthers the University's mission by disseminating knowledge in the pursuit of education, learning and research at the highest international levels of excellence.

www.cambridge.org
Information on this title: www.cambridge.org/9781107005587

© Cambridge University Press 2012

This publication is in copyright. Subject to statutory exception and to the provisions of relevant collective licensing agreements, no reproduction of any part may take place without the written permission of Cambridge University Press.

First published 2012
4th printing 2014

A catalogue record for this publication is available from the British Library

Library of Congress Cataloguing in Publication data
   Compressed sensing : theory and applications / edited by Yonina C. Eldar, Gitta Kutyniok.
     p.   cm.
   Includes bibliographical references and index.
   ISBN 978-1-107-00558-7
   1. Signal processing.   2. Wavelets (Mathematics)   I. Eldar, Yonina C.   II. Kutyniok, Gitta.
   QA601.C638 2012
   621.382'2–dc23   2011040519

ISBN 978-1-107-00558-7 Hardback

Cambridge University Press has no responsibility for the persistence or accuracy of URLs for external or third-party internet websites referred to in this publication, and does not guarantee that any content on such websites is, or will remain, accurate or appropriate.

Contents

	List of contributors	*page* vii
	Preface	ix
1	**Introduction to compressed sensing** MARK A. DAVENPORT, MARCO F. DUARTE, YONINA C. ELDAR, AND GITTA KUTYNIOK	1
2	**Second-generation sparse modeling: structured and collaborative signal analysis** ALEXEY CASTRODAD, IGNACIO RAMIREZ, GUILLERMO SAPIRO, PABLO SPRECHMANN, AND GUOSHEN YU	65
3	**Xampling: compressed sensing of analog signals** MOSHE MISHALI AND YONINA C. ELDAR	88
4	**Sampling at the rate of innovation: theory and applications** JOSE ANTONIO URIGÜEN, YONINA C. ELDAR, PIER LUIGI DRAGOTTI, AND ZVIKA BEN-HAIM	148
5	**Introduction to the non-asymptotic analysis of random matrices** ROMAN VERSHYNIN	210
6	**Adaptive sensing for sparse recovery** JARVIS HAUPT AND ROBERT NOWAK	269
7	**Fundamental thresholds in compressed sensing: a high-dimensional geometry approach** WEIYU XU AND BABAK HASSIBI	305
8	**Greedy algorithms for compressed sensing** THOMAS BLUMENSATH, MICHAEL E. DAVIES, AND GABRIEL RILLING	348

9	**Graphical models concepts in compressed sensing** ANDREA MONTANARI	394
10	**Finding needles in compressed haystacks** ROBERT CALDERBANK AND SINA JAFARPOUR	439
11	**Data separation by sparse representations** GITTA KUTYNIOK	485
12	**Face recognition by sparse representation** ARVIND GANESH, ANDREW WAGNER, ZIHAN ZHOU, ALLEN Y. YANG, YI MA, AND JOHN WRIGHT	515
	Index	540

Contributors

Zvika Ben-Haim
Technion-Israel Institute of Technology, Haifa, Israel

Thomas Blumensath
University of Oxford, UK

Robert Calderbank
Duke University, Durham, USA

Alexey Castrodad
University of Minnesota, Minneapolis, USA

Mark A. Davenport
Stanford University, USA

Michael E. Davies
University of Edinburgh, UK

Pier Luigi Dragotti
Imperial College London, UK

Marco F. Duarte
Duke University, Durham, USA

Yonina C. Eldar
Technion-Israel Institute of Technology, Haifa, Israel
Visiting Professor at Stanford University, USA

Arvind Ganesh
University of Illinois, USA

Babak Hassibi
California Institute of Technology, Pasadena, USA

Jarvis Haupt
University of Minnesota, Minneapolis, USA

Sina Jafarpour
Princeton University, USA

Gitta Kutyniok
Technische Universität Berlin, Germany

Yi Ma
Microsoft Research Asia, Beijing, China

Moshe Mishali
Technion-Israel Institute of Technology, Haifa, Israel

Andrea Montanari
Stanford University, USA

Robert Nowak
University of Wisconsin, Madison, USA

Ignacio Ramirez
University of Minnesota, Minneapolis, USA

Gabriel Rilling
University of Edinburgh, UK

Guillermo Sapiro
University of Minnesota, Minneapolis, USA

Pablo Sprechmann
University of Minnesota, Minneapolis, USA

Jose Antonio Urigüen
Imperial College London, UK

Roman Vershynin
University of Michigan, Ann Arbor, USA

Andrew Wagner
University of Illinois, USA

John Wright
Microsoft Research Asia, Beijing, China

Weiyu Xu
Cornell University, USA

Allen Y. Yang
University of California, Berkeley, USA

Guoshen Yu
University of Minnesota, Minneapolis, USA

Zihan Zhou
University of Illinois, USA

Preface

Compressed sensing (CS) is an exciting, rapidly growing field that has attracted considerable attention in electrical engineering, applied mathematics, statistics, and computer science. Since its initial introduction several years ago, an avalanche of results have been obtained, both of theoretical and practical nature, and various conferences, workshops, and special sessions have been dedicated to this growing research field. This book provides the first comprehensive introduction to the subject, highlighting recent theoretical advances and a range of applications, as well as outlining numerous remaining research challenges.

CS offers a framework for simultaneous sensing and compression of finite-dimensional vectors, that relies on linear dimensionality reduction. Quite surprisingly, it predicts that sparse high-dimensional signals can be recovered from highly incomplete measurements by using efficient algorithms. To be more specific, let x be a length-n vector. In CS we do not measure x directly, but rather acquire $m < n$ linear measurements of the form $y = Ax$ using an $m \times n$ CS matrix A. Ideally, the matrix is designed to reduce the number of measurements as much as possible while allowing for recovery of a wide class of signals from their measurement vectors y. Thus, we would like to choose $m \ll n$. However, this renders the matrix A rank-deficient, meaning that it has a nonempty nullspace. This implies that for any particular signal x_0, an infinite number of signals x will yield the same measurements $y = Ax = Ax_0$ for the chosen CS matrix. In order to enable recovery, we must therefore limit ourselves to a special class of input signals x.

The most prevalent signal structure used in CS is that of sparsity. In its simplest form, sparsity implies that x has only a small number of nonzero values. More generally, CS ideas can be applied when a suitable representation of x is sparse. The surprising result at the heart of CS is that if x (or a suitable representation of x) is k-sparse, i.e., it has at most k nonzero elements, then it can be recovered from $y = Ax$ using a number of measurements m that is on the order of $k \log n$. Furthermore, recovery is possible using simple, polynomial-time algorithms. In addition, these methods can be shown to be robust to noise and mismodelling of x. Many of the first research papers in CS were devoted to the analysis of theoretical guarantees on the CS matrix A in order to enable stable recovery, as well as the development of accompanying efficient algorithms.

This basic discovery has led to a fundamentally new approach to signal processing, image recovery, and compression algorithms, to name a few areas that have benefited from CS. Interestingly, the research field of CS draws from a variety of other areas such as approximation theory, Banach space theory, convex optimization, frame theory, numerical linear algebra, random matrix theory, and signal processing. The combined efforts of mathematicians, computer scientists, and engineers have led to a deluge of significant contributions to theory and applications of CS. This includes various constructions of efficient sensing matrices, fast algorithms for sparse recovery, extension of the notion of sparsity to more general signal structures including low-rank matrices and analog signal models, hardware designs of sub-Nyquist converters that rely on ideas of CS, as well as applications to radar analysis, face recognition, image processing, biomedical imaging, and many more. CS also holds promise for increasing resolution by exploiting the signal structure. This can potentially revolutionize many applications such as radar and microscopy by making efficient use of the available degrees of freedom in these settings. Consumer electronics, microscopy, civilian and military surveillance, medical imaging, radar and many other applications rely on efficient sampling and are resolution-limited. Reducing the sampling rate in these applications and increasing resolution can improve the user experience, increase data transfer, improve imaging quality and reduce exposure time.

This book is the first monograph in the literature to provide a comprehensive survey of compressed sensing. The potential reader of this book could be a researcher in the areas of applied mathematics, computer science, and electrical engineering, or a related research area, or a graduate student seeking to learn about CS. The particular design of this volume ensures that it can serve as both a state-of-the-art reference for researchers as well as a textbook for students.

The book contains 12 diverse chapters written by recognized leading experts from all over the world covering a large variety of topics. The book begins with a comprehensive introduction to CS which serves as a background for the remaining chapters, and also sets the notation to be used throughout the book. It does not assume any prior knowledge in the field. The following chapters are then organized into 4 categories: Extended signal models (Chapters 2–4), sensing matrix design (Chapters 5–6), recovery algorithms and performance guarantees (Chapters 7–9), and applications (Chapters 10–12). The chapters are self-contained, covering the most recent research results in the respective topic, and can all be treated independent of the others. A brief summary of each chapter is given next.

Chapter 1 provides a comprehensive introduction to the basics of CS. After a brief historical overview, the chapter begins with a discussion of sparsity and other low-dimensional signal models. The authors then treat the central question of how to accurately recover a high-dimensional signal from a small set of measurements and provide performance guarantees for a variety of sparse recovery algorithms. The chapter concludes with a discussion of some extensions of the sparse recovery framework.

Chapter 2 goes beyond traditional sparse modeling, and addresses collaborative structured sparsity to add stability and prior information to the representation. In structured sparse modeling, instead of considering the dictionary atoms as singletons, the atoms are partitioned in groups, and a few groups are selected at a time for the signal encoding. Further structure is then added via collaboration, where multiple signals, which are known to follow the same model, are allowed to collaborate in the coding. The authors discuss applications of these models to image restoration and source separation.

Chapter 3 generalizes CS to reduced-rate sampling of analog signals. It introduces Xampling, a unified framework for low rate sampling and processing of signals lying in a union of subspaces. A hardware-oriented viewpoint is advocated throughout, addressing practical constraints and exemplifying hardware realizations of sub-Nyquist systems. A variety of analog CS applications are reviewed within the unified Xampling framework including multiband communications with unknown carrier frequencies, ultrasound imaging, and wideband radar.

Chapter 4 considers reduced-rate sampling of finite rate of innovation (FRI) analog signals such as streams of pulses from discrete measurements. Exploiting the fact that only a small number of parameters per unit of time are needed to fully describe FRI signals allows to sample them at rates below Nyquist. The authors provide an overview of the theory and algorithms along with a diverse set of applications in areas such as superresolution, radar and ultrasound.

Chapter 5 considers constructions of random CS matrices with proven performance guarantees. The author provides an overview of basic non-asymptotic methods and concepts in random matrix theory. Several tools from geometric functional analysis and probability theory are put together in order to analyze the extreme singular values of random matrices. This then allows deducing results on random matrices used for sensing in CS.

Chapter 6 investigates the advantages of sequential measurement schemes that adaptively focus sensing using information gathered throughout the measurement process. This is in contrast to most theory and methods for sparse recovery which are based on an assumption of non-adaptive measurements. In particular, the authors show that adaptive sensing can be significantly more powerful when the measurements are contaminated with additive noise.

Chapter 7 introduces a unified high dimensional geometric framework for analyzing the phase transition phenomenon of ℓ_1 minimization in sparse recovery. This framework connects studying the phase transitions of ℓ_1 minimization with computing the Grassmann angles in high dimensional convex geometry. The authors further demonstrate the broad applications of this Grassmann angle framework by giving sharp phase transitions for related recovery methods.

Chapter 8 presents an overview of several greedy methods and explores their theoretical properties. Greedy algorithms are very fast and easy to implement and often have similar theoretical performance guarantees to convex methods. The authors detail some of the leading greedy approaches for sparse recovery, and consider extensions of these methods to more general signal structures.

Chapter 9 surveys recent work in applying ideas from graphical models and message passing algorithms to solve large scale regularized regression problems. In particular, the focus is on CS reconstruction via ℓ_1 penalized least-squares. The author discusses how to derive fast approximate message passing algorithms to solve this problem and shows how the analysis of such algorithms allows to prove exact high-dimensional limit results on the recovery error.

Chapter 10 considers compressed learning, where learning is performed directly in the compressed domain. The authors provide tight bounds demonstrating that the linear kernel SVM's classifier in the measurement domain, with high probability, has true accuracy close to the accuracy of the best linear threshold classifier in the data domain. It is also shown that for a family of well-known CS matrices, compressed learning is provided on the flight. The authors then demonstrate these results in the context of texture analysis.

Chapter 11 surveys methods for data separation by sparse representations. The author considers the use of sparsity in problems in which the data is composed of two or more morphologically distinct constituents. The key idea is to choose a deliberately overcomplete representation made of several frames, each one providing a sparse expansion of one of the components to be extracted. The morphological difference between the components is then encoded as incoherence conditions of those frames which allows for separation using CS algorithms.

Chapter 12 applies CS to the classical problem of face recognition. The authors consider the problem of recognizing human faces in the presence of real-world nuisances such as occlusion and variabilities in pose and illumination. The main idea behind the proposed approach is to explain any query image using a small number of training images from a single subject category. This core idea is then generalized to account for various physical variabilities encountered in face recognition. The authors demonstrate how the resulting system is capable of accurately recognizing subjects out of a database of several hundred subjects with state-of-the-art accuracy.

1 Introduction to compressed sensing

Mark A. Davenport, Marco F. Duarte, Yonina C. Eldar, and Gitta Kutyniok

Compressed sensing (CS) is an exciting, rapidly growing, field that has attracted considerable attention in signal processing, statistics, and computer science, as well as the broader scientific community. Since its initial development only a few years ago, thousands of papers have appeared in this area, and hundreds of conferences, workshops, and special sessions have been dedicated to this growing research field. In this chapter, we provide an up-to-date review of the basics of the theory underlying CS. This chapter should serve as a review to practitioners wanting to join this emerging field, and as a reference for researchers. We focus primarily on the theory and algorithms for sparse recovery in finite dimensions. In subsequent chapters of the book, we will see how the fundamentals presented in this chapter are expanded and extended in many exciting directions, including new models for describing structure in both analog and discrete-time signals, new sensing design techniques, more advanced recovery results and powerful new recovery algorithms, and emerging applications of the basic theory and its extensions.

1.1 Introduction

We are in the midst of a digital revolution that is driving the development and deployment of new kinds of sensing systems with ever-increasing fidelity and resolution. The theoretical foundation of this revolution is the pioneering work of Kotelnikov, Nyquist, Shannon, and Whittaker on sampling continuous-time bandlimited signals [162, 195, 209, 247]. Their results demonstrate that signals, images, videos, and other data can be exactly recovered from a set of uniformly spaced samples taken at the so-called *Nyquist rate* of twice the highest frequency present in the signal of interest. Capitalizing on this discovery, much of signal processing has moved from the analog to the digital domain and ridden the wave of Moore's law. Digitization has enabled the creation of sensing and processing systems that are more robust, flexible, cheaper and, consequently, more widely used than their analog counterparts.

As a result of this success, the amount of data generated by sensing systems has grown from a trickle to a torrent. Unfortunately, in many important and emerging applications,

Compressed Sensing: Theory and Applications, ed. Yonina C. Eldar and Gitta Kutyniok. Published by Cambridge University Press. © Cambridge University Press 2012.

the resulting Nyquist rate is so high that we end up with far too many samples. Alternatively, it may simply be too costly, or even physically impossible, to build devices capable of acquiring samples at the necessary rate [146,241]. Thus, despite extraordinary advances in computational power, the acquisition and processing of signals in application areas such as imaging, video, medical imaging, remote surveillance, spectroscopy, and genomic data analysis continues to pose a tremendous challenge.

To address the logistical and computational challenges involved in dealing with such high-dimensional data, we often depend on compression, which aims at finding the most concise representation of a signal that is able to achieve a target level of acceptable distortion. One of the most popular techniques for signal compression is known as *transform coding*, and typically relies on finding a basis or frame that provides *sparse* or *compressible* representations for signals in a class of interest [31,77,106]. By a sparse representation, we mean that for a signal of length n, we can represent it with $k \ll n$ nonzero coefficients; by a compressible representation, we mean that the signal is well-approximated by a signal with only k nonzero coefficients. Both sparse and compressible signals can be represented with high fidelity by preserving only the values and locations of the largest coefficients of the signal. This process is called *sparse approximation*, and forms the foundation of transform coding schemes that exploit signal sparsity and compressibility, including the JPEG, JPEG2000, MPEG, and MP3 standards.

Leveraging the concept of transform coding, *compressed sensing* has emerged as a new framework for signal acquisition and sensor design that enables a potentially large reduction in the sampling and computation costs for sensing signals that have a sparse or compressible representation. While the Nyquist–Shannon sampling theorem states that a certain minimum number of samples is required in order to perfectly capture an arbitrary bandlimited signal, when the signal is sparse in a known basis we can vastly reduce the number of measurements that need to be stored. Consequently, when sensing sparse signals we might be able to do better than suggested by classical results. This is the fundamental idea behind CS: rather than first sampling at a high rate and then compressing the sampled data, we would like to find ways to *directly* sense the data in a compressed form – i.e., at a lower sampling rate. The field of CS grew out of the work of Candès, Romberg, and Tao and of Donoho, who showed that a finite-dimensional signal having a sparse or compressible representation can be recovered from a small set of linear, non-adaptive measurements [3, 33, 40–42, 44, 82]. The design of these measurement schemes and their extensions to practical data models and acquisition systems are central challenges in the field of CS.

While this idea has only recently gained significant attention in the signal processing community, there have been hints in this direction dating back as far as the eighteenth century. In 1795, Prony proposed an algorithm for the estimation of the parameters associated with a small number of complex exponentials sampled in the presence of noise [201]. The next theoretical leap came in the early 1900s, when Carathéodory showed that a positive linear combination of *any* k sinusoids is uniquely determined by its value at $t = 0$ and at *any* other $2k$ points in time [46,47]. This represents far fewer samples than the number of Nyquist-rate samples when k is small and the range of possible frequencies is large. In the

1990s, this work was generalized by George, Gorodnitsky, and Rao, who studied sparsity in biomagnetic imaging and other contexts [134–136, 202]. Simultaneously, Bresler, Feng, and Venkataramani proposed a sampling scheme for acquiring certain classes of signals consisting of k components with nonzero bandwidth (as opposed to pure sinusoids) under restrictions on the possible spectral supports, although exact recovery was not guaranteed in general [29, 117, 118, 237]. In the early 2000s Blu, Marziliano, and Vetterli developed sampling methods for certain classes of parametric signals that are governed by only k parameters, showing that these signals can be sampled and recovered from just $2k$ samples [239].

A related problem focuses on recovery of a signal from partial observation of its Fourier transform. Beurling proposed a method for extrapolating these observations to determine the entire Fourier transform [22]. One can show that if the signal consists of a finite number of impulses, then Beurling's approach will correctly recover the entire Fourier transform (of this non-bandlimited signal) from *any* sufficiently large piece of its Fourier transform. His approach – to find the signal with smallest ℓ_1 norm among all signals agreeing with the acquired Fourier measurements – bears a remarkable resemblance to some of the algorithms used in CS.

More recently, Candès, Romberg, Tao [33, 40–42, 44], and Donoho [82] showed that a signal having a sparse representation can be recovered *exactly* from a small set of linear, non-adaptive measurements. This result suggests that it may be possible to sense sparse signals by taking far fewer measurements, hence the name *compressed* sensing. Note, however, that CS differs from classical sampling in three important respects. First, sampling theory typically considers infinite-length, continuous-time signals. In contrast, CS is a mathematical theory focused on measuring finite-dimensional vectors in $\mathbb{R}^n$. Second, rather than sampling the signal at specific points in time, CS systems typically acquire measurements in the form of inner products between the signal and more general test functions. This is in fact in the spirit of modern sampling methods which similarly acquire signals by more general linear measurements [113, 230]. We will see throughout this book that *randomness* often plays a key role in the design of these test functions. Third, the two frameworks differ in the manner in which they deal with *signal recovery*, i.e., the problem of recovering the original signal from the compressive measurements. In the Nyquist–Shannon framework, signal recovery is achieved through sinc interpolation – a linear process that requires little computation and has a simple interpretation. In CS, however, signal recovery is typically achieved using highly nonlinear methods.[1] See Section 1.6 as well as the survey in [226] for an overview of these techniques.

Compressed sensing has already had a notable impact on several applications. One example is medical imaging [178–180, 227], where it has enabled speedups by a factor of seven in pediatric MRI while preserving diagnostic quality [236]. Moreover, the broad applicability of this framework has inspired research that extends

---

[1] It is also worth noting that it has recently been shown that nonlinear methods can be used in the context of traditional sampling as well, when the sampling mechanism is nonlinear [105].

the CS framework by proposing practical implementations for numerous applications, including sub-Nyquist sampling systems [125, 126, 186–188, 219, 224, 225, 228], compressive imaging architectures [99, 184, 205], and compressive sensor networks [7, 72, 141].

The aim of this book is to provide an up-to-date review of some of the important results in CS. Many of the results and ideas in the various chapters rely on the fundamental concepts of CS. Since the focus of the remaining chapters is on more recent advances, we concentrate here on many of the basic results in CS that will serve as background material to the rest of the book. Our goal in this chapter is to provide an overview of the field and highlight some of the key technical results, which are then more fully explored in subsequent chapters. We begin with a brief review of the relevant mathematical tools, and then survey many of the low-dimensional models commonly used in CS, with an emphasis on sparsity and the union of subspaces models. We next focus attention on the theory and algorithms for sparse recovery in finite dimensions. To facilitate our goal of providing both an elementary introduction as well as a comprehensive overview of many of the results in CS, we provide proofs of some of the more technical lemmas and theorems in the Appendix.

1.2 Review of vector spaces

For much of its history, signal processing has focused on signals produced by physical systems. Many natural and man-made systems can be modeled as linear. Thus, it is natural to consider signal models that complement this kind of linear structure. This notion has been incorporated into modern signal processing by modeling signals as *vectors* living in an appropriate *vector space*. This captures the linear structure that we often desire, namely that if we add two signals together then we obtain a new, physically meaningful signal. Moreover, vector spaces allow us to apply intuitions and tools from geometry in $\mathbb{R}^3$, such as lengths, distances, and angles, to describe and compare signals of interest. This is useful even when our signals live in high-dimensional or infinite-dimensional spaces. This book assumes that the reader is relatively comfortable with vector spaces. We now provide a brief review of some of the key concepts in vector spaces that will be required in developing the CS theory.

1.2.1 Normed vector spaces

Throughout this book, we will treat signals as real-valued functions having domains that are either continuous or discrete, and either infinite or finite. These assumptions will be made clear as necessary in each chapter. We will typically be concerned with *normed vector spaces*, i.e., vector spaces endowed with a *norm*.

In the case of a discrete, finite domain, we can view our signals as vectors in an n-dimensional Euclidean space, denoted by $\mathbb{R}^n$. When dealing with vectors in $\mathbb{R}^n$, we will

Introduction to compressed sensing

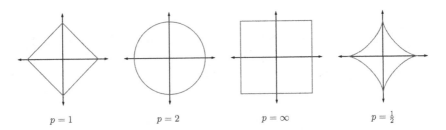

Figure 1.1 Unit spheres in $\mathbb{R}^2$ for the ℓ_p norms with $p = 1, 2, \infty$, and for the ℓ_p quasinorm with $p = \frac{1}{2}$.

make frequent use of the ℓ_p norms, which are defined for $p \in [1, \infty]$ as

$$\|x\|_p = \begin{cases} \left(\sum_{i=1}^n |x_i|^p\right)^{\frac{1}{p}}, & p \in [1, \infty); \\ \max_{i=1,2,\ldots,n} |x_i|, & p = \infty. \end{cases} \quad (1.1)$$

In Euclidean space we can also consider the standard *inner product* in $\mathbb{R}^n$, which we denote

$$\langle x, z \rangle = z^T x = \sum_{i=1}^n x_i z_i.$$

This inner product leads to the ℓ_2 norm: $\|x\|_2 = \sqrt{\langle x, x \rangle}$.

In some contexts it is useful to extend the notion of ℓ_p norms to the case where $p < 1$. In this case, the "norm" defined in (1.1) fails to satisfy the triangle inequality, so it is actually a quasinorm. We will also make frequent use of the notation $\|x\|_0 := |\operatorname{supp}(x)|$, where $\operatorname{supp}(x) = \{i : x_i \neq 0\}$ denotes the support of x and $|\operatorname{supp}(x)|$ denotes the cardinality of $\operatorname{supp}(x)$. Note that $\|\cdot\|_0$ is not even a quasinorm, but one can easily show that

$$\lim_{p \to 0} \|x\|_p^p = |\operatorname{supp}(x)|,$$

justifying this choice of notation. The ℓ_p (quasi-)norms have notably different properties for different values of p. To illustrate this, in Figure 1.1 we show the unit sphere, i.e., $\{x : \|x\|_p = 1\}$, induced by each of these norms in $\mathbb{R}^2$.

We typically use norms as a measure of the strength of a signal, or the size of an error. For example, suppose we are given a signal $x \in \mathbb{R}^2$ and wish to approximate it using a point in a one-dimensional affine space A. If we measure the approximation error using an ℓ_p norm, then our task is to find the $\widehat{x} \in A$ that minimizes $\|x - \widehat{x}\|_p$. The choice of p will have a significant effect on the properties of the resulting approximation error. An example is illustrated in Figure 1.2. To compute the closest point in A to x using each ℓ_p norm, we can imagine growing an ℓ_p sphere centered on x until it intersects with A. This will be the point $\widehat{x} \in A$ that is closest to x in the corresponding ℓ_p norm. We observe that larger p tends to spread out the error more evenly among the two coefficients, while smaller p leads to an error that is more unevenly distributed and tends to be sparse. This intuition generalizes to higher dimensions, and plays an important role in the development of CS theory.

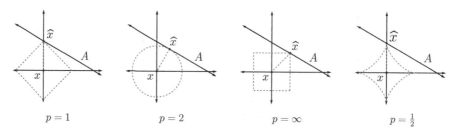

Figure 1.2 Best approximation of a point in $\mathbb{R}^2$ by a one-dimensional subspace using the ℓ_p norms for $p = 1, 2, \infty$, and the ℓ_p quasinorm with $p = \frac{1}{2}$.

1.2.2 Bases and frames

A set $\{\phi_i\}_{i=1}^n$ is called a basis for $\mathbb{R}^n$ if the vectors in the set span $\mathbb{R}^n$ and are linearly independent.[2] This implies that each vector in the space has a unique representation as a linear combination of these basis vectors. Specifically, for any $x \in \mathbb{R}^n$, there exist (unique) coefficients $\{c_i\}_{i=1}^n$ such that

$$x = \sum_{i=1}^n c_i \phi_i.$$

Note that if we let Φ denote the $n \times n$ matrix with columns given by ϕ_i and let c denote the length-n vector with entries c_i, then we can represent this relation more compactly as

$$x = \Phi c.$$

An important special case of a basis is an orthonormal basis, defined as a set of vectors $\{\phi_i\}_{i=1}^n$ satisfying

$$\langle \phi_i, \phi_j \rangle = \begin{cases} 1, & i = j; \\ 0, & i \neq j. \end{cases}$$

An orthonormal basis has the advantage that the coefficients c can be easily calculated as

$$c_i = \langle x, \phi_i \rangle,$$

or

$$c = \Phi^T x$$

in matrix notation. This can easily be verified since the orthonormality of the columns of Φ means that $\Phi^T \Phi = I$, where I denotes the $n \times n$ identity matrix.

It is often useful to generalize the concept of a basis to allow for sets of possibly linearly dependent vectors, resulting in what is known as a *frame* [48, 55, 65, 163, 164, 182]. More

---

[2] In any n-dimensional vector space, a basis will always consist of exactly n vectors. Fewer vectors are not sufficient to span the space, while additional vectors are guaranteed to be linearly dependent.

formally, a frame is a set of vectors $\{\phi_i\}_{i=1}^n$ in $\mathbb{R}^d$, $d < n$ corresponding to a matrix $\Phi \in \mathbb{R}^{d \times n}$, such that for all vectors $x \in \mathbb{R}^d$,

$$A \|x\|_2^2 \leq \|\Phi^T x\|_2^2 \leq B \|x\|_2^2$$

with $0 < A \leq B < \infty$. Note that the condition $A > 0$ implies that the rows of Φ must be linearly independent. When A is chosen as the largest possible value and B as the smallest for these inequalities to hold, then we call them the *(optimal) frame bounds*. If A and B can be chosen as $A = B$, then the frame is called *A-tight*, and if $A = B = 1$, then Φ is a *Parseval frame*. A frame is called *equal-norm*, if there exists some $\lambda > 0$ such that $\|\phi_i\|_2 = \lambda$ for all $i = 1, \ldots, n$, and it is *unit-norm* if $\lambda = 1$. Note also that while the concept of a frame is very general and can be defined in infinite-dimensional spaces, in the case where Φ is a $d \times n$ matrix A and B simply correspond to the smallest and largest eigenvalues of $\Phi\Phi^T$, respectively.

Frames can provide richer representations of data due to their redundancy [26]: for a given signal x, there exist infinitely many coefficient vectors c such that $x = \Phi c$. In order to obtain a set of feasible coefficients we exploit the *dual frame* $\widetilde{\Phi}$. Specifically, any frame satisfying

$$\Phi \widetilde{\Phi}^T = \widetilde{\Phi} \Phi^T = I$$

is called an (alternate) dual frame. The particular choice $\widetilde{\Phi} = (\Phi\Phi^T)^{-1}\Phi$ is referred to as the *canonical dual frame*. It is also known as the Moore–Penrose pseudoinverse. Note that since $A > 0$ requires Φ to have linearly independent rows, this also ensures that $\Phi\Phi^T$ is invertible, so that $\widetilde{\Phi}$ is well-defined. Thus, one way to obtain a set of feasible coefficients is via

$$c_d = \widetilde{\Phi}^T x = \Phi^T (\Phi\Phi^T)^{-1} x.$$

One can show that this sequence is the smallest coefficient sequence in ℓ_2 norm, i.e., $\|c_d\|_2 \leq \|c\|_2$ for all c such that $x = \Phi c$.

Finally, note that in the sparse approximation literature, it is also common for a basis or frame to be referred to as a *dictionary* or *overcomplete dictionary* respectively, with the dictionary elements being called *atoms*.

1.3 Low-dimensional signal models

At its core, signal processing is concerned with efficient algorithms for acquiring, processing, and extracting information from different types of signals or data. In order to design such algorithms for a particular problem, we must have accurate models for the signals of interest. These can take the form of generative models, deterministic classes, or probabilistic Bayesian models. In general, models are useful for incorporating *a priori* knowledge to help distinguish classes of interesting or probable signals from uninteresting or improbable signals. This can help in efficiently and accurately acquiring, processing, compressing, and communicating data and information.

As noted in the introduction, much of classical signal processing is based on the notion that signals can be modeled as vectors living in an appropriate vector space (or subspace). To a large extent, the notion that any possible vector is a valid signal has driven the explosion in the dimensionality of the data we must sample and process. However, such simple linear models often fail to capture much of the structure present in many common classes of signals – while it may be reasonable to model signals as vectors, in many cases not all possible vectors in the space represent valid signals. In response to these challenges, there has been a surge of interest in recent years, across many fields, in a variety of *low-dimensional signal models* that quantify the notion that the number of degrees of freedom in high-dimensional signals is often quite small compared to their ambient dimensionality.

In this section we provide a brief overview of the most common low-dimensional structures encountered in the field of CS. We will begin by considering the traditional sparse models for finite-dimensional signals, and then discuss methods for generalizing these classes to infinite-dimensional (continuous-time) signals. We will also briefly discuss low-rank matrix and manifold models and describe some interesting connections between CS and some other emerging problem areas.

1.3.1 Sparse models

Signals can often be well-approximated as a linear combination of just a few elements from a known basis or dictionary. When this representation is exact we say that the signal is *sparse*. Sparse signal models provide a mathematical framework for capturing the fact that in many cases these high-dimensional signals contain relatively little information compared to their ambient dimension. Sparsity can be thought of as one incarnation of *Occam's razor* — when faced with many possible ways to represent a signal, the simplest choice is the best one.

Sparsity and nonlinear approximation

Mathematically, we say that a signal x is *k-sparse* when it has at most k nonzeros, i.e., $\|x\|_0 \leq k$. We let

$$\Sigma_k = \{x : \|x\|_0 \leq k\}$$

denote the set of all k-sparse signals. Typically, we will be dealing with signals that are not themselves sparse, but which admit a sparse representation in some basis Φ. In this case we will still refer to x as being k-sparse, with the understanding that we can express x as $x = \Phi c$ where $\|c\|_0 \leq k$.

Sparsity has long been exploited in signal processing and approximation theory for tasks such as compression [77, 199, 215] and denoising [80], and in statistics and learning theory as a method for avoiding overfitting [234]. Sparsity also figures prominently in the theory of statistical estimation and model selection [139, 218], in the study of the human visual system [196], and has been exploited heavily in image processing tasks, since the multiscale wavelet transform [182] provides nearly sparse representations for natural images. An example is shown in Figure 1.3.

As a traditional application of sparse models, we consider the problems of image compression and image denoising. Most natural images are characterized by large smooth or

Figure 1.3 Sparse representation of an image via a multiscale wavelet transform. (a) Original image. (b) Wavelet representation. Large coefficients are represented by light pixels, while small coefficients are represented by dark pixels. Observe that most of the wavelet coefficients are close to zero.

textured regions and relatively few sharp edges. Signals with this structure are known to be very nearly sparse when represented using a multiscale wavelet transform [182]. The wavelet transform consists of recursively dividing the image into its low- and high-frequency components. The lowest frequency components provide a coarse scale approximation of the image, while the higher frequency components fill in the detail and resolve edges. What we see when we compute a wavelet transform of a typical natural image, as shown in Figure 1.3, is that most coefficients are very small. Hence, we can obtain a good approximation of the signal by setting the small coefficients to zero, or *thresholding* the coefficients, to obtain a k-sparse representation. When measuring the approximation error using an ℓ_p norm, this procedure yields the *best k-term approximation* of the original signal, i.e., the best approximation of the signal using only k basis elements.[3]

Figure 1.4 shows an example of such an image and its best k-term approximation. This is the heart of nonlinear approximation [77] – nonlinear because the choice of which coefficients to keep in the approximation depends on the signal itself. Similarly, given the knowledge that natural images have approximately sparse wavelet transforms, this same thresholding operation serves as an effective method for rejecting certain common types of noise, which typically do *not* have sparse wavelet transforms [80].

---

[3] Thresholding yields the best k-term approximation of a signal with respect to an orthonormal basis. When redundant frames are used, we must rely on sparse approximation algorithms like those described in Section 1.6 [106, 182].

Figure 1.4 Sparse approximation of a natural image. (a) Original image. (b) Approximation of image obtained by keeping only the largest 10% of the wavelet coefficients.

Geometry of sparse signals

Sparsity is a highly nonlinear model, since the choice of which dictionary elements are used can change from signal to signal [77]. This can be seen by observing that given a pair of k-sparse signals, a linear combination of the two signals will in general no longer be k-sparse, since their supports may not coincide. That is, for any $x, z \in \Sigma_k$, we do not necessarily have that $x + z \in \Sigma_k$ (although we do have that $x + z \in \Sigma_{2k}$). This is illustrated in Figure 1.5, which shows Σ_2 embedded in $\mathbb{R}^3$, i.e., the set of all 2-sparse signals in $\mathbb{R}^3$.

The set of sparse signals Σ_k does not form a linear space. Instead it consists of the union of all possible $\binom{n}{k}$ canonical subspaces. In Figure 1.5 we have only $\binom{3}{2} = 3$ possible subspaces, but for larger values of n and k we must consider a potentially huge number of subspaces. This will have significant algorithmic consequences in the development of the algorithms for sparse approximation and sparse recovery described in Sections 1.5 and 1.6.

Compressible signals

An important point in practice is that few real-world signals are *truly* sparse; rather they are compressible, meaning that they can be well-approximated by sparse signals. Such signals have been termed compressible, approximately sparse, or relatively sparse in various contexts. Compressible signals are well approximated by sparse signals in the same way that signals living close to a subspace are well approximated by the first few principal components [139]. In fact, we can quantify the compressibility by calculating the error incurred by approximating a signal x by some $\widehat{x} \in \Sigma_k$:

$$\sigma_k(x)_p = \min_{\widehat{x} \in \Sigma_k} \|x - \widehat{x}\|_p. \tag{1.2}$$

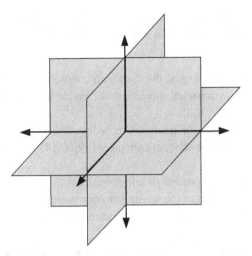

Figure 1.5 Union of subspaces defined by $\Sigma_2 \subset \mathbb{R}^3$, i.e., the set of all 2-sparse signals in $\mathbb{R}^3$.

If $x \in \Sigma_k$, then clearly $\sigma_k(x)_p = 0$ for any p. Moreover, one can easily show that the thresholding strategy described above (keeping only the k largest coefficients) results in the optimal approximation as measured by (1.2) for all ℓ_p norms [77].

Another way to think about compressible signals is to consider the rate of decay of their coefficients. For many important classes of signals there exist bases such that the coefficients obey a power law decay, in which case the signals are highly compressible. Specifically, if $x = \Phi c$ and we sort the coefficients c_i such that $|c_1| \geq |c_2| \geq \cdots \geq |c_n|$, then we say that the coefficients obey a power law decay if there exist constants $C_1, q > 0$ such that

$$|c_i| \leq C_1 i^{-q}.$$

The larger q is, the faster the magnitudes decay, and the more compressible a signal is. Because the magnitudes of their coefficients decay so rapidly, compressible signals can be represented accurately by $k \ll n$ coefficients. Specifically, for such signals there exist constants $C_2, r > 0$ depending only on C_1 and q such that

$$\sigma_k(x)_2 \leq C_2 k^{-r}.$$

In fact, one can show that $\sigma_k(x)_2$ will decay as k^{-r} if and only if the sorted coefficients c_i decay as $i^{-r+1/2}$ [77].

1.3.2 Finite unions of subspaces

In certain applications, the signal has a structure that cannot be completely expressed using sparsity alone. For instance, when only certain sparse support patterns are allowable in the signal, it is possible to leverage such constraints to formulate more concise signal models. We give a few representative examples below; see Chapters 2 and 8 for more detail on structured sparsity.

- For piecewise-smooth signals and images, the dominant coefficients in the wavelet transform tend to cluster into a connected rooted subtree inside the wavelet parent–child binary tree [79, 103, 104, 167, 168].
- In applications such as surveillance or neuronal recording, the coefficients might appear clustered together, or spaced apart from each other [49, 50, 147]. See Chapter 11 for more details.
- When multiple sparse signals are recorded simultaneously, their supports might be correlated according to the properties of the sensing environment [7, 63, 76, 114, 121, 185]. One possible structure leads to the multiple measurement vector problem; see Section 1.7 for more details.
- In certain cases the small number of components of a sparse signal correspond not to vectors (columns of a matrix Φ), but rather to points known to lie in particular subspaces. If we construct a frame by concatenating bases for such subspaces, the nonzero coefficients of the signal representations form block structures at known locations [27, 112, 114]. See Chapters 3, 11, and 12 for further description and potential applications of this model.

Such examples of additional structure can be captured in terms of restricting the feasible signal supports to a small subset of the possible $\binom{n}{k}$ selections of nonzero coefficients for a k-sparse signal. These models are often referred to as structured sparsity models [4, 25, 102, 114, 177]. In cases where nonzero coefficients appear in clusters, the structure can be expressed in terms of a sparse union of subspaces [102, 114]. The structured sparse and union of subspaces models extend the notion of sparsity to a much broader class of signals that can incorporate both finite-dimensional and infinite-dimensional representations.

In order to define these models, recall that for canonically sparse signals, the union Σ_k is composed of canonical subspaces $\mathcal{U}_i$ that are aligned with k out of the n coordinate axes of $\mathbb{R}^n$. See, for example, Figure 1.5, which illustrates this for the case where $n = 3$ and $k = 2$. Allowing for more general choices of $\mathcal{U}_i$ leads to powerful representations that accommodate many interesting signal priors. Specifically, given the knowledge that x resides in one of M possible subspaces $\mathcal{U}_1, \mathcal{U}_2, \ldots, \mathcal{U}_M$, we have that x lies in the *union* of M subspaces [114, 177]:

$$x \in \mathcal{U} = \bigcup_{i=1}^{M} \mathcal{U}_i.$$

It is important to note that, as in the generic sparse setting, union models are nonlinear: the sum of two signals from a union $\mathcal{U}$ is generally no longer in $\mathcal{U}$. This nonlinear behavior of the signal set renders any processing that exploits these models more intricate. Therefore, instead of attempting to treat all unions in a unified way, we focus our attention on some specific classes of union models, in order of complexity.

The simplest class of unions arises when the number of subspaces comprising the union is finite and each subspace has finite dimensions. We call this setup a finite union of subspaces model. Under the finite-dimensional framework, we revisit the two types of models described above:

- *Structured sparse supports*: This class consists of sparse vectors that meet additional restrictions on the support (i.e., the set of indices for the vector's nonzero entries). This corresponds to only certain subspaces $\mathcal{U}_i$ out of the $\binom{n}{k}$ subspaces present in Σ_k being allowed [4].
- *Sparse union of subspaces*: This class consists of vectors where each subspace $\mathcal{U}_i$ comprising the union is a direct sum of k low-dimensional subspaces [114].

$$\mathcal{U}_i = \bigoplus_{j=1}^{k} \mathcal{A}_{i_j}. \qquad (1.3)$$

Here $\{\mathcal{A}_i\}$ are a given set of subspaces with dimensions $\dim(\mathcal{A}_i) = d_i$, and $i_1, i_2, \ldots, i_k$ select k of these subspaces. Thus, each subspace $\mathcal{U}_i$ corresponds to a different choice of k out of M subspaces $\mathcal{A}_i$ that comprise the sum. This framework can model standard sparsity by letting $\mathcal{A}_j$ be the one-dimensional subspace spanned by the jth canonical vector. It can be shown that this model leads to block sparsity, in which certain blocks in a vector are zero and others are not [112].

These two cases can be combined to allow for only certain sums of k subspaces to be part of the union $\mathcal{U}$. Both models can be leveraged to further reduce sampling rate and allow for the application of CS to a broader class of signals.

1.3.3 Unions of subspaces for analog signal models

One of the primary motivations for CS is to design new sensing systems for acquiring continuous-time, analog signals or images. In contrast, the finite-dimensional sparse model described above inherently assumes that the signal x is discrete. It is sometimes possible to extend this model to continuous-time signals using an intermediate discrete representation. For example, a bandlimited, periodic signal can be perfectly represented by a finite-length vector consisting of its Nyquist-rate samples. However, it will often be more useful to extend the concept of sparsity to provide union of subspaces models for analog signals [97, 109, 114, 125, 186–188, 239]. Two of the broader frameworks that treat sub-Nyquist sampling of analog signals are Xampling and finite-rate of innovation, which are discussed in Chapters 3 and 4, respectively.

In general, when treating unions of subspaces for analog signals there are three main cases to consider, as elaborated further in Chapter 3 [102]:

- finite unions of infinite-dimensional spaces;
- infinite unions of finite-dimensional spaces;
- infinite unions of infinite-dimensional spaces.

In each of the three settings above there is an element that can take on infinite values, which is a result of the fact that we are considering analog signals: either the underlying subspaces are infinite-dimensional, or the number of subspaces is infinite.

There are many well-known examples of analog signals that can be expressed as a union of subspaces. For example, an important signal class corresponding to a finite union of infinite-dimensional spaces is the multiband model [109]. In this model, the analog signal consists of a finite sum of bandlimited signals, where typically the signal components have a relatively small bandwidth but are distributed across a comparatively large frequency range [117,118,186,237,238]. Sub-Nyquist recovery techniques for this class of signals can be found in [186–188].

Another example of a signal class that can often be expressed as a union of subspaces is the class of signals having a finite rate of innovation [97,239]. Depending on the specific structure, this model corresponds to an infinite or finite union of finite-dimensional subspaces [19,125,126], and describes many common signals having a small number of degrees of freedom. In this case, each subspace corresponds to a certain choice of parameter values, with the set of possible values being infinite dimensional, and thus the number of subspaces spanned by the model being infinite as well. The eventual goal is to exploit the available structure in order to reduce the sampling rate; see Chapters 3 and 4 for more details. As we will see in Chapter 3, by relying on the analog union of subspaces model we can design efficient hardware that samples analog signals at sub-Nyquist rates, thus moving the analog CS framework from theory to practice.

1.3.4 Low-rank matrix models

Another model closely related to sparsity is the set of low-rank matrices:

$$\mathcal{L} = \{M \in \mathbb{R}^{n_1 \times n_2} : \text{rank}(M) \leq r\}.$$

The set $\mathcal{L}$ consists of matrices M such that $M = \sum_{k=1}^{r} \sigma_k u_k v_k^*$ where $\sigma_1, \sigma_2, \ldots, \sigma_r \geq 0$ are the nonzero singular values, and $u_1, u_2, \ldots, u_r \in \mathbb{R}^{n_1}$, $v_1, v_2, \ldots, v_r \in \mathbb{R}^{n_2}$ are the corresponding singular vectors. Rather than constraining the number of elements used to construct the signal, we are constraining the number of nonzero singular values. One can easily observe that the set $\mathcal{L}$ has $r(n_1 + n_2 - r)$ degrees of freedom by counting the number of free parameters in the singular value decomposition. For small r this is significantly less than the number of entries in the matrix – $n_1 n_2$. Low-rank matrices arise in a variety of practical settings. For example, low-rank (Hankel) matrices correspond to low-order linear, time-invariant systems [198]. In many data-embedding problems, such as sensor geolocation, the matrix of pairwise distances will typically have rank 2 or 3 [172,212]. Finally, approximately low-rank matrices arise naturally in the context of collaborative filtering systems such as the now-famous Netflix recommendation system [132] and the related problem of *matrix completion*, where a low-rank matrix is recovered from a small sample of its entries [39,151,204]. While we do not focus in-depth on matrix completion or the more general problem of low-rank matrix recovery, we note that many of the concepts and tools treated in this book are highly relevant to this emerging field, both from a theoretical and algorithmic perspective [36,38,161,203].

1.3.5 Manifold and parametric models

Parametric or manifold models form another, more general class of low-dimensional signal models. These models arise in cases where (i) a k-dimensional continuously valued parameter θ can be identified that carries the relevant information about a signal and (ii) the signal $f(\theta) \in \mathbb{R}^n$ changes as a continuous (typically nonlinear) function of these parameters. Typical examples include a one-dimensional (1-D) signal shifted by an unknown time delay (parameterized by the translation variable), a recording of a speech signal (parameterized by the underlying phonemes being spoken), and an image of a 3-D object at an unknown location captured from an unknown viewing angle (parameterized by the 3-D coordinates of the object and its roll, pitch, and yaw) [90, 176, 240]. In these and many other cases, the signal class forms a nonlinear k-dimensional manifold in $\mathbb{R}^n$, i.e.,

$$\mathcal{M} = \{f(\theta) : \theta \in \Theta\},$$

where Θ is the k-dimensional parameter space. Manifold-based methods for image processing have attracted considerable attention, particularly in the machine learning community. They can be applied to diverse applications including data visualization, signal classification and detection, parameter estimation, systems control, clustering, and machine learning [14, 15, 58, 61, 89, 193, 217, 240, 244]. Low-dimensional manifolds have also been proposed as approximate models for a number of nonparametric signal classes such as images of human faces and handwritten digits [30, 150, 229].

Manifold models are closely related to all of the models described above. For example, the set of signals x such that $\|x\|_0 = k$ forms a k-dimensional Riemannian manifold. Similarly, the set of $n_1 \times n_2$ matrices of rank r forms an $r(n_1 + n_2 - r)$-dimensional Riemannian manifold [233].[4]

A number of the signal models used in this book are closely related to manifold models. For example, the union of subspaces models in Chapter 3, the finite rate of innovation models considered in Chapter 4, and the continuum models in Chapter 11 can all be viewed from a manifold perspective. For the most part we will not explicitly exploit this structure in the book. However, low-dimensional manifolds have a close connection to many of the key results in CS. In particular, many of the randomized sensing matrices used in CS can also be shown to preserve the structure in low-dimensional manifolds [6]. For details and further applications see [6, 71, 72, 101].

1.4 Sensing matrices

In order to make the discussion more concrete, for the remainder of this chapter we will restrict our attention to the standard finite-dimensional CS model. Specifically, given a

---

[4] Note that in the case where we allow signals with sparsity less than or equal to k, or matrices of rank less than or equal to r, these sets fail to satisfy certain technical requirements of a topological manifold (due to the behavior where the sparsity/rank changes). However, the manifold viewpoint can still be useful in this context [68].

signal $x \in \mathbb{R}^n$, we consider measurement systems that acquire m linear measurements. We can represent this process mathematically as

$$y = Ax, \qquad (1.4)$$

where A is an $m \times n$ matrix and $y \in \mathbb{R}^m$. The matrix A represents a *dimensionality reduction*, i.e., it maps $\mathbb{R}^n$, where n is generally large, into $\mathbb{R}^m$, where m is typically much smaller than n. Note that in the standard CS framework we assume that the measurements are *non-adaptive*, meaning that the rows of A are fixed in advance and do not depend on the previously acquired measurements. In certain settings adaptive measurement schemes can lead to significant performance gains. See Chapter 6 for further details.

As noted earlier, although the standard CS framework assumes that x is a finite-length vector with a discrete-valued index (such as time or space), in practice we will often be interested in designing measurement systems for acquiring continuously indexed signals such as continuous-time signals or images. It is sometimes possible to extend this model to continuously indexed signals using an intermediate discrete representation. For a more flexible approach, we refer the reader to Chapters 3 and 4. For now we will simply think of x as a finite-length window of Nyquist-rate samples, and we temporarily ignore the issue of how to directly acquire compressive measurements without first sampling at the Nyquist rate.

There are two main theoretical questions in CS. First, how should we design the sensing matrix A to ensure that it preserves the information in the signal x? Second, how can we recover the original signal x from measurements y? In the case where our data is sparse or compressible, we will see that we can design matrices A with $m \ll n$ that ensure that we will be able to recover the original signal accurately and efficiently using a variety of practical algorithms.

We begin in this section by first addressing the question of how to design the sensing matrix A. Rather than directly proposing a design procedure, we instead consider a number of desirable properties that we might wish A to have. We then provide some important examples of matrix constructions that satisfy these properties.

1.4.1 Null space conditions

A natural place to begin is by considering the null space of A, denoted

$$\mathcal{N}(A) = \{z : Az = 0\}.$$

If we wish to be able to recover *all* sparse signals x from the measurements Ax, then it is immediately clear that for any pair of distinct vectors $x, x' \in \Sigma_k$, we must have $Ax \neq Ax'$, since otherwise it would be impossible to distinguish x from x' based solely on the measurements y. More formally, by observing that if $Ax = Ax'$ then $A(x-x') = 0$ with $x - x' \in \Sigma_{2k}$, we see that A uniquely represents all $x \in \Sigma_k$ if and only if $\mathcal{N}(A)$ contains no vectors in Σ_{2k}. While there are many equivalent ways of characterizing this property, one of the most common is known as the *spark* [86].

DEFINITION 1.1 *The spark of a given matrix A is the smallest number of columns of A that are linearly dependent.*

This definition allows us to pose the following straightforward guarantee.

THEOREM 1.1 (Corollary 1 of [86]) *For any vector $y \in \mathbb{R}^m$, there exists at most one signal $x \in \Sigma_k$ such that $y = Ax$ if and only if* spark$(A) > 2k$.

Proof. We first assume that, for any $y \in \mathbb{R}^m$, there exists at most one signal $x \in \Sigma_k$ such that $y = Ax$. Now suppose for the sake of a contradiction that spark$(A) \leq 2k$. This means that there exists some set of at most $2k$ columns that are linearly independent, which in turn implies that there exists an $h \in \mathcal{N}(A)$ such that $h \in \Sigma_{2k}$. In this case, since $h \in \Sigma_{2k}$ we can write $h = x - x'$, where $x, x' \in \Sigma_k$. Thus, since $h \in \mathcal{N}(A)$ we have that $A(x - x') = 0$ and hence $Ax = Ax'$. But this contradicts our assumption that there exists at most one signal $x \in \Sigma_k$ such that $y = Ax$. Therefore, we must have that spark$(A) > 2k$.

Now suppose that spark$(A) > 2k$. Assume that for some y there exist $x, x' \in \Sigma_k$ such that $y = Ax = Ax'$. We therefore have that $A(x - x') = 0$. Letting $h = x - x'$, we can write this as $Ah = 0$. Since spark$(A) > 2k$, all sets of up to $2k$ columns of A are linearly independent, and therefore $h = 0$. This in turn implies $x = x'$, proving the theorem. □

It is easy to see that spark$(A) \in [2, m+1]$. Therefore, Theorem 1.1 yields the requirement $m \geq 2k$.

When dealing with *exactly* sparse vectors, the spark provides a complete characterization of when sparse recovery is possible. However, when dealing with *approximately* sparse signals we must consider somewhat more restrictive conditions on the null space of A [57]. Roughly speaking, we must also ensure that $\mathcal{N}(A)$ does not contain any vectors that are too compressible in addition to vectors that are sparse. In order to state the formal definition we define the following notation that will prove to be useful throughout much of this book. Suppose that $\Lambda \subset \{1, 2, \ldots, n\}$ is a subset of indices and let $\Lambda^c = \{1, 2, \ldots, n\} \setminus \Lambda$. By x_Λ we typically mean the length n vector obtained by setting the entries of x indexed by Λ^c to zero. Similarly, by A_Λ we typically mean the $m \times n$ matrix obtained by setting the columns of A indexed by Λ^c to zero.[5]

DEFINITION 1.2 *A matrix A satisfies the* null space property (NSP) *of order k if there exists a constant $C > 0$ such that,*

$$\|h_\Lambda\|_2 \leq C \frac{\|h_{\Lambda^c}\|_1}{\sqrt{k}} \qquad (1.5)$$

holds for all $h \in \mathcal{N}(A)$ and for all Λ such that $|\Lambda| \leq k$.

The NSP quantifies the notion that vectors in the null space of A should not be too concentrated on a small subset of indices. For example, if a vector h is exactly k-sparse,

---

[5] We note that this notation will occasionally be abused to refer to the length $|\Lambda|$ vector obtained by keeping only the entries corresponding to Λ, or the $m \times |\Lambda|$ matrix obtained by only keeping the columns corresponding to Λ, respectively. The usage should be clear from the context, but in most cases there is no substantive difference between the two.

then there exists a Λ such that $\|h_{\Lambda^c}\|_1 = 0$ and hence (1.5) implies that $h_\Lambda = 0$ as well. Thus, if a matrix A satisfies the NSP then the only k-sparse vector in $\mathcal{N}(A)$ is $h = 0$.

To fully illustrate the implications of the NSP in the context of sparse recovery, we now briefly discuss how we will measure the performance of sparse recovery algorithms when dealing with general non-sparse x. Towards this end, let $\Delta : \mathbb{R}^m \to \mathbb{R}^n$ represent our specific recovery method. We will focus primarily on guarantees of the form

$$\|\Delta(Ax) - x\|_2 \leq C \frac{\sigma_k(x)_1}{\sqrt{k}} \tag{1.6}$$

for all x, where $\sigma_k(x)_1$ is as defined in (1.2). This guarantees exact recovery of all possible k-sparse signals, but also ensures a degree of robustness to non-sparse signals that directly depends on how well the signals are approximated by k-sparse vectors. Such guarantees are called *instance-optimal* since they guarantee optimal performance for each instance of x [57]. This distinguishes them from guarantees that only hold for some subset of possible signals, such as sparse or compressible signals – the quality of the guarantee adapts to the particular choice of x. These are also commonly referred to as *uniform guarantees* since they hold uniformly for all x.

Our choice of norms in (1.6) is somewhat arbitrary. We could easily measure the reconstruction error using other ℓ_p norms. The choice of p, however, will limit what kinds of guarantees are possible, and will also potentially lead to alternative formulations of the NSP. See, for instance, [57]. Moreover, the form of the right-hand side of (1.6) might seem somewhat unusual in that we measure the approximation error as $\sigma_k(x)_1/\sqrt{k}$ rather than simply something like $\sigma_k(x)_2$. However, we will see in Section 1.5.3 that such a guarantee is actually not possible without taking a prohibitively large number of measurements, and that (1.6) represents the best possible guarantee we can hope to obtain.

We will see in Section 1.5 (Theorem 1.8) that the NSP of order $2k$ is sufficient to establish a guarantee of the form (1.6) for a practical recovery algorithm (ℓ_1 minimization). Moreover, the following adaptation of a theorem in [57] demonstrates that if there exists *any* recovery algorithm satisfying (1.6), then A must necessarily satisfy the NSP of order $2k$.

THEOREM 1.2 (Theorem 3.2 of [57]) *Let $A : \mathbb{R}^n \to \mathbb{R}^m$ denote a sensing matrix and $\Delta : \mathbb{R}^m \to \mathbb{R}^n$ denote an arbitrary recovery algorithm. If the pair (A, Δ) satisfies (1.6) then A satisfies the NSP of order $2k$.*

Proof. Suppose $h \in \mathcal{N}(A)$ and let Λ be the indices corresponding to the $2k$ largest entries of h. We next split Λ into Λ_0 and Λ_1, where $|\Lambda_0| = |\Lambda_1| = k$. Set $x = h_{\Lambda_1} + h_{\Lambda^c}$ and $x' = -h_{\Lambda_0}$, so that $h = x - x'$. Since by construction $x' \in \Sigma_k$, we can apply (1.6) to obtain $x' = \Delta(Ax')$. Moreover, since $h \in \mathcal{N}(A)$, we have

$$Ah = A(x - x') = 0$$

so that $Ax' = Ax$. Thus, $x' = \Delta(Ax)$. Finally, we have that

$$\|h_\Lambda\|_2 \leq \|h\|_2 = \|x - x'\|_2 = \|x - \Delta(Ax)\|_2 \leq C\frac{\sigma_k(x)_1}{\sqrt{k}} = \sqrt{2}C\frac{\|h_{\Lambda^c}\|_1}{\sqrt{2k}},$$

where the last inequality follows from (1.6). □

1.4.2 The restricted isometry property

While the NSP is both necessary and sufficient for establishing guarantees of the form (1.6), these guarantees do not account for *noise*. When the measurements are contaminated with noise or have been corrupted by some error such as quantization, it will be useful to consider somewhat stronger conditions. In [43], Candès and Tao introduced the following isometry condition on matrices A and established its important role in CS.

DEFINITION 1.3 *A matrix A satisfies the* restricted isometry property *(RIP) of order k if there exists a $\delta_k \in (0,1)$ such that*

$$(1 - \delta_k)\|x\|_2^2 \leq \|Ax\|_2^2 \leq (1 + \delta_k)\|x\|_2^2 \tag{1.7}$$

holds for all $x \in \Sigma_k$.

If a matrix A satisfies the RIP of order $2k$, then we can interpret (1.7) as saying that A approximately preserves the distance between any pair of k-sparse vectors. This will clearly have fundamental implications concerning robustness to noise. Moreover, the potential applications of such *stable embeddings* range far beyond acquisition for the sole purpose of signal recovery. See Chapter 10 for examples of additional applications.

It is important to note that while in our definition of the RIP we assume bounds that are symmetric about 1, this is merely for notational convenience. In practice, one could instead consider arbitrary bounds

$$\alpha\|x\|_2^2 \leq \|Ax\|_2^2 \leq \beta\|x\|_2^2$$

where $0 < \alpha \leq \beta < \infty$. Given any such bounds, one can always scale A so that it satisfies the symmetric bound about 1 in (1.7). Specifically, multiplying A by $\sqrt{2/(\beta+\alpha)}$ will result in an $\widetilde{A}$ that satisfies (1.7) with constant $\delta_k = (\beta - \alpha)/(\beta + \alpha)$. While we will not explicitly show this, one can check that all of the theorems in this chapter based on the assumption that A satisfies the RIP actually hold as long as there exists some scaling of A that satisfies the RIP. Thus, since we can always scale A to satisfy (1.7), we lose nothing by restricting our attention to this simpler bound.

Note also that if A satisfies the RIP of order k with constant δ_k, then for any $k' < k$ we automatically have that A satisfies the RIP of order k' with constant $\delta_{k'} \leq \delta_k$. Moreover, in [190] it is shown that if A satisfies the RIP of order k with a sufficiently small constant, then it will also automatically satisfy the RIP of order γk for certain $\gamma > 1$, albeit with a somewhat worse constant.

LEMMA 1.1 (Corollary 3.4 of [190]) *Suppose that A satisfies the RIP of order k with constant δ_k. Let γ be a positive integer. Then A satisfies the RIP of order $k' = \gamma \lfloor \frac{k}{2} \rfloor$ with constant $\delta_{k'} < \gamma \cdot \delta_k$, where $\lfloor \cdot \rfloor$ denotes the floor operator.*

This lemma is trivial for $\gamma = 1, 2$, but for $\gamma \geq 3$ (and $k \geq 4$) this allows us to extend from RIP of order k to higher orders. Note however, that δ_k must be sufficiently small in order for the resulting bound to be useful.

The RIP and stability

We will see in Sections 1.5 and 1.6 that if a matrix A satisfies the RIP, then this is sufficient for a variety of algorithms to be able to successfully recover a sparse signal from noisy measurements. First, however, we will take a closer look at whether the RIP is actually necessary. It should be clear that the lower bound in the RIP is a necessary condition if we wish to be able to recover all sparse signals x from the measurements Ax for the same reasons that the NSP is necessary. We can say even more about the necessity of the RIP by considering the following notion of stability [67].

DEFINITION 1.4 *Let $A : \mathbb{R}^n \to \mathbb{R}^m$ denote a sensing matrix and $\Delta : \mathbb{R}^m \to \mathbb{R}^n$ denote a recovery algorithm. We say that the pair (A, Δ) is C-stable if for any $x \in \Sigma_k$ and any $e \in \mathbb{R}^m$ we have that*

$$\|\Delta(Ax + e) - x\|_2 \leq C \|e\|_2.$$

This definition simply says that if we add a small amount of noise to the measurements, then the impact of this on the recovered signal should not be arbitrarily large. Theorem 1.3 below demonstrates that the existence of any decoding algorithm (potentially impractical) that can stably recover from noisy measurements requires that A satisfy the lower bound of (1.7) with a constant determined by C.

THEOREM 1.3 (Theorem 3.1 of [67]) *If the pair (A, Δ) is C-stable, then*

$$\frac{1}{C} \|x\|_2 \leq \|Ax\|_2 \qquad (1.8)$$

for all $x \in \Sigma_{2k}$.

Proof. Pick any $x, z \in \Sigma_k$. Define

$$e_x = \frac{A(z-x)}{2} \quad \text{and} \quad e_z = \frac{A(x-z)}{2},$$

and note that

$$Ax + e_x = Az + e_z = \frac{A(x+z)}{2}.$$

Let $\widehat{x} = \Delta(Ax + e_x) = \Delta(Az + e_z)$. From the triangle inequality and the definition of C-stability, we have that

$$\begin{aligned}\|x - z\|_2 &= \|x - \widehat{x} + \widehat{x} - z\|_2 \\ &\leq \|x - \widehat{x}\|_2 + \|\widehat{x} - z\|_2 \\ &\leq C\|e_x\|_2 + C\|e_z\|_2 \\ &= C\|Ax - Az\|_2.\end{aligned}$$

Since this holds for any $x, z \in \Sigma_k$, the result follows. $\square$

Note that as $C \to 1$, we have that A must satisfy the lower bound of (1.7) with $\delta_{2k} = 1 - 1/C^2 \to 0$. Thus, if we desire to reduce the impact of noise in our recovered signal then we must adjust A so that it satisfies the lower bound of (1.7) with a tighter constant.

One might respond to this result by arguing that since the upper bound is not necessary, we can avoid redesigning A simply by rescaling A so that as long as A satisfies the RIP with $\delta_{2k} < 1$, the rescaled version αA will satisfy (1.8) for any constant C. In settings where the size of the noise is independent of our choice of A, this is a valid point – by scaling A we are essentially adjusting the gain on the "signal" part of our measurements, and if increasing this gain does not impact the noise, then we can achieve arbitrarily high signal-to-noise ratios, so that eventually the noise is negligible compared to the signal.

However, in practice we will typically not be able to rescale A to be arbitrarily large. Moreover, in many practical settings the noise is not independent of A. For example, consider the case where the noise vector e represents quantization noise produced by a finite dynamic range quantizer with B bits. Suppose the measurements lie in the interval $[-T, T]$, and we have adjusted the quantizer to capture this range. If we rescale A by α, then the measurements now lie between $[-\alpha T, \alpha T]$, and we must scale the dynamic range of our quantizer by α. In this case the resulting quantization error is simply αe, and we have achieved *no reduction* in the reconstruction error.

Measurement bounds

We can also consider how many measurements are necessary to achieve the RIP. If we ignore the impact of δ_{2k} and focus only on the dimensions of the problem (n, m, and k) then we can establish a simple lower bound, which is proven in Section A.1.

THEOREM 1.4 (Theorem 3.5 of [67]) *Let A be an $m \times n$ matrix that satisfies the RIP of order $2k$ with constant $\delta_{2k} \in (0, \frac{1}{2}]$. Then*

$$m \geq Ck \log\left(\frac{n}{k}\right)$$

where $C = 1/2 \log(\sqrt{24} + 1) \approx 0.28$.

Note that the restriction to $\delta_{2k} \leq 1/2$ is arbitrary and is made merely for convenience – minor modifications to the argument establish bounds for $\delta_{2k} \leq \delta_{\max}$ for any $\delta_{\max} < 1$.

Moreover, although we have made no effort to optimize the constants, it is worth noting that they are already quite reasonable.

While the proof is somewhat less direct, one can establish a similar result (in terms of its dependence on n and k) by examining the *Gelfand width* of the ℓ_1 ball [124]. However, both this result and Theorem 1.4 fail to capture the precise dependence of m on the desired RIP constant δ_k. In order to quantify this dependence, we can exploit recent results concerning the *Johnson–Lindenstrauss lemma*, which relates to embeddings of finite sets of points in low-dimensional spaces [158]. Specifically, it is shown in [156] that if we are given a point cloud with p points and wish to embed these points in $\mathbb{R}^m$ such that the squared ℓ_2 distance between any pair of points is preserved up to a factor of $1 \pm \epsilon$, then we must have that

$$m \geq \frac{c_0 \log(p)}{\epsilon^2},$$

where $c_0 > 0$ is a constant.

The Johnson-Lindenstrauss lemma is closely related to the RIP. In [5] it is shown that any procedure that can be used for generating a linear, distance-preserving embedding for a point cloud can also be used to construct a matrix that satisfies the RIP. Moreover, in [165] it is shown that if a matrix A satisfies the RIP of order $k = c_1 \log(p)$ with constant δ_k, then A can be used to construct a distance-preserving embedding for p points with $\epsilon = \delta_k/4$. Combining these we obtain

$$m \geq \frac{c_0 \log(p)}{\epsilon^2} = \frac{16 c_0 k}{c_1 \delta^2}.$$

Thus, for very small δ the number of measurements required to ensure that A satisfies the RIP of order k will be proportional to k/δ_k^2, which may be significantly higher than $k \log(n/k)$. See [165] for further details.

The relationship between the RIP and the NSP

Finally, we will now show that if a matrix satisfies the RIP, then it also satisfies the NSP. Thus, the RIP is strictly stronger than the NSP.

THEOREM 1.5 *Suppose that A satisfies the RIP of order $2k$ with $\delta_{2k} < \sqrt{2} - 1$. Then A satisfies the NSP of order $2k$ with constant*

$$C = \frac{2}{1 - (1 + \sqrt{2})\delta_{2k}}.$$

The proof of this theorem involves two useful lemmas. The first of these follows directly from standard norm inequalities by relating a k-sparse vector to a vector in $\mathbb{R}^k$. We include a simple proof for the sake of completeness.

LEMMA 1.2 *Suppose $u \in \Sigma_k$. Then*

$$\frac{\|u\|_1}{\sqrt{k}} \leq \|u\|_2 \leq \sqrt{k}\|u\|_\infty.$$

Proof. For any u, $\|u\|_1 = |\langle u, \mathrm{sgn}(u)\rangle|$. By applying the Cauchy–Schwarz inequality we obtain $\|u\|_1 \leq \|u\|_2 \|\mathrm{sgn}(u)\|_2$. The lower bound follows since $\mathrm{sgn}(u)$ has exactly k nonzero entries all equal to ± 1 (since $u \in \Sigma_k$) and thus $\|\mathrm{sgn}(u)\|_2 = \sqrt{k}$. The upper bound is obtained by observing that each of the k nonzero entries of u can be upper bounded by $\|u\|_\infty$. □

Below we state the second key lemma that we will need in order to prove Theorem 1.5. This result is a general result which holds for arbitrary h, not just vectors $h \in \mathcal{N}(A)$. It should be clear that when we do have $h \in \mathcal{N}(A)$, the argument could be simplified considerably. However, this lemma will prove immensely useful when we turn to the problem of sparse recovery from noisy measurements in Section 1.5, and thus we establish it now in its full generality. The intuition behind this bound will become more clear after reading Section 1.5. We state the lemma here, which is proven in Section A.2.

LEMMA 1.3 *Suppose that A satisfies the RIP of order $2k$, and let $h \in \mathbb{R}^n$, $h \neq 0$ be arbitrary. Let Λ_0 be any subset of $\{1,2,\ldots,n\}$ such that $|\Lambda_0| \leq k$. Define Λ_1 as the index set corresponding to the k entries of $h_{\Lambda_0^c}$ with largest magnitude, and set $\Lambda = \Lambda_0 \cup \Lambda_1$. Then*

$$\|h_\Lambda\|_2 \leq \alpha \frac{\|h_{\Lambda_0^c}\|_1}{\sqrt{k}} + \beta \frac{|\langle Ah_\Lambda, Ah\rangle|}{\|h_\Lambda\|_2},$$

where

$$\alpha = \frac{\sqrt{2}\delta_{2k}}{1-\delta_{2k}}, \quad \beta = \frac{1}{1-\delta_{2k}}.$$

Again, note that Lemma 1.3 holds for arbitrary h. In order to prove Theorem 1.5, we merely need to apply Lemma 1.3 to the case where $h \in \mathcal{N}(A)$.

Proof of Theorem 1.5. Suppose that $h \in \mathcal{N}(A)$. It is sufficient to show that

$$\|h_\Lambda\|_2 \leq C \frac{\|h_{\Lambda^c}\|_1}{\sqrt{2k}} \tag{1.9}$$

holds for the case where Λ is the index set corresponding to the $2k$ largest entries of h. Thus, we can take Λ_0 to be the index set corresponding to the k largest entries of h and apply Lemma 1.3.

The second term in Lemma 1.3 vanishes since $Ah = 0$, and thus we have

$$\|h_\Lambda\|_2 \leq \alpha \frac{\|h_{\Lambda_0^c}\|_1}{\sqrt{k}}.$$

Using Lemma 1.2,

$$\|h_{\Lambda_0^c}\|_1 = \|h_{\Lambda_1}\|_1 + \|h_{\Lambda^c}\|_1 \leq \sqrt{k}\|h_{\Lambda_1}\|_2 + \|h_{\Lambda^c}\|_1$$

resulting in

$$\|h_\Lambda\|_2 \leq \alpha \left(\|h_{\Lambda_1}\|_2 + \frac{\|h_{\Lambda^c}\|_1}{\sqrt{k}}\right).$$

Since $\|h_{\Lambda_1}\|_2 \leq \|h_\Lambda\|_2$, we have that

$$(1-\alpha)\|h_\Lambda\|_2 \leq \alpha \frac{\|h_{\Lambda^c}\|_1}{\sqrt{k}}.$$

The assumption $\delta_{2k} < \sqrt{2} - 1$ ensures that $\alpha < 1$, and thus we may divide by $1 - \alpha$ without changing the direction of the inequality to establish (1.9) with constant

$$C = \frac{\sqrt{2}\alpha}{1-\alpha} = \frac{2\delta_{2k}}{1-(1+\sqrt{2})\delta_{2k}},$$

as desired. □

1.4.3 Coherence

While the spark, NSP, and RIP all provide guarantees for the recovery of k-sparse signals, verifying that a general matrix A satisfies any of these properties would typically require a combinatorial search over all $\binom{n}{k}$ submatrices. In many cases it is preferable to use properties of A that are easily computable to provide more concrete recovery guarantees. The *coherence* of a matrix is one such property [86, 222].

DEFINITION 1.5 *The coherence of a matrix A, $\mu(A)$, is the largest absolute inner product between any two columns a_i, a_j of A:*

$$\mu(A) = \max_{1 \leq i < j \leq n} \frac{|\langle a_i, a_j \rangle|}{\|a_i\|_2 \|a_j\|_2}.$$

It is possible to show that the coherence of a matrix is always in the range $\mu(A) \in \left[\sqrt{\frac{n-m}{m(n-1)}}, 1\right]$; the lower bound is known as the Welch bound [207, 214, 245]. Note that when $n \gg m$, the lower bound is approximately $\mu(A) \geq 1/\sqrt{m}$. The concept of coherence can also be extended to certain structured sparsity models and specific classes of analog signals [27, 111, 112].

One can sometimes relate coherence to the spark, NSP, and RIP. For example, the coherence and spark properties of a matrix can be related by employing the Geršgorin circle theorem [127, 235].

THEOREM 1.6 (Theorem 2 of [127]) *The eigenvalues of an $n \times n$ matrix M with entries m_{ij}, $1 \leq i, j \leq n$, lie in the union of n discs $d_i = d_i(c_i, r_i)$, $1 \leq i \leq n$, centered at $c_i = m_{ii}$ and with radius $r_i = \sum_{j \neq i} |m_{ij}|$.*

Applying this theorem on the Gram matrix $G = A_\Lambda^T A_\Lambda$ leads to the following straightforward result.

LEMMA 1.4 *For any matrix A,*

$$\text{spark}(A) \geq 1 + \frac{1}{\mu(A)}.$$

Proof. Since spark(A) does not depend on the scaling of the columns, we can assume without loss of generality that A has unit-norm columns. Let $\Lambda \subseteq \{1,\ldots,n\}$ with $|\Lambda| = p$ determine a set of indices. We consider the restricted Gram matrix $G = A_\Lambda^T A_\Lambda$, which satisfies the following properties:

- $g_{ii} = 1, 1 \leq i \leq p$;
- $|g_{ij}| \leq \mu(A), 1 \leq i, j \leq p, i \neq j$.

From Theorem 1.6, if $\sum_{j \neq i} |g_{ij}| < |g_{ii}|$ then the matrix G is positive definite, so that the columns of A_Λ are linearly independent. Thus, the spark condition implies $(p-1)\mu(A) < 1$ or, equivalently, $p < 1 + 1/\mu(A)$ for all $p < \text{spark}(A)$, yielding $\text{spark}(A) \geq 1 + 1/\mu(A)$. □

By merging Theorem 1.1 with Lemma 1.4, we can pose the following condition on A that guarantees uniqueness.

THEOREM 1.7 (Theorem 12 of [86]) *If*

$$k < \frac{1}{2}\left(1 + \frac{1}{\mu(A)}\right),$$

then for each measurement vector $y \in \mathbb{R}^m$ *there exists at most one signal* $x \in \Sigma_k$ *such that* $y = Ax$.

Theorem 1.7, together with the Welch bound, provides an upper bound on the level of sparsity k that guarantees uniqueness using coherence: $k = O(\sqrt{m})$. Another straightforward application of the Geršgorin circle theorem (Theorem 1.6) connects the RIP to the coherence property.

LEMMA 1.5 *If A has unit-norm columns and coherence $\mu = \mu(A)$, then A satisfies the RIP of order k with $\delta_k = (k-1)\mu$ for all $k < 1/\mu$.*

The proof of this lemma is similar to that of Lemma 1.4.

1.4.4 Sensing matrix constructions

Now that we have defined the relevant properties of a matrix A in the context of CS, we turn to the question of how to construct matrices that satisfy these properties. To begin, it is straightforward to show that an $m \times n$ Vandermonde matrix V constructed from m distinct scalars has $\text{spark}(V) = m+1$ [57]. Unfortunately, these matrices are poorly conditioned for large values of n, rendering the recovery problem numerically unstable. Similarly, there are known matrices A of size $m \times m^2$ that achieve the coherence lower bound $\mu(A) = 1/\sqrt{m}$, such as the Gabor frame generated from the Alltop sequence [148] and more general equiangular tight frames [214]. These constructions restrict the number of measurements needed to recover a k-sparse signal to be $m = O(k^2 \log n)$. It is also possible to deterministically construct matrices of size $m \times n$ that satisfy the RIP of order k, but such constructions also require m to be relatively large [28, 78, 140, 152]. For example, the construction in [78] requires $m = O(k^2 \log n)$ while the construction in [152] requires $m = O(kn^\alpha)$ for some constant α. In many real-world settings, these results would lead to an unacceptably large requirement on m.

Fortunately, these limitations can be overcome by randomizing the matrix construction. For example, random matrices A of size $m \times n$ whose entries are independent and identically distributed (i.i.d.) with continuous distributions have $\mathrm{spark}(A) = m + 1$ with probability one. More significantly, it can also be shown that random matrices will satisfy the RIP with high probability if the entries are chosen according to a Gaussian, Bernoulli, or more generally any sub-gaussian distribution. See Chapter 5 for details, and in particular, Theorem 5.65. This theorem states that if a matrix A is chosen according to a sub-gaussian distribution with $m = O\left(k \log(n/k)/\delta_{2k}^2\right)$, then A will satisfy the RIP of order $2k$ with probability at least $1 - 2\exp(-c_1 \delta_{2k}^2 m)$. Note that in light of the measurement bounds in Section 1.4.2 we see that this achieves the optimal number of measurements up to a constant. It also follows from Theorem 1.5 that these random constructions provide matrices satisfying the NSP. Furthermore, it can be shown that if the distribution used has zero mean and finite variance, then the coherence converges to $\mu(A) = \sqrt{(2 \log n)/m}$ in the asymptotic regime (as m and n grow) [32, 37, 83].

Using random matrices to construct A has a number of additional benefits. To illustrate these, we will focus on the RIP. First, one can show that for random constructions the measurements are *democratic*, meaning that it is possible to recover a signal using any sufficiently large subset of the measurements [73, 169]. Thus, by using random A one can be robust to the loss or corruption of a small fraction of the measurements. Second, and perhaps more significantly, in practice we are often more interested in the setting where x is sparse with respect to some basis Φ. In this case what we actually require is that the product $A\Phi$ satisfies the RIP. If we were to use a deterministic construction then we would need to explicitly take Φ into account in our construction of A, but when A is chosen randomly we can avoid this consideration. For example, if A is chosen according to a Gaussian distribution and Φ is an orthonormal basis then one can easily show that $A\Phi$ will also have a Gaussian distribution, and so provided that m is sufficiently high $A\Phi$ will satisfy the RIP with high probability, just as before. Although less obvious, similar results hold for sub-gaussian distributions as well [5]. This property, sometimes referred to as *universality*, constitutes a significant advantage of using random matrices to construct A. See Chapter 5 for further details on random matrices and their role in CS.

Finally, we note that since the fully random matrix approach is sometimes impractical to build in hardware, several hardware architectures have been implemented and/or proposed that enable random measurements to be acquired in practical settings. Examples include the random demodulator [224], random filtering [225], the modulated wideband converter [187], random convolution [1, 206], and the compressive multiplexer [211]. These architectures typically use a reduced amount of randomness and are modeled via matrices A that have significantly more structure than a fully random matrix. Perhaps somewhat surprisingly, while it is typically not quite as easy as in the fully random case, one can prove that many of these constructions also satisfy the RIP and/or have low coherence. Furthermore, one can analyze the effect of inaccuracies in the matrix A implemented by the system [54, 149]; in the simplest cases, such sensing matrix errors can be addressed through system calibration.

1.5 Signal recovery via ℓ_1 minimization

While there now exist a wide variety of approaches to recover a sparse signal x from a small number of linear measurements, as we will see in Section 1.6, we begin by considering a natural first approach to the problem of sparse recovery.

Given measurements y and the knowledge that our original signal x is sparse or compressible, it is natural to attempt to recover x by solving an optimization problem of the form

$$\widehat{x} = \arg\min_{z} \|z\|_0 \quad \text{subject to} \quad z \in \mathcal{B}(y), \tag{1.10}$$

where $\mathcal{B}(y)$ ensures that $\widehat{x}$ is consistent with the measurements y. For example, in the case where our measurements are exact and noise-free, we can set $\mathcal{B}(y) = \{z : Az = y\}$. When the measurements have been contaminated with a small amount of bounded noise, we could instead consider $\mathcal{B}(y) = \{z : \|Az - y\|_2 \le \epsilon\}$. In both cases, (1.10) finds the sparsest x that is consistent with the measurements y.

Note that in (1.10) we are inherently assuming that x itself is sparse. In the more common setting where $x = \Phi c$, we can easily modify the approach and instead consider

$$\widehat{c} = \arg\min_{z} \|z\|_0 \quad \text{subject to} \quad z \in \mathcal{B}(y) \tag{1.11}$$

where $\mathcal{B}(y) = \{z : A\Phi z = y\}$ or $\mathcal{B}(y) = \{z : \|A\Phi z - y\|_2 \le \epsilon\}$. By considering $\widetilde{A} = A\Phi$ we see that (1.10) and (1.11) are essentially identical. Moreover, as noted in Section 1.4.4, in many cases the introduction of Φ does not significantly complicate the construction of matrices A such that $\widetilde{A}$ will satisfy the desired properties. Thus, for the remainder of this chapter we will restrict our attention to the case where $\Phi = I$. It is important to note, however, that this restriction does impose certain limits in our analysis when Φ is a general dictionary and not an orthonormal basis. For example, in this case $\|\widehat{x} - x\|_2 = \|\Phi\widehat{c} - \Phi c\|_2 \ne \|\widehat{c} - c\|_2$, and thus a bound on $\|\widehat{c} - c\|_2$ cannot directly be translated into a bound on $\|\widehat{x} - x\|_2$, which is often the metric of interest. For further discussion of these and related issues see [35].

While it is possible to analyze the performance of (1.10) under the appropriate assumptions on A (see [56, 144] for details), we do not pursue this strategy since the objective function $\|\cdot\|_0$ is non-convex, and hence (1.10) is potentially very difficult to solve. In fact, one can show that for a general matrix A, even finding a solution that approximates the true minimum is NP-hard [189].

One avenue for translating this problem into something more tractable is to replace $\|\cdot\|_0$ with its convex approximation $\|\cdot\|_1$. Specifically, we consider

$$\widehat{x} = \arg\min_{z} \|z\|_1 \quad \text{subject to} \quad z \in \mathcal{B}(y). \tag{1.12}$$

Provided that $\mathcal{B}(y)$ is convex, (1.12) is computationally feasible. In fact, when $\mathcal{B}(y) = \{z : Az = y\}$, the resulting problem can be posed as a linear program [53].

While it is clear that replacing (1.10) with (1.12) transforms a computationally intractable problem into a tractable one, it may not be immediately obvious that the solution to (1.12) will be at all similar to the solution to (1.10). However, there are certainly intuitive reasons to expect that the use of ℓ_1 minimization will indeed promote sparsity. As an example, recall that in Figure 1.2, the solutions to the ℓ_1 minimization problem coincided exactly with the solution to the ℓ_p minimization problem for any $p < 1$, and notably, was sparse. Moreover, the use of ℓ_1 minimization to promote or exploit sparsity has a long history, dating back at least to the work of Beurling on Fourier transform extrapolation from partial observations [22].

Additionally, in a somewhat different context, in 1965 Logan [91, 174] showed that a bandlimited signal can be perfectly recovered in the presence of *arbitrary* corruptions on a small interval (see also extensions of these conditions in [91]). Again, the recovery method consists of searching for the bandlimited signal that is closest to the observed signal in the ℓ_1 norm. This can be viewed as further validation of the intuition gained from Figure 1.2 – the ℓ_1 norm is well-suited to sparse errors.

Historically, the use of ℓ_1 minimization on large problems finally became practical with the explosion of computing power in the late 1970s and early 1980s. In one of its first applications, it was demonstrated that geophysical signals consisting of spike trains could be recovered from only the high-frequency components of these signals by exploiting ℓ_1 minimization [171, 216, 242]. Finally, in the 1990s there was renewed interest in these approaches within the signal processing community for the purpose of finding sparse approximations to signals and images when represented in overcomplete dictionaries or unions of bases [53, 182]. Separately, ℓ_1 minimization received significant attention in the statistics literature as a method for variable selection in regression, known as the Lasso [218].

Thus, there are a variety of reasons to suspect that ℓ_1 minimization will provide an accurate method for sparse signal recovery. More importantly, this also constitutes a computationally tractable approach to sparse signal recovery. In this section we provide an overview of ℓ_1 minimization from a theoretical perspective. We discuss algorithms for ℓ_1 minimization in Section 1.6.

1.5.1 Noise-free signal recovery

In order to analyze ℓ_1 minimization for various specific choices of $\mathcal{B}(y)$, we require the following general result which builds on Lemma 1.3 and is proven in Section A.3.

LEMMA 1.6 *Suppose that A satisfies the RIP of order $2k$ with $\delta_{2k} < \sqrt{2} - 1$. Let $x, \widehat{x} \in \mathbb{R}^n$ be given, and define $h = \widehat{x} - x$. Let Λ_0 denote the index set corresponding to the k entries of x with largest magnitude and Λ_1 the index set corresponding to the k entries of $h_{\Lambda_0^c}$ with largest magnitude. Set $\Lambda = \Lambda_0 \cup \Lambda_1$. If $\|\widehat{x}\|_1 \leq \|x\|_1$, then*

$$\|h\|_2 \leq C_0 \frac{\sigma_k(x)_1}{\sqrt{k}} + C_1 \frac{|\langle Ah_\Lambda, Ah \rangle|}{\|h_\Lambda\|_2},$$

where
$$C_0 = 2\frac{1-(1-\sqrt{2})\delta_{2k}}{1-(1+\sqrt{2})\delta_{2k}}, \quad C_1 = \frac{2}{1-(1+\sqrt{2})\delta_{2k}}.$$

Lemma 1.6 establishes an error bound for the class of ℓ_1 minimization problems described by (1.12) when combined with a measurement matrix A satisfying the RIP. In order to obtain specific bounds for concrete examples of $\mathcal{B}(y)$, we must examine how requiring $\widehat{x} \in \mathcal{B}(y)$ affects $|\langle Ah_\Lambda, Ah\rangle|$. As an example, in the case of noise-free measurements we obtain the following theorem.

THEOREM 1.8 (Theorem 1.1 of [34]) *Suppose that A satisfies the RIP of order $2k$ with $\delta_{2k} < \sqrt{2}-1$ and we obtain measurements of the form $y = Ax$. Then when $\mathcal{B}(y) = \{z : Az = y\}$, the solution $\widehat{x}$ to (1.12) obeys*

$$\|\widehat{x} - x\|_2 \le C_0 \frac{\sigma_k(x)_1}{\sqrt{k}}.$$

Proof. Since $x \in \mathcal{B}(y)$ we can apply Lemma 1.6 to obtain that for $h = \widehat{x} - x$,

$$\|h\|_2 \le C_0 \frac{\sigma_k(x)_1}{\sqrt{k}} + C_1 \frac{|\langle Ah_\Lambda, Ah\rangle|}{\|h_\Lambda\|_2}.$$

Furthermore, since $x, \widehat{x} \in \mathcal{B}(y)$ we also have that $y = Ax = A\widehat{x}$ and hence $Ah = 0$. Therefore the second term vanishes, and we obtain the desired result. □

Theorem 1.8 is rather remarkable. By considering the case where $x \in \Sigma_k$ we can see that provided A satisfies the RIP – which as shown in Section 1.4.4 allows for as few as $O(k\log(n/k))$ measurements – we can recover any k-sparse x *exactly*. This result seems improbable on its own, and so one might expect that the procedure would be highly sensitive to noise, but we will see below that Lemma 1.6 can also be used to demonstrate that this approach is actually stable.

Note that Theorem 1.8 assumes that A satisfies the RIP. One could easily modify the argument to replace this with the assumption that A satisfies the NSP instead. Specifically, if we are only interested in the noiseless setting, in which case h lies in the nullspace of A, then Lemma 1.6 simplifies and its proof could essentially be broken into two steps: (i) show that if A satisfies the RIP then it satisfies the NSP (as shown in Theorem 1.5), and (ii) the NSP implies the simplified version of Lemma 1.6. This proof directly mirrors that of Lemma 1.6. Thus, by the same argument as in the proof of Theorem 1.8, it is straightforward to show that if A satisfies the NSP then it will obey the same error bound.

1.5.2 Signal recovery in noise

The ability to perfectly reconstruct a sparse signal from noise-free measurements represents a very promising result. However, in most real-world systems the measurements are likely to be contaminated by some form of noise. For instance, in order to process data in a computer we must be able to represent it using a finite number of bits, and hence the measurements will typically be subject to quantization error. Moreover, systems which are implemented in physical hardware will be subject to a variety of different types of noise depending on the setting. Another important noise source is on the signal itself. In many settings the signal x to be estimated is contaminated by some form of random noise. The implications of this type of noise on the achievable sampling rates has been recently analyzed in [19,67,219]. Here we focus on measurement noise, which has received much more attention in the literature.

Perhaps somewhat surprisingly, one can show that it is possible to stably recover sparse signals under a variety of common noise models [18,42,87,88,144,169,170]. As might be expected, both the RIP and coherence are useful in establishing performance guarantees in noise. We begin our discussion below with robustness guarantees for matrices satisfying the RIP. We then turn to results for matrices with low coherence.

Bounded noise

We first provide a bound on the worst-case performance for uniformly bounded noise, as first investigated in [42].

THEOREM 1.9 (Theorem 1.2 of [34]) *Suppose that A satisfies the RIP of order $2k$ with $\delta_{2k} < \sqrt{2} - 1$ and let $y = Ax + e$ where $\|e\|_2 \leq \epsilon$. Then when $\mathcal{B}(y) = \{z : \|Az - y\|_2 \leq \epsilon\}$, the solution $\widehat{x}$ to (1.12) obeys*

$$\|\widehat{x} - x\|_2 \leq C_0 \frac{\sigma_k(x)_1}{\sqrt{k}} + C_2 \epsilon,$$

where

$$C_0 = 2\frac{1 - (1 - \sqrt{2})\delta_{2k}}{1 - (1 + \sqrt{2})\delta_{2k}}, \quad C_2 = 4\frac{\sqrt{1 + \delta_{2k}}}{1 - (1 + \sqrt{2})\delta_{2k}}.$$

Proof. We are interested in bounding $\|h\|_2 = \|\widehat{x} - x\|_2$. Since $\|e\|_2 \leq \epsilon$, $x \in \mathcal{B}(y)$, and therefore we know that $\|\widehat{x}\|_1 \leq \|x\|_1$. Thus we may apply Lemma 1.6, and it remains to bound $|\langle Ah_\Lambda, Ah \rangle|$. To do this, we observe that

$$\|Ah\|_2 = \|A(\widehat{x} - x)\|_2 = \|A\widehat{x} - y + y - Ax\|_2 \leq \|A\widehat{x} - y\|_2 + \|y - Ax\|_2 \leq 2\epsilon$$

where the last inequality follows since $x, \widehat{x} \in \mathcal{B}(y)$. Combining this with the RIP and the Cauchy–Schwarz inequality we obtain

$$|\langle Ah_\Lambda, Ah \rangle| \leq \|Ah_\Lambda\|_2 \|Ah\|_2 \leq 2\epsilon\sqrt{1 + \delta_{2k}} \|h_\Lambda\|_2.$$

Thus,

$$\|h\|_2 \leq C_0 \frac{\sigma_k(x)_1}{\sqrt{k}} + C_1 2\epsilon\sqrt{1 + \delta_{2k}} = C_0 \frac{\sigma_k(x)_1}{\sqrt{k}} + C_2 \epsilon,$$

completing the proof. □

In order to place this result in context, consider how we would recover a sparse vector x if we happened to already know the k locations of the nonzero coefficients, which we denote by Λ_0. This is referred to as the *oracle estimator*. In this case a natural approach is to reconstruct the signal using a simple pseudoinverse:[6]

$$\widehat{x}_{\Lambda_0} = A^\dagger_{\Lambda_0} y = (A^T_{\Lambda_0} A_{\Lambda_0})^{-1} A^T_{\Lambda_0} y$$
$$\widehat{x}_{\Lambda_0^c} = 0. \tag{1.13}$$

The implicit assumption in (1.13) is that A_{Λ_0} has full column-rank (and hence we are considering the case where A_{Λ_0} is the $m \times k$ matrix with the columns indexed by Λ_0^c removed) so that there is a unique solution to the equation $y = A_{\Lambda_0} x_{\Lambda_0}$. With this choice, the recovery error is given by

$$\|\widehat{x} - x\|_2 = \|(A^T_{\Lambda_0} A_{\Lambda_0})^{-1} A^T_{\Lambda_0}(Ax + e) - x\|_2 = \|(A^T_{\Lambda_0} A_{\Lambda_0})^{-1} A^T_{\Lambda_0} e\|_2.$$

We now consider the worst-case bound for this error. Using standard properties of the singular value decomposition, it is straightforward to show that if A satisfies the RIP of order $2k$ (with constant δ_{2k}), then the largest singular value of $A^\dagger_{\Lambda_0}$ lies in the range $[1/\sqrt{1+\delta_{2k}}, 1/\sqrt{1-\delta_{2k}}]$. Thus, if we consider the worst-case recovery error over all e such that $\|e\|_2 \leq \epsilon$, then the recovery error can be bounded by

$$\frac{\epsilon}{\sqrt{1+\delta_{2k}}} \leq \|\widehat{x} - x\|_2 \leq \frac{\epsilon}{\sqrt{1-\delta_{2k}}}.$$

Therefore, in the case where x is exactly k-sparse, the guarantee for the pseudoinverse recovery method, which is given *perfect knowledge of the true support of x*, cannot improve upon the bound in Theorem 1.9 by more than a constant value.

We now consider a slightly different noise model. Whereas Theorem 1.9 assumed that the noise norm $\|e\|_2$ was small, the theorem below analyzes a different recovery algorithm known as the *Dantzig selector* in the case where $\|A^T e\|_\infty$ is small [45]. We will see below that this will lead to a simple analysis of the performance of this algorithm in Gaussian noise.

THEOREM 1.10 *Suppose that A satisfies the RIP of order $2k$ with $\delta_{2k} < \sqrt{2} - 1$ and we obtain measurements of the form $y = Ax + e$ where $\|A^T e\|_\infty \leq \lambda$. Then when $\mathcal{B}(y) = \{z : \|A^T(Az - y)\|_\infty \leq \lambda\}$, the solution $\widehat{x}$ to (1.12) obeys*

$$\|\widehat{x} - x\|_2 \leq C_0 \frac{\sigma_k(x)_1}{\sqrt{k}} + C_3 \sqrt{k} \lambda,$$

where

$$C_0 = 2 \frac{1 - (1 - \sqrt{2})\delta_{2k}}{1 - (1 + \sqrt{2})\delta_{2k}}, \quad C_3 = \frac{4\sqrt{2}}{1 - (1 + \sqrt{2})\delta_{2k}}.$$

---

[6] Note that while the pseudoinverse approach can be improved upon (in terms of ℓ_2 error) by instead considering alternative biased estimators [16, 108, 155, 159, 213], this does not fundamentally change the above conclusions.

Proof. The proof mirrors that of Theorem 1.9. Since $\|A^T e\|_\infty \leq \lambda$, we again have that $x \in \mathcal{B}(y)$, so $\|\widehat{x}\|_1 \leq \|x\|_1$ and thus Lemma 1.6 applies. We follow a similar approach as in Theorem 1.9 to bound $|\langle Ah_\Lambda, Ah\rangle|$. We first note that

$$\|A^T Ah\|_\infty \leq \|A^T(A\widehat{x} - y)\|_\infty + \|A^T(y - Ax)\|_\infty \leq 2\lambda$$

where the last inequality again follows since $x, \widehat{x} \in \mathcal{B}(y)$. Next, note that $Ah_\Lambda = A_\Lambda h_\Lambda$. Using this we can apply the Cauchy–Schwarz inequality to obtain

$$|\langle Ah_\Lambda, Ah\rangle| = |\langle h_\Lambda, A_\Lambda^T Ah\rangle| \leq \|h_\Lambda\|_2 \|A_\Lambda^T Ah\|_2.$$

Finally, since $\|A^T Ah\|_\infty \leq 2\lambda$, we have that every coefficient of $A^T Ah$ is at most 2λ, and thus $\|A_\Lambda^T Ah\|_2 \leq \sqrt{2k}(2\lambda)$. Thus,

$$\|h\|_2 \leq C_0 \frac{\sigma_k(x)_1}{\sqrt{k}} + C_1 2\sqrt{2k}\lambda = C_0 \frac{\sigma_k(x)_1}{\sqrt{k}} + C_3 \sqrt{k}\lambda,$$

as desired. □

Gaussian noise

Finally, we also consider the performance of these approaches in the presence of Gaussian noise. The case of Gaussian noise was first considered in [144], which examined the performance of ℓ_0 minimization with noisy measurements. We now see that Theorems 1.9 and 1.10 can be leveraged to provide similar guarantees for ℓ_1 minimization. To simplify our discussion we will restrict our attention to the case where $x \in \Sigma_k$, so that $\sigma_k(x)_1 = 0$ and the error bounds in Theorems 1.9 and 1.10 depend only on the noise e.

To begin, suppose that the coefficients of $e \in \mathbb{R}^m$ are i.i.d. according to a Gaussian distribution with mean zero and variance σ^2. By using standard properties of the Gaussian distribution, one can show (see, for example, Corollary 5.17 of Chapter 5) that there exists a constant $c_0 > 0$ such that for any $\epsilon > 0$,

$$\mathbb{P}\left(\|e\|_2 \geq (1+\epsilon)\sqrt{m}\sigma\right) \leq \exp\left(-c_0 \epsilon^2 m\right), \quad (1.14)$$

where $\mathbb{P}(E)$ denotes the probability that the event E occurs. Applying this result to Theorem 1.9 with $\epsilon = 1$, we obtain the following result for the special case of Gaussian noise.

COROLLARY 1.1 *Suppose that A satisfies the RIP of order $2k$ with $\delta_{2k} < \sqrt{2} - 1$. Furthermore, suppose that $x \in \Sigma_k$ and that we obtain measurements of the form $y = Ax + e$ where the entries of e are i.i.d. $\mathcal{N}(0, \sigma^2)$. Then when $\mathcal{B}(y) = \{z : \|Az - y\|_2 \leq 2\sqrt{m}\sigma\}$, the solution $\widehat{x}$ to (1.12) obeys*

$$\|\widehat{x} - x\|_2 \leq 8 \frac{\sqrt{1 + \delta_{2k}}}{1 - (1 + \sqrt{2})\delta_{2k}} \sqrt{m}\sigma$$

with probability at least $1 - \exp(-c_0 m)$.

We can similarly consider Theorem 1.10 in the context of Gaussian noise. If we assume that the columns of A have unit norm, then each coefficient of $A^T e$ is a Gaussian random variable with mean zero and variance σ^2. Using standard tail bounds for the Gaussian distribution (see, for example, (5.5) of Chapter 5), we have that

$$\mathbb{P}\left(\left|[A^T e]_i\right| \geq t\sigma\right) \leq \exp\left(-t^2/2\right)$$

for $i = 1, 2, \ldots, n$. Thus, using the union bound over the bounds for different i, we obtain

$$\mathbb{P}\left(\left\|A^T e\right\|_\infty \geq 2\sqrt{\log n}\,\sigma\right) \leq n\exp(-2\log n) = \frac{1}{n}.$$

Applying this to Theorem 1.10, we obtain the following result, which is a simplified version of Theorem 1.1 of [45].

COROLLARY 1.2 *Suppose that A has unit-norm columns and satisfies the RIP of order $2k$ with $\delta_{2k} < \sqrt{2} - 1$. Furthermore, suppose that $x \in \Sigma_k$ and that we obtain measurements of the form $y = Ax + e$ where the entries of e are i.i.d. $\mathcal{N}(0, \sigma^2)$. Then when $\mathcal{B}(y) = \{z : \|A^T(Az - y)\|_\infty \leq 2\sqrt{\log n}\,\sigma\}$, the solution $\widehat{x}$ to (1.12) obeys*

$$\|\widehat{x} - x\|_2 \leq 4\sqrt{2}\frac{\sqrt{1 + \delta_{2k}}}{1 - (1 + \sqrt{2})\delta_{2k}}\sqrt{k\log n}\,\sigma$$

with probability at least $1 - \frac{1}{n}$.

Ignoring the precise constants and the probabilities with which the stated bounds hold (which we have made no effort to optimize), we observe that in the case when $m = O(k\log n)$ these results appear to be essentially the same. However, there is a subtle difference. Specifically, if m and n are fixed and we consider the effect of varying k, we can see that Corollary 1.2 yields a bound that is adaptive to this change, providing a stronger guarantee when k is small, whereas the bound in Corollary 1.1 does not improve as k is reduced. Thus, while they provide very similar guarantees, there are certain circumstances where the Dantzig selector is preferable. See [45] for further discussion of the comparative advantages of these approaches.

It can also be seen that results such as Corollary 1.2 guarantee that the Dantzig selector achieves an error $\|\widehat{x} - x\|_2^2$ which is bounded by a constant times $k\sigma^2\log n$, with high probability. Note that since we typically require $m > k\log n$, this can be substantially lower than the expected noise power $\mathbb{E}\|e\|_2^2 = m\sigma^2$, illustrating the fact that sparsity-based techniques are highly successful in reducing the noise level.

The value $k\sigma^2\log n$ is nearly optimal in several respects. First, an "oracle" estimator which knows the locations of the nonzero components and uses a least-squares technique to estimate their values achieves an estimation error on the order of $k\sigma^2$. For this reason, guarantees such as Corollary 1.2 are referred to as near-oracle results. The Cramér–Rao bound (CRB) for estimating x is also on the order of $k\sigma^2$ [17]. This is of practical interest since the CRB is achieved by the maximum likelihood estimator at high SNR, implying that for low-noise settings, an error of $k\sigma^2$ is achievable. However, the maximum likelihood estimator is NP-hard to compute, so that near-oracle results are still of interest. Interestingly, the $\log n$ factor seems to be an unavoidable result of the fact that the locations of the nonzero elements are unknown.

Coherence guarantees

Thus far, we have examined performance guarantees based on the RIP. As noted in Section 1.4.3, in practice it is typically impossible to verify that a matrix A satisfies the RIP or calculate the corresponding RIP constant δ. In this respect, results based on coherence are appealing, since they can be used with arbitrary dictionaries.

One quick route to coherence-based performance guarantees is to combine RIP-based results such as Corollaries 1.1 and 1.2 with coherence bounds such as Lemma 1.5. This technique yields guarantees based only on the coherence, but the results are often overly pessimistic. It is typically more enlightening to instead establish guarantees by directly exploiting coherence [18, 37, 87, 88]. In order to illustrate the types of guarantees that this approach can yield, we provide the following representative examples.

THEOREM 1.11 (Theorem 3.1 of [88]) *Suppose that A has coherence μ and that $x \in \Sigma_k$ with $k < (1/\mu + 1)/4$. Furthermore, suppose that we obtain measurements of the form $y = Ax + e$. Then when $\mathcal{B}(y) = \{z : \|Az - y\|_2 \leq \epsilon\}$, the solution $\widehat{x}$ to (1.12) obeys*

$$\|x - \widehat{x}\|_2 \leq \frac{\|e\|_2 + \epsilon}{\sqrt{1 - \mu(4k - 1)}}.$$

Note that this theorem holds for the case where $\epsilon = 0$ as well as where $\|e\|_2 = 0$. Thus, it also applies to the noise-free setting as in Theorem 1.8. Furthermore, there is no requirement that $\|e\|_2 \leq \epsilon$. In fact, this theorem is valid even when $\epsilon = 0$ but $\|e\|_2 \neq 0$. This constitutes a significant difference between this result and Theorem 1.9, and might cause us to question whether we actually need to pose alternative algorithms to handle the noisy setting. However, as noted in [88], Theorem 1.11 is the result of a worst-case analysis and will typically overestimate the actual error. In practice, the performance of (1.12) where $\mathcal{B}(y)$ is modified to account for the noise can lead to significant improvements.

In order to describe an additional type of coherence-based guarantee, we must consider an alternative, but equivalent, formulation of (1.12). Specifically, consider the optimization problem

$$\widehat{x} = \arg\min_z \frac{1}{2} \|Az - y\|_2^2 + \lambda \|z\|_1. \tag{1.15}$$

This formulation is exploited in the following result, which provides guarantees for (1.15) that go beyond what we have seen so far by providing explicit results concerning the recovery of the original support of x.

THEOREM 1.12 (Corollary 1 of [18]) *Suppose that A has coherence μ and that $x \in \Sigma_k$ with $k \leq 1/(3\mu)$. Furthermore, suppose that we obtain measurements of the form $y = Ax + e$ where the entries of e are i.i.d. $\mathcal{N}(0, \sigma^2)$. Set*

$$\lambda = \sqrt{8\sigma^2 (1 + \alpha) \log(n - k)}$$

for some fairly small value $\alpha > 0$. Then with probability exceeding

$$\left(1 - \frac{1}{(n-k)^\alpha}\right)(1 - \exp(-k/7)),$$

the solution $\widehat{x}$ to (1.15) is unique, $\mathrm{supp}(\widehat{x}) \subset \mathrm{supp}(x)$, and

$$\|\widehat{x} - x\|_2^2 \leq \left(\sqrt{3} + 3\sqrt{2(1+\alpha)\log(n-k)}\right)^2 k\sigma^2.$$

In this case we see that we are guaranteed that any nonzero of $\widehat{x}$ corresponds to a true nonzero of x. Note that this analysis allows for the worst-case signal x. It is possible to improve upon this result by instead assuming that the signal x has a limited amount of randomness. Specifically, in [37] it is shown that if $\mathrm{supp}(x)$ is chosen uniformly at random and that the signs of the nonzero entries of x are independent and equally likely to be ± 1, then it is possible to significantly relax the assumption on μ. Moreover, by requiring the nonzeros of x to exceed some minimum magnitude one can also guarantee perfect recovery of the true support.

1.5.3 Instance-optimal guarantees revisited

We now briefly return to the noise-free setting to take a closer look at instance-optimal guarantees for recovering non-sparse signals. To begin, recall that in Theorem 1.8 we bounded the ℓ_2-norm of the reconstruction error $\|\widehat{x} - x\|_2$ by a constant C_0 times $\sigma_k(x)_1/\sqrt{k}$. One can generalize this result to measure the reconstruction error using the ℓ_p-norm for any $p \in [1,2]$. For example, by a slight modification of these arguments, one can also show that $\|\widehat{x} - x\|_1 \leq C_0 \sigma_k(x)_1$ (see [34]). This leads us to ask whether we might replace the bound for the ℓ_2 error with a result of the form $\|\widehat{x} - x\|_2 \leq C\sigma_k(x)_2$. Unfortunately, obtaining such a result requires an unreasonably large number of measurements, as quantified by the following theorem of [57], proven in Section A.4.

THEOREM 1.13 (Theorem 5.1 of [57]) *Suppose that A is an $m \times n$ matrix and that $\Delta : \mathbb{R}^m \to \mathbb{R}^n$ is a recovery algorithm that satisfies*

$$\|x - \Delta(Ax)\|_2 \leq C\sigma_k(x)_2 \tag{1.16}$$

for some $k \geq 1$, then $m > \left(1 - \sqrt{1 - 1/C^2}\right)n$.

Thus, if we want a bound of the form (1.16) that holds for *all* signals x with a constant $C \approx 1$, then regardless of what recovery algorithm we use we will need to take $m \approx n$ measurements. However, in a sense this result is overly pessimistic, and we will now see that the results from Section 1.5.2 can actually allow us to overcome this limitation by essentially treating the approximation error as noise.

Towards this end, notice that all the results concerning ℓ_1 minimization stated thus far are deterministic instance-optimal guarantees that apply simultaneously to all x given any matrix that satisfies the RIP. This is an important theoretical property, but as noted in Section 1.4.4, in practice it is very difficult to obtain a deterministic guarantee that the matrix A satisfies the RIP. In particular, constructions that rely on randomness are only known to satisfy the RIP with high probability. As an example, Theorem 5.65 of Chapter 5 states that if a matrix A is chosen according to a sub-gaussian distribution with $m = O\left(k \log(n/k)/\delta_{2k}^2\right)$, then A will satisfy the RIP of order $2k$ with probability at least $1 - 2\exp(-c_1 \delta_{2k}^2 m)$. Results of this kind open the door to slightly weaker results that hold only with high probability.

Even within the class of probabilistic results, there are two distinct flavors. The typical approach is to combine a probabilistic construction of a matrix that will satisfy the RIP with high probability with the previous results in this chapter. This yields a procedure that, with high probability, will satisfy a deterministic guarantee applying to all possible signals x. A weaker kind of result is one that states that given a signal x, we can draw a random matrix A and with high probability expect certain performance *for that signal* x. This type of guarantee is sometimes called *instance-optimal in probability*. The distinction is essentially whether or not we need to draw a new random A for each signal x. This may be an important distinction in practice, but if we assume for the moment that it is permissible to draw a new matrix A for each x, then we can see that Theorem 1.13 may be somewhat pessimistic, exhibited by the following result.

THEOREM 1.14 *Let $x \in \mathbb{R}^n$ be fixed. Set $\delta_{2k} < \sqrt{2} - 1$. Suppose that A is an $m \times n$ sub-gaussian random matrix with $m = O\left(k \log(n/k)/\delta_{2k}^2\right)$. Suppose we obtain measurements of the form $y = Ax$. Set $\epsilon = 2\sigma_k(x)_2$. Then with probability exceeding $1 - 2\exp(-c_1 \delta_{2k}^2 m) - \exp(-c_0 m)$, when $\mathcal{B}(y) = \{z : \|Az - y\|_2 \leq \epsilon\}$, the solution $\widehat{x}$ to (1.12) obeys*

$$\|\widehat{x} - x\|_2 \leq \frac{8\sqrt{1 + \delta_{2k}} - (1 + \sqrt{2})\delta_{2k}}{1 - (1 + \sqrt{2})\delta_{2k}} \sigma_k(x)_2.$$

Proof. First we recall that, as noted above, from Theorem 5.65 of Chapter 5 we have that A will satisfy the RIP of order $2k$ with probability at least $1 - 2\exp(-c_1 \delta_{2k}^2 m)$. Next, let Λ denote the index set corresponding to the k entries of x with largest magnitude and write $x = x_\Lambda + x_{\Lambda^c}$. Since $x_\Lambda \in \Sigma_k$, we can write $Ax = Ax_\Lambda + Ax_{\Lambda^c} = Ax_\Lambda + e$. If A is sub-gaussian then Ax_{Λ^c} is also sub-gaussian (see Chapter 5 for details), and one can apply a similar result to (1.14) to obtain that with probability at least $1 - \exp(-c_0 m)$, $\|Ax_{\Lambda^c}\|_2 \leq 2\|x_{\Lambda^c}\|_2 = 2\sigma_k(x)_2$. Thus, applying the union bound we have that with probability exceeding $1 - 2\exp(-c_1 \delta_{2k}^2 m) - \exp(-c_0 m)$, we satisfy the necessary conditions to apply Theorem 1.9 to x_Λ, in which case $\sigma_k(x_\Lambda)_1 = 0$ and hence

$$\|\widehat{x} - x_\Lambda\|_2 \leq 2C_2 \sigma_k(x)_2.$$

From the triangle inequality we thus obtain

$$\|\widehat{x} - x\|_2 = \|\widehat{x} - x_\Lambda + x_\Lambda - x\|_2 \leq \|\widehat{x} - x_\Lambda\|_2 + \|x_\Lambda - x\|_2 \leq (2C_2 + 1)\sigma_k(x)_2$$

which establishes the theorem. □

Thus, while it is not possible to achieve a deterministic guarantee of the form in (1.16) without taking a prohibitively large number of measurements, it *is* possible to show that such performance guarantees can hold with high probability while simultaneously taking far fewer measurements than would be suggested by Theorem 1.13. Note that the above result applies only to the case where the parameter is selected correctly, which requires some limited knowledge of x, namely $\sigma_k(x)_2$. In practice this limitation can easily be overcome through a parameter selection technique such as cross-validation [243], but there also exist more intricate analyses of ℓ_1 minimization that show it is possible to obtain similar performance without requiring an oracle for parameter selection [248]. Note that Theorem 1.14 can also be generalized to handle other measurement matrices and to the case where x is compressible rather than sparse. Moreover, this proof technique is applicable to a variety of the greedy algorithms described in Chapter 8 that do not require knowledge of the noise level to establish similar results [56, 190].

1.5.4 The cross-polytope and phase transitions

While the RIP-based analysis of ℓ_1 minimization allows us to establish a variety of guarantees under different noise settings, one drawback is that the analysis of how many measurements are actually required for a matrix to satisfy the RIP is relatively loose. An alternative approach to analyzing ℓ_1 minimization algorithms is to examine them from a more geometric perspective. Towards this end, we define the closed ℓ_1 ball, also known as the *cross-polytope*:

$$C^n = \{x \in \mathbb{R}^n : \|x\|_1 \leq 1\}.$$

Note that C^n is the convex hull of $2n$ points $\{p_i\}_{i=1}^{2n}$. Let $AC^n \subseteq \mathbb{R}^m$ denote the convex polytope defined as either the convex hull of $\{Ap_i\}_{i=1}^{2n}$ or equivalently as

$$AC^n = \{y \in \mathbb{R}^m : y = Ax, x \in C^n\}.$$

For any $x \in \Sigma_k$, we can associate a k-face of C^n with the support and sign pattern of x. One can show that the number of k-faces of AC^n is precisely the number of index sets of size k for which signals supported on them can be recovered by (1.12) with $\mathcal{B}(y) = \{z : Az = y\}$. Thus, ℓ_1 minimization yields the same solution as ℓ_0 minimization for all $x \in \Sigma_k$ if and only if the number of k-faces of AC^n is identical to the number of k-faces of C^n. Moreover, by counting the number of k-faces of AC^n, we can quantify exactly what fraction of sparse vectors can be recovered using ℓ_1 minimization with A as our sensing matrix. See [81, 84, 92–94] for more details and [95] for an overview of the implications of this body of work. Note also that by replacing the cross-polytope with certain other polytopes (the simplex and the hypercube), one can apply the same technique to obtain results concerning the recovery of more limited signal classes, such as sparse signals with non-negative or bounded entries [95].

Given this result, one can then study random matrix constructions from this perspective to obtain probabilistic bounds on the number of k-faces of AC^n with A generated at random, such as from a Gaussian distribution. Under the assumption that $k = \rho m$ and $m = \gamma n$, one can obtain asymptotic results as $n \to \infty$. This analysis shows that there is a *phase transition*, i.e., for very large problem sizes there are sharp thresholds dictating that the fraction of k-faces preserved will tend to either one or zero with very high probability, depending on ρ and γ [95]. For the precise values of ρ and γ which will enable successful recovery and for further discussion of similar results, see Chapters 7 and 9.

These results provide sharp bounds on the minimum number of measurements required in the noiseless case. In general, these bounds are significantly stronger than the corresponding measurement bounds obtained within the RIP-based framework, which tend to be extremely loose in terms of the constants involved. However, these sharper bounds also require somewhat more intricate analysis and typically more restrictive assumptions on A (such as it being Gaussian). Thus, one of the main strengths of the RIP-based analysis presented in this chapter is that it gives results for a very broad class of matrices that can also be extended to noisy settings.

1.6 Signal recovery algorithms

We now discuss a number of algorithmic approaches to the problem of signal recovery from CS measurements. While this problem has received significant attention in recent years in the context of CS, many of these techniques pre-date the field of CS. There are a variety of algorithms that have been used in applications such as sparse approximation, statistics, geophysics, and theoretical computer science that were developed to exploit sparsity in other contexts and can be brought to bear on the CS recovery problem. We briefly review some of these, and refer the reader to later chapters as well as the overview in [226] for further details.

Note that we restrict our attention here to algorithms that actually reconstruct the original signal x. In some settings the end goal is to solve some kind of inference problem such as detection, classification, or parameter estimation, in which case a full reconstruction may not be necessary [69–71, 74, 100, 101, 143, 145].

ℓ_1 minimization algorithms

The ℓ_1 minimization approach analyzed in Section 1.5 provides a powerful framework for recovering sparse signals. The power of ℓ_1 minimization is that not only will it lead to a provably accurate recovery, but the formulations described in Section 1.5 are also convex optimization problems for which there exist efficient and accurate numerical solvers [194]. For example, (1.12) with $\mathcal{B}(y) = \{z : Az = y\}$ can be posed as a linear program. In the cases where $\mathcal{B}(y) = \{z : \|Az - y\|_2 \leq \epsilon\}$ or $\mathcal{B}(y) = \{z : \|A^T(Az - y)\|_\infty \leq \lambda\}$, the minimization problem (1.12) becomes a convex program with a conic constraint.

While these optimization problems could all be solved using general-purpose convex optimization software, there now also exist a tremendous variety of algorithms designed to explicitly solve these problems in the context of CS. This body of literature has primarily focused on the case where $\mathcal{B}(y) = \{z : \|Az - y\|_2 \leq \epsilon\}$. However, there exist multiple equivalent formulations of this program. For instance, the majority of ℓ_1 minimization algorithms in the literature have actually considered the unconstrained version

Introduction to compressed sensing

of this problem, i.e.,

$$\widehat{x} = \arg\min_{z} \frac{1}{2} \|Az - y\|_2^2 + \lambda \|z\|_1.$$

See, for example, [11, 120, 122, 138, 175, 197, 246, 249–251]. Note that for some choice of the parameter λ this optimization problem will yield the same result as the constrained version of the problem given by

$$\widehat{x} = \arg\min_{z} \|z\|_1 \quad \text{subject to} \quad \|Az - y\|_2 \leq \epsilon.$$

However, in general the value of λ which makes these problems equivalent is unknown a priori. Several approaches for choosing λ are discussed in [110, 123, 133]. Since in many settings ϵ is a more natural parameterization (being determined by the noise or quantization level), it is also useful to have algorithms that directly solve the latter formulation. While there are fewer efforts in this direction, there also exist some excellent solvers for this problem [12, 13, 231]. Note that [13] also provides solvers for a variety of other ℓ_1 minimization problems, such as for the Dantzig selector.

Greedy algorithms

While convex optimization techniques are powerful methods for computing sparse representations, there is also a variety of greedy/iterative methods for solving such problems [21, 23, 24, 56, 64, 66, 75, 85, 96, 153, 182, 183, 190–192, 220, 222, 223]. Greedy algorithms rely on iterative approximation of the signal coefficients and support, either by iteratively identifying the support of the signal until a convergence criterion is met, or alternatively by obtaining an improved estimate of the sparse signal at each iteration that attempts to account for the mismatch to the measured data. Some greedy methods can actually be shown to have performance guarantees that match those obtained for convex optimization approaches. In fact, some of the more sophisticated greedy algorithms are remarkably similar to those used for ℓ_1 minimization described above. However, the techniques required to prove performance guarantees are substantially different.

We refer the reader to Chapter 8 for a more detailed overview of greedy algorithms and their performance. Here we briefly highlight some of the most common methods and their theoretical guarantees. Two of the oldest and simplest greedy approaches are *Orthogonal Matching Pursuit* (OMP) and *iterative thresholding*. We first consider OMP [183], which begins by finding the column of A most correlated with the measurements. The algorithm then repeats this step by correlating the columns with the signal residual, which is obtained by subtracting the contribution of a partial estimate of the signal from the original measurement vector. The algorithm is formally defined as Algorithm 1.1, where $H_k(x)$ denotes the *hard thresholding* operator on x that sets all entries to zero except for the k entries of x with largest magnitude. The stopping criterion can consist of either a limit on the number of iterations, which also limits the number of nonzeros in $\widehat{x}$, or a requirement that $y \approx A\widehat{x}$ in some sense. Note that in either case, if OMP runs for m iterations then it will always produce an estimate $\widehat{x}$ such that $y = A\widehat{x}$. Iterative thresholding algorithms are often even more straightforward. For an overview see [107]. As an example, we consider *Iterative Hard Thresholding* (IHT) [24], which is described

---
Algorithm 1.1 Orthogonal Matching Pursuit
Inputs: CS matrix/dictionary A, measurement vector y
Initialize: $\widehat{x}_0 = 0$, $r_0 = y$, $\Lambda_0 = \emptyset$.
for $i = 1; i := i+1$ until stopping criterion is met **do**
    $g_i \leftarrow A^T r_{i-1}$ {form signal estimate from residual}
    $\Lambda_i \leftarrow \Lambda_{i-1} \cup \mathrm{supp}(H_1(g_i))$ {add largest residual entry to support}
    $\widehat{x}_i|_{\Lambda_i} \leftarrow A^\dagger_{\Lambda_i} y$, $\widehat{x}_i|_{\Lambda_i^c} \leftarrow 0$ {update signal estimate}
    $r_i \leftarrow y - A\widehat{x}_i$ {update measurement residual}
end for
Output: Sparse representation $\widehat{x}$

Algorithm 1.2 Iterative Hard Thresholding
Inputs: CS matrix/dictionary A, measurement vector y, sparsity level k
Initialize: $\widehat{x}_0 = 0$.
for $i = 1; i := i+1$ until stopping criterion is met **do**
    $\widehat{x}_i = H_k\left(\widehat{x}_{i-1} + A^T(y - A\widehat{x}_{i-1})\right)$
end for
Output: Sparse representation $\widehat{x}$
---

in Algorithm 1.2. Starting from an initial signal estimate $\widehat{x}_0 = 0$, the algorithm iterates a gradient descent step followed by hard thresholding until a convergence criterion is met.

OMP and IHT both satisfy many of the same guarantees as ℓ_1 minimization. For example, under a slightly stronger assumption on the RIP constant, iterative hard thresholding satisfies a very similar guarantee to that of Theorem 1.9. We refer the reader to Chapter 8 for further details on the theoretical properties of thresholding algorithms, and focus here on OMP.

The simplest guarantees for OMP state that for exactly k-sparse x with noise-free measurements $y = Ax$, OMP will recover x exactly in k iterations. This analysis has been performed for both matrices satisfying the RIP [75] and matrices with bounded coherence [220]. In both results, however, the required constants are relatively small, so that the results only apply when $m = O(k^2 \log(n))$.

There have been many efforts to improve upon these basic results. As one example, in [173] the required number of measurements is reduced to $m = O(k^{1.6} \log(n))$ by allowing OMP to run for more than k iterations. More recently, it has been shown that this can be even further relaxed to the more familiar $m = O(k \log(n))$ and that OMP is stable with respect to bounded noise, yielding a guarantee along the lines of Theorem 1.9 but only for exactly sparse signals [254]. Both of these analyses have exploited the RIP. There has also been recent progress in using the RIP to analyze the performance of OMP on non-sparse signals [10]. At present, however, RIP-based analysis of OMP remains a topic of ongoing work.

Note that all of the above efforts have aimed at establishing uniform guarantees (although often restricted to exactly sparse signals). In light of our discussion of probabilistic guarantees in Section 1.5.3, one might expect to see improvements by considering less restrictive guarantees. As an example, it has been shown that by considering random matrices for A OMP can recover k-sparse signals in k iterations with high probability using only $m = O(k \log(n))$ measurements [222]. Similar improvements are also possible by placing restrictions on the smallest nonzero value of the signal, as in [88]. Furthermore, such restrictions also enable near-optimal recovery guarantees when the measurements are corrupted by Gaussian noise [18].

Combinatorial algorithms

In addition to ℓ_1 minimization and greedy algorithms, there is another important class of sparse recovery algorithms that we will refer to as *combinatorial algorithms*. These algorithms, mostly developed by the theoretical computer science community, in many cases pre-date the compressive sensing literature but are highly relevant to the sparse signal recovery problem.

The historically oldest of these algorithms were developed in the context of *combinatorial group testing* [98, 116, 160, 210]. In this problem we suppose that there are n total items and k anomalous elements that we wish to find. For example, we might wish to identify defective products in an industrial setting, or identify a subset of diseased tissue samples in a medical context. In both of these cases the vector x indicates which elements are anomalous, i.e., $x_i \neq 0$ for the k anomalous elements and $x_i = 0$ otherwise. Our goal is to design a collection of tests that allow us to identify the support (and possibly the values of the nonzeros) of x while also minimizing the number of tests performed. In the simplest practical setting these tests are represented by a binary matrix A whose entries a_{ij} are equal to 1 if and only if the jth item is used in the ith test. If the output of the test is linear with respect to the inputs, then the problem of recovering the vector x is essentially the same as the standard sparse recovery problem in CS.

Another application area in which combinatorial algorithms have proven useful is computation on *data streams* [59, 189]. As an example of a typical data streaming problem, suppose that x_i represents the number of packets passing through a network router with destination i. Simply storing the vector x is typically infeasible since the total number of possible destinations (represented by a 32-bit IP address) is $n = 2^{32}$. Thus, instead of attempting to store x directly, one can store $y = Ax$ where A is an $m \times n$ matrix with $m \ll n$. In this context the vector y is often called a *sketch*. Note that in this problem y is computed in a different manner than in the compressive sensing context. Specifically, in the network traffic example we do not ever observe x_i directly, rather we observe increments to x_i (when a packet with destination i passes through the router). Thus we construct y iteratively by adding the ith column to y each time we observe an increment to x_i, which we can do since $y = Ax$ is linear. When the network traffic is dominated by traffic to a small number of destinations, the vector x is compressible, and thus the problem of recovering x from the sketch Ax is again essentially the same as the sparse recovery problem in CS.

Despite the fact that in both of these settings we ultimately wish to recover a sparse signal from a small number of linear measurements, there are also some important differences between these settings and CS. First, in these settings it is natural to assume that the designer of the reconstruction algorithm also has full control over A, and is thus free to choose A in a manner that reduces the amount of computation required to perform recovery. For example, it is often useful to design A so that it has very few nonzeros, i.e., the sensing matrix itself is also sparse [8, 128, 154]. In general, most methods involve careful construction of the sampling matrix A (although some schemes do involve "generic" sparse matrices, for example, see [20]). This is in contrast with the optimization and greedy methods that work with any matrix satisfying the conditions described in Section 1.4. Of course, this additional structure can often lead to significantly faster algorithms [51, 60, 129, 130].

Second, note that the computational complexity of all the convex methods and greedy algorithms described above is always at least linear in terms of n, since in order to recover x we must at least incur the computational cost of reading out all n entries of x. While this may be acceptable in most typical CS applications, this becomes impractical when n is extremely large, as in the network monitoring example. In this context, one may seek to develop algorithms whose complexity is linear only in the *length of the representation* of the signal, i.e., its sparsity k. In this case the algorithm does not return a complete reconstruction of x but instead returns only its k largest elements (and their indices). As surprising as it may seem, such algorithms are indeed possible. See [60, 129, 130] for examples.

1.7 Multiple measurement vectors

Many potential applications of CS involve distributed acquisition of multiple correlated signals. The multiple signal case where all l signals involved are sparse and exhibit the same indices for their nonzero coefficients is well known in sparse approximation literature, where it has been termed the multiple measurement vector (MMV) problem [52, 63, 134, 185, 221, 223, 232]. In the MMV setting, rather than trying to recover each single sparse vector x_i independently for $i = 1, \ldots, l$, the goal is to jointly recover the set of vectors by exploiting their common sparse support. Stacking these vectors into the columns of a matrix X, there will be at most k non-zero rows in X. That is, not only is each vector k-sparse, but the non-zero values occur on a common location set. We therefore say that X is *row-sparse* and use the notation $\Lambda = \mathrm{supp}(X)$ to denote the index set corresponding to non-zero rows.[7]

MMV problems appear quite naturally in many different application areas. Early work on MMV algorithms focused on magnetoencephalography, which is a modality for imaging the brain [134, 135, 200]. Similar ideas were also developed in the context

---

[7] The MMV problem can be converted into a block-sparse recovery problem through appropriate rasterizing of the matrix X and the construction of a single matrix $A' \in \mathbb{R}^{lm \times ln}$ dependent on the matrix $A \in \mathbb{R}^{m \times n}$ used for each of the signals.

of array processing [135, 157, 181], equalization of sparse communication channels [2, 62, 119, 142], and more recently cognitive radio and multiband communications [9, 114, 186–188, 252].

Conditions on measurement matrices

As in standard CS, we assume that we are given measurements $\{y_i\}_{i=1}^{l}$ where each vector is of length $m < n$. Letting Y be the $m \times l$ matrix with columns y_i, our problem is to recover X assuming a known measurement matrix A so that $Y = AX$. Clearly, we can apply any CS method to recover x_i from y_i as before. However, since the vectors $\{x_i\}$ all have a common support, we expect intuitively to improve the recovery ability by exploiting this joint information. In other words, we should in general be able to reduce the number of measurements ml needed to represent X below sl, where s is the number of measurements required to recover one vector x_i for a given matrix A.

Since $|\Lambda| = k$, the rank of X satisfies $\mathrm{rank}(X) \leq k$. When $\mathrm{rank}(X) = 1$, all the sparse vectors x_i are multiples of each other, so that there is no advantage to their joint processing. However, when $\mathrm{rank}(X)$ is large, we expect to be able to exploit the diversity in its columns in order to benefit from joint recovery. This essential result is captured nicely by the following necessary and sufficient uniqueness condition:

THEOREM 1.15 (Theorem 2 of [76]) *A necessary and sufficient condition for the measurements $Y = AX$ to uniquely determine the row sparse matrix X is that*

$$|\mathrm{supp}(X)| < \frac{\mathrm{spark}(A) - 1 + \mathrm{rank}(X)}{2}. \tag{1.17}$$

As shown in [76], we can replace $\mathrm{rank}(X)$ by $\mathrm{rank}(Y)$ in (1.17). The sufficient direction of this condition was shown in [185] to hold even in the case where there are infinitely many vectors x_i. A direct consequence of Theorem 1.15 is that matrices X with larger rank can be recovered from fewer measurements. Alternatively, matrices X with larger support can be recovered from the same number of measurements. When $\mathrm{rank}(X) = k$ and $\mathrm{spark}(A)$ takes on its largest possible value equal to $m + 1$, condition (1.17) becomes $m \geq k + 1$. Therefore, in this best-case scenario, only $k + 1$ measurements per signal are needed to ensure uniqueness. This is much lower than the value of $2k$ obtained in standard CS via the spark (cf. Theorem 1.7), which we refer to here as the single measurement vector (SMV) setting. Furthermore, when X is full rank, it can be recovered by a simple algorithm, in contrast to the combinatorial complexity needed to solve the SMV problem from $2k$ measurements for general matrices A. See Chapter 8 for more details.

Recovery algorithms

A variety of algorithms have been proposed that exploit the joint sparsity in different ways when X is not full rank. As in the SMV setting, two main approaches to solving MMV problems are based on convex optimization and greedy methods. The analog of (1.10) in the MMV case is

$$\widehat{X} = \arg \min_{X \in \mathbb{R}^{n \times l}} \|X\|_{p,0} \text{ subject to } Y = AX, \tag{1.18}$$

where we define the $\ell_{p,q}$ norms for matrices as

$$\|X\|_{p,q} = \left(\sum_i \|x^i\|_p^q \right)^{1/q}$$

with x^i denoting the ith row of X. With a slight abuse of notation, we also consider the $q=0$ case where $\|X\|_{p,0} = |\mathrm{supp}(X)|$ for any p. Optimization-based algorithms relax the ℓ_0 norm in (1.18) and attempt to recover X by mixed norm minimization:

$$\widehat{X} = \arg \min_{X \in \mathbb{R}^{n \times l}} \|X\|_{p,q} \text{ subject to } Y = AX$$

for some $p, q \geq 1$; values for p and q of 1, 2, and ∞ have been advocated [52, 63, 114, 121, 221, 223]. The standard greedy approaches in the SMV setting have also been extended to the MMV case; see Chapter 8 for more details. Furthermore, one can also reduce the MMV problem into an SMV problem and solve using standard CS recovery algorithms [185]. This reduction can be particularly beneficial in large-scale problems, such as those resulting from analog sampling.

Multiple measurement vector models can also be used to perform blind CS, in which the sparsifying basis is learned together with the representation coefficients [131]. While all standard CS algorithms assume that the sparsity basis is known in the recovery process, blind CS does not require this knowledge. When multiple measurements are available it can be shown that under certain conditions on the sparsity basis, blind CS is possible thus avoiding the need to know the sparsity basis in both the sampling and the recovery process.

In terms of theoretical guarantees, it can be shown that MMV extensions of SMV algorithms will recover X under similar conditions to the SMV setting in the worst-case scenario [4, 52, 114, 115] so that theoretical equivalence results for arbitrary values of X do not predict any performance gain with joint sparsity. In practice, however, multichannel reconstruction techniques perform much better than recovering each channel individually. The reason for this discrepancy is that these results apply to all possible input signals, and are therefore worst-case measures. Clearly, if we input the same signal to each channel, namely when $\mathrm{rank}(X) = 1$, no additional information on the joint support is provided from multiple measurements. However, as we have seen in Theorem 1.15, higher ranks of the input X improve the recovery ability.

Another way to improve performance guarantees is by considering random values of X and developing conditions under which X is recovered with high probability [7, 115, 137, 208]. Average case analysis can be used to show that fewer measurements are needed in order to recover X exactly [115]. In addition, under a mild condition on the sparsity and on the matrix A, the failure probability decays exponentially in the number of channels l [115].

Finally, we note that algorithms similar to those used for MMV recovery can also be adapted to block-sparse reconstruction [112, 114, 253].

1.8 Summary

Compressed sensing is an exciting, rapidly growing field that has attracted considerable attention in signal processing, statistics, and computer science, as well as the broader scientific community. Since its initial development, only a few years ago, thousands of papers have appeared in this area, and hundreds of conferences, workshops, and special sessions have been dedicated to this growing research field. In this chapter, we have reviewed some of the basics of the theory underlying CS. We have also aimed, throughout our summary, to highlight new directions and application areas that are at the frontier of CS research. This chapter should serve as a review to practitioners wanting to join this emerging field, and as a reference for researchers. Our hope is that this presentation will attract the interest of both mathematicians and engineers in the desire to encourage further research into this new frontier as well as promote the use of CS in practical applications. In subsequent chapters of the book, we will see how the fundamentals presented in this chapter are expanded and extended in many exciting directions, including new models for describing structure in both analog and discrete-time signals, new sensing design techniques, more advanced recovery results and powerful new recovery algorithms, and emerging applications of the basic theory and its extensions.

Acknowledgements

The authors would like to thank Ewout van den Berg, Piotr Indyk, Yaniv Plan, and the authors contributing to this book for their valuable feedback on a preliminary version of this manuscript.

A Appendix: proofs for Chapter 1

A.1 Proof of Theorem 1.4

To prove Theorem 1.4 we first provide a preliminary lemma. The proof of this result is based on techniques from [166].

LEMMA A.1 *Let k and n satisfying $k < n/2$ be given. There exists a set $X \subset \Sigma_k$ such that for any $x \in X$ we have $\|x\|_2 \leq \sqrt{k}$ and for any $x, z \in X$ with $x \neq z$*

$$\|x - z\|_2 \geq \sqrt{k/2} \tag{A.1}$$

and

$$\log |X| \geq \frac{k}{2} \log\left(\frac{n}{k}\right).$$

Proof. We will begin by considering the set

$$U = \{x \in \{0, +1, -1\}^n : \|x\|_0 = k\}.$$

By construction, $\|x\|_2^2 = k$ for all $x \in U$. Thus if we construct X by picking elements from U then we automatically have $\|x\|_2 \leq \sqrt{k}$.

Next, observe that $|U| = \binom{n}{k} 2^k$. Note also that if $x, z \in U$ then $\|x - z\|_0 \leq \|x - z\|_2^2$, and thus if $\|x - z\|_2^2 \leq k/2$ then $\|x - z\|_0 \leq k/2$. From this we observe that for any fixed $x \in U$,

$$\left| \{z \in U : \|x - z\|_2^2 \leq k/2\} \right| \leq |\{z \in U : \|x - z\|_0 \leq k/2\}| \leq \binom{n}{k/2} 3^{k/2}.$$

Thus, suppose we construct the set X by iteratively choosing points that satisfy (A.1). After adding j points to the set, there are at least

$$\binom{n}{k} 2^k - j \binom{n}{k/2} 3^{k/2}$$

points left to pick from. Thus, we can construct a set of size $|X|$ provided that

$$|X| \binom{n}{k/2} 3^{k/2} \leq \binom{n}{k} 2^k. \quad (A.2)$$

Next, observe that

$$\frac{\binom{n}{k}}{\binom{n}{k/2}} = \frac{(k/2)!(n - k/2)!}{k!(n - k)!} = \prod_{i=1}^{k/2} \frac{n - k + i}{k/2 + i} \geq \left(\frac{n}{k} - \frac{1}{2}\right)^{k/2},$$

where the inequality follows from the fact that $(n - k + i)/(k/2 + i)$ is decreasing as a function of i. Thus, if we set $|X| = (n/k)^{k/2}$ then we have

$$|X| \left(\frac{3}{4}\right)^{k/2} = \left(\frac{3n}{4k}\right)^{k/2} = \left(\frac{n}{k} - \frac{n}{4k}\right)^{k/2} \leq \left(\frac{n}{k} - \frac{1}{2}\right)^{k/2} \leq \frac{\binom{n}{k}}{\binom{n}{k/2}}.$$

Hence, (A.2) holds for $|X| = (n/k)^{k/2}$, which establishes the lemma. $\square$

Using this lemma, we can establish Theorem 1.4.

THEOREM 1.4 (Theorem 3.5 of [67]) *Let A be an $m \times n$ matrix that satisfies the RIP of order $2k$ with constant $\delta_{2k} \in (0, 1/2]$. Then*

$$m \geq Ck \log\left(\frac{n}{k}\right)$$

where $C = 1/2 \log(\sqrt{24} + 1) \approx 0.28$.

Proof. We first note that since A satisfies the RIP, then for the set of points X in Lemma A.1 we have,

$$\|Ax - Az\|_2 \geq \sqrt{1 - \delta_{2k}} \|x - z\|_2 \geq \sqrt{k/4}$$

for all $x, z \in X$, since $x - z \in \Sigma_{2k}$ and $\delta_{2k} \leq 1/2$. Similarly, we also have

$$\|Ax\|_2 \leq \sqrt{1+\delta_{2k}} \|x\|_2 \leq \sqrt{3k/2}$$

for all $x \in X$.

From the lower bound we can say that for any pair of points $x, z \in X$, if we center balls of radius $\sqrt{k/4}/2 = \sqrt{k/16}$ at Ax and Az, then these balls will be disjoint. In turn, the upper bound tells us that the entire set of balls is itself contained within a larger ball of radius $\sqrt{3k/2} + \sqrt{k/16}$. If we let $B^m(r) = \{x \in \mathbb{R}^m : \|x\|_2 \leq r\}$, then

$$\mathrm{Vol}\left(B^m\left(\sqrt{3k/2} + \sqrt{k/16}\right)\right) \geq |X| \cdot \mathrm{Vol}\left(B^m\left(\sqrt{k/16}\right)\right),$$

$$\Leftrightarrow \left(\sqrt{3k/2} + \sqrt{k/16}\right)^m \geq |X| \cdot \left(\sqrt{k/16}\right)^m,$$

$$\Leftrightarrow \left(\sqrt{24} + 1\right)^m \geq |X|,$$

$$\Leftrightarrow m \geq \frac{\log |X|}{\log\left(\sqrt{24} + 1\right)}.$$

The theorem follows by applying the bound for $|X|$ from Lemma A.1. □

A.2 Proof of Lemma 1.3

To begin, we establish the following preliminary lemmas.

LEMMA A.2 *Suppose u, v are orthogonal vectors. Then*

$$\|u\|_2 + \|v\|_2 \leq \sqrt{2} \|u+v\|_2.$$

Proof. We begin by defining the 2×1 vector $w = [\|u\|_2, \|v\|_2]^T$. By applying Lemma 1.2 with $k = 2$, we have $\|w\|_1 \leq \sqrt{2} \|w\|_2$. From this we obtain

$$\|u\|_2 + \|v\|_2 \leq \sqrt{2} \sqrt{\|u\|_2^2 + \|v\|_2^2}.$$

Since u and v are orthogonal, $\|u\|_2^2 + \|v\|_2^2 = \|u+v\|_2^2$, which yields the desired result. □

LEMMA A.3 *If A satisfies the RIP of order $2k$, then for any pair of vectors $u, v \in \Sigma_k$ with disjoint support,*

$$|\langle Au, Av \rangle| \leq \delta_{2k} \|u\|_2 \|v\|_2.$$

Proof. Suppose $u, v \in \Sigma_k$ with disjoint support and that $\|u\|_2 = \|v\|_2 = 1$. Then, $u \pm v \in \Sigma_{2k}$ and $\|u \pm v\|_2^2 = 2$. Using the RIP we have

$$2(1-\delta_{2k}) \leq \|Au \pm Av\|_2^2 \leq 2(1+\delta_{2k}).$$

Finally, applying the parallelogram identity

$$|\langle Au, Av\rangle| \le \frac{1}{4}\left|\|Au+Av\|_2^2 - \|Au-Av\|_2^2\right| \le \delta_{2k}$$

establishes the lemma. □

LEMMA A.4 *Let Λ_0 be an arbitrary subset of $\{1,2,\dots,n\}$ such that $|\Lambda_0| \le k$. For any vector $u \in \mathbb{R}^n$, define Λ_1 as the index set corresponding to the k largest entries of $u_{\Lambda_0^c}$ (in absolute value), Λ_2 as the index set corresponding to the next k largest entries, and so on. Then*

$$\sum_{j\ge 2}\|u_{\Lambda_j}\|_2 \le \frac{\|u_{\Lambda_0^c}\|_1}{\sqrt{k}}.$$

Proof. We begin by observing that for $j \ge 2$,

$$\|u_{\Lambda_j}\|_\infty \le \frac{\|u_{\Lambda_{j-1}}\|_1}{k}$$

since the Λ_j sort u to have decreasing magnitude. Applying Lemma 1.2 we have

$$\sum_{j\ge 2}\|u_{\Lambda_j}\|_2 \le \sqrt{k}\sum_{j\ge 2}\|u_{\Lambda_j}\|_\infty \le \frac{1}{\sqrt{k}}\sum_{j\ge 1}\|u_{\Lambda_j}\|_1 = \frac{\|u_{\Lambda_0^c}\|_1}{\sqrt{k}},$$

proving the lemma. □

We are now in a position to prove Lemma 1.3. The key ideas in this proof follow from [34].

LEMMA 1.3 *Suppose that A satisfies the RIP of order $2k$. Let Λ_0 be an arbitrary subset of $\{1,2,\dots,n\}$ such that $|\Lambda_0| \le k$, and let $h \in \mathbb{R}^n$ be given. Define Λ_1 as the index set corresponding to the k entries of $h_{\Lambda_0^c}$ with largest magnitude, and set $\Lambda = \Lambda_0 \cup \Lambda_1$. Then*

$$\|h_\Lambda\|_2 \le \alpha\frac{\|h_{\Lambda_0^c}\|_1}{\sqrt{k}} + \beta\frac{|\langle Ah_\Lambda, Ah\rangle|}{\|h_\Lambda\|_2},$$

where

$$\alpha = \frac{\sqrt{2}\delta_{2k}}{1-\delta_{2k}}, \quad \beta = \frac{1}{1-\delta_{2k}}.$$

Proof. Since $h_\Lambda \in \Sigma_{2k}$, the lower bound of the RIP immediately yields

$$(1-\delta_{2k})\|h_\Lambda\|_2^2 \le \|Ah_\Lambda\|_2^2. \tag{A.3}$$

Define Λ_j as in Lemma A.4, then since $Ah_\Lambda = Ah - \sum_{j\ge 2}Ah_{\Lambda_j}$, we can rewrite (A.3) as

$$(1-\delta_{2k})\|h_\Lambda\|_2^2 \le \langle Ah_\Lambda, Ah\rangle - \left\langle Ah_\Lambda, \sum_{j\ge 2}Ah_{\Lambda_j}\right\rangle. \tag{A.4}$$

In order to bound the second term of (A.4), we use Lemma A.3, which implies that

$$|\langle Ah_{\Lambda_i}, Ah_{\Lambda_j}\rangle| \leq \delta_{2k}\|h_{\Lambda_i}\|_2\|h_{\Lambda_j}\|_2, \qquad (A.5)$$

for any i,j. Furthermore, Lemma A.2 yields $\|h_{\Lambda_0}\|_2+\|h_{\Lambda_1}\|_2 \leq \sqrt{2}\|h_\Lambda\|_2$. Substituting into (A.5) we obtain

$$\left|\langle Ah_\Lambda, \sum_{j\geq 2} Ah_{\Lambda_j}\rangle\right| = \left|\sum_{j\geq 2}\langle Ah_{\Lambda_0}, Ah_{\Lambda_j}\rangle + \sum_{j\geq 2}\langle Ah_{\Lambda_1}, Ah_{\Lambda_j}\rangle\right|$$

$$\leq \sum_{j\geq 2}|\langle Ah_{\Lambda_0}, Ah_{\Lambda_j}\rangle| + \sum_{j\geq 2}|\langle Ah_{\Lambda_1}, Ah_{\Lambda_j}\rangle|$$

$$\leq \delta_{2k}\|h_{\Lambda_0}\|_2 \sum_{j\geq 2}\|h_{\Lambda_j}\|_2 + \delta_{2k}\|h_{\Lambda_1}\|_2 \sum_{j\geq 2}\|h_{\Lambda_j}\|_2$$

$$\leq \sqrt{2}\delta_{2k}\|h_\Lambda\|_2 \sum_{j\geq 2}\|h_{\Lambda_j}\|_2.$$

From Lemma A.4, this reduces to

$$\left|\langle Ah_\Lambda, \sum_{j\geq 2} Ah_{\Lambda_j}\rangle\right| \leq \sqrt{2}\delta_{2k}\|h_\Lambda\|_2 \frac{\|h_{\Lambda_0^c}\|_1}{\sqrt{k}}. \qquad (A.6)$$

Combining (A.6) with (A.4) we obtain

$$(1-\delta_{2k})\|h_\Lambda\|_2^2 \leq \left|\langle Ah_\Lambda, Ah\rangle - \langle Ah_\Lambda, \sum_{j\geq 2} Ah_{\Lambda_j}\rangle\right|$$

$$\leq |\langle Ah_\Lambda, Ah\rangle| + \left|\langle Ah_\Lambda, \sum_{j\geq 2} Ah_{\Lambda_j}\rangle\right|$$

$$\leq |\langle Ah_\Lambda, Ah\rangle| + \sqrt{2}\delta_{2k}\|h_\Lambda\|_2 \frac{\|h_{\Lambda_0^c}\|_1}{\sqrt{k}},$$

which yields the desired result upon rearranging. □

A.3 Proof of Lemma 1.6

We now return to the proof of Lemma 1.6. The key ideas in this proof follow from [34].

LEMMA 1.6 *Suppose that A satisfies the RIP of order $2k$ with $\delta_{2k} < \sqrt{2}-1$. Let $x,\widehat{x} \in \mathbb{R}^n$ be given, and define $h = \widehat{x} - x$. Let Λ_0 denote the index set corresponding to the k entries of x with largest magnitude and Λ_1 the index set corresponding to the k*

entries of $h_{\Lambda_0^c}$ with largest magnitude. Set $\Lambda = \Lambda_0 \cup \Lambda_1$. If $\|\widehat{x}\|_1 \leq \|x\|_1$, then

$$\|h\|_2 \leq C_0 \frac{\sigma_k(x)_1}{\sqrt{k}} + C_1 \frac{|\langle Ah_\Lambda, Ah\rangle|}{\|h_\Lambda\|_2},$$

where

$$C_0 = 2\frac{1-(1-\sqrt{2})\delta_{2k}}{1-(1+\sqrt{2})\delta_{2k}}, \quad C_1 = \frac{2}{1-(1+\sqrt{2})\delta_{2k}}.$$

Proof. We begin by observing that $h = h_\Lambda + h_{\Lambda^c}$, so that from the triangle inequality

$$\|h\|_2 \leq \|h_\Lambda\|_2 + \|h_{\Lambda^c}\|_2. \tag{A.7}$$

We first aim to bound $\|h_{\Lambda^c}\|_2$. From Lemma A.4 we have

$$\|h_{\Lambda^c}\|_2 = \left\|\sum_{j\geq 2} h_{\Lambda_j}\right\|_2 \leq \sum_{j\geq 2} \|h_{\Lambda_j}\|_2 \leq \frac{\|h_{\Lambda_0^c}\|_1}{\sqrt{k}}, \tag{A.8}$$

where the Λ_j are defined as in Lemma A.4, i.e., Λ_1 is the index set corresponding to the k largest entries of $h_{\Lambda_0^c}$ (in absolute value), Λ_2 as the index set corresponding to the next k largest entries, and so on.

We now wish to bound $\|h_{\Lambda_0^c}\|_1$. Since $\|x\|_1 \geq \|\widehat{x}\|_1$, by applying the triangle inequality we obtain

$$\|x\|_1 \geq \|x+h\|_1 = \|x_{\Lambda_0} + h_{\Lambda_0}\|_1 + \|x_{\Lambda_0^c} + h_{\Lambda_0^c}\|_1$$
$$\geq \|x_{\Lambda_0}\|_1 - \|h_{\Lambda_0}\|_1 + \|h_{\Lambda_0^c}\|_1 - \|x_{\Lambda_0^c}\|_1.$$

Rearranging and again applying the triangle inequality,

$$\|h_{\Lambda_0^c}\|_1 \leq \|x\|_1 - \|x_{\Lambda_0}\|_1 + \|h_{\Lambda_0}\|_1 + \|x_{\Lambda_0^c}\|_1$$
$$\leq \|x - x_{\Lambda_0}\|_1 + \|h_{\Lambda_0}\|_1 + \|x_{\Lambda_0^c}\|_1.$$

Recalling that $\sigma_k(x)_1 = \|x_{\Lambda_0^c}\|_1 = \|x - x_{\Lambda_0}\|_1$,

$$\|h_{\Lambda_0^c}\|_1 \leq \|h_{\Lambda_0}\|_1 + 2\sigma_k(x)_1. \tag{A.9}$$

Combining this with (A.8) we obtain

$$\|h_{\Lambda^c}\|_2 \leq \frac{\|h_{\Lambda_0}\|_1 + 2\sigma_k(x)_1}{\sqrt{k}} \leq \|h_{\Lambda_0}\|_2 + 2\frac{\sigma_k(x)_1}{\sqrt{k}}$$

where the last inequality follows from Lemma 1.2. By observing that $\|h_{\Lambda_0}\|_2 \leq \|h_\Lambda\|_2$ this combines with (A.7) to yield

$$\|h\|_2 \leq 2\|h_\Lambda\|_2 + 2\frac{\sigma_k(x)_1}{\sqrt{k}}. \tag{A.10}$$

We now turn to establishing a bound for $\|h_\Lambda\|_2$. Combining Lemma 1.3 with (A.9) and applying Lemma 1.2 we obtain

$$\|h_\Lambda\|_2 \leq \alpha \frac{\|h_{\Lambda_0^c}\|_1}{\sqrt{k}} + \beta \frac{|\langle Ah_\Lambda, Ah\rangle|}{\|h_\Lambda\|_2}$$

$$\leq \alpha \frac{\|h_{\Lambda_0}\|_1 + 2\sigma_k(x)_1}{\sqrt{k}} + \beta \frac{|\langle Ah_\Lambda, Ah\rangle|}{\|h_\Lambda\|_2}$$

$$\leq \alpha \|h_{\Lambda_0}\|_2 + 2\alpha \frac{\sigma_k(x)_1}{\sqrt{k}} + \beta \frac{|\langle Ah_\Lambda, Ah\rangle|}{\|h_\Lambda\|_2}.$$

Since $\|h_{\Lambda_0}\|_2 \leq \|h_\Lambda\|_2$,

$$(1-\alpha)\|h_\Lambda\|_2 \leq 2\alpha \frac{\sigma_k(x)_1}{\sqrt{k}} + \beta \frac{|\langle Ah_\Lambda, Ah\rangle|}{\|h_\Lambda\|_2}.$$

The assumption that $\delta_{2k} < \sqrt{2}-1$ ensures that $\alpha < 1$. Dividing by $(1-\alpha)$ and combining with (A.10) results in

$$\|h\|_2 \leq \left(\frac{4\alpha}{1-\alpha} + 2\right) \frac{\sigma_k(x)_1}{\sqrt{k}} + \frac{2\beta}{1-\alpha} \frac{|\langle Ah_\Lambda, Ah\rangle|}{\|h_\Lambda\|_2}.$$

Plugging in for α and β yields the desired constants. □

A.4 Proof of Theorem 1.13

THEOREM 1.13 (Theorem 5.1 of [57]) *Suppose that A is an $m \times n$ matrix and that $\Delta : \mathbb{R}^m \to \mathbb{R}^n$ is a recovery algorithm that satisfies*

$$\|x - \Delta(Ax)\|_2 \leq C\sigma_k(x)_2 \tag{A.11}$$

for some $k \geq 1$, then $m > \left(1 - \sqrt{1-1/C^2}\right)n$.

Proof. We begin by letting $h \in \mathbb{R}^n$ denote any vector in $\mathcal{N}(A)$. We write $h = h_\Lambda + h_{\Lambda^c}$ where Λ is an arbitrary set of indices satisfying $|\Lambda| \leq k$. Set $x = h_{\Lambda^c}$, and note that $Ax = Ah_{\Lambda^c} = Ah - Ah_\Lambda = -Ah_\Lambda$, since $h \in \mathcal{N}(A)$. Since $h_\Lambda \in \Sigma_k$, (A.11) implies that $\Delta(Ax) = \Delta(-Ah_\Lambda) = -h_\Lambda$. Hence, $\|x - \Delta(Ax)\|_2 = \|h_{\Lambda^c} - (-h_\Lambda)\|_2 = \|h\|_2$. Furthermore, we observe that $\sigma_k(x)_2 \leq \|x\|_2$, since by definition $\sigma_k(x)_2 \leq \|x - \widetilde{x}\|_2$ for all $\widetilde{x} \in \Sigma_k$, including $\widetilde{x} = 0$. Thus $\|h\|_2 \leq C \|h_{\Lambda^c}\|_2$. Since $\|h\|_2^2 = \|h_\Lambda\|_2^2 + \|h_{\Lambda^c}\|_2^2$, this yields

$$\|h_\Lambda\|_2^2 = \|h\|_2^2 - \|h_{\Lambda^c}\|_2^2 \leq \|h\|_2^2 - \frac{1}{C^2}\|h\|_2^2 = \left(1 - \frac{1}{C^2}\right)\|h\|_2^2.$$

This must hold for any vector $h \in \mathcal{N}(A)$ and for any set of indices Λ such that $|\Lambda| \leq k$. In particular, let $\{v_i\}_{i=1}^{n-m}$ be an orthonormal basis for $\mathcal{N}(A)$, and define the vectors

$\{h_i\}_{i=1}^n$ as follows:

$$h_j = \sum_{i=1}^{n-m} v_i(j) v_i. \qquad (A.12)$$

We note that $h_j = \sum_{i=1}^{n-m} \langle e_j, v_i \rangle v_i$ where e_j denotes the vector of all zeros except for a 1 in the jth entry. Thus we see that $h_j = P_\mathcal{N} e_j$ where $P_\mathcal{N}$ denotes an orthogonal projection onto $\mathcal{N}(A)$. Since $\|P_\mathcal{N} e_j\|_2^2 + \|P_{\mathcal{N}^\perp} e_j\|_2^2 = \|e_j\|_2^2 = 1$, we have that $\|h_j\|_2 \leq 1$. Thus, by setting $\Lambda = \{j\}$ for h_j we observe that

$$\left| \sum_{i=1}^{n-m} |v_i(j)|^2 \right|^2 = |h_j(j)|^2 \leq \left(1 - \frac{1}{C^2}\right) \|h_j\|_2^2 \leq 1 - \frac{1}{C^2}.$$

Summing over $j = 1, 2, \ldots, n$, we obtain

$$n\sqrt{1 - 1/C^2} \geq \sum_{j=1}^{n} \sum_{i=1}^{n-m} |v_i(j)|^2 = \sum_{i=1}^{n-m} \sum_{j=1}^{n} |v_i(j)|^2 = \sum_{i=1}^{n-m} \|v_i\|_2^2 = n - m,$$

and thus $m \geq \left(1 - \sqrt{1 - 1/C^2}\right) n$, as desired. $\qquad \square$

References

[1] W. Bajwa, J. Haupt, G. Raz, S. Wright, and R. Nowak. Toeplitz-structured compressed sensing matrices. *Proc IEEE Work Stat Sig Proce*, Madison, WI, 2007.

[2] W. Bajwa, J. Haupt, A. Sayeed, and R. Nowak. Compressed channel sensing: A new approach to estimating sparse multipath channels. *Proc IEEE*, 98(6):1058–1076, 2010.

[3] R. Baraniuk. Compressive sensing. *IEEE Signal Proc Mag*, 24(4):118–120, 124, 2007.

[4] R. Baraniuk, V. Cevher, M. Duarte, and C. Hegde. Model-based compressive sensing. *IEEE Trans. Inform Theory*, 56(4):1982–2001, 2010.

[5] R. Baraniuk, M. Davenport, R. DeVore, and M. Wakin. A simple proof of the restricted isometry property for random matrices. *Const Approx*, 28(3):253–263, 2008.

[6] R. Baraniuk and M. Wakin. Random projections of smooth manifolds. *Found Comput Math*, 9(1):51–77, 2009.

[7] D. Baron, M. Duarte, S. Sarvotham, M. Wakin, and R. Baraniuk. Distributed compressed sensing of jointly sparse signals. *Proc Asilomar Conf Sign, Syst Comput*, Pacific Grove, CA, 2005.

[8] D. Baron, S. Sarvotham, and R. Baraniuk. Sudocodes – Fast measurement and reconstruction of sparse signals. *Proc IEEE Int Symp Inform Theory (ISIT)*, Seattle, WA, 2006.

[9] J. Bazerque and G. Giannakis. Distributed spectrum sensing for cognitive radio networks by exploiting sparsity. *IEEE Trans Sig Proc*, 58(3):1847–1862, 2010.

[10] P. Bechler and P. Wojtaszczyk. Error estimates for orthogonal matching pursuit and random dictionaries. *Const Approx*, 33(2):273–288, 2011.

[11] A. Beck and M. Teboulle. A fast iterative shrinkage-thresholding algorithm for linear inverse problems. *SIAM J Imag Sci*, 2(1):183–202, 2009.

[12] S. Becker, J. Bobin, and E. Candès. NESTA: A fast and accurate first-order method for sparse recovery. *SIAM J Imag Sci*, 4(1):1–39, 2011.

[13] S. Becker, E. Candès, and M. Grant. Templates for convex cone problems with applications to sparse signal recovery. *Math Prog Comp*, 3(3):165–218, 2011.

[14] M. Belkin and P. Niyogi. Laplacian eigenmaps for dimensionality reduction and data representation. *Neural Comput*, 15(6):1373–1396, 2003.

[15] M. Belkin and P. Niyogi. Semi-supervised learning on Riemannian manifolds. *Mach Learning*, 56:209–239, 2004.

[16] Z. Ben-Haim and Y. C. Eldar. Blind minimax estimation. *IEEE Trans. Inform Theory*, 53(9):3145–3157, 2007.

[17] Z. Ben-Haim and Y. C. Eldar. The Cramer-Rao bound for estimating a sparse parameter vector. *IEEE Trans Sig Proc*, 58(6):3384–3389, 2010.

[18] Z. Ben-Haim, Y. C. Eldar, and M. Elad. Coherence-based performance guarantees for estimating a sparse vector under random noise. *IEEE Trans Sig Proc*, 58(10):5030–5043, 2010.

[19] Z. Ben-Haim, T. Michaeli, and Y. C. Eldar. Performance bounds and design criteria for estimating finite rate of innovation signals. Preprint, 2010.

[20] R. Berinde, A. Gilbert, P. Indyk, H. Karloff, and M. Strauss. Combining geometry and combinatorics: A unified approach to sparse signal recovery. *Proc Allerton Conf Commun. Control, Comput*, Monticello, IL, Sept. 2008.

[21] R. Berinde, P. Indyk, and M. Ruzic. Practical near-optimal sparse recovery in the ℓ_1 norm. In *Proc Allerton Conf Comm Control, Comput*, Monticello, IL, Sept. 2008.

[22] A. Beurling. Sur les intégrales de Fourier absolument convergentes et leur application à une transformation fonctionelle. In *Proc Scandi Math Congr*, Helsinki, Finland, 1938.

[23] T. Blumensath and M. Davies. Gradient pursuits. *IEEE Trans Sig Proc*, 56(6):2370–2382, 2008.

[24] T. Blumensath and M. Davies. Iterative hard thresholding for compressive sensing. *Appl Comput Harmon Anal*, 27(3):265–274, 2009.

[25] T. Blumensath and M. Davies. Sampling theorems for signals from the union of finite-dimensional linear subspaces. *IEEE Trans Inform Theory*, 55(4):1872–1882, 2009.

[26] B. Bodmann, P. Cassaza, and G. Kutyniok. A quantitative notion of redundancy for finite frames. *Appl Comput Harmon Anal*, 30(3):348–362, 2011.

[27] P. Boufounos, H. Rauhut, and G. Kutyniok. Sparse recovery from combined fusion frame measurements. *IEEE Trans Inform Theory*, 57(6):3864–3876, 2011.

[28] J. Bourgain, S. Dilworth, K. Ford, S. Konyagin, and D. Kutzarova. Explicit constructions of RIP matrices and related problems. *Duke Math J*, 159(1):145–185, 2011.

[29] Y. Bresler and P. Feng. Spectrum-blind minimum-rate sampling and reconstruction of 2-D multiband signals. *Proc IEEE Int Conf Image Proc (ICIP)*, Zurich, Switzerland, 1996.

[30] D. Broomhead and M. Kirby. The Whitney reduction network: A method for computing autoassociative graphs. *Neural Comput*, 13:2595–2616, 2001.

[31] A. Bruckstein, D. Donoho, and M. Elad. From sparse solutions of systems of equations to sparse modeling of signals and images. *SIAM Rev*, 51(1):34–81, 2009.

[32] T. Cai and T. Jiang. Limiting laws of coherence of random matrices with applications to testing covariance structure and construction of compressed sensing matrices. *Ann Stat*, 39(3):1496–1525, 2011.

[33] E. Candès. Compressive sampling. *Proc Int Congre Math*, Madrid, Spain, 2006.

[34] E. Candès. The restricted isometry property and its implications for compressed sensing. *C R Acad Sci, Sér I*, 346(9-10):589–592, 2008.

[35] E. Candès, Y. C. Eldar, D. Needell, and P. Randall. Compressed sensing with coherent and redundant dictionaries. *Appl Comput Harmon Anal*, 31(1):59–73, 2010.

[36] E. Candès, X. Li, Y. Ma, and J. Wright. Robust principal component analysis? *J ACM*, 58(1):1–37, 2009.

[37] E. Candès and Y. Plan. Near-ideal model selection by ℓ_1 minimization. *Ann Stat*, 37(5A):2145–2177, 2009.

[38] E. Candès and Y. Plan. Matrix completion with noise. *Proc IEEE*, 98(6):925–936, 2010.

[39] E. Candès and B. Recht. Exact matrix completion via convex optimization. *Found Comput Math*, 9(6):717–772, 2009.

[40] E. Candès and J. Romberg. Quantitative robust uncertainty principles and optimally sparse decompositions. *Found Comput Math*, 6(2):227–254, 2006.

[41] E. Candès, J. Romberg, and T. Tao. Robust uncertainty principles: Exact signal reconstruction from highly incomplete frequency information. *IEEE Trans Inform Theory*, 52(2):489–509, 2006.

[42] E. Candès, J. Romberg, and T. Tao. Stable signal recovery from incomplete and inaccurate measurements. *Comm Pure Appl Math*, 59(8):1207–1223, 2006.

[43] E. Candès and T. Tao. Decoding by linear programming. *IEEE Trans Inform Theory*, 51(12):4203–4215, 2005.

[44] E. Candès and T. Tao. Near optimal signal recovery from random projections: Universal encoding strategies? *IEEE Trans Inform Theory*, 52(12):5406–5425, 2006.

[45] E. Candès and T. Tao. The Dantzig selector: Statistical estimation when p is much larger than n. *Ann Stat*, 35(6):2313–2351, 2007.

[46] C. Carathéodory. Über den Variabilitätsbereich der Koeffizienten von Potenzreihen, die gegebene Werte nicht annehmen. *Math Ann*, 64:95–115, 1907.

[47] C. Carathéodory. Über den Variabilitätsbereich der Fourierschen Konstanten von positiven harmonischen Funktionen. *Rend Circ Mat Palermo*, 32:193–217, 1911.

[48] P. Casazza and G. Kutyniok. *Finite Frames*. Birkhäuser, Boston, MA, 2012.

[49] V. Cevher, M. Duarte, C. Hegde, and R. Baraniuk. Sparse signal recovery using Markov random fields. In *Proc Adv Neural Proc Syst(NIPS)*, Vancouver, BC, 2008.

[50] V. Cevher, P. Indyk, C. Hegde, and R. Baraniuk. Recovery of clustered sparse signals from compressive measurements. *Proc Sampling Theory Appl (SampTA)*, Marseilles, France, 2009.

[51] M. Charikar, K. Chen, and M. Farach-Colton. Finding frequent items in data streams. In *Proc Int Coll Autom Lang Programm*, Málaga, Spain, 2002.

[52] J. Chen and X. Huo. Theoretical results on sparse representations of multiple-measurement vectors. *IEEE Trans Sig Proc*, 54(12):4634–4643, 2006.

[53] S. Chen, D. Donoho, and M. Saunders. Atomic decomposition by basis pursuit. *SIAM J Sci Comp*, 20(1):33–61, 1998.

[54] Y. Chi, L. Scharf, A. Pezeshki, and R. Calderbank. Sensitivity to basis mismatch in compressed sensing. *IEEE Trans Sig Proc*, 59(5):2182–2195, 2011.

[55] O. Christensen. *An Introduction to Frames and Riesz Bases*. Birkhäuser, Boston, MA, 2003.

[56] A. Cohen, W. Dahmen, and R. DeVore. Instance optimal decoding by thresholding in compressed sensing. *Int Conf Harmonic Analysis and Partial Differential Equations*, Madrid, Spain, 2008.

[57] A. Cohen, W. Dahmen, and R. DeVore. Compressed sensing and best k-term approximation. *J Am Math Soc*, 22(1):211–231, 2009.

[58] R. Coifman and M. Maggioni. Diffusion wavelets. *Appl Comput Harmon Anal*, 21(1): 53–94, 2006.

[59] G. Cormode and M. Hadjieleftheriou. Finding the frequent items in streams of data. *Comm ACM*, 52(10):97–105, 2009.

[60] G. Cormode and S. Muthukrishnan. Improved data stream summaries: The count-min sketch and its applications. *J Algorithms*, 55(1):58–75, 2005.

[61] J. Costa and A. Hero. Geodesic entropic graphs for dimension and entropy estimation in manifold learning. *IEEE Trans Sig Proc*, 52(8):2210–2221, 2004.

[62] S. Cotter and B. Rao. Sparse channel estimation via matching pursuit with application to equalization. *IEEE Trans Commun*, 50(3):374–377, 2002.

[63] S. Cotter, B. Rao, K. Engan, and K. Kreutz-Delgado. Sparse solutions to linear inverse problems with multiple measurement vectors. *IEEE Trans Sig Proc*, 53(7):2477–2488, 2005.

[64] W. Dai and O. Milenkovic. Subspace pursuit for compressive sensing signal reconstruction. *IEEE Trans Inform Theory*, 55(5):2230–2249, 2009.

[65] I. Daubechies. *Ten Lectures on Wavelets*. SIAM, Philadelphia, PA, 1992.

[66] I. Daubechies, M. Defrise, and C. De Mol. An iterative thresholding algorithm for linear inverse problems with a sparsity constraint. *Comm Pure Appl Math*, 57(11):1413–1457, 2004.

[67] M. Davenport. Random observations on random observations: Sparse signal acquisition and processing. PhD thesis, Rice University, 2010.

[68] M. Davenport and R. Baraniuk. Sparse geodesic paths. *Proc AAAI Fall Symp Manifold Learning*, Arlington, VA, 2009.

[69] M. Davenport, P. Boufounos, and R. Baraniuk. Compressive domain interference cancellation. *Proc Work Struc Parc Rep Adap Signaux (SPARS)*, Saint-Malo, France, 2009.

[70] M. Davenport, P. Boufounos, M. Wakin, and R. Baraniuk. Signal processing with compressive measurements. *IEEE J Sel Top Sig Proc*, 4(2):445–460, 2010.

[71] M. Davenport, M. Duarte, M. Wakin, *et al*. The smashed filter for compressive classification and target recognition. In *Proc IS&T/SPIE Symp Elec Imag: Comp Imag*, San Jose, CA, 2007.

[72] M. Davenport, C. Hegde, M. Duarte, and R. Baraniuk. Joint manifolds for data fusion. *IEEE Trans Image Proc*, 19(10):2580–2594, 2010.

[73] M. Davenport, J. Laska, P. Boufounos, and R. Baraniuk. A simple proof that random matrices are democratic. Technical Report TREE 0906, Rice Univ., ECE Dept, 2009.

[74] M. Davenport, S. Schnelle, J. P. Slavinsky, *et al*. A wideband compressive radio receiver. *Proc IEEE Conf Mil Comm (MILCOM)*, San Jose, CA, 2010.

[75] M. Davenport and M. Wakin. Analysis of orthogonal matching pursuit using the restricted isometry property. *IEEE Trans Inform Theory*, 56(9):4395–4401, 2010.

[76] M. Davies and Y. C. Eldar. Rank awareness in joint sparse recovery. To appear in *IEEE Trans Inform Theory*, 2011.

[77] R. DeVore. Nonlinear approximation. *Acta Numer*, 7:51–150, 1998.

[78] R. DeVore. Deterministic constructions of compressed sensing matrices. *J Complex*, 23(4):918–925, 2007.

[79] M. Do and C. La. Tree-based majorize-minimize algorithm for compressed sensing with sparse-tree prior. *Int Workshop Comput Adv Multi-Sensor Adapt Proc (CAMSAP)*, Saint Thomas, US Virgin Islands, 2007.

[80] D. Donoho. Denoising by soft-thresholding. *IEEE Trans Inform Theory*, 41(3):613–627, 1995.

[81] D. Donoho. Neighborly polytopes and sparse solutions of underdetermined linear equations. Technical Report 2005-04, Stanford Univ., Stat. Dept, 2005.

[82] D. Donoho. Compressed sensing. *IEEE Trans Inform Theory*, 52(4):1289–1306, 2006.

[83] D. Donoho. For most large underdetermined systems of linear equations, the minimal ℓ_1-norm solution is also the sparsest solution. *Comm Pure Appl Math*, 59(6):797–829, 2006.

[84] D. Donoho. High-dimensional centrally symmetric polytopes with neighborliness proportional to dimension. *Discrete Comput Geom*, 35(4):617–652, 2006.

[85] D. Donoho, I. Drori, Y. Tsaig, and J.-L. Stark. Sparse solution of underdetermined linear equations by stagewise orthogonal matching pursuit. *Tech Report Stanford Univ.*, 2006.

[86] D. Donoho and M. Elad. Optimally sparse representation in general (nonorthogonal) dictionaries via ℓ_1 minimization. *Proc Natl Acad Sci*, 100(5):2197–2202, 2003.

[87] D. Donoho and M. Elad. On the stability of basis pursuit in the presence of noise. *EURASIP Sig Proc J*, 86(3):511–532, 2006.

[88] D. Donoho, M. Elad, and V. Temlyahov. Stable recovery of sparse overcomplete representations in the presence of noise. *IEEE Trans Inform Theory*, 52(1):6–18, 2006.

[89] D. Donoho and C. Grimes. Hessian eigenmaps: Locally linear embedding techniques for high-dimensional data. *Proc Natl Acad. Sci*, 100(10):5591–5596, 2003.

[90] D. Donoho and C. Grimes. Image manifolds which are isometric to Euclidean space. *J Math Imag Vision*, 23(1):5–24, 2005.

[91] D. Donoho and B. Logan. Signal recovery and the large sieve. *SIAM J Appl Math*, 52(6):577–591, 1992.

[92] D. Donoho and J. Tanner. Neighborliness of randomly projected simplices in high dimensions. *Proc Natl Acad Sci*, 102(27):9452–9457, 2005.

[93] D. Donoho and J. Tanner. Sparse nonnegative solutions of undetermined linear equations by linear programming. *Proc Natl Acad Sci*, 102(27):9446–9451, 2005.

[94] D. Donoho and J. Tanner. Counting faces of randomly-projected polytopes when the projection radically lowers dimension. *J Am Math Soc*, 22(1):1–53, 2009.

[95] D. Donoho and J. Tanner. Precise undersampling theorems. *Proc IEEE*, 98(6):913–924, 2010.

[96] D. Donoho and Y. Tsaig. Fast solution of ℓ_1 norm minimization problems when the solution may be sparse. *IEEE Trans Inform Theory*, 54(11):4789–4812, 2008.

[97] P. Dragotti, M. Vetterli, and T. Blu. Sampling moments and reconstructing signals of finite rate of innovation: Shannon meets Strang-Fix. *IEEE Trans Sig Proc*, 55(5):1741–1757, 2007.

[98] D. Du and F. Hwang. *Combinatorial Group Testing and Its Applications*. World Scientific, Singapore, 2000.

[99] M. Duarte, M. Davenport, D. Takhar, *et al*. Single-pixel imaging via compressive sampling. *IEEE Sig Proc Mag*, 25(2):83–91, 2008.

[100] M. Duarte, M. Davenport, M. Wakin, and R. Baraniuk. Sparse signal detection from incoherent projections. *Proc IEEE Int Conf Acoust, Speech, Sig Proc (ICASSP)*, Toulouse, France, 2006.

[101] M. Duarte, M. Davenport, M. Wakin, *et al*. Multiscale random projections for compressive classification. *Proc IEEE Int Conf Image Proc (ICIP)*, San Antonio, TX, Sept. 2007.

[102] M. Duarte and Y. C. Eldar. Structured compressed sensing: Theory and applications. *IEEE Trans Sig Proc*, 59(9):4053–4085, 2011.

[103] M. Duarte, M. Wakin, and R. Baraniuk. Fast reconstruction of piecewise smooth signals from random projections. *Proc Work Struc Parc Rep Adap Signaux (SPARS)*, Rennes, France, 2005.

[104] M. Duarte, M. Wakin, and R. Baraniuk. Wavelet-domain compressive signal reconstruction using a hidden Markov tree model. *Proc IEEE Int Conf Acoust, Speech, Signal Proc (ICASSP)*, Las Vegas, NV, Apr. 2008.

[105] T. Dvorkind, Y. C. Eldar, and E. Matusiak. Nonlinear and non-ideal sampling: Theory and methods. *IEEE Trans Sig Proc*, 56(12):471–481, 2009.

[106] M. Elad. *Sparse and Redundant Representations: From Theory to Applications in Signal and Image Processing*. Springer, New York, NY, 2010.

[107] M. Elad, B. Matalon, J. Shtok, and M. Zibulevsky. A wide-angle view at iterated shrinkage algorithms. *Proc SPIE Optics Photonics: Wavelets*, San Diego, CA, 2007.

[108] Y. C. Eldar. Rethinking biased estimation: Improving maximum likelihood and the Cramer-Rao bound Found. *Trends Sig Proc*, 1(4):305–449, 2008.

[109] Y. C. Eldar. Compressed sensing of analog signals in shift-invariant spaces. *IEEE Trans Sig Proc*, 57(8):2986–2997, 2009.

[110] Y. C. Eldar. Generalized SURE for exponential families: Applications to regularization. *IEEE Trans Sig Proc*, 57(2):471–481, 2009.

[111] Y. C. Eldar. Uncertainty relations for shift-invariant analog signals. *IEEE Trans Inform Theory*, 55(12):5742–5757, 2009.

[112] Y. C. Eldar, P. Kuppinger, and H. Bölcskei. Block-sparse signals: Uncertainty relations and efficient recovery. *IEEE Trans Sig Proc*, 58(6):3042–3054, 2010.

[113] Y. C. Eldar and T. Michaeli. Beyond bandlimited sampling. *IEEE Sig Proc Mag*, 26(3): 48–68, 2009.

[114] Y. C. Eldar and M. Mishali. Robust recovery of signals from a structured union of subspaces. *IEEE Trans Inform Theory*, 55(11):5302–5316, 2009.

[115] Y. C. Eldar and H. Rauhut. Average case analysis of multichannel sparse recovery using convex relaxation. *IEEE Trans Inform Theory*, 6(1):505–519, 2010.

[116] Y. Erlich, N. Shental, A. Amir, and O. Zuk. Compressed sensing approach for high throughput carrier screen. *Proc Allerton Conf Commun Contr Comput*, Monticello, IL, Sept. 2009.

[117] P. Feng. Universal spectrum blind minimum rate sampling and reconstruction of multiband signals. PhD thesis, University of Illinois at Urbana-Champaign, Mar. 1997.

[118] P. Feng and Y. Bresler. Spectrum-blind minimum-rate sampling and reconstruction of multiband signals. *Proc IEEE Int Conf Acoust, Speech, Sig Proc (ICASSP)*, Atlanta, GA, May 1996.

[119] I. Fevrier, S. Gelfand, and M. Fitz. Reduced complexity decision feedback equalization for multipath channels with large delay spreads. *IEEE Trans Communi*, 47(6):927–937, 1999.

[120] M. Figueiredo, R. Nowak, and S. Wright. Gradient projections for sparse reconstruction: Application to compressed sensing and other inverse problems. *IEEE J Select Top Sig Proc*, 1(4):586–597, 2007.

[121] M. Fornassier and H. Rauhut. Recovery algorithms for vector valued data with joint sparsity constraints. *SIAM J Numer Anal*, 46(2):577–613, 2008.

[122] J. Friedman, T. Hastie, and R. Tibshirani. Regularization paths for generalized linear models via coordinate descent. *J Stats Software*, 33(1):1–22, 2010.

[123] N. Galatsanos and A. Katsaggelos. Methods for choosing the regularization parameter and estimating the noise variance in image restoration and their relation. *IEEE Trans Image Proc*, 1(3):322–336, 1992.

[124] A. Garnaev and E. Gluskin. The widths of Euclidean balls. *Dokl An SSSR*, 277:1048–1052, 1984.

[125] K. Gedalyahu and Y. C. Eldar. Time-delay estimation from low-rate samples: A union of subspaces approach. *IEEE Trans Sig Proc*, 58(6):3017–3031, 2010.

[126] K. Gedalyahu, R. Tur, and Y. C. Eldar. Multichannel sampling of pulse streams at the rate of innovation. *IEEE Trans Sig Proc*, 59(4):1491–1504, 2011.

[127] S. Geršgorin. Über die Abgrenzung der Eigenwerte einer Matrix. *Izv. Akad Nauk SSSR Ser Fiz.-Mat*, 6:749–754, 1931.

[128] A. Gilbert and P. Indyk. Sparse recovery using sparse matrices. *Proc IEEE*, 98(6):937–947, 2010.

[129] A. Gilbert, Y. Li, E. Porat, and M. Strauss. Approximate sparse recovery: Optimizing time and measurements. *Proc ACM Symp Theory Comput*, Cambridge, MA, Jun. 2010.

[130] A. Gilbert, M. Strauss, J. Tropp, and R. Vershynin. One sketch for all: Fast algorithms for compressed sensing. *Proc ACM Symp Theory Comput*, San Diego, CA, Jun. 2007.

[131] S. Gleichman and Y. C. Eldar. Blind compressed sensing. To appear in *IEEE Trans Inform Theory*, 57(10):6958–6975, 2011.

[132] D. Goldberg, D. Nichols, B. Oki, and D. Terry. Using collaborative filtering to weave an information tapestry. *Commun ACM*, 35(12):61–70, 1992.

[133] G. Golub and M. Heath. Generalized cross-validation as a method for choosing a good ridge parameter. *Technometrics*, 21(2):215–223, 1970.

[134] I. Gorodnitsky, J. George, and B. Rao. Neuromagnetic source imaging with FOCUSS: A recursive weighted minimum norm algorithm. *Electroen Clin Neuro*, 95(4):231–251, 1995.

[135] I. Gorodnitsky and B. Rao. Sparse signal reconstruction from limited data using FOCUSS: A re-weighted minimum norm algorithm. *IEEE Trans Sig Proc*, 45(3):600–616, 1997.

[136] I. Gorodnitsky, B. Rao, and J. George. Source localization in magnetoencephalography using an iterative weighted minimum norm algorithm. *Proc Asilomar Conf Sig, Syst and Comput*, Pacific Grove, CA, Oct. 1992.

[137] R. Gribonval, H. Rauhut, K. Schnass, and P. Vandergheynst. Atoms of all channels, unite! Average case analysis of multi-channel sparse recovery using greedy algorithms. *J Fourier Anal Appl*, 14(5):655–687, 2008.

[138] E. Hale, W. Yin, and Y. Zhang. A fixed-point continuation method for ℓ_1-regularized minimization with applications to compressed sensing. Technical Report TR07-07, Rice Univ., CAAM Dept, 2007.

[139] T. Hastie, R. Tibshirani, and J. Friedman. *The Elements of Statistical Learning*. Springer, New York, NY, 2001.

[140] J. Haupt, L. Applebaum, and R. Nowak. On the restricted isometry of deterministically subsampled Fourier matrices. *Conf Inform Sci Syste (CISS)*, Princeton, NJ, 2010.

[141] J. Haupt, W. Bajwa, M. Rabbat, and R. Nowak. Compressed sensing for networked data. *IEEE Sig Proc Mag*, 25(2):92–101, 2008.

[142] J. Haupt, W. Bajwa, G. Raz, and R. Nowak. Toeplitz compressed sensing matrices with applications to sparse channel estimation. *IEEE Trans Inform Theory*, 56(11):5862–5875, 2010.

[143] J. Haupt, R. Castro, R. Nowak, G. Fudge, and A. Yeh. Compressive sampling for signal classification. *Proc Asilomar Conf Sig, Syste, Comput*, Pacific Grove, CA, 2006.

[144] J. Haupt and R. Nowak. Signal reconstruction from noisy random projections. *IEEE Trans Inform Theory*, 52(9):4036–4048, 2006.

[145] J. Haupt and R. Nowak. Compressive sampling for signal detection. In *Proc. IEEE Int. Conf Acoust, Speech, Sig Proc (ICASSP)*, Honolulu, HI, 2007.

[146] D. Healy. Analog-to-information: BAA #05-35, 2005. Available online at http://www.darpa.mil/mto/solicitations/baa05-35/s/index.html.

[147] C. Hegde, M. Duarte, and V. Cevher. Compressive sensing recovery of spike trains using a structured sparsity model. *Proc Work Struc Parc Rep Adap Signaux (SPARS)*, Saint-Malo, France, 2009.

[148] M. Herman and T. Strohmer. High-resolution radar via compressed sensing. *IEEE Trans Sig Proc*, 57(6):2275–2284, 2009.

[149] M. Herman and T. Strohmer. General deviants: An analysis of perturbations in compressed sensing. *IEEE J Select Top Sig Proc*, 4(2):342–349, 2010.

[150] G. Hinton, P. Dayan, and M. Revow. Modelling the manifolds of images of handwritten digits. *IEEE Trans Neural Networks*, 8(1):65–74, 1997.

[151] L. Hogben. *Handbook of Linear Algebra. Discrete Mathematics and its Applications.* Chapman & Hall / CRC, Boca Raton, FL, 2007.

[152] P. Indyk. Explicit constructions for compressed sensing of sparse signals. *Proc. ACM-SIAM Symp Discrete Algorithms (SODA)*, San Franciso, CA, 2008.

[153] P. Indyk and M. Ruzic. Near-optimal sparse recovery in the ℓ_1 norm. In *Proc IEEE Symp Found Comp Science (FOCS)*, Philadelphia, PA, 2008.

[154] S. Jafarpour, W. Xu, B. Hassibi, and R. Calderbank. Efficient and robust compressed sensing using optimized expander graphs. *IEEE Trans Inform Theory*, 55(9):4299–4308, 2009.

[155] W. James and C. Stein. Estimation of quadratic loss. In *Proc 4th Berkeley Symp Math Statist Prob*, 1: 361–379. University of California Press, Berkeley, 1961.

[156] T. Jayram and D. Woodruff. Optimal bounds for Johnson-Lindenstrauss transforms and streaming problems with sub-constant error. *Proc ACM-SIAM Symp Discrete Algorithms (SODA)*, San Francisco, CA, 2011.

[157] B. Jeffs. Sparse inverse solution methods for signal and image processing applications. In *Proc IEEE Int Conf Acoust, Speech Sig Proce (ICASSP)*, Seattle, WA, 1998.

[158] W. Johnson and J. Lindenstrauss. Extensions of Lipschitz mappings into a Hilbert space. *Proc Conf Modern Anal Prob*, New Haven, CT, 1982.

[159] G. Judge and M. Bock. *The Statistical Implications of Pre-Test and Stein-Rule Estimators in Econometrics*. North-Holland, Amsterdam, 1978.

[160] R. Kainkaryam, A. Breux, A. Gilbert, P. Woolf, and J. Schiefelbein. poolMC: Smart pooling of mRNA samples in microarray experiments. *BMC Bioinformatics*, 11(1):299, 2010.

[161] R. Keshavan, A. Montanari, and S. Oh. Matrix completion from a few entries. *IEEE Trans Inform Theory*, 56(6):2980–2998, 2010.

[162] V. Kotelnikov. On the carrying capacity of the ether and wire in telecommunications. *Izd Red Upr Svyazi RKKA*, Moscow, Russia, 1933.

[163] J. Kovačević and A. Chebira. Life beyond bases: The advent of frames (Part I). *IEEE Sig Proc Mag*, 24(4):86–104, 2007.

[164] J. Kovačević and A. Chebira. Life beyond bases: The advent of frames (Part II). *IEEE Sig Proc Mag*, 24(4):115–125, 2007.

[165] F. Krahmer and R. Ward. New and improved Johnson-Lindenstrauss embeddings via the restricted isometry property. *SIAM J Math Anal*, 43(3):1269–1281, 2011.

[166] T. Kühn. A lower estimate for entropy numbers. *J Approx Theory*, 110(1):120–124, 2001.

[167] C. La and M. Do. Signal reconstruction using sparse tree representation. *Proc SPIE Optics Photonics: Wavelets*, San Diego, CA, 2005.

[168] C. La and M. Do. Tree-based orthogonal matching pursuit algorithm for signal reconstruction. *IEEE Int Conf Image Proc (ICIP)*, Atlanta, GA, 2006.

[169] J. Laska, P. Boufounos, M. Davenport, and R. Baraniuk. Democracy in action: Quantization, saturation, and compressive sensing. *Appl Comput Harman Anal*, 31(3):429–443, 2011.

[170] J. Laska, M. Davenport, and R. Baraniuk. Exact signal recovery from corrupted measurements through the pursuit of justice. *Proc Asilomar Conf Sign, Syste Compute*, Pacific Grove, CA, 2009.

[171] S. Levy and P. Fullagar. Reconstruction of a sparse spike train from a portion of its spectrum and application to high-resolution deconvolution. *Geophysics*, 46(9):1235–1243, 1981.

[172] N. Linial, E. London, and Y. Rabinovich. The geometry of graphs and some of its algorithmic applications. *Combinatorica*, 15(2):215–245, 1995.

[173] E. Livshitz. On efficiency of Orthogonal Matching Pursuit. Preprint, 2010.

[174] B. Logan. Properties of high-pass signals. PhD thesis, Columbia University, 1965.

[175] I. Loris. On the performance of algorithms for the minimization of ℓ_1-penalized functions. *Inverse Problems*, 25(3):035008, 2009.

[176] H. Lu. Geometric theory of images. PhD thesis, University of California, San Diego, 1998.

[177] Y. Lu and M. Do. Sampling signals from a union of subspaces. *IEEE Sig Proce Mag*, 25(2):41–47, 2008.

[178] M. Lustig, D. Donoho, and J. Pauly. Rapid MR imaging with compressed sensing and randomly under-sampled 3DFT trajectories. *Proc Ann Meeting of ISMRM*, Seattle, WA, 2006.

[179] M. Lustig, J. Lee, D. Donoho, and J. Pauly. Faster imaging with randomly perturbed, under-sampled spirals and ℓ_1 reconstruction. *Proc Ann Meeting ISMRM*, Miami, FL, 2005.

[180] M. Lustig, J. Santos, J. Lee, D. Donoho, and J. Pauly. Application of compressed sensing for rapid MR imaging. *Proc Work Struc Parc Rep Adap Sign (SPARS)*, Rennes, France, 2005.

[181] D. Malioutov, M. Cetin, and A. Willsky. A sparse signal reconstruction perspective for source localization with sensor arrays. *IEEE Trans Sign Proce*, 53(8):3010–3022, 2005.

[182] S. Mallat. *A Wavelet Tour of Signal Processing*. Academic Press, San Diego, CA, 1999.

[183] S. Mallat and Z. Zhang. Matching pursuits with time-frequency dictionaries. *IEEE Trans Sig Proc*, 41(12):3397–3415, 1993.

[184] R. Marcia, Z. Harmany, and R. Willett. Compressive coded aperture imaging. *Proc IS&T/SPIE Symp Elec Imag: Comp Imag*, San Jose, CA, 2009.

[185] M. Mishali and Y. C. Eldar. Reduce and boost: Recovering arbitrary sets of jointly sparse vectors. *IEEE Trans Sig Proc*, 56(10):4692–4702, 2008.

[186] M. Mishali and Y. C. Eldar. Blind multi-band signal reconstruction: Compressed sensing for analog signals. *IEEE Trans Sig Proc*, 57(3):993–1009, 2009.

[187] M. Mishali and Y. C. Eldar. From theory to practice: Sub-Nyquist sampling of sparse wideband analog signals. *IEEE J Select Top Sig Proc*, 4(2):375–391, 2010.

[188] M. Mishali, Y. C. Eldar, O. Dounaevsky, and E. Shoshan. Xampling: Analog to digital at sub-Nyquist rates. *IET Circ Dev Syst*, 5(1):8–20, 2011.

[189] S. Muthukrishnan. *Data Streams: Algorithms and Applications, Foundations and Trends in Theoretical Computer Science*. Now Publishers, Boston, MA, 2005.

[190] D. Needell and J. Tropp. CoSaMP: Iterative signal recovery from incomplete and inaccurate samples. *Appl Comput Harmon Anal*, 26(3):301–321, 2009.

[191] D. Needell and R. Vershynin. Uniform uncertainty principle and signal recovery via regularized orthogonal matching pursuit. *Found Comput Math*, 9(3):317–334, 2009.

[192] D. Needell and R. Vershynin. Signal recovery from incomplete and inaccurate measurements via regularized orthogonal matching pursuit. *IEEE J Select Top Sig Proc*, 4(2):310–316, 2010.

[193] P. Niyogi. Manifold regularization and semi-supervised learning: Some theoretical analyses. Technical Report TR-2008-01, Univ. of Chicago, Comput Sci Dept., 2008.

[194] J. Nocedal and S. Wright. *Numerical Optimization*. Springer-Verlag, 1999.

[195] H. Nyquist. Certain topics in telegraph transmission theory. *Trans AIEE*, 47:617–644, 1928.

[196] B. Olshausen and D. Field. Emergence of simple-cell receptive field properties by learning a sparse representation. *Nature*, 381:607–609, 1996.

[197] S. Osher, Y. Mao, B. Dong, and W. Yin. Fast linearized Bregman iterations for compressive sensing and sparse denoising. *Commun Math Sci*, 8(1):93–111, 2010.

[198] J. Partington. *An Introduction to Hankel Operators*. Cambridge University Press, Cambridge, 1988.

[199] W. Pennebaker and J. Mitchell. *JPEG Still Image Data Compression Standard*. Van Nostrand Reinhold, 1993.

[200] J. Phillips, R. Leahy, and J. Mosher. MEG-based imaging of focal neuronal current sources. *IEEE Trans Med Imaging*, 16(3):338–348, 1997.

[201] R. Prony. Essai expérimental et analytique sur les lois de la Dilatabilité des fluides élastiques et sur celles de la Force expansive de la vapeur de l'eau et de la vapeur de l'alkool, à différentes températures. *J de l'École Polytechnique*, Floréal et Prairial III, 1(2):24–76, 1795. R. Prony is Gaspard Riche, baron de Prony.

[202] B. Rao. Signal processing with the sparseness constraint. *Proc IEEE Int Conf Acoust, Speech, Sig Proc (ICASSP)*, Seattle, WA, 1998.

[203] B. Recht. A simpler approach to matrix completion. To appear in *J. Machine Learning Rese*, 12(12):3413–3430, 2011.

[204] B. Recht, M. Fazel, and P. Parrilo. Guaranteed minimum rank solutions of matrix equations via nuclear norm minimization. *SIAM Rev*, 52(3):471–501, 2010.

[205] R. Robucci, L. Chiu, J. Gray, *et al.* Compressive sensing on a CMOS separable transform image sensor. *Proc IEEE Int Conf Acoust, Speech, Sig Proc (ICASSP)*, Las Vegas, NV, 2008.

[206] J. Romberg. Compressive sensing by random convolution. *SIAM J Imag Sci*, 2(4):1098–1128, 2009.

[207] M. Rosenfeld. In praise of the Gram matrix, *The Mathematics of Paul Erdős II*, 318–323. Springer, Berlin, 1996.

[208] K. Schnass and P. Vandergheynst. Average performance analysis for thresholding. *IEEE Sig Proc Letters*, 14(11):828–831, 2007.

[209] C. Shannon. Communication in the presence of noise. *Proc Inst Radio Engineers*, 37(1):10–21, 1949.

[210] N. Shental, A. Amir, and O. Zuk. Identification of rare alleles and their carriers using compressed se(que)nsing. *Nucl Acids Res*, 38(19):e179, 2009.

[211] J. P. Slavinsky, J. Laska, M. Davenport, and R. Baraniuk. The compressive multiplexer for multi-channel compressive sensing. *Proc IEEE Int Conf Acoust, Speech, Sig Proc (ICASSP)*, Prague, Czech Republic, 2011.

[212] A. So and Y. Ye. Theory of semidefinite programming for sensor network localization. *Math Programming, Series A B*, 109(2):367–384, 2007.

[213] C. Stein. Inadmissibility of the usual estimator for the mean of a multivariate normal distribution. *Proc 3rd Berkeley Symp Math Statist Prob*, 1: 197–206. University of California Press, Berkeley, 1956.

[214] T. Strohmer and R. Heath. Grassmanian frames with applications to coding and communication. *Appl Comput Harmon Anal*, 14(3):257–275, 2003.

[215] D. Taubman and M. Marcellin. *JPEG 2000: Image Compression Fundamentals, Standards and Practice*. Kluwer, 2001.

[216] H. Taylor, S. Banks, and J. McCoy. Deconvolution with the ℓ_1 norm. *Geophysics*, 44(1):39–52, 1979.

[217] J. Tenenbaum, V. de Silva, and J. Landford. A global geometric framework for nonlinear dimensionality reduction. *Science*, 290:2319–2323, 2000.

[218] R. Tibshirani. Regression shrinkage and selection via the Lasso. *J Roy Statist Soc B*, 58(1):267–288, 1996.

[219] J. Treichler, M. Davenport, and R. Baraniuk. Application of compressive sensing to the design of wideband signal acquisition receivers. *Proc US/Australia Joint Work. Defense Apps of Signal Processing (DASP)*, Lihue, HI, 2009.

[220] J. Tropp. Greed is good: Algorithmic results for sparse approximation. *IEEE Trans Inform Theory*, 50(10):2231–2242, 2004.

[221] J. Tropp. Algorithms for simultaneous sparse approximation. Part II: Convex relaxation. *Sig Proc*, 86(3):589–602, 2006.

[222] J. Tropp and A. Gilbert. Signal recovery from partial information via orthogonal matching pursuit. *IEEE Trans Inform. Theory*, 53(12):4655–4666, 2007.

[223] J. Tropp, A. Gilbert, and M. Strauss. Algorithms for simultaneous sparse approximation. Part I: Greedy pursuit. *Sig Proc*, 86(3):572–588, 2006.

[224] J. Tropp, J. Laska, M. Duarte, J. Romberg, and R. Baraniuk. Beyond Nyquist: Efficient sampling of sparse, bandlimited signals. *IEEE Trans Inform Theory*, 56(1):520–544, 2010.

[225] J. Tropp, M. Wakin, M. Duarte, D. Baron, and R. Baraniuk. Random filters for compressive sampling and reconstruction. *Proc IEEE Int Conf Acoust, Speech, Sig Proc (ICASSP)*, Toulouse, France, 2006.

[226] J. Tropp and S. Wright. Computational methods for sparse solution of linear inverse problems. *Proc IEEE*, 98(6):948–958, 2010.

[227] J. Trzasko and A. Manduca. Highly undersampled magnetic resonance image reconstruction via homotopic ℓ_0-minimization. *IEEE Trans Med Imaging*, 28(1):106–121, 2009.

[228] R. Tur, Y. C. Eldar, and Z. Friedman. Innovation rate sampling of pulse streams with application to ultrasound imaging. *IEEE Trans Sig Proc*, 59(4):1827–1842, 2011.

[229] M. Turk and A. Pentland. Eigenfaces for recognition. *J Cogni Neurosci*, 3(1):71–86, 1991.

[230] M. Unser. Sampling – 50 years after Shannon. *Proc IEEE*, 88(4):569–587, 2000.

[231] E. van den Berg and M. Friedlander. Probing the Pareto frontier for basis pursuit solutions. *SIAM J Sci Comp*, 31(2):890–912, 2008.

[232] E. van den Berg and M. Friedlander. Theoretical and empirical results for recovery from multiple measurements. *IEEE Trans Inform Theory*, 56(5):2516–2527, 2010.

[233] B. Vandereycken and S. Vandewalle. Riemannian optimization approach for computing low-rank solutions of Lyapunov equations. *Proc SIAM Conf Optimization*, Boston, MA, 2008.

[234] V. Vapnik. *The Nature of Statistical Learning Theory*. Springer-Verlag, New York, 1999.

[235] R. Varga. *Geršgorin and His Circles*. Springer, Berlin, 2004.

[236] S. Vasanawala, M. Alley, R. Barth, *et al*. Faster pediatric MRI via compressed sensing. *Proc Ann Meeting Soc Pediatric Radiology (SPR)*, Carlsbad, CA, 2009.

[237] R. Venkataramani and Y. Bresler. Further results on spectrum blind sampling of 2-D signals. *Proc IEEE Int Conf Image Proc (ICIP)*, Chicago, IL, 1998.

[238] R. Venkataramani and Y. Bresler. Perfect reconstruction formulas and bounds on aliasing error in sub-Nyquist nonuniform sampling of multiband signals. *IEEE Trans Inform Theory*, 46(6):2173–2183, 2000.

[239] M. Vetterli, P. Marziliano, and T. Blu. Sampling signals with finite rate of innovation. *IEEE Trans Sig Proc*, 50(6):1417–1428, 2002.

[240] M. Wakin, D. Donoho, H. Choi, and R. Baraniuk. The multiscale structure of non-differentiable image manifolds. *Proc SPIE Optics Photonics: Wavelets*, San Diego, CA, 2005.

[241] R. Walden. Analog-to-digital converter survey and analysis. *IEEE J Sel Areas Commun*, 17(4):539–550, 1999.

[242] C. Walker and T. Ulrych. Autoregressive recovery of the acoustic impedance. *Geophysics*, 48(10):1338–1350, 1983.

[243] R. Ward. Compressive sensing with cross validation. *IEEE Trans Inform Theory*, 55(12):5773–5782, 2009.

[244] K. Weinberger and L. Saul. Unsupervised learning of image manifolds by semidefinite programming. *Int J Computer Vision*, 70(1):77–90, 2006.

[245] L. Welch. Lower bounds on the maximum cross correlation of signals. *IEEE Trans Inform Theory*, 20(3):397–399, 1974.

[246] Z. Wen, W. Yin, D. Goldfarb, and Y. Zhang. A fast algorithm for sparse reconstruction based on shrinkage, subspace optimization and continuation. *SIAM J Sci Comput*, 32(4):1832–1857, 2010.

[247] E. Whittaker. On the functions which are represented by the expansions of the interpolation theory. *Proc Roy Soc Edin Sec A*, 35:181–194, 1915.

[248] P. Wojtaszczyk. Stability and instance optimality for Gaussian measurements in compressed sensing. *Found Comput Math*, 10(1):1–13, 2010.

[249] S. Wright, R. Nowak, and M. Figueiredo. Sparse reconstruction by separable approximation. *IEEE Trans Sig Proc*, 57(7):2479–2493, 2009.

[250] A. Yang, S. Sastray, A. Ganesh, and Y. Ma. Fast ℓ_1-minimization algorithms and an application in robust face recognition: A review. *Proc IEEE Int Conf Image Proc (ICIP)*, Hong Kong, 2010.

[251] W. Yin, S. Osher, D. Goldfarb, and J. Darbon. Bregman iterative algorithms for ℓ_1-minimization with applications to compressed sensing. *SIAM J Imag Sci*, 1(1):143–168, 2008.

[252] Z. Yu, S. Hoyos, and B. Sadler. Mixed-signal parallel compressed sensing and reception for cognitive radio. *Proc IEEE Int Conf Acoust, Speech, Sig Proc (ICASSP)*, Las Vegas, NV, 2008.

[253] M. Yuan and Y. Lin. Model selection and estimation in regression with grouped variables. *J Roy Stat Soc Ser B*, 68(1):49–67, 2006.

[254] T. Zhang. Sparse recovery with Orthogonal Matching Pursuit under RIP. *IEEE Trans Inform Theory*, 59(9):6125–6221, 2011.

2 Second-generation sparse modeling: structured and collaborative signal analysis

Alexey Castrodad, Ignacio Ramirez, Guillermo Sapiro, Pablo Sprechmann, and Guoshen Yu

In this chapter the authors go beyond traditional sparse modeling, and address collaborative structured sparsity to add stability and prior information to the representation. In structured sparse modeling, instead of considering the dictionary atoms as singletons, the atoms are partitioned in groups, and a few groups are selected at a time for the signal encoding. A complementary way of adding structure, stability, and prior information to a model is via collaboration. Here, multiple signals, which are known to follow the same model, are allowed to collaborate in the coding. The first studied framework connects sparse modeling with Gaussian Mixture Models and leads to state-of-the-art image restoration. The second framework derives a hierarchical structure on top of the collaboration and is well fitted for source separation. Both models enjoy very important theoretical virtues as well.

2.1 Introduction

In traditional sparse modeling, it is assumed that a signal can be accurately represented by a sparse linear combination of atoms from a (learned) dictionary. A large class of signals, including most natural images and sounds, is well described by this model, as demonstrated by numerous state-of-the-art results in various signal processing applications.

From a data modeling point of view, sparsity can be seen as a form of *regularization*, that is, as a device to restrict or control the set of coefficient values which are allowed in the model to produce an estimate of the data. The idea is that, by reducing the *flexibility* of the model (that is, the ability of a model to fit given data), one gains robustness by ruling out unrealistic estimates of the coefficients. In traditional sparse models, regularization translates to the requirement that "a few nonzero coefficients should be enough to represent the data well."

However, a great deal of flexibility in sparse models is still present in the usually very large possible number of nonzero coefficients subsets, a number which grows exponentially with the number of atoms in the dictionary. It turns out that, in certain applications (for example, inverse filtering), where the problem of estimating the model coefficients is very ill-posed, the reduction in the flexibility of the model (sometimes also referred to as *degree of freedom*) provided by traditional sparse models may not be enough to produce both stable and accurate estimates. Note that each set of nonzero coefficients defines one subspace, and therefore standard sparse models permit an often very large number of subspaces for data representation.

To further reduce the flexibility of the model, and obtain more stable estimates (which at the same time are often faster to compute), structured sparse models impose further restrictions on the possible subsets of the estimated active coefficients. For example, groups of atoms are forced to be simultaneously selected, thereby reducing the number of subspaces utilized to represent the data.

Another reason to add structure to sparse models is to incorporate prior knowledge about the signal of interest and how it can be represented. For example, in factor analysis problems, where one is interested in finding those factors which play a role in predicting response variables, an explanatory factor is often known to be represented by groups of input variables. In those cases, selecting the input variables in groups rather than individually produces better results [43].

Another complementary way of adding structure, stability, and prior information, to a model is via collaboration. Here, if multiple signals, which are known to follow the same model, are allowed to collaborate in the coding, further stability is obtained in the overall sparse representation, leading to more accurate estimates. In audio signal processing, for example, sounds are typically locally stationary, and dependency can be added between coefficients in consecutive short-time frames to improve the signal's estimate [39].

This chapter presents two concrete structured and collaborative sparse models that are shown to be effective for various applications in image and audio processing, and pattern recognition. Incorporating prior knowledge about the underlying signals, both models account for structure by constructing the dictionary with groups of atoms and calculating the coding in a collaborative way.

Section 1.2 presents a structured sparse model for image restoration problems such as inpainting, zooming, and deblurring [42]. Since these inverse problems are ill-posed, stabilizing the estimation is indispensable to obtain accurate estimates. The structured dictionary is composed of a number of PCA (Principal Component Analysis) bases, each capturing one class of signals, for example local patterns of different directions for images. A signal is estimated with a collaborative linear filter using a single PCA from the collection, and the best estimation is then used as the corresponding representation. This method is less flexible, more stable, and computationally much less expensive than traditional sparse modeling, leading to state-of-the-art results in a number of ill-posed problems. The model follows a Gaussian mixture model interpretation as well, thereby connecting these classical linear models with sparse coding.

Section 1.3 describes a collaborative hierarchical sparse model for a number of source separation and pattern recognition tasks [32]. The source separation problem naturally arises in a variety of signal processing applications such as audio segmentation and material identification in hyperspectral imaging. The structured dictionary in this case is composed by a set of sub-dictionaries, each of them learned to sparsely model one of a set of possible classes. The coding is performed by efficiently (and collaboratively) solving a convex optimization problem that combines standard sparsity with group sparsity. After coding the mixed signal, the sources can be recovered by reconstructing the sub-signals associated to the used sub-dictionaries. Having stable atom selection is thus particularly important for these tasks of source identification. The collaborative filtering in the coding is critical to further stabilize the sparse representation, letting the samples collaborate in identifying the classes (sources), while allowing different samples to have different internal representations (inside the group). As an important special case, one can address pattern recognition problems by characterizing the signals based on the group selection, similarly to the above model, where each PCA represents a class of signal. Comparing with the classic sparse modeling, again this collaborative and structured sparse model improves both the signal recognition and reconstruction.

The rest of this chapter provides additional details on these examples of structured and collaborative sparse modeling, further connects between them, and presents a number of examples in image restoration, source separation in audio, image classification, and hyperspectral image segmentation. For additional details and theoretical foundations, as well as the full color figures here reproduced in black and white, see [9, 31, 32, 42].

2.2 Image restoration and inverse problems

Image restoration often requires to solve an inverse problem, where the image $\mathbf{u}$ to be recovered is estimated from a measurement

$$\mathbf{y} = \mathbf{A}\mathbf{u} + \epsilon,$$

obtained through a non-invertible linear degradation operator $\mathbf{A}$, and contaminated by an additive noise ϵ. Typical degradation operators include masking, sub-sampling in a uniform grid and convolution, the corresponding inverse problems often named inpainting or interpolation, zooming, and deblurring. Estimating $\mathbf{u}$ requires some prior information on the image, or equivalently, image models. Finding good image models is therefore at the heart of image estimation.

As mentioned in Section 1.1, sparse models have been shown to be very effective for some of the aforementioned image processing applications. First, because the sparse representation hypothesis applies very well to image data (in particular when image patches are considered instead of entire images), and second, due to the noise-rejection properties of sparsity-regularized regression models.

However, for certain applications, the degree of freedom of classical sparse models is too large for the estimates to be stable, unless further structure (prior information) is added to the system. This section describes a structured and collaborative sparse modeling framework introduced in [42] that provides general and computationally efficient solutions for image restoration problems. Preliminary theoretical results supporting this model can be found at [40], while applications for non-imaging types of signals are reported in [21].

2.2.1 Traditional sparse modeling

Traditional sparse estimation using learned dictionaries provides the foundation to several effective algorithms for solving inverse problems. However, issues such as highly correlated atoms in the dictionary (high mutual coherence), or excessive degree of freedom, often lead to instability and errors in the estimation. Some of these algorithms [13, 18, 23, 38] are briefly reviewed below.

A signal $\mathbf{u} \in \mathbb{R}^m$ is estimated by taking advantage of prior information specified via a dictionary $\Phi \in \mathbb{R}^{m \times n}$, having n columns corresponding to a set of atoms $\{\phi_i : i = 1, \ldots, n\}$, in which an approximation to $\mathbf{u}$ can be sparsely represented. This dictionary may be a basis (e.g., DCT, Wavelet) or some redundant frame with $n \geq m$. Sparsity means that $\mathbf{u}$ can be well approximated by a representation $\mathbf{u}_T$ over a subspace $\mathbb{V}_T$ generated by a small number $|T| \ll n$ of column vectors $\{\phi_i\}_{i \in T}$ of Φ:

$$\mathbf{u} = \mathbf{u}_T + \epsilon_T = \Phi \mathbf{x} + \epsilon_T, \tag{2.1}$$

where $\mathbf{x} \in \mathbb{R}^n$ is the transform coefficient vector with support $T = \{i : x_i \neq 0\}$, and $\|\epsilon_T\|^2 \ll \|\mathbf{u}\|^2$ is a small approximation error. The support set T is also commonly referred to as the *active set*. A sparse representation of $\mathbf{u}$ in terms of a redundant dictionary Φ can be calculated for example by solving the following ℓ_1-regularized regression problem [11],

$$\hat{\mathbf{x}} = \frac{1}{2} \arg\min_{\mathbf{x}} \|\Phi \mathbf{x} - \mathbf{u}\|^2 + \lambda \|\mathbf{x}\|_1, \tag{2.2}$$

which is also known as the Lagrangian version of the Lasso [33]. Other alternatives include for example the greedy Matching Pursuit algorithm [24]. The parameter λ controls the trade-off between sparsity and approximation accuracy. See [27] for more on this parameter and [45, 44] for non-parametric sparse models.

Sparse inversion algorithms try to estimate, from the degraded signal $\mathbf{y} = \mathbf{A}\mathbf{u} + \epsilon$, the support T and the coefficients $\mathbf{x}$ in T that specify the representation of $\mathbf{u}$ in the approximation subspace $\mathbb{V}_T$. It results from (2.1) that

$$\mathbf{y} = \mathbf{A}\Phi\mathbf{x} + \epsilon', \text{ with } \epsilon' = \mathbf{A}\epsilon_T + \epsilon. \tag{2.3}$$

This means that $\mathbf{y}$ is well approximated by the same sparse set T of atoms and the same coefficients $\mathbf{x}$ in the transformed dictionary $\Psi = \mathbf{A}\Phi$, whose columns are the transformed vectors $\{\psi_i = \mathbf{A}\phi_i\}_{i=1,\ldots,n}$. Therefore, a sparse approximation $\hat{\mathbf{y}} = \Psi\hat{\mathbf{x}}$ of $\mathbf{y}$ could in principle be calculated with the ℓ_1 minimization (2.2) using the transformed dictionary $\mathbf{A}\Phi$

$$\hat{\mathbf{x}} = \arg\min_{\mathbf{x}} \frac{1}{2} \|\Psi \mathbf{x} - \mathbf{y}\|^2 + \lambda \|\mathbf{x}\|_1. \tag{2.4}$$

The resulting sparse estimation of $\mathbf{u}$ is

$$\hat{\mathbf{u}} = \Phi\hat{\mathbf{x}}. \tag{2.5}$$

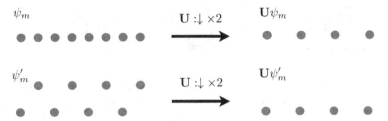

Figure 2.1 Coherence of the transformed dictionary $\mathbf{A}\Phi$. Here ϕ_i and $\phi_{i'}$ are respectively a constant atom and an oscillatory one in Φ, and $\mathbf{A}$ is the regular sub-sampling operator. Transformed by the operator, $\mathbf{A}\phi_i$ and $\mathbf{A}\phi_{i'}$ in $\mathbf{A}\Phi$ become identical.

This simple approach is often ineffective, motivating the introduction of structure and collaboration for inverse problems. This is because the geometric properties of non-invertible operators $\mathbf{A}$ such as sub-sampling on a uniform grid and convolution, do not guarantee correct recovery of representation coefficients via penalized ℓ_1 based estimation algorithms. For example, conditions based on the *restricted isometry property* [8, 12], are violated by such operators even for very low sparsity levels $|T|$. The same occurs with the recovery guarantees which are based on the *mutual coherence* of the transformed dictionary $\Psi = \mathbf{A}\Phi$, which measures the orthogonality of its atoms [34]. Stable and accurate sparse inverse problem estimation requires that the transformed dictionary Ψ be incoherent enough (that is, close to orthogonal). The estimation may be unstable if some of the columns in Ψ are too similar. This can be seen in the following toy example. Let ϕ_i and $\phi_{i'}$ be respectively a constant atom and an oscillatory one, and let $\mathbf{A}$ be a sub-sampling operator, as illustrated in Figure 2.1. After sub-sampling $\psi_i = \mathbf{A}\phi_i$ and $\psi_{i'} = \mathbf{A}\phi_{i'}$ coincide. Therefore the sparse estimation (1.4) can't distinguish between them, which results in an unstable inverse problem estimate (2.5). The coherence of Ψ depends on Φ as well as on the operator $\mathbf{A}$. Operators $\mathbf{A}$ such as sub-sampling on a uniform grid and convolution, usually lead to a coherent Ψ, which makes accurate inverse problem estimation difficult.

In the unstable scenario of inverse filtering, the degree of freedom of traditional sparse models is another source of instability. Without further constraints, for a dictionary with n atoms, the number of possible subspaces $\mathbb{V}_T$ where the signal can be represented is $\binom{n}{|T|}$, usually very large. For example, in a local patch-based sparse estimation setting with patches of size 8×8, typical values of $n = 256$ and $|T| = 8$ result in a huge $\binom{256}{8} \sim 10^{14}$ number of possible subspaces to choose from, further stressing the inaccuracy (and relatively high computational complexity) of estimating the correct $\mathbf{x}$ given the transformed dictionary $\mathbf{A}\Phi$. Adding structure to the dictionary, where atoms are selected in groups, reduces the number of possible subspaces, thereby stabilizing the selection. We describe next a particular case of this, where atoms are grouped and a single group is selected at a time, with a drastic reduction in the number of possible subspaces, and at the same time improvement in computational cost and restoration accuracy. For additional structured sparse models and results, see for example [4, 15, 20] as well as the model in the next section. This collection of works clearly show the advantages of adding structure to sparse models, in particular when the structure can be naturally derived or learned from

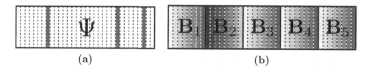

Figure 2.2 (a) Unstructured standard overcomplete dictionary. Each column represents an atom in the dictionary. In traditional sparse estimation it is allowed to select any combination of atoms (marked by the darkest columns). (b) The underlying structured sparse dictionary of the proposed approach. The dictionary is composed of a family of PCA bases whose atoms are pre-ordered by their associated eigenvalues. For each image patch, an optimal *linear* estimator is calculated in each PCA basis, and the best linear estimate among the bases is selected ($\mathbf{B}_2$).

the data. This has been shown to have advantages at all levels, from the theoretical results, with improved bounds when compared to standard sparsity, to the practical applications.

2.2.2 Structured sparse modeling

Figure 2.2(b) illustrates the proposed structured sparse modeling dictionary, compare with standard sparse modeling as represented in Figure 2.2(a). Structure is introduced to constrain the atoms selection and thus stabilize the sparse estimation. The dictionary is composed of a union of c blocks, each block being a PCA basis whose atoms are pre-ordered by their associated eigenvalues. To estimate a signal, a collaborative *linear* estimator is calculated in each PCA basis, and the best linear estimate among these bases is selected with a *non-linear* model selection at the block level. The resulting piecewise linear estimator (PLE) sharply reduces the degree of freedom of the model (only c options) with respect to the traditional sparse estimation, thus stabilizing the estimate.

The proposed structured sparse modeling is based on the Gaussian mixture model. This follows from the straightforward relationship between Gaussian functions and PCAs, basically the latter being the eigenvectors of the covariance matrix of the former.

Gaussian mixture model (GMM)

Natural images include rich and non-stationary content, whereas, when restricted to local windows, image structures appear to be simpler and are therefore easier to model. To this end, an image is first decomposed into overlapping $\sqrt{m} \times \sqrt{m}$ patches,

$$\mathbf{y}_j = \mathbf{A}_j \mathbf{u}_j + \epsilon_j, \quad \text{with} \quad 1 \leq j \leq p, \tag{2.6}$$

where p is the total number of patches, $\mathbf{A}_j$ is the degradation operator restricted to the patch j, $\mathbf{y}_j$ and $\mathbf{u}_j$ are respectively the degraded and original image patches, and ϵ_j is the noise restricted to the patch. The noise is assumed to be white Gaussian, $\epsilon_j \sim \mathcal{N}(\mathbf{0}, \sigma^2 \mathbf{I})$, where $\mathbf{I}$ is the identity matrix. Treated as a signal, each of the patches is estimated, and their corresponding estimates are finally combined and averaged, leading to the estimate of the image.

The GMM describes local image patches with a mixture of Gaussian distributions. Assume there exist c Gaussian distributions, $\{\mathcal{N}(\mu_r, \Sigma_r)\}_{1 \leq r \leq c}$, parameterized by their means μ_r and covariances Σ_r,

$$f(\mathbf{u}_j) = \frac{1}{(2\pi)^{m/2}|\Sigma_{r_j}|^{1/2}} \exp\left(-\frac{1}{2}(\mathbf{u}_j - \mu_{r_j})^T \Sigma_{r_j}^{-1}(\mathbf{u}_j - \mu_{r_j})\right). \quad (2.7)$$

Each image patch $\mathbf{u}_j$ is independently drawn from one of these Gaussian distributions. This is represented by an unknown index $r_j \in [1, c], 1 \leq j \leq p$. In this framework, estimating $\{\mathbf{u}_j\}_{1 \leq j \leq p}$ from $\{\mathbf{y}_j\}_{1 \leq j \leq p}$ can then be cast into the following subproblems:

- Estimate the Gaussian parameters $\{(\mu_r, \Sigma_r)\}_{1 \leq r \leq c}$, from the degraded data $\{\mathbf{y}_j\}_{1 \leq j \leq p}$.
- Identify the Gaussian distribution r_j that generates $\mathbf{u}_j, \forall 1 \leq j \leq p$.
- Estimate $\mathbf{u}_j$ from its generating Gaussian distribution $(\mu_{r_j}, \Sigma_{r_j}), \forall 1 \leq j \leq p$.

This problem is overall non-convex. The next section will present a maximum a posteriori expectation-maximization (MAP-EM) algorithm that calculates a local-minimum solution [2].

MAP-EM algorithm

Following initialization, which is addressed in detail in [42], the MAP-EM algorithm iterations alternate between two steps, an E-step and an M-step, described next.

E-step: signal estimation and model selection

In the E-step, the estimates of the Gaussian parameters $\{(\hat{\mu}_r, \hat{\Sigma}_r)\}_{1 \leq r \leq c}$ are assumed to be known. To simplify the notation, we assume without loss of generality that the Gaussian distributions have zero means, $\hat{\mu}_r = \mathbf{0}$, as one can always center the image patches with respect to the means.

Each image patch $\mathbf{u}_j$ is calculated to maximize the log a-posteriori probability (MAP) $\log f(\mathbf{u}_j | \mathbf{y}_j, \hat{\Sigma}_{r_j})$. It can be shown that this maximization can be calculated in two steps: first, the MAP estimate of $\mathbf{u}_j$ is computed in terms of each Gaussian model, and then the model with largest MAP probability is selected. More specifically:

1. **Signal estimation with each Gaussian model**
   Given a Gaussian signal model $\mathbf{u} \sim \mathcal{N}(\mathbf{0}, \hat{\Sigma}_r)$, the MAP estimate

   $$\hat{\mathbf{u}}_j^r = \arg\min_{\mathbf{u}} \left(\|\mathbf{A}_j \mathbf{u} - \mathbf{y}_j\|^2 + \sigma^2 \mathbf{u}^T \hat{\Sigma}_r^{-1} \mathbf{u} \right) \quad (2.8)$$

   is calculated. One can verify, [42], that the solution to (1.8) can be obtained with linear filtering

   $$\hat{\mathbf{u}}_j^r = \mathbf{H}_{r,j} \mathbf{y}_j, \quad (2.9)$$

   where

   $$\mathbf{H}_{r,j} = (\mathbf{A}_j^T \mathbf{A}_j + \sigma^2 \Sigma_r^{-1})^{-1} \mathbf{A}_j^T \quad (2.10)$$

   is a Wiener filter matrix. Assuming Σ_r is positive definite, since $\mathbf{A}_j^T \mathbf{A}_j$ is semi-definite positive, $\mathbf{A}_j^T \mathbf{A}_j + \sigma^2 \Sigma_r^{-1}$ is also positive definite and its inverse is well defined.

2. **Best model selection**
The best Gaussian model $\hat{r}_j$ that generates the maximum MAP probability $\log f(\mathbf{u}_j | \mathbf{y}_j, \hat{\Sigma}_{r_j})$ among all the models is then selected with the estimated $\hat{\mathbf{u}}_j^r$

$$\hat{r}_j = \arg\min_r \left(\|\mathbf{A}_j \hat{\mathbf{u}}_j^r - \mathbf{y}\|^2 + \sigma^2 (\hat{\mathbf{u}}_j^r)^T \Sigma_r^{-1} \hat{\mathbf{u}}_j^r + \sigma^2 \log \left| \hat{\Sigma}_r \right| \right). \tag{2.11}$$

The signal estimate is obtained by plugging in the best model $\hat{r}_j$ in the MAP estimate,

$$\hat{\mathbf{u}}_j = \hat{\mathbf{u}}_j^{\hat{r}_j}. \tag{2.12}$$

M-step: model estimation
In the M-step, the Gaussian model selection $\hat{r}_j$ and the signal estimate $\hat{\mathbf{u}}_j$ of all the patches are assumed to be known. Let $\mathcal{C}_r$ be the ensemble of the patch indices j that are assigned to the rth Gaussian model, i.e., $\mathcal{C}_r = \{j : \hat{r}_j = r\}$, and let $|\mathcal{C}_r|$ be its cardinality. The parameters of each Gaussian model are collaboratively found via maximum likelihood, using all the patches assigned to that Gaussian cluster,

$$\hat{\mu}_r = \frac{1}{|\mathcal{C}_r|} \sum_{j \in \mathcal{C}_r} \hat{\mathbf{u}}_j \quad \text{and} \quad \hat{\Sigma}_r = \frac{1}{|\mathcal{C}_r|} \sum_{j \in \mathcal{C}_r} (\hat{\mathbf{u}}_j - \hat{\mu}_r)(\hat{\mathbf{u}}_j - \hat{\mu}_r)^T. \tag{2.13}$$

The empirical covariance estimate may be improved through regularization when there is lack of data [29]. In this case, a simple eigenvalue-based regularization is used, $\hat{\Sigma}_r \leftarrow \hat{\Sigma}_r + \varepsilon \mathbf{I}$, where ϵ is a small constant [42].

As the MAP-EM algorithm described above iterates, the joint MAP probability of the observed signals, $f(\{\hat{\mathbf{u}}_j\}_{1 \leq j \leq p} | \{\mathbf{y}_j\}_{1 \leq j \leq p}, \{\hat{\mu}_r, \hat{\Sigma}_r\}_{1 \leq r \leq c})$, always increases. This can be observed by interpreting the E- and M-steps as a block-coordinate descent optimization algorithm [19]. In the experiments reported below, both the patch clustering and resulting PSNR converge after a few iterations.

The computational cost of the EM-MAP algorithm is dominated by the E-step, which consists basically of a set of linear filtering operations. For typical applications such as zooming and deblurring where the degradation operators $\mathbf{A}_j$ are translation-invariant and do not depend on the patch index j, i.e., $\mathbf{A}_j \equiv \mathbf{A}, \forall 1 \leq j \leq p$, the Wiener filter matrices $\mathbf{H}_{r,j} \equiv \mathbf{H}_r$ (1.10) can be precomputed for each of the c Gaussian distributions. Calculating (1.9) thus requires only $2m^2$ floating-point operations (flops), where m is the image patch size. For a translation-variant degradation $\mathbf{A}_j$, random masking for example, $\mathbf{H}_{r,i}$ needs to be calculated at each position where $\mathbf{A}_j$ changes. In this case, the calculation is dominated by the matrix inversion in (1.10) which can be implemented with $m^3/3$ flops through a Cholesky factorization [6].

Structured sparse estimation in PCA bases
The PCA bases bridge the GMM/MAP-EM framework presented above with the sparse estimation model described in Section 2.2.1, the former leading to structured sparse estimation.

Given data $\{\mathbf{u}_j\}$, the PCA basis is defined as the orthonormal matrix that diagonalizes its empirical covariance matrix $\Sigma_r = E[\mathbf{u}_j \mathbf{u}_j^T]$,

$$\Sigma_r = \mathbf{B}_r \mathbf{S}_r \mathbf{B}_r^T, \qquad (2.14)$$

where $\mathbf{B}_r$ is the PCA basis and $\mathbf{S}_r = \text{diag}(\lambda_1^r, \ldots, \lambda_m^r)$ is a diagonal matrix, whose diagonal elements $\lambda_1^r \geq \lambda_2^r \geq \ldots \geq \lambda_m^r$ are the sorted eigenvalues.

Transforming $\mathbf{u}_j$ from the canonical basis to the PCA basis $\mathbf{x}_j^r = \mathbf{B}_r^T \mathbf{u}_j$, one can verify that the MAP estimate (2.8)–(2.10) can be equivalently calculated as

$$\hat{\mathbf{u}}_j^r = \mathbf{B}_r \hat{\mathbf{x}}_j^r, \qquad (2.15)$$

where, following simple calculus, the MAP estimate of the PCA coefficients $\hat{\mathbf{x}}_j^r$ is obtained by the linear problem

$$\hat{\mathbf{x}}_j^r = \arg\min_{\mathbf{x}} \left(\|\mathbf{A}_j \mathbf{B}_r \mathbf{x} - \mathbf{y}_j\|^2 + \sigma^2 \sum_{i=1}^{m} \frac{|x_i|^2}{\lambda_i^r} \right). \qquad (2.16)$$

Comparing (2.16) with (2.4), the MAP-EM estimation can thus be interpreted as a structured sparse estimation. Note also the collaboration obtained via the eigenvalues weighting of the norm. The fast decreasing eigenvalues provide additional "sparsity" inside the blocks, noting that almost identical results are obtained when keeping only the largest eigenvalues and setting the rest to zero.

As illustrated in Figure 2.2, the proposed dictionary Φ is overcomplete, composed of a family of PCAs, and is adapted to the image of interest thanks to the Gaussian model estimation in the M-step (which is equivalent to updating the PCAs), thus enjoying the same advantages as traditional sparse models. However, the PLE estimation is more structured than the one obtained using traditional nonlinear sparse models. The PLE is calculated with a *nonlinear* best basis selection and a *linear* estimation in each basis:

- **Nonlinear block sparsity** The dictionary is composed of a union of c PCA bases. To represent an image patch, the *nonlinear* model selection (2.11) in the E-step restricts the estimation to only one basis (m atoms out of cm selected in group). In this way, the number of possible subspaces is only c, compared to the $\binom{cm}{m}$ that a traditional sparse estimation method would allow, resulting in a very sharp reduction in the degree of freedom of the estimation.
- **Linear collaborative filtering** Inside each PCA basis, the atoms are pre-ordered by their associated eigenvalues, in decreasing order (which typically decay very fast, leading to sparsity inside the block as well). In contrast to the nonlinear sparse ℓ_1 estimation (2.4), the MAP estimate (2.16) implements the regularization with the ℓ_2 norm of the coefficients weighted by the eigenvalues $\{\lambda_i^r : 1 \leq i \leq m\}$, and is calculated with *linear* filtering, (2.9) and (2.10). The eigenvalues λ_i^r are computed from all the signals that were assigned to the same Gaussian distribution $\{\mathbf{u}_j : \hat{r}_j = r\}$. The resulting estimation is therefore *collaboratively* incorporating the information from all the signals in the same cluster [1]. The weighting scheme privileges the coefficients x_i

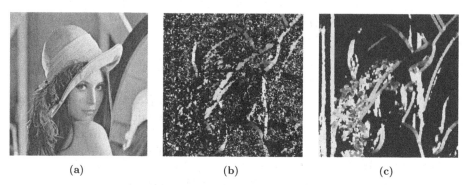

Figure 2.3 (a) Lena image. (b) Patch clustering obtained with the initial directional PCAs. The patches are densely overlapped and each pixel represents the model r_j selected for the 8×8 patch around it, different gray levels encoding different direction values of $\hat{r}_j$, from 1 to $c = 19$. (c) Patch clustering at the 2nd iteration.

corresponding to the principal directions with large eigenvalues λ_i, where the energy is likely to be high, and penalizes the others. For ill-posed inverse problems, the collaborative prior information incorporated in the eigenvalues $\{\lambda_i^r\}_{1 \leq i \leq m}$ further stabilizes the estimate. Note that this weighting scheme, which comes directly from the mixture of Gaussians/PCA model, is fundamentally different from standard and popular re-weighted schemes, e.g., [7, 10], where the weights are signal and coefficient dependent.

2.2.3 Experimental results

We now present a number of experimental results obtained with the above described model. For additional results and comparisons, see [42]. In particular, the reader is referred to this extended report showing how standard sparsity fails in challenging inverse problems, while PLE, at a significantly lower computational cost, succeeds in obtaining accurate reconstructions. In fact, the structured sparsity PLE model leads to state-of-the-art results in numerous image inverse problems at a fraction of the computational cost of comparable (in reconstruction accuracy) algorithms.

First, Figure 2.3 illustrates the Lena image and the corresponding patch clustering, i.e., (evolution of) the model selection $\hat{r}_j$. The patches are densely overlapped and each pixel in Figure 2.3(b) represents the model r_j selected for the 8×8 patch around it, different gray levels encoding different values of $\hat{r}_j$. For example, on the edge of the hat, patches where the image patterns follow similar directions are clustered together, as expected. On uniform regions such as the background, where there is no directional preference, all the bases provide equally sparse representations. As the $\log|\Sigma_r| = \sum_{i=1}^{m} \log \lambda_i^r$ term in the model selection (1.11) is initially 0 for all the Gaussian models, the clustering is random in these regions. As the MAP-EM algorithm evolves, the clustering improves.

Figure 2.4 illustrates, in an inpainting context on Barbara's cloth, which is rich in texture, the evolution of the patch clustering as well as that of typical PCA bases as

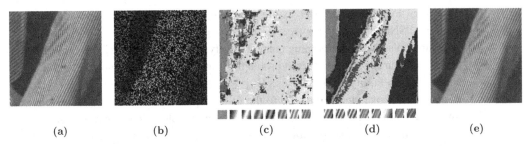

Figure 2.4 Evolution of the representations. (a) The original image cropped from Barbara. (b) The image masked with 30 percent available data. (c) Bottom: The first few atoms of an initial PCA basis corresponding to the texture on the right of the image. Top: The resulting patch clustering after the 1st iteration. Different gray levels represent different clusters. (d) Bottom: The first few atoms of the PCA basis updated after the 2nd iteration. Top: The resulting patch clustering after the 2nd iteration. (e) The inpainting estimate after the 2nd iteration (32.30 dB).

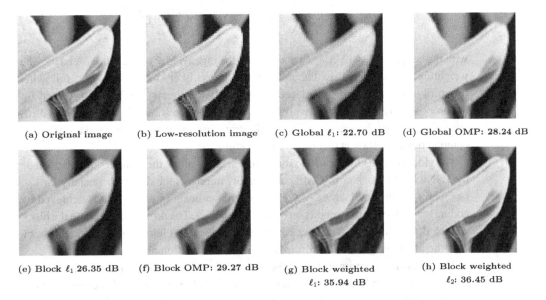

(a) Original image (b) Low-resolution image (c) Global ℓ_1: 22.70 dB (d) Global OMP: 28.24 dB

(e) Block ℓ_1 26.35 dB (f) Block OMP: 29.27 dB (g) Block weighted ℓ_1: 35.94 dB (h) Block weighted ℓ_2: 36.45 dB

Figure 2.5 Comparison of different estimation methods on super-resolution zooming. (a) The original image cropped from Lena. (b) The low-resolution image, shown at the same scale by pixel duplication. From (c) to (h) are the super-resolution results obtained with different estimation methods. See text for more details.

the MAP-EM algorithm iterates. The clustering becomes cleaner as the algorithm iterates. Some high-frequency atoms are promoted to better capture the oscillatory patterns, resulting in a significant PSNR improvement of more than 3 dB. On contour images such as Lena's hat illustrated in Figure 2.5, on the contrary, although the patch clustering becomes cleaner as the algorithm iterates, the resulting local PSNR evolves little after the initialization, which already produces an accurate estimation, since the initial directional PCA bases themselves are calculated over synthetic contour images. The resulting PSNRs typically converge in three to five iterations.

Figure 2.5 shows, in a zooming context on a typical region of Lena, the gain of introducing structure in sparse inverse problem estimation. An overcomplete dictionary Φ composed of a family of PCA bases $\{\mathbf{B}_r\}_{1 \leq r \leq c}$, illustrated in Figure 2.2(b), is learned as described above, and is then fed to the following estimation schemes. (i) **Global ℓ_1 and OMP**: the ensemble of Φ is used as an overcomplete dictionary, and the zooming estimation is calculated with the sparse estimate (1.4) through, respectively, an ℓ_1 minimization or an orthogonal matching pursuit (OMP). (ii) **Block ℓ_1 and OMP**: the sparse estimate is calculated in each PCA basis $\mathbf{B}_r$ through ℓ_1 minimization and OMP respectively, and the best estimate is selected with a model selection procedure similar to (2.11), thereby reducing the degree of freedom in the estimation with respect to the global ℓ_1 and OMP [41]. (iii) **Block weighted ℓ_1** : on top of the block ℓ_1, weights are included for each coefficient amplitude in the regularizer,

$$\hat{\mathbf{x}}_j^r = \arg\min_{\mathbf{x}} \left(\|\mathbf{A}_j \mathbf{B}_r \mathbf{x} - \mathbf{y}_j\|^2 + \sigma^2 \sum_{i=1}^{m} \frac{|\mathbf{x}_i|}{\tau_i^r} \right), \qquad (2.17)$$

with the weights $\tau_i^r = (\lambda_i^r)^{1/2}$, where λ_i^r are the eigenvalues of the rth PCA basis. The weighting scheme penalizes the atoms that are less likely to be important, following the spirit of the weighted ℓ_2 deduced from the MAP estimate. (iv) **Block weighted ℓ_2**: the proposed PLE. Comparing with (2.17), the difference is that the weighted ℓ_2 (2.16) takes the place of the weighted ℓ_1, thereby transforming the problem into a stable and computationally efficient piecewise linear estimation.

The global ℓ_1 and OMP produce some clear artifacts along the contours, which degrade the PSNRs. The block ℓ_1 or OMP considerably improves the results (especially for ℓ_1). Comparing with the block ℓ_1 or OMP, a very significant improvement is achieved by adding the collaborative weights on top of the block ℓ_1. The proposed PLE with the block weighted ℓ_2, computed with linear filtering, further improves the estimation accuracy over the block weighted ℓ_1, with a much lower computational cost.

These illustrative examples show the high quality of the results achieved with the proposed structured and collaborative sparse model. In [42] we show that this extremely simple and computationally efficient approach leads to state-of-the-art results in a number of applications. We now move to a different way of introducing structure and collaboration in sparse modeling, this time tailored to source identification and separation instead of signal reconstruction. Note that the PLE model can also be extended to consider more than one PCA (dictionary block) active at a time, coming closer to the model presented next.

2.3 Source identification and separation via structured and collaborative models

In the previous section it was shown that imposing structure in a model is an effective way of stabilizing the solutions of otherwise highly unstable inverse problems such as image deblurring. In general, adding structure to a model further restricts the span of a model to those solutions that are compatible with our prior knowledge.

In this section, another set of applications is presented where prior information naturally translates into structural constraints on the solutions of sparse models, leading to a new family of structured sparse models. The model now presented was first introduced in [32], where the reader is referred to for additional details as well as connections with the literature such as [17, 20]. The applications for source identification, in particular in music, was introduced in [31], and the pattern detection classification in [28].

Problem statement

In the *source separation* problem, an observed signal $\mathbf{y}$ is assumed to be a linear superposition (mixture) of several (say c) sources, plus additive noise, $\mathbf{y} = \sum_{r=1}^{c} \alpha_r \mathbf{y}_r + \epsilon$, and the primary task is to estimate each of the unmixed sources $\mathbf{y}_r$ out of it. If the task is only to identify the active sources, the problem is called *source identification*. In turn, the source identification problem includes as a special case the *pattern classification* problem, in which only one out of the c sources is assumed to be present at a time, that is, α_r is nonzero for only one index r_0. The E-step of the MAP-EM algorithm presented in Section 1.2.2 is an example of the latter case.

Note that, in the source identification setting (including pattern classification), since the original sources do not need to be recovered, the modeling can be done in terms of features extracted from the original signals in a non-bijective way.

In all the above problems, a sparsity assumption arises naturally when one can assume that, out of the c sources, only a few $k \ll c$ of them are actually active (nonzero). This can happen for example in a piece of music, where only a few instruments are playing simultaneously.

Continuing with the music example, traditional sparse modeling tools can be used to learn a dictionary $\mathbf{B}_r$ for each one of the c possible instruments [22], such that the sound produced by each instrument at each instant is efficiently represented by a few atoms from its corresponding dictionary. Concatenating the dictionaries (as shown in Figure 2.2), $\Phi = [\mathbf{B}_1|\mathbf{B}_2|\ldots|\mathbf{B}_c]$, any mixture signal produced by that ensemble will be represented accurately as a sparse linear combination of the atoms of this larger dictionary Φ. However, contrary to a general sparse model, in this case one expects the resulting sparsity patterns to have a particular structure: only a few blocks are active at a time, and inside these blocks, only a few atoms participate in the representation of the corresponding class.

Translated to the coefficients vector $\mathbf{x}$, this constitutes a hierarchical sparsity assumption, where only a few out of many possible groups are active (each group of coefficients is associated to an instrument sub-dictionary), and, within each active group, only a few atoms are required to represent the sound produced by the corresponding instrument accurately. The first model presented in this section imposes these assumptions explicitly in the model, promoting a hierarchical sparsity pattern in the sparse codes. In cases where these assumptions hold, it can be shown that the success in recovering the unmixed sources can only improve using this model.

The second model presented, also designed for source separation and identification problems, further exploits typical prior knowledge available in such problems. Continuing with the music example, one can assume by temporal continuity that, over short contiguous passages of a piece, the few instruments that are playing simultaneously will remain approximately the same, whereas the sound each of them produces can vary more rapidly. This new prior knowledge can be exploited by imposing, during coding, that all samples in a given time window have the same few active groups (same few instruments playing), thus resulting in a collaborative coding scheme. However, it cannot be expected that the sound produced in each of these samples by each instrument will be the same, thus the collaboration has to address only the detection of the active groups, while allowing the in-group sparsity patterns (representing the actual sound at each instant) to vary from sample to sample.

We begin this section by briefly introducing the Lasso and Group Lasso sparse regression models, which form the foundation to the models described above. The hierarchical and collaborative hierarchical models are then presented, along with examples of their application to tasks from image and audio analysis.

2.3.1 The Group Lasso

Given a data sample $\mathbf{y}$, and a dictionary Φ, the ℓ_1 regularizer in the Lasso formulation (2.4) induces sparsity in the resulting code $\mathbf{x}$. This is desirable not only from a regularization point of view, but also from a model selection perspective, where one wants to identify the features or factors (atoms) that play an active part in generating each sample $\mathbf{y}$.

In many situations, however, one knows that certain groups of features can become active or inactive only simultaneously (e.g. measurements of gene expression levels). In that case, structure can be added to the model so that the relevant factors are represented not as singletons, but as predefined groups of atoms. This particular case of structured sparsity was introduced in [43] and is referred to as the Group Lasso model.

In the setting presented here, the coefficients vector $\mathbf{x}$ is partitioned into blocks, $\mathbf{x} = (\mathbf{x}_1, \mathbf{x}_2, \ldots, \mathbf{x}_c)$, such that each block corresponds to the sub-dictionary with the same index in $\Phi = [\mathbf{B}_1 | \mathbf{B}_2 | \ldots | \mathbf{B}_c]$. To perform group-selection according to this structure, the Group Lasso solves the following problem,

$$\min_{\mathbf{x} \in \mathbb{R}^n} \frac{1}{2} \|\mathbf{y} - \Phi \mathbf{x}\|_2^2 + \lambda \sum_{r=1}^{c} \|\mathbf{x}_r\|_2. \tag{2.18}$$

The Group Lasso regularizer $\sum_{r=1}^{n} \|\mathbf{x}_r\|_2$ can be seen as an ℓ_1 norm on Euclidean norms of the sub-vectors of coefficients from the same group $\mathbf{x}_r$. This is a generalization of the ℓ_1 regularizer, as the latter arises from the special case where each coefficient in $\mathbf{x}$ is its own group (the groups are singletons), and as such, its effect on the groups of $\mathbf{x}$ is also a natural generalization of the one obtained with the Lasso: it "turns on/off" coefficients in groups. Note that in contrast with the PLE model presented before, and since now the number of active blocks is unknown, an optimization of this type is needed. When only one block is active at a time as in the PLE model, it is more efficient to simply try all blocks in the search for the best one.

The model selection properties of the Group Lasso have been studied both from the statistical (consistency, oracle properties) and compressive sensing (exact signal recovery) points of view, in both cases as extensions of the corresponding results for the Lasso [3, 14, 43].

2.3.2 The Hierarchical Lasso

The Group Lasso trades sparsity at the single-coefficient level with sparsity at a group level, while, inside each group, the solution is generally dense. If Group Lasso is applied to analyze a mixture of signals with a dictionary Φ constructed by concatenating class-specific sub-dictionaries $\mathbf{B}_r$ (as explained before), it will be unable to recover the hierarchical sparsity pattern in the coefficient vector. It will produce a vector of coefficients where, hopefully, only the groups of coefficients corresponding to the sources active in the mixture are nonzero. However, due to the properties of the Group Lasso regularizer, the solutions within each group will be generally dense, that is, all or most of the coefficients in an active group will be nonzero.

Since each sub-dictionary $\mathbf{B}_r$ is learned for representing signals from its class in a sparse manner, the entire dictionary Φ is appropriate to sparsely represent signals of all the classes, as well as mixtures of them. Therefore, a sparsity-aware method such as the Lasso is consistent with that model, and will produce efficient representations of such signals. However, Lasso will not be aware of the group structure, and will thus be unable to perform model selection.

The following model, referred to as Hierarchical Lasso (HiLasso hereafter), combines the group-selection properties of Group Lasso while being consistent with the in-group sparsity assumptions of the Lasso by simply combining the regularization terms from those two formulations (again, see [32] for the original developments on this and relations to the literature),

$$\min_{\mathbf{x} \in \mathbb{R}^n} \frac{1}{2} \|\mathbf{y}_j - \Phi \mathbf{x}\|_2^2 + \lambda_2 \sum_{r=1}^{c} \|\mathbf{x}_r\|_2 + \lambda_1 \|\mathbf{x}\|_1. \tag{2.19}$$

The hierarchical sparsity pattern produced by the solutions of (2.19) is depicted in Figure 2.6(a). For simplicity of the description all the groups are assumed to have the same number of elements. The extension to the general case is obtained by multiplying each group norm by the square root of the number of atoms inside the corresponding group. This model then achieves the desired effect of promoting sparsity at the group/class level while at the same time leading to overall sparse feature selection.

The selection of λ_1 and λ_2 has an important influence on the sparsity of the obtained solution. Intuitively as λ_2/λ_1 increases, the group constraint becomes dominant and the solution tends to be more sparse at a group level but less sparse within groups.

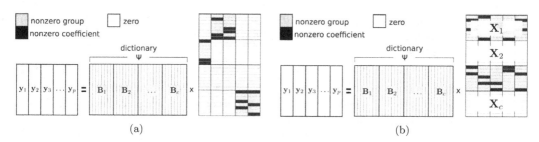

Figure 2.6 Sparsity patterns induced by the described model selection programs. (a) HiLasso. (b) Collaborative HiLasso. Notice that the C-HiLasso imposes the same group-sparsity pattern in all the samples (same class), whereas the in-group sparsity patterns can vary between samples (samples themselves are different).

As a note of warning, the HiLasso model should not be confused with the Elastic Net (EN) of Zou and Hastie [46]. Although superficially very similar in the formulation, the "only" difference is that the ℓ_2 term is squared in EN, the resulting model selection properties are radically different. For instance, EN is incapable, just as the Lasso, of performing group model selection.

2.3.3 Collaborative Hierarchical Lasso

In numerous applications, one expects that certain collections of samples y_j share the same active components from the dictionary, that is, the indices of the nonzero coefficients in x_j are the same for all the samples in the collection. Imposing such dependency in the ℓ_1 regularized regression problem gives rise to the so-called collaborative (also called "multitask" or "simultaneous") sparse coding problem [16, 25, 35, 36]. Note that PLE imposed the collaboration differently, by jointly designing the PCA and using the corresponding eigenvalues in the weighted optimization metric (this all coming from the MAP estimators).

More specifically, considering the matrix of coefficients $\mathbf{X} = [\mathbf{x}_1, \ldots, \mathbf{x}_p] \in \mathbb{R}^{n \times p}$ associated with the reconstruction of the samples $\mathbf{Y} = [\mathbf{y}_1, \ldots, \mathbf{y}_p] \in \mathbb{R}^{m \times p}$, the collaborative sparse coding model is given by

$$\min_{\mathbf{X} \in \mathbb{R}^{n \times p}} \frac{1}{2} \|\mathbf{Y} - \Phi \mathbf{X}\|_F^2 + \lambda \sum_{i=1}^{n} \|\mathbf{x}^i\|_2, \qquad (2.20)$$

where $\mathbf{x}^i \in \mathbb{R}^p$ is the ith row of $\mathbf{X}$, that is, the vector of the p different values that the coefficient associated to the ith atom takes for each sample $j = 1, \ldots, p$. Extending this idea to the Group Lasso, a *collaborative Group Lasso* (C-GLasso) formulation can be obtained,

$$\min_{\mathbf{X} \in \mathbb{R}^{n \times p}} \frac{1}{2} \|\mathbf{Y} - \Phi \mathbf{X}\|_F^2 + \lambda \sum_{r=1}^{c} \|\mathbf{X}_r\|_2, \qquad (2.21)$$

where $\mathbf{X}_r$ indicates the sub-matrix of rows of $\mathbf{X}$ which belong to group r. This regularizer is the natural extension of the regularizer in (1.18) for the collaborative case.

The second model described in this section, called Collaborative Hierarchical Lasso (C-HiLasso), combines the collaborative coding idea with the hierarchical sparse model presented in the previous section. However, in this case, *collaboration is performed only at the group level*, leaving the ℓ_1 part decoupled between samples. The formulation is as follows [32],

$$\min_{\mathbf{X}\in\mathbb{R}^{n\times p}} \frac{1}{2}\|\mathbf{Y}-\Phi\mathbf{X}\|_F^2 + \lambda_2\sum_{r=1}^{c}\|\mathbf{X}_r\|_2 + \sum_{j=1}^{p}\lambda_1\|\mathbf{x}_j\|_1. \quad (2.22)$$

The sparsity patterns obtained with solutions to (2.22) is shown in Figure 2.6(b). This model contains the collaborative Group Lasso as a particular case when $\lambda_1 = 0$, and the *non-collaborative* Lasso (effectively decomposing into p Lasso problems, one for each signal $\mathbf{u}_j, j = 1,\ldots,p$) by setting $\lambda_2 = 0$. Additional levels of hierarchy and collaboration can be easily included in the model.

In this way, the model encourages all the signals to share the same groups (classes), while allowing the active sets inside each group to vary between signals. The idea behind this is to model mixture signals such as the music example presented in the introduction to this section, where it is reasonable to assume that the same instruments (groups) are active during several consecutive signal samples, but the actual sound produced by a given instrument varies from sample to sample.

The coding of C-HiLasso is described in detail in [32]. The algorithm is an iterative procedure that decouples the overall problem into two simpler sub-problems: one that breaks the multi-signal problem into p single-signal Lasso-like sparse coding problems, and another that treats the multi-signal case as a collaborative Group Lasso-like problem. The scheme is guaranteed to converge to the global optimum and it consists of a series of simple vector soft-thresholding operations derived from a combination of the Alternating Direction Method of Multipliers (ADMOM) [5] with SPARSA [37].

As with the traditional sparse regression problems, Lasso and Group Lasso, it is of special interest to know under which conditions the HiLasso will be able to recover the true underlying sparse vector of coefficients $\mathbf{x}$ given an observation $\mathbf{y}$. This is of paramount importance to the source identification and classification problems, where one is interested more in $\mathbf{x}$ and its active set rather than in recovering $\mathbf{x}$. This has been carefully studied in [32], where we provide conditions for the existence of unique solutions of the HiLasso model and for the correct recovery of the active set.

2.3.4 Experimental results

This section presents four different examples to show how C-HiLasso can be applied in signal processing tasks. First, three different examples on source separation and identification, covering a wide range of signal types, are presented. Then, an object recognition application is presented to illustrate how the ideas behind C-HiLasso can be used for pattern recognition tasks.

Digits separation under missing information

This example addresses the separation of digit images. Although this is a rather artificial example, it is useful to clearly show the concepts explained in the previous sections. Hand-written digit images are an example where sparse models have been shown to be effective [28]. In this case, each sample vector contains the gray level intensities of an $\sqrt{m} \times \sqrt{m}$ image of a hand-written digit. The dataset used is the USPS dataset, which contains several samples of each digit from 0 to 9, and is divided, as usual, into a training and a testing subset.

In this example, each sub-dictionary $\mathbf{B}_r$ is learned from the training samples to sparsely represent a single digit. The set of mixtures, $\mathbf{Y}$, is then simulated by drawing random digits "3" and "5" from the testing dataset, and then setting 60 percent of the pixels to 0. The simulated input was then encoded using C-HiLasso and Lasso. Figure 2.7 shows the recovered coefficients for each method. One can see that only C-HiLasso can successfully detect which digits were present in the mixture. The collaborative coding technique, that considers jointly all signals to perform the model selection, is crucial

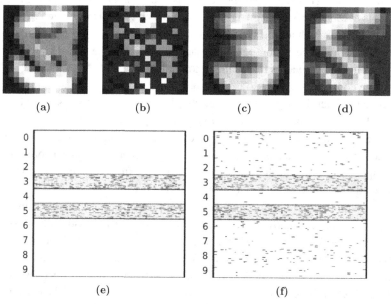

Figure 2.7 Example of recovered digits (3 and 5) from a mixture with 60 percent of missing components. (a) Noiseless mixture. (b) Observed mixture with missing pixels highlighted in dark gray. (c) and (d) Recovered digits 3 and 5 respectively. (e) and (f) Active set recovered for all samples using the C-HiLasso and Lasso respectively. The coefficients corresponding to the sub-dictionaries for digits 3 and 5 are marked as gray bands. Active sets of the recovered coefficients matrix $\mathbf{X}$ are shown as a binary matrix the same size as $\mathbf{X}$ (atom indices in the vertical axis and sample indices in the horizontal axis), with black dots indicating nonzero coefficients. Notice that the C-HiLasso exploits the hypothesis of collaborative group-sparsity, succeeding in recovering the correct active groups in all the samples. The Lasso, which lacks this prior knowledge, is clearly not capable of doing so, and active sets spread all over the groups.

to overcome the challenges of missing information. One can further support this by a quantitative measure of recovery performance by looking at the "separation error" [30], $\frac{1}{pc}\sum_{r=1}^{c}\sum_{j=1}^{p}\left\|\mathbf{u}_j^r - \hat{\mathbf{u}}_j^r\right\|_2^2$, where $\mathbf{u}_j^r$ is the component corresponding to source r in the signal j, and $\hat{\mathbf{u}}_j^r$ is the recovered one. See [32] for further details.

Source identification in audio

In this section C-HiLasso is used to automatically identify the sources present in a mixture of audio signals. The goal is to identify the speakers talking simultaneously on a single recording.

Audio signals have in general very rich structures and their properties rapidly change over time. A natural approach is to decompose them into a set of overlapping local windows, where the properties of the signal remain stable. There is a straightforward analogy with the approach explained in Section 2.2.2, where images were decomposed into collections of patches.

A challenging aspect when identifying audio sources is to obtain features that are specific to each source and at the same time invariant to changes in the fundamental frequency (tone) of the sources. In the case of speech, a common choice is to use the short-term power spectrum envelopes as feature vectors (refer to [31] for details on the feature extraction process and implementation). The spectral envelope in human speech varies with time, producing different patterns for each phoneme. Thus, a speaker does not produce a unique spectral envelope, but a set of spectral envelopes that live in a union of manifolds. Since such manifolds are well represented by sparse models, the problem of speaker identification is well suited for a sparse modeling framework.

For this experiment, the dataset consists of recordings of five different German radio speakers, two female and three male. Each recording is six minutes long. One quarter of the samples were used for training, and the rest for testing. For each speaker, a sub-dictionary was learned from the training dataset. For testing, 10 non-overlapping frames of 15 seconds each were extracted (including silences made by the speakers while talking), and encoded using C-HiLasso. The experiment was repeated for all possible combinations of two speakers, and all the speakers talking alone. The results are presented in Figure 2.8. C-HiLasso manages to automatically detect the number of sources very accurately.

Hyperspectral images

Hyperspectral imaging (HSI) is a classical example where class mixtures naturally appear. Low spatial resolution distorts the geometric features in the scene, and introduces the possibility of having multiple materials inside a pixel. In addition, partial occlusions caused by elevation differences will also cause such mixtures. For example, if there are tree branches over a road in the scene, the measured pixels are a combination of the energy reflected from the tree leaves and from the partially occluded road. Therefore, in general, the pixels in the acquired scene are *not* pure. This effect is known as *spectral mixing*. In general, one may assume that the pixels in the scene contain mixtures of multiple materials. Therefore, a realistic approach to HSI classification is to allow for a pixel to have one or more labels, each corresponding to a material class. In

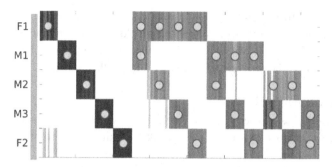

Figure 2.8 Speaker identification results obtained with C-HiLasso. Each column corresponds to the sources identified for a specific time frame, with the true ones marked by light gray circles. The vertical axis indicates the estimated activity of the different sources, where darker shades indicate higher energy. For each possible combination of speakers, 10 frames (15 seconds of audio) were evaluated.

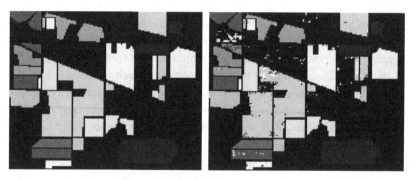

Figure 2.9 Ground-truth classification of the standard Indian Pines HSI, followed by the classification obtained via a collaborative and hierarchical structured sparse model.

[9] we developed ideas related to the C-HiLasso model, including the incorporation of block cross-incoherence and spatial regularization in the collaborative coding, to address this challenge, obtaining state-of-the-art results for HSI classification from significantly under-sampled data. An example is shown in Figure 2.9.

Pattern classification

The task in this example is to detect the presence of an object belonging to a specific class in an image, by classifying each patch in the image as either belonging to that object class (in this case, "bike") or "background." The algorithm consists of an offline stage in which a dictionary is learned for each class ("bike" and "background") using training images, and an online stage where it is validated on another separate set of images. The images were taken from the Graz02 "bikes" dataset [26]. Example results are shown in Figure 2.10, see [28] for details. Note that here, as in the PLE model, a single block is active at a time per patch/pixel, making the optimization of the hierarchical model even simpler.

(a)   (b)   (c)   (d)

Figure 2.10   Bike detection on the Graz dataset. Two example images are shown, with their corresponding detection maps to the right. Lighter shades indicate that the corresponding patches are more "bike-like."

2.4 Concluding remarks

In this chapter we introduced models for collaborative and structured sparse representations, leading to state-of-the-art results in classical applications and expanding the reach of sparse modeling to new arenas. We motivated such models by the need both to stabilize the coding and to introduce available prior knowledge about the signals and tasks at hand. This second generation of sparse models is still in infancy (while standard sparse models are already in adolescence), opening the door to numerous new developments at all levels, from the theory to the applications into old and new problems.

Acknowledgements

We thank our collaborators on the topics described in this chapter: S. Mallat, Y. Eldar, L. Carin, Z. Xing, J. Greer, P. Cancela, and E. Bosch. We have learned from them a lot, and they make this research much more fun. Work partially supported by NSF, ONR, NGA, ARO, and NSSEFF.

References

[1] J. Abernethy, F. Bach, T. Evgeniou, and J.P. Vert. A new approach to collaborative filtering: Operator estimation with spectral regularization. *J Mach Learn Res*, 10:803–826, 2009.

[2] S. Allassonniere, Y. Amit, and A. Trouvé. Towards a coherent statistical framework for dense deformable template estimation. *J R Statist Soc B*, 69(1):3–29, 2007.

[3] F. Bach. Consistency of the group lasso and multiple kernel learning. *J Mach Learn Res*, 9:1179–1225, 2008.

[4] R. G. Baraniuk, V. Cevher, M. F. Duarte, and C. Hegde. Model-based compressive sensing. *IEEE Trans IT*, 56(4):1982–2001, 2010.

[5] D. Bertsekas and J. Tsitsiklis. *Parallel and Distributed Comptutation: Numerical Methods*. Prentice Hall, 1989.

[6] S. P. Boyd and L. Vandenberghe. *Convex Optimization*. Cambridge University Press, 2004.

[7] E. J. Candès, M. Wakin, and S. Boyd. Enhancing sparsity by reweighted ℓ_1 minimization. *J Fourier Anal Appl*, 14(5):877–905, 2008.

[8] E.J. Candés and T. Tao. Near-optimal signal recovery from random projections: Universal encoding strategies? *IEEE Trans Inform Theory*, 52(12):5406–5425, 2006.

[9] A. Castrodad, Z. Xing, J. Greer, et al. Learning discriminative sparse models for source separation and mapping of hyperspectral imagery. Submitted, 2010.

[10] R. Chartrand and W. Yin. Iteratively reweighted algorithms for compressive sensing. 2008.

[11] S. S. Chen, D. L. Donoho, and M. A. Saunders. Atomic decomposition by basis pursuit. *SIAM J Sci Comp*, 20:33, 1999.

[12] D. L. Donoho. Compressed sensing. *IEEE Trans Inform Theory*, 52(4):1289–1306, 2006.

[13] M. Elad, J. L. Starck, P. Querre, and D. L. Donoho. Simultaneous cartoon and texture image inpainting using morphological component analysis (MCA). *Appl Comput Harmon Anal*, 19(3):340–358, 2005.

[14] Y. C. Eldar, P. Kuppinger, and H. Bölcskei. Compressed sensing of block-sparse signals: Uncertainty relations and efficient recovery. *IEEE Trans Sig Proc*, 58:3042–3054, 2010.

[15] Y. C. Eldar and M. Mishali. Robust recovery of signals from a structured union of subspaces. *IEEE Trans Inform Theory*, 55(11):5302–5316, 2009.

[16] Y. C. Eldar and H. Rauhut. Average case analysis of multichannel sparse recovery using convex relaxation. *IEEE Trans Inform Theory*, 56:505–519, 2010.

[17] J. Friedman, T. Hastie, and R. Tibshirani. A note on the group lasso and a sparse group lasso. Preprint, 2010.

[18] O. G. Guleryuz. Nonlinear approximation based image recovery using adaptive sparse reconstructions and iterated denoising–Part II: Adaptive algorithms. *IEEE Trans Image Proc*, 15(3):555–571, 2006.

[19] R. J. Hathaway. Another interpretation of the EM algorithm for mixture distributions. *Stat Prob Letters*, 4(2):53–56, 1986.

[20] R. Jenatton, J. Audibert, and F. Bach. Structured variable selection with sparsity-inducing norms. Technical Report arXiv:0904.3523v1, INRIA, 2009.

[21] F. Leger, G. Yu, and G. Sapiro. Efficient matrix completion with Gaussian models. In http://arxiv.org/abs/1010.4050, 2010.

[22] J. Mairal, F. Bach, J. Ponce, and G. Sapiro. Online dictionary learning for sparse coding. In *ICML '09: Proc 26th Ann Inte Conf Mach Learning*, 689–696, New York, 2009.

[23] J. Mairal, M. Elad, and G. Sapiro. Sparse representation for color image restoration. *IEEE Trans Image Proc*, 17, 2008.

[24] S. G. Mallat and Z. Zhang. Matching pursuits with time-frequency dictionaries. *IEEE Trans Sig Proc*, 41(12):3397–3415, 1993.

[25] M. Mishali and Y. C. Eldar. Reduce and boost: Recovering arbitrary sets of jointly sparse vectors. *IEEE Trans Sig Proc*, 56(10):4692–4702, 2008.

[26] A. Opelt, A. Pinz, M. Fussenegger, and P. Auer. Generic object recognition with boosting. *IEEE Trans PAMI*, 28(3), 2006.

[27] I. Ramirez and G. Sapiro. Universal regularizers for robust sparse coding and modeling. Submitted, http://arxiv.org/abs/1003.2941, 2010.

[28] I. Ramirez, P. Sprechmann, and G. Sapiro. Classification and clustering via dictionary learning with structured incoherence. *CVPR*, 2010.

[29] J. Schafer and K. Strimmer. A shrinkage approach to large-scale covariance matrix estimation and implications for functional genomics. *Statist Appl Genet Mol Biol*, 4(1):1175, 2005.

[30] N. Shoham and M. Elad. Alternating KSVD-denoising for texture separation. *IEEE 25th Convention Electr Electron Eng Israel*, 2008.

[31] P. Sprechmann, I. Ramirez, P. Cancela, and G. Sapiro. Collaborative sources identification in mixed signals via hierarchical sparse modeling. http://arxiv.org/abs/1010.4893, 2010.

[32] P. Sprechmann, I. Ramirez, G. Sapiro, and Y. C. Eldar. C-HiLasso: A collaborative hierarchical sparse modeling framework. http://arxiv.org/abs/1003.0400, 2010.

[33] R. Tibshirani. Regression shrinkage and selection via the lasso. *J Roy Stat Soci*: 267–288, 1996.

[34] J. Tropp. Greed is good: Algorithmic results for sparse approximation. *IEEE Trans Inform Theory*, 50(10):2231–2242, 2004.

[35] J. A. Tropp. Algorithms for simultaneous sparse approximation. Part ii: Convex relaxation. *Signal Processing, special issue "Sparse approximations in signal and image processing,"* 86:589–602, 2006.

[36] B. Turlach, W. Venables, and S. Wright. Simultaneous variable selection. *Technometrics*, 27:349–363, 2004.

[37] S. Wright, R. Nowak, and M. Figueiredo. Sparse reconstruction by separable approximation. *IEEE Trans Sig Proc*, 57(7):2479–2493, 2009.

[38] J. Yang, J. Wright, T. Huang, and Y. Ma. Image super-resolution via sparse representation. Accepted *IEEE Trans Image Proc*, 2010.

[39] G. Yu, S. Mallat, and E. Bacry. Audio denoising by time-frequency block thresholding. *IEEE Trans Sig Proc*, 56(5):1830–1839, 2008.

[40] G. Yu and G. Sapiro. Statistical compressive sensing of Gaussian mixture models. In http://arxiv.org/abs/1010.4314, 2010.

[41] G. Yu, G. Sapiro, and S. Mallat. Image modeling and enhancement via structured sparse model selection. *ICIP, Hong Kong*, 2010.

[42] G. Yu, G. Sapiro, and S. Mallat. Solving inverse problems with piecewise linear estimators: From Gaussian mixture models to structured sparsity. Submitted, http://arxiv.org/abs/1006.3056, 2010.

[43] M. Yuan and Y. Lin. Model selection and estimation in regression with grouped variables. *J Roy Stat Soc, Ser B*, 68:49–67, 2006.

[44] M. Zhou, H. Chen, J. Paisley, *et al*. Nonparametric Bayesian dictionary learning for analysis of noisy and incomplete images. IMA Preprint, Apr. 2010, http://www.ima.umn.edu/preprints/apr2010/2307.pdf.

[45] M. Zhou, H. Chen, J. Paisley, *et al*. Non-parametric Bayesian dictionary learning for sparse image representations. *Adv. NIPS*, 2009.

[46] H. Zou and T. Hastie. Regularization and variable selection via the elastic net. *J Roy Stat Soc Ser B*, 67:301–320, 2005.

3 Xampling: compressed sensing of analog signals

Moshe Mishali and Yonina C. Eldar

This chapter generalizes compressed sensing (CS) to reduced-rate sampling of analog signals. It introduces Xampling, a unified framework for low-rate sampling and processing of signals lying in a union of subspaces. Xampling consists of two main blocks: analog compression that narrows down the input bandwidth prior to sampling with commercial devices followed by a nonlinear algorithm that detects the input subspace prior to conventional signal processing. A variety of analog CS applications are reviewed within the unified Xampling framework including a general filter-bank scheme for sparse shift-invariant spaces, periodic nonuniform sampling and modulated wideband conversion for multiband communications with unknown carrier frequencies, acquisition techniques for finite rate of innovation signals with applications to medical and radar imaging, and random demodulation of sparse harmonic tones. A hardware-oriented viewpoint is advocated throughout, addressing practical constraints and exemplifying hardware realizations where relevant.

3.1 Introduction

Analog-to-digital conversion (ADC) technology constantly advances along the route that was delineated in the last century by the celebrated Shannon–Nyquist [1, 2] theorem, essentially requiring the sampling rate to be at least twice the highest frequency in the signal. This basic principle underlies almost all digital signal processing (DSP) applications such as audio, video, radio receivers, wireless communications, radar applications, medical devices, optical systems and more. The ever growing demand for data, as well as advances in radio frequency (RF) technology, have promoted the use of high-bandwidth signals, for which the rates dictated by the Shannon–Nyquist theorem impose demanding challenges on the acquisition hardware and on the subsequent storage and DSP processors. A holy grail of compressed sensing is to build acquisition devices that exploit signal structure in order to reduce the sampling rate, and subsequent demands on storage and DSP. In such an approach, the actual information contents should dictate the sampling

---

Compressed Sensing: Theory and Applications, ed. Yonina C. Eldar and Gitta Kutyniok. Published by Cambridge University Press. © Cambridge University Press 2012.

rate, rather than the ambient signal bandwidth. Indeed, CS was motivated in part by the desire to sample wideband signals at rates far below the Shannon–Nyquist rate, while still maintaining the vital information encoded in the underlying signal [3,4].

At its core, CS is a mathematical framework that studies rate reduction in a discrete setup. A vector $\mathbf{x}$ of length n represents a signal of interest. A measurement vector $\mathbf{y} = \mathbf{A}\mathbf{x}$ is computed using an $m \times n$ matrix $\mathbf{A}$. In a typical CS setup $m \ll n$, so that there are fewer measurements in $\mathbf{y}$ than the ambient dimension of $\mathbf{x}$. Since $\mathbf{A}$ is non-invertible in this setting, recovery must incorporate some prior knowledge on $\mathbf{x}$. The structure that is widely assumed in CS is sparsity, namely that $\mathbf{x}$ has only a few nonzero entries. Convex programming, e.g., ℓ_1 minimization, and various greedy methods have been shown to be successful in reconstructing sparse signals $\mathbf{x}$ from short measurement vectors $\mathbf{y}$.

The discrete machinery nicely captures the notion of reduced-rate sampling by the choice $m \ll n$ and affirms robust recovery from incomplete measurements. Nevertheless, since the starting point is of a finite-dimensional vector $\mathbf{x}$, one important aspect is not clearly addressed – how to actually acquire an analog input $x(t)$ at a low-rate. In many applications, our interest is to process and represent signals which arrive from the physical domain and are therefore naturally represented as continuous-time functions rather than discrete vectors. A conceptual route to implementing CS in these real-world problems is to first obtain a discrete high-rate representation using standard hardware, and then apply CS to reduce dimensionality. This, however, contradicts the motivation at the heart of CS: reducing acquisition rate as much as possible. Achieving the holy grail of compressive ADCs requires a broader framework which can treat more general signal models including analog signals with various types of structure, as well as practical measurement schemes that can be implemented in hardware. To further gain advantage from the sampling rate decrease, processing speed in the digital domain should also be reduced. Our goal therefore is to develop an end-to-end system, consisting of sampling, processing, and reconstruction, where all operations are performed at a low-rate, below the Nyquist-rate of the input.

The key to developing low-rate analog sensing methods is relying on structure in the input. Signal processing algorithms have a long history of leveraging structure for various tasks. As an example, MUSIC [5] and ESPRIT [6] are popular techniques for spectrum estimation that exploit signal structure. Model-order selection methods in estimation [7], parametric estimation and parametric feature detection [8] are further examples where structure is heavily exploited. In our context, we are interested in utilizing signal models in order to reduce sampling rate. Classic approaches to sub-Nyquist sampling include carrier demodulation [9], undersampling [10], and nonuniform methods [11–13], which all assume a linear model corresponding to a bandlimited input with predefined frequency support and fixed carrier frequencies. In the spirit of CS, where unknown nonzero locations result in a nonlinear model, we would like to extend the classical treatment to analog inputs with unknown frequency support, as well as more broadly to scenarios that involve nonlinear input structures. The approach we take in this chapter follows the recently proposed Xampling framework [14], which treats a nonlinear model of union of subspaces (UoS). In this structure, originally introduced by Lu and Do [15], the input signal belongs to a single subspace out of multiple, possibly even infinitely many, candidate subspaces. The exact subspace to which the signal belongs is unknown a priori.

In Section 3.2, we motivate the use of UoS modeling by considering two example sampling problems of analog signals: an RF receiver which intercepts multiple narrowband transmissions, termed multiband communication, but is not provided with their carrier frequencies, and identification of a fading channel which creates echoes of its input at several unknown delays and attenuations. The latter example belongs to a broad model of signals with finite rate of innovation (FRI), discussed in detail in Chapter 4 of this book. FRI models also include other interesting problems in radar and sonar. As we show throughout this chapter, union modeling is a key to savings in acquisition and processing resources.

In Section 3.3, we study a high-level architecture of Xampling systems [14]. The proposed architecture consists of two main functions: low-rate analog to digital conversion (X-ADC) and low-rate digital signal processing (X-DSP). The X-ADC block compresses $x(t)$ in the analog domain, by generating a version of the input that contains all vital information but with relatively lower bandwidth, often substantially below the Nyquist-rate of $x(t)$. The important point is that the chosen analog compression can be efficiently realized with existing hardware components. The compressed version is then sampled at a low-rate. X-DSP is responsible for reducing processing rates in the digital domain. To accomplish this goal, the exact signal subspace within the union is detected digitally, using either CS techniques or comparable methods for subspace identification, such as MUSIC [5] or ESPRIT [6]. Identifying the input's subspace allows one to execute existing DSP algorithms and interpolation techniques at the low rate of the streaming measurements, that is without going through reconstruction of the Nyquist-rate samples of $x(t)$. Together, when applicable, X-ADC and X-DSP alleviate the Nyquist-rate burden from the entire signal path. Pronounced as CS-Sampling (phonetically /k'sæmplm), the nomenclature Xampling symbolizes the combination between recent developments in CS and the successful machinery of analog sampling theory developed in the past century.

The main body of this chapter is dedicated to studying low-rate sampling of various UoS signal models in light of Xampling, capitalizing on the underlying analog model, compressive sensing hardware, and digital recovery algorithms. Section 3.4 introduces a framework for sampling sparse shift-invariant (SI) subspaces [16], which extends the classic notion of SI sampling developed for inputs lying in a single subspace [17, 18]. Multiband models [11–13, 19–21] are considered in Section 3.5 with applications to wideband carrier-unaware reception [22] and cognitive radio communication [23]. In particular, this section achieves the X-DSP goal, by considering multiband inputs consisting of a set of digital transmissions whose information bits are recovered and processed at the low-rate of the streaming samples. Sections 3.6 and 3.7 address FRI signals [24,25] and sequences of innovation [26], respectively, with applications to pulse stream acquisition and ultrasonic imaging [27,28]. In radar imaging [29], the Xampling viewpoint not only offers a reduced-rate sampling method, but also allows the authors to increase resolution in target identification and decrease the overall time–bandwidth product of the radar system (when the noise is not too large). Section 3.8 describes sampling strategies that are based on application of CS on discretized analog models, e.g., sampling a sparse sum of harmonic tones [30] and works on quantized CS radar [31–33].

Besides reviewing sampling strategies, we provide some insights into analog sensing. In Section 3.5, we use the context of multiband sampling to exemplify a full development cycle of analog CS systems, from theory to hardware. The cycle begins with a nonuniform method [19] that is derived from the sparse-SI framework. Analyzing this approach in a more practical perspective, reveals that nonuniform acquisition requires ADC devices with Nyquist-rate front end since they are connected directly to the wideband input. We next review the hardware-oriented design of the modulated wideband converter (MWC) [20,22], which incorporates RF preprocessing to compress the wideband input, so that actual sampling is carried out by commercial low-rate and low-bandwidth ADC devices. To complete the cycle, we take a glimpse at circuit challenges and solutions as reported in the design of an MWC hardware prototype [22]. The MWC appears to be the first reported wideband technology borrowing CS ideas with provable hardware that samples and processes wideband signals at a rate that is directly proportional to the actual bandwidth occupation and not the highest frequency (280 MHz sampling of 2 GHz Nyquist-rate inputs in [22]).

Xampling advocates use of traditional tools from sampling theory for modeling analog signals, according to which a continuous-time signal $x(t)$ is determined by a countable sequence $c[n]$ of numbers, e.g., a bandlimited input $x(t)$ and its equally spaced pointwise values $c[n] = x(nT)$. The UoS approach is instrumental in capturing similar infinite structures by taking to infinity either the dimensions of the individual subspaces, the number of subspaces in the union or both. In Section 3.8 we review alternative analog CS methods which treat continuous signals that are determined by a finite set of parameters. This approach was taken, for example, in the development of the random demodulator (RD) [30] and works on quantized CS radar [31–33]. Whilst effective in the finite scenarios for which they were developed, the application of these methods to general analog models (which possess a countable representation) can lead to performance degradation. We exemplify differences when comparing hardware and software complexities of the RD and MWC systems. Visualizing radar performance of quantized [33] vs. analog [29] approaches further demonstrates the possible differences. Based on the insights gained throughout this chapter, several operative conclusions are suggested in Section 3.9 for extending CS to general analog signals.

3.2 From subspaces to unions

The traditional paradigm in sampling theory assumes that $x(t)$ lies in a single subspace. Bandlimited sampling is undoubtedly the most studied example. Subspace modeling is quite powerful, as it allows perfect recovery of the signal from its linear and nonlinear samples under very broad conditions [17, 18, 34–36]. Furthermore, recovery can be achieved by digital and analog filtering. This is a very appealing feature of the subspace model, which generalizes the Shannon–Nyquist theorem to a broader set of input classes.

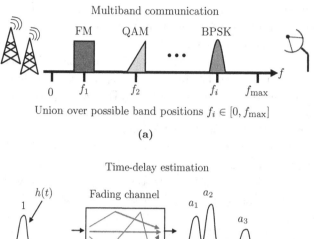

Figure 3.1 Example applications of UoS modeling. Adapted from [38], © [2011] IEEE.

Despite the simplicity and intuitive appeal of subspace modeling, in modern applications many signals are characterized by parameters which are not necessarily known to the sampler. As we will show now via several examples, we can often still describe the signal by a subspace model. However, in order to include all possible parameter choices, the subspace has to have large dimension with enough degrees of freedom to capture the uncertainty, leading to extremely high sampling rates. The examples below build the motivation for low-rate sampling solutions which we discuss in the rest of this chapter.

Consider first the scenario of a multiband input $x(t)$, which has sparse spectra, such that its continuous-time Fourier transform (CTFT) $X(f)$ is supported on N frequency intervals, or bands, with individual widths not exceeding B Hz. Figure 3.1 illustrates a typical multiband spectra. When the band positions are known and fixed, the signal model is linear, since the CTFT of any combination of two inputs is supported on the same frequency bands. This scenario is typical in communication, when a receiver intercepts several RF transmissions, each modulated on a different high carrier frequency f_i. Knowing the band positions, or the carriers f_i, allows the receiver to demodulate a transmission of interest to baseband, that is to shift the contents from the relevant RF band to the origin. Several demodulation topologies are reviewed in [37]. Subsequent sampling and processing are carried out at the low rate corresponding to the individual band of interest. When the input consists of a single transmission, an alternative approach to shift contents to baseband is by uniform undersampling at a properly chosen sub-Nyquist rate [10]. Nonuniform sampling methods that can treat more than a single transmission were developed in [12, 13], under the assumption that the digital recovery algorithm is provided with knowledge of the spectral support.

When the carrier frequencies f_i are unknown, we are interested in the set of all possible multiband signals that occupy up to NB Hz of the spectrum. In this scenario, the transmissions can lie anywhere below $f_{\max}$. At first sight, it may seem that sampling at the Nyquist rate

$$f_{\text{NYQ}} = 2f_{\max}, \tag{3.1}$$

is necessary, since every frequency interval below $f_{\max}$ appears in the support of some multiband $x(t)$. On the other hand, since each specific $x(t)$ in this model fills only a portion of the Nyquist range (only NB Hz), we intuitively expect to be able to reduce the sampling rate below f_{NYQ}. Standard demodulation cannot be used since f_i are unknown, which makes this sampling problem challenging.

Another interesting application is estimation of time delays from observation of a signal of the following form

$$x(t) = \sum_{\ell=1}^{L} a_\ell h(t - t_\ell), \quad t \in [0, \tau]. \tag{3.2}$$

For fixed time delays t_ℓ, (3.2) defines a linear space of inputs with L degrees of freedom, one per each amplitude a_ℓ. In this case, L samples of $x(t)$ can be used to reconstruct the input $x(t)$. In practice, however, there are many interesting situations with unknown t_ℓ. Inputs of this type belong to the broader family of FRI signals [24, 25], and are treated in detail in Chapter 4 of this book. For example, when a communication channel introduces multipath fading, the transmitter can assist the receiver in channel identification by sending a short probing pulse $h(t)$. Since the receiver knows the shape of $h(t)$, it can resolve the delays t_ℓ and use this information to decode the following information messages. Another example is radar, where the delays t_ℓ correspond to target locations, while the amplitudes a_ℓ encode Doppler shifts indicating target speeds. Medical imaging techniques, e.g., ultrasound, use signals of the form (3.2) to probe density changes in human tissues as a vital tool in medical diagnosis. Underwater acoustics also conform with (3.2). Since in all these applications, the pulse $h(t)$ is short in time, sampling $x(t)$ according to its Nyquist bandwidth, which is effectively that of $h(t)$, results in unnecessarily large sampling rates. In contrast, it follows intuitively from (3.2), that only $2L$ unknowns determine $x(t)$, namely $t_\ell, a_\ell, 1 \leq \ell \leq L$. Since with unknown delays, (3.2) describes a nonlinear model, subspace modeling cannot achieve the optimal sampling rate of $2L/\tau$, which in all the above applications can be substantially lower than Nyquist.

The example applications above motivate the need for signal modeling that is more sophisticated than the conventional single subspace approach. In order to capture real-world scenarios within a convenient mathematical formulation without unnecessarily increasing the rate, we introduce in the next section the Xampling framework which treats UoS signal classes and is applicable to many interesting applications. Using the Xampling framework, we will analyze sampling strategies for several union models in detail, and show that although sampling can still be obtained by linear filtering, recovery becomes more involved and requires nonlinear algorithms, following the spirit of CS.

3.3 Xampling

In this section, we introduce Xampling – our proposed framework for acquisition and digital processing of UoS signal models [14].

3.3.1 Union of subspaces

As motivated earlier, the key to reduced-rate sampling of analog signals is based on UoS modeling of the input set. The concept of allowing more than a single input subspace was first suggested by Lu and Do in [15]. We denote by $x(t)$ an analog signal in the Hilbert space $\mathcal{H} = L_2(\mathbb{R})$, which lies in a parameterized family of subspaces

$$x(t) \in \mathcal{U} \triangleq \bigcup_{\lambda \in \Lambda} \mathcal{A}_\lambda, \qquad (3.3)$$

where Λ is an index set, and each individual $\mathcal{A}_\lambda$ is a subspace of $\mathcal{H}$. The key property of the UoS model (3.3) is that the input $x(t)$ resides within $\mathcal{A}_{\lambda^*}$ for some $\lambda^* \in \Lambda$, but a priori, the exact subspace index λ^* is unknown. For example, multiband signals with unknown carriers f_i can be described by (3.3), where each $\mathcal{A}_\lambda$ corresponds to signals with specific carrier positions and the union is taken over all possible $f_i \in [0, f_{\max}]$. Pulses with unknown time delays of the form (3.2) also obey UoS modeling, where each $\mathcal{A}_\lambda$ is an L-dimensional subspace that captures the coefficients a_ℓ, whereas the union over all possible delays $t_\ell \in [0, \tau]$ provides an efficient way to group these subspaces to a single set $\mathcal{U}$.

Union of subspaces modeling enables treating $x(t)$ directly in its analog formulation. This approach is fundamentally different than previous attempts to treat similar problems, which rely on discretization of the analog input to finite representations. Namely, models in which both cardinalities, Λ and each $\mathcal{A}_\lambda$, are finite. Standard CS which treats vectors in $\mathbb{R}^n$ having at most k nonzeros is a special case of a finite representation. Each individual subspace has dimensions k, defined by the locations of the nonzeros, and the union is over $\binom{n}{k}$ possibilities of choosing the nonzero locations. In Section 3.8, we discuss in detail the difference between union modeling and discretization. As we show, the major consequences of imposing a finite representation on an analog signal that does not inherently conform to a finite model are twofold: model sensitivity and high computational loads. Therefore, the main core of this chapter focuses on the theory and applications developed for general UoS modeling (3.3). We note that there are examples of continuous-time signals that naturally possess finite representations. One such example are trigonometric polynomials. However, our interest here is in signals of the form described in Section 3.2, that do not readily admit a finite representation.

The union (3.3) over all possible signal locations forms a nonlinear signal set $\mathcal{U}$, where its nonlinearity refers to the fact that the sum (or any linear combination) of $x_1(t), x_2(t) \in \mathcal{U}$ does not lie in $\mathcal{U}$, in general. Consequently, $\mathcal{U}$ is a true subset of the linear affine space

$$\Sigma = \left\{ x(t) = \sum_{\lambda \in \Lambda} \alpha_\lambda x_\lambda(t) : \alpha_\lambda \in \mathbb{R}, x_\lambda(t) \in \mathcal{A}_\lambda \right\}, \quad (3.4)$$

which we refer to as the Nyquist subspace of $\mathcal{U}$. Since every $x(t) \in \mathcal{U}$ also belongs to Σ, one can in principle apply conventional sampling strategies with respect to the single subspace Σ [18]. However, this technically correct approach often leads to practically infeasible sampling systems with a tremendous waste of expensive hardware and software resources. For example, in multiband sampling, Σ is the $f_{\max}$-bandlimited space, for which no rate reduction is possible. Similarly, in time-delay estimation problems, Σ has the high bandwidth of $h(t)$, and again no rate reduction can be achieved.

We define the sampling problem for the union set (3.3) as the design of a system that provides:

1. **ADC:** an acquisition operator which converts the analog input $x(t) \in \mathcal{U}$ to a sequence $y[n]$ of measurements,
2. **DSP:** a toolbox of processing algorithms, which uses $y[n]$ to perform classic tasks, e.g., estimation, detection, data retrieval etc., and
3. **DAC:** a method for reconstructing $x(t)$ from the samples $y[n]$.

In order to exclude from consideration inefficient solutions, such as those treating the Nyquist subspace Σ and not exploiting the union structure, we adopt as a general design constraint that the above goals should be accomplished with minimum use of resources. Minimizing the sampling rate, for example, excludes inefficient Nyquist-rate solutions and promotes potential approaches to wisely incorporate the union structure to stand this resource constraint. For reference, this requirement is outlined as

$$\textbf{ADC + DSP + DAC} \rightarrow \textbf{minimum use of resources.} \quad (3.5)$$

In practice, besides constraining the sampling rate, (3.5) translates to the minimization of several other resources of interest, including the number of devices in the acquisition stage, design complexity, processing speed, memory requirements, power dissipation, system cost, and more.

In essence, the UoS model follows the spirit of classic sampling theory by assuming that $x(t)$ belongs to a single underlying subspace $\mathcal{A}_{\lambda^*}$. However, in contrast to the traditional paradigm, the union setting permits uncertainty in the exact signal subspace, opening the door to interesting sampling problems. The challenge posed in (3.5) is to treat the uncertainty of the union model at an overall complexity (of hardware and software) that is comparable with a system which knows the exact $\mathcal{A}_{\lambda^*}$. In Section 3.5, we describe strategies which acquire and process signals from the multiband union at a low rate, proportional to NB. Sections 3.6 and 3.7 describe variants of FRI unions, including (3.2), and their low-rate sampling solutions, which approach the rate of innovation $2L/\tau$. A line of other UoS applications that are described throughout this chapter exhibit similar rationale – the sampling rate is reduced by exploiting the fact that the input belongs to

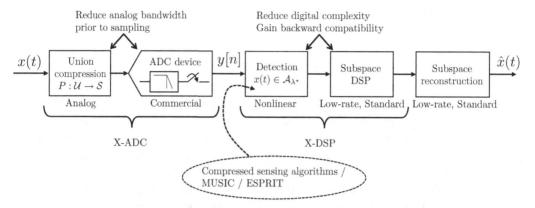

Figure 3.2 Xampling – A pragmatic framework for signal acquisition and processing in union of subspaces. Adapted from [14], © [2011] IEEE.

a single subspace $\mathcal{A}_{\lambda^*}$, even though the exact subspace index λ^* is unknown. The next subsection proposes a systematic architecture for the design of sampling systems for UoS signal classes. As we show in the ensuing sections, this architecture unifies a variety of sampling strategies developed for different instances of UoS models.

3.3.2 Architecture

The Xampling system we propose has the high-level architecture presented in Figure 3.2 [14]. The first two blocks, termed X-ADC, perform the conversion of $x(t)$ to digital. An operator P compresses the high-bandwidth input $x(t)$ into a signal with lower bandwidth, effectively capturing the entire union $\mathcal{U}$ by a subspace $\mathcal{S}$ with substantially lower sampling requirements. A commercial ADC device then takes pointwise samples of the compressed signal, resulting in the sequence of samples $y[n]$. The role of P in Xampling is to narrow down the analog bandwidth, so that low-rate ADC devices can subsequently be used. As in digital compression, the goal is to capture all vital information of the input in the compressed version, though here this functionality is achieved by hardware rather than software. The design of P therefore needs to wisely exploit the union structure, in order not to lose any essential information while reducing the bandwidth.

In the digital domain, Xampling consists of three computational blocks. A nonlinear step detects the signal subspace $\mathcal{A}_{\lambda^*}$ from the low-rate samples. Compressed sensing algorithms, e.g., those described in the relevant chapters of this book, as well as comparable methods for subspace identification, e.g., MUSIC [5] or ESPRIT [6], can be used for that purpose. Once the index λ^* is determined, we gain backward compatibility, meaning standard DSP methods apply and commercial DAC devices can be used for signal reconstruction. The combination of nonlinear detection and standard DSP is referred to as X-DSP. As we demonstrate, besides backward compatibility, the nonlinear detection decreases computational loads, since the subsequent DSP and DAC stages need to treat only the single subspace $\mathcal{A}_{\lambda^*}$, complying with (3.5). The important point is that the detection stage can be performed efficiently at the low acquisition rate, without requiring Nyquist-rate processing.

Xampling is a generic template architecture. It does not specify the exact acquisition operator P or nonlinear detection method to be used. These are application-dependent functions. Our goal in introducing Xampling is to propose a high-level system architecture and a basic set of guidelines:

1. an analog pre-processing unit to compress the input bandwidth,
2. commercial low-rate ADC devices for actual acquisition at a low-rate,
3. subspace detection in software, and
4. standard DSP and DAC methods.

The Xampling framework is developed in [14] based on two basic assumptions:

(**A1**) DSP is the main purpose of signal acquisition, and
(**A2**) the ADC device has limited bandwidth.

The DSP assumption (**A1**) highlights the ultimate use of many sampling systems – substituting analog processing by modern software algorithms. Digital signal processing is perhaps the most profound reason for signal acquisition: hardware development can rarely compete with the convenience and flexibilities that software environments provide. In many applications, therefore, DSP is what essentially motivates the ADC and decreasing processing speeds can sometimes be an important requirement, regardless of whether the sampling rate is reduced as well. In particular, the digital flow proposed in Figure 3.2 is beneficial even when a high ADC rate is acceptable. In this case, $x(t)$ can be acquired directly without narrowing down its bandwidth prior to ADC, but we would still like to reduce computational loads and storage requirements in the digital domain. This can be accomplished by imitating rate reduction in software, detecting the signal subspace, and processing at the actual information bandwidth. The compounded usage of both X-ADC and X-DSP is for mainstream applications, where reducing the rate of both signal acquisition and processing is of interest.

Assumption (**A2**) basically says that we expect the conversion device to have limited front-end bandwidth. The X-ADC can be realized on a circuit board, chip design, optical system, or other appropriate hardware. In all these platforms, the front-end has certain bandwidth limitations which obey (**A2**), thereby motivating the use of a preceding analog compression step P in order to capture all vital information within a narrow range of frequencies that the acquisition device can handle. Section 3.5 elaborates on this property.

Considering the architecture of Figure 3.2 in conjunction with requirement (3.5) reveals an interesting aspect of Xampling. In standard CS, most of the system complexity concentrates in digital reconstruction, since sensing is as simple as applying $\mathbf{y} = \mathbf{A}\mathbf{x}$. In Xampling, we attempt to balance between analog and digital complexities. As discussed in Section 3.8, a properly chosen analog preprocessing operator P can lead to substantial savings in digital complexities and vice versa.

We next describe sampling solutions for UoS models according to the Xampling paradigm. In general, when treating unions of analog signals, there are three main cases to consider:

- finite unions of infinite-dimensional spaces;
- infinite unions of finite-dimensional spaces;
- infinite unions of infinite-dimensional spaces.

In each one of the three settings above there is an element that can take on infinite values, which is a result of the fact that we are considering general analog signals: either the underlying subspaces $\mathcal{A}_\lambda$ are infinite dimensional, or the number of subspaces $|\Lambda|$ is infinite. In the next sections, we present general theory and results behind each of these cases, and focus in additional detail on a representative example application for each class. Sections 3.4 and 3.5 cover the first scenario, introducing the sparse-SI framework and reviewing multiband sampling strategies, respectively. Sections 3.6 and 3.7 discuss variants of innovation rate sampling and cover the other two cases. Methods that are based on completely finite unions, when both $|\Lambda|$ and $\mathcal{A}_\lambda$ are finite, are discussed in Section 3.8. While surveying these different cases, we will attempt to shed light into pragmatic considerations that underlie Xampling, and hint on possible routes to promote these compressive methods to actual hardware realizations.

3.4 Sparse shift-invariant framework

3.4.1 Sampling in shift-invariant subspaces

We first briefly introduce the notion of sampling in SI subspaces, which plays a key role in the development of standard (subspace) sampling theory [17, 18]. We then discuss how to incorporate the union structure into SI settings.

Shift invariant signals are characterized by a set of generators $\{h_\ell(t), 1 \leq \ell \leq N\}$ where in principle N can be finite or infinite (as is the case in Gabor or wavelet expansions of L_2). Here we focus on the case in which N is finite. Any signal in such an SI space can be written as

$$x(t) = \sum_{\ell=1}^{N} \sum_{n \in \mathbb{Z}} d_\ell[n] h_\ell(t - nT), \qquad (3.6)$$

for some set of sequences $\{d_\ell[n] \in \ell_2, 1 \leq \ell \leq N\}$ and period T. This model encompasses many signals used in communication and signal processing including bandlimited functions, splines [39], multiband signals (with known carrier positions) [11, 12], and pulse amplitude modulation signals.

The subspace of signals described by (3.6) has infinite dimensions, since every signal is associated with infinitely many coefficients $\{d_\ell[n], 1 \leq \ell \leq N\}$. Any such signal can be recovered from samples at a rate of N/T; one possible sampling paradigm at the minimal rate is given in Figure 3.3 [16, 34].

Here $x(t)$ is filtered with a bank of N filters, each with impulse response $s_\ell(t)$ which can be almost arbitrary. The outputs are uniformly sampled with period T, resulting in the sample sequences $c_\ell[n]$. Denote by $\mathbf{c}(\omega)$ a vector collecting the frequency responses of $c_\ell[n], 1 \leq \ell \leq N$, and similarly $\mathbf{d}(\omega)$ for the frequency responses of $d_\ell[n], 1 \leq \ell \leq N$.

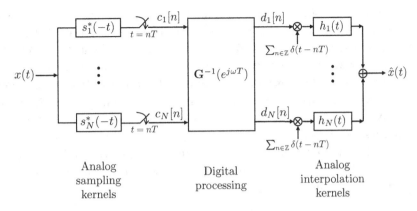

Figure 3.3 Sampling and reconstruction in shift-invariant spaces [16, 34]. Adapted from [16], © [2009] IEEE.

Then, it can be shown that [16]

$$\mathbf{c}(\omega) = \mathbf{G}(e^{j\omega T})\mathbf{d}(\omega), \qquad (3.7)$$

where $\mathbf{G}(e^{j\omega T})$ is an $N \times N$ matrix, with entries

$$[\mathbf{G}(e^{j\omega T})]_{i\ell} = \frac{1}{T}\sum_{k \in \mathbb{Z}} S_i^*\left(\frac{\omega}{T} - \frac{2\pi}{T}k\right) H_\ell\left(\frac{\omega}{T} - \frac{2\pi}{T}k\right). \qquad (3.8)$$

The notations $S_i(\omega), H_\ell(\omega)$ stand for the CTFT of $s_i(t), h_\ell(t)$, respectively. To allow recovery, the condition on the sampling filters $s_i(t)$ is that (3.8) results in an invertible frequency response $\mathbf{G}(e^{j\omega T})$. The signal is then recovered by processing the samples with a filter bank with frequency response $\mathbf{G}^{-1}(e^{j\omega T})$. In this way, we invert (3.7) and obtain the vectors

$$\mathbf{d}(\omega) = \mathbf{G}^{-1}(e^{j\omega T})\mathbf{c}(\omega). \qquad (3.9)$$

Each output sequence $d_\ell[n]$ is then modulated by a periodic impulse train $\sum_{n \in \mathbb{Z}} \delta(t - nT)$ with period T, followed by filtering with the corresponding analog filter $h_\ell(t)$. In practice, interpolation with finitely many samples gives sufficiently accurate reconstruction, provided that $h_\ell(t)$ decay fast enough [40], similar to finite interpolation in the Shannon–Nyquist theorem.

3.4.2 Sparse union of SI subspaces

In order to incorporate further structure into the generic SI model (3.6), we treat signals of the form (3.6) involving a small number K of generators, chosen from a finite set Λ of N generators. Specifically, we consider the input model

$$x(t) = \sum_{|\ell|=K} \sum_{n \in \mathbb{Z}} d_\ell[n] h_\ell(t - nT), \qquad (3.10)$$

where $|\ell| = K$ means a sum over at most K elements. If the K active generators are known, then according to Figure 3.3 it suffices to sample at a rate of K/T corresponding to uniform samples with period T at the output of K appropriate filters. A more difficult question is whether the rate can be reduced if we know that only K of the generators are active, but do not know in advance which ones. In terms of (3.10) this means that only K of the sequences $d_\ell[n]$ have nonzero energy. Consequently, for each value n, $\|\mathbf{d}[n]\|_0 \leq K$, where $\mathbf{d}[n] = [d_1[n], \ldots, d_N[n]]^T$ collects the unknown generator coefficients for time instance n.

For this model, it is possible to reduce the sampling rate to as low as $2K/T$ [16] as follows. We target a compressive sampling system that produces a vector of low-rate samples $\mathbf{y}[n] = [y_1[n], \ldots, y_p[n]]^T$ at $t = nT$ which satisfies a relation

$$\mathbf{y}[n] = \mathbf{A}\mathbf{d}[n], \quad \|\mathbf{d}[n]\|_0 \leq K, \tag{3.11}$$

with a sensing matrix $\mathbf{A}$ that allows recovery of sparse vectors. The choice $p < N$ reduces the sampling rate below Nyquist. In principle, a parameterized family of underdetermined systems, by the time index n in the case of (3.11), can be treated by applying CS recovery algorithms independently for each n. A more robust and efficient technique which exploits the joint sparsity over n is described in the next section. The question is therefore how to design a sampling scheme which would boil down to a relation such as (3.11) in the digital domain. Figure 3.4 provides a system for obtaining $\mathbf{y}[n]$, where the following theorem gives the expression for its sampling filters $w_\ell(t)$ [16].

THEOREM 3.1 *Let $s_\ell(t)$ be a set of N filters and $\mathbf{G}(e^{j\omega T})$ the response matrix defined in (3.9) (so that $s_\ell(t)$ can be used in the Nyquist-rate scheme of Figure 3.3), and let $\mathbf{A}$ be a given $p \times N$ sensing matrix. Sampling $x(t)$ with a bank of filters $w_\ell(t), 1 \leq \ell \leq p$ defined by*

$$\mathbf{w}(\omega) = \mathbf{A}^* \mathbf{G}^{-*}(e^{j\omega T})\mathbf{s}(\omega), \tag{3.12}$$

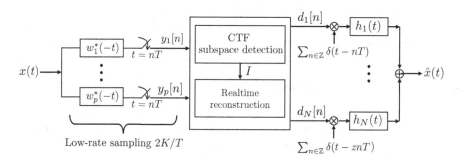

Figure 3.4 Compressive sensing acquisition for sparse union of shift-invariant subspaces. Adapted from [16], © [2009] IEEE.

gives a set of compressed measurements $y_\ell[n], 1 \leq \ell \leq p$ that satisfies (3.11). In (3.12), the vectors $\mathbf{w}(\omega), \mathbf{s}(\omega)$ have ℓth elements $W_\ell(\omega), S_\ell(\omega)$, denoting CTFTs of the corresponding filters, and $(\cdot)^{-*}$ denotes the conjugate of the inverse.

The filters $w_\ell(t)$ of Figure 3.4 form an analog compression operator P as suggested in the X-ADC architecture. The sampling rate is effectively reduced by taking linear combinations of the outputs $c_\ell[n]$ of the Nyquist scheme of Figure 3.3, with combination coefficients defined by the sensing matrix $\mathbf{A}$. This structure is revealed by examining (3.12) – sampling by $w_\ell[n]$ is tantamount to filtering $x(t)$ by $s_\ell(t)$, applying $\mathbf{G}^{-1}(e^{j\omega T})$ to obtain the sparse set of sequences $d_\ell[n]$, and then combining these sequences by an underdetermined matrix $\mathbf{A}$. A more general result of [16] enables further flexibility in choosing the sampling filters by letting $\mathbf{w}(\omega) = \mathbf{P}^*(e^{j\omega T})\mathbf{A}^*\mathbf{G}^*(e^{j\omega T})\mathbf{s}(\omega)$, for some arbitrary invertible $p \times p$ matrix $\mathbf{P}^*(e^{j\omega T})$. In this case, (3.11) holds with respect to sequences obtained by post-processing the compressive measurements $y_\ell[n]$ by $\mathbf{P}^{-1}(e^{j\omega T})$.

The sparse-SI model (3.10) can be generalized to a sparse sum of arbitrary subspaces, where each subspace $\mathcal{A}_\lambda$ of the union (3.3) consists of a direct sum of K low-dimensional subspaces [41]

$$\mathcal{A}_\lambda = \bigoplus_{|j|=K} \mathcal{V}_j. \qquad (3.13)$$

Here $\{\mathcal{V}_j, 1 \leq j \leq N\}$ are a given set of subspaces with dimensions $\dim(\mathcal{V}_j) = v_j$, and as before $|j| = K$ denotes a sum over K indices. Thus, each subspace $\mathcal{A}_\lambda$ corresponds to a different choice of K subspaces $\mathcal{V}_j$ that comprise the sum. The sparse-SI model is a special case of (3.13), in which each $\mathcal{V}_j$ is an SI subspace with a single shift kernel $h_j(t)$. In [41], sampling and reconstruction algorithms are developed for the case of finite Λ and finite-dimensional $\mathcal{A}_\lambda$. The approach utilizes the notion of set transforms to cast the sampling problem into an underdetermined system with an unknown block-sparse solution, which is found via a polynomial-time mixed-norm optimization program. Block-sparsity is studied in more detail in [41–44].

3.4.3 Infinite measurement model and continuous to finite

In the sparse-SI framework, the acquisition scheme is mapped into the system (3.11). Reconstruction of $x(t)$ therefore depends on our ability to resolve $d_\ell[n]$ from this underdetermined system. More generally, we are interested in solving a parameterized underdetermined linear system with sensing matrix dimensions $p \times N, p < N$

$$\mathbf{y}(\theta) = \mathbf{A}\mathbf{x}(\theta), \quad \theta \in \Theta, \qquad (3.14)$$

where Θ is a set whose cardinality can be infinite. In particular, Θ may be uncountable, such as the frequencies $\omega \in [-\pi, \pi)$ of (3.12), or countable as in (3.11). The system (3.14) is referred to as an infinite measurement vector (IMV) model with sparsity K, if the vectors $\mathbf{x}(\Theta) = \{\mathbf{x}(\theta)\}$ share a joint sparsity pattern [45]. That is, the nonzero elements are supported within a fixed location set I of size K.

The IMV model includes as a special case standard CS, when taking $\Theta = \{\theta^*\}$ to be a single element set. It also includes the case of a finite set Θ, termed multiple measurement vectors (MMV) in the CS literature [45–50]. In the finite cases it is easy to see that if $\sigma(\mathbf{A}) \geq 2k$, where $\sigma(\mathbf{A}) = \text{spark}(\mathbf{A}) - 1$ is the Kruskal-rank of $\mathbf{A}$, then $\mathbf{x}(\Theta)$ is the unique K-sparse solution of (3.14) [48]. A simple necessary and sufficient condition in terms of $\text{rank}(\mathbf{y}(\Theta))$ is derived in [51], which improves upon earlier (sufficient only) conditions in [48]. Similar conditions hold for a jointly K-sparse IMV system [45].

The major difficulty with the IMV model is how to recover the solution set $\mathbf{x}(\Theta)$ from the infinitely many equations (3.14). One strategy is to solve (3.14) independently for each θ. However, this strategy may be computationally intensive in practice, since it would require to execute a CS solver for each individual θ; for example, in the context of (3.11), this amounts to solving a sparse recovery problem for each time instance n. A more efficient strategy exploits the fact that $\mathbf{x}(\Theta)$ are jointly sparse, so that the index set

$$I = \{l : x_l(\theta) \neq 0\} \tag{3.15}$$

is independent of θ. Therefore, I can be estimated from several instances of $\mathbf{y}(\Theta)$, which increases the robustness of the estimate. Once I is found, recovery of the entire set $\mathbf{x}(\Theta)$ is straightforward. To see this, note that using I, (3.14) can be written as

$$\mathbf{y}(\theta) = \mathbf{A}_I \mathbf{x}_I(\theta), \quad \theta \in \Theta, \tag{3.16}$$

where $\mathbf{A}_I$ denotes the matrix containing the columns of $\mathbf{A}$ whose indices belong to I, and $\mathbf{x}_I(\theta)$ is the vector consisting of entries of $\mathbf{x}(\theta)$ in locations I. Since $\mathbf{x}(\Theta)$ is K-sparse, $|I| \leq K$. Therefore, the columns of $\mathbf{A}_I$ are linearly independent (because $\sigma(\mathbf{A}) \geq 2K$), implying that $\mathbf{A}_I^\dagger \mathbf{A}_I = \mathbf{I}$, where $\mathbf{A}_I^\dagger = \left(\mathbf{A}_I^H \mathbf{A}_I\right)^{-1} \mathbf{A}_I^H$ is the pseudo-inverse of $\mathbf{A}_I$ and $(\cdot)^H$ denotes the Hermitian conjugate. Multiplying (3.16) by $\mathbf{A}_I^\dagger$ on the left gives

$$\mathbf{x}_I(\theta) = \mathbf{A}_I^\dagger \mathbf{y}(\theta), \quad \theta \in \Theta. \tag{3.17}$$

The components in $\mathbf{x}(\theta)$ not supported on S are all zero. In contrast to applying a CS solver for each θ, (3.17) requires only one matrix-vector multiplication per $\mathbf{y}(\theta)$, typically requiring far fewer computations.

It remains to determine I efficiently. In [45] it was shown that I can be found exactly by solving a finite MMV. The steps used to formulate this MMV are grouped under a block referred to as continuous-to-finite (CTF). The essential idea is that every finite collection of vectors spanning the subspace $\text{span}(\mathbf{y}(\Theta))$ contains sufficient information to recover I, as incorporated in the following theorem [45]:

THEOREM 3.2 *Suppose that $\sigma(\mathbf{A}) \geq 2K$, and let $\mathbf{V}$ be a matrix with column span equal to $\text{span}(\mathbf{y}(\Theta))$. Then, the linear system*

$$\mathbf{V} = \mathbf{A}\mathbf{U} \tag{3.18}$$

has a unique K-sparse solution $\mathbf{U}$ whose support equals I.

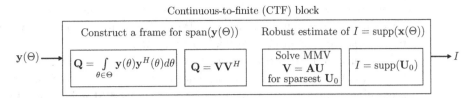

Figure 3.5 The fundamental stages for the recovery of the nonzero location set I in an IMV model using only one finite-dimensional program. Adapted from [45], © [2008] IEEE.

The advantage of Theorem 3.2 is that it allows us to avoid the infinite structure of (3.14) and instead find the finite set I by solving a single MMV system of the form (3.18).

For example, in the sparse SI model, such a frame can be constructed by

$$\mathbf{Q} = \sum_n \mathbf{y}[n]\mathbf{y}^H[n], \qquad (3.19)$$

where typically $2K$ snapshots $\mathbf{y}[n]$ are sufficient [20]. Optionally, $\mathbf{Q}$ is decomposed to another frame $\mathbf{V}$, such that $\mathbf{Q} = \mathbf{V}\mathbf{V}^H$, allowing removal of the noise space [20]. Applying the CTF in this setting provides a robust estimate of $I = \text{supp}(d_\ell[n])$, namely the indices of the active generators that comprise $x(t)$. This is essentially the subspace detection part of X-DSP, where the joint support set I determines the signal subspace $\mathcal{A}_{\lambda^*}$. The crux of the CTF now becomes apparent – the indices of the nonidentically zero rows of the matrix $\mathbf{U}_0$ that solves the finite underdetermined system (3.19) coincide with the index set $I = \text{supp}(d_\ell[n])$ that is associated with the continuous signal $x(t)$ [45]. Once I is found, (3.11) can be inverted on the column subset I by (3.17), where the time index n takes the role of θ. Reconstruction from that point on is carried out in real time; one matrix-vector multiplication (3.17) per incoming vector of samples $\mathbf{y}[n]$ recovers $\mathbf{d}_I[n]$, denoting the entries of $\mathbf{d}[n]$ indicated by I.

Figure 3.5 summarizes the CTF steps for identifying the nonzero location set of an IMV system. In the figure, the summation (3.19) is formulated as integration over $\theta \in \Theta$ for the general IMV setting (3.14). The additional requirement of Theorem 3.2 is to construct a frame matrix $\mathbf{V}$ having column span equal to $\text{span}(\mathbf{y}(\Theta))$, which, in practice, is computed efficiently from the samples.

The mapping of Figure 3.4 to an IMV system (3.11) and the CTF recovery create a nice connection to results of standard CS. The number of branches p is the number of rows in $\mathbf{A}$, and the choice of sampling filters $w_\ell(t)$ translate to its entries via Theorem 3.1. Since recovery boils down to solving an MMV system with sensing matrix $\mathbf{A}$, we should design the hardware so that the resulting matrix $\mathbf{A}$ in (3.12) has "nice" CS properties.[1] Precisely, an MMV system of size $p \times N$ and joint sparsity of order K needs to be solved correctly with that $\mathbf{A}$. In practice, to solve the MMV (3.18), we can make use of existing algorithms from the CS literature, cf. [45–49]. The Introduction and relevant chapters

---

[1] We comment that most known constructions of "nice" CS matrices involve randomness. In practice, the X-ADC hardware is fixed and defines a deterministic sensing matrix $\mathbf{A}$ for the corresponding IMV system.

of this book describe various conditions on CS matrices to ensure stable recovery. The dimension requirements of the specific MMV solver in use will impact the number of branches p, and consequently the total sampling rate.

The sparse-SI framework can be used, in principle, to reduce the rate of any signal of the form (3.10). In the next section, we treat multiband signals and derive a sub-Nyquist acquisition strategy for this model from the general sparse-SI architecture of Figure 3.4.

3.5 From theory to hardware of multiband sampling

The prime goal of Xampling is to enable theoretical ideas to develop from the math to hardware, to real-world applications. In this section, we study sub-Nyquist sampling of multiband signals in the eyes of a practitioner, aiming to design low-rate sampling hardware. We define the multiband model and propose a union formulation that fits the sparse-SI framework introduced in the previous section. A periodic nonuniform sampling (PNS) solution [19] is then derived from Figure 3.4. Moving on to practical aspects, we examine front-end bandwidth specifications of commercial ADC devices, and conclude that devices with Nyquist-rate bandwidth are required whenever the ADC is directly connected to a wideband input. Consequently, although PNS as well as the general architecture of Figure 3.4, enable in principle sub-Nyquist sampling, in practice, high analog bandwidth is necessary, which can be limiting in high-rate applications. To overcome this possible limitation, an alternative scheme, the MWC [20], is presented and analyzed. We conclude our study with a glimpse at circuit aspects that are unique to Xampling systems, as were reported in the circuit design of an MWC prototype hardware [22].

3.5.1 Signal model and sparse-SI formulation

The class of multiband signals models a scenario in which $x(t)$ consists of several concurrent RF transmissions. A receiver that intercepts a multiband $x(t)$ sees the typical spectral support that is depicted in Figure 3.1. We assume that the multiband spectrum contains at most N (symmetric) frequency bands with carriers f_i, each of maximal width B. The carriers are limited to a maximal frequency $f_{\max}$. The information bands represent analog messages or digital bits transmitted over a shared channel.

When the carrier frequencies f_i are fixed, the resulting signal model can be described as a subspace, and standard demodulation techniques may be used to sample each of the bands at a low-rate. A more challenging scenario is when the carriers f_i are unknown. This situation arises, for example, in spectrum sensing for mobile cognitive radio (CR) receivers [23, 52], which aim at utilizing unused frequency regions on an opportunistic basis. Commercialization of CR technology necessitates a spectrum sensing mechanism that can sense a wideband spectrum which consists of several narrowband transmissions, and determines in real time which frequency bands are active.

Since each combination of carrier frequencies determines a single subspace, a multiband signal can be described in terms of a union of subspaces. In principle, f_i lies in

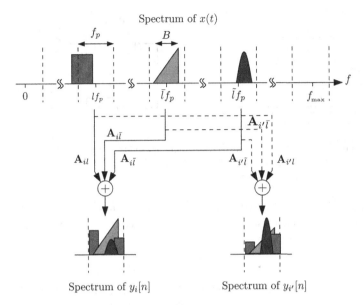

Figure 3.6 Spectrum slices of $x(t)$ are overlaid in the spectrum of the output sequences $y_i[n]$. In the example, channels i and i' realize different linear combinations of the spectrum slices centered around $lf_p, \bar{l}f_p, \tilde{l}f_p$. For simplicity, the aliasing of the negative frequencies is not drawn. Adapted from [22], © [2011] IET.

the continuum $f_i \in [0, f_{\max}]$, so that the union contains infinitely many subspaces. To utilize the sparse-SI framework with finitely many SI generators, a different viewpoint can be used, which treats the multiband model as a finite union of bandpass subspaces, termed spectrum slices [20]. To obtain the finite union viewpoint, the Nyquist range $[-f_{\max}, f_{\max}]$ is conceptually divided into $M = 2L + 1$ consecutive, non-overlapping, slices of individual widths $f_p = 1/T$, such that $M/T \geq f_{\text{NYQ}}$, as depicted in Figure 3.6. Each spectrum slice represents an SI subspace $\mathcal{V}_i$ of a single bandpass slice. By choosing $f_p \geq B$, we ensure that no more than $2N$ spectrum slices are active, namely contain signal energy. Thus, (3.13) holds with $\mathcal{A}_\lambda$ being the sum over $2N$ SI bandpass subspaces $\mathcal{V}_i$. Consequently, instead of enumerating over the unknown carriers f_i, the union is defined over the active bandpass subspaces [16, 19, 20], which can be written in the form (3.10). Note that the conceptual division to spectrum slices does not restrict the band positions; a single band can split between adjacent slices.

Formulating the multiband model with unknown carriers as a sparse-SI problem, we can now apply the sub-Nyquist sampling scheme of Figure 3.4 to develop an analog CS system for this setting.

3.5.2 Analog compressed sensing via nonuniform sampling

One way to realize the sampling scheme of Figure 3.4 is through PNS [19]. This strategy is derived from Figure 3.4 when choosing

$$w_i(t) = \delta(t - c_i T_{\text{NYQ}}), \quad 1 \leq i \leq p, \quad (3.20)$$

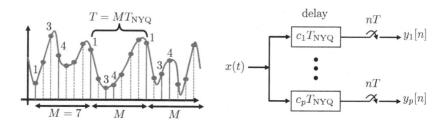

Figure 3.7 Periodic nonuniform sampling for sub-Nyquist sensing. In the example, out of $M = 7$ points, only $p = 3$ are active, with time shifts $c_i = 1, 3, 4$.

where $T_{\text{NYQ}} = 1/f_{\text{NYQ}}$ is the Nyquist period, and using a sampling period of $T = MT_{\text{NYQ}}$. Here c_i are integers which select part of the uniform Nyquist grid, resulting in p uniform sequences

$$y_i[n] = x((nM + c_i)T_{\text{NYQ}}). \tag{3.21}$$

The sampling sequences are illustrated in Figure 3.7. It can be shown that the PNS sequences $y_i[n]$ satisfy an IMV system of the form (3.11) with $d_\ell[n]$ representing the contents of the ℓth bandpass slice. The sensing matrix $\mathbf{A}$ in this setting has $i\ell$th entry

$$\mathbf{A}_{i\ell} = e^{j\frac{2\pi}{M}c_i \ell}, \tag{3.22}$$

that is a partial discrete Fourier transform (DFT), obtained by taking only the row indices c_i from the full $M \times M$ DFT matrix. The CS properties of partial-DFT matrices are studied in [4], for example.

To recover $x(t)$, we can apply the CTF framework and obtain spectrum blind reconstruction (SBR) of $x(t)$ [19]. Specifically, a frame $\mathbf{Q}$ is computed with (3.19) and is optionally decomposed to another frame $\mathbf{V}$ (to combat noise). Solving (3.18) then indicates the active sequences $d_\ell[n]$, and equivalently estimates the frequency support of $x(t)$ at a coarse resolution of slice width f_p. Continuous reconstruction is then obtained by standard lowpass interpolation of the active sequences $d_\ell[n]$ and modulation to the corresponding positions on the spectrum. This procedure is termed SBR4 in [19], where 4 designates that under the choice of $p \geq 4N$ sampling sequences (and additional conditions), this algorithm guarantees perfect reconstruction of a multiband $x(t)$. With the earlier choice $f_p = 1/T \geq B$, the average sampling rate can be as low as $4NB$.

The rate can be further reduced by a factor of 2 exploiting the way a multiband spectra is arranged in spectrum slices. Using several CTF instances, an algorithm reducing the required rate was developed in [19] under the name SBR2, leading to $p \geq 2N$ sampling branches, so that the sampling rate can approach $2NB$. This is essentially the provable optimal rate [19], since regardless of the sampling strategy, theoretic arguments show that $2NB$ is the lowest possible sampling rate for multiband signals with unknown spectrum support [19]. Figure 3.8 depicts recovery performance in Monte Carlo simulations of a (complex-valued) multiband model with $N = 3$ bands, widths $B = 1$ GHz and $f_{\text{NYQ}} = 20$ GHz. Recovery of noisy signals is also simulated in [19]. We demonstrate robustness to

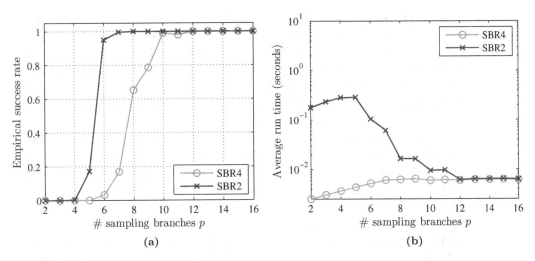

Figure 3.8 Comparing algorithms SBR4 and SBR2. (a) Empirical recovery rate for different sampling rates and (b) digital complexity as measured by average run time. Adapted from [19], © [2009] IEEE.

noise later on in this section in the context of MWC sampling. The robustness follows from that of the MMV system used for SBR.

We note that PNS was utilized for multiband sampling already in classic studies, though the traditional goal was to approach a rate of NB samples/second. This rate is optimal according to the Landau theorem [53], though achieving it for all input signals is possible only when the spectral support is known and fixed. When the carrier frequencies are unknown, the optimal rate is $2NB$ [19]. Indeed, [11,54] utilized knowledge of the band positions to design a PNS grid and the required interpolation filters for reconstruction. The approaches in [12, 13] were semi-blind: a sampler design independent of band positions combined with the reconstruction algorithm of [11] which requires exact support knowledge. Other techniques targeted the rate NB by imposing alternative constraints on the input spectrum [21]. Here we demonstrate how analog CS tools [16,45] can lead to a fully blind sampling system of multiband inputs with unknown spectra at the appropriate optimal rate [19]. A more thorough discussion in [19] studies the differences between the analog CS method presented here based on [16, 19, 45] and earlier approaches.

3.5.3 Modeling practical ADC devices

Analog CS via PNS results in a simple acquisition strategy, which consists of p delay components and p uniform ADC devices. Furthermore, if high sampling rate is not an obstacle and only low processing rates are of interest, then PNS can be simulated by first sampling $x(t)$ at its Nyquist rate and then reducing the rate digitally by discarding some of the samples. Nonuniform topologies of this class are also popular in the design of Nyquist-rate time-interleaved ADC devices, in which case $p = M$ [55,56].

Realization of a PNS grid with standard ADCs remains simple as long as the input bandwidth is not too high. For high bandwidth signals, PNS is potentially limited, as we now explain by zooming into the drawing of the ADC device of Figure 3.2. In the signal processing community, an ADC is often modeled as an ideal pointwise sampler that takes snapshots of $x(t)$ at a constant rate of r samples/second. The sampling rate r is the main parameter that is highlighted in the datasheets of popular ADC devices; see online catalogs [57,58] for many examples.

For most analysis purposes, the first-order model of pointwise acquisition approximates the true ADC operation sufficiently well. Another property of practical devices, also listed in datasheets, is about to play a major role in the UoS settings – the analog bandwidth power b. The parameter b measures the -3 dB point in the frequency response of the ADC device, which stems from the responses of all circuitries comprising the internal front-end. See the datasheet quote of AD9057 in Figure 3.9. Consequently, inputs with frequencies up to b Hz can be reliably converted. Any information beyond b is attenuated and distorted. Figure 3.9 depicts an ADC model in which the pointwise sampler is preceded by a lowpass filter with cutoff b, in order to take into account the bandwidth limitation [20]. In Xampling, the input signal $x(t)$ belongs to a union set $\mathcal{U}$ which typically has high bandwidth, e.g., multiband signals whose spectrum reaches up to $f_{\max}$ or FRI signals with wideband pulse $h(t)$. This explains the necessity of an analog compression operator P to reduce the bandwidth prior to the actual ADC. The next stage can then employ commercial devices with low analog bandwidth b.

The Achilles heel of nonuniform sampling is the pointwise acquisition of a wideband input. While the rate of each sequence $y_i[n]$ is low, namely f_{NYQ}/M, the ADC device still needs to capture a snapshot of a wideband input with frequencies possibly reaching up to $f_{\max}$. In practice, this requires an ADC with front-end bandwidth that reaches the Nyquist rate, which can be challenging in wideband scenarios.

3.5.4 Modulated wideband converter

To circumvent analog bandwidth issues, an alternative to PNS sensing referred to as the modulated wideband converter (MWC) was developed in [20]. The MWC combines the spectrum slices $d_\ell[n]$ according to the scheme depicted in Figure 3.10. This architecture allows one to implement an effective demodulator without the carrier frequencies being known to the receiver. A nice feature of the MWC is a modular design so that for known carrier frequencies the same receiver can be used with fewer channels or lower sampling rate. Furthermore, by increasing the number of channels or the rate on each channel the same realization can be used for sampling full band signals at the Nyquist rate.

The MWC consists of an analog front-end with p channels. In the ith channel, the input signal $x(t)$ is multiplied by a periodic waveform $p_i(t)$ with period T, lowpass filtered by an analog filter with impulse response $h(t)$ and cutoff $1/2T$, and then sampled at rate $f_s = 1/T$. The mixing operation scrambles the spectrum of $x(t)$, such that a portion of the energy of all bands appears in baseband. Specifically, since $p_i(t)$ is periodic, it has a

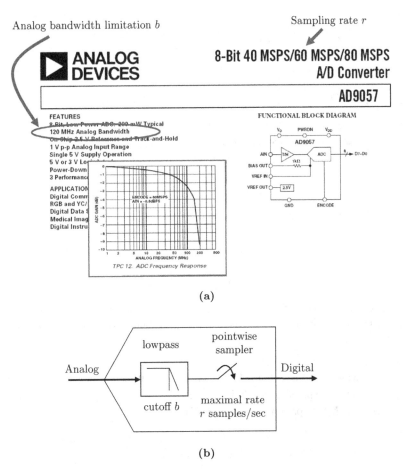

Figure 3.9   (a) Datasheet of AD9057 (with permission, source: www.analog.com/static/imported-files/data_sheets/AD9057.pdf). (b) Modeling the inherent bandwidth limitation of the ADC front-end as a lowpass filter preceding pointwise acquisition. Adapted from [20], © [2010] IEEE.

Fourier expansion

$$p_i(t) = \sum_{\ell=-\infty}^{\infty} c_{i\ell} e^{j\frac{2\pi}{T}\ell t}. \qquad (3.23)$$

In the frequency domain, mixing by $p_i(t)$ is tantamount to convolution between $X(f)$ and the Fourier transform of $p_i(t)$. The latter is a weighted Dirac-comb, with Dirac locations on $f = l/T$ and weights $c_{i\ell}$. Thus, as before, the spectrum is conceptually divided into slices of width $1/T$, represented by the unknown sequences $d_\ell[n]$, and a weighted-sum of these slices is shifted to the origin [20]. The lowpass filter $h(t)$ transfers only the narrowband frequencies up to $f_s/2$ from that mixture to the output sequence $y_i[n]$. The output has the same aliasing pattern that was illustrated in Figure 3.6. Sensing

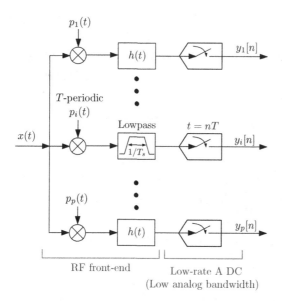

Figure 3.10 Block diagram of the modulated wideband converter. The input passes through p parallel branches, where it is mixed with a set of periodic functions $p_i(t)$, lowpass filtered and sampled at a low rate. Adapted from [22], © [2011] IET.

with the MWC results in the IMV system (3.11) with a sensing matrix $\mathbf{A}$ whose entries are the Fourier expansion coefficients $c_{i\ell}$.

The basic MWC parameter setting is [20]

$$p \geq 4N, \quad f_s = \frac{1}{T} \geq B. \qquad (3.24)$$

Using the SBR2 algorithm of [19], the required number of branches is $p \geq 2N$ so that the sampling rate is reduced by a factor of 2 and can approach the minimal rate of $2NB$. Advanced configurations enable additional hardware savings by collapsing the number of branches p by a factor of q at the expense of increasing the sampling rate of each channel by the same factor, ultimately enabling a single-channel sampling system [20]. This property is unique to MWC sensing, since it decouples the aliasing from the actual acquisition.

The periodic functions $p_i(t)$ define a sensing matrix $\mathbf{A}$ with entries $c_{i\ell}$. Thus, as before, $p_i(t)$ need to be chosen such that the resulting $\mathbf{A}$ has "nice" CS properties. In principle, any periodic function with high-speed transitions within the period T can satisfy this requirement. One possible choice for $p_i(t)$ is a sign-alternating function, with $M = 2L + 1$ sign intervals within the period T [20]. Popular binary patterns, e.g., Gold or Kasami sequences, are especially suitable for the MWC [59]. Imperfect sign alternations are allowed as long as periodicity is maintained [22]. This property is crucial since precise sign alternations at high speeds are extremely difficult to maintain, whereas simple hardware wirings ensure that $p_i(t) = p_i(t + T)$ for every $t \in \mathbb{R}$ [22]. Another important practical design aspect is that a filter $h(t)$ with non-flat frequency response can be used since a non-ideal response can be compensated for in the digital domain, using an algorithm developed in [60].

In practical scenarios, $x(t)$ is contaminated by wideband analog noise $e_{\text{analog}}(t)$ and measurement noise $e_{\ell,\text{meas.}}[n]$ that is added to the compressive sequences $y_\ell[n]$. This results in a noisy IMV system

$$\mathbf{y}[n] = \mathbf{A}(\mathbf{d}[n] + \mathbf{e}_{\text{analog}}[n]) + \mathbf{e}_{\text{meas.}}[n] = \mathbf{A}\mathbf{d}[n] + \mathbf{e}_{\text{eff.}}[n], \quad (3.25)$$

with an effective error term $\mathbf{e}_{\text{eff.}}[n]$. This means that noise has the same effects in analog CS as it has in the standard CS framework with an increase in variance due to the term $\mathbf{A}\mathbf{e}_{\text{analog}}[n]$. Therefore, existing algorithms can be used to try to combat the noise. Furthermore, we can translate known results and error guarantees developed in the context of CS to handle noisy analog environments. In particular, as is known in standard CS, the total noise, i.e., in both zero and nonzero locations, is what dictates the behavior of various algorithms and recovery guarantees. Similarly, analog CS systems, such as sparse-SI [16], PNS [19], or MWC [20], aggregate wideband noise power from the entire Nyquist range $[-f_{\max}, f_{\max}]$ into their samples. This is different from standard demodulation that aggregates only in-band noise, since only a specific range of frequencies is shifted to baseband. Nonetheless, as demonstrated below, analog CS methods exhibit robust recovery performance which degrades gracefully as noise levels increase.

Numerical simulations were used in [59] to evaluate the MWC performance in noisy environments. A multiband model with $N = 6$, $B = 50$ MHz and $f_{\text{NYQ}} = 10$ GHz was used to generate inputs $x(t)$, which were contaminated by additive wideband Gaussian noise. An MWC system with $f_p = 51$ MHz and a varying number p of branches was considered, with sign alternating waveforms of length $M = 195$. Performance of support recovery using CTF is depicted in Figure 3.11 for various (wideband) signal-to-noise

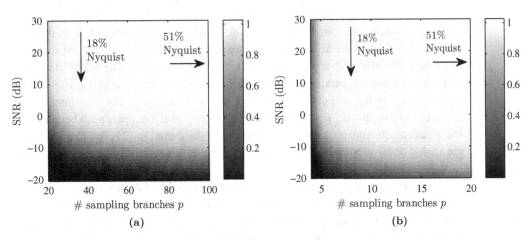

Figure 3.11 Image intensity represents percentage of correct recovery of the active slices set I, for different number of sampling branches p and under several SNR levels. The collapsing factors are (a) $q = 1$ and (b) $q = 5$. The markers indicate reference points with same total sampling rate pf_s as a fraction of $f_{\text{NYQ}} = 10$ GHz. Adapted from [20], © [2010] IEEE.

ratio (SNR) levels. Two MWC configurations were tested: a basic version with sampling rate f_p per branch, and an advanced setup with a collapsing factor $q = 5$, in which case each branch samples at rate qf_p. The results affirm saving in hardware branches by a factor of 5 while maintaining comparable recovery performance. Signal reconstruction is demonstrated in the next subsection using samples obtained by a hardware MWC prototype.

Note that the MWC achieves a similar effect of aliasing bandpass slices to the origin as does the PNS system. However, in contrast to PNS, the MWC accomplishes this goal with analog pre-processing prior to sampling, as proposed in Xampling, which allows the use of standard low-rate ADCs. In other words, the practical aspects of front-end bandwidth motivate a solution which departs from the generic scheme of Figure 3.4. This is analogous to the advantage of standard demodulation over plain undersampling; both demodulation and undersampling can shift a single bandpass subspace to the origin. However, while undersampling requires an ADC with Nyquist-rate front-end bandwidth, demodulation uses RF technology to interact with the wideband input, thereby requiring only low-rate and low-bandwidth ADC devices.

3.5.5 Hardware design

The MWC has been implemented as a board-level hardware prototype [22]. The hardware specifications cover inputs with 2 GHz Nyquist rate and $NB = 120$ MHz spectrum occupation. The prototype has $p = 4$ sampling branches, with total sampling rate of 280 MHz, far below the 2 GHz Nyquist rate. In order to save analog components, the hardware realization incorporates the advanced configuration of the MWC [20] with a collapsing factor $q = 3$. In addition, a single shift-register provides a basic periodic pattern, from which p periodic waveforms are derived using delays, that is, by tapping p different locations of the register. Photos of the hardware are presented in Figure 3.12.

Several nonordinary RF blocks in the MWC prototype are highlighted in Figure 3.12. These non-ordinary circuitries stem from the unique application of sub-Nyquist sampling as described in detail in [22]. For instance, ordinary analog mixers are specified and commonly used with a pure sinusoid in their oscillator port. The MWC, however, requires simultaneous mixing with the many sinusoids comprising $p_i(t)$. This results in attenuation of the output and substantial nonlinear distortion not accounted for in datasheet specifications. To address this challenge, power control, special equalizers, and local adjustments on datasheet specifications were used in [22] in order to design the analog acquisition, taking into account the non-ordinary mixer behavior due to the periodic mixing.

Another circuit challenge pertains to generating $p_i(t)$ with 2 GHz alternation rates. The waveforms can be generated either by analog or digital means. Analog waveforms, such as sinusoid, square, or sawtooth waveforms, are smooth within the period, and therefore do not have enough transients at high frequencies which is necessary to ensure sufficient aliasing. On the other hand, digital waveforms can be programmed to any desired number of alternations within the period, but are required to meet timing constraints on the order of the clock period. For 2 GHz transients, the clock interval $1/f_{\text{NYQ}} = 480$ picosecs leads

Xampling: compressed sensing of analog signals

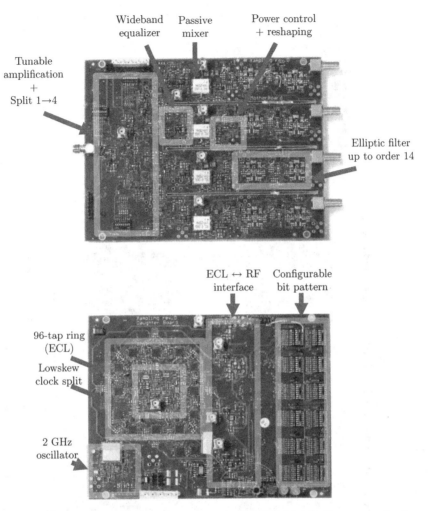

Figure 3.12 Hardware realization of the MWC consisting of two circuit boards. The top pane implements $m = 4$ sampling channels, whereas the bottom pane provides four sign-alternating periodic waveforms of length $M = 108$, derived from different taps of a single shift-register. Adapted from [61], © [2011] IEEE.

to tight timing constraints that are difficult to satisfy with existing digital devices. The timing constraints involved in this logic are overcome in [22] by operating commercial devices beyond their datasheet specifications. The reader is referred to [22] for further technical details.

Correct support detection and signal reconstruction in the presence of three narrowband transmissions was verified in [22]. Figure 3.13 depicts the setup of three signal generators that were combined at the input terminal of the MWC prototype: an amplitude-modulated (AM) signal at 807.8 MHz with 100 kHz envelope, a frequency-modulation (FM) source at 631.2 MHz with 1.5 MHz frequency deviation and 10 kHz modulation rate, and a pure sine waveform at 981.9 MHz. Signal powers were set to about 35 dB

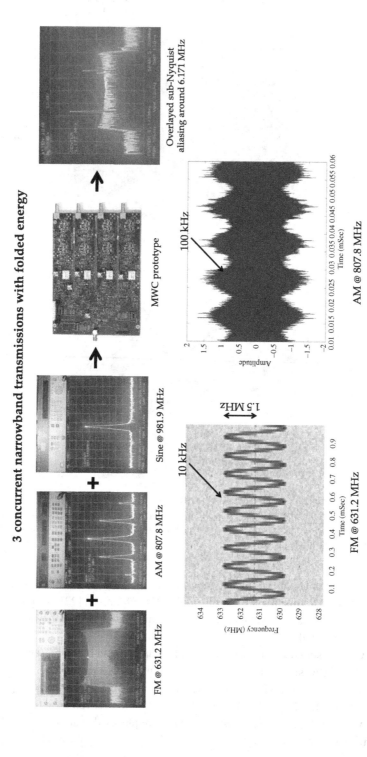

Figure 3.13 Three signal generators are combined to the system input terminal. The spectrum of the low-rate samples (first channel) reveals overlapped aliasing at baseband. The recovery algorithm finds the correct carriers and reconstructs the original individual signals. Adapted from [22], © [2011] IET.

SNR with respect to the wideband noise that folded to baseband. The carrier positions were chosen so that their aliases overlay at baseband, as the photos in Figure 3.13 demonstrate. The CTF was executed and detected the correct support set I. The unknown carrier frequencies were estimated up to 10 kHz accuracy. In addition, the figure demonstrates correct reconstruction of the AM and FM signal contents. Our lab experiments also indicate an average of 10 millisecond duration for the digital computations, including CTF support detection and carrier estimation. The small dimensions of $\mathbf{A}$ (12×100 in the prototype configuration) is what makes the MWC practically feasible from a computational perspective.

The results of Figure 3.13 connect between theory and practice. The same digital algorithms that were used in the numerical simulations of [20] are successfully applied in [22] on real data, acquired by the hardware. This demonstrates that the theoretical principles are sufficiently robust to accommodate circuit non-idealities, which are inevitable in practice. A video recording of these experiments and additional documentation for the MWC hardware are available at http://webee.technion.ac.il/Sites/People/YoninaEldar/Info/hardware.html. A graphical package demonstrating the MWC numerically is available at http://webee.technion.ac.xil/Sites/People/YoninaEldar/Info/software/GUI/MWC_GUI.htm.

The MWC board appears to be the first reported hardware example borrowing ideas from CS to realize a sub-Nyquist sampling system for wideband signals, where the sampling and processing rates are directly proportional to the actual bandwidth occupation and not the highest frequency. Alternative approaches which employ discretization of the analog input are discussed in Section 3.8. The realization of these methods recover signals with Nyquist-rates below 1 MHz, falling outside of the class of wideband samplers. Additionally, the signal representations that result from discretization have size proportional to the Nyquist frequency, leading to recovery problems in the digital domain that are much larger than those posed by the MWC.

3.5.6 Sub-Nyquist signal processing

A nice feature of the MWC recovery stage is that it interfaces seamlessly with standard DSPs by providing (samples of) the narrowband quadrature information signals $I_i(t), Q_i(t)$ which build the ith band of interest

$$s_i(t) = I_i(t)\cos(2\pi f_i t) + Q_i(t)\sin(2\pi f_i t). \tag{3.26}$$

The signals $I_i(t), Q_i(t)$ could have been obtained by classic demodulation had the carriers f_i been known. In the union settings, with unknown carrier frequencies f_i, this capability is provided by a digital algorithm, named Back-DSP, that is developed in [14] and illustrated in Figure 3.14. The Back-DSP algorithm[2] translates the sequences $\mathbf{d}[n]$ to the narrowband signals $I_i(t), Q_i(t)$ that standard DSP packages expect to receive,

---

[2] Matlab code is available online at http://webee.technion.ac.il/Sites/People/YoninaEldar/Info/software/FR/FR.htm.

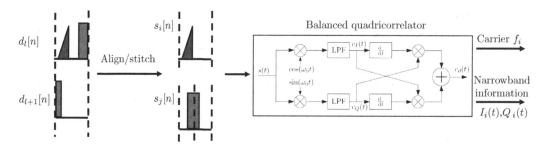

Figure 3.14 The flow of information extractions begins with detecting the band edges. The slices are filtered, aligned and stitched appropriately to construct distinct quadrature sequences $s_i[n]$ per information band. The balanced quadricorrelator finds the carrier f_i and extracts the narrowband information signals. Adapted from [38], © [2011] IEEE.

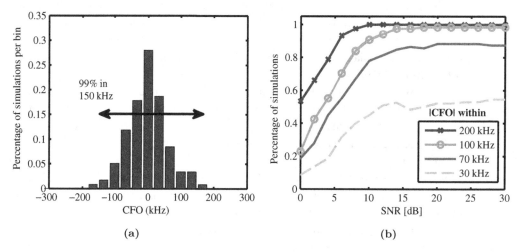

Figure 3.15 The distribution of CFO for fixed SNR=10 dB (a). The curves (b) represent the percentage of simulations in which the CFO magnitude is within the specified range. Adapted from [14], © [2011] IEEE.

thereby providing backward compatibility. Only low-rate computations, proportional to the rate of $I_i(t), Q_i(t)$, are used. Back-DSP first detects the band edges, then separates bands occupying the same slice to distinct sequences and stitches together energy that was split between adjacent slices. Finally, the balanced quadricorrelator [62] is applied in order to estimate the carrier frequencies.

Numerical simulations of the Back-DSP algorithm, in a wideband setup similar to the one of Figure 3.11, evaluated the Back-DSP performance in two aspects. The carrier frequency offset (CFO), estimated vs. true value of f_i, is plotted in Figure 3.15. In most cases, algorithm Back-DSP approaches the true carriers as close as 150 kHz. For reference, the 40 part-per-million (ppm) CFO specifications of IEEE 802.11 standards tolerate 150 kHz offsets for transmissions located around 3.75 GHz [63]. To verify data retrieval, a binary phase-shift keying (BPSK) transmission was generated, such that the band energy splits between two adjacent spectrum slices. A Monte Carlo simulation was used to compute bit error rate (BER) at the output of Back-DSP. Estimated BERs for 3 dB and 5 dB SNR, respectively, are better than $0.77 \cdot 10^{-6}$ and $0.71 \cdot 10^{-6}$. No erroneous bits were detected for SNR of 7 and 9 dB. See [14] for full results.

3.6 Finite rate of innovation signals

The second class we consider are analog signals in infinite unions of finite-dimensional spaces; these are continuous-time signals that can be characterized by a finite number of coefficients, also termed finite rate of innovation (FRI) signals as coined by Vetterli et al. [24,25]. One important problem that is studied in this framework is that of time-delay estimation, in which the input contains several, say L, echoes of a known pulse shape $h(t)$, though the echo positions t_ℓ and amplitudes a_ℓ are unknown [64]. Time-delay estimation is analogous to estimation of frequencies and amplitudes in a mixture of sinusoids. Both problems were widely studied in the classic literature [5, 65–69], with parametric estimation techniques that date back to methods developed by Rife and Boorstyn in 1974 [70] and earlier by David Slepian in the 1950s. The classic approaches focused on improving estimation performance in the digital domain, so that the error in estimating the time delays, or equivalently the sinusoid frequencies, approaches the optimal defined by the relevant Cramér–Rao bounds. The starting point, however, is discrete samples at the Nyquist rate of the input. The concept of FRI is fundamentally different, as it aims to obtain similar estimates from samples taken at the rate of innovation, namely proportional to $2L$ samples per observation interval, rather than at the typically much higher rate corresponding to the Nyquist bandwidth of $h(t)$. Chapter 4 in this book provides a comprehensive review of the FRI field. In the present chapter, we focus on Xampling-related aspects, with emphasis on possible hardware configurations for sub-Nyquist FRI acquisition. Recovery algorithms are briefly reviewed for the chapter to be self-contained.

3.6.1 Analog signal model

As we have seen in Section 3.4, the SI model (3.6) is a convenient way to describe analog signals in infinite-dimensional spaces. We can use a similar approach to describe analog signals that lie within finite-dimensional spaces by restricting the number of unknown gains $a_\ell[n]$ to be finite, leading to the parameterization

$$x(t) = \sum_{\ell=1}^{L} a_\ell h_\ell(t). \qquad (3.27)$$

In order to incorporate infiniteness into this model, we assume that each generator $h_\ell(t)$ has an unknown parameter α_ℓ associated with it, which can take on values in a continuous interval, resulting in the model

$$x(t) = \sum_{\ell=1}^{L} a_\ell h_\ell(t, \alpha_\ell). \tag{3.28}$$

Each possible choice of the set $\{\alpha_\ell\}$ leads to a different L-dimensional subspace of signals $\mathcal{A}_\lambda$, spanned by the functions $\{h(t, \alpha_\ell)\}$. Since α_ℓ can take on any value in a given interval, the model (3.28) corresponds to an infinite union of finite-dimensional subspaces (i.e., $|\Lambda| = \infty$), where each subspace $\mathcal{A}_\lambda$ in (3.3) contains those analog signals corresponding to a particular configuration of $\{\alpha_\ell\}_{\ell=1}^{L}$.

An important example of (3.28) is when $h_\ell(t, \alpha_\ell) = h(t - t_\ell)$ for some unknown time delay t_ℓ, leading to a stream of pulses

$$x(t) = \sum_{\ell=1}^{L} a_\ell h(t - t_\ell). \tag{3.29}$$

Here $h(t)$ is a known pulse shape and $\{t_\ell, a_\ell\}_{\ell=1}^{L}$, $t_\ell \in [0, \tau)$, $a_\ell \in \mathbb{C}$, $\ell = 1 \ldots L$ are unknown delays and amplitudes. This model was introduced by Vetterli *et al.* [24, 25] as a special case of signals having a finite number of degrees of freedom per unit time, termed FRI signals. Our goal is to sample $x(t)$ and reconstruct it from a minimal number of samples. Since in FRI applications the primary interest is in pulses which have small time-support, the required Nyquist rate can be very high. Bearing in mind that the pulse shape $h(t)$ is known, there are only $2L$ degrees of freedom in $x(t)$, and therefore we expect the minimal number of samples to be $2L$, much lower than the number of samples resulting from Nyquist rate sampling.

3.6.2 Compressive signal acquisition

To date, there are no general acquisition methods for signals of the form (3.28), while there are known solutions to various instances of (3.29). We begin by focusing on a simpler version of the problem, in which the signal $x(t)$ of (3.29) is repeated periodically leading to the model

$$x(t) = \sum_{m \in \mathbb{Z}} \sum_{\ell=1}^{L} a_\ell h(t - t_\ell - m\tau), \tag{3.30}$$

where τ is a known period. This periodic setup is easier to treat because we can exploit the properties of the Fourier series representation of $x(t)$ due to the periodicity. The dimensionality and number of subspaces included in the model (3.3) remain unchanged.

The key to designing an efficient X-ADC stage for this model is in identifying the connection to a standard problem in signal processing: the retrieval of the frequencies and amplitudes of a sum of sinusoids. The Fourier series coefficients $X[k]$ of the periodic pulse stream $x(t)$ are actually a sum of complex exponentials, with amplitudes $\{a_\ell\}$, and frequencies directly related to the unknown time-delays [24]:

$$X[k] = \frac{1}{\tau} H(2\pi k/\tau) \sum_{\ell=1}^{L} a_\ell e^{-j2\pi k t_\ell/\tau}, \qquad (3.31)$$

where $H(\omega)$ is the CTFT of the pulse $h(t)$. Therefore, once the Fourier coefficients are known, the unknown delays and amplitudes can be found using standard tools developed in the context of array processing and spectral estimation [24,71]. For further details see Chapter 4 in this book. Our focus here is on how to obtain the Fourier coefficients $X[k]$ efficiently from $x(t)$.

There are several X-ADC operators P which can be used to obtain the Fourier coefficients from time-domain samples of the signal. One choice is to set P to be a lowpass filter, as suggested in [24]. The resulting reconstruction requires $2L+1$ samples and therefore presents a near-critical sampling scheme. A general condition on the sampling kernel $s(t)$ that enables obtaining the Fourier coefficients was derived in [27]: its CTFT $S(\omega)$ should satisfy

$$S(\omega) = \begin{cases} 0, & \omega = 2\pi k/\tau, k \notin \mathcal{K} \\ \text{nonzero}, & \omega = 2\pi k/\tau, k \in \mathcal{K} \\ \text{arbitrary}, & \text{otherwise}, \end{cases} \qquad (3.32)$$

where $\mathcal{K}$ is a set of $2L$ consecutive indices such that $H(2\pi k/\tau) \neq 0$ for all $k \in \mathcal{K}$. The resulting X-ADC consists of a filter with a suitable impulse response $s(t)$ followed by a uniform sampler.

A special class of filters satisfying (3.32) are Sum of Sincs (SoS) in the frequency domain [27], which lead to compactly supported filters in the time domain. These filters are given in the Fourier domain by

$$G(\omega) = \frac{\tau}{\sqrt{2\pi}} \sum_{k \in \mathcal{K}} b_k \operatorname{sinc}\left(\frac{\omega}{2\pi/\tau} - k\right), \qquad (3.33)$$

where $b_k \neq 0, k \in \mathcal{K}$. It is easy to see that this class of filters satisfies (3.32) by construction. Switching to the time domain leads to

$$g(t) = \operatorname{rect}\left(\frac{t}{\tau}\right) \sum_{k \in \mathcal{K}} b_k e^{j2\pi k t/\tau}. \qquad (3.34)$$

For the special case in which $\mathcal{K} = \{-p, \ldots, p\}$ and $b_k = 1$,

$$g(t) = \operatorname{rect}\left(\frac{t}{\tau}\right) \sum_{k=-p}^{p} e^{j2\pi k t/\tau} = \operatorname{rect}\left(\frac{t}{\tau}\right) D_p(2\pi t/\tau), \qquad (3.35)$$

where $D_p(t)$ denotes the Dirichlet kernel.

While periodic streams are mathematically convenient, finite pulse streams of the form (3.29) are ubiquitous in real-world applications. A finite pulse stream can be viewed as a restriction of a periodic FRI signal to a single period. As long as the analog preprocessing P does not involve values of $x(t)$ outside the observation interval $[0, \tau]$, this implies that sampling and reconstruction methods developed for the periodic case also apply to finite settings. Treating time-limited signals with lowpass P, however, may be difficult since it has infinite time support, beyond the interval $[0, \tau]$ containing the finite pulse stream. Instead, we can choose fast-decaying sampling kernels or SoS filters such as (3.34) that have compact time support τ by construction.

To treat the finite case, a Gaussian sampling kernel was proposed in [24]; however, this method is numerically unstable since the samples are multiplied by a rapidly diverging or decaying exponent. As an alternative, we may use compactly supported sampling kernels for certain classes of pulse shapes based on splines [25]; this enables obtaining moments of the signal rather than its Fourier coefficients. These kernels have several advantages in practice as detailed in the next chapter. The moments are then processed in a similar fashion (see the next subsection for details). However, this approach is unstable for high values of L [25]. To improve robustness, the SoS class is extended to the finite case by exploiting the compact support of the filters [27]. This approach exhibits superior noise robustness when compared to the Gaussian and spline methods, and can be used for stable reconstruction even for very high values of L, e.g., $L = 100$.

The model of (3.29) can be further extended to the infinite stream case, in which

$$x(t) = \sum_{\ell \in \mathbb{Z}} a_\ell h(t - t_\ell), \quad t_\ell, a_\ell \in \mathbb{R}. \qquad (3.36)$$

Both [25] and [27] exploit the compact support of their sampling filters, and show that under certain conditions the infinite stream may be divided into a series of finite problems, which are solved independently with the existing finite algorithm. However, both approaches operate far from the rate of innovation, since proper spacing is required between the finite streams in order to allow the reduction stage, mentioned earlier. In the next section we consider a special case of (3.36) in which the time delays repeat periodically (but not the amplitudes). As we will show in this special case, efficient sampling and recovery is possible even using a single filter, and without requiring the pulse $h(t)$ to be time limited.

An alternative choice of analog compression operator P to enable recovery of infinite streams of pulses is to introduce multichannel sampling schemes. This approach was first considered for Dirac streams, where moments of the signal were obtained by a successive chain of integrators [72]. Unfortunately, the method is highly sensitive to noise. A simple sampling and reconstruction scheme consisting of two channels, each with an RC circuit, was presented in [73] for the special case where there is no more than one Dirac per sampling period. A more general multichannel architecture that can treat a broader class of pulses, while being much more stable, is depicted in Figure 3.16 [28]. The system is very similar to the MWC presented in the previous section, and as such it also complies with the general Xampling architecture. In each channel of this X-ADC, the signal is mixed with a modulating waveform $s_\ell(t)$, followed by an integrator, resulting in a mixture of the Fourier coefficients of the signal. By correct choice of the mixing coefficients, the

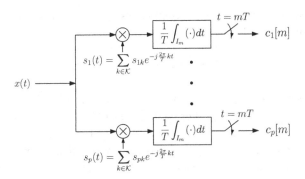

Figure 3.16 Extended sampling scheme using modulating waveforms for an infinite pulse stream. Adapted from [28], © [2011] IEEE.

Fourier coefficients may be extracted from the samples by a simple matrix inversion. This method exhibits superior noise robustness over the integrator chain method [72] and allows for more general compactly supported pulse-shapes. A recent method studied multichannel sampling for analog signals comprised of several, possibly overlapping, finite duration pulses with unknown shapes and time positions [74].

From a practical hardware perspective it is often more convenient to implement the multichannel scheme rather than a single-channel acquisition with an analog filter that satisfies the SoS structure. It is straightforward to show that the SoS filtering approach can also be implemented in the form of Figure 3.16 with coefficient matrix $\mathbf{S} = \mathbf{Q}$ where $\mathbf{Q}$ is chosen according to the definition following (3.37), for the SoS case. We point out that the multichannel architecture of Figure 3.16 can be readily implemented using the MWC prototype hardware. Mixing functions $s_\ell(t)$ comprised of finitely many sinusoids can be obtained by properly filtering a general periodic waveform. Integration over T is a first-order lowpass filter which can be assembled in place of the typically higher-order filter of the MWC system [61].

3.6.3 Recovery algorithms

In both the single-channel and multichannel approaches, recovery of the unknown delays and amplitudes proceeds according to Xampling by detecting the parameters t_ℓ that identify the signal subspace. The approach consists of two steps. First, the vector of samples $\mathbf{c}$ is related to the Fourier coefficients vector $\mathbf{x}$ through a $p \times |\mathcal{K}|$ mixing matrix $\mathbf{Q}$, as

$$\mathbf{c} = \mathbf{Q}\mathbf{x}. \tag{3.37}$$

Here $p \geq 2L$ represents the number of samples. When using the SoS approach with a filter $S(\omega)$, $\mathbf{Q} = \mathbf{V}\mathbf{S}$ where $\mathbf{S}$ is a $p \times p$ diagonal matrix with diagonal elements $S^*(-2\pi\ell/\tau)$, $1 \leq \ell \leq p$, and $\mathbf{V}$ is a $p \times |\mathcal{K}|$ Vandermonde matrix with ℓth element given by $e^{j2\pi\ell T/\tau}$, $1 \leq \ell \leq p$, where T denotes the sampling period. For the multichannel architecture of

Figure 3.16, $\mathbf{Q}$ consists of the modulation coefficients $s_{\ell k}$. The Fourier coefficients $\mathbf{x}$ can be obtained from the samples as

$$\mathbf{x} = \mathbf{Q}^\dagger \mathbf{c}. \qquad (3.38)$$

The unknown parameters $\{t_\ell, a_\ell\}_{\ell=1}^{L}$ are then recovered from $\mathbf{x}$ using standard spectral estimation tools, e.g. the annihilating filter method (see [24, 71] and the next chapter for details). These techniques can operate with as low as $2L$ Fourier coefficients. When a larger number of samples is available, alternative techniques that are more robust to noise can be used, such as the matrix-pencil method [75], and the Tufts and Kumaresan technique [76]. In Xampling terminology, these methods detect the input subspace, analogous to the role that CS plays in the CTF block for sparse-SI or multiband unions.

Reconstruction results for the sampling scheme using an SoS filter (3.33) with $b_k = 1$ are depicted in Figure 3.17. The original signal consists of $L = 5$ Gaussian pulses, and $N = 11$ samples were used for reconstruction. The reconstruction is exact to numerical

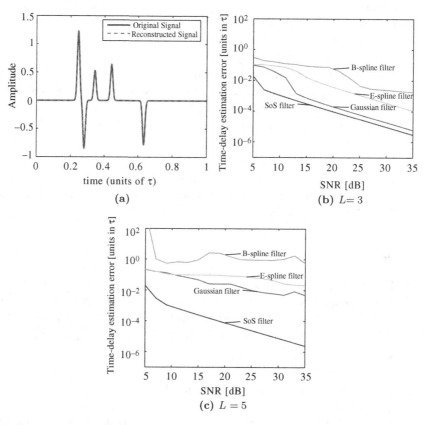

Figure 3.17 Performance comparison of finite pulse stream recovery using Gaussian, B-spline, E-spline, and SoS sampling kernels. (a) Reconstructed signal using SoS filters vs. original one. The reconstruction is exact to numerical precision. (b) $L = 3$ Dirac pulses are present, (c) $L = 5$ pulses. Adapted from [27], © [2011] IEEE.

precision. A comparison of the performance of various methods in the presence of noise is depicted in Figure 3.17 for a finite stream consisting of 3 and 5 pulses. The pulse-shape is a Dirac delta, and white Gaussian noise is added to the samples with a proper level in order to reach the desired SNR for all methods. All approaches operate using $2L+1$ samples. The results affirm stable recovery when using SoS filters. Chapter 4 of this book reviews in detail FRI recovery in the presence of noise [77] and outlines potential applications in super-resolution imaging [78], ultrasound [27], and radar imaging [29].

3.7 Sequences of innovation signals

The conventional SI setting (3.6) treats a single input subspace spanned by the shifts of N given generators $h_\ell(t)$. Combining the SI setting (3.6) and the time uncertainties of Section 3.6, we now incorporate structure by assuming that each generator $h_\ell(t)$ is given up to some unknown parameter α_ℓ associated with it, leading to an infinite union of infinite-dimensional spaces. As with its finite counterpart, there is currently no general sampling framework available to treat such signals. Instead, we focus on a special time-delay scenario of this model for which efficient sampling techniques have been developed.

3.7.1 Analog signal model

An interesting special case of the general model (3.28) is when $h_\ell(t) = h(t)$ and $\alpha_\ell = t_\ell$ represent unknown delays, leading to [26, 28, 29]

$$x(t) = \sum_{n \in \mathbb{Z}} \sum_{\ell=1}^{L} a_\ell[n] h(t - t_\ell - nT), \qquad (3.39)$$

where $\mathbf{t} = \{t_\ell\}_{\ell=1}^{L}$ is a set of unknown time delays contained in the time interval $[0, T)$, $\{a_\ell[n]\}$ are arbitrary bounded energy sequences, presumably representing low-rate streams of information, and $h(t)$ is a known pulse shape. For a given set of delays $\mathbf{t}$, each signal of the form (3.39) lies in an SI subspace $\mathcal{A}_\lambda$, spanned by L generators $\{h(t - t_\ell)\}_{\ell=1}^{L}$. Since the delays can take on any values in the continuous interval $[0, T)$, the set of all signals of the form (3.39) constitutes an infinite union of SI subspaces, i.e., $|\Lambda| = \infty$. Additionally, since any signal has parameters $\{a_\ell[n]\}_{n \in \mathbb{Z}}$, each of the $\mathcal{A}_\lambda$ subspaces has infinite cardinality. This model generalizes (3.36) with time delays that repeat periodically, where (3.39) allows the pulse shapes to have infinite support.

3.7.2 Compressive signal acquisition

To obtain a Xampling system, we follow a similar approach to that in Section 3.4, which treats a structured SI setting where there are N possible generators. The difference though is that in this current case there are infinitely many possibilities. Therefore, we replace the

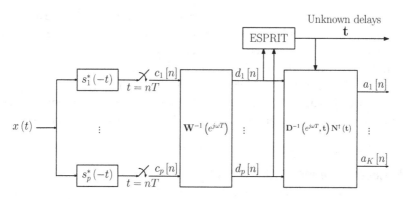

Figure 3.18 Sampling and reconstruction scheme for signals of the form (3.39). Adapted from [26], © [2010] IEEE.

CTF detection in the X-DSP of Figure 3.4 with a detection technique that supports this continuity: we will see that the ESPRIT method essentially replaces the CTF block [6].

A sampling and reconstruction scheme for signals of the form (3.39) is depicted in Figure 3.18 [26]. The analog compression operator P is comprised of p parallel sampling channels, where $p = 2L$ is possible under mild conditions on the sampling filters [26]. In each channel, the input signal $x(t)$ is filtered by a bandlimited sampling kernel $s_\ell^*(-t)$ with frequency support contained in an interval of width $2\pi p/T$, followed by a uniform sampler operating at a rate of $1/T$, thus providing the sampling sequence $c_\ell[n]$. Note that just as in the MWC (Section 3.5.4), the sampling filters can be collapsed to a single filter whose output is sampled at p times the rate of a single channel. In particular, acquisition can be as simple as a single channel with a lowpass filter followed by a uniform sampler. Analog compression of (3.39) is obtained by spreading out the energy of the signal in time, in order to capture all vital information with the narrow range $2\pi p/T$ of frequencies. To understand the importance of this stage, consider the case where $g(t) = \delta(t)$ and there are $L = 2$ Diracs per period of $T = 1$, as illustrated in Figure 3.19(a). We use a sampling scheme consisting of a complex bandpass filter-bank with four channels, each with width $2\pi/T$. In Figure 3.19(b) to (d), the outputs of the first three sampling channels are shown. It can be seen that the sampling kernels "smooth" the short pulses (Diracs in this example) in the time domain so that even when the sampling rate is low, the samples contain signal information. In contrast, if the input signal was sampled directly, then most of the samples would be zero.

3.7.3 Recovery algorithms

To recover the signal from the samples, a properly designed digital filter correction bank, whose frequency response in the DTFT domain is given by $\mathbf{W}^{-1}(e^{j\omega T})$, is applied to the sampling sequences in a manner similar to (3.9). The matrix $\mathbf{W}(e^{j\omega T})$ depends on the choice of the sampling kernels $s_\ell^*(-t)$ and the pulse shape $h(t)$. Its entries are defined

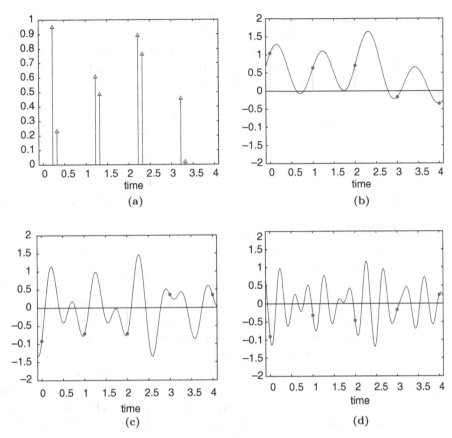

Figure 3.19 Stream of Diracs. (a) $L = 2$ Diracs per period $T = 1$. (b)–(d) The outputs of the first three sampling channels, the dashed lines denote the sampling instances. Adapted from [26], © [2010] IEEE.

for $1 \leq \ell, m \leq p$ as

$$\mathbf{W}\left(e^{j\omega T}\right)_{\ell,m} = \frac{1}{T} S_\ell^*(\omega + 2\pi m/T) H(\omega + 2\pi m/T). \tag{3.40}$$

After the digital correction stage, it can be shown that the corrected sample vector $\mathbf{d}[n]$ is related to the unknown amplitudes vector $\mathbf{a}[n] = \{a_\ell[n]\}$ by a Vandermonde matrix which depends on the unknown delays [26]. Therefore, subspace detection can be performed by exploiting known tools from the direction of arrival [79] and spectral estimation [71] literature to recover the delays $\mathbf{t} = \{t_1, \ldots, t_L\}$, such as the well-known ESPRIT algorithm [6]. Once the delays are determined, additional filtering operations are applied on the samples to recover the information sequences $a_\ell[n]$. In particular, referring to Figure 3.18, the matrix $\mathbf{D}$ is a diagonal matrix with diagonal elements equal to $e^{-j\omega t_k}$, and $\mathbf{N}(\mathbf{t})$ is a Vandermonde matrix with elements $e^{-j2\pi m t_k/T}$.

In our setting, the ESPRIT algorithm consists of the following steps:

1. Construct the correlation matrix $\mathbf{R}_{dd} = \sum_{n \in \mathbb{Z}} \mathbf{d}[n] \mathbf{d}^H[n]$.
2. Perform an SVD decomposition of $\mathbf{R}_{dd}$ and construct the matrix $\mathbf{E}_s$ consisting of the L singular vectors associated with the non-zero singular values in its columns.
3. Compute the matrix $\mathbf{\Phi} = \mathbf{E}_{s\downarrow}^{\dagger} \mathbf{E}_{s\uparrow}$. The notations $\mathbf{E}_{s\downarrow}$ and $\mathbf{E}_{s\uparrow}$ denote the sub matrices extracted from $\mathbf{E}_s$ by deleting its last/first row respectively.
4. Compute the eigenvalues of $\mathbf{\Phi}$, $\lambda_i, i = 1, 2, \ldots, L$.
5. Retrieve the unknown delays by $t_i = -\frac{T}{2\pi} \arg(\lambda_i)$.

In general, the number of sampling channels p required to ensure unique recovery of the delays and sequences using the proposed scheme has to satisfy $p \geq 2L$ [26]. This leads to a minimal sampling rate of $2L/T$. For certain signals, the sampling rate can be reduced even further to $(L+1)/T$ [26]. Interestingly, the minimal sampling rate is not related to the Nyquist rate of the pulse $h(t)$. Therefore, for wideband pulse shapes, the reduction in rate can be quite substantial. As an example, consider the setup in [80], used for characterization of ultra-wide band wireless indoor channels. Under this setup, pulses with bandwidth of $W = 1$ GHz are transmitted at a rate of $1/T = 2$ MHz. Assuming that there are 10 significant multipath components, we can reduce the sampling rate down to 40 MHz compared with the 2 GHz Nyquist rate.

We conclude by noting that the approach of [26] imposes only minimal conditions on the possible generator $h(t)$ in (3.39), so that in principle almost arbitrary generators can be treated according to Figure 3.18, including $h(t)$ with unlimited time support. As mentioned earlier, implementing this sampling strategy can be as simple as collapsing the entire system to a single channel that consists of a lowpass filter and a uniform sampler. Reconstruction, however, involves a $p \times p$ bank of digital filters $\mathbf{W}^{-1}(e^{j\omega T})$, which can be computationally demanding. In scenarios with time-limited $h(t)$ sampling with the multichannel scheme of Figure 3.16 can be more convenient, since digital filtering is not required so that ESPRIT is applied directly on the samples [28].

3.7.4 Applications

Problems of the form (3.39) appear in a variety of different settings. For example, the model (3.39) can describe multipath medium identification problems, which arise in applications such as radar [81], underwater acoustics [82], wireless communications [83], and more. In this context, pulses with known shape are transmitted through a multipath medium, which consists of several propagation paths, at a constant rate. As a result the received signal is composed of delayed and weighted replicas of the transmitted pulses. The delays t_ℓ represent the propagation delays of each path, while the sequences $a_\ell[n]$ describe the time-varying gain coefficient of each multipath component.

An example of multipath channel identification is shown in Figure 3.20. The channel consists of four propagation paths and is probed by pulses at a rate of $1/T$. The output is sampled at a rate of $5/T$, with white Gaussian noise with SNR of 20 dB added to the samples. Figure 3.20 demonstrates recovery of the propagations delays, and the time-varying gain coefficients, from low-rate samples corrupted by noise. This is essentially

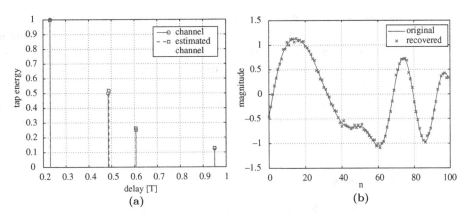

Figure 3.20 Channel estimation with $p = 5$ sampling channels, and SNR $= 20$ dB. (a) Delay recovery. (b) Recovery of the time-varying gain coefficient of the first path. Adapted from [26], © [2010] IEEE.

a combination of X-ADC and X-DSP, where the former is used to reduce the sampling rate, while the latter is responsible for translating the compressed sample sequences $c_\ell[n]$ to the set of low-rate streams $d_\ell[n]$, which convey the actual information to the receiver. In this example the scheme of Figure 3.18 was used with a bank of ideal band-pass filters covering consecutive frequency bands:

$$S_\ell(\omega) = \begin{cases} T, & \omega \in \left[(\ell-1)\frac{2\pi}{T}, \ell\frac{2\pi}{T}\right] \\ 0, & \text{otherwise.} \end{cases} \quad (3.41)$$

As can be seen, even in the presence of noise, the channel is recovered almost perfectly from low-rate samples. Applications to radar are explored in Chapter 4 and later on in Section 3.8.5.

3.8 Union modeling vs. finite discretization

The approach we have been describing so far treats analog signals by taking advantage of a UoS model, where the inherent infiniteness of the analog signal enters either through the dimensions of the underlying subspaces $\mathcal{A}_\lambda$, the cardinality of the union $|\Lambda|$, or both. An alternative strategy is to assume that the analog signal has some finite representation to begin with, i.e., that both Λ and $\mathcal{A}_\lambda$ are finite. Sampling in this case can be readily mapped to a standard underdetermined CS system $\mathbf{y} = \mathbf{A}\mathbf{x}$ (that is with a single vector of unknowns rather than infinitely many as in the IMV setting).

The methods we review in this section treat continuous signals that have an underlying finite parameterization: the RD [30] and quantized CS radar [31–33]. In addition to surveying [30–33], we examine the option of applying sampling strategies developed for finite settings on general analog models with infinite cardinalities. To address this option, we compare hardware and digital complexities of the RD and MWC systems when treating multiband inputs, and imaging performance of quantized [31–33] vs. analog radar [29]. In order to obtain a close approximation to union modeling, a sufficiently dense discretization of the input is required, which in turn can degrade performance in various practical metrics. Thus, whilst methods such as [30–33] are effective for the models for which they were developed, their application to general analog signals, presumably by discretization, may limit the range of signal classes that can be treated.

3.8.1 Random demodulator

The RD approach treats signals consisting of a discrete set of harmonic tones with the system that is depicted in Figure 3.21 [30].

Signal model A multitone signal $f(t)$ consists of a sparse combination of integral frequencies:

$$f(t) = \sum_{\omega \in \Omega} a_\omega e^{j2\pi\omega t}, \quad (3.42)$$

where Ω is a finite set of K out of an even number Q of possible harmonics

$$\Omega \subset \{0, \pm\Delta, \pm 2\Delta, \ldots, \pm(0.5Q-1)\Delta, 0.5Q\Delta\}. \quad (3.43)$$

The model parameters are the tone spacing Δ, number of active tones K, and grid length Q. The Nyquist rate is $Q\Delta$. Whenever normalized, Δ is omitted from formulae under the convention that all variables take nominal values (e.g., $R = 10$ instead of $R = 10$ Hz).

Sampling The input signal $f(t)$ is mixed by a pseudorandom chipping sequence $p_c(t)$ which alternates at a rate of W. The mixed output is then integrated and dumped at a constant rate R, resulting in the sequence $y[n]$, $1 \leq n \leq R$. The development in [30] uses the following parameter setup

$$\Delta = 1, \quad W = Q, \quad R \in \mathbb{Z} \text{ such that } \frac{W}{R} \in \mathbb{Z}. \quad (3.44)$$

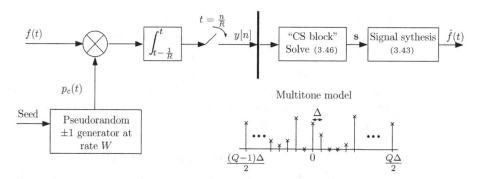

Figure 3.21 Block diagram of the random demodulator. Adapted from [30], © [2010] IEEE.

It was proven in [30] that if W/R is an integer and (3.44) holds, then the vector of samples $\mathbf{y} = [y[1], \ldots, y[R]]^T$ can be written as

$$\mathbf{y} = \mathbf{\Phi}\mathbf{x}, \quad \mathbf{x} = \mathbf{F}\mathbf{s}, \quad \|\mathbf{s}\|_0 \leq K. \tag{3.45}$$

The matrix $\mathbf{\Phi}$ has dimensions $R \times W$, effectively capturing the mechanism of integration over W/R Nyquist intervals, where the polarity of the input is flipped on each interval according to the chipping function $p_c(t)$. See Figure 3.23(a) in the sequel for further details on $\mathbf{\Phi}$. The W-squared DFT matrix $\mathbf{F}$ accounts for the sparsity in the frequency domain. The vector $\mathbf{s}$ has Q entries s_ω which are up to a constant scaling from the corresponding tone amplitudes a_ω. Since the signal has only K active tones, $\|\mathbf{s}\|_0 \leq K$.

Reconstruction Equation (3.45) is an underdetermined system that can be solved with existing CS algorithms, e.g., ℓ_1 minimization or greedy methods. As before, a "nice" CS matrix $\mathbf{\Phi}$ is required in order to solve (3.45) with sparsity order K efficiently with existing polynomial-time algorithms. In Figure 3.21, the CS block refers to solving (3.45) with a polynomial-time CS algorithm and a "nice" $\mathbf{\Phi}$, which requires a sampling rate on the order of [30]

$$R \approx 1.7K \log(W/K + 1). \tag{3.46}$$

Once the sparse $\mathbf{s}$ is found, the amplitudes a_ω are determined from s_ω by constant scaling, and the output $\hat{f}(t)$ is synthesized according to (3.42).

3.8.2 Finite-model sensitivity

The RD system is sensitive to inputs with tones slightly displaced from the theoretical grid, as was indicated by several studies [14,84,85]. For example, [14] repeated the developments of [30] for an unnormalized multitone model, with Δ as a free parameter and W, R that are not necessarily integers. The measurements still obey the underdetermined system (3.45) as before, where now [14]

$$W = Q\Delta, \quad R = N_R\Delta, \quad \frac{W}{R} \in \mathbb{Z}, \tag{3.47}$$

and N_R is the number of samples taken by the RD. The equalities in (3.47) imply that the rates W, R need to be perfectly synchronized with the tones spacing Δ. If (3.47) does not hold, either due to hardware imperfections so that the rates W, R deviate from their nominal values, or due to model mismatch so that the actual spacing Δ is different than what was assumed, then the reconstruction error grows high.

The following toy-example demonstrates this sensitivity. Let $W = 1000, R = 100$ Hz, with $\Delta = 1$ Hz. Construct $f(t)$ by drawing $K = 30$ locations uniformly at random on the tones grid and normally distributed amplitudes a_ω. Basis pursuit gave exact recovery $\hat{f}(t) = f(t)$ for $\Delta = 1$. For 5 part-per-million (ppm) deviation in Δ the squared-error reached 37%:

$$\Delta = 1 + 0.000005 \quad \rightarrow \quad \frac{\|f(t) - \hat{f}(t)\|^2}{\|f(t)\|^2} = 37\%. \tag{3.48}$$

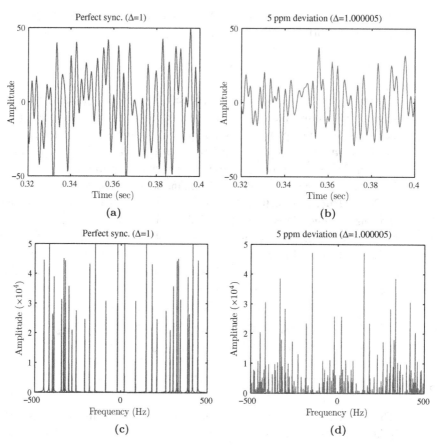

Figure 3.22 Effects of non-integral tones on the output of the random demodulator. Panels (a),(b) plot the recovered signal in the time domain. The frequency contents are compared in panels (c),(d). Adapted from [14], © [2011] IEEE.

Figure 3.22 plots $f(t)$ and $\hat{f}(t)$ in time and frequency, revealing many spurious tones due to the model mismatch. The equality $W = Q$ in the normalized setup (3.44) hints at the required synchronization, though the dependency on the tones spacing is implicit since $\Delta = 1$. With $\Delta \neq 1$, this issue appears explicitly.

The sensitivity that is demonstrated in Figure 3.22 is a source of error already in the finite multitone setting (3.42). The implication is that utilizing the RD for the counterpart problem of sampling multiband signals with continuous spectra requires a sufficiently dense grid of tones. Otherwise, a non-negligible portion of the multiband energy resides off the grid, which can lead to recovery errors due to the model mismatch. As discussed below, a dense grid of tones translates to high computational loads in the digital domain.

The MWC is less sensitive to model mismatches in comparison. Since only inequalities are used in (3.24), the number of branches p and aliasing rate f_p can be chosen with some safeguards with respect to the specified number of bands N and individual widths B. Thus, the system can handle inputs with more than N bands and widths larger than

B, up to the safeguards that were set. The band positions are not restricted to any specific displacement with respect to the spectrum slices; a single band can split between slices, as depicted in Figure 3.6. Nonetheless, both the PNS [19] and MWC [20] approaches require specifying the multiband spectra by a pair of maximal quantities (N, B). This modeling can be inefficient (in terms of resulting sampling rate) when the individual band widths are significantly different from each other. For example, a multiband model with N_1 bands of lengths $B_1 = k_1 b$ and N_2 bands of lengths $B_2 = k_2 b$ is described by a pair $(N_1 + N_2, \max(B_1, B_2))$, with spectral occupation potentially larger than actually used. A more flexible modeling in this scenario would assume only the total actual bandwidth being occupied, i.e., $N_1 B_1 + N_2 B_2$. This issue can partially be addressed at the expense of hardware size by designing the system (PNS/MWC) to accommodate $N_1 k_1 + N_2 k_2$ bands of lengths b.

3.8.3 Hardware complexity

We next compare the hardware complexity of the RD/MWC systems. In both approaches, the acquisition stage is mapped into an underdetermined CS system: Figure 3.21 leads to a standard sparse recovery problem (3.45) in the RD system, while in the MWC approach, Figure 3.10 results in an IMV problem (3.11). A crucial point is that the hardware needs to be sufficiently accurate for that mapping to hold, since this is the basis for reconstruction. While the RD and MWC sampling stages seem similar, they rely on different analog properties of the hardware to ensure accurate mapping to CS, which in turn imply different design complexities.

To better understand this issue, we examine Figure 3.23. The figure depicts the Nyquist-equivalent of each method, which is the system that samples the input at its Nyquist rate and then computes the relevant sub-Nyquist samples by applying the sensing matrix digitally. The RD-equivalent integrates and dumps the input at rate W, and then applies $\boldsymbol{\Phi}$ on Q serial measurements, $\mathbf{x} = [x[1], \ldots, x[Q]]^T$. To coincide with the sub-Nyquist samples of Figure 3.21, $\boldsymbol{\Phi} = \mathbf{HD}$ is used, where $\mathbf{D}$ is diagonal with ± 1 entries, according to the values $p_c(t)$ takes on $t = n/W$, and $\mathbf{H}$ sums over W/R entries [30]. The MWC-equivalent has M channels, with the ℓth channel demodulating the relevant spectrum slice to the origin and sampling at rate $1/T$, which results in $d_\ell[n]$. The sensing matrix $\mathbf{A}$ is applied on $\mathbf{d}[n]$. While sampling according to the equivalent systems of Figure 3.23 is a clear waste of resources, it enables us to view the internal mechanism of each strategy. Note that the reconstruction algorithms remain the same; it does not matter whether the samples were actually obtained at a sub-Nyquist rate, according to Figures 3.21 or 3.10, or if they were computed after sampling according to Figure 3.23.

Analog compression In the RD approach, time-domain properties of the hardware dictate the necessary accuracy. For example, the impulse-response of the integrator needs to be a square waveform with a width of $1/R$ seconds, so that $\mathbf{H}$ has exactly W/R consecutive 1's in each row. For a diagonal $\mathbf{D}$, the sign alternations of $p_c(t)$ need to be sharply aligned on $1/W$ time intervals. If either of these properties is non-ideal, then the

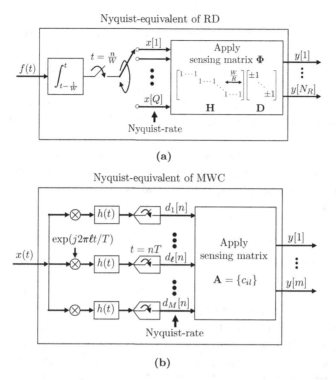

Figure 3.23 The Nyquist-equivalents of the RD (a) and MWC (b) sample the input at its Nyquist rate and apply the sensing matrix digitally. Adapted from [14], © [2011] IEEE.

mapping to CS becomes nonlinear and signal dependent. Precisely, (3.45) becomes [30]

$$\mathbf{y} = \mathbf{H}(\mathbf{x})\mathbf{D}(\mathbf{x})\mathbf{x}. \qquad (3.49)$$

A non-integer ratio W/R affects both $\mathbf{H}$ and $\mathbf{D}$ [30]. Since $f(t)$ is unknown, $\mathbf{x}$, $\mathbf{H}(\mathbf{x})$ and $\mathbf{D}(\mathbf{x})$ are also unknown. It is suggested in [30] to train the system on example signals, so as to approximate a linear system. Note that if (3.47) is not satisfied, then the DFT expansion also becomes nonlinear and signal-dependent $\mathbf{x} = \mathbf{F}(\Delta)\mathbf{s}$. The *form factor* of the RD is therefore the time-domain accuracy that can be achieved in practice.

The MWC requires periodicity of the waveforms $p_i(t)$ and lowpass response for $h(t)$, which are both frequency-domain properties. The sensing matrix $\mathbf{A}$ is constant as long as $p_i(t)$ are periodic, regardless of the time-domain appearance of these waveforms. Therefore, non-ideal time-domain properties of $p_i(t)$ have no effect on the MWC. The consequence is that stability in the frequency domain dictates the form factor of the MWC. For example, 2 GHz periodic functions were demonstrated in a circuit prototype of the MWC [22]. More broadly, circuit publications report the design of high-speed sequence generators up to 23 and even 80 GHz speeds [86, 87], where stable frequency properties are verified experimentally. Accurate time-domain appearance is not considered a design factor in [86, 87], and is in fact not maintained in practice as shown in

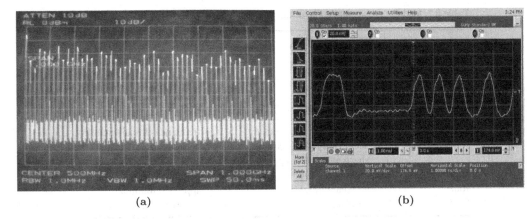

Figure 3.24 The spectrum (a) and the time-domain appearance (b) of a 2 GHz sign-alternating periodic waveform. Adapted from [22], © [2011] IET.

[22, 86, 87]. For example, Figure 3.24 demonstrates frequency stability vs. inaccurate time-domain appearance [22].

The MWC scheme requires an ideal lowpass filter $h(t)$ with rectangular frequency response, which is difficult to implement due to its sharp edges. This problem appears as well in Nyquist sampling, where it is addressed by alternative sampling kernels with smoother edges at the expense of oversampling. Similar edge-free filters $h(t)$ can be used in the MWC system with slight oversampling [74]. Ripples in the passband and non-smooth transitions in the frequency response can be compensated for digitally using the algorithm in [60].

Sampling rate In theory, both the RD and MWC approach the minimal rate for their model. The RD system, however, requires in addition an integer ratio W/R; see (3.44) and (3.47). In general, a substantial rate increase may be needed to meet this requirement. The MWC does not limit the rate granularity; see a numerical comparison in the next subsection.

Continuous reconstruction The RD synthesizes $\hat{f}(t)$ using (3.42). Realizing (3.42) in hardware can be excessive, since it requires K oscillators, one per each active tone. Computing (3.42) digitally needs a processing rate of W, and then a DAC device at the same rate. Thus, the synthesis complexity scales with the Nyquist rate. The MWC reconstructs $\hat{x}(t)$ using commercial DAC devices, running at the low-rate $f_s = 1/T$. It needs N branches. Wideband continuous inputs require prohibitively large K, W to be adequately represented on a discrete grid of tones. In contrast, despite the infinitely many frequencies that comprise a multiband input, N is typically small. We note however that the MWC may incur difficulties in reconstructing contents around the frequencies $(\ell + 0.5)f_p$, $-L \leq \ell \leq L$, since these are irregular points of transitions between spectrum slices. Reconstruction accuracy of these irregular points depends on the cutoff curvature of $h(t)$ and relative amplitudes of consecutive $c_{i\ell}$. Reconstruction of an input consisting of pure tones at these specific frequencies may be imperfect. In practice, the bands encode information signals, which can be reliably decoded, even when signal energy

Table 3.1. Model and hardware comparison

	RD (multitone)	**MWC** (multiband)
Model parameters	K, Q, Δ	$N, B, f_{\max}$
System parameters	R, W, N_R	$m, 1/T$
Setup	(3.44)	(3.24)
	Sensitive, Eq. (3.47), Figure 3.22	Robust
Form factor	time-domain appearance	frequency-domain stability
Requirements	accurate $1/R$ integration	periodic $p_i(t)$
	sharp alternations $p_c(t)$	
ADC topology	integrate-and-dump	commercial
Rate	gap due to (3.44)	approach minimal
DAC	1 device at rate W	N devices at rate f_s

is located around the frequencies $(l + 0.5)fp$. As discussed in Section 3.5.6, when the bands contain digital transmissions and the SNR is sufficiently high, algorithm Back-DSP [14] enables recovery of the underlying information bits, and in turn allows DSP at a low rate, even when a band energy is split between adjacent slices. This algorithm also allows reconstruction of $x(t)$ with only N DAC devices instead of $2N$ that are required for arbitrary multiband reconstruction. Table 3.1 summarizes the model and hardware comparison.

3.8.4 Computational loads

In this subsection, we compare computational loads when treating multiband signals, either using the MWC system or in the RD framework by discretizing the continuous frequency axis to a grid of $Q = f_{\text{NYQ}}$ tones, out of which only $K = NB$ are active [30]. We emphasize that the RD system was designed for multitone inputs, though for the study of computational loads we examine the RD on multiband inputs by considering a comparable grid of tones of the same Nyquist bandwidth. Table 3.2 compares between the RD and MWC for an input with 10 GHz Nyquist rate and 300 MHz spectral occupancy. For the RD we consider two discretization configurations, $\Delta = 1$ Hz and $\Delta = 100$ Hz. The table reveals high computational loads that stem from the dense discretization that is required to represent an analog multiband input. We also included the sampling rate and DAC speeds to complement the previous section. The notation in the table is self-explanatory, though a few aspects are emphasized below.

The sensing matrix $\boldsymbol{\Phi} = \mathbf{HD}$ of the RD has dimensions

$$\boldsymbol{\Phi}: R \times W \propto K \times Q \quad \text{(huge)}. \tag{3.50}$$

The dimension scales with the Nyquist rate; already for $Q = 1$ MHz Nyquist-rate input, there are 1 million unknowns in (3.45). The sensing matrix $\mathbf{A}$ of the MWC has dimensions

$$\mathbf{A}: m \times M \propto N \times \frac{f_{\text{NYQ}}}{B} \quad \text{(small)}. \tag{3.51}$$

Table 3.2. Discretization impact on computational loads

		RD		MWC
	Discretization spacing	$\Delta = 1$ Hz	$\Delta = 100$ Hz	
Model	K tones	$300 \cdot 10^6$	$3 \cdot 10^6$	N bands
	out of Q tones	$10 \cdot 10^9$	$10 \cdot 10^7$	width B
	alternation speed W	10 GHz	10 GHz	m channels[§]
Sampling setup				M Fourier coefficients
	rate R, Eq. (3.46), theory	2.9 GHz	2.9 GHz	f_s per channel
	Eq. (3.44), practice	5 GHz	5 GHz	total rate
Underdetermined system	(3.45): $\mathbf{y} = \mathbf{HDF}\mathbf{s}$, $\|\mathbf{s}\|_0 \leq K$			(3.18): $\mathbf{V} = \mathbf{AU}$, $\|\mathbf{U}\|_0 \leq 2N$
Preparation				
Collect samples	Num. of samples N_R	$5 \cdot 10^9$	$5 \cdot 10^7$	$2N$ snapshots of $\mathbf{y}[n]$
Delay	N_R/R	1 sec	10 msec	$2N/f_s$
Complexity				
Matrix dimensions	$\mathbf{\Phi} = \mathbf{HDF} = N_R \times Q$		$5 \cdot 10^7 \times 10^8$	$\mathbf{A} = m \times M$
Apply matrix[#]	$\mathcal{O}(W \log W)$			$\mathcal{O}(mM)$
Storage[#]	$\mathcal{O}(W)$			$\mathcal{O}(mM)$
Realtime (fixed support)	$\mathbf{s}_\Omega = (\mathbf{\Phi}\mathbf{F})^\dagger_\Omega \mathbf{y}$			$\mathbf{d}_\lambda[n] = \mathbf{A}^\dagger_\lambda \mathbf{y}[n]$
Memory length	N_R	$5 \cdot 10^9$	$5 \cdot 10^7$	1 snapshot of $\mathbf{y}[n]$
Delay	N_R/R	1 sec	10 msec	$1/f_s$
Mult.-ops. (per window)	$K N_R$	$1.5 \cdot 10^{18}$	$1.5 \cdot 10^{14}$	$2Nm$
(100 MHz cycle)	$K N_R/((N_R/R) \cdot 100\mathrm{M})$	$1.5 \cdot 10^{10}$	$1.5 \cdot 10^6$	$2Nm f_s/100\mathrm{M}$
Reconstruction	1 DAC at rate $W = 10$ GHz			$N = 6$ DACs at individual rates $f_s = 51$ MHz
Technology barrier (estimated)	CS algorithms (~10 MHz)			Waveform generator (~23 GHz)

		MWC
		6
		50 MHz
		35
		195
		51 MHz
		1.8 GHz
		$12 \cdot 35 = 420$
		235 nsec
		35×195
		35
		19.5 nsec
		420
		214

[§] with $q = 1$; in practice, hardware size is collapsed with $q > 1$ [22].   [#] for the RD, taking into account the structure **HDF**.

For the comparable spectral occupancy we consider, Φ has dimensions that are six to eight orders of magnitude higher, in both the row and column dimensions, than the MWC sensing matrix $\mathbf{A}$. The size of the sensing matrix is a prominent factor since it affects many digital complexities: the delay and memory length that are associated with collecting the measurements, the number of multiplications when applying the sensing matrix on a vector, and the storage requirement of the matrix. See the table for a numerical comparison of these factors.

We also compare the reconstruction complexity, in the more simple scenario that the support is fixed. In this setting, the recovery is merely a matrix-vector multiplication with the relevant pseudo-inverse. As before, the size of Φ results in long delay and huge memory length for collecting the samples. The number of scalar multiplications (Mult.-ops.) for applying the pseudo-inverse reveals again orders of magnitude differences. We expressed the Mult.-ops. per block of samples, and in addition scaled them to operations per clock cycle of a 100 MHz DSP processor.

We conclude the table with our estimation of the technology barrier of each approach. Computational loads and memory requirements in the digital domain are the bottleneck of the RD approach. Therefore the size of CS problems that can be solved with available processors limits the recovery. We estimate that $W \approx 1$ MHz may be already quite demanding using convex solvers, whereas $W \approx 10$ MHz is probably the barrier using greedy methods.[3] The MWC is limited by the technology for generating the periodic waveforms $p_i(t)$, which depends on the specific choice of waveform. The estimated barrier of 23 GHz refers to implementation of the periodic waveforms according to [86, 87], though realizing a full MWC system at these high rates can be a challenging task. Our barrier estimates are roughly consistent with the hardware publications of these systems: [89, 90] report the implementation of (single, parallel) RD for Nyquist-rate $W = 800$ kHz. An MWC prototype demonstrates faithful reconstruction of $f_{\text{NYQ}} = 2$ GHz wideband inputs [22].

3.8.5 Analog vs. discrete CS radar

The question of whether finite modeling can be used to treat general analog scenarios was also studied in [29] in the context of radar imaging. Here, rate reduction can be translated to increased resolution and decreased time–bandwidth product of the radar system.

An intercepted radar signal $x(t)$ has the form

$$x(t) = \sum_{k=1}^{K} \alpha_k h(t - t_k) e^{j2\pi \nu_k t} \tag{3.52}$$

with each triplet (t_k, ν_k, α_k) corresponding to an echo of the radar waveform $h(t)$ from a distinct target [91]. Equation (3.52) represents an infinite union, parameterized by

---

[3] A bank of RD channels was studied in [88], the parallel system duplicates the analog issues and its computational complexity is not improved by much.

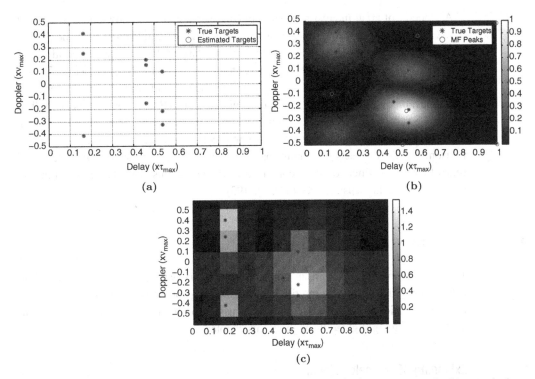

Figure 3.25 Recovery of the delay–Doppler plane using (a) a union of subspaces approach, (b) a standard matched filter, and (c) a discretized delay–Doppler plane. Adapted from [29], © [2011] IEEE.

$\lambda = (t_k, \nu_k)$, of K-dimensional subspaces $\mathcal{A}_\lambda$ which capture the amplitudes α_k within the chosen subspace. The UoS approach was taken in [29], where reconstruction is obtained by the general scheme for time delay recovery of [26], with subspace estimation that uses standard spectral estimation tools [71]. A finite modeling approach to radar assumes that the delays t_k and frequencies ν_k lie on a grid, effectively quantizing the delay–Doppler space (t, ν) [32,33,92]. Compressed sensing algorithms are then used for reconstruction of the targets scene.

An example of identification of nine targets (in a noiseless setting) is illustrated in Figure 3.25 for three approaches: the union-based approach [29] with a simple lowpass acquisition, classic matched filtering, and quantized-CS recovery. The discretization approach causes energy leakage in the quantized space into adjacent grid points. As the figure shows, union modeling is superior with respect to both alternative approaches. Identification results in the presence of noise appear in [29] and affirm imaging performance that degrades gracefully as noise levels increase, as long as the noise is not too large. These results affirm that UoS modeling not only offers a reduced-rate sampling method, but allows one to increase the resolution in target identification, as long as the noise is not too high. At high noise levels, match-filtering is superior. We refer to [77] for rigorous analysis of noise effects in general FRI models.

A property of great interest in radar applications is the time–bandwidth WT product of the system, where W refers to the bandwidth of the transmitted pulse $h(t)$ and T indicates the round-trip time between radar station and targets. Ultimately we would like to minimize both quantities, since W impacts antenna size and sampling rate, while T poses a physical limitation on latency, namely the time it takes to identify targets. Uncertainty principles, though, imply that we cannot minimize both W and T simultaneously. The analog CS radar approach results in minimal time–bandwidth product, much lower than that obtained using standard matched-filtering techniques; see [29] for a precise comparison. Practical aspects of sparsity-based radar imaging, such as improved decoding time of target identification from compressive measurements as well as efficient matrix structures for radar sensing, are studied in [93].

3.9 Discussion

Table 3.3 summarizes the various applications we surveyed, suggesting that Xampling is broad enough to capture a multitude of engineering solutions, under the same logical flow of operations. We conclude with a discussion on the properties and insights into analog sensing highlighted throughout this chapter.

3.9.1 Extending CS to analog signals

The influential works by Donoho [3] and Candès *et al.* [4] coined the CS terminology, in which the goal is to reduce the sampling rate below Nyquist. These pioneering works established CS via a study of underdetermined systems, where the sensing matrix abstractly replaces the role of the sampling operator, and the ambient dimensions represent the high Nyquist rate. In practice, however, the study of underdetermined systems does not hint at the actual sub-Nyquist sampling of analog signals. One cannot apply a sensing matrix on a set of Nyquist rate samples, as performed in the conceptual systems in Figure 3.23, since that would contradict the whole idea of reducing the sampling rate. The previous sections demonstrate how extensions of CS to continuous signals can be significantly different in many practical aspects. Based on the insights gained, we draw several operative conclusions in Table 3.4 regarding the choice of analog compression P in continuous-time CS systems. The first point follows immediately from Figure 3.22 and basically implies that model and sampler parameters should not be tightly related, implicitly or explicitly. We elaborate below on the other two suggestions.

Input signals are eventually generated by some source, which has its own accuracy specifications. Therefore, if designing P imposes constraints on the hardware that are not stricter than those required to generate the input signal, then there are no essential limitations on the input range. We support this conclusion by several examples. The MWC requires accuracy that is achieved with RF technology, which also defines the possible range of multiband transmissions. The same principle of shifting spectral slices to the origin with different weights can be achieved by PNS [19]. This strategy, however, can result in a narrower input range that can be treated, since current RF technology

Table 3.3. Applications of union of subspaces

Application	Signal model	Cardinality		Analog compression	Subspace detection
		union	subspaces		
Sparse-SI [16]	see (3.10)	finite	∞	filter-bank, Figure 3.4	CTF
PNS [19]	multiband, Figure 3.6	finite	∞	time shifts	CTF [45]
MWC [20]	multiband, Figure 3.6	finite	∞	periodic mixing + lowpass	CTF [45]
RD [30]	$f(t) = \sum_\omega a_\omega e^{j2\pi\omega t}$ $\omega \in$ discrete grid Ω	finite	finite	sign flipping + integrate-dump	CS
FRI	$x(t) = \sum_{\ell=1}^{L} d_\ell g(t - t_\ell)$				
periodic [24, 94]	$x(t) = x(t+T)$	∞	finite	lowpass	annihilating filter [24, 94]
finite [25]	$0 \leq t \leq T$	∞	finite	splines	moments factoring [25]
periodic/finite [27, 28]	either of the above	∞	finite	SoS filtering	annihilating filter
Sequences of innovation [26, 28]	see (3.39)	∞	∞	lowpass, or periodic mixing + integrate-dump	MUSIC [5] or ESPRIT [6]
NYFR [95]	multiband	finite	∞	jittered undersampling	n/a

Table 3.4. Suggested guidelines for extending CS to analog signals

#1	set system parameters with safeguards to accommodate possible model mismatches
#2	incorporate design constraints on P that suit the technology generating the source signals
#3	balance between nonlinear (subspace detection) and linear (interpolation) reconstruction complexities

can generate source signals at frequencies that exceed front-end bandwidths of existing ADC devices [20]. Multiband inputs generated by optical sources, however, may require a different compression stage P than that of the RF-based MWC system.

Along the same line, time-domain accuracy constraints may limit the range of multitone inputs that can be treated in the RD approach, if these signals are generated by RF sources. On the other hand, consider a model of piecewise constant inputs, with knots at the integers and only K non-identically zero pieces out of Q. Sampling these signals with the RD system would map to (3.45), but with an identity basis instead of the DFT matrix $\mathbf{F}$. In this setting, the time-domain accuracy required to ensure that the mapping to (3.45) holds is within the tolerances of the input source.

Moving on to our third suggestion, we attempt to reason the computational loads encountered in Table 3.2. Over 1 second, both approaches reconstruct their inputs from a comparable set of numbers; $K = 300 \cdot 10^6$ tone coefficients or $2Nf_s = 612 \cdot 10^6$ amplitudes of active sequences $d_\ell[n]$. The difference is, however, that the RD recovers all these unknowns by a single execution of a nonlinear CS algorithm on the system (3.45), which has large dimensions. In contrast, the MWC splits the recovery task to a small-size nonlinear part (i.e., CTF) and real-time linear interpolation. This distinction can be traced back to model assumptions. The nonlinear part of a multitone model, namely the number of subspaces $|\Lambda| = \binom{Q}{K}$, is exponentially larger than $\binom{M}{2N}$ which specifies a multiband union of the same Nyquist bandwidth. Clearly, a prerequisite for balancing computation loads is an input model with as many unknowns as possible in its linear part (subspaces $\mathcal{A}_\lambda$), so as to decrease the nonlinear cardinality $|\Lambda|$ of the union. The important point is that in order to benefit from such modeling, P must be properly designed to incorporate this structure and reduce computational loads.

For example, consider a block-sparse multitone model with K out of Q tones, such that the active tones are clustered in K/d blocks of length d. A plain RD system which does not incorporate this block structure would still result in a large $R \times W$ sensing matrix with its associated digital complexities. Block-sparse recovery algorithms, e.g., [43], can be used to partially decrease the complexity, but the bottleneck remains the fact that the hardware compression is mapped to a large sensing matrix.[4] A potential analog

---

[4] Note that simply modifying the chipping and integrate-dumping intervals, in the existing scheme of Figure 3.21, to d times larger results in a sensing matrix smaller by the same factor, though (3.45) in this setting would force reconstructing each block of tones by a single tone, presumably corresponding to a model of K/d active tones out of Q/d at spacing $d\Delta$.

compression for this block-sparse model can be an MWC system designed for $N = K/d$ and $B = d\Delta$ specifications.

Our conclusions here stem from the study of the RD and MWC systems, and are therefore mainly relevant for choosing P in Xampling systems that map their hardware to underdetermined systems and incorporate CS algorithms for recovery. Nonetheless, our suggestions above do not necessitate such a relation to CS, and may hold more generally with regard to other compression techniques.

Finally, we point out the Nyquist-folding receiver (NYFR) of [95] which suggests an interesting alternative route towards sub-Nyquist sampling. This method introduces a deliberate jitter in an undersampling grid, which results in induced phase modulations at baseband such that the modulation magnitudes depend on the unknown carrier positions. This strategy is exceptional as it relies on a time-varying acquisition effect, which departs from the linear time-invariant P that unifies all the works we surveyed herein. In principle, to enable recovery, one would need to infer the magnitudes of the phase modulations. A reconstruction algorithm was not reported yet for this class of sampling, which is why we do not elaborate further on this method. Nonetheless, this is an interesting venue for developing sub-Nyquist strategies and opens a wide range of possibilities to explore.

3.9.2 Is CS a universal sampling scheme?

The discussion on extending CS to analog signals draws an interesting connection to the notion of CS universality. In the discrete setup of sensing, the measurement model is $\mathbf{y} = \mathbf{\Phi}\mathbf{x}$ and the signal is sparse in some given transform basis $\mathbf{x} = \mathbf{\Psi}\mathbf{s}$. The concept of CS universality refers to the attractive property of sensing with $\mathbf{\Phi}$ without knowledge of $\mathbf{\Psi}$, so that $\mathbf{\Psi}$ enters only in the reconstruction algorithm. This notion is further emphasized with the default choice of the identity basis $\mathbf{\Psi} = \mathbf{I}$ in many CS publications, which is justified by no loss of generality, since $\mathbf{\Psi}$ is conceptually absorbed into the sensing matrix $\mathbf{\Phi}$.

In contrast, in many analog CS systems, the hardware design benefits from incorporating knowledge on the sparsity basis of the input. Refer to the Nyquist-equivalent system of the MWC in Figure 3.23(b), for example. The input $x(t)$ is conceptually first preprocessed into a set of high-rate streams of measurements $\mathbf{d}[n]$, and then a sensing matrix $\mathbf{A} = \{c_{i\ell}\}$ is applied to reduce the rate. In PNS [20], the same set of streams $\mathbf{d}[n]$ is sensed by the partial DFT matrix (3.22), which depends on the time shifts c_i of the PNS sequences. This sensing structure also appears in Theorem 3.1, where the term $\mathbf{G}^{-*}(e^{j\omega T})\mathbf{s}(\omega)$ in (3.12) first generates $\mathbf{d}[n]$, and then a sensing matrix $\mathbf{A}$ is applied. In all these scenarios, the intermediate sequences $\mathbf{d}[n]$ are sparse for all n, so that the sensing hardware effectively incorporates knowledge on the (continuous) sparsity basis of the input.

Figure 3.26 generalizes this point. The analog compression stage P in Xampling systems can be thought of as a two-stage sampling system. First, a sparsifying stage which generates a set of high-rate streams of measurements, out of which only a few are nonidentically zero. Second, a sensing matrix is applied, where in principle, any

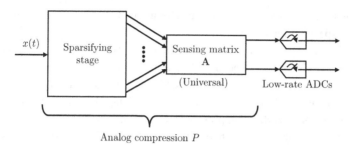

Figure 3.26 Analog compression operator P in X-ADC architecture consists of a sparsifying stage and sensing matrix, which are combined into one efficient analog preprocessing stage.

sensing matrix can be used in that stage. In practice, however, the trick is to choose a sensing matrix which can be combined with the sparsifying part into a single hardware mechanism, so that the system does not actually go through Nyquist-rate sampling. This combination is achieved by periodic mixing in the MWC system, time delays in the case of PNS, and the filters $w_\ell(t)$ in the sparse-SI framework of Theorem 3.1. We can therefore suggest a slightly different interpretation of the universality concept for analog CS systems, which is the flexibility to choose any sensing matrix $\mathbf{A}$ in the second stage of P, provided that it can be efficiently combined with the given sparsifying stage.

3.9.3 Concluding remarks

Starting from the work in [15], union of subspaces models appear at the frontier of research on sampling methods. The ultimate goal is to build a complete sampling theory for UoS models of the general form (3.3) and then derive specific sampling solutions for applications of interest. Although several promising advances have already been made [15,16,26,41,44], this esteemed goal is yet to be accomplished.

In this chapter we described a line of works which extend CS ideas to the analog domain based on UoS modeling. The Xampling framework of [14] unifies the treatment of several classes of UoS signals, by leveraging insights and pragmatic considerations into the generic architecture of Figure 3.2.

Our hope is that the template scheme of Figure 3.2 can serve as a substrate for developing future sampling strategies for UoS models, and inspire future developments that will eventually lead to a complete generalized sampling theory in unions of subspaces.

Acknowledgements

The authors would like to thank Kfir Gedalyahu and Ronen Tur for their collaboration on many topics related to this review and for authorizing the use of their figures, and Waheed Bajwa for many useful comments.

References

[1] C. E. Shannon. Communication in the presence of noise. *Proc IRE*, 37:10–21, 1949.

[2] H. Nyquist. Certain topics in telegraph transmission theory. *Trans AIEE*, 47(2):617–644, 1928.

[3] D. L. Donoho. Compressed sensing. *IEEE Trans Inf Theory*, 52(4):1289–1306, 2006.

[4] E. J. Candès, J. Romberg, and T. Tao. Robust uncertainty principles: Exact signal reconstruction from highly incomplete frequency information. *IEEE Trans Inf Theory*, 52(2):489–509, 2006.

[5] R. Schmidt. Multiple emitter location and signal parameter estimation. *IEEE Trans. Antennas Propag*, 34(3):276–280, 1986. First presented at RADC Spectrum Estimation Workshop, Griffiss AFB, NY, 1979.

[6] R. Roy and T. Kailath. ESPRIT-estimation of signal parameters via rotational invariance techniques. *IEEE Trans Acoust, Speech Sig Proc*, 37(7):984–995, 1989.

[7] P. Stoica and Y. Selen. Model-order selection: A review of information criterion rules. *IEEE Sig Proc Mag*, 21(4):36–47, 2004.

[8] S. Baker, S. K. Nayar, and H. Murase. Parametric feature detection. *Int J Comp Vision*, 27(1):27–50, 1998.

[9] J. Crols and M. S. J. Steyaert. Low-IF topologies for high-performance analog front ends of fully integrated receivers. *IEEE Trans Circ Syst II: Analog Digital Sig Proc*, 45(3):269–282, 1998.

[10] R. G. Vaughan, N. L. Scott, and D. R. White. The theory of bandpass sampling. *IEEE Trans Sig Proc*, 39(9):1973–1984, 1991.

[11] Y. P. Lin and P. P. Vaidyanathan. Periodically nonuniform sampling of bandpass signals. *IEEE Trans Circu Syst II*, 45(3):340–351, 1998.

[12] C. Herley and P. W. Wong. Minimum rate sampling and reconstruction of signals with arbitrary frequency support. *IEEE Trans Inf Theory*, 45(5):1555–1564, 1999.

[13] R. Venkataramani and Y. Bresler. Perfect reconstruction formulas and bounds on aliasing error in sub-Nyquist nonuniform sampling of multiband signals. *IEEE Trans Inf Theory*, 46(6):2173–2183, 2000.

[14] M. Mishali, Y. C. Eldar, and A. Elron. Xampling: Signal acquisition and processing in union of subspaces. *IEEE Trans Sig Proc*, 59(10):4719–4734, 2011.

[15] Y. M. Lu and M. N. Do. A theory for sampling signals from a union of subspaces. *IEEE Trans Sig Proc*, 56(6):2334–2345, 2008.

[16] Y. C. Eldar. Compressed sensing of analog signals in shift-invariant spaces. *IEEE Trans Sig Proc*, 57(8):2986–2997, 2009.

[17] M. Unser. Sampling – 50 years after Shannon. *Proc IEEE*, 88(4):569–587, 2000.

[18] Y. C. Eldar and T. Michaeli. Beyond bandlimited sampling. *IEEE Sig Proc Mag*, 26(3):48–68, 2009.

[19] M. Mishali and Y. C. Eldar. Blind multi-band signal reconstruction: compressed sensing for analog signals. *IEEE Trans Sig Proc*, 57(3):993–1009, 2009.

[20] M. Mishali and Y. C. Eldar. From theory to practice: Sub-Nyquist sampling of sparse wideband analog signals. *IEEE J Sel Topics Sig Proc*, 4(2):375–391, 2010.

[21] P. Feng and Y. Bresler. Spectrum-blind minimum-rate sampling and reconstruction of multiband signals. *Proc IEEE Int Conf ASSP*, 3:1688–1691, 1996.

[22] M. Mishali, Y. C. Eldar, O. Dounaevsky, and E. Shoshan. Xampling: Analog to digital at sub-Nyquist rates. *IET Circ Dev Syst*, 5(1):8–20, 2011.

[23] M. Mishali and Y. C. Eldar. Wideband spectrum sensing at sub-Nyquist rates. *IEEE Sig Proc Mag*, 28(4):102–135, 2011.

[24] M. Vetterli, P. Marziliano, and T. Blu. Sampling signals with finite rate of innovation. *IEEE Trans Sig Proc*, 50(6):1417–1428, 2002.

[25] P. L. Dragotti, M. Vetterli, and T. Blu. Sampling moments and reconstructing signals of finite rate of innovation: Shannon meets Strang fix. *IEEE Trans Sig Proc*, 55(5):1741–1757, 2007.

[26] K. Gedalyahu and Y. C. Eldar. Time delay estimation from low-rate samples: A union of subspaces approach. *IEEE Trans Sig Proc*, 58(6):3017–3031, 2010.

[27] R. Tur, Y. C. Eldar, and Z. Friedman. Innovation rate sampling of pulse streams with application to ultrasound imaging. *IEEE Trans Sig Proc*, 59(4):1827–1842, 2011.

[28] K. Gedalyahu, R. Tur, and Y. C. Eldar. Multichannel sampling of pulse streams at the rate of innovation. *IEEE Trans Sig Proc*, 59(4):1491–1504, 2011.

[29] W. U. Bajwa, K. Gedalyahu, and Y. C. Eldar. Identification of parametic underspread linear systems super-resolution radar. *IEEE Trans Sig Proc*, 59(6):2548–2561, 2011.

[30] J. A. Tropp, J. N. Laska, M. F. Duarte, J. K. Romberg, and R. G. Baraniuk. Beyond Nyquist: Efficient sampling of sparse bandlimited signals. *IEEE Trans Inf Theory*, 56(1):520–544, 2010.

[31] A. W. Habboosh, R. J. Vaccaro, and S. Kay. An algorithm for detecting closely spaced delay/Doppler components. *ICASSP 1997*: 535–538, 1997.

[32] W. U. Bajwa, A. M. Sayeed, and R. Nowak. Learning sparse doubly-selective channels. *Allerton Conf Commun Contr Comput*, 575–582, 2008.

[33] M. A. Herman and T. Strohmer. High-resolution radar via compressed sensing. *IEEE Trans Sig Proc*, 57(6):2275–2284, 2009.

[34] M. Unser and A. Aldroubi. A general sampling theory for nonideal acquisition devices. *IEEE Trans Sig Proc*, 42(11):2915–2925, 1994.

[35] A. Aldroubi and K. Gröchenig. Non-uniform sampling and reconstruction in shift-invariant spaces. *SIAM Rev*, 43(4):585–620, 2001.

[36] T. G. Dvorkind, Y. C. Eldar, and E. Matusiak. Nonlinear and non-ideal sampling: Theory and methods. *IEEE Trans Sig Proc*, 56(12):5874–5890, 2008.

[37] M. A. Davenport, P. T. Boufounos, M. B. Wakin, and R. G. Baraniuk. Signal processing with compressive measurements. *IEEE J Sel Topics Sig Proc*, 4(2):445–460, 2010.

[38] M. Mishali and Y. C. Eldar. Sub-Nyquist sampling: Bridging theory and practice. *IEEE Sig Proc Mag*, 2011.

[39] M. Unser. Splines: A perfect fit for signal and image processing. *IEEE Sig Proc Mag*, 16(6): 22–38, 1999.

[40] T. Blu and M. Unser. Quantitative Fourier analysis of approximation techniques: Part I – Interpolators and projectors. *IEEE Trans Sig Proc*, 47(10):2783–2795, 1999.

[41] Y. C. Eldar and M. Mishali. Robust recovery of signals from a structured union of subspaces. *IEEE Trans Inf Theory*, 55(11):5302–5316, 2009.

[42] M. Yuan and Y. Lin. Model selection and estimation in regression with grouped variables. *J Roy Stat Soc Ser B Stat Methodol*, 68(1):49–67, 2006.

[43] Y. C. Eldar, P. Kuppinger, and H. Bölcskei. Block-sparse signals: Uncertainty relations and efficient recovery. *IEEE Trans Sig Proc*, 58(6):3042–3054, 2010.

[44] R. G. Baraniuk, V. Cevher, M. F. Duarte, and C. Hegde. Model-based compressive sensing. *IEEE Trans Inf Theory*, 56(4):1982–2001, 2010.

[45] M. Mishali and Y. C. Eldar. Reduce and boost: Recovering arbitrary sets of jointly sparse vectors. *IEEE Trans Sig Proc*, 56(10):4692–4702, 2008.

[46] J. A. Tropp. Algorithms for simultaneous sparse approximation. Part I: Greedy pursuit. *Sig Proc* (Special Issue on Sparse Approximations in Signal and Image Processing). 86:572–588, 2006.

[47] J. A. Tropp. Algorithms for simultaneous sparse approximation. Part II: Convex relaxation. *Sig Proc* (Special Issue on Sparse Approximations in Signal and Image Processing). 86:589–602, 2006.

[48] S. F. Cotter, B. D. Rao, K. Engan, and K. Kreutz-Delgado. Sparse solutions to linear inverse problems with multiple measurement vectors. *IEEE Trans Sig Proc*, 53(7):2477–2488, 2005.

[49] J. Chen and X. Huo. Theoretical results on sparse representations of multiple-measurement vectors. *IEEE Trans Sig Proc*, 54(12):4634–4643, 2006.

[50] Y. C. Eldar and H. Rauhut. Average case analysis of multichannel sparse recovery using convex relaxation. *IEEE Trans Inf Theory*, 56(1):505–519, 2010.

[51] M. E. Davies and Y. C. Eldar. Rank awareness in joint sparse recovery. To appear in IEEE *Trans Inf Theory*.

[52] J. Mitola III. Cognitive radio for flexible mobile multimedia communications. *Mobile Netw. Appl*, 6(5):435–441, 2001.

[53] H. J. Landau. Necessary density conditions for sampling and interpolation of certain entire functions. *Acta Math*, 117:37–52, 1967.

[54] A. Kohlenberg. Exact interpolation of band-limited functions. *J Appl Phys*, 24:1432–1435, 1953.

[55] W. Black and D. Hodges. Time interleaved converter arrays. In: *Solid-State Circuits Conference. Digest of Technical Papers*. 1980 IEEE Int, XXIII: 14–15.

[56] P. Nikaeen and B. Murmann. Digital compensation of dynamic acquisition errors at the front-end of high-performance A/D converters. *IEEE Trans Sig Proc*, 3(3):499–508, 2009.

[57] A/D Converters. Analog Devices Corp:[Online]. Available: www.analog.com/en/analog-to-digital-converters/ad-converters/products/index.html, 2009.

[58] Data converters. Texas Instruments Corp. 2009:[Online]. Available: http://focus.ti.com/analog/docs/dataconvertershome.tsp.

[59] M. Mishali and Y. C. Eldar. Expected-RIP: Conditioning of the modulated wideband converter. *Inform Theory Workshop*, IEEE: 343–347, 2009.

[60] Y. Chen, M. Mishali, Y. C. Eldar, and A. O. Hero III. Modulated wideband converter with non-ideal lowpass filters. *ICASSP*: 3630–3633, 2010.

[61] M. Mishali, R. Hilgendorf, E. Shoshan, I. Rivkin, and Y. C. Eldar. Generic sensing hardware and real-time reconstruction for structured analog signals. *ISCAS*: 1748–1751, 2011.

[62] F. Gardner. Properties of frequency difference detectors. *IEEE Trans Commun*, 33(2):131–138, 1985.

[63] Wireless LAN Medium Access Control (MAC) and Physical Layer (PHY) specifications: High-speed physical layer in the 5 GHz band. IEEE Std 80211a-1999.

[64] A. Quazi. An overview on the time delay estimate in active and passive systems for target localization. *IEEE Trans Acoust, Speech Sig Proc*, 29(3):527–533, 1981.

[65] A. Bruckstein, T. J. Shan, and T. Kailath. The resolution of overlapping echos. *IEEE Trans Acoust, Speech, Sig Proc*, 33(6):1357–1367, 1985.

[66] M. A. Pallas and G. Jourdain. Active high resolution time delay estimation for large BT signals. *IEEE Trans Sig Proc*, 39(4):781–788, 1991.

[67] Z. Q. Hou and Z. D. Wu. A new method for high resolution estimation of time delay. *ICASSP '82*. 7:420–423, 1982.

[68] H. Saarnisaari. TLS-ESPRIT in a time delay estimation. In: IEEE 47th Vehic Techn Conf, 3:1619–1623, 1997.

[69] F. X. Ge, D. Shen, Y. Peng, and V. O. K. Li. Super-resolution time delay estimation in multipath environments. *IEEE Trans Circ Syst, I*: 54(9):1977–1986, 2007.

[70] D. Rife and R. Boorstyn. Single tone parameter estimation from discrete-time observations. *IEEE Trans Inf Theory*, 20(5):591–598, 1974.

[71] P. Stoica and R. Moses. *Introduction to Spectral Analysis*. Upper Saddle River, NJ: Prentice-Hall; 1997.

[72] J. Kusuma and V. K. Goyal. Multichannel sampling of parametric signals with a successive approximation property. *IEEE Int. Conf Image Proc (ICIP)*: 1265–1268, 2006.

[73] C. S. Seelamantula, and M. Unser. A generalized sampling method for finite-rate-of-innovation-signal reconstruction. *IEEE Sig Proc Letters*, 15:813–816, 2008.

[74] E. Matusiak and Y. C. Eldar. Sub-Nyquist sampling of short pulses: Theory. Submitted to *IEEE Trans Sig Theory*; [Online] arXivorg 10103132. 2010 Oct;.

[75] Y. Hua and T. K. Sarkar. Matrix pencil method for estimating parameters of exponentially damped/undamped sinusoids in noise. *IEEE Trans Acoust, Speech, Sig Proc*, 38(5):814–824, 1990.

[76] D. W. Tufts and R. Kumaresan. Estimation of frequencies of multiple sinusoids: Making linear prediction perform like maximum likelihood. *Proc IEEE*, 70(9):975–989, 1982.

[77] Z. Ben-Haim, T. Michaeli, and Y. C. Eldar. Performance bounds and design criteria for estimating finite rate of innovation signals. Submitted to *IEEE Trans Inf Theory*; [Online] arXivorg 10092221. 2010 Sep;.

[78] L. Baboulaz and P. L. Dragotti. Exact feature extraction using finite rate of innovation principles with an application to image super-resolution. *IEEE Trans Image Proc*, 18(2):281–298, 2009.

[79] H. Krim and M. Viberg. Two decades of array signal processing research: The parametric approach. *IEEE Sig Proc Mag*, 13(4):67–94, 1996.

[80] M. Z. Win and R. A. Scholtz. Characterization of ultra-wide bandwidth wireless indoor channels: A communication-theoretic view. *IEEE J Sel Areas Commun*, 20(9):1613–1627, 2002.

[81] A. Quazi. An overview on the time delay estimate in active and passive systems for target localization. *IEEE Trans Acoust, Speech, Sig Proc*, 29(3):527–533, 1981.

[82] R. J. Urick. *Principles of Underwater Sound*. New York: McGraw-Hill; 1983.

[83] G. L. Turin. Introduction to spread-spectrum antimultipath techniques and their application to urban digital radio. *Proc IEEE*, 68(3):328–353, 1980.

[84] M. F. Duarte and R. G. Baraniuk. Spectral compressive sensing;[Online]. Available: www.math.princeton.edu/~mduarte/images/SCS-TSP.pdf, 2010.

[85] Y. Chi, A. Pezeshki, L. Scharf, and R. Calderbank. Sensitivity to basis mismatch in compressed sensing. *ICASSP 2010*; 3930–3933, 2010.

[86] E. Laskin and S. P. Voinigescu. A 60 mW per Lane, 4×23-Gb/s 2^7-1 PRBS Generator. *IEEE J Solid-State Circ*, 41(10):2198–2208, 2006.

[87] T. O. Dickson, E. Laskin, I. Khalid, *et al*. An 80-Gb/s $2^{31}-1$ pseudorandom binary sequence generator in SiGe BiCMOS technology. *IEEE J Solid-State Circ*, 40(12):2735–2745, 2005.

[88] Z. Yu, S. Hoyos, and B. M. Sadler. Mixed-signal parallel compressed sensing and reception for cognitive radio. *ICASSP 2008*: 3861–3864, 2008.

[89] T. Ragheb, J. N. Laska, H. Nejati, *et al*. A prototype hardware for random demodulation based compressive analog-to-digital conversion. 51st Midwest Symp *Circ Syst, 2008. MWSCAS*: 37–40, 2008.

[90] Z. Yu, X. Chen, S. Hoyos, *et al*. Mixed-signal parallel compressive spectrum sensing for cognitive radios. *Int J Digit Multimedia Broadcast*, 2010.

[91] M. I. Skolnik. *Introduction to Radar Systems*. 3rd edn. *New York: McGraw-Hill*; 2001.

[92] X. Tan, W. Roberts, J. Li, and P. Stoica. Range-Doppler imaging via a train of probing pulses. *IEEE Trans Sig Proc*, 57(3):1084–1097, 2009.

[93] L. Applebaum, S. D. Howard, S. Searle, and R. Calderbank. Chirp sensing codes: Deterministic compressed sensing measurements for fast recovery. *Appl Comput Harmon Anal*, 26(2): 283–290, 2009.

[94] I. Maravic and M. Vetterli. Sampling and reconstruction of signals with finite rate of innovation in the presence of noise. *IEEE Trans Sig Proc*, 53(8):2788–2805, 2005.

[95] G. L. Fudge, R. E. Bland, M. A. Chivers, *et al*. A Nyquist folding analog-to-information receiver. *Proc 42nd Asilomar Conf Sig, Syst Comput*: 541–545, 2008.

4 Sampling at the rate of innovation: theory and applications

Jose Antonio Urigüen, Yonina C. Eldar, Pier Luigi Dragotti, and Zvika Ben-Haim

Parametric signals, such as streams of short pulses, appear in many applications including bio-imaging, radar, and spread-spectrum communication. The recently developed finite rate of innovation (FRI) framework has paved the way to low-rate sampling of such signals, by exploiting the fact that only a small number of parameters per unit of time are needed to fully describe them. For example, a stream of pulses can be uniquely defined by the time delays of the pulses and their amplitudes, which leads to far fewer degrees of freedom than the signal's Nyquist rate samples. This chapter provides an overview of FRI theory, algorithms, and applications. We begin by discussing theoretical results and practical algorithms allowing perfect reconstruction of FRI signals from a minimal number of samples. We then turn to treat recovery from noisy measurements. Finally, we overview a diverse set of applications of FRI theory, in areas such as super-resolution, radar, and ultrasound.

4.1 Introduction

We live in an analog world, but we would like our digital computers to interact with it. For example, sound is a continuous-time phenomenon, which can be characterized by the variations in air pressure as a function of time. For digital processing of such real-world signals to be possible, we require a sampling mechanism which converts continuous signals to discrete sequences of numbers, while preserving the information present in those signals.

In classical sampling theory, which dates back to the beginning of the twentieth century [1–3], a bandlimited signal whose maximum frequency is $f_{\max}$ is sampled at or above the Nyquist rate $2f_{\max}$. It is well known that the signal can then be perfectly reconstructed from its samples. Unfortunately, real-world signals are rarely truly bandlimited, if only because most signals have finite duration in time. Even signals which are approximately bandlimited often have to be sampled at a fairly high Nyquist rate, requiring expensive sampling hardware and high-throughput digital machinery.

Compressed Sensing: Theory and Applications, ed. Yonina C. Eldar and Gitta Kutyniok. Published by Cambridge University Press. © Cambridge University Press 2012.

Classical sampling theory necessitates a high sampling rate whenever a signal has a high bandwidth, even if the actual information content in the signal is low. For instance, a piecewise linear signal is non-differentiable; it is therefore not bandlimited, and moreover, its Fourier transform decays at the fairly low-rate $O(1/f^2)$. However, the signal is completely described by the positions of knots (transitions between linear segments) and the signal values at those positions. Thus, as long as the knots are known to have a minimum separation, this signal has a finite information rate. It seems wasteful to sample such signals at the Nyquist rate. It would be more efficient to have a variety of sampling techniques, tailored to different signal models, such as bandlimited or piecewise linear signals. Such an approach echoes the fundamental quest of compressive sampling, which is to capture only the essential information embedded in a signal. This chapter, together with Chapter 3, on Xampling, applies the idea of compressed sensing to certain classes of analog signals. While the focus of Xampling is on signals lying in unions of subspaces and on developing a unified architecture for efficient sampling of various classes of signals, here we concentrate on a comprehensive review of finite rate of innovation (FRI) theory.

To be specific, suppose that a function $x(t)$ has the property that any finite duration segment of length τ is completely determined by no more than K parameters. In this case, the function $x(t)$ is said to have a local rate of innovation equal to K/τ [4], because it has no more than K degrees of freedom every τ seconds. In general, a signal is said to have FRI if its local rate of innovation is finite for a sufficiently large τ. For example, the aforementioned piecewise linear signal has this property. Many important signal models, such as splines and pulse streams, also satisfy the FRI property, and will be explored in depth later in this chapter.

An elegant and powerful result is that, in many cases, certain types of FRI signals can be reconstructed without error from samples taken at the rate of innovation [4]. The advantage of this result is self-evident: FRI signals need not be bandlimited, and even if they are, the Nyquist frequency can be much higher than the rate of innovation. Thus, by using FRI techniques, the sampling rate required for perfect reconstruction can be lowered substantially. However, exploiting these capabilities requires careful design of the sampling mechanism and of the digital post-processing. The purpose of this chapter is to review the theory, recovery techniques, and applications of the FRI model.

4.1.1 The sampling scheme

Consider the sampling setup shown in Figure 4.1, where the original continuous-time signal $x(t)$ is filtered before being uniformly sampled at a rate of $f_s = 1/T$. The filtering may be a design choice or may be due to the acquisition device. If we denote the filtered version of $x(t)$ by $y(t) = h(t) * x(t)$, then the samples $\{y_n\}$ are given by

$$y_n = y(nT) = \left\langle x(t), \varphi\left(\frac{t}{T} - n\right)\right\rangle = \int_{-\infty}^{\infty} x(t)\varphi\left(\frac{t}{T} - n\right) dt, \qquad (4.1)$$

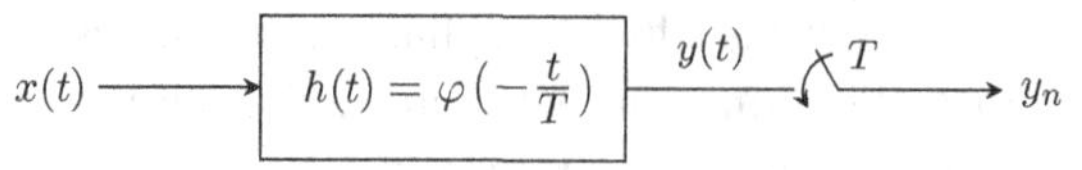

$$x(t) \longrightarrow \boxed{h(t) = \varphi\left(-\frac{t}{T}\right)} \xrightarrow{y(t)} \overset{T}{\swarrow} \longrightarrow y_n$$

Figure 4.1 Traditional sampling scheme. The continuous-time input signal $x(t)$ is filtered with $h(t)$ and sampled every T seconds. The samples are then given by $y_n = (x * h)(t)|_{t=nT}$.

where the *sampling kernel* $\varphi(t)$ is the scaled and time-reversed version of $h(t)$. For example, the previously discussed classical sampling setup often incorporates an anti-aliasing lowpass filter $h(t) = \text{sinc}(t)$, which eliminates any signal components having frequencies above $f_s/2$.

Changing the sampling kernel $\varphi(t)$ provides considerable flexibility in the information transferred to the samples $\{y_n\}$. Indeed, many modern sampling techniques, such as sampling in shift-invariant spaces, rely on an appropriate choice of the sampling kernel [5, 6]. As we will see, the model of Figure 4.1 with adequate sampling kernels also provides the basis for most FRI sampling techniques. On the other hand, FRI recovery methods are typically more elaborate, and involve nonlinear digital processing of the samples. This is an important practical aspect of FRI techniques: the sampling hardware is simple, linear, and easy to implement, but it is followed by nonlinear algorithms in the digital stage, since this is typically easier and cheaper to customize.

Two basic questions arise in the context of the sampling scheme of Figure 4.1. First, under what conditions is there a one-to-one mapping between the measurements $\{y_n\}$ and the original signal $x(t)$? Second, assuming such a mapping exists and given the samples $\{y_n\}$, how can a practical algorithm recover the original signal?

Sampling is a typical ill-posed problem in that one can construct an infinite number of signals that lead to the same samples $\{y_n\}$. To make the problem tractable one then has to impose some constraints on the choice of $x(t)$. Bandlimited signals are the prototypical example of such a constraint, and yield both a one-to-one mapping and a practical recovery technique. The set of band-limited signals also happens to form a shift-invariant subspace of the space of continuous-time functions. As it turns out, the classical sampling theorem can be extended to signals belonging to arbitrary shift-invariant subspaces, such as splines having uniformly spaced knots [5, 6].

In many cases, however, requiring that a signal belongs to a subspace is too strong a restriction. Consider the example of piecewise linear functions. Is the set of all such functions a subspace? Indeed it is, since the sum of any two piecewise linear signals is again piecewise linear, as is the product of a piecewise linear function with a scalar. However, the sum of two such functions will usually contain a knot wherever either of the summands has a knot. Repeatedly summing piecewise linear signals will therefore lead to functions containing an infinite number of infinitesimally spaced knots; these contain an infinite amount of information per time unit and clearly cannot be recovered from samples taken at a finite rate.

To avoid this difficulty, we could consider uniform piecewise linear signals, i.e., we could allow knots only at predetermined, equally spaced locations. This leads to the shift-invariant subspace setting mentioned above, for which stable recovery techniques exist [6]. However, instead of forcing fixed knot positions, one could merely require, for example, a combination of a finite number of piecewise linear signals with arbitrary known locations. In many cases, such a restriction better characterizes real-world signals, although it can no longer be modeled as a linear subspace. Rather, this is an instance of a union of subspaces [7, 8]: each choice of valid knot positions forms a subspace, and the class of allowed signals is the union of such subspaces. The minimum separation model also satisfies the FRI property, and can be recovered efficiently from samples taken at the rate of innovation. The union of subspaces structure, which is explored in more detail in Chapter 3, is useful in developing a geometrical intuition of FRI recovery techniques. There is, however, a distinction between the union of subspaces and FRI models. In particular, there exist FRI settings which cannot be described in terms of unions of subspaces, for example, when the signal parameters do not include an amplitude component. There are also unions of subspaces which do not conform to the FRI scenario, in particular when the parameters affect the signal in a non-local manner, so that finite-duration segments are not determined by a finite number of parameters.

Our discussion thus far has concentrated on perfect recovery of FRI signals in the absence of noise. However, empirical observations indicate that, for some noisy FRI signals, substantial performance improvements are achievable when the sampling rate is increased beyond the rate of innovation [9–12]. This leads to two areas of active research on FRI: first, the development of algorithms with improved noise robustness [9–11, 13–17], and, second, the derivation of bounds on the best possible performance at a given noise level [12, 16]. By comparing FRI techniques with performance bounds, we will demonstrate that while noise treatment has improved in recent years, there remain cases in which state-of-the-art techniques can still be enhanced.

4.1.2 History of FRI

The idea of analyzing FRI signals was first proposed by Vetterli *et al.* [4]. Although the minimal sampling rate required for such settings has been derived, no generic reconstruction scheme exists for the general problem. Nonetheless, some special cases have been treated in previous work, including streams of pulses, which will be our focus in this chapter.

A stream of pulses can be viewed as a parametric signal, uniquely defined by the time delays of the pulses and their amplitudes. An efficient sampling scheme for *periodic* streams of impulses, having K impulses in each period, was proposed in [4]. Using this technique, one obtains a set of Fourier series coefficients of the periodic signal. Once these coefficients are known, the problem of determining the time delays and amplitudes of the pulses becomes that of finding the frequencies and amplitudes of a sum of sinusoids. The latter is a standard problem in spectral analysis [18] which can be solved using conventional approaches, such as the annihilating filter method [18, 19], as long as the number of samples is no smaller than $2K$. This result is intuitive since $2K$ is the number of degrees of freedom in each period: K time delays and K amplitudes.

Periodic streams of pulses are mathematically convenient to analyze, but not very practical. By contrast, *finite* streams of pulses are prevalent in applications such as ultrasound imaging [10]. The first treatment of finite Dirac streams appears in [4], in which a Gaussian sampling kernel was proposed. The time delays and amplitudes are then estimated from the samples. However, this approach is subject to numerical instability, caused by the exponential decay of the kernel. A different approach, based on moments of the signal, was developed in [9], where the sampling kernels have compact time support. This method treats streams of Diracs, differentiated Diracs, and short pulses with compact support. The moments characterize the input akin to the Fourier coefficients used in [4]. In fact, the time delays and pulses can again be determined from the moments by using standard spectral estimation tools. Another technique that utilizes finite-support sampling kernels, was proposed in [10]. This approach has improved numerical stability, thanks to the choice of the sampling kernel, especially for high rates of innovation. The method was then generalized in [11].

Infinite streams of pulses arise in applications such as ultra-wideband (UWB) communications, where the communicated data changes frequently. Using a compactly supported filter [9], and under certain limitations on the signal, the infinite stream can be divided into a sequence of separate finite problems. The individual finite cases may be treated using methods for the finite setting; however, this leads to a sampling rate that is higher than the rate of innovation. A technique achieving the rate of innovation was proposed in [11], based on a multichannel sampling scheme which uses a number of sampling kernels in parallel.

In related work, a *semi-periodic* pulse model was proposed in [17], wherein the pulse time delays do not change from period to period, but the amplitudes vary. This is a hybrid case in which the number of degrees of freedom in the time delays is finite, but there is an infinite number of degrees of freedom in the amplitudes. Therefore, the proposed recovery scheme generally requires an infinite number of samples.

The effect of digital noise on the recovery procedure was first analyzed by Maravic and Vetterli [13], where an improved model-based approach was proposed. In this technique, known as the subspace estimator, proper use of the algebraic structure of the signal subspace is exploited, leading to improved noise robustness. An iterative version of the subspace estimator was later proposed by Blu *et al.* [19]. This approach is optimal for the sinc sampling kernel of [4], but can also be adapted to compactly supported kernels. Finite rate of innovation recovery in the presence of noise was also examined from a stochastic modeling perspective by Tan and Goyal [14] and by Erdozain and Crespo [15]. The performance in the presence of analog noise has been recently examined in [12]. Treating analog noise allows the interaction between FRI techniques and the underlying sampling methods to be analyzed. In particular, bounds are obtained which are independent of the sampling method. For different classes of FRI signals, this allows one to identify an optimal sampling approach that achieves the bound. In addition, it is shown that under certain scenarios the sampling schemes of [11] are optimal in the presence of analog noise. This framework can also be used to identify FRI settings in which noise-free recovery techniques deteriorate substantially under slight noise levels.

There has also been some work on FRI setups departing from the simple one-dimensional scheme of Figure 4.1. We have already mentioned multichannel setups, in which sampling is performed simultaneously using several distinct kernels, but with a lower total sampling rate [11, 17, 20]. The problem of recovering an FRI pulse stream in which the pulse shape is unknown was examined in [21]. Some forms of distributed sampling have been studied in [22]. There has also been work on multidimensional FRI signals, i.e., signals which are a function of two or more parameters (such as images) [23, 24]. The many applications of FRI theory include image super-resolution [25, 26], ultrasound imaging [10], radar [27], multipath identification [17], and wideband communications [28, 29].

4.1.3 Chapter outline

Throughout the rest of the chapter, we treat the basic concepts underlying FRI theory in greater detail. We mainly focus on FRI pulse streams, and consider in particular the cases of periodic, finite, infinite, and semi-periodic pulse streams. In Section 4.2, we provide a general definition and some examples of FRI signals. In Section 4.3, we treat the problem of recovering FRI signals from noiseless samples taken at the rate of innovation. Specifically, we concentrate on a pulse stream input signal and develop recovery procedures for various types of sampling kernels. Modifications of these techniques when noise is present in the system are discussed in Section 4.4. Simulations illustrating the ability to recover FRI signals are provided in Section 4.5. We conclude the chapter in Section 4.6 with several extensions of the FRI model and a brief discussion of some of its practical application areas.

4.1.4 Notation and conventions

The following notation will be used throughout the chapter. $\mathbb{R}$, $\mathbb{C}$ and $\mathbb{Z}$ denote the sets of real, complex, and integer numbers, respectively. Boldface uppercase letters $\mathbf{M}$ denote matrices, while boldface lowercase letters $\mathbf{v}$ indicate vectors. The identity matrix is denoted $\mathbf{I}$. The notation $\mathbf{M}_{a \times b}$ explicitly indicates that the matrix is of dimensions $a \times b$. The superscripts $(\cdot)^T$, $(\cdot)^*$, $(\cdot)^{-1}$, and $(\cdot)^\dagger$, when referring to operations on matrices or vectors, mean the transpose, Hermitian conjugate, inverse, and Moore–Penrose pseudoinverse respectively. Continuous-time functions are denoted $x(t)$, whereas discrete-time sequences are denoted x_n or $x[n]$. The expectation operator is $\mathbb{E}(\cdot)$. The box function $\text{rect}(t)$ equals 1 in the range $[-1/2, 1/2]$ and 0 elsewhere. The Heaviside or step function $u(t)$ is 0 for $t < 0$ and 1 for $t \geq 0$.

The continuous-time Fourier transform $\hat{x}(\omega)$ of the function $x(t)$ is defined as

$$\hat{x}(\omega) \triangleq \int_{-\infty}^{\infty} x(t) e^{-jt\omega} dt, \quad (4.2)$$

while the discrete-time Fourier transform (DTFT) of a sequence $a[n]$ is given by

$$\hat{a}(e^{j\omega T}) \triangleq \sum_{n \in \mathbb{Z}} a[n] e^{-j\omega n T}. \quad (4.3)$$

The Fourier series $\{\hat{x}_m\}_{m\in\mathbb{Z}}$ of a τ-periodic function is defined as

$$\hat{x}_m \triangleq \frac{1}{\tau}\int_0^\tau x(t)e^{-j2\pi m\frac{t}{\tau}}\,dt. \tag{4.4}$$

We will also use the Fourier series (4.4) for finite-duration signals, i.e., signals whose support is contained in $[0,\tau]$.

We conclude the section with some identities which will be used in several proofs throughout the chapter. These are the Poisson summation formula [30]

$$\sum_{n\in\mathbb{Z}} x(t+nT) = \frac{1}{T}\sum_{k\in\mathbb{Z}} \hat{x}\left(\frac{2\pi k}{T}\right)e^{j2\pi k\frac{t}{T}} \tag{4.5}$$

and Parseval's theorem for the equivalence of the inner product [30, 31]

$$\langle x(t), y(t)\rangle = \frac{1}{2\pi}\langle \hat{x}(\omega), \hat{y}(\omega)\rangle, \tag{4.6}$$

where $\langle x(t), y(t)\rangle = \int_{-\infty}^{\infty} x^*(t)y(t)\,dt$.

4.2 Signals with finite rate of innovation

As explained at the beginning of this chapter, FRI signals are those that can be described by a finite number of parameters per unit time. In this section we introduce the original definition as stated by Vetterli *et al.* in [4]. In addition, we provide some examples of FRI signals that can be sampled and perfectly reconstructed at their rate of innovation using the techniques of [4, 9–11, 17]. We also formally define periodic, semi-periodic, and finite duration signals.

4.2.1 Definition of signals with FRI

The concept of FRI is intimately related to parametric signal modeling. If a signal variation depends on a few unknown parameters, then we can see it as having a limited number of degrees of freedom per unit time.

More precisely, given a set of known functions $\{g_r(t)\}_{r=0}^{R-1}$, arbitrary shifts t_k, and amplitudes $\gamma_{k,r}$, consider a signal of the form:

$$x(t) = \sum_{k\in\mathbb{Z}}\sum_{r=0}^{R-1} \gamma_{k,r} g_r(t-t_k). \tag{4.7}$$

Since the set of functions $\{g_r(t)\}_{r=0}^{R-1}$ is known, the only free parameters of the signal are the coefficients $\gamma_{k,r}$ and the time shifts t_k. Consider a counting function $C_x(t_a, t_b)$ that is able to compute the number of parameters over a time interval $[t_a, t_b]$. The rate of innovation is defined as follows

$$\rho = \lim_{\tau\to\infty} \frac{1}{\tau} C_x\left(-\frac{\tau}{2}, \frac{\tau}{2}\right). \tag{4.8}$$

DEFINITION 4.1 [4] *A signal with Finite Rate of Innovation can be defined as a signal with a parametric representation such as that given by (4.7), and with a finite ρ given by (4.8).*

Another useful concept is that of a *local* rate of innovation over a window of size τ, defined as:

$$\rho_\tau(t) = \frac{1}{\tau} C_x\left(t - \frac{\tau}{2}, t + \frac{\tau}{2}\right). \tag{4.9}$$

Note that $\rho_\tau(t)$ clearly tends to ρ as τ tends to infinity.

Given an FRI signal with a rate of innovation ρ, we expect to be able to recover $x(t)$ from ρ samples (or parameters) per unit time. The rate of innovation turns out to have another interesting interpretation in the presence of noise: it is a lower bound on the ratio between the average mean squared error (MSE) achievable by any unbiased estimator of $x(t)$ and the noise variance, regardless of the sampling method [12].

4.2.2 Examples of FRI signals

It is well known from classical sampling theory that a signal bandlimited to $[-B/2, B/2]$ can be expressed as an infinite sum of properly weighted and shifted versions of the sinc function:

$$x(t) = \sum_{n \in \mathbb{Z}} x[n] \operatorname{sinc}(Bt - n), \tag{4.10}$$

where $x[n] = \langle x(t), B\operatorname{sinc}(Bt - n)\rangle$. Comparing Equations (4.10) and (4.7) immediately reveals that a bandlimited signal can be interpreted as having finite rate of innovation. In this case, we can say that the signal $x(t)$ has B degrees of freedom per second, since it is exactly defined by a sequence of numbers $\{x[n]\}_{n \in \mathbb{Z}}$ spaced $T = B^{-1}$ seconds apart, given that the basis function sinc is known.

This idea can be generalized by replacing the sinc basis function with any other function $\varphi(t)$. The set of signals

$$x(t) = \sum_{n \in \mathbb{Z}} x[n] \varphi(Bt - n), \tag{4.11}$$

defines a shift-invariant subspace, which is not necessarily bandlimited, but that again has a rate of innovation $\rho = B$. Such functions can be efficiently sampled and reconstructed using linear methods [5,6], and thus typically do not require the more elaborate techniques of FRI theory. However, many FRI families of signals form a union of subspaces [7,8], rather than a subspace, and can still be sampled and perfectly reconstructed at the rate of innovation. As a motivation for the forthcoming analysis, several examples of such signals are plotted in Figure 4.2 and described below. For simplicity, these examples describe finite-duration FRI signals defined over the range $[0, 1]$, but the extension to infinite or periodic FRI models is straightforward.

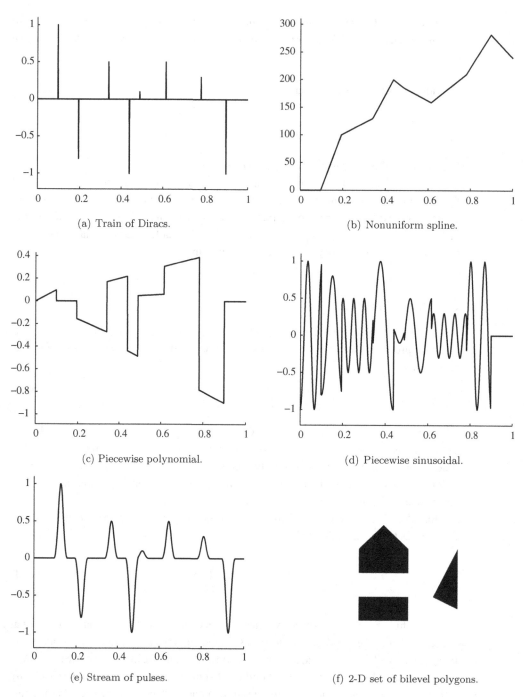

Figure 4.2 Examples of FRI signals that can be sampled and perfectly reconstructed at the rate of innovation.

(i) The first signal of interest is a *stream of K Diracs* with amplitudes $\{a_k\}_{k=0}^{K-1}$ and time locations $\{t_k\}_{k=0}^{K-1}$. Mathematically, $x(t)$ can be written as

$$x(t) = \sum_{k=0}^{K-1} a_k \delta(t - t_k). \qquad (4.12)$$

The signal has $2K$ degrees of freedom, because it has K amplitudes and K locations that are unknown. A typical realization of such a signal can be seen in Figure 4.2(a).

(ii) A signal $x(t)$ is a *nonuniform spline of order R* with amplitudes $\{a_k\}_{k=0}^{K-1}$ and knots at $\{t_k\}_{k=0}^{K-1} \in [0,1]$ if and only if its $(R+1)$th derivative is a stream of K weighted Diracs. Equivalently, such a signal consists of $K+1$ segments, each of which is a polynomial of degree R, such that the entire function is differentiable R times. This signal also has $2K$ degrees of freedom, because it is only the K amplitudes and K locations of the Diracs that are unknown. An example is the piecewise linear signal described in Section 4.1 and shown in Figure 4.2(b). The second derivative of this signal is the train of Diracs shown in (a).

(iii) A *stream of K differentiated Diracs* with amplitudes $\{a_{kr}\}_{k=0,r=0}^{K-1,R_k-1}$ and time locations $\{t_k\}_{k=0}^{K-1}$ is similar to the stream of Diracs, but combining linearly a set of properly displaced and weighted differentiated Diracs, $\delta^{(r)}(t)$. Mathematically, we can write:

$$x(t) = \sum_{k=0}^{K-1} \sum_{r=0}^{R_k-1} a_{kr} \delta^{(r)}(t - t_k). \qquad (4.13)$$

In this case, the number of degrees of freedom of the signal is determined by K locations and $\tilde{K} = \sum_{k=0}^{K-1} R_k$ different weights.

(iv) A signal $x(t)$ is a *piecewise polynomial* with K segments of maximum degree $R-1$ ($R > 0$) if and only if its Rth derivative is a stream of differentiated Diracs. The signal again has $K + \tilde{K}$ degrees of freedom. An example is shown in Figure 4.2(c). The difference between a piecewise polynomial and a spline is that the former is not differentiable at the knots.

(v) Another family of signals, considered in [16], are *piecewise sinusoidal* functions. These are a linear combination of truncated sinusoids, with unknown amplitudes a_{kd}, angular frequencies ω_{kd}, and phases θ_{kd}, so that

$$x(t) = \sum_{k=0}^{K-1} \sum_{d=0}^{D-1} a_{kd} \cos(\omega_{kd} t - \theta_{kd}) \xi_d(t), \qquad (4.14)$$

with $\xi_d(t) = u(t - t_d) - u(t - t_{d+1})$, where t_d are locations to be determined, and $u(t)$ is the Heaviside step function. Figure 4.2(d) shows an example of such a signal.

(vi) An important example we focus on in this chapter is a *stream of pulses*, which is uniquely defined by a known pulse shape $p(t)$ and the unknown locations $\{t_k\}_{k=0}^{K-1}$ and amplitudes $\{a_k\}_{k=0}^{K-1}$ that characterize the pulses. The signal can thus be expressed mathematically as

$$x(t) = \sum_{k=0}^{K-1} a_k p(t - t_k). \tag{4.15}$$

The stream of pulses has $2K$ degrees of freedom. A realization of a train of pulses is shown in Figure 4.2(e).

(vii) Finally, it is also possible to consider FRI signals in higher dimensions. For instance, a *2-D-stream of Diracs* can be written as

$$f(x, y) = \sum_{k=0}^{K-1} a_k \delta(x - x_k, y - y_k). \tag{4.16}$$

In Figure 4.2(f) we show another type of two-dimensional signal, a *2-D set of bilevel polygons*.

We conclude this section by focusing on streams of pulses, which are the prototypical signals we use from now on in the remainder of the chapter. We thus assume for simplicity a single pulse shape $p(t)$ in (4.7), and describe an infinite-length stream of pulses as

$$x(t) = \sum_{k \in \mathbb{Z}} a_k p(t - t_k). \tag{4.17}$$

Periodic FRI signals turn out to be particularly convenient for analysis, and will be discussed in depth in Section 4.3.1. If we assume that there are only K different time locations $\{t_k\}$ and amplitudes $\{a_k\}$ in (4.17), and that they are repeated every τ, we have

$$x(t) = \sum_{m \in \mathbb{Z}} \sum_{k=0}^{K-1} a_k p(t - t_k - m\tau). \tag{4.18}$$

The total number of parameters determining the signal for each period is thus $2K$, leading to a rate of innovation given by $2K/\tau$.

Another variant is a finite-duration pulse stream, which consists of K pulses, whose shifts $\{t_k\}$ are known to be located within a finite segment of length τ. Under the assumption of a single pulse shape $p(t)$, we can express finite-duration FRI signals as

$$x(t) = \sum_{k=0}^{K-1} a_k p(t - t_k). \tag{4.19}$$

Such signals are of practical relevance, since it is unrealistic to expect any measured signal to continue indefinitely. Here again, a finite number of parameters determines $x(t)$ entirely. In this case we are interested in the local rate of innovation $\rho_\tau = 2K/\tau$.

We will also consider semi-periodic signals, which we define as signals of the form

$$x(t) = \sum_{k=0}^{K-1} \sum_{m \in \mathbb{Z}} a_k[m] p(t - t_k - m\tau). \tag{4.20}$$

Such signals are similar to the periodic pulse stream (4.18), with amplitudes that vary from period to period. Signals from this class can be used, for example, to describe the propagation of a pulse with known shape $p(t)$ which is transmitted at a constant rate $1/\tau$ through a medium consisting of K paths. Each path has a constant delay t_k and a time-varying gain $a_k[m]$ [17]. Due to the delays being repeated over the subsequent periods of the signal, estimation in this model is simpler than in the finite or infinite signal cases [11, 12].

4.3 Sampling and recovery of FRI signals in the noise-free setting

In this section, we present the basic mechanisms for reconstruction of pulse stream FRI signals from their low-rate samples. Recovery is achieved by first linearly combining the samples in order to obtain a new set of measurements $\{\hat{x}_m\}$, which represent the Fourier transform of $x(t)$, and then recovering the FRI signal parameters from $\{\hat{x}_m\}$. The latter stage is equivalent to the problem of determining the frequencies of a signal formed by a sum of complex exponentials. This problem has been treated extensively in the array processing literature, and can be solved using conventional tools from spectral estimation theory [18] such as the matrix pencil [32], subspace-based estimators [33, 34], and the annihilating filter [19].

Recovery of FRI signals is most readily understood in the setting of a periodic stream of pulses given by (4.18), and this is therefore the first scenario we explore. We later discuss recovery techniques that use finite-support sampling kernels. These can be used in the finite setting of Equation (4.19) as well as the original infinite FRI model of Equation (4.7). Finally, we also discuss a technique for recovering semi-periodic signals of the form (4.20).

4.3.1 Sampling using the sinc kernel

Consider a τ-periodic stream of K pulses $p(t)$ at locations $\{t_k\}_{k=0}^{K-1}$ and with amplitudes $\{a_k\}_{k=0}^{K-1}$, as defined in (4.18). The pulse shape is known a priori, and therefore the signal has only $2K$ degrees of freedom per period.

Since $x(t)$ is periodic it can be represented in terms of its Fourier series coefficients $\hat{x}_m$ as

$$x(t) = \sum_{k=0}^{K-1} a_k \sum_{m \in \mathbb{Z}} p(t - t_k - m\tau) \qquad (4.21)$$

$$\stackrel{(a)}{=} \sum_{k=0}^{K-1} a_k \frac{1}{\tau} \sum_{m \in \mathbb{Z}} \hat{p}\left(\frac{2\pi m}{\tau}\right) e^{j2\pi m \frac{t-t_k}{\tau}}$$

$$= \sum_{m \in \mathbb{Z}} \hat{x}_m e^{j2\pi m \frac{t}{\tau}},$$

where in (a) we used Poisson summation formula (4.5), and

$$\hat{x}_m = \frac{1}{\tau}\hat{p}\left(\frac{2\pi m}{\tau}\right) \sum_{k=0}^{K-1} a_k e^{-j2\pi m \frac{t_k}{\tau}}, \qquad (4.22)$$

are the Fourier series coefficients of $x(t)$.

If we have direct access to a set $\mathcal{K}$ of M consecutive Fourier coefficients for which $\hat{p}(2\pi m/\tau) \neq 0$, and $M \geq 2K$, then it is possible to retrieve the $2K$ free parameters $\{a_k, t_k\}$, $k = 0, 1, \ldots, K-1$ by using conventional tools from spectral analysis [18] such as Prony's method or the annihilating filter method [18, 19]. To show this fact we first write (4.22) as

$$\hat{x}_m \hat{p}^{-1}\left(\frac{2\pi m}{\tau}\right) = \frac{1}{\tau}\sum_{k=0}^{K-1} a_k u_k^m, \qquad (4.23)$$

where $u_k = e^{-j2\pi \frac{t_k}{\tau}}$ and $\hat{p}^{-1}$ denotes the multiplicative inverse of p. Since $p(t)$ is known a priori, we assume for simplicity of notation that $\hat{p}(2\pi m/\tau) = 1$ for $m \in \mathcal{K}$; this happens for example when $x(t)$ is a stream of Diracs. Otherwise one must simply divide each measurement by the corresponding value of $\hat{p}(2\pi m/\tau)$.

In order to find the values u_k in (4.23), let $\{h_m\}_{m=0}^{K}$ denote the filter whose z-transform is

$$\hat{h}(z) = \sum_{m=0}^{K} h_m z^{-m} = \prod_{m=0}^{K-1} \left(1 - u_k z^{-1}\right). \qquad (4.24)$$

That is, the roots of $\hat{h}(z)$ equal the values u_k to be found. Then, it follows that:

$$h_m * \hat{x}_m = \sum_{i=0}^{K} h_i \hat{x}_{m-i} = \sum_{i=0}^{K}\sum_{k=0}^{K-1} a_k h_i u_k^{m-i} = \sum_{k=0}^{K-1} a_k u_k^m \underbrace{\sum_{i=0}^{K} h_i u_k^{-i}}_{=0} = 0 \qquad (4.25)$$

where the last equality is due to the fact that $\hat{h}(u_k) = 0$. The filter $\{h_m\}$ is called an annihilating filter, since it zeroes the signal $\hat{x}_m$. Its roots uniquely define the set of values u_k, provided that the locations t_k are distinct.

Assuming without loss of generality that $h_0 = 1$, the identity in (4.25) can be written in matrix/vector form as

$$\begin{pmatrix} \hat{x}_{-1} & \hat{x}_{-2} & \cdots & \hat{x}_{-K} \\ \hat{x}_0 & \hat{x}_{-1} & \cdots & \hat{x}_{-K+1} \\ \vdots & \vdots & \ddots & \vdots \\ \hat{x}_{K-2} & \hat{x}_{K-3} & \cdots & \hat{x}_{-1} \end{pmatrix} \begin{pmatrix} h_1 \\ h_2 \\ \vdots \\ h_K \end{pmatrix} = -\begin{pmatrix} \hat{x}_0 \\ \hat{x}_1 \\ \vdots \\ \hat{x}_{K-1} \end{pmatrix} \qquad (4.26)$$

which reveals that we need at least $2K$ consecutive values of $\hat{x}_m$ to solve the above system. Once the filter has been found, the locations t_k are retrieved from the zeros u_k of the z-transform in (4.24). Given the locations, the weights a_k can then be obtained by considering for instance K consecutive Fourier series coefficients in (4.23). For example, if we use the coefficients for $k = 0, 1, \ldots, K-1$, then we can write (4.23) in matrix/vector form as follows:

$$\frac{1}{\tau} \begin{pmatrix} 1 & 1 & \cdots & 1 \\ u_0 & u_1 & \cdots & u_{K-1} \\ \vdots & \vdots & \ddots & \vdots \\ u_0^{K-1} & u_1^{K-1} & \cdots & u_{K-1}^{K-1} \end{pmatrix} \begin{pmatrix} a_0 \\ a_1 \\ \vdots \\ a_{K-1} \end{pmatrix} = \begin{pmatrix} \hat{x}_0 \\ \hat{x}_1 \\ \vdots \\ \hat{x}_{K-1} \end{pmatrix}. \quad (4.27)$$

This is a Vandermonde system of equations that yields a unique solution for the weights a_k since the u_ks are distinct. We thus conclude that the original signal $x(t)$ is completely determined by the knowledge of $2K$ consecutive Fourier coefficients.

However, the Fourier coefficients are not readily available, rather they need to be determined from the samples $y_n = \langle x(t), \varphi(\frac{t}{T} - n) \rangle$ (see also Figure 4.1). In [4], the sampling kernel considered is the sinc function of bandwidth B, where $B\tau$ is assumed to be an *odd* integer. We denote this kernel by $\phi_B(t)$. In this case, the Fourier coefficients can be related to the samples as follows:

$$y_n = \langle x(t), \phi_B(nT - t) \rangle \quad (4.28)$$

$$\overset{(a)}{=} \sum_{m \in \mathbb{Z}} \hat{x}_m \left\langle e^{j2\pi m \frac{t}{\tau}}, \phi_B(nT - t) \right\rangle$$

$$\overset{(b)}{=} \frac{1}{2\pi} \sum_{m \in \mathbb{Z}} \hat{x}_m \left\langle \delta\left(\omega - \frac{2\pi m}{\tau}\right), \hat{\phi}_B(\omega) e^{j\omega nT} \right\rangle$$

$$= \sum_{m \in \mathbb{Z}} \hat{x}_m \hat{\phi}_B\left(\frac{2\pi m}{\tau}\right) e^{j2\pi n \frac{T}{N} \frac{m}{\tau}}$$

$$= \frac{1}{B} \sum_{|m| \leq M = \lfloor \frac{B\tau}{2} \rfloor} \hat{x}_m e^{j2\pi \frac{mn}{N}}$$

where in (a), (4.21) and the linearity of the inner product have been used, and for (b) Parseval's theorem (4.6) has been applied. Equation (4.28) relates the samples y_n and the Fourier series coefficients $\hat{x}_m$ by means of the inverse discrete Fourier transform (IDFT). Thus, calculating the DFT of the samples would directly yield $\hat{x}_m$ for $|m| \leq M$. Since we need $2K$ consecutive Fourier coefficients and we require $B\tau$ to be an odd number we obtain the requirement $B\tau \geq 2K + 1$.

We summarize the above sampling and recovery discussion by highlighting the main steps necessary for the retrieval of $x(t)$:

(1) Obtain the Fourier series coefficients $\hat{x}_m$ for $|m| \leq M$. This can be done by calculating the DFT coefficients of the samples using $\hat{y}_m = \sum_{n=0}^{N-1} y_n e^{-j2\pi \frac{nm}{N}}$ and the fact that they relate through $\hat{x}_m = B\hat{y}_m, |m| \leq M$.
(2) Retrieve the coefficients of the filter that annihilates $\hat{x}_m$. These coefficients can be found by writing down (4.25) as a linear system of equations of the form (4.26),

which has K equations and K unknowns. There is only one solution to the system, since the filter h_m is unique for the given signal $\hat{x}_m$.

(3) Obtain the roots of the filter $\hat{h}(z)$, which yield the values u_k and, therefore, the locations t_k.

(4) Find the amplitudes a_k using the first K consecutive equations in (4.23). This yields the Vandermonde system of equations (4.27), which also has a unique solution for different values of the locations t_k.

We note that while the mechanism described above correctly identifies the signal parameters in the present setting, it becomes inaccurate if noise is added to the system. Techniques which are better suited to dealing with noise will be discussed in Section 4.4.

For the sake of brevity we have concentrated on the annihilating filter method for retrieving the signal parameters. However, other techniques exist such as the matrix pencil method [32] as well as subspace-based estimators [33, 34]. In the presence of noise the latter methods can provide improved performance compared to the annihilating filter approach [12, 13].

4.3.2 Sampling using the sum of sincs kernel

While the above procedure has shown that it is indeed possible to reconstruct exactly a periodic stream of pulses, it has the disadvantage that it uses a sampling kernel of infinite support and slow decay. It is thus natural to investigate whether a similar procedure can be used with alternative, possibly compactly supported, kernels. As we will see shortly, another important advantage of compactly supported kernels is that the resulting methods can be used in conjunction with finite and infinite FRI signals, rather than periodic signals as was the case in Section 4.3.1. Essentially, we are looking for alternative kernels that can still be used to relate the samples y_n to the Fourier coefficients of $x(t)$. This is because we have seen that, given the Fourier coefficients, $x(t)$ can be retrieved using spectral estimation techniques.

Consider for now a periodic FRI signal (4.18). Assuming a generic sampling kernel $g(t)$, we have that [10]

$$y_n = \langle x(t), g(t - nT) \rangle \quad (4.29)$$

$$= \left\langle \sum_{m \in \mathbb{Z}} \hat{x}_m e^{j2\pi m \frac{t}{\tau}}, g(t - nT) \right\rangle$$

$$\stackrel{(a)}{=} \sum_{m \in \mathbb{Z}} \hat{x}_m e^{j2\pi m \frac{nT}{\tau}} \left\langle e^{j2\pi m \frac{t}{\tau}}, g(t) \right\rangle$$

$$\stackrel{(b)}{=} \sum_{m \in \mathbb{Z}} \hat{x}_m e^{j2\pi m \frac{nT}{\tau}} \hat{g}^* \left(\frac{2\pi m}{\tau} \right),$$

where (a) follows from the linearity of the inner product and a change of variable, and (b) is due to the definition (4.2) of the Fourier transform.

Having control over the filter $g(t)$, we now impose the following condition on its Fourier transform:

$$\hat{g}^*(\omega) = \begin{cases} 0, & \omega = \frac{2\pi m}{\tau}, \quad m \notin \mathcal{K}, \\ \text{nonzero}, & \omega = \frac{2\pi m}{\tau}, \quad m \in \mathcal{K}, \\ \text{arbitrary}, & \text{otherwise}, \end{cases} \quad (4.30)$$

where $\mathcal{K}$ is a set of coefficients which will be determined shortly. Then, we have

$$y_n = \sum_{m \in \mathcal{K}} \hat{x}_m e^{j2\pi m \frac{nT}{\tau}} \hat{g}^*\left(\frac{2\pi m}{\tau}\right). \quad (4.31)$$

In general, the system in (4.31) has a unique solution provided the number of samples N is no smaller than the cardinality of $\mathcal{K}$, which we will call $M = |\mathcal{K}|$. The reason is that, in this case, the matrix defined by the elements $e^{j2\pi m \frac{nT}{\tau}}$ is left-invertible. The idea is that each sample y_n is a combination of the elements $\hat{x}_m$, and the kernel $g(t)$ is designed to pass the coefficients for $m \in \mathcal{K}$ and suppress those for $m \notin \mathcal{K}$. Note that for any real filter satisfying (4.30), we have that if $m \in \mathcal{K}$, then $-m \in \mathcal{K}$, since by conjugate symmetry $\hat{g}(2\pi m/\tau) = \hat{g}^*(-2\pi m/\tau)$.

In the particular situation in which the number of samples N equals M, and when the sampling period T is related to the total period τ by $T = \tau/N$, we can write

$$y_n = \sum_{m \in \mathcal{K}} \hat{g}_m^* \hat{x}_m e^{j\frac{2\pi mn}{N}} \quad (4.32)$$

where $\hat{g}_m^* = \hat{g}^*(2\pi m/\tau)$. This equation relates the samples y_n and the Fourier coefficients of the input $\hat{x}_m$ through a "weighted" IDFT. This means that calculating the DFT of the samples yields each of the weighted Fourier series coefficients $\text{DFT}\{y_n\} = \hat{y}_m = \hat{g}_m^* \hat{x}_m$ or, equivalently, the coefficients themselves by inversion of each equation, $\hat{x}_m = \hat{g}_m^{*-1} \hat{y}_m$. Thus, sampling with a filter that satisfies (4.30) allows us to obtain the Fourier coefficients $\hat{x}_m$ in a simple manner.

It is straightforward to see that one particular case of a filter obeying (4.30) is the sinc function $g(t) = \text{sinc}(Bt)$ with $B = M/\tau$. A family of alternative kernels satisfying (4.30) was introduced in [10] and is known as the family of Sum of Sincs (SoS). This class of kernels is defined in the frequency domain as

$$\hat{g}(\omega) = \tau \sum_{m \in \mathcal{K}} b_m \text{sinc}\left(\frac{\omega}{\frac{2\pi}{\tau}} - m\right), \quad (4.33)$$

where $b_m \neq 0$ for $m \in \mathcal{K}$. The resulting filter is real valued if $m \in \mathcal{K}$ implies $-m \in \mathcal{K}$ and $b_m = b_{-m}^*$. In the time domain, the sampling kernel is of compact support, and can be written as

$$g(t) = \text{rect}\left(\frac{t}{\tau}\right) \sum_{m \in \mathcal{K}} b_m e^{j2\pi m \frac{t}{\tau}}. \quad (4.34)$$

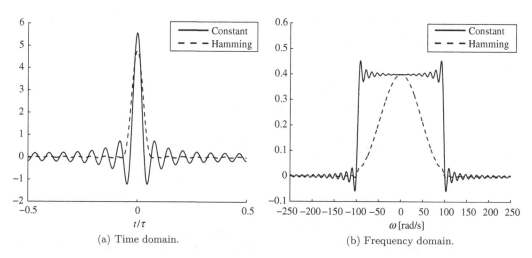

Figure 4.3 SoS sampling kernels. The figures show the time and frequency domain representations of the SoS family of kernels given by (4.34) and (4.33) for $b_m = 1, \forall m$ and when the coefficients follow a Hamming window pattern.

The filter can be further generalized when using a function $\phi(t)$ instead of the sinc in (4.33). This could be useful when we need a smoother time implementation than the one involving the rect function as in (4.34). A key feature of $g(t)$ is that it is compactly supported in time. This will become important when sampling finite-length FRI signals.

One interesting set of coefficients is $b_m = 1$ for $m = -p, \ldots, p$, so that the filter in (4.34) becomes:

$$g(t) = \text{rect}\left(\frac{t}{\tau}\right) \sum_{m=-p}^{p} e^{j2\pi m \frac{t}{\tau}} = \text{rect}\left(\frac{t}{\tau}\right) D_p\left(\frac{2\pi t}{\tau}\right) \quad (4.35)$$

where $D_p(t)$ is the Dirichlet kernel. It is shown in [10] that under certain conditions this choice is optimal in the presence of noise. Figure 4.3 shows this kernel together with the one obtained when the coefficients b_m form a Hamming window [10]. Here M is the cardinality of the set $\mathcal{K}$ and $M \geq 2K$. In general, the free parameters $\{b_k\}_{k \in \mathcal{K}}$ may be optimized for different goals.

To summarize, given the samples y_n, we need to obtain their DFT, and the resulting sequence is related to the Fourier series coefficients $\hat{x}_m$ through (4.32) (we use $N = M$ and $\tau = NT$). We can then build a system of equations as in (4.26) to determine the annihilating filter coefficients, from which the locations t_k are found by calculating its roots. Finally we build another system of equations like (4.27) to determine the amplitudes a_k, using (4.32).

The fact that the SoS kernels have compact support allows us to depart from the case of periodic signals, facilitating sampling finite- and infinite-length FRI signals, as discussed below.

Sampling finite streams of pulses

Finite streams of pulses can be processed based on the above analysis for the periodic case. For the finite-length scenario, we need to relate the samples obtained from the finite stream of pulses to those of the periodic stream. Let $\tilde{x}(t)$ be a finite FRI signal of the form (4.19). It is shown in [10] that

$$y_n = \langle \tilde{x}(t), \tilde{g}(t-nT) \rangle = \langle x(t), g(t-nT) \rangle \tag{4.36}$$

where $x(t)$ is the periodic continuation of the finite stream $\tilde{x}(t)$, and where we have defined the periodic extension of the filter $g(t)$ as $\tilde{g}(t) = \sum_{m \in \mathbb{Z}} g(t - m\tau)$. Therefore, the set of samples $y_n = \langle x(t), g(t-nT) \rangle$, which uniquely represent a τ-periodic stream of pulses, are equivalent to those that could be obtained by sampling the finite-length signal $\tilde{x}(t)$ with the τ-periodic extension of the filter, $\tilde{g}(t-nT)$.

However, it is not practical to use an infinitely long sampling kernel. Assume the pulse $p(t)$ is equal to zero for any $|t| \geq R/2$. Then, the samples have the form [10]

$$y_n = \left\langle \tilde{x}(t), \sum_{m=-r}^{r} g(t - nT - m\tau) \right\rangle, \tag{4.37}$$

where $r = \left\lceil \frac{R+3}{2} \right\rceil - 1$. The advantage of this approach is that we can immediately follow the same retrieval procedure as with the periodic stream of pulses. The reason is that now we obtain the same set of samples given by Equation (4.36) sampling the finite-length signal $\tilde{x}(t)$ with the finite support kernel

$$g_r(t) = \sum_{m=-r}^{r} g(t - nT - m\tau). \tag{4.38}$$

Moreover, if the support of $p(t)$ satisfies $R \leq \tau$, then $r = 1$, and the extension of $g(t)$ will contain only three repetitions, i.e. $g_r(t) = g(t) + g(t+\tau) + g(t-\tau)$.

The multichannel sampling scheme of [11] can also be used to sample finite FRI signals. As we will see in Section 4.3.4, the use of a filter (or modulator) bank allows us to avoid forming a delayed pulse as in $g_r(t)$. In cases in which such delays are difficult to implement in hardware, it may be advantageous to use multiple channels without the need for delays.

Sampling infinite-length streams of pulses

A similar technique may also be used to sample and recover infinite-length FRI pulse streams of the form

$$x(t) = \sum_{k \in \mathbb{Z}} a_k p(t - t_k). \tag{4.39}$$

Concretely, in this case, we assume the signal is characterized by bursts of maximal duration τ which contain at most K pulses, separated by quiet phases of a certain length. This separation depends on the support of the sampling kernel which, in turn, is related to the pulse shape $p(t)$. For example, in order to sample a finite-length stream of Diracs we showed that the filter $g_{3p}(t) = g(t) + g(t \pm \tau)$ was capable of sampling the signal leading to its perfect reconstruction. The support of the filter is 3τ and we then know that, if we want to use a sequential retrieval algorithm for the infinite-length input signal case, the separation of consecutive bursts has to be at least $3\tau/2$. However, this technique requires a sampling rate which is higher than the rate of innovation. Achieving perfect reconstruction for infinite FRI signals from samples taken at the rate of innovation requires a multichannel sampling scheme, and is the subject of Section 4.3.4.

4.3.3 Sampling using exponential reproducing kernels

Another important class of compact support kernels that can be used to sample FRI signals is given by the family of exponential reproducing kernels.

An exponential reproducing kernel is any function $\varphi(t)$ that, together with its shifted versions, can generate complex exponentials of the form $e^{\alpha_m t}$. Specifically,

$$\sum_{n \in \mathbb{Z}} c_{m,n} \varphi(t-n) = e^{\alpha_m t} \qquad (4.40)$$

where $m = 0, 1, \ldots, P$ and $\alpha_0, \lambda \in \mathbb{C}$. The coefficients are given by $c_{m,n} = \langle e^{\alpha_m t}, \tilde{\varphi}(t-n) \rangle$, where $\tilde{\varphi}(t)$ is the dual of $\varphi(t)$, that is, $\langle \varphi(t-n), \tilde{\varphi}(t-k) \rangle = \delta_{n,k}$. When we use these kernels in the FRI process, the choice of the exponents in (4.40) is restricted to $\alpha_m = \alpha_0 + m\lambda$ with $\alpha_0, \lambda \in \mathbb{C}$ and $m = 0, 1, \ldots, P$. This is done to allow the use of the annihilating filter method at the reconstruction stage. This point will be more evident later on.

The theory related to the reproduction of exponentials relies on the concept of E-splines [35]. A function $\beta_\alpha(t)$ with Fourier transform $\hat{\beta}_\alpha(\omega) = \frac{1-e^{\alpha-j\omega}}{j\omega - \alpha}$ is called an E-spline of first order, with $\alpha \in \mathbb{C}$. The time domain representation of such a function is $\beta_\alpha(t) = e^{\alpha t} \text{rect}(t - 1/2)$. The function $\beta_\alpha(t)$ is of compact support, and a linear combination of its shifted versions $\beta_\alpha(t-n)$ reproduces the exponential $e^{\alpha t}$. Higher-order E-splines can be obtained through convolution of first-order ones, e.g., $\beta_{\vec{\alpha}}(t) = (\beta_{\alpha_0} * \beta_{\alpha_1} * \ldots * \beta_{\alpha_P})(t)$, where $\vec{\alpha} = (\alpha_0, \alpha_1, \ldots, \alpha_P)$. This can also be written in the Fourier domain as follows:

$$\hat{\beta}_{\vec{\alpha}}(\omega) = \prod_{k=0}^{P} \frac{1 - e^{\alpha_k - j\omega}}{j\omega - \alpha_k}. \qquad (4.41)$$

Higher-order E-splines are also of compact support and, combined with their shifted versions, $\beta_{\vec{\alpha}}(t-n)$, can reproduce any exponential in the subspace spanned by $\{e^{\alpha_0}, e^{\alpha_1}, \ldots, e^{\alpha_P}\}$ [9, 35]. Notice that the exponent α_m can be complex, which indicates that E-splines need not be real. However, this can be avoided by choosing complex conjugate exponents. Figure 4.4 shows examples of real E-spline functions of orders one to four. Finally, note that the exponential reproduction property is preserved through convolution [9, 35] and, therefore, any function $\varphi(t) = \psi(t) * \beta_{\vec{\alpha}}(t)$, combined with its

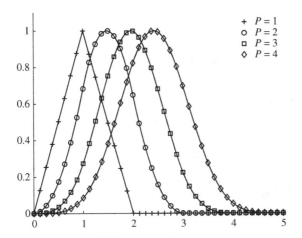

Figure 4.4 Example of exponential reproducing kernels. The shortest function shown is obtained by convolving two first-order splines with complex parameters $\pm j\omega_0 = \pm j\frac{2\pi}{N}$ and $N = 32$ samples, resulting in a real function. The successive E-splines, shown in order from left to right, are obtained by convolving kernels with parameters $\alpha_m = j\omega_0(2m - P)$, $m = 0, \ldots, P$.

shifted versions, is also able to reproduce the exponentials in the subspace spanned by $\{e^{\alpha_0}, e^{\alpha_1}, \ldots, e^{\alpha_P}\}$.

Reconstruction of FRI signals using exponential reproducing kernels is better understood in the time domain. For simplicity, we assume that $p(t)$ is a Dirac function, even though other types of pulses can be sampled and perfectly recovered. In fact, any pulse satisfying $\hat{p}(\omega) \neq 0$ for $\omega = \alpha_m$ can be used. Here $\alpha_m, m = 0, 1, \ldots, P$ are the exponents of the exponentials reproduced by the kernel. This is due to the fact that sampling a stream of pulses with the kernel $\varphi(t)$ is equivalent to sampling a stream of Diracs with the kernel $p(t) * \varphi(t)$. The above condition guarantees that $p(t) * \varphi(t)$ is still able to reproduce exponentials.

Consider a finite-duration FRI signal of length τ:

$$x(t) = \sum_{k=0}^{K-1} a_k \delta(t - t_k). \tag{4.42}$$

Assuming a sampling period of $T = \tau/N$, the measurements are

$$y_n = \left\langle x(t), \varphi\left(\frac{t}{T} - n\right)\right\rangle = \sum_{k=0}^{K-1} a_k \varphi\left(\frac{t_k}{T} - n\right), \tag{4.43}$$

for $n = 0, 1, \ldots, N - 1$. The E-spline reconstruction scheme, first proposed in [9], operates as follows. The samples are first linearly combined with the coefficients $c_{m,n}$ of (4.40) to obtain the new measurements

$$s_m = \sum_{n=0}^{N-1} c_{m,n} y_n, \quad m = 0, 1, \ldots, P. \quad (4.44)$$

Then, using (4.43), we have that

$$s_m = \left\langle x(t), \sum_n c_{m,n} \varphi\left(\frac{t}{T} - n\right) \right\rangle = \int_{-\infty}^{\infty} x(t) e^{\alpha_m t} dt \quad (4.45)$$

$$= \sum_{k=0}^{K-1} \hat{a}_k u_k^m, \quad m = 0, 1, \ldots, P$$

where $\hat{a}_k = a_k e^{\alpha_0 \frac{t_k}{T}}$ and $u_k = e^{\lambda \frac{t_k}{T}}$. Here we have used the fact that $\alpha_m = \alpha_0 + m\lambda$. Note that the new measurements s_m represent the bilateral Laplace transform of $x(t)$ at locations α_m, $m = 0, 1, \ldots, P$. These measurements are again in a power sum series form as those discussed in the previous sections. Therefore the pairs of unknowns $\{\hat{a}_k, u_k\}$ can be retrieved from $s_m = \sum_{k=0}^{K-1} \hat{a}_k u_k^m$ using the annihilating filter method. Consequently, the main steps in the reconstruction of FRI signals with exponential reproducing kernels are the same as those discussed previously. The only difference is that the samples were previously combined using a weighted DFT, whereas in this case the linear combination is dictated by the coefficients $c_{m,n}$. Since $2K$ consecutive coefficients s_m are needed to run the annihilating filter method, we have the condition $P \geq 2K - 1$.

We conclude by highlighting the generality of exponential reproducing kernels. First, when the exponent α_m is purely imaginary, that is, when $\alpha_m = j\omega_m$, then $s_m = \hat{x}(\omega_m)$ is precisely the Fourier transform of $x(t)$ at ω_m. Since $x(t)$ is time-limited, this can be thought of as the Fourier series coefficients of the signal. In this case, and for a proper choice of the parameters N and P, it can be shown [36] that the coefficients $c_{m,n}$ constitute a DFT. For this situation the above analysis converges to the one of Section 4.3.1. Moreover, the SoS sampling kernel introduced in Section 4.3.2 is an exponential reproducing kernel of this type [36]. Second, when $\alpha_m = 0$, $m = 0, 1, \ldots, P$, the E-spline becomes a polynomial spline (or B-spline). In general, when $\alpha_m = 0$, any exponential reproducing kernel reduces to a kernel satisfying the Strang–Fix conditions [37]. These are still valid sampling kernels but reproduce polynomials rather than exponentials. Functions satisfying Strang–Fix conditions are extensively used in wavelet theory and the above result provides an intriguing connection between sampling of FRI signals and wavelets. This connection allows us to combine FRI theory with wavelets to develop efficient centralized and distributed algorithms for the compression of piecewise smooth functions [38,39]. Finally, it is possible to show that any device whose input and output are related by linear differential equations can be turned into an exponential reproducing kernel and can therefore be used to sample FRI signals [9]. This includes, for example, any linear electrical circuit. Given the ubiquity of such devices and the fact that in many cases the sampling kernel is given and cannot be modified, FRI theory with exponential reproducing kernels becomes even more relevant in practical scenarios.

4.3.4 Multichannel sampling

The techniques discussed so far were based on uniform sampling of the signal $x(t)$ convolved with a single kernel $h(t)$ (see Figure 4.1). While this is the simplest possible sampling scheme, improved performance and lower sampling rates can be achieved at the cost of slightly more complex hardware. In particular, one can consider a multichannel sampling setup, in which the signal $x(t)$ is convolved with P different kernels $s_1^*(-t), \ldots, s_P^*(-t)$, and the output of each channel is sampled at a rate $1/T$ [11, 17]. The set of samples in this case is given by

$$c_\ell[m] = \langle s_\ell(t - mT), x(t) \rangle, \quad \ell = 1, \ldots, P, \quad m \in \mathbb{Z}. \tag{4.46}$$

The system is said to have a total sampling rate of P/T. Note that the standard (single-channel) scenario is a special case of this scheme, which can be obtained either by choosing $P = 1$ sampling channels, or with $P > 1$ copies of the sampling kernel $h(t)$ which are shifted in time.

An alternative multichannel structure can be obtained in which the filter is replaced by a modulator (i.e. multiplier) followed by an integrator. In this case the output of each branch is given by

$$c_\ell[m] = \int_{(m-1)T}^{mT} x(t) s_\ell(t), \quad \ell = 1, \ldots, P, \quad m \in \mathbb{Z}, \tag{4.47}$$

where $s_\ell(t)$ is the modulating function on the ℓth branch. This scheme is particularly simple, and as we show below, can be used to treat all classes of FRI signals: periodic, finite, infinite, and semi-periodic, under the assumption that the pulse $p(t)$ is compactly supported. In contrast, the filterbank approach is beneficial in particular for semi-periodic pulse streams and can accommodate arbitrary pulse shapes $p(t)$, including infinite-length functions. Furthermore, the multichannel filter bank structure can often be collapsed to a single sampling channel followed by a serial to parallel converter, in order to produce the parallel sampling sequences in (4.46). Thus, when applicable, this scheme may lead to savings in hardware over the modulator-based approach, while still retaining the benefits of low sampling rate.

Due to its generality and simplicity, we begin by discussing the modulator-based multichannel structure. The merits of this approach are best exposed by first considering a τ-periodic stream of K pulses.

Before proceeding we note that alternative multichannel systems have been proposed in the literature. In [20] a multichannel extension of the method in [9] was presented. This scheme allows reduced sampling rate in each channel, but the overall sampling rate is similar to [9] and therefore does not achieve the rate of innovation. Two alternative multichannel methods were proposed in [40] and [41]. These approaches, which are based on a chain of integrators [40] and exponential filters [41], allow only sampling of infinite streams of Diracs at the rate of innovation. In addition, we show in the simulation section, that these methods are unstable, especially for high rates of innovation.

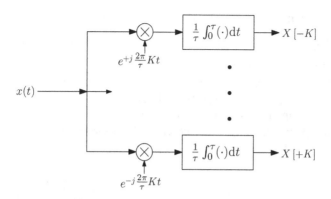

Figure 4.5 Multichannel sampling scheme for periodic FRI signals. The resulting samples are the Fourier series coefficients of $x(t)$. Note that we only sample once every period, thus $T = \tau$.

Periodic FRI signals

Consider a τ-periodic stream of K pulses, as in (4.18). Recall from Section 4.3.1 that if the Fourier coefficients of this signal are available, then standard techniques of spectral analysis can be used to recover the unknown pulse shifts and amplitudes. The multichannel setup, shown in Figure 4.5, provides a simple and intuitive method for obtaining these Fourier coefficients by correlating the signal $x(t)$ with the Fourier basis functions

$$s_\ell(t) = \begin{cases} e^{j\frac{2\pi}{\tau}\ell t}, & t \in [0,\tau], \\ 0, & \text{elsewhere,} \end{cases} \quad (4.48)$$

for $\ell \in \mathcal{L}$, where $\mathcal{L}$ is a set of $2K$ contiguous integers. We set the sampling interval T to be equal to the signal period τ, yielding a total sampling rate of $2K/T$ for all channels. Thus we have a sampling scheme functioning at the rate of innovation, and yielding $2K$ Fourier coefficients of $x(t)$. These can then be used to recover the original signal, for example using the annihilating filter method discussed in Section 4.3.1. An additional advantage of this approach is that the kernels have compact support; indeed, the support corresponds to precisely one period of the FRI signal, which is smaller than the support of the kernel proposed in Section 4.3.2. This property will facilitate the extension of the multichannel system to infinite FRI signals.

Instead of functions of the form (4.48), one can just as well use sampling kernels which are a linear combination of these sinusoids, as in Figure 4.6. This can be advantageous from a hardware point of view, since it may be difficult in practice to implement accurate sinusoids. On the other hand, by allowing such linear combinations, the modulating functions $s_\ell(t)$ can be chosen to have a simple form, such as lowpassed versions of binary sequences [11]. These sequences were shown to be advantageous in other sub-Nyquist configurations as well, such as the modulated wideband converter, designed to sample wideband signals at sub-Nyquist rates [42, 43], and sampling of streams of pulses with unknown shapes [21]. In addition, in real-life scenarios one or more channels might fail, due to malfunction or noise corruption, and therefore we lose the information stored

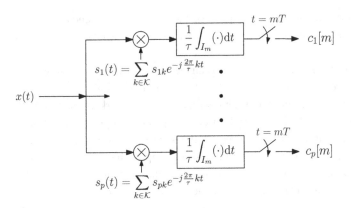

Figure 4.6 Multichannel sampling scheme for infinite FRI signals. Here $T = \tau$.

in that channel. By mixing the coefficients we distribute the information about each Fourier coefficient among several sampling channels. Consequently, when one or more channels fail, the required Fourier coefficients may still be recovered from the remaining operating channels.

When using a mixture of sinusoids, a linear operation is needed to recover the Fourier coefficients from the resulting samples. Specifically, denoting by **x** the vector of Fourier coefficients of $x(t)$, the output of Figure 4.6 is given by **Sx** where **S** is the matrix of elements s_{ik}. As long as **S** has full column rank, we can recover **x** from the samples and then proceed using, e.g., the annihilating method to recover the delays and amplitudes. The new kernels retain the desirable property of compact support with length equal to a single signal period. It is also interesting to note that by proper choice of these linear combinations, the modulator bank can implement the SoS filters [11]. This offers an alternative implementation for finite-length FRI signals that avoids the need to form delayed versions of the SoS kernel at the expense of more complex hardware.

Connection to the modulated wideband converter

The concept of using modulation waveforms, is based on ideas which were presented in [42–44]. We now briefly review the sampling problem treated in [43] and its relation to our setup. We also show that the practical hardware implementation of both systems is similar. For a more detailed description of this scheme see Chapter 3.

The model in [43] is of multiband signals: signals whose CTFT is concentrated on N_{bands} frequency bands, and the width of each band is no greater than B. The location of the bands is unknown in advance. A low-rate sampling scheme allowing recovery of such signals at a rate of $4BN_{\text{bands}}$ was proposed in [45]. This scheme exploits the sparsity of multiband signals in the frequency domain, to reduce the sampling rate well below the Nyquist rate. In [42, 43], this approach was extended to a more practical sampling scheme, which uses a modulation stage and is referred to as the modulated wideband converter (MWC). In each channel of the MWC, the input is modulated with some periodic waveform, and then sampled using a lowpass filter (LPF) followed by a low-rate uniform sampler. The main idea is that in each channel, the spectrum of the signal is shuffled, such that a portion of the energy of all bands appears at baseband. Mixing the frequency bands in [43] is analogous to mixing the Fourier coefficients in Figure 4.6.

We note here some differences between the methods. First, following the mixing stage, we use an integrator in contrast to the LPF used in [43]. This is a result of the different signal quantities measured: Fourier coefficients in our work as opposed to the frequency bands content in [43]. The second difference is in the purpose of the mixing procedure. In [43] mixing is performed to reduce the sampling rate relative to the Nyquist rate. In our setting, the mixing is used in order to simplify hardware implementation and to improve robustness to failure in one of the sampling channels.

Nonetheless, the hardware considerations in the mixing stage in both systems are similar. Recently, a prototype of the MWC has been implemented in hardware [42]. This design is composed of $P = 4$ sampling channels, where the repetition rate of the modulating waveforms is $1/T \approx 20$ MHz. In each period there are 108 rectangular pulses. This prototype, with certain modifications, can be used to implement our sampling scheme as well. These modifications mainly include adding shaping filters on modulating waveforms lines, and reducing the number of rectangular pulses in each period.

Infinite FRI signals

Consider an infinite-duration FRI signal of the form (4.39), where we use $T = \tau$. Furthermore, suppose that the T-local rate of innovation is $2K/T$, for some specified value T. Thus, there are no more than K pulses in any interval of size T, i.e. $I_m = [(m-1)T, mT]$. Assume further that the pulses do not overlap interval boundaries, i.e., if $t_k \in I_m$ then $p(t - t_k) = 0$ for all $t \notin I_m$. Such a requirement automatically holds if $p(t)$ is a Dirac, and will hold with high probability as long as the support of $p(t)$ is substantially smaller than T.

The signal parameters in each interval can now be treated separately. Specifically, consider the T-periodic signal obtained by periodic continuation of the values of $x(t)$ within a particular interval I_m. This periodic signal can be recovered by obtaining $2K$ of its Fourier coefficients. As explained above, these coefficients can be determined using sampling kernels of the form (4.48), whose support is limited to the interval I_m itself (rather than its periodic continuation).

This precise technique can thus be used directly on the non-periodic signal $x(t)$, since the portion of the periodic signal which is sampled includes only the interval I_m [11]. Specifically, this requires obtaining a sample from each of the channels once every T seconds, and using $P \geq 2K$ channels. The resulting procedure is equivalent to a multichannel sampling scheme with rate $1/T$, as depicted in Figure 4.6. Observe that the success of this technique hinges on the availability of sampling kernels whose support is limited to a single period of the periodic waveform. The output of the channel is equal to $\mathbf{c}[m] = \mathbf{S}\mathbf{x}[m]$ where $\mathbf{S}$ is the matrix of elements s_{ik}, and $\mathbf{x}[m]$ are the Fourier coefficients of $x(t)$ over the interval I_m. We can then invert $\mathbf{S}$ to obtain the Fourier coefficients over each interval.

Semi-periodic FRI signals

The multichannel scheme is also effective for reconstructing FRI signals having the semi-periodic structure of (4.20). That is, signals consisting of K pulses occurring at repeated intervals T, with amplitudes $a_k[m]$ which vary from one period to the next. The modulator approach can be used as in the infinite case, with the difference that now the samples from different periods can be jointly processed to improve performance.

Specifically, as before, we can recover $\mathbf{x}[m]$ from the output of the modulator bank. Since the delays are constant for each interval I_m, it can be shown (after normalizing the Fourier coefficients by the Fourier coefficients of the pulse if necessary) that in the frequency domain

$$\mathbf{x}[m] = \mathbf{N}(\mathbf{t})\mathbf{a}[m], \quad m \in \mathbb{Z}, \qquad (4.49)$$

where $\mathbf{a}[m]$ is the vector of coefficients $a_k[m]$, and $\mathbf{N}(\mathbf{t})$ is the $P \times K$ Vandermonde matrix with $k\ell$th element $e^{-j2\pi k \frac{t_\ell}{T}}$. When only one time instant m is available, we can solve (4.49) by using the annihilating filter method to recover the delays t_ℓ, and then the coefficients $a_k[m]$. However, now we have many vectors $\mathbf{x}[m]$ that share the same delays, namely, use the same matrix $\mathbf{N}$. This allows the use of robust methods that recover the delays more reliably, by jointly processing the samples for all m. Examples include the ESPRIT [46] or MUSIC [47] algorithms. These approaches, known as subspace methods, are far more robust than techniques based on a single set of samples. They proceed by computing the correlation matrix $\sum_{m \in \mathbb{Z}} \mathbf{x}[m]\mathbf{x}^T[m]$, and then separate the range of this matrix into two subspaces, the signal and noise subspaces. The delays associated with the matrix $\mathbf{N}$ are then found by exploiting this decomposition.

Clearly, the condition for the general infinite model $P \geq 2K$ is a sufficient condition here as well in order to ensure recovery of $x(t)$. However the additional prior on the signal's structure can be used to reduce the number of sampling channels. In particular it is sufficient to use

$$P \geq 2K - \eta + 1 \qquad (4.50)$$

channels, where η is the dimension of the minimal subspace containing the vector set $\{\mathbf{a}[m], m \in \mathbb{Z}\}$. This condition implies that in some cases the number of channels P can be reduced beyond the lower limit $2K$ for the general model.

An alternative scheme for the semi-periodic setting is the filterbank system. The advantage of this technique is that one need not assume the existence of distinct pulse intervals, nor is it necessary for the pulse shape to have compact support [17]. Here as well we will exploit the periodicity to jointly process the samples by using subspace methods.

When the pulse shape $p(t)$ is arbitrary, the derivation departs somewhat from the canonical technique presented in Section 4.3.1. This is a result of the fact that the signal is not periodic and cannot be divided into distinct intervals, so that one can no longer speak of its Fourier series. Instead, assume that the sampling interval T equals the signal period τ. The DTFT of the samples (4.46) is then

$$\hat{c}_\ell\left(e^{j\omega T}\right) = \frac{1}{T} \sum_{m \in \mathbb{Z}} \hat{s}_\ell^*\left(\omega - \frac{2\pi}{T}m\right) \hat{x}\left(\omega - \frac{2\pi}{T}m\right) \qquad (4.51)$$

where the Fourier transform of a function $f(t)$ is denoted $\hat{f}(\omega)$. By computing the Fourier transform of the semi-periodic signal $x(t)$ of (4.20), we have

$$\hat{c}_\ell\left(e^{j\omega T}\right) = \sum_{k=0}^{K-1} \hat{a}_k\left(e^{j\omega T}\right) e^{-j\omega t_k} \frac{1}{T} \sum_{m \in \mathbb{Z}} \hat{s}_\ell^*\left(\omega - \frac{2\pi}{T}m\right) \hat{p}\left(\omega - \frac{2\pi}{T}m\right) e^{j\frac{2\pi}{T}m t_k}, \tag{4.52}$$

where we used the fact that $\hat{a}_k\left(e^{j\omega T}\right)$ is $2\pi/T$-periodic.

Let us restrict our attention to $\omega \in [0, 2\pi/T)$, which can be done without loss of information since expressions in the DTFT domain are $2\pi/T$-periodic. Denote by $\hat{\mathbf{c}}(e^{j\omega T})$ the length-P column vector whose ℓth element is $\hat{c}_\ell(e^{j\omega T})$, and by $\hat{\mathbf{a}}(e^{j\omega T})$ the length-K column vector whose kth element is $\hat{a}_k(e^{j\omega T})$. Also define the vector $\mathbf{t} = (t_0, \ldots, t_{K-1})^T$. We can then write (4.52) in matrix form as

$$\hat{\mathbf{c}}\left(e^{j\omega T}\right) = \mathbf{M}\left(e^{j\omega T}, \mathbf{t}\right) \mathbf{D}\left(e^{j\omega T}, \mathbf{t}\right) \hat{\mathbf{a}}\left(e^{j\omega T}\right). \tag{4.53}$$

Here $\mathbf{M}\left(e^{j\omega T}, \mathbf{t}\right)$ is a $P \times K$ matrix whose ℓkth element is

$$\mathbf{M}_{\ell k}\left(e^{j\omega T}, \mathbf{t}\right) = \frac{1}{T} \sum_{m \in \mathbb{Z}} \hat{s}_\ell^*\left(\omega - \frac{2\pi}{T}m\right) \hat{p}\left(\omega - \frac{2\pi}{T}m\right) e^{j\frac{2\pi}{T}m t_k}, \tag{4.54}$$

and $\mathbf{D}\left(e^{j\omega T}, \mathbf{t}\right)$ is a diagonal matrix whose kth diagonal element equals $e^{-j\omega t_k}$.

Defining the vector $\mathbf{b}\left(e^{j\omega T}\right)$ as

$$\mathbf{b}\left(e^{j\omega T}\right) = \mathbf{D}\left(e^{j\omega T}, \mathbf{t}\right) \hat{\mathbf{a}}\left(e^{j\omega T}\right), \tag{4.55}$$

we can rewrite (4.53) in the form

$$\hat{\mathbf{c}}\left(e^{j\omega T}\right) = \mathbf{M}\left(e^{j\omega T}, \mathbf{t}\right) \mathbf{b}\left(e^{j\omega T}\right). \tag{4.56}$$

Our problem can then be reformulated as that of recovering $\mathbf{b}\left(e^{j\omega T}\right)$ and the unknown delay set $\mathbf{t}$ from the vectors $\hat{\mathbf{c}}\left(e^{j\omega T}\right)$, for all $\omega \in [0, 2\pi/T)$. Once these are known, the vectors $\hat{\mathbf{a}}\left(e^{j\omega T}\right)$ can be recovered using the relation in (4.55).

To proceed, we focus our attention on sampling filters $\hat{s}_\ell(\omega)$ with finite support in the frequency domain, contained in the frequency range

$$\mathcal{F} = \left[\frac{2\pi}{T}\gamma, \frac{2\pi}{T}(P+\gamma)\right], \tag{4.57}$$

where $\gamma \in \mathbb{Z}$ is an index which determines the working frequency band $\mathcal{F}$. This choice should be such that it matches the frequency occupation of $p(t)$ (although $p(t)$ does not have to be bandlimited). This freedom allows our sampling scheme to support both complex and real-valued signals. For simplicity, we assume here that $\gamma = 0$. Under this choice of filters, each element $\mathbf{M}_{\ell k}\left(e^{j\omega T}, \mathbf{t}\right)$ of (4.54) can be expressed as

$$\mathbf{M}_{\ell k}\left(e^{j\omega T}, \mathbf{t}\right) = \sum_{m=1}^{P} \mathbf{W}_{\ell m}\left(e^{j\omega T}\right) \mathbf{N}_{mk}(\mathbf{t}), \tag{4.58}$$

where $\mathbf{W}(e^{j\omega T})$ is a $P \times P$ matrix whose ℓmth element is given by

$$\mathbf{W}_{\ell m}(e^{j\omega T}) = \frac{1}{T}\hat{s}_\ell^*\left(\omega + \frac{2\pi}{T}(m-1+\gamma)\right)\hat{p}\left(\omega + \frac{2\pi}{T}(m-1+\gamma)\right),$$

and $\mathbf{N}(\mathbf{t})$ is a $P \times K$ Vandermonde matrix. Substituting (4.58) into (4.56), we have

$$\hat{\mathbf{c}}(e^{j\omega T}) = \mathbf{W}(e^{j\omega T})\mathbf{N}(\mathbf{t})\mathbf{b}(e^{j\omega T}). \tag{4.59}$$

If $\mathbf{W}(e^{j\omega T})$ is stably invertible, then we can define the modified measurement vector $\mathbf{d}(e^{j\omega T})$ as $\mathbf{d}(e^{j\omega T}) = \mathbf{W}^{-1}(e^{j\omega T})\hat{\mathbf{c}}(e^{j\omega T})$. This vector satisfies

$$\mathbf{d}(e^{j\omega T}) = \mathbf{N}(\mathbf{t})\mathbf{b}(e^{j\omega T}). \tag{4.60}$$

Since $\mathbf{N}(\mathbf{t})$ is not a function of ω, from the linearity of the DTFT, we can express (4.60) in the time domain as

$$\mathbf{d}[n] = \mathbf{N}(\mathbf{t})\mathbf{b}[n], \quad n \in \mathbb{Z}. \tag{4.61}$$

The elements of the vectors $\mathbf{d}[n]$ and $\mathbf{b}[n]$ are the discrete time sequences, obtained from the inverse DTFT of the elements of the vectors $\mathbf{b}(e^{j\omega T})$ and $\mathbf{d}(e^{j\omega T})$ respectively.

Equation (4.61) has the same structure as (4.49) and can therefore be treated in a similar fashion. Relying on methods such as ESPRIT and MUSIC one can first recover $\mathbf{t}$ from the measurements [17]. After $\mathbf{t}$ is known, the vectors $\mathbf{b}(e^{j\omega T})$ and $\hat{\mathbf{a}}(e^{j\omega T})$ can be found using linear filtering relations by

$$\mathbf{b}(e^{j\omega T}) = \mathbf{N}^\dagger(\mathbf{t})\mathbf{d}(e^{j\omega T}). \tag{4.62}$$

Since $\mathbf{N}(\mathbf{t})$ is a Vandermonde matrix, its columns are linearly independent, and consequently $\mathbf{N}^\dagger \mathbf{N} = \mathbf{I}_K$. Using (4.55),

$$\hat{\mathbf{a}}(e^{j\omega T}) = \mathbf{D}^{-1}(e^{j\omega T},\mathbf{t})\mathbf{N}^\dagger(\mathbf{t})\mathbf{d}(e^{j\omega T}). \tag{4.63}$$

The resulting sampling and reconstruction scheme is depicted in Figure 4.7.

Our last step is to derive conditions on the filters $s_1^*(-t),\ldots,s_P^*(-t)$ and the function $p(t)$ such that the matrix $\mathbf{W}(e^{j\omega T})$ will be stably invertible. To this end, we decompose the matrix $\mathbf{W}(e^{j\omega T})$ as

$$\mathbf{W}(e^{j\omega T}) = \mathbf{S}(e^{j\omega T})\mathbf{P}(e^{j\omega T}) \tag{4.64}$$

where $\mathbf{S}(e^{j\omega T})$ is a $P \times P$ matrix whose ℓmth element is

$$\mathbf{S}_{\ell m}(e^{j\omega T}) = \frac{1}{T}\hat{s}_\ell^*\left(\omega + \frac{2\pi}{T}(m-1+\gamma)\right) \tag{4.65}$$

and $\mathbf{P}(e^{j\omega T})$ is a $P \times P$ diagonal matrix with mth diagonal element

$$\mathbf{P}_{mm}(e^{j\omega T}) = \hat{p}\left(\omega + \frac{2\pi}{T}(m-1+\gamma)\right). \tag{4.66}$$

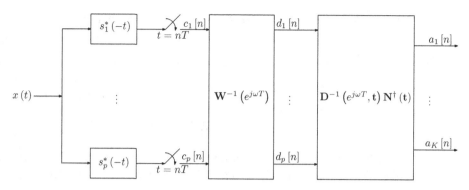

Figure 4.7 Sampling and reconstruction scheme for a semi-periodic signal.

We can guarantee stable invertibility of $\mathbf{W}(e^{j\omega T})$ by ensuring that both $\mathbf{S}(e^{j\omega T})$ and $\mathbf{P}(e^{j\omega T})$ are stably invertible. From (4.66), it is readily seen that the matrix $\mathbf{P}(e^{j\omega T})$ is stably invertible if there exist constants $a, b \in \mathbb{R}$ such that

$$0 < a \leq |\hat{p}(\omega)| \leq b < \infty \text{ almost everywhere } \omega \in \mathcal{F}. \quad (4.67)$$

In addition, the filters $s_\ell^*(-t)$ should be chosen in such a way that they form a stably invertible matrix $\mathbf{S}(e^{j\omega T})$. One example of a set of sampling kernels satisfying this requirement is the ideal bandpass filterbank given by

$$\hat{s}_\ell(\omega) = \begin{cases} T, & \omega \in \left[(\ell-1)\frac{2\pi}{T}, \ell\frac{2\pi}{T}\right], \\ 0, & \text{otherwise.} \end{cases} \quad (4.68)$$

Another example is an LPF with cutoff $\pi P/T$ followed by a uniform sampler at a rate of P/T. The samples can then be converted into P parallel streams to mimic the output of P branches. Further discussion of sampling kernels satisfying these requirements can be found in [17].

To summarize, we derived a general technique for the recovery of pulse parameters from a semi-periodic pulse stream. The technique is outlined in Figure 4.7. This method is guaranteed to perfectly recover the signal parameters from samples taken at a total rate of $2K/T$ or higher, provided that the pulse shape satisfies the stability condition (4.67) and the sampling kernels are chosen so as to yield a stable recovery matrix, for example by using the bandpass filterbank (4.68).

4.4 The effect of noise on FRI recovery

Real-world signals are often contaminated by noise and thus do not conform precisely to the FRI scheme. Furthermore, like any mathematical model, the FRI framework is an approximation which does not precisely hold in practical scenarios, an effect known as mismodeling error. It is therefore of interest to design noise-robust FRI recovery techniques.

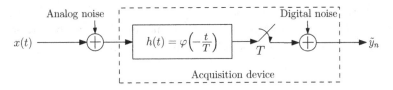

Figure 4.8 Noise perturbations in a "real-world" sampling set up. The continuous signal $x(t)$ can be corrupted both in the analog and the digital paths.

Noise may arise both in the analog and digital domains, i.e., before and after sampling, as illustrated in Figure 4.8. The resulting samples can then be written as

$$\tilde{y}_n = \langle x(t), h(t - nT) \rangle + \epsilon_n \qquad (4.69)$$

with ϵ_n being the overall noise introduced in the process.

When noise is present, it is no longer possible to perfectly recover the original signal from its samples. However, one can sometimes mitigate the effects of noise by oversampling, i.e., by increasing the sampling rate beyond the rate of innovation. In Section 4.4.3 we describe several modifications of the recovery techniques of Section 4.3 designed for situations in which a larger number of measurements is available. These are based on the noise model we introduce in Section 4.4.2.

Oversampling increases the number of measurements of the signal, and it is consequently not surprising that this technique can sometimes be used to improve performance under noise. However, the degree to which improvement is possible depends on the setting under consideration. Indeed, in some cases sampling at the rate of innovation is optimal even in the presence of noise, and cannot be improved by oversampling. Reaching such conclusions requires a theoretical analysis of the effects of noise on the ability to recover FRI signals. This issue will be discussed in Sections 4.4.1 and 4.4.2.

4.4.1 Performance bounds under continuous-time noise

In the next two sections, we analyze the effect of noise on the accuracy with which FRI signals can be recovered. A standard tool for accomplishing this is the Cramér–Rao bound (CRB), which is a lower bound on the MSE achievable by any unbiased estimator [48]. As such, it provides a measure of the difficulty of a given estimation problem, and can indicate whether or not existing techniques come close to optimal. It can also be used to measure the relative merit of different types of measurements. Thus, we will see that the CRB can identify which of the sampling kernels proposed in Section 4.3 provides more robustness to noise, as well as quantify the benefit achievable by oversampling.

As we have already mentioned, in practical applications two types of noise may arise, namely, continuous-time noise which corrupts the signal prior to sampling, and discrete noise contributed by the sampling system (see Figure 4.8). To simplify the discussion, we separately examine each of these models: we begin below with continuous-time noise and discuss sampling noise in Section 4.4.2. Further details concerning the combined effect of the two sources of noise can be found in [12].

For the purpose of the performance analysis, we focus on finite-duration FRI signals of the form (4.19). Thus, our signal $x(t)$ is determined by a finite number $2K$ of parameters $\{a_k, t_k\}_{k=0}^{K-1}$. For future use, we define the parameter vector

$$\boldsymbol{\theta} = (t_0, \ldots, t_{K-1}, a_0, \ldots, a_{K-1})^T. \tag{4.70}$$

An important aspect of continuous-time noise is that it is independent of the sampling process. This noise model can thus be used to identify ultimate limits on the achievable estimation accuracy of a given signal. To be specific, suppose we sample the signal

$$y(t) = x(t) + w(t) \tag{4.71}$$

where $x(t)$ is the finite-duration FRI signal (4.19) and $w(t)$ is continuous-time white Gaussian noise with variance σ^2.

Sampling-indifferent bound

To bound the MSE that can be achieved by any sampling method, it is of interest to derive the CRB for estimating $x(t)$ directly from the continuous-time process $y(t)$. Clearly, no sampling mechanism can do better than exhausting all of the information contained in $y(t)$.

This bound turns out to have a particularly simple closed-form expression which depends on the number of pulses in the signal (or, equivalently, on the rate of innovation) – but not on the class of FRI signals being estimated. Indeed, for a signal $x(t)$ of duration τ, it can be shown that the MSE of any unbiased, finite-variance estimator $\hat{x}(t)$ satisfies [12]

$$\frac{1}{\tau} \int \mathbb{E}\left(|x(t) - \hat{x}(t)|^2\right) dt \geq \rho_\tau \sigma^2 \tag{4.72}$$

where we recall that the τ-local rate of innovation satisfies $\rho_\tau = 2K/\tau$ for finite FRI signals.

Thus, in the noisy setting, the rate of innovation can be given a new interpretation as the ratio between the best achievable MSE and the noise variance σ^2. This is to be contrasted with the characterization of the rate of innovation in the noise-free case as the lowest sampling rate allowing for perfect recovery of the signal; indeed, when noise is present, perfect recovery is no longer possible.

Bound for sampled measurements

We next consider a lower bound for estimating $x(t)$ from *samples* of the signal $y(t)$ of (4.71). To keep the discussion general, we consider samples of the form

$$\tilde{y}_n = \langle y(t), \varphi_n(t) \rangle, \quad n = 0, \ldots, N-1 \tag{4.73}$$

where $\{\varphi_n(t)\}$ is a set of sampling kernels. For example, pointwise sampling at the output of an anti-aliasing filter $\varphi(-t)$ corresponds to the sampling kernels $\varphi_n(t) = \varphi(t - nT)$. We denote by Φ the subspace spanned by the sampling kernels. In this setting, the samples inherit the noise $w(t)$ embedded in the signal $y(t)$. Note that unless the sampling kernels

$\{\varphi_n(t)\}$ happen to be orthogonal, the resulting measurements will not be statistically independent. This is a crucial difference with respect to the sampling noise model of Section 4.4.2 below.

We assume that there exists a Fréchet derivative $\partial x/\partial\theta$ which quantifies the sensitivity of $x(t)$ to changes in θ. Informally, $\partial x/\partial\theta$ is an operator from $\mathbb{R}^{2K}$ to the space of square-integrable functions L_2 such that

$$x(t)|_{\theta+\delta} \approx x(t)|_\theta + \frac{\partial x}{\partial\theta}\delta. \tag{4.74}$$

Suppose for a moment that there exist elements in the range space of $\partial x/\partial\theta$ which are orthogonal to Φ. This implies that one can perturb $x(t)$ without changing the distribution of the measurements $\tilde{y}_0, \ldots, \tilde{y}_{N-1}$. This situation occurs, for example, when the number of measurements N is smaller than the number $2K$ of parameters defining $x(t)$. While it may still be possible to reconstruct some of the information concerning $x(t)$ from these measurements, this is an undesirable situation from an estimation point of view. Thus we will assume that

$$\frac{\partial x}{\partial\theta} \cap \Phi^\perp = \{\mathbf{0}\}. \tag{4.75}$$

Under these assumptions, it can be shown that any unbiased, finite-variance estimator $\hat{x}(t)$ of $x(t)$ from the samples (4.73) satisfies [12]

$$\frac{1}{\tau}\int \mathbb{E}\left(|x(t)-\hat{x}(t)|^2\right)\mathrm{d}t \geq \frac{\sigma^2}{\tau}\mathrm{Tr}\left[\left(\frac{\partial x}{\partial\theta}\right)^*\left(\frac{\partial x}{\partial\theta}\right)\left(\left(\frac{\partial x}{\partial\theta}\right)^*\mathbf{P}_\Phi\left(\frac{\partial x}{\partial\theta}\right)\right)^{-1}\right] \tag{4.76}$$

where $\mathbf{P}_\Phi$ is the projection onto the subspace Φ.

Note that despite the involvement of continuous-time operators, the expression within the trace in (4.76) is a $2K \times 2K$ matrix and can therefore be computed numerically. Also observe that, in contrast to the continuous-time bound of (4.72), the sampled bound depends on the value of θ. Thus, for a specific sampling scheme, some signals can potentially be more difficult to estimate than others.

As expected, the sampled bound (4.76) is never lower than the ultimate (sample-indifferent) bound (4.72). However, the two bounds can sometimes coincide. If this occurs, then at least in terms of the performance bounds, estimators based on the samples (4.73) will suffer no degradation compared with the "ideal" estimator based on the entire set of continuous-time measurements. Such a situation occurs if $x(t) \in \Phi$ for any feasible value of $x(t)$, a situation which we refer to as "Nyquist-equivalent" sampling. In this case, $\mathbf{P}_\Phi \frac{\partial x}{\partial\theta} = \frac{\partial x}{\partial\theta}$, so that (4.76) reduces to

$$\frac{1}{\tau}\int \mathbb{E}\left(|x(t)-\hat{x}(t)|^2\right)\mathrm{d}t \geq \frac{\sigma^2}{\tau}\mathrm{Tr}(\mathbf{I}_{2K\times 2K}) = \sigma^2\rho_\tau \tag{4.77}$$

and the two bounds coincide.

Many practical FRI signal models are not contained in any finite-dimensional subspace, and in these cases, any increase in the sampling rate can improve estimation performance. Even if there exists a subspace containing the entire family of FRI signals, its dimension is often much larger than the number of parameters $2K$ defining the signal; consequently, fully exploiting the information in the signal requires sampling at the Nyquist-equivalent rate, which is potentially much higher than the rate of innovation. This fact provides an analytical explanation of the empirically observed phenomena that oversampling often provides improvement over sampling at the rate of innovation in the presence of noise. A practical example of the benefit of oversampling is described in Section 4.5.

It is interesting to examine this phenomenon from a union of subspaces viewpoint. Suppose that the set of feasible signals $\mathcal{X}$ can be described as a union of an infinite number of subspaces $\{\mathcal{U}_\alpha\}$ indexed by the continuous parameter α, so that

$$\mathcal{X} = \bigcup_\alpha \mathcal{U}_\alpha. \tag{4.78}$$

In this case, a finite sampling rate captures all of the information present in the signal if and only if

$$\dim\left(\sum_\alpha \mathcal{U}_\alpha\right) < \infty \tag{4.79}$$

where $\dim(\mathcal{M})$ is the dimension of the subspace $\mathcal{M}$. By contrast, in the noise-free case, it has been previously shown [7] that the number of samples required to recover $x(t)$ is given by

$$\max_{\alpha_1,\alpha_2} \dim(\mathcal{U}_{\alpha_1} + \mathcal{U}_{\alpha_2}), \tag{4.80}$$

i.e., the largest dimension among sums of *two* subspaces belonging to the union. In general, the dimension of (4.79) will be much higher than (4.80), illustrating the qualitative difference between the noisy and noise-free settings. For example, if the subspaces $\mathcal{U}_\alpha$ are finite dimensional, then (4.80) is also necessarily finite, whereas (4.79) need not be.

The bounds developed for analog noise can also be used to optimize the sampling kernels for a given fixed rate. Under certain assumptions, it can be shown that for the case of finite pulse streams, using exponential functions, or Fourier samples, as in the schemes of [10, 11], is optimal. However, in some cases of pulse shapes the bounds demonstrate that there is room for substantial improvement in the reconstruction stage of these algorithms. Another insight gained from these bounds is that estimation in the semi-periodic setting is far more robust than in the infinite case. As we have discussed, this is because joint processing of the samples is possible. As a rule of thumb, it appears that for union of subspace signals, performance is improved at low-rates if most of the parameters identify the position within the subspace, rather than the subspace itself.

4.4.2 Performance bounds under sampling noise

In this section, we derive the CRB for estimating the parameters of a finite-duration FRI signal (4.19) from samples, in the presence of discrete sampling noise. Specifically, we

consider unbiased estimators of the parameters $\boldsymbol{\theta}$, as given in (4.70), from the noisy samples
$$\tilde{\mathbf{y}} = (\tilde{y}_0, \ldots, \tilde{y}_{N-1})^T, \qquad (4.81)$$
which are given by
$$\tilde{y}_n = \left\langle x(t), \varphi\left(\frac{t}{T} - n\right)\right\rangle + \epsilon_n. \qquad (4.82)$$

We assume throughout that ϵ_n is white Gaussian noise with variance σ^2.

This setting is distinct from the scenario discussed in Section 4.4.1 in two respects. First, we now consider noise introduced after the sampling process, rather than continuous-time noise. It is therefore possible to discuss performance bounds only in the context of a given sampling scheme, so that a sampling-indifferent lower bound such as (4.72) is not available in this case. Another implication is that since the noise originates from the sampling process, it is reasonable to assume that the noise in different samples is independent. Second, we consider in this section the problem of estimating the *parameters* $\boldsymbol{\theta}$ defining the signal $x(t)$, rather than the signal itself. Bounds on the recovery of $x(t)$ from samples corrupted by discrete noise can be found in [12].

For concreteness, we focus in this section on the problem of estimating a τ-periodic stream of Diracs, given by
$$x(t) = \sum_{m \in \mathbb{Z}} \sum_{k=0}^{K-1} a_k \delta(t - t_k - m\tau). \qquad (4.83)$$

The samples (4.82) then become
$$\tilde{y}_n = \sum_{m \in \mathbb{Z}} \sum_{k=0}^{K-1} a_k \varphi(nT - t_k - m\tau) + \epsilon_n = f(\boldsymbol{\theta}, n) + \epsilon_n. \qquad (4.84)$$

Thus, the measurement vector $\tilde{\mathbf{y}}$ has a Gaussian distribution with mean $(f(\boldsymbol{\theta}, 0), \ldots, f(\boldsymbol{\theta}, N-1))^T$ and covariance $\sigma^2 \mathbf{I}_{N \times N}$. The CRB is given by [48]
$$\text{CRB}(\boldsymbol{\theta}) = (\mathbf{J}(\boldsymbol{\theta}))^{-1} \qquad (4.85)$$

where $\mathbf{J}(\boldsymbol{\theta})$ is the Fisher information matrix
$$\mathbf{J}(\boldsymbol{\theta}) = \frac{1}{\sigma^2} \sum_{n=0}^{N-1} \nabla f(\boldsymbol{\theta}, n) \nabla f(\boldsymbol{\theta}, n)^T. \qquad (4.86)$$

It follows that the MSE of any unbiased estimator $\hat{\boldsymbol{\theta}}$ of $\boldsymbol{\theta}$ satisfies
$$\mathbb{E}\left\{\|\hat{\boldsymbol{\theta}} - \boldsymbol{\theta}\|^2\right\} \geq \text{Tr}\left[(\mathbf{J}(\boldsymbol{\theta}))^{-1}\right]. \qquad (4.87)$$

Note that a very similar technique can be used to obtain bounds on FRI signals composed of arbitrary pulse shapes, as well as periodic FRI signals. The only difference is that the expression for $f(\boldsymbol{\theta}, n)$ becomes more cumbersome.

Table 4.1. Summary of the uncertainties on the locations and amplitudes for various sampling kernels. The uncertainties are obtained from the Cramér–Rao Bounds derived in the Appendix.

Kernel	$\frac{\Delta t_0}{\tau} \geq$	$\frac{\Delta a_0}{|a_0|} \geq$
sinc	$\frac{1}{\pi}\sqrt{\frac{\tau}{N}\frac{3}{(B^2\tau^2-1)}}\mathrm{PSNR}^{-\frac{1}{2}}$	$\sqrt{\frac{\tau}{N}}\mathrm{PSNR}^{-\frac{1}{2}}$
B-spline	$\frac{2}{3}\frac{1}{N}\sqrt{\frac{\tau}{N}}\mathrm{PSNR}^{-\frac{1}{2}}$	$\frac{2}{\sqrt{3}}\sqrt{\frac{\tau}{N}}\mathrm{PSNR}^{-\frac{1}{2}}$
E-spline	$\frac{\omega_0-\cos\omega_0\sin\omega_0}{\omega_0\sin\omega_0}\frac{1}{\omega_0 N}\sqrt{\frac{\tau}{N}}\mathrm{PSNR}^{-\frac{1}{2}}$	$\frac{1}{\omega_0}\sqrt{\frac{\omega_0^2-\cos^2\omega_0\sin^2\omega_0}{\omega_0\sin^2\omega_0}}\sqrt{\frac{\tau}{N}}\mathrm{PSNR}^{-\frac{1}{2}}$
SoS	$\frac{1}{2\pi}\sqrt{\frac{\tau}{N}\frac{\sum_{k\in\mathcal{K}}|b_k|^2}{\sum_{k\in\mathcal{K}}k^2|b_k|^2}}\mathrm{PSNR}^{-\frac{1}{2}}$	$\sqrt{\frac{\tau}{N}}\mathrm{PSNR}^{-\frac{1}{2}}$

Comparing sampling kernels in the presence of noise

As an example for which closed-form expressions of the CRB can be obtained, we now consider the special case of estimating the parameters of a periodic stream of Diracs in which each period contains a single pulse. We thus have $K=1$, and the unknown parameters are $\boldsymbol{\theta}=(t_0,a_0)^T$. While this is a very simple case, the ability to derive a closed form will enable us to reach conclusions about the relative merit of various sampling schemes. In particular, we will compare the bounds obtained using the sinc, B-spline, E-spline, and SoS sampling kernels.

The CRBs for estimating this periodic FRI signal using various kernels are derived in the Appendix. The square root of the resulting MSE is then used to bound the uncertainties in the locations and amplitudes. The expressions obtained for B-splines and E-splines depend on t_0. We remove this dependency by assuming that t_0 is uniformly distributed over τ and then compute the expected values of the uncertainties. We restrict our analysis to cardinal and trigonometric exponential splines [35]. For all the derivations and the summary given in this section we define the peak signal-to-noise ratio (SNR) as $\mathrm{PSNR}=(a_0/\sigma)^2$. To obtain a fair comparison between the sampling kernels under consideration the kernels are normalized to have unit norm.

To compare the different kernels, assume that the sinc kernel is chosen with $B\tau=N$ for odd N and $B\tau=N-1$ for even N, as is commonly accepted [19]. Also assume that the SoS kernel has $b_k=1$ for all k, which yields optimal results for this kernel [10]. Under these assumptions, it can be seen from Table 4.1 that the uncertainties in the location for all the kernels follow the same trend, up to a constant factor: they are proportional to $\frac{1}{N}\sqrt{\frac{\tau}{N}}\mathrm{PSNR}^{-\frac{1}{2}}$. Thus, performance improves considerably with an increase in the sampling rate (corresponding to larger values of N), and also improves as the square root of the SNR. Interestingly, it can easily be shown that the SoS kernel has precisely the same uncertainty as that of the sinc kernel. To see this, note that $|\mathcal{K}|=2M+1$ and that the number of samples has to satisfy $N\geq|\mathcal{K}|\geq 2K$.

Using typical values for the parameters of the results given in Table 4.1 we can compare the performance of the kernels. For instance, assume a fixed interval $\tau=1$, and constant number of samples $N=32$, with sampling period $T=\tau/N$ for all the kernels, $b_k=1, \forall k$

for the SoS, $P = 1$ for the B-spline and $P = 1$ and $\omega_0 = 2\pi/N$ for the E-spline, with only $K = 1$ Diracs. In this situation, the sinc and SoS kernels have the best behavior, both in terms of uncertainty in the location and amplitude. For the B-spline and E-spline kernels of lowest possible order ($P = 1$), the uncertainties are almost identical, and slightly worse than optimal. For any support larger than the minimum, the uncertainties achieved by these latter kernels increase.

4.4.3 FRI techniques improving robustness to sampling noise

A central step in each of the reconstruction algorithms examined in Section 4.3 was the search for an annihilating filter $\{h_m\}_{m=0}^{K}$ which satisfies a given system of linear equations (4.26). This annihilation equation was obtained by observing that $\hat{x} * h = 0$ for any filter $\{h_m\}$ whose z-transform has roots (zeros) at the values $\{u_k = e^{-j2\pi \frac{t_k}{\tau}}\}_{k=0}^{K-1}$. In the noise-free setting it was sensible to choose $\{h_m\}$ having degree K, the lowest possible degree for a filter with K zeros. However, any filter of degree $L \geq K$ can also be chosen, as long as $\{u_k\}_{k=0}^{K-1}$ are among its L zeros. Conversely, any filter which annihilates the coefficients $\{\hat{x}_m\}$ is also such that the values u_k are among its zeros.

When noise is present in the system, we can no longer compute the sequence $\{\hat{x}_m\}$ precisely; instead, we have access only to a noisy version $\{\tilde{\hat{x}}_m\}$. On the other hand, since the annihilating equation is satisfied for any contiguous sequence within $\{\hat{x}_m\}$, we can choose to increase the number of measurements, and consequently obtain the sequence $\{\tilde{\hat{x}}_m\}$ in the range $-M \leq m \leq M$, for some $M > L/2$. The annihilating equation can then be written as

$$\begin{pmatrix} \tilde{\hat{x}}_{-M+L} & \tilde{\hat{x}}_{-M+L-1} & \cdots & \tilde{\hat{x}}_{-M} \\ \tilde{\hat{x}}_{-M+L+1} & \tilde{\hat{x}}_{-M+L} & \cdots & \tilde{\hat{x}}_{-M+1} \\ \vdots & \vdots & \ddots & \vdots \\ \tilde{\hat{x}}_{M} & \tilde{\hat{x}}_{M-1} & \cdots & \tilde{\hat{x}}_{M-L} \end{pmatrix} \begin{pmatrix} h_0 \\ h_1 \\ \vdots \\ h_L \end{pmatrix} \approx \begin{pmatrix} 0 \\ 0 \\ \vdots \\ 0 \end{pmatrix} \quad (4.88)$$

which has $2M - L + 1$ equations and $L + 1$ unknowns. The equation is not satisfied exactly due to the presence of noise in the measurements $\{\tilde{\hat{x}}_m\}$. Equivalently, we can write the same equation more compactly as

$$\tilde{\mathbf{A}} \mathbf{h} \approx \mathbf{0} \quad (4.89)$$

where the tilde sign in $\tilde{\mathbf{A}}$ serves to remind us of the fact that this matrix contains noisy measurements. We will denote by $\mathbf{A}$ the matrix obtained when we form the same system of equations with noiseless measurements.

Note that we do not require $h_0 = 1$. Indeed, there exist $L - K + 1$ linearly independent polynomials of degree L with zeros at u_k. Thus, there are $L - K + 1$ independent vectors h that satisfy (4.88). In other words, the rank of $\tilde{\mathbf{A}}$ never exceeds K. This is a key point which forms the basis for many of the methods for signal reconstruction in the presence of noise. We now review two such techniques, namely the total least-squares approach and Cadzow iterative algorithm introduced in [19]. Note that for these techniques to work as explained next, the sampled noise ϵ_n has to be a set of additive, white and Gaussian measurements.

Total least-squares approach

In the presence of noise, the measurements $\{\hat{x}_m\}$ are not known precisely, and one therefore has access only to a noisy version $\tilde{\mathbf{A}}$ of matrix $\mathbf{A}$, so that the modified annihilating equation (4.89) is true. However, it is reasonable to seek an approximate solution to (4.89) by using the method of total least-squares (TLS) [19], which is defined as the solution to the minimization problem

$$\min_{\mathbf{h}} \|\tilde{\mathbf{A}}\mathbf{h}\|^2 \quad \text{subject to } \|\mathbf{h}\|^2 = 1. \tag{4.90}$$

It is not difficult to show that the filter $\mathbf{h}$ solving (4.90) is given by the singular vector corresponding to the smallest singular value of $\tilde{\mathbf{A}}$. Once the filter $\mathbf{h}$ is found, one can determine its roots and hence identify the time delays, as explained in Section 4.3.

Cadzow iterative denoising algorithm

When the level of noise increases, the TLS approach becomes unreliable. Therefore, it is necessary to use a technique that reduces the noise prior to applying TLS. The idea of the Cadzow technique is to exploit the fact that the noise-free matrix $\mathbf{A}$ is Toeplitz with rank K. Our goal is therefore to find a rank-K Toeplitz matrix $\mathbf{A}'$ which is closest to the noisy matrix $\tilde{\mathbf{A}}$, in the sense of a minimal Frobenius norm. Thus, we would like to solve the optimization problem

$$\min_{\mathbf{A}'} \|\tilde{\mathbf{A}} - \mathbf{A}'\|_F^2 \quad \text{such that } \text{rank}(\mathbf{A}') \leq K \text{ and } \mathbf{A}' \text{ is Toeplitz}. \tag{4.91}$$

To solve (4.91), we employ an algorithm that iteratively updates a target matrix $\mathbf{B}$ until convergence. The iterations alternate between finding the best rank-K approximation and finding the best Toeplitz approximation to $\mathbf{B}$. Thus, we must independently solve the two optimization problems

$$\min_{\mathbf{A}'} \|\mathbf{B} - \mathbf{A}'\|_F^2 \quad \text{such that } \text{rank}(\mathbf{A}') \leq K \tag{4.92}$$

and

$$\min_{\mathbf{A}'} \|\mathbf{B} - \mathbf{A}'\|_F^2 \quad \text{such that } \mathbf{A}' \text{ is Toeplitz}. \tag{4.93}$$

The solution to (4.93) is easily obtained by averaging the diagonals of $\mathbf{B}$. To solve (4.92), we compute the singular value decomposition (SVD) $\mathbf{B} = \mathbf{U}\mathbf{S}\mathbf{V}^*$ of $\mathbf{B}$, where $\mathbf{U}$ and $\mathbf{V}$ are unitary and $\mathbf{S}$ is a diagonal matrix whose diagonal entries are the singular values of $\mathbf{B}$. We then discard all but the K largest singular values in $\mathbf{S}$. In other words, we construct a diagonal matrix $\mathbf{S}'$ whose diagonal contains the K largest entries in $\mathbf{S}$, and zero elsewhere. The rank-K matrix closest to $\mathbf{B}$ is then given by $\mathbf{U}\mathbf{S}'\mathbf{V}^*$.

The entire iterative algorithm for solving (4.91) can be summarized as follows:

1. Let **B** equal the original (noisy) measurement matrix $\tilde{\mathbf{A}}$.
2. Compute the SVD decomposition of **B** such that $\mathbf{B} = \mathbf{USV}^*$, where **U** and **V** are unitary and **S** is diagonal.
3. Build the diagonal matrix **S'** consisting of the K largest elements in **S**, and zero elsewhere.
4. Update **B** to its best rank-K approximation $\mathbf{B} = \mathbf{US'V}^*$.
5. Update **B** to its best Toeplitz approximation by averaging over the diagonals of **B**.
6. Repeat from step (2.) until convergence or until a specified number of iterations has been performed.

Applying even a small number of iterations of Cadzow's algorithm will yield a matrix $\mathbf{A}'$ whose error $\|\mathbf{A}' - \mathbf{A}\|_F^2$ is much lower than the error of the original measurement matrix $\tilde{\mathbf{A}}$. This procedure works best when $\tilde{\mathbf{A}}$ is as close as possible to a square matrix [19], and so a good choice would be to use $L = M = \lfloor \frac{B\tau}{2} \rfloor$. The denoised matrix $\mathbf{A}'$ can then be used in conjunction with the TLS technique, as described previously.

4.5 Simulations

In this section we provide some results obtained from implementation of the FRI methods we described. We first show how perfect reconstruction is possible using the proposed kernels in the absence of noise. We then demonstrate the performance of the various kernels when samples are corrupted by additive i.i.d. Gaussian noise. In all simulations we consider only real-valued sampling kernels.

4.5.1 Sampling and reconstruction in the noiseless setting

Figure 4.9 shows an example of the sampling and reconstruction process of Section 4.3.1 for periodic inputs consisting of Diracs. Note that in this setting, using a sinc sampling kernel, or an SoS filter with $b_k = 1$ is equivalent. In Figure 4.9(a) we show the original and reconstructed signals plotted together, while in Figure 4.9(b) we plot the filtered input and the samples taken at a uniform interval. The reconstruction of the signal is exact to numerical precision.

Figure 4.10 shows perfect reconstruction of $K = 4$ closely spaced Diracs using a real-valued E-spline. Here again, reconstruction is exact to numerical precision.

As a final example, consider a periodic input $x(t)$ in which each period consists of $K = 5$ delayed and weighted versions of a Gaussian pulse, with $\tau = 1$. We select the amplitudes and locations at random. Sampling is performed using an SoS kernel with indices $\mathcal{K} = -K, \ldots, K$ and cardinality $M = |\mathcal{K}| = 2K + 1 = 11$. We filter $x(t)$ with $g(t)$ defined in (4.34), and set the coefficients $b_k, k \in \mathcal{K}$ to be a length-M symmetric Hamming window. The output of the filter is sampled uniformly N times, with sampling period $T = \tau/N$, where $N = M = 11$. The sampling process is depicted in Figure 4.11(b). The reconstructed and original signals are depicted in Figure 4.11(a). Once again the estimation and reconstruction are exact to numerical precision.

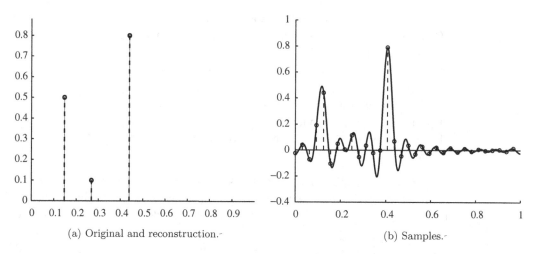

Figure 4.9 Example of sampling and reconstruction of a stream of Diracs with a *sinc* kernel. (a) The original signal along with its reconstruction, exact to numerical precision. (b) Convolution of the sinc kernel with the input. The samples, taken at uniform intervals of T seconds, are also indicated.

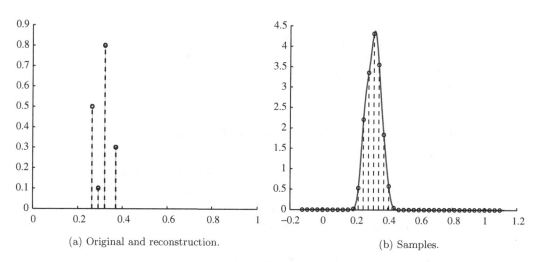

Figure 4.10 Sampling and reconstruction of $K = 4$ closely spaced Diracs with the E-spline kernel. (a) The original signal along with its reconstruction, exact to numerical precision. (b) Convolution of the E-spline kernel with the input. The samples, taken at uniform intervals of T seconds, are also indicated.

4.5.2 Sampling and reconstruction in the presence of noise

In the presence of noise, exact retrieval of the input signal is no longer possible. In order to obtain reasonable recovery, it is necessary to employ some denoising strategies, such as those explained in Section 4.4.

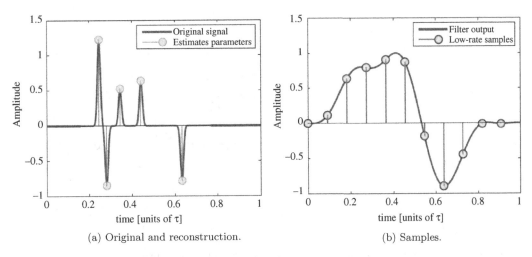

Figure 4.11 Example of sampling and reconstruction of a stream of pulses with an SoS kernel. (a) The train of pulses, with Gaussian shape, and the estimated parameters. (b) Convolution of the input with the SoS kernel, and the samples taken at uniform intervals.

Periodic pulse streams

We start by showing that the proposed robust reconstruction strategies can achieve the CRBs on digital noise given in Section 4.4.2 for a wide range of SNRs. We concentrate on the SoS kernel with coefficients $b_k = 1$. Notice that in this case, the SoS is the Dirichlet function and is therefore equivalent to the periodic sinc of [19]. Figure 4.12 shows the results of the SoS kernel when the input is a periodic train of $K = 3$ Diracs, and the samples are corrupted by i.i.d. Gaussian noise with SNR = 10 dB, where we define the SNR as $\text{SNR} = \frac{\|\mathbf{y}\|_2^2}{N\sigma^2}$, for a single realization. We use $M = \lfloor \frac{B\tau}{2} \rfloor$, and 20 iterations of Cadzow. The result shows that there is a very small error in the estimated locations despite the fairly low SNR. There is, however, a bigger error when estimating the amplitudes. This happens because the kernel is optimized to minimize the error in the estimation of the location of the Diracs rather than in the amplitude.

Now we consider a periodic stream with a single Dirac (e.g. $K = 1$). In the simulations, the amplitude of the Dirac is fixed. The samples are corrupted by i.i.d. Gaussian noise with variance σ^2 such that the SNR ranges from -10 dB to 30 dB. We define the error in time-delay estimation as the average over all experiments of $\|\mathbf{t} - \hat{\mathbf{t}}\|_2^2$, where $\mathbf{t}$ and $\hat{\mathbf{t}}$ denote the true and estimated time delays, respectively, sorted in increasing order. We then calculate the square root of the average to obtain the MSE, which equals the standard deviation for unbiased estimators. Figure 4.13 shows the results obtained from averaging 10 000 realizations and using 10 iterations of Cadzow's algorithm. More specifically, 4.13(a) shows the estimated positions with respect to the real location and 4.13(b) the estimation error compared to the deviation predicted by the CRB. The retrieval of the FRI signal made of one Dirac is almost optimal for SNR levels above 5 dB since the uncertainty on these locations reaches the (unbiased) theoretical minimum given by CRBs.

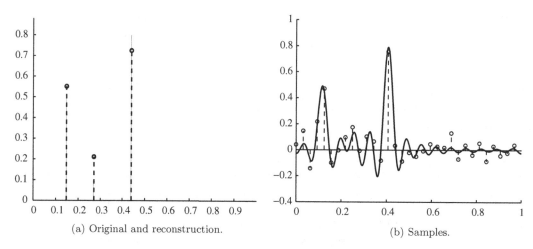

Figure 4.12 Example of sampling and reconstruction of a stream of Diracs with an SoS kernel. (a) The original signal to be sampled and its reconstruction, overlapping the input. (b) Convolution of the kernel with the input. The noisy samples are also shown.

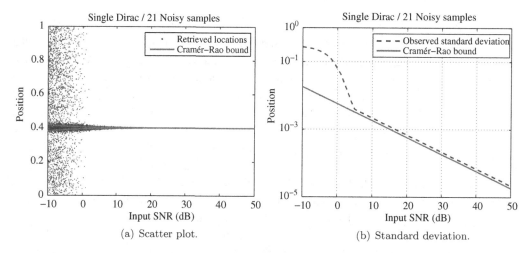

Figure 4.13 Retrieval of the locations of a FRI signal. (a) Scatterplot of the locations. (b) Standard deviation (averaged over 10 000 realizations) compared to the Cramér–Rao lower bound.

The reconstruction quality can be further improved at the expense of oversampling. This is illustrated in Figure 4.14 where two Diracs are reconstructed. Here we show recovery performance for oversampling factors of 2, 4, and 8.

In the following simulation, we consider exponential reproducing kernels and analyze their performance in the presence of noise. Any exponential reproducing kernel is of the form $\varphi(t) = \psi(t) * \beta_{\vec{\alpha}}(t)$ where $\beta_{\vec{\alpha}}(t)$ is the E-spline with exponents $\vec{\alpha} = \{\alpha_0, \alpha_1, \ldots, \alpha_P\}$ and $\psi(t)$ can essentially be any function, even a distribution. The aim here is to understand how to set both $\vec{\alpha}$ and $\psi(t)$ in order to have maximum resilience

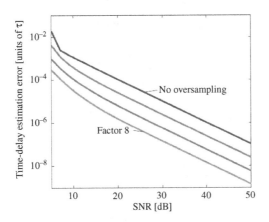

Figure 4.14 Effect of oversampling. The performance of the recovery improves for all SNR when more samples are available.

to noise when noise is additive i.i.d. Gaussian as assumed so far. It turns out that the best choice of the exponents is $\alpha_m = j2\pi \frac{m}{N}$ [36]. The choice of $\psi(t)$ is not unique and depends on the desired support of $\varphi(t)$. If $\varphi(t)$ has the same support as the SoS kernel, then the best choice of $\psi(t)$ leads to an exponential reproducing kernel with the property that its coefficients $c_{m,n}$ constitute a DFT. Moreover, when the order P of the resulting exponential reproducing kernel equals $P = N - 1$ then the kernel behaves like the Dirichlet function [36]. The simulation results of Figure 4.15 confirm the above analysis. Here we retrieve two Diracs in the presence of noise using an E-spline with arbitrary exponents ($P = 9$, d), an E-spline with the correct exponents $\alpha_m = j2\pi \frac{m}{N}$ ($P = 9$, o), and finally using two of the most stable exponential reproducing kernels ($P = 15, 30$, n) (the best being the Dirichlet function). We use the notation "d" to indicate default kernel, "o" orthogonal rows of coefficients, and "n" orthonormal rows.

Here, we have $N = 31$ samples and the input $x(t)$ is a τ-periodic stream of Diracs, where $\tau = 1$ second. We run 1000 experiments contaminating the samples with i.i.d. Gaussian noise of desired SNR by controlling its variance, and we denoise the calculated moments doing 30 iterations of Cadzow's denoising algorithm. We can see the improvement in performance by going from the first to the last type of exponential reproducing kernel. In fact, as expected, proper choice of the exponents α_m improves the estimation of the locations, and the appropriate choice of $\psi(t)$ enhances the results further. Interestingly, if we use pure E-splines $\beta_{\vec{\alpha}}(t)$ then there is an order from which the performance declines. In Figure 4.15 we plot the optimum order ($P = 9$). In contrast, when we design the optimum exponential reproducing kernel the performance improves constantly until it matches that of the Dirichlet kernel.

Finite pulse streams

We now turn to demonstrate FRI recovery methods when using finite pulse streams. We examine four scenarios, in which the signal consists of $K = 2, 3, 5, 20$ Diracs.[1] In our

---

[1] Due to computational complexity of calculating the time-domain expression for high-order E-splines, the functions were simulated up to order 9, which allows for $K = 5$ pulses.

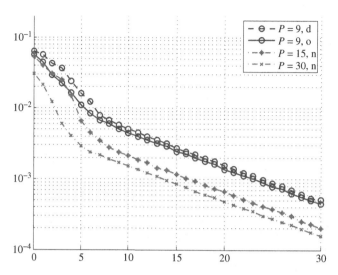

Figure 4.15 Sampling with exponential reproducing kernels. Results of the estimation of the location of $K = 2$ Diracs in the presence of noise. The performance of exponential reproducing kernels can be enhanced by proper selection of the parameter α_m (solid line with -o-) and depending on the choice of $\psi(t)$ (dashed-dotted lines).

setup, the time delays are equally distributed in the window $[0, \tau)$, with $\tau = 1$, and remain constant throughout the experiments. All amplitudes are set to one. The index set of the SoS filter is $\mathcal{K} = \{-K, \ldots, K\}$. Both B-splines and E-splines are taken of order $2K - 1$, and for E-splines we use purely imaginary exponents, equally distributed around the complex unit circle. The sampling period for all methods is $T = \tau/N$, where the number of samples is $N = 2K + 1 = 5, 7, 11, 41$ for the SoS and $N = 2K + 1 + S = 9, 13, 21, 81$ for the spline-based methods, where S is the spline support. Hard thresholding was implemented in order to improve the spline methods. The threshold was chosen to be 3σ, where σ is the standard deviation of the noise. For the Gaussian sampling kernel the parameter σ_g was optimized and took on the value of $\sigma_g = 0.25, 0.28, 0.32, 0.9$, respectively.

The results are given in Figure 4.16. For $K = 2$ all methods are stable, where E-splines exhibit better performance than B-splines, and Gaussian and SoS approaches demonstrate the lowest errors. As the value of K grows, the advantage of the SoS filter becomes more prominent, where for $K \geq 5$, the performance of Gaussian and both spline methods deteriorate and have errors approaching the order of τ. In contrast, the SoS filter retains its performance nearly unchanged even up to $K = 20$, where the B-spline and Gaussian methods are unstable.

Infinite pulse streams

We now demonstrate the performance of FRI methods for infinite pulse streams in the presence of white Gaussian noise, when working at the rate of innovation. We compare three methods that can achieve the innovation rate in the infinite case: an integrator-based approach detailed in [40], exponential filters [41], and the multichannel approach described in Section 4.3.4 based on modulators. For the modulators, we examine three

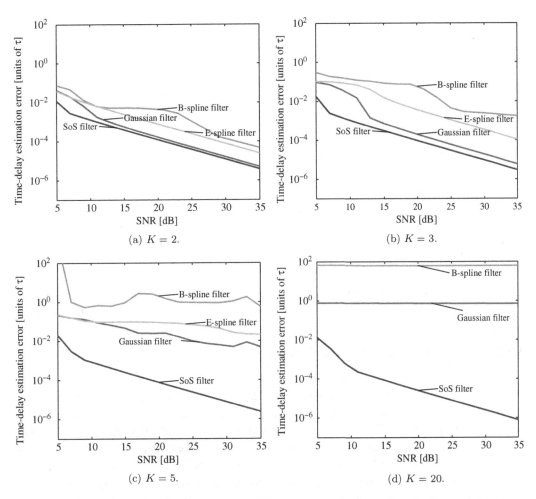

Figure 4.16 Performance in the presence of noise: finite stream case. SoS, B-spline, E-spline, and Gaussian sampling kernels. (a) $K = 2$ Dirac pulses are present, (b) $K = 3$ pulses, (c) high value of $K = 5$ pulses, and (d) the performance for a very high value of $K = 20$ (without E-spline simulation, due to computational complexity of calculating the time-domain expression for high values of K).

waveforms: cosine and sine waveform (tones), filtered rectangular alternating pulses (rectangular), and waveforms obtained from delayed versions of the SoS filter (SoS). Following [41], the parameters defining the impulse response of the exponential filters are chosen as $\alpha = 0.2T$ and $\beta = 0.8T$.

We focus on one period of the input signal, which consists of $K = 10$ Diracs with times chosen in the interval $[0, T)$ and amplitudes equal one, and $P = 21$ channels. The estimation error of the time delays versus the SNR is depicted in Figure 4.17, for the various approaches. The instability of the integrators and exponential filters based methods becomes apparent for these high orders. The SoS approach, in contrast, achieves

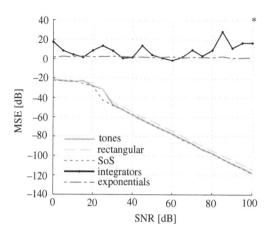

Figure 4.17 Performance in the presence of noise at the rate of innovation. The signal consists of $K = 10$ Diracs.

good estimation results. There is a slight advantage for the schemes based on tones and SoS, over alternating pulses, where the first two configurations have similar performance.

4.5.3 Periodic vs. semi-periodic FRI signals

As we have seen above, the reconstruction of signals of the form (4.18) in the presence of noise is often severely hampered when sampled at or slightly above the rate of innovation. Rather than indicating a lack of appropriate algorithms, in many cases this phenomenon results from fundamental limits on the ability to recover such signals from noisy measurements. A similar effect was demonstrated [10] in the finite pulse stream model (4.19). On the other hand, some types of FRI signals exhibit remarkable noise resilience, and do not appear to require substantial oversampling in the presence of noise [17]. As we now show, the CRB for analog noise can be used to verify that such phenomena arise from a fundamental difference between families of FRI signals.

As an example, we compare the CRB for reconstructing the periodic signal (4.18) with the semi-periodic signal (4.20). Recall that in the former case, each period consists of pulses having unknown amplitudes and time shifts. By contrast, in the latter signal, the time delays are identical throughout all periods, but the amplitudes can change from one period to the next.

While these are clearly different types of signals, an effort was made to form a fair comparison between the reconstruction capabilities in the two cases. To this end, we chose an identical pulse $g(t)$ in both cases. We selected the signal segment $[0, \tau]$, where $\tau = 1$, and chose the signal parameters so as to guarantee an identical τ-local rate of innovation. We also used identical sampling kernels in both settings: specifically, we chose the kernels which measure the N lowest frequency components of the signal.

To simplify the analysis and focus on the fundamental differences between these settings, we will assume in this section that the pulses $p(t)$ are compactly supported, and that the time delays are chosen such that pulses from one period do not overlap with

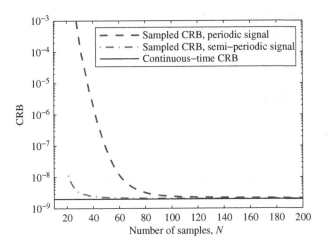

Figure 4.18 Comparison between the CRB for a periodic signal (4.18) and a semi-periodic signal (4.20).

other periods. For the periodic signal, we chose $K = 10$ pulses with random delays and amplitudes. A period of $\tau = 1$ was selected. This implies that the signal of interest is determined by $2K = 20$ parameters (K amplitudes and K time delays). To construct a semi-periodic signal with the same number of parameters, we chose a period of $T = 1/9$ containing $K = 2$ pulses. The segment $[0, \tau]$ then contains precisely $M = 9$ periods, for a total of 20 parameters. While it may seem plausible to require the same number of periods for both signals, this would actually disadvantage the periodic approach, as it would require the estimation of much more closely spaced pulses.

Note that since the number of parameters to be estimated is identical in both signal models, the continuous-time CRBs for the two settings coincide (see Section 4.4.1). Consequently, for a large number of measurements, the sampled bounds also converge to the same values. However, when the number of samples is closer to the rate of innovation, the bound on the reconstruction error for the semi-periodic signal is much lower than that of the periodic signal, as shown in Figure 4.18. As mentioned above, this is in agreement with previously reported findings for the two types of signals [4, 11, 17].

To find an explanation for this difference, it is helpful to recall that both signals can be described using the union of subspaces viewpoint. Each of the signals in this experiment is defined by precisely 20 parameters, which determine the subspace to which the signal belongs and the position within this subspace. Specifically, the values of the time delays select the subspace, and the pulse amplitudes define a point within this subspace. Thus, in the above setting, the periodic signal contains 10 parameters for selecting the subspace and 10 additional parameters determining the position within it; whereas for the semi-periodic signal, only 2 parameters determine the subspace while the remaining 18 parameters set the location in the subspace. Evidently, identification of the subspace is challenging, especially in the presence of noise, but once the subspace is determined, the remaining parameters can be estimated using a simple linear operation (a projection onto the chosen subspace). Consequently, if many of the unknown parameters identify the position within a subspace, estimation can be performed more accurately. This may provide an explanation for the difference between the two examined signal models.

4.6 Extensions and applications

4.6.1 Sampling piecewise sinusoidal signals

While most of the previous sections have concentrated on sampling streams of pulses, the theory of FRI extends beyond this class of signals and can be applied, for instance, to sample piecewise polynomial signals or classes of 2-D signals. In this section we demonstrate that piecewise sinusoidal signals can also be sampled and perfectly reconstructed using FRI theory [16].

The signals we consider can be written as follows:

$$x(t) = \sum_{d=1}^{D} \sum_{n=1}^{N} A_{d,n} \cos(\omega_{d,n} t + \theta_{d,n}) \xi_d(t),$$

where $\xi_d(t) = u(t - t_d) - u(t - t_{d+1})$ and $-\infty < t_1 < \ldots < t_d < \ldots < t_{D+1} < \infty$. Namely, we consider piecewise sinusoidal signals with a maximum of D pieces and with a maximum of N sinusoids per piece. Piecewise sinusoidal signals are traditionally difficult to sample because they are not bandlimited, have information concentrated both in time and frequency (e.g., time location of the switching points, frequency of each sine wave), and finally cannot be sparsely described in a basis or a frame. However, they are completely specified by a finite number of parameters and are therefore FRI signals.

We assume, for simplicity, that the signal $x(t)$ is acquired using an exponential reproducing kernel; however, similar analysis applies to the sinc and SoS kernels. We have seen in (4.44) and (4.45) that given the samples y_n the new measurements $s_m = \sum_{n=0}^{N-1} c_{m,n} y_n$, $m = 0, 1, \ldots, P$ correspond to the Laplace transform of $x(t)$ evaluated at $\alpha_m = \alpha_0 + m\lambda$. In the piecewise sinusoidal case the Laplace transform is given by:

$$s_m = \sum_{d=1}^{D} \sum_{n=1}^{2N} \bar{A}_{d,n} \frac{[e^{t_{d+1}(j\omega_{d,n} + \alpha_m)} - e^{t_d(j\omega_{d,n} + \alpha_m)}]}{(j\omega_{d,n} + \alpha_m)}, \quad (4.94)$$

where $\bar{A}_{d,n} = A_{d,n} e^{j\theta_{d,n}}$. We now define the polynomial $Q(\alpha_m)$ as follows:

$$Q(\alpha_m) = \prod_{d=1}^{D} \prod_{n=1}^{2N} (j\omega_{d,n} + \alpha_m) = \sum_{j=0}^{J} r_j \alpha_m^j. \quad (4.95)$$

Multiplying both sides of the equation by $Q(\alpha_m)$ we obtain:

$$Q(\alpha_m) s_m = \sum_{d=1}^{D} \sum_{n=1}^{2N} \bar{A}_{d,n} R(\alpha_m) [e^{t_{d+1}(j\omega_{d,n} + \alpha_m)} - e^{t_d(j\omega_{d,n} + \alpha_m)}], \quad (4.96)$$

where $R(\alpha_m)$ is a polynomial. Since $\alpha_m = \alpha_0 + \lambda m$ the right-hand side of (4.96) is a power-sum series and can be annihilated:

$$Q(\alpha_m)s_m * h_m = 0. \qquad (4.97)$$

More precisely the right-hand side of (4.96) is equivalent to $\sum_{d=1}^{D} \sum_{r=0}^{2DN-1} b_{r,d} m^r e^{\lambda t_d m}$ where $b_{r,d}$ are weights that depend on α_m but do not need to be computed here. Therefore a filter of the type:

$$\hat{h}(z) = \prod_{d=1}^{D}(1 - e^{\lambda t_d} z^{-1})^{2DN} = \sum_{k=0}^{K} h_k z^{-k}$$

will annihilate $Q(\alpha_m)s_m$. In matrix/vector form (4.97) can be written as

$$\begin{pmatrix} s_K & \alpha_K^J s_K & \cdots & s_0 & \cdots & \alpha_0^J s_0 \\ s_{K+1} & \alpha_{K+1}^J s_{K+1} & \cdots & s_1 & \cdots & \alpha_1^J s_1 \\ \vdots & \vdots & \vdots & \ddots & \vdots & \vdots \\ s_P & \alpha_P^J s_P & \cdots & s_{p-k} & \cdots & \alpha_{P-K}^J s_{P-K} \end{pmatrix} \begin{pmatrix} h_0 r_0 \\ \vdots \\ h_0 r_J \\ \vdots \\ h_K r_0 \\ \vdots \\ h_K r_J \end{pmatrix} = \begin{pmatrix} 0 \\ 0 \\ \vdots \\ 0 \end{pmatrix}.$$

Solving the system for $h_0 = 1$ enables finding the coefficients r_j, from which we can obtain the coefficients h_k. The roots of the filter $\hat{h}(z)$ and of the polynomial $Q(\alpha_m)$ give the locations of the switching points and the frequencies of the sine waves respectively. The number of values s_m required to build a system with enough equations to find the parameters of the signal is $P \geq 4D^3N^2 + 4D^2N^2 + 4D^2N + 6DN$.

An illustration of the sampling and reconstruction of a piecewise sinusoidal signal is shown in Figure 4.19. For more details about the sampling of these signals we refer the reader to [16].

4.6.2 Signal compression

We have seen that specific classes of signals can be parsimoniously sampled using FRI sampling theory. Moreover, the sampling kernels involved include scaling functions used in the construction of wavelet bases such as, for example, B-splines or Daubechies scaling function.

We are now going to concentrate on this type of kernel and investigate the potential impact of such sampling schemes in compression where samples are also quantized and represented with a bit stream. In this context, the best way to analyze the compression algorithm is by using standard rate-distortion (R-D) theory since this gives the best achievable trade-off between the number of bits used and the reconstruction fidelity. It is often assumed that the error due to quantization can be modeled as additive noise.

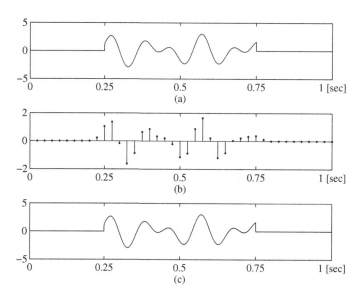

Figure 4.19 Sampling a piecewise sinusoidal signal. (a) The original continuous-time waveform, this is made of two truncated sinewaves. (b) The observed samples. (c) The reconstructed signal, where the retrieval of the two switch points and of the sine waves parameters is exact to machine precision.

While this assumption is normally not accurate, it allows us to connect R-D theory with the CRB discussed in the previous section and therefore relate the theory of sampling FRI signals with compression.

The classes of signals we consider here are piecewise smooth functions, that is, functions which are made of regular pieces. The regularity of a function is normally measured using the Lipschitz coefficients [49]. We thus assume that the signals we consider are made of pieces with Lipschitz regularity α.

The FRI-based compression algorithm we propose is characterized by a simple linear encoding strategy and a more complex decoding. This is in contrast with standard wavelet-based compression algorithms that involve a fairly sophisticated encoding strategy, but simple decoding. There might be situations, however, where it is important to have simple encoders. In our setup, at the encoder the signal is decomposed using a standard wavelet transform and the resulting coefficients are quantized linearly. This means that the lowpass coefficients (equivalent to the samples in the FRI framework) are quantized first followed by the wavelet coefficients from the coarse to the finest scale.

At the decoder, the FRI reconstruction strategy is used to estimate the discontinuities in the signal using the scaling coefficients, while the other coefficients are used to reconstruct the smooth parts of the signals. By modeling the quantization error and any model mismatch as additive noise, one can use the CRB to estimate the performance of this compression strategy. The rate-distortion behavior of this FRI-based algorithm is [38, 39]:

$$D_{\text{FRI}}(R) \leq c_1 R^{-2\alpha} + c_2 \qquad (4.98)$$

where c_2 is a systematic estimation error due to the model mismatch. Standard wavelet-based compression algorithms instead are characterized by a complex encoder and a simple decoder and can achieve the optimal rate-distortion behavior [50]:

$$D_{\text{wave}}(R) \leq c_3 R^{-2\alpha}. \tag{4.99}$$

This indicates that if the systematic error in (4.98) is sufficiently small the FRI-based algorithm, which shifts the complexity from the encoder to the decoder, can achieve the same performance of the best wavelet-based compression algorithms for a wide range of bit rates.

4.6.3 Super-resolution imaging

An image super-resolution algorithm aims at creating a single detailed image, called a super-resolved image (SR) from a set of low-resolution input images of the same scene [51]. If different images from the same scene have been taken such that their relative shifts are not integer-multiple of the pixel size, then sub-pixel information exists among the set. This allows us to obtain higher resolution accuracy of the scene once the images have been properly registered.

Image registration involves any group of transformations that removes the disparity between any two low-resolution (LR) images. This is followed by image fusion, which blends the properly aligned LR images into a higher resolution output, possibly removing blur and noise introduced by the system [26].

The registration step is crucial in order to obtain a good quality SR image. The theory of FRI can be extended to provide super-resolution imaging, combined with B-spline or E-spline processing. The key idea of this approach is that, using a proper model for the point-spread function of the scene acquisition system, it is possible to retrieve the underlying "continuous geometric moments" of the irradiance light-field. From this information, and assuming the disparity between any two images can be characterized by a global affine transformation, the set of images can be exactly registered.

Concretely, if the smoothing kernel that models the 2-D image acquisition is considered to be a B-spline or a more generic function such as a spline, then the continuous moments of the image can be found using a proper linear combination of the samples [25, 26]. From them, it is possible to find the central and complex moments of the signal, from which the disparities between any two LR images can be estimated. Thus, this allows for proper registration of the set of input images, which can now be combined into a super-resolved output. Figure 4.20 shows an example of the results obtained using the method presented in [26].

4.6.4 Ultrasound imaging

Another application of the stream of pulses FRI framework is ultrasound imaging [10]. In this application, an acoustic pulse is transmitted into the scanned tissue. Echoes of the pulse bounce off scatterers within the tissue, and create a signal consisting of a

(a) HR. (b) LR. (c) SR.

Figure 4.20 Image super-resolution from translated images with registration from the extracted edges and detected corners. (a) Original high-resolution image (512 × 512 pixels). (b) One of the 20 low-resolution images (64 × 64 pixels) used in the super-resolution simulation. (c) Super-resolved image with the proposed edge detector and Wiener Filter, 512 × 512 pixels, PSNR = 15.6 dB.

stream of pulses at the receiver. The time delays and amplitudes of the echoes indicate the position and strength of the various scatterers, respectively. Therefore, determining these parameters from low-rate samples of the received signal is an important problem. Reducing the rate allows more efficient processing which can translate to power and size reduction of the ultrasound imaging system.

The stream of pulses is finite since the pulse energy decays within the tissue. In order to demonstrate the viability of an FRI framework, we model the multiple echo signal recorded at the receiver as a finite stream of pulses, like (4.15). The unknown time delays correspond to the locations of the various scatterers, whereas the amplitudes are the reflection coefficients. The pulse shape in this case is Gaussian, due to the physical characteristics of the electro-acoustic transducer (mechanical damping).

As an example, we chose a phantom consisting of uniformly spaced pins, mimicking point scatterers, and scanned it by GE Healthcare's Vivid-i portable ultrasound imaging system, using a 3S-RS probe. We use the data recorded by a single element in the probe, which is modeled as a 1-D stream of pulses. The center frequency of the probe is $f_c = 1.7021$ MHz, the width of the transmitted Gaussian pulse in this case is $\sigma = 3 \cdot 10^{-7}$ seconds, and the depth of imaging is $R_{\max} = 0.16$ m corresponding to a time window of $\tau = 2.08 \cdot 10^{-4}$ seconds.[2] We carried out our sampling and reconstruction scheme on the data. We set $K = 4$, looking for the strongest four echoes. Since the data is corrupted by strong noise we oversampled the signal, obtaining twice the minimal number of samples. In addition, hard-thresholding of the samples was implemented, where we set the threshold to 10 percent of the maximal value. Figure 4.21 depicts the reconstructed signal together with the full demodulated signal. Clearly, the time delays were estimated with high precision. The amplitudes were estimated as well, but the amplitude of the second pulse has a large error. However, the exact locations of the scatterers is typically more important than the accurate reflection coefficients. This is because the

---

[2] The speed of sound within the tissue is 1550 m/s.

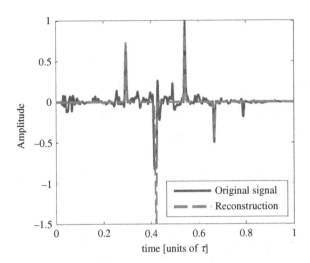

Figure 4.21 Example of sampling and reconstruction of real ultrasound imaging data. The input signal, in continuous line, is sampled assuming there exist $K = 4$ pulses, and using an oversampling factor of 2. The output is a stream of Gaussian pulses, where the unknown locations and amplitudes have been estimated from the $N = 17$ samples obtained from the input, denoising with hard-thresholding.

time of arrival indicates the scatterer's location within the tissue. Accurate estimation of tissue boundaries and scatterer locations allows for reliable detection of certain illnesses, and is therefore of major clinical importance. The location of the boundaries is often more important than the power of the reflection which is incorporated in the received amplitudes.

Current ultrasound imaging technology operates at the high rate sampled data, e.g., $f_s = 20$ MHz in our setting. Since there are usually 100 different elements in a single ultrasonic probe each sampled at a very high rate, data throughput becomes very high, and imposes high computational complexity to the system, limiting its capabilities. Therefore, there is a demand for lowering the sampling rate, which in turn will reduce the complexity of reconstruction. Exploiting the parametric point of view, our sampling scheme reduces the sampling rate by over two orders of magnitude, while estimating the locations of the scatterers with high accuracy.

4.6.5 Multipath medium identification

Another nice application of the FRI model is to the problem of time-varying channel estimation in wireless communication [17]. In such an application the aim of the receiver is to estimate the channel's parameters from the samples of the received signal [52].

We consider a baseband communication system operating in a multipath fading environment with pulse amplitude modulation (PAM). The data symbols are transmitted at a symbol rate of $1/T$, modulated by a known pulse $p(t)$. The transmitted signal $x_t(t)$ is given by

$$x_T(t) = \sum_{n=1}^{N_{sym}} d[n]p(t-nT) \qquad (4.100)$$

where $d[n]$ are the data symbols taken from a finite alphabet, and N_{sym} is the total number of transmitted symbols.

The transmitted signal $x_T(t)$ passes through a baseband time-varying multipath channel whose impulse response is modeled as

$$h(\tau,t) = \sum_{k=1}^{K} \alpha_k(t)\delta(\tau-\tau_k) \qquad (4.101)$$

where $\alpha_k(t)$ is the path time-varying complex gain for the kth multipath propagation path and τ_k is the corresponding time delay. The total number of paths is denoted by K. We assume that the channel is slowly varying relative to the symbol rate, so that the path gains are considered to be constant over one symbol period:

$$\alpha_k(t) = \alpha_k[nT] \text{ for } t \in [nT, (n+1)T]. \qquad (4.102)$$

In addition, we assume that the propagation delays are confined to one symbol, i.e. $\tau_k \in [0,T)$. Under these assumptions, the received signal at the receiver is given by

$$x_R(t) = \sum_{k=1}^{K}\sum_{n=1}^{N_{sym}} a_k[n]p(t-\tau_k-nT) + n(t) \qquad (4.103)$$

where $a_k[n] = \alpha_k[nT]d[n]$ and $n(t)$ denotes the channel noise.

The received signal $x_R(t)$ fits the semi-periodic FRI signal model. Therefore, we can use the methods we described to recover the time delays of the propagation paths. In addition, if the transmitted symbols are known to the receiver, then the time-varying path gains can be recovered from the sequences $a_k[n]$. As a result our sampling scheme can estimate the channel's parameters from samples of the output at a low-rate, proportional to the number of paths.

As an example, we can look at the channel estimation problem in code division multiple access (CDMA) communication. This problem was handled using subspace techniques in [53,54]. In these works the sampling is done at the chip rate $1/T_c$ or above, where T_c is the chip duration given by $T_c = T/N$ and N is the spreading factor which is usually high (1023, for example, in GPS applications). In contrast, our sampling scheme can provide recovery of the channel's parameters at a sampling rate of $2K/T$. For a channel with a small number of paths, this sampling rate can be significantly lower than the chip rate.

4.6.6 Super-resolution radar

We end with an application of the semi-periodic model (4.20) to super-resolution radar [27].

In this context, we can translate the rate reduction to increased resolution, thus enabling super-resolution radar from low-rate samples. Here the goal is to identify the range and velocity of a set of targets. The delay in this case captures the range while the time-varying coefficients are a result of the Doppler delay related to the target velocity. More specifically, we assume that several targets can have the same delays but possibly different Doppler shifts so that $\{t_\ell\}_{\ell=1}^{K}$ denote the set of distinct delays. For each delay value t_ℓ there are K_ℓ values of associated Doppler shifts $\nu_{\ell k}$ and reflection coefficients $\alpha_{\ell k}$. It is further assumed that the system is highly underspread, namely $\nu_{\max}T \ll 1$, where $\nu_{\max}$ denotes the maximal Doppler shift, and T denotes the maximal delay. To identify the

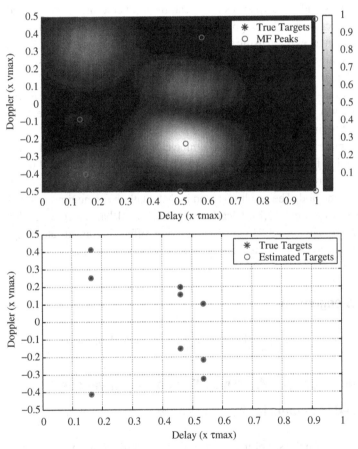

Figure 4.22 Comparison between the target-detection performance of matched-filtering and the procedure described in [27] for the case of nine targets (represented by $*$) in the delay–Doppler space with $\tau_{max} = 10\mu s$, $\nu_{max} = 10$ kHz, $\mathcal{W} = 1.2$ MHz, and $\mathcal{T} = 0.48$ ms. The probing sequence $\{x_n\}$ corresponds to a random binary (± 1) sequence with $N = 48$, the pulse $p(t)$ is designed to have a nearly flat frequency response and the pulse repetition interval is $T = 10\ \mu s$. (a) Target detection by matched-filtering. (b) Target detection using the proposed procedure with $P = 12$.

targets we transmit the signal

$$x_T = \sum_{n=0}^{N-1} x_n p(t - nT),\qquad(4.104)$$

where x_n is a known N-length probing sequence, and $p(t)$ is a known pulse shape. The received signal can then be described in the form (4.20), where the sequences $a_\ell[n]$ satisfy

$$a_\ell[n] = x_n \sum_{k=1}^{K_\ell} \alpha_{\ell k} e^{j2\pi \nu_{\ell k} nT}.\qquad(4.105)$$

The delays and the sequences $a_\ell[n]$ can be recovered using the general scheme for time-delay recovery. The Doppler shifts and reflection coefficients are then determined from the sequences $a_\ell[n]$ using standard spectral estimation tools [18]. The targets can be exactly identified as long as the bandwidth of the transmitted pulse satisfies $W \geq 4\pi K/T$, and the length of the probing sequence satisfies $N \geq 2\max K_\ell$ [27]. This leads to a minimal time–bandwidth product of the input signal of $WT \geq 8\pi K \max K_\ell$, which is much lower than that obtained using standard radar processing techniques, such as matched-filtering (MF).

An example of the identification of nine close targets is illustrated in Figure 4.22(a). The sampling filter used is a simple LPF. The original and recovered targets are shown on the Doppler–delay plane. Evidently all the targets were correctly identified using our FRI-based method. The result obtained by MF, with the same time–bandwidth product, is shown in Figure 4.22(b). Clearly the FRI method has superior resolution than the standard MF. Thus, the FRI viewpoint not only offers a reduced-rate sampling method, but allows us to increase the resolution in target identification.

4.7 Acknowledgement

Y. Eldar would like to thank Kfir Gedalyahu for many helpful comments.

Appendix to Chapter 4: Cramér–Rao bound derivations
Cramér–Rao bounds for the sinc kernel

Here we focus on the simplified case in which the input is a single τ-periodic Dirac, for which we can obtain a closed-form expression for (4.85).

In the absence of noise, the samples taken at uniform intervals of time T with $K = 1$ Dirac, can be expressed as:

$$y_n = a_0 \psi(nT - t_0) = f(\boldsymbol{\theta}, n)\qquad(4.106)$$

where $\psi(t)$ is the Dirichlet kernel

$$\psi(t) = \sum_{m \in \mathbb{Z}} \phi_B(t - m\tau) = \frac{1}{B\tau} \frac{\sin\left(\frac{\pi(2M+1)t}{\tau}\right)}{\sin\left(\frac{\pi t}{\tau}\right)} = \frac{1}{B\tau} \frac{\sin(\pi B t)}{\sin\left(\frac{\pi t}{\tau}\right)} \quad (4.107)$$

and $\boldsymbol{\theta} = (t_0, a_0)^T$. The Fisher information matrix is the following square and size 2×2 matrix:

$$\mathbf{I}(\boldsymbol{\theta}) = \sigma^{-2} \begin{pmatrix} \sum_{n=0}^{N-1} (a_0 \psi'(nT - t_0))^2 & \sum_{n=0}^{N-1} a_0 \psi'(nT - t_0) \psi(nT - t_0) \\ \sum_{n=0}^{N-1} \psi(nT - t_0) a_0 \psi'(nT - t_0) & \sum_{n=0}^{N-1} (\psi(nT - t_0))^2 \end{pmatrix}. \quad (4.108)$$

In order to evaluate the summations it is convenient to use the Fourier series representations of the signals $\psi(t)$ and $\psi'(t)$ because the following holds [55]:

$$\sum_{n=0}^{N-1} f(nT) g^*(nT) \stackrel{(a)}{=} \sum_{n=0}^{N-1} \left(\sum_k \hat{f}_k e^{j2\pi k n \frac{T}{\tau}}\right) \left(\sum_{k'} \hat{g}_{k'}^* e^{-j2\pi k' n \frac{T}{\tau}}\right) \quad (4.109)$$

$$= \sum_k \hat{f}_k \sum_{k'} \hat{g}_{k'}^* \frac{1 - e^{j2\pi(k-k') N \frac{T}{\tau}}}{e^{-j2\pi(k-k') \frac{T}{\tau}}}$$

$$\stackrel{(b)}{=} \sum_k \hat{f}_k \sum_{k'} \hat{g}_{k'}^* N \delta_{k,k'} = N \sum_k \hat{f}_k \hat{g}_k^*,$$

where in (a) we have used the fact that $f(t)$ and $g(t)$ are assumed to be periodic, and in (b) the fact that for $\tau = NT$ the sum is only nonzero when $k = k'$.

Furthermore, if we call $\hat{\psi}_k$ the coefficients for $\psi(t)$, then $\hat{\psi}'_k = j2\pi \frac{k}{\tau} \hat{\psi}_k$ would be the coefficients for its derivative and $\hat{\psi}_k^{(t_0)} = e^{-j2\pi k \frac{t_0}{\tau}} \hat{\psi}_k$ the coefficients for its shifted version by t_0.

These last equivalences and (4.109) simplify the calculations of the sums in (4.108), because the function $\psi(t)$ is characterized by the Fourier series coefficients $\hat{\psi}_k = \frac{1}{B\tau} |k| \leq M$ and $\hat{\psi}_k = 0$ otherwise. Element (1,1) in (4.108) can therefore be calculated as

$$\sigma^{-2} \sum_{n=0}^{N-1} (a_0 \psi'(nT - t_0))^2 = \sigma^{-2} a_0^2 N \sum_{|k| \leq M} \psi'_k e^{-j2\pi k \frac{t_0}{\tau}} \psi_k'^* e^{j2\pi k \frac{t_0}{\tau}} \quad (4.110)$$

$$[\mathbf{I}(\boldsymbol{\theta})]_{11} = N \left(\frac{a_0}{\sigma} \frac{2\pi}{B\tau^2}\right)^2 \frac{M(M+1)(2M+1)}{3}.$$

The other elements in (4.108) can be calculated likewise. Due to the fact that the elements in the anti-diagonal are zero, the inverse of the Fisher information matrix can be computed by just inverting the diagonal elements, yielding

$$\mathrm{CRB}(\boldsymbol{\theta}) = \begin{pmatrix} \frac{1}{N} \left(\frac{\sigma}{a_0} \frac{B\tau^2}{2\pi}\right)^2 \frac{3}{M(M+1)(2M+1)} & 0 \\ 0 & \frac{1}{N} \sigma^2 B^2 \tau^2 \frac{1}{2M+1} \end{pmatrix}, \quad (4.111)$$

where $M = \lfloor \frac{B\tau}{2} \rfloor$. Note that, for this M, it always holds that $2M+1 = B\tau$.

To end, we can determine the uncertainties in the location and the amplitude from the values derived in the CRB. We know that the diagonal values in (4.111) are lower bounds for $\text{var}\{t_0\}$ and $\text{var}\{a_0\}$ respectively. And since we are interested in unbiased estimators, the variances equal the MSE for each unknown. Therefore, we can write the uncertainty in the location as follows:

$$\frac{\Delta t_0}{\tau} \geq \sqrt{\frac{1}{N}\left(\frac{\sigma}{a_0}\frac{B\tau}{2\pi}\right)^2 \frac{3}{M(M+1)(2M+1)}} \qquad (4.112)$$

$$\stackrel{(a)}{=} \frac{1}{\pi}\sqrt{\frac{3B\tau}{N(B^2\tau^2-1)}} \text{PSNR}^{-\frac{1}{2}},$$

where we have defined the peak signal-to-noise ratio as $\text{PSRN} = (a_0/\sigma)^2$, and in (a) we have used the fact that $4M(M+1) = (2M+1)^2 - 1 = B^2\tau^2 - 1$. We can also write the uncertainty in the amplitude as:

$$\frac{\Delta a_0}{|a_0|} \geq \sqrt{\frac{1}{N}\frac{\sigma^2}{a_0^2}B^2\tau^2\frac{1}{2M+1}} = \sqrt{\frac{B\tau}{N}}\text{PSNR}^{-\frac{1}{2}}. \qquad (4.113)$$

Cramér–Rao bounds for the SoS kernel

The derivation of the CRB for the SoS kernel follows exactly the same steps as in the previous section for the sinc. First, we express the samples as $y_n = a_0\eta(t_0 - nT) = f(\boldsymbol{\theta}, n)$, where $\eta(t) = \sum_{m \in \mathbf{Z}} g(t - m\tau)$, and $g(t)$ is the filter defined in (4.34). Now, we can once more rely on the Fourier series representation of the signals to calculate the summations. From (4.33), the Fourier series coefficients of the periodic expansion of the kernel $\eta(t)$ are $\eta_k = \frac{1}{\tau}\hat{g}\left(\frac{2\pi k}{\tau}\right) = b_k$, for $k \in \mathcal{K}$.

The elements of the Fisher information matrix are found as in the sinc case, using the equivalence shown in (4.109) and the properties for the coefficients of the derivative of $\eta(t)$ and of its shifted version by t_0. The only additional consideration is that, when computing the elements of the anti-diagonal, we encounter a term of the form $\sum_{k \in \mathcal{K}} k|b_k|^2$. This is always equal to zero as long as $|b_k| = |b_{-k}|$ which is true, for instance, if we want to design real filters, for which $b_k = b^*_{-k}$. Thus

$$\text{CRB}(\boldsymbol{\theta}) = \begin{pmatrix} \frac{1}{N}\left(\frac{\sigma}{a_0}\right)^2\left(\frac{\tau}{2\pi}\right)^2\frac{1}{\sum_{k \in \mathcal{K}} k^2|b_k|^2} & 0 \\ 0 & \sigma^2\frac{1}{N}\frac{1}{\sum_{k \in \mathcal{K}}|b_k|^2} \end{pmatrix}. \qquad (4.114)$$

The uncertainty in the location is

$$\frac{\Delta t_0}{\tau} \geq \frac{1}{2\pi}\sqrt{\frac{1}{N}\frac{1}{\sum_{k \in \mathcal{K}} k^2|b_k|^2}}\text{PSNR}^{-\frac{1}{2}}, \qquad (4.115)$$

and the uncertainty in the amplitude

$$\frac{\Delta a_0}{|a_0|} \geq \sqrt{\frac{1}{N} \frac{1}{\sum_{k \in \mathcal{K}} |b_k|^2}} \text{PSNR}^{-\frac{1}{2}}. \quad (4.116)$$

Cramér–Rao bounds for B-splines

We now derive the lower bounds on the variances when estimating the location t_0 and amplitude a_0 of a single τ-periodic Dirac when the sampling kernel is a B-spline. We restrict the analysis to the shortest possible B-spline capable of sampling one Dirac, i.e. the first-order B-spline ($P=1$) obtained as the convolution of two box functions. It has the following form:

$$\beta_1(t) = \begin{cases} t, & 0 \leq t < 1, \\ 2-t, & 1 \leq t < 2. \end{cases} \quad (4.117)$$

In the absence of noise, the samples taken at uniform intervals of time T can be expressed as $y_n = a_0 \sum_{m \in \mathbb{Z}} \beta_1(t_0/T - n - mN) = f(\boldsymbol{\theta}, n)$ where we have used $\tau = NT$.

If we want to sample an infinitely long signal with a finite support kernel, we need to have zero samples in between blocks of nonzero samples. For a kernel of size $L = P+1$ we need at least 1 zero per period $\tau \geq (L+1)T \leftrightarrow N \geq L+1$. We assume the only nonzero samples are located in the positions $n = 0, \ldots, L-1$ (or we would do a circular shift otherwise).

We are working with a finite support kernel of length $L = P+1 = 2$. This allows us to remove the dependence of the Fisher information matrix on m, since fixing $t_0 \in [PT, (P+1)T) = [T, 2T)$, which is equivalent to $n = 0, \ldots, L-1$, makes the only possible m value to be equal to zero.

We can now evaluate the terms of the Fisher information matrix, which has a form identical to (4.108). Contrary to the previous sections, now we have to work in the time domain, using the definition of the B-spline (4.117) and of its derivative. We have finite-length sums over n, so it is possible to derive closed-form results. For example, the first element of the diagonal can be calculated as

$$\sigma^{-2} \sum_{n=0}^{1} \left(\frac{a_0}{T} \beta_1'\left(\frac{t_0}{T} - n\right) \right)^2 = \sigma^{-2} \left(\frac{a_0}{T}\right)^2 [1^2 + 1^2] \quad (4.118)$$

$$[\mathbf{I}(\boldsymbol{\theta})]_{11} = 2\sigma^{-2} \left(\frac{a_0}{T}\right)^2.$$

Once we obtain all the terms, the CRB can be found by inverting the Fisher information matrix. In this scenario, the bounds depend on t_0:

$$\text{CRB}(\boldsymbol{\theta}) = \begin{pmatrix} (2t_0^2 - 6Tt_0 + 5T^2)\left(\frac{\sigma}{a_0}\right)^2 & (3T - 2t_0)\frac{\sigma^2}{a_0} \\ (3T - 2t_0)\frac{\sigma^2}{a_0} & 2\sigma^2 \end{pmatrix}. \quad (4.119)$$

In order to remove the dependence on t_0, we can consider various options. For instance, we may calculate the expected value of $\mathrm{CRB}(\boldsymbol{\theta})$ assuming that t_0 is uniformly distributed over τ. This leads to:

$$\frac{\Delta t_0}{\tau} \geq \frac{1}{N}\sqrt{\frac{2}{3}}\mathrm{PSNR}^{-\frac{1}{2}}, \tag{4.120}$$

$$\frac{\Delta a_0}{|a_0|} \geq \sqrt{2}\mathrm{PSNR}^{-\frac{1}{2}}. \tag{4.121}$$

Cramér–Rao bounds for E-splines

To conclude, we derive lower bounds on the variances of the estimated location and amplitude for a τ-periodic single Dirac when the sampling kernel is an E-spline. The method is the same as that explained for B-splines, but requires further assumptions.

We restrict the analysis to cardinal exponential splines, which we also assume to be trigonometric [35]. The first property means that the exponential splines are defined on a uniform grid, and the second property that the complex parameters are purely imaginary and equally spaced around the origin, which yields a real-valued function.

We need to be careful when working with E-splines for two main reasons. The first is that the periodicity of the complex exponentials causes the moments to be periodic too. This imposes a limit in the locations that can be retrieved. The second is that E-splines are no longer a basis for certain combinations of the complex parameters [35]. These conditions, plus the fact that we want that the exponential reproduction formula coefficients form an orthogonal basis with $\omega_0 = 2\pi/N$, translate into a bound for t_0 which has to satisfy $t_0 < NT/2 = \tau/2$, and for the number of samples, which requires $N > \max(P+1, 2P)$. For a more detailed explanation of these conditions, we refer to [56].

If we focus on the first-order real E-spline, then it is possible to derive a closed-form expression for the CRB. Note that this function is obtained through convolution of the zero-order components having complex parameters $\pm j\omega_0$. The obtained kernel has the following form:

$$e_1(t) = \begin{cases} \dfrac{\sin(\omega_0 t)}{\omega_0}, & 0 \leq t < 1, \\ -\dfrac{\sin(\omega_0(t-2))}{\omega_0}, & 1 \leq t < 2. \end{cases} \tag{4.122}$$

The CRB can be obtained by inverting the Fisher information matrix, derived similarly to the B-spline case. In this situation, again the bounds depend on t_0. Calculating the average values leads to:

$$\frac{\Delta t_0}{\tau} \geq \frac{1}{N}\sqrt{\frac{\omega_0 - \cos\omega_0 \sin\omega_0}{\omega_0 \sin^2\omega_0}}\mathrm{PSNR}^{-\frac{1}{2}}, \tag{4.123}$$

$$\frac{\Delta a_0}{|a_0|} \geq \sqrt{\omega_0\frac{\omega_0 + \cos\omega_0 \sin\omega_0}{\sin^2\omega_0}}\mathrm{PSNR}^{-\frac{1}{2}}. \tag{4.124}$$

References

[1] E. T. Whittaker. On the functions which are represented by the expansions of the interpolation theory. *Proc. Roy Soc Edin A*, 35:181–194, 1915.

[2] H. Nyquist, Certain topics in telegraph transmission theory. *Trans AIEE*, 47:617–644, 1928.

[3] C. Shannon. A mathematical theory of communication. *Bell Syst Tech J*. 27:379–423, 623–656, 1948.

[4] M. Vetterli, P. Marziliano, and T. Blu. Sampling signals with finite rate of innovation. *IEEE Trans Signal Proc*, 50:1417–1428, 2002.

[5] Y. C. Eldar and T. Michaeli. Beyond bandlimited sampling. *IEEE Sig Proc Mag*, 26:48–68, 2009.

[6] M. Unser. Sampling – 50 years after Shannon, in *Proc IEEE*, pp. 569–587, 2000.

[7] Y. M. Lu and M. N. Do. A theory for sampling signals from a union of subspaces. *IEEE Trans Sig Proc*, 56(6):2334–2345, 2008.

[8] M. Mishali and Y. C. Eldar. Robust recovery of signals from a structured union of subspaces. *IEEE Trans Inf Theory*, 55:5302–5316, 2009.

[9] P. L. Dragotti, M. Vetterli, and T. Blu. Sampling moments and reconstructing signals of finite rate of innovation: Shannon meets Strang-Fix. *IEEE Tran Sig Proc*, 55(5):1741–1757, 2007.

[10] R. Tur, Y. C. Eldar, and Z. Friedman. Innovation rate sampling of pulse streams with application to ultrasound imaging. *IEEE Trans Sig Proc*, 59(4):1827–1842, 2011.

[11] K. Gedalyahu, R. Tur, and Y. C. Eldar. Multichannel sampling of pulse streams at the rate of innovation. *IEEE Trans Sig Proc*, 59(4):1491–1504, 2011.

[12] Z. Ben-Haim, T. Michaeli, and Y. C. Eldar. Performance bounds and design criteria for estimating finite rate of innovation signals. Sensor Array and Multichannel Processing Workshop (SAM), Jerusalem, October 4–7, 2010.

[13] I. Maravic and M. Vetterli. Sampling and reconstruction of signals with finite rate of innovation in the presence of noise. *IEEE Trans Sig Proc*, 53:2788–2805, 2005.

[14] V. Y. F. Tan and V. K. Goyal. Estimating signals with finite rate of innovation from noisy samples: A stochastic algorithm. *IEEE Trans Sig Proc*, 56(10):5135–5146, 2008.

[15] A. Erdozain and P. M. Crespo. A new stochastic algorithm inspired on genetic algorithms to estimate signals with finite rate of innovation from noisy samples. *Sig Proc*, 90:134–144, 2010.

[16] J. Berent, P. L. Dragotti, and T. Blu. "Sampling piecewise sinusoidal signals with finite rate of innovation methods." *IEEE Trans Sig Proc*, 58(2):613–625, 2010.

[17] K. Gedalyahu and Y. C. Eldar, Time delay estimation from low rate samples: A union of subspaces approach. *IEEE Trans Sig Proc*, 58:3017–3031, 2010.

[18] P. Stoica and R. L. Moses. *Introduction to Spectral Analysis*. Englewood Cliffs, NJ: Prentice-Hall, 2000.

[19] T. Blu, P. L. Dragotti, M. Vetterli, P. Marziliano, and L. Coulot. Sparse sampling of signal innovations. *IEEE Sig Proc Mag*, 25(2):31–40, 2008.

[20] H. Akhondi Asl, P. L. Dragotti, and L. Baboulaz. Multichannel sampling of signals with finite rate of innovation. *IEEE Sig Proc Letters*, 17:762–765, 2010.

[21] E. Matusiak and Y. C. Eldar, Sub-Nyquist sampling of short pulses: Part I, to appear in *IEEE Trans Sig Proc* arXiv:1010.3132v1.

[22] A. Hormati, O. Roy, Y. M. Lu, and M. Vetterli. Distributed sampling of signals linked by sparse filtering: Theory and applications. *Trans Sig Proc*, 58(3):1095–1109, 2010.

[23] I. Maravic and M. Vetterli. Exact sampling results for some classes of parametric non-bandlimited 2-D signals. *IEEE Trans Sig Proc*, 52(1):175–189, 2004.

[24] P. Shukla and P. L. Dragotti, Sampling schemes for multidimensional signals with finite rate of innovation. *IEEE Trans Sig Proc*, 2006.

[25] L. Baboulaz and P. L. Dragotti. Distributed acquisition and image super-resolution based on continuous moments from samples. *Proc IEEE Int Conf Image Proc (ICIP)*, pp. 3309–3312, 2006.

[26] L. Baboulaz and P. L. Dragotti. Exact feature extraction using finite rate of innovation principles with an application to image super-resolution. *IEEE Trans Image Proc*, 18(2):281–298, 2009.

[27] W. U. Bajwa, K. Gedalyahu, and Y. C. Eldar. Identification of parametric underspread linear systems and super-resolution radar, to appear in *IEEE Trans Sig Proc*, 59(6):2548–2561, 2011.

[28] I. Maravic, J. Kusuma, and M. Vetterli. Low-sampling rate UWB channel characterization and synchronization, *J Commu Networks KOR, special issue on ultra-wideband systems*, 5(4):319–327, 2003.

[29] I. Maravic, M. Vetterli, and K. Ramchandran. Channel estimation and synchronization with sub-Nyquist sampling and application to ultra-wideband systems. *Proc IEEE Int Symp Circuits Syst*, 5:381–384, 2004.

[30] M. Vetterli and J. Kovacevic. *Wavelets and Subband Coding*. Englewood Cliffs, NJ: Prentice Hall, 1995.

[31] I. Gohberg and S. Goldberg. *Basic Operator Theory*. Boston, MA: Birkhäuser, 1981.

[32] Y. Hua and T. K. Sarkar, Matrix pencil method for estimating parameters of exponentially damped/undamped sinusoids in noise. *IEEE Trans Acoust, Speech, Sig Proc*, 70:1272–1281, 1990.

[33] S. Y. Kung, K. S. Arun, and D. V. B. Rao. State-space and singular-value decomposition-based approximation methods for the harmonic retrieval problem. *J Opt Soci Amer*, 73:1799–1811, 1983.

[34] B. D. Rao and K. S. Arun. Model based processing of signals: A state space approach, *Proc IEEE*, 80(2):283–309, 1992.

[35] M. Unser and T. Blu. Cardinal exponential splines: Part I – Theory and filtering algorithms. *IEEE Trans Sig Proc*, 53:1425–1438, 2005.

[36] J. A. Urigüen, P. L. Dragotti, and T. Blu. On the exponential reproducing kernels for sampling signals with finite rate of innovation. *Proc 9th Int Workshop Sampling Theory Appl (SampTA'11)*, Singapore, May 2–6, 2011.

[37] G. Strang and G. Fix, Fourier analysis of the finite element variational method. *Construc Asp Funct Anal*; 796–830, 1971.

[38] V. Chaisinthop and P. L. Dragotti. Semi-parametric compression of piecewise smooth functions. *Proc. Europ Sig Proc Conf (EUSIPCO)*, 2009.

[39] V. Chaisinthop and P. L. Dragotti, Centralized and distributed semi-parametric compression of piecewise smooth functions. *IEEE Trans Sig Proc*, 59(7):3071–3085, 2011.

[40] J. Kusuma and V. Goyal, Multichannel sampling of parametric signals with a successive approximation property. *IEEE Int Conf Image Proc (ICIP2006)*: 1265–1268, 2006.

[41] H. Olkkonen and J. Olkkonen. Measurement and reconstruction of impulse train by parallel exponential filters. *IEEE Sig Proc Letters*, 15:241–244, 2008.

[42] M. Mishali, Y. C. Eldar, O. Dounaevsky, and E. Shoshan. Xampling: Analog to digital at sub-Nyquist rates. *IET Circ, Devices Syst*, 5:8–20, 2011.

[43] M. Mishali and Y. C. Eldar. From theory to practice: Sub-Nyquist sampling of sparse wideband analog signals. *IEEE J Sel Top Sig Proc*, 4:375–391, 2010.

[44] Y. C. Eldar and V. Pohl. Recovering signals from lowpass data. *IEEE Trans Sig Proc*, 58(5):2636–2646, 2010.

[45] M. Mishali and Y. C. Eldar. Blind multiband signal reconstruction: Compressed sensing for analog signals, *IEEE Trans Sig Proc*, 57(3):993–1009, 2009.

[46] R. Roy and T. Kailath. ESPRIT-estimation of signal parameters via rotational invariance techniques. *IEEE Trans Acoust, Speech Sig Proc*, 37:984–995, 1989.

[47] R. Schmidt. Multiple emitter location and signal parameter estimation. *IEEE Trans Antennas Propagation*, 34:276–280, 1986.

[48] S. M. Kay. *Fundamentals of Statistical Sig Processing: Estimation Theory*. Englewood Cliffs, NJ: Prentice Hall, 1993.

[49] S. Mallat. *A Wavelet Tour of Signal Processing*. Academic Press, 1998.

[50] A. Cohen, I. Daubechies, O. G. Guleryuzz, and M. T. Orchard. On the importance of combining wavelet-based nonlinear approximation with coding strategies. *IEEE Trans Inf Theory*, 48:1895–1921, 2002.

[51] S. Farsiu, D. Robinson, M. Elad, and P. Milanfar. Advances and challenges in super-resolution. *Int J Imaging Syst Technol*, 13(10):1327–1344, 2004.

[52] H. Meyr, M. Moeneclaey, and S. A. Fechtel. *Digital Communication Receivers: Synchronization, Channel Estimation, and Signal Processing*. New York: Wiley-Interscience, 1997.

[53] S. E. Bensley and B. Aazhang. Subspace-based channel estimation for code division multiple access communication systems. *IEEE Trans Commun*, 44(8):1009–1020, 1996.

[54] E. G. Strom, S. Parkvall, S. L. Miller, and B. E. Ottersten. DS-CDMA synchronization in time-varying fading channels. *IEEE J Selected Areas in Commun*, 14(8):1636–1642, 1996.

[55] L. Coulot, M. Vetterli, T. Blu, and P. L. Dragotti, Sampling signals with finite rate of innovation in the presence of noise. Tech. rep., École Polytechnique Federale de Lausanne, Switzerland, 2007.

[56] F. J. Homann and P. L. Dragotti. Robust sampling of 'almost' sparse signals. Tech. rep., Imperial College London, 2008.

5 Introduction to the non-asymptotic analysis of random matrices

Roman Vershynin

This is a tutorial on some basic non-asymptotic methods and concepts in random matrix theory. The reader will learn several tools for the analysis of the extreme singular values of random matrices with independent rows or columns. Many of these methods sprung off from the development of geometric functional analysis since the 1970s. They have applications in several fields, most notably in theoretical computer science, statistics and signal processing. A few basic applications are covered in this text, particularly for the problem of estimating covariance matrices in statistics and for validating probabilistic constructions of measurement matrices in compressed sensing. This tutorial is written particularly for graduate students and beginning researchers in different areas, including functional analysts, probabilists, theoretical statisticians, electrical engineers, and theoretical computer scientists.

5.1 Introduction

Asymptotic and non-asymptotic regimes

Random matrix theory studies properties of $N \times n$ matrices A chosen from some distribution on the set of all matrices. As dimensions N and n grow to infinity, one observes that the spectrum of A tends to stabilize. This is manifested in several *limit laws*, which may be regarded as random matrix versions of the central limit theorem. Among them is Wigner's semicircle law for the eigenvalues of symmetric Gaussian matrices, the circular law for Gaussian matrices, the Marchenko–Pastur law for Wishart matrices $W = A^*A$ where A is a Gaussian matrix, the Bai–Yin and Tracy–Widom laws for the extreme eigenvalues of Wishart matrices W. The books [51, 5, 23, 6] offer thorough introduction to the classical problems of random matrix theory and its fascinating connections.

The asymptotic regime where the dimensions $N, n \to \infty$ is well suited for the purposes of statistical physics, e.g. when random matrices serve as finite-dimensional models of infinite-dimensional operators. But in some other areas including statistics, geometric functional analysis, and compressed sensing, the limiting regime may not be very useful [69]. Suppose, for example, that we ask about the largest singular value $s_{\max}(A)$ (i.e.

Compressed Sensing: Theory and Applications, ed. Yonina C. Eldar and Gitta Kutyniok. Published by Cambridge University Press. © Cambridge University Press 2012.

the largest eigenvalue of $(A^*A)^{1/2}$); to be specific assume that A is an $n \times n$ matrix whose entries are independent standard normal random variables. The asymptotic random matrix theory answers this question as follows: the Bai–Yin law (see Theorem 5.31) states that

$$s_{\max}(A)/2\sqrt{n} \to 1 \quad \text{almost surely}$$

as the dimension $n \to \infty$. Moreover, the limiting distribution of $s_{\max}(A)$ is known to be the Tracy–Widom law (see [71, 27]). In contrast to this, a non-asymptotic answer to the same question is the following: in *every* dimension n, one has

$$s_{\max}(A) \le C\sqrt{n} \quad \text{with probability at least } 1 - e^{-n},$$

here C is an absolute constant (see Theorems 5.32 and 5.39). The latter answer is less precise (because of an absolute constant C) but more quantitative because for fixed dimensions n it gives an exponential probability of success.[1] This is the kind of answer we will seek in this text – guarantees up to absolute constants in all dimensions, and with large probability.

Tall matrices are approximate isometries

The following heuristic will be our guideline: *tall random matrices should act as approximate isometries*. So, an $N \times n$ random matrix A with $N \gg n$ should act almost like an isometric embedding of ℓ_2^n into ℓ_2^N:

$$(1-\delta)K\|x\|_2 \le \|Ax\|_2 \le (1+\delta)K\|x\|_2 \quad \text{for all } x \in \mathbb{R}^n$$

where K is an appropriate normalization factor and $\delta \ll 1$. Equivalently, this says that all the singular values of A are close to each other:

$$(1-\delta)K \le s_{\min}(A) \le s_{\max}(A) \le (1+\delta)K,$$

where $s_{\min}(A)$ and $s_{\max}(A)$ denote the smallest and the largest singular values of A. Yet equivalently, this means that tall matrices are well conditioned: the *condition number* of A is $\kappa(A) = s_{\max}(A)/s_{\min}(A) \le (1+\delta)/(1-\delta) \approx 1$.

In the asymptotic regime and for random matrices with independent entries, our heuristic is justified by Bai–Yin's law, which is Theorem 5.31 below. Loosely speaking, it states that as the dimensions N, n increase to infinity while the aspect ratio N/n is fixed, we have

$$\sqrt{N} - \sqrt{n} \approx s_{\min}(A) \le s_{\max}(A) \approx \sqrt{N} + \sqrt{n}. \tag{5.1}$$

In these notes, we study $N \times n$ random matrices A with independent rows or independent columns, but not necessarily independent entries. We develop non-asymptotic versions

---

[1] For this specific model (Gaussian matrices), Theorems 5.32 and 5.35 even give a sharp absolute constant $C \approx 2$ here. But the result mentioned here is much more general as we will see later; it only requires independence of rows or columns of A.

of (5.1) for such matrices, which should hold for all dimensions N and n. The desired results should have the form

$$\sqrt{N} - C\sqrt{n} \le s_{\min}(A) \le s_{\max}(A) \le \sqrt{N} + C\sqrt{n} \qquad (5.2)$$

with large probability, e.g. $1 - e^{-N}$, where C is an absolute constant.[2] For tall matrices, where $N \gg n$, both sides of this inequality would be close to each other, which would guarantee that A is an approximate isometry.

Models and methods
We shall study quite general models of random matrices – those with independent rows or independent columns that are sampled from high-dimensional distributions. We will place either strong moment assumptions on the distribution (sub-gaussian growth of moments), or no moment assumptions at all (except finite variance). This leads us to four types of main results:

1. Matrices with independent sub-gaussian rows: Theorem 5.39
2. Matrices with independent heavy-tailed rows: Theorem 5.41
3. Matrices with independent sub-gaussian columns: Theorem 5.58
4. Matrices with independent heavy-tailed columns: Theorem 5.62

These four models cover many natural classes of random matrices that occur in applications, including random matrices with independent entries (Gaussian and Bernoulli in particular) and random sub-matrices of orthogonal matrices (random Fourier matrices in particular).

The analysis of these four models is based on a variety of tools of probability theory and geometric functional analysis, most of which have not been covered in the texts on the "classical" random matrix theory. The reader will learn basics on sub-gaussian and sub-exponential random variables, isotropic random vectors, large deviation inequalities for sums of independent random variables, extensions of these inequalities to random matrices, and several basic methods of high-dimensional probability such as symmetrization, decoupling, and covering (ε-net) arguments.

Applications
In these notes we shall emphasize two applications, one in statistics and one in compressed sensing. Our analysis of random matrices with independent rows immediately applies to a basic problem in statistics – *estimating covariance matrices* of high-dimensional distributions. If a random matrix A has i.i.d. rows A_i, then $A^*A = \sum_i A_i \otimes A_i$ is the *sample covariance matrix*. If A has independent columns A_j, then $A^*A = (\langle A_j, A_k \rangle)_{j,k}$ is the *Gram matrix*. Thus our analysis of the row-independent and column-independent models can be interpreted as a study of sample covariance matrices and Gram matrices of high-dimensional distributions. We will see in Section 5.4.3 that

---

[2] More accurately, we should expect $C = O(1)$ to depend on easily computable quantities of the distribution, such as its moments. This will be clear from the context.

for a general distribution in $\mathbb{R}^n$, its covariance matrix can be estimated from a sample of size $N = O(n \log n)$ drawn from the distribution. Moreover, for sub-gaussian distributions we have an even better bound $N = O(n)$. For low-dimensional distributions, much fewer samples are needed – if a distribution lies close to a subspace of dimension r in $\mathbb{R}^n$, then a sample of size $N = O(r \log n)$ is sufficient for covariance estimation.

In compressed sensing, the best known measurement matrices are random. A sufficient condition for a matrix to succeed for the purposes of compressed sensing is given by the *restricted isometry property*. Loosely speaking, this property demands that all submatrices of given size be well-conditioned. This fits well in the circle of problems of the non-asymptotic random matrix theory. Indeed, we will see in Section 5.6 that all basic models of random matrices are nice restricted isometries. These include Gaussian and Bernoulli matrices, more generally all matrices with sub-gaussian independent entries, and even more generally all matrices with sub-gaussian independent rows or columns. Also, the class of restricted isometries includes random Fourier matrices, more generally random sub-matrices of bounded orthogonal matrices, and even more generally matrices whose rows are independent samples from an isotropic distribution with uniformly bounded coordinates.

Related sources

This chapter is a tutorial rather than a survey, so we focus on explaining methods rather than results. This forces us to make some concessions in our choice of the subjects. *Concentration of measure* and its applications to random matrix theory are only briefly mentioned. For an introduction into concentration of measure suitable for a beginner, see [9] and [49, Chapter 14]; for a thorough exposition see [56, 43]; for connections with random matrices see [21, 44]. The monograph [45] also offers an introduction into concentration of measure and related probabilistic methods in analysis and geometry, some of which we shall use in these notes.

We completely avoid the important (but more difficult) model of *symmetric random matrices* with independent entries on and above the diagonal. Starting from the work of Füredi and Komlos [29], the largest singular value (the spectral norm) of symmetric random matrices has been a subject of study in many works; see e.g. [50, 83, 58] and the references therein.

We also did not even attempt to discuss sharp small *deviation inequalities* (of Tracy–Widom type) for the extreme eigenvalues. Both these topics and much more are discussed in the surveys [21, 44, 69], which serve as bridges between asymptotic and non-asymptotic problems in random matrix theory.

Because of the absolute constant C in (5.2), our analysis of the smallest singular value (the *"hard edge"*) will only be useful for sufficiently tall matrices, where $N \geq C^2 n$. For square and almost square matrices, the hard edge problem will be only briefly mentioned in Section 5.3. The surveys [76, 69] discuss this problem at length, and they offer a glimpse of connections to other problems of random matrix theory and additive combinatorics.

Many of the results and methods presented in these notes are known in one form or another. Some of them are published while some others belong to the folklore of probability in Banach spaces, geometric functional analysis, and related areas. When available, historic references are given in Section 5.7.

Acknowledgements

The author is grateful to the colleagues who made a number of improving suggestions for the earlier versions of the manuscript, in particular to Richard Chen, Alexander Litvak, Deanna Needell, Holger Rauhut, S. V. N. Vishwanathan, and the anonymous referees. Special thanks are due to Ulas Ayaz and Felix Krahmer who thoroughly read the entire text, and whose numerous comments led to significant improvements of this tutorial.

5.2 Preliminaries

5.2.1 Matrices and their singular values

The main object of our study will be an $N \times n$ matrix A with real or complex entries. We shall state all results in the real case; the reader will be able to adjust them to the complex case as well. Usually but not always one should think of tall matrices A, those for which $N \geq n > 1$. By passing to the adjoint matrix A^*, many results can be carried over to "flat" matrices, those for which $N \leq n$.

It is often convenient to study A through the $n \times n$ symmetric positive semidefinite matrix, the matrix A^*A. The eigenvalues of $|A| := \sqrt{A^*A}$ are therefore non-negative real numbers. Arranged in a non-decreasing order, they are called the *singular values*[3] of A and denoted $s_1(A) \geq \cdots \geq s_n(A) \geq 0$. Many applications require estimates on the extreme singular values

$$s_{\max}(A) := s_1(A), \quad s_{\min}(A) := s_n(A).$$

The smallest singular value is only of interest for tall matrices, since for $N < n$ one automatically has $s_{\min}(A) = 0$.

Equivalently, $s_{\max}(A)$ and $s_{\min}(A)$ are respectively the smallest number M and the largest number m such that

$$m\|x\|_2 \leq \|Ax\|_2 \leq M\|x\|_2 \quad \text{for all } x \in \mathbb{R}^n. \tag{5.3}$$

In order to interpret this definition geometrically, we look at A as a linear operator from $\mathbb{R}^n$ into $\mathbb{R}^N$. The Euclidean distance between any two points in $\mathbb{R}^n$ can increase by at most the factor $s_{\max}(A)$ and decrease by at most the factor $s_{\max}(A)$ under the action of A. Therefore, the extreme singular values control the distortion of the Euclidean geometry under the action of A. If $s_{\max}(A) \approx s_{\min}(A) \approx 1$ then A acts as an *approximate isometry*, or more accurately an approximate isometric embedding of ℓ_2^n into ℓ_2^N.

---

[3] In the literature, singular values are also called *s-numbers*.

The extreme singular values can also be described in terms of the *spectral norm of A*, which is by definition

$$\|A\| = \|A\|_{\ell_2^n \to \ell_2^N} = \sup_{x \in \mathbb{R}^n \setminus \{0\}} \frac{\|Ax\|_2}{\|x\|_2} = \sup_{x \in S^{n-1}} \|Ax\|_2. \qquad (5.4)$$

Equation (5.3) gives a link between the extreme singular values and the spectral norm:

$$s_{\max}(A) = \|A\|, \quad s_{\min}(A) = 1/\|A^\dagger\|$$

where $A^\dagger$ denotes the pseudoinverse of A; if A is invertible then $A^\dagger = A^{-1}$.

5.2.2 Nets

Nets are convenient means to discretize compact sets. In our study we will mostly need to discretize the unit Euclidean sphere S^{n-1} in the definition of the spectral norm (5.4). Let us first recall a general definition of an ε-net.

DEFINITION 5.1 (Nets, covering numbers) *Let (X,d) be a metric space and let $\varepsilon > 0$. A subset $\mathcal{N}_\varepsilon$ of X is called an ε-net of X if every point $x \in X$ can be approximated to within ε by some point $y \in \mathcal{N}_\varepsilon$, i.e. so that $d(x,y) \le \varepsilon$. The minimal cardinality of an ε-net of X, if finite, is denoted $\mathcal{N}(X,\varepsilon)$ and is called the* covering number *of X (at scale ε).*[4]

From a characterization of compactness we remember that X is compact if and only if $\mathcal{N}(X,\varepsilon) < \infty$ for each $\varepsilon > 0$. A quantitative estimate on $\mathcal{N}(X,\varepsilon)$ would give us a *quantitative version of compactness* of X.[5] Let us therefore take a simple example of a metric space, the unit Euclidean sphere S^{n-1} equipped with the Euclidean metric[6] $d(x,y) = \|x-y\|_2$, and estimate its covering numbers.

LEMMA 5.2 (Covering numbers of the sphere) *The unit Euclidean sphere S^{n-1} equipped with the Euclidean metric satisfies for every $\varepsilon > 0$ that*

$$\mathcal{N}(S^{n-1}, \varepsilon) \le \left(1 + \frac{2}{\varepsilon}\right)^n.$$

Proof. This is a simple *volume argument*. Let us fix $\varepsilon > 0$ and choose $\mathcal{N}_\varepsilon$ to be a maximal ε-separated subset of S^{n-1}. In other words, $\mathcal{N}_\varepsilon$ is such that $d(x,y) \ge \varepsilon$ for all $x,y \in \mathcal{N}_\varepsilon$, $x \ne y$, and no subset of S^{n-1} containing $\mathcal{N}_\varepsilon$ has this property.[7]

---

[4] Equivalently, $\mathcal{N}(X,\varepsilon)$ is the minimal number of balls with radii ε and with centers in X needed to cover X.

[5] In statistical learning theory and geometric functional analysis, $\log \mathcal{N}(X,\varepsilon)$ is called *the metric entropy of X*. In some sense it measures the "complexity" of metric space X.

[6] A similar result holds for the geodesic metric on the sphere, since for small ε these two distances are equivalent.

[7] One can in fact construct $\mathcal{N}_\varepsilon$ inductively by first selecting an arbitrary point on the sphere, and at each next step selecting a point that is at distance at least ε from those already selected. By compactness, this algorithm will terminate after finitely many steps and it will yield a set $\mathcal{N}_\varepsilon$ as we required.

The maximality property implies that $\mathcal{N}_\varepsilon$ is an ε-net of S^{n-1}. Indeed, otherwise there would exist $x \in S^{n-1}$ that is at least ε-far from all points in $\mathcal{N}_\varepsilon$. So $\mathcal{N}_\varepsilon \cup \{x\}$ would still be an ε-separated set, contradicting the minimality property.

Moreover, the separation property implies via the triangle inequality that the balls of radii $\varepsilon/2$ centered at the points in $\mathcal{N}_\varepsilon$ are disjoint. On the other hand, all such balls lie in $(1+\varepsilon/2)B_2^n$ where B_2^n denotes the unit Euclidean ball centered at the origin. Comparing the volume gives $\text{vol}\left(\frac{\varepsilon}{2}B_2^n\right) \cdot |\mathcal{N}_\varepsilon| \leq \text{vol}\left((1+\frac{\varepsilon}{2})B_2^n\right)$. Since $\text{vol}\left(rB_2^n\right) = r^n \text{vol}(B_2^n)$ for all $r \geq 0$, we conclude that $|\mathcal{N}_\varepsilon| \leq (1+\frac{\varepsilon}{2})^n/(\frac{\varepsilon}{2})^n = (1+\frac{2}{\varepsilon})^n$ as required. □

Nets allow us to reduce the complexity of computations with linear operators. One such example is the computation of the spectral norm. To evaluate the spectral norm by definition (5.4) one needs to take the supremum over the whole sphere S^{n-1}. However, one can essentially replace the sphere by its ε-net:

LEMMA 5.3 (Computing the spectral norm on a net) *Let A be an $N \times n$ matrix, and let $\mathcal{N}_\varepsilon$ be an ε-net of S^{n-1} for some $\varepsilon \in [0,1)$. Then*

$$\max_{x \in \mathcal{N}_\varepsilon} \|Ax\|_2 \leq \|A\| \leq (1-\varepsilon)^{-1} \max_{x \in \mathcal{N}_\varepsilon} \|Ax\|_2.$$

Proof. The lower bound in the conclusion follows from the definition. To prove the upper bound let us fix $x \in S^{n-1}$ for which $\|A\| = \|Ax\|_2$, and choose $y \in \mathcal{N}_\varepsilon$ which approximates x as $\|x - y\|_2 \leq \varepsilon$. By the triangle inequality we have $\|Ax - Ay\|_2 \leq \|A\|\|x - y\|_2 \leq \varepsilon\|A\|$. It follows that

$$\|Ay\|_2 \geq \|Ax\|_2 - \|Ax - Ay\|_2 \geq \|A\| - \varepsilon\|A\| = (1-\varepsilon)\|A\|.$$

Taking the maximum over all $y \in \mathcal{N}_\varepsilon$ in this inequality, we complete the proof. □

A similar result holds for symmetric $n \times n$ matrices A, whose spectral norm can be computed via the associated quadratic form: $\|A\| = \sup_{x \in S^{n-1}} |\langle Ax, x \rangle|$. Again, one can essentially replace the sphere by its ε-net:

LEMMA 5.4 (Computing the spectral norm on a net) *Let A be a symmetric $n \times n$ matrix, and let $\mathcal{N}_\varepsilon$ be an ε-net of S^{n-1} for some $\varepsilon \in [0,1)$. Then*

$$\|A\| = \sup_{x \in S^{n-1}} |\langle Ax, x \rangle| \leq (1-2\varepsilon)^{-1} \sup_{x \in \mathcal{N}_\varepsilon} |\langle Ax, x \rangle|.$$

Proof. Let us choose $x \in S^{n-1}$ for which $\|A\| = |\langle Ax, x \rangle|$, and choose $y \in \mathcal{N}_\varepsilon$ which approximates x as $\|x - y\|_2 \leq \varepsilon$. By the triangle inequality we have

$$|\langle Ax, x \rangle - \langle Ay, y \rangle| = |\langle Ax, x - y \rangle + \langle A(x-y), y \rangle|$$
$$\leq \|A\|\|x\|_2\|x-y\|_2 + \|A\|\|x-y\|_2\|y\|_2 \leq 2\varepsilon\|A\|.$$

It follows that $|\langle Ay, y \rangle| \geq |\langle Ax, x \rangle| - 2\varepsilon\|A\| = (1-2\varepsilon)\|A\|$. Taking the maximum over all $y \in \mathcal{N}_\varepsilon$ in this inequality completes the proof. □

5.2.3 Sub-gaussian random variables

In this section we introduce the class of sub-gaussian random variables,[8] those whose distributions are dominated by the distribution of a centered gaussian random variable. This is a convenient and quite wide class, which contains in particular the standard normal and all bounded random variables.

Let us briefly recall some of the well-known properties of the *standard normal random variable* X. The distribution of X has density $\frac{1}{\sqrt{2\pi}}e^{-x^2/2}$ and is denoted $N(0,1)$. Estimating the integral of this density between t and ∞ one checks that the tail of a standard normal random variable X decays super-exponentially:

$$\mathbb{P}\{|X| > t\} = \frac{2}{\sqrt{2\pi}} \int_t^\infty e^{-x^2/2}\, dx \leq 2e^{-t^2/2}, \quad t \geq 1, \tag{5.5}$$

see e.g. [26, Theorem 1.4] for a more precise two-sided inequality. The absolute moments of X can be computed as

$$(\mathbb{E}|X|^p)^{1/p} = \sqrt{2}\left[\frac{\Gamma((1+p)/2)}{\Gamma(1/2)}\right]^{1/p} = O(\sqrt{p}), \quad p \geq 1. \tag{5.6}$$

The moment generating function of X equals

$$\mathbb{E}\exp(tX) = e^{t^2/2}, \quad t \in \mathbb{R}. \tag{5.7}$$

Now let X be a general random variable. We observe that these three properties are equivalent – a super-exponential tail decay like in (5.5), the moment growth (5.6), and the growth of the moment generating function like in (5.7). We will then focus on the class of random variables that satisfy these properties, which we shall call sub-gaussian random variables.

LEMMA 5.5 (Equivalence of sub-gaussian properties) *Let X be a random variable. Then the following properties are equivalent with parameters $K_i > 0$ differing from each other by at most an absolute constant factor.*[9]

1. *Tails:* $\mathbb{P}\{|X| > t\} \leq \exp(1 - t^2/K_1^2)$ *for all* $t \geq 0$;
2. *Moments:* $(\mathbb{E}|X|^p)^{1/p} \leq K_2\sqrt{p}$ *for all* $p \geq 1$;
3. *Super-exponential moment:* $\mathbb{E}\exp(X^2/K_3^2) \leq e$.

Moreover, if $\mathbb{E}X = 0$ then properties 1–3 are also equivalent to the following one:

4. *Moment generating function:* $\mathbb{E}\exp(tX) \leq \exp(t^2 K_4^2)$ *for all* $t \in \mathbb{R}$.

---

[8] It would be more rigorous to say that we study *sub-gaussian probability distributions*. The same concerns some other properties of random variables and random vectors we study later in this text. However, it is convenient for us to focus on random variables and vectors because we will form random matrices out of them.

[9] The precise meaning of this equivalence is the following. There exists an absolute constant C such that property i implies property j with parameter $K_j \leq CK_i$ for any two properties $i, j = 1, 2, 3$.

Proof. **1 ⇒ 2** Assume property 1 holds. By homogeneity, rescaling X to X/K_1 we can assume that $K_1 = 1$. Recall that for every non-negative random variable Z, integration by parts yields the identity $\mathbb{E}Z = \int_0^\infty \mathbb{P}\{Z \geq u\}\,du$. We apply this for $Z = |X|^p$. After change of variables $u = t^p$, we obtain using property 1 that

$$\mathbb{E}|X|^p = \int_0^\infty \mathbb{P}\{|X| \geq t\}\,pt^{p-1}\,dt \leq \int_0^\infty e^{1-t^2} pt^{p-1}\,dt = \left(\frac{ep}{2}\right)\Gamma\left(\frac{p}{2}\right) \leq \left(\frac{ep}{2}\right)\left(\frac{p}{2}\right)^{p/2}.$$

Taking the pth root yields property 2 with a suitable absolute constant K_2.

2 ⇒ 3 Assume property 2 holds. As before, by homogeneity we may assume that $K_2 = 1$. Let $c > 0$ be a sufficiently small absolute constant. Writing the Taylor series of the exponential function, we obtain

$$\mathbb{E}\exp(cX^2) = 1 + \sum_{p=1}^\infty \frac{c^p \mathbb{E}(X^{2p})}{p!} \leq 1 + \sum_{p=1}^\infty \frac{c^p (2p)^p}{p!} \leq 1 + \sum_{p=1}^\infty (2c/e)^p.$$

The first inequality follows from property 2; in the second one we use $p! \geq (p/e)^p$. For small c this gives $\mathbb{E}\exp(cX^2) \leq e$, which is property 3 with $K_3 = c^{-1/2}$.

3 ⇒ 1 Assume property 3 holds. As before we may assume that $K_3 = 1$. Exponentiating and using Markov's inequality[10] and then property 3, we have

$$\mathbb{P}\{|X| > t\} = \mathbb{P}\{e^{X^2} \geq e^{t^2}\} \leq e^{-t^2}\mathbb{E}e^{X^2} \leq e^{1-t^2}.$$

This proves property 1 with $K_1 = 1$.

2 ⇒ 4 Let us now assume that $\mathbb{E}X = 0$ and property 2 holds; as usual we can assume that $K_2 = 1$. We will prove that property 4 holds with an appropriately large absolute constant $C = K_4$. This will follow by estimating Taylor series for the exponential function

$$\mathbb{E}\exp(tX) = 1 + t\mathbb{E}X + \sum_{p=2}^\infty \frac{t^p \mathbb{E}X^p}{p!} \leq 1 + \sum_{p=2}^\infty \frac{t^p p^{p/2}}{p!} \leq 1 + \sum_{p=2}^\infty \left(\frac{e|t|}{\sqrt{p}}\right)^p. \quad (5.8)$$

The first inequality here follows from $\mathbb{E}X = 0$ and property 2; the second one holds since $p! \geq (p/e)^p$. We compare this with Taylor's series for

$$\exp(C^2 t^2) = 1 + \sum_{k=1}^\infty \frac{(C|t|)^{2k}}{k!} \geq 1 + \sum_{k=1}^\infty \left(\frac{C|t|}{\sqrt{k}}\right)^{2k} = 1 + \sum_{p \in 2\mathbb{N}} \left(\frac{C|t|}{\sqrt{p/2}}\right)^p. \quad (5.9)$$

The first inequality here holds because $p! \leq p^p$; the second one is obtained by substitution $p = 2k$. One can show that the series in (5.8) is bounded by the series in (5.9). We conclude that $\mathbb{E}\exp(tX) \leq \exp(C^2 t^2)$, which proves property 4.

---

[10] This simple argument is sometimes called exponential Markov's inequality.

4 ⇒ 1 Assume property 4 holds; we can also assume that $K_4 = 1$. Let $\lambda > 0$ be a parameter to be chosen later. By exponential Markov inequality, and using the bound on the moment generating function given in property 4, we obtain

$$\mathbb{P}\{X \geq t\} = \mathbb{P}\{e^{\lambda X} \geq e^{\lambda t}\} \leq e^{-\lambda t}\mathbb{E}e^{\lambda X} \leq e^{-\lambda t + \lambda^2}.$$

Optimizing in λ and thus choosing $\lambda = t/2$ we conclude that $\mathbb{P}\{X \geq t\} \leq e^{-t^2/4}$. Repeating this argument for $-X$, we also obtain $\mathbb{P}\{X \leq -t\} \leq e^{-t^2/4}$. Combining these two bounds we conclude that $\mathbb{P}\{|X| \geq t\} \leq 2e^{-t^2/4} \leq e^{1-t^2/4}$. Thus property 1 holds with $K_1 = 2$. The lemma is proved. □

REMARK 5.6 1. The constants 1 and e in properties 1 and 3 respectively are chosen for convenience. Thus the value 1 can be replaced by any positive number and the value e can be replaced by any number greater than 1.
2. The assumption $\mathbb{E}X = 0$ is only needed to prove the necessity of property 4; the sufficiency holds without this assumption.

DEFINITION 5.7 (Sub-gaussian random variables) *A random variable X that satisfies one of the equivalent properties 1–3 in Lemma 5.5 is called a* sub-gaussian random variable. *The* sub-gaussian norm *of X, denoted $\|X\|_{\psi_2}$, is defined to be the smallest K_2 in property 2. In other words,*[11]

$$\|X\|_{\psi_2} = \sup_{p \geq 1} p^{-1/2}(\mathbb{E}|X|^p)^{1/p}.$$

The class of sub-gaussian random variables on a given probability space is thus a normed space. By Lemma 5.5, every sub-gaussian random variable X satisfies:

$$\mathbb{P}\{|X| > t\} \leq \exp(1 - ct^2/\|X\|_{\psi_2}^2) \quad \text{for all } t \geq 0; \quad (5.10)$$

$$(\mathbb{E}|X|^p)^{1/p} \leq \|X\|_{\psi_2}\sqrt{p} \quad \text{for all } p \geq 1; \quad (5.11)$$

$$\mathbb{E}\exp(cX^2/\|X\|_{\psi_2}^2) \leq e;$$

if $\mathbb{E}X = 0$ then $\mathbb{E}\exp(tX) \leq \exp(Ct^2\|X\|_{\psi_2}^2) \quad \text{for all } t \in \mathbb{R}, \quad (5.12)$

where $C, c > 0$ are absolute constants. Moreover, up to absolute constant factors, $\|X\|_{\psi_2}$ is the smallest possible number in each of these inequalities.

Example 5.8 Classical examples of sub-gaussian random variables are Gaussian, Bernoulli and all bounded random variables.

1. **(Gaussian):** A standard normal random variable X is sub-gaussian with $\|X\|_{\psi_2} \leq C$ where C is an absolute constant. This follows from (5.6). More generally, if X is a centered normal random variable with variance σ^2, then X is sub-gaussian with $\|X\|_{\psi_2} \leq C\sigma$.

---

[11] The sub-gaussian norm is also called ψ_2 norm in the literature.

2. **(Bernoulli):** Consider a random variable X with distribution $\mathbb{P}\{X = -1\} = \mathbb{P}\{X = 1\} = 1/2$. We call X a *symmetric Bernoulli random variable*. Since $|X| = 1$, it follows that X is a sub-gaussian random variable with $\|X\|_{\psi_2} = 1$.
3. **(Bounded):** More generally, consider any bounded random variable X, thus $|X| \leq M$ almost surely for some M. Then X is a sub-gaussian random variable with $\|X\|_{\psi_2} \leq M$. We can write this more compactly as $\|X\|_{\psi_2} \leq \|X\|_\infty$.

A remarkable property of the normal distribution is *rotation invariance*. Given a finite number of independent centered normal random variables X_i, their sum $\sum_i X_i$ is also a centered normal random variable, obviously with $\mathrm{Var}(\sum_i X_i) = \sum_i \mathrm{Var}(X_i)$. Rotation invariance passes onto sub-gaussian random variables, although approximately:

LEMMA 5.9 (Rotation invariance) *Consider a finite number of independent centered sub-gaussian random variables X_i. Then $\sum_i X_i$ is also a centered sub-gaussian random variable. Moreover,*

$$\Big\|\sum_i X_i\Big\|_{\psi_2}^2 \leq C \sum_i \|X_i\|_{\psi_2}^2$$

where C is an absolute constant.

Proof. The argument is based on estimating the moment generating function. Using independence and (5.12) we have for every $t \in \mathbb{R}$:

$$\mathbb{E}\exp\Big(t\sum_i X_i\Big) = \mathbb{E}\prod_i \exp(tX_i) = \prod_i \mathbb{E}\exp(tX_i) \leq \prod_i \exp(Ct^2\|X_i\|_{\psi_2}^2)$$

$$= \exp(t^2 K^2) \quad \text{where } K^2 = C\sum_i \|X_i\|_{\psi_2}^2.$$

Using the equivalence of properties 2 and 4 in Lemma 5.5 we conclude that $\|\sum_i X_i\|_{\psi_2} \leq C_1 K$ where C_1 is an absolute constant. The proof is complete. □

The rotation invariance immediately yields a *large deviation inequality* for sums of independent sub-gaussian random variables:

PROPOSITION 5.10 (Hoeffding-type inequality) *Let $X_1, \ldots, X_N$ be independent centered sub-gaussian random variables, and let $K = \max_i \|X_i\|_{\psi_2}$. Then for every $a = (a_1, \ldots, a_N) \in \mathbb{R}^N$ and every $t \geq 0$, we have*

$$\mathbb{P}\Big\{\Big|\sum_{i=1}^N a_i X_i\Big| \geq t\Big\} \leq e \cdot \exp\Big(-\frac{ct^2}{K^2\|a\|_2^2}\Big)$$

where $c > 0$ is an absolute constant.

Proof. The rotation invariance (Lemma 5.9) implies the bound $\|\sum_i a_i X_i\|_{\psi_2}^2 \leq C\sum_i a_i^2 \|X_i\|_{\psi_2}^2 \leq CK^2\|a\|_2^2$. Property (5.10) yields the required tail decay. □

REMARK 5.11 One can interpret these results (Lemma 5.9 and Proposition 5.10) as one-sided *non-asymptotic manifestations of the central limit theorem*. For example, consider the normalized sum of independent symmetric Bernoulli random variables $S_N = \frac{1}{\sqrt{N}} \sum_{i=1}^{N} \varepsilon_i$. Proposition 5.10 yields the tail bounds $\mathbb{P}\{|S_N| > t\} \leq e \cdot e^{-ct^2}$ for any number of terms N. Up to the absolute constants e and c, these tails coincide with those of the standard normal random variable (5.5).

Using moment growth (5.11) instead of the tail decay (5.10), we immediately obtain from Lemma 5.9 a general form of the well-known Khintchine inequality:

COROLLARY 5.12 (Khintchine inequality) *Let X_i be a finite number of independent sub-gaussian random variables with zero mean, unit variance, and $\|X_i\|_{\psi_2} \leq K$. Then, for every sequence of coefficients a_i and every exponent $p \geq 2$ we have*

$$\Big(\sum_i a_i^2\Big)^{1/2} \leq \Big(\mathbb{E} \Big|\sum_i a_i X_i\Big|^p\Big)^{1/p} \leq CK\sqrt{p}\Big(\sum_i a_i^2\Big)^{1/2}$$

where C is an absolute constant.

Proof. The lower bound follows by independence and Hölder's inequality: indeed, $\big(\mathbb{E}|\sum_i a_i X_i|^p\big)^{1/p} \geq \big(\mathbb{E}|\sum_i a_i X_i|^2\big)^{1/2} = \big(\sum_i a_i^2\big)^{1/2}$. For the upper bound, we argue as in Proposition 5.10, but use property (5.11). □

5.2.4 Sub-exponential random variables

Although the class of sub-gaussian random variables is natural and quite wide, it leaves out some useful random variables which have tails heavier than gaussian. One such example is a standard exponential random variable – a non-negative random variable with exponential tail decay

$$\mathbb{P}\{X \geq t\} = e^{-t}, \quad t \geq 0. \tag{5.13}$$

To cover such examples, we consider a class of *sub-exponential random variables*, those with at least an exponential tail decay. With appropriate modifications, the basic properties of sub-gaussian random variables hold for sub-exponentials. In particular, a version of Lemma 5.5 holds with a similar proof for sub-exponential properties, except for property 4 of the moment generating function. Thus for a random variable X the following properties are equivalent with parameters $K_i > 0$ differing from each other by at most an absolute constant factor:

$$\mathbb{P}\{|X| > t\} \leq \exp(1 - t/K_1) \quad \text{for all } t \geq 0; \tag{5.14}$$

$$(\mathbb{E}|X|^p)^{1/p} \leq K_2 p \quad \text{for all } p \geq 1; \tag{5.15}$$

$$\mathbb{E}\exp(X/K_3) \leq e. \tag{5.16}$$

DEFINITION 5.13 (Sub-exponential random variables) *A random variable X that satisfies one of the equivalent properties (5.14)–(5.16) is called a* sub-exponential random variable. *The sub-exponential norm of X, denoted $\|X\|_{\psi_1}$, is defined to be the smallest parameter K_2. In other words,*

$$\|X\|_{\psi_1} = \sup_{p\geq 1} p^{-1}(\mathbb{E}|X|^p)^{1/p}.$$

LEMMA 5.14 (Sub-exponential is sub-gaussian squared) *A random variable X is sub-gaussian if and only if X^2 is sub-exponential. Moreover,*

$$\|X\|_{\psi_2}^2 \leq \|X^2\|_{\psi_1} \leq 2\|X\|_{\psi_2}^2.$$

Proof. This follows easily from the definition. □

The moment generating function of a sub-exponential random variable has a similar upper bound as in the sub-gaussian case (property 4 in Lemma 5.5). The only real difference is that the bound only holds in a neighborhood of zero rather than on the whole real line. This is inevitable, as the moment generating function of an exponential random variable (5.13) does not exist for $t \geq 1$.

LEMMA 5.15 (Mgf of sub-exponential random variables) *Let X be a centered sub-exponential random variable. Then, for t such that $|t| \leq c/\|X\|_{\psi_1}$, one has*

$$\mathbb{E}\exp(tX) \leq \exp(Ct^2\|X\|_{\psi_1}^2)$$

where $C, c > 0$ are absolute constants.

Proof. The argument is similar to the sub-gaussian case. We can assume that $\|X\|_{\psi_1} = 1$ by replacing X with $X/\|X\|_{\psi_1}$ and t with $t\|X\|_{\psi_1}$. Repeating the proof of the implication $2 \Rightarrow 4$ of Lemma 5.5 and using $\mathbb{E}|X|^p \leq p^p$ this time, we obtain that $\mathbb{E}\exp(tX) \leq 1 + \sum_{p=2}^\infty (e|t|)^p$. If $|t| \leq 1/2e$ then the right-hand side is bounded by $1+2e^2t^2 \leq \exp(2e^2t^2)$. This completes the proof. □

Sub-exponential random variables satisfy a *large deviation inequality* similar to the one for sub-gaussians (Proposition 5.10). The only significant difference is that *two tails* have to appear here – a gaussian tail responsible for the central limit theorem, and an exponential tail coming from the tails of each term.

PROPOSITION 5.16 (Bernstein-type inequality) *Let $X_1, \ldots, X_N$ be independent centered sub-exponential random variables, and $K = \max_i \|X_i\|_{\psi_1}$. Then for every $a = (a_1, \ldots, a_N) \in \mathbb{R}^N$ and every $t \geq 0$, we have*

$$\mathbb{P}\Big\{\Big|\sum_{i=1}^N a_i X_i\Big| \geq t\Big\} \leq 2\exp\Big[-c\min\Big(\frac{t^2}{K^2\|a\|_2^2}, \frac{t}{K\|a\|_\infty}\Big)\Big]$$

where $c > 0$ is an absolute constant.

Proof. Without loss of generality, we assume that $K = 1$ by replacing X_i with X_i/K and t with t/K. We use the exponential Markov inequality for the sum $S = \sum_i a_i X_i$ and with a parameter $\lambda > 0$:

$$\mathbb{P}\{S \geq t\} = \mathbb{P}\{e^{\lambda S} \geq e^{\lambda t}\} \leq e^{-\lambda t} \mathbb{E} e^{\lambda S} = e^{-\lambda t} \prod_i \mathbb{E} \exp(\lambda a_i X_i).$$

If $|\lambda| \leq c/\|a\|_\infty$ then $|\lambda a_i| \leq c$ for all i, so Lemma 5.15 yields

$$\mathbb{P}\{S \geq t\} \leq e^{-\lambda t} \prod_i \exp(C\lambda^2 a_i^2) = \exp(-\lambda t + C\lambda^2 \|a\|_2^2).$$

Choosing $\lambda = \min(t/2C\|a\|_2^2, c/\|a\|_\infty)$, we obtain that

$$\mathbb{P}\{S \geq t\} \leq \exp\left[-\min\left(\frac{t^2}{4C\|a\|_2^2}, \frac{ct}{2\|a\|_\infty}\right)\right].$$

Repeating this argument for $-X_i$ instead of X_i, we obtain the same bound for $\mathbb{P}\{-S \geq t\}$. A combination of these two bounds completes the proof. $\square$

COROLLARY 5.17 *Let $X_1, \ldots, X_N$ be independent centered sub-exponential random variables, and let $K = \max_i \|X_i\|_{\psi_1}$. Then, for every $\varepsilon \geq 0$, we have*

$$\mathbb{P}\left\{\left|\sum_{i=1}^N X_i\right| \geq \varepsilon N\right\} \leq 2\exp\left[-c\min\left(\frac{\varepsilon^2}{K^2}, \frac{\varepsilon}{K}\right)N\right]$$

where $c > 0$ is an absolute constant.

Proof. This follows from Proposition 5.16 for $a_i = 1$ and $t = \varepsilon N$. $\square$

REMARK 5.18 (Centering) The definitions of sub-gaussian and sub-exponential random variables X do not require them to be centered. In any case, one can always center X using the simple fact that if X is sub-gaussian (or sub-exponential), then so is $X - \mathbb{E}X$. Moreover,

$$\|X - \mathbb{E}X\|_{\psi_2} \leq 2\|X\|_{\psi_2}, \quad \|X - \mathbb{E}X\|_{\psi_1} \leq 2\|X\|_{\psi_1}.$$

This follows by the triangle inequality $\|X - \mathbb{E}X\|_{\psi_2} \leq \|X\|_{\psi_2} + \|\mathbb{E}X\|_{\psi_2}$ along with $\|\mathbb{E}X\|_{\psi_2} = |\mathbb{E}X| \leq \mathbb{E}|X| \leq \|X\|_{\psi_2}$, and similarly for the sub-exponential norm.

5.2.5 Isotropic random vectors

Now we carry our work over to higher dimensions. We will thus be working with random vectors X in $\mathbb{R}^n$, or equivalently probability distributions in $\mathbb{R}^n$.

While the concept of the mean $\mu = \mathbb{E}Z$ of a random variable Z remains the same in higher dimensions, the second moment $\mathbb{E}Z^2$ is replaced by the $n \times n$ *second moment matrix* of a random vector X, defined as

$$\Sigma = \Sigma(X) = \mathbb{E}X \otimes X = \mathbb{E}XX^T$$

where $\otimes$ denotes the outer product of vectors in $\mathbb{R}^n$. Similarly, the concept of variance $\text{Var}(Z) = \mathbb{E}(Z-\mu)^2 = \mathbb{E}Z^2 - \mu^2$ of a random variable is replaced in higher dimensions with the *covariance matrix* of a random vector X, defined as

$$\text{Cov}(X) = \mathbb{E}(X-\mu) \otimes (X-\mu) = \mathbb{E}X \otimes X - \mu \otimes \mu$$

where $\mu = \mathbb{E}X$. By translation, many questions can be reduced to the case of centered random vectors, for which $\mu = 0$ and $\text{Cov}(X) = \Sigma(X)$. We will also need a higher-dimensional version of unit variance:

DEFINITION 5.19 (Isotropic random vectors) *A random vector X in $\mathbb{R}^n$ is called isotropic if $\Sigma(X) = I$. Equivalently, X is isotropic if*

$$\mathbb{E}\langle X, x \rangle^2 = \|x\|_2^2 \quad \text{for all } x \in \mathbb{R}^n. \tag{5.17}$$

Suppose $\Sigma(X)$ is an invertible matrix, which means that the distribution of X is not essentially supported on any proper subspace of $\mathbb{R}^n$. Then $\Sigma(X)^{-1/2}X$ is an isotropic random vector in $\mathbb{R}^n$. Thus every non-degenerate random vector can be made isotropic by an appropriate linear transformation.[12] This allows us to mostly focus on studying isotropic random vectors in the future.

LEMMA 5.20 *Let X, Y be independent isotropic random vectors in $\mathbb{R}^n$. Then $\mathbb{E}\|X\|_2^2 = n$ and $\mathbb{E}\langle X, Y \rangle^2 = n$.*

Proof. The first part follows from $\mathbb{E}\|X\|_2^2 = \mathbb{E}\,\text{tr}(X \otimes X) = \text{tr}(\mathbb{E}X \otimes X) = \text{tr}(I) = n$. The second part follows by conditioning on Y, using isotropy of X and using the first part for Y: this way we obtain $\mathbb{E}\langle X, Y \rangle^2 = \mathbb{E}\|Y\|_2^2 = n$. □

Example 5.21

1. **(Gaussian):** The (standard) *Gaussian random vector* X in $\mathbb{R}^n$ chosen according to the standard normal distribution $N(0, I)$ is isotropic. The coordinates of X are independent standard normal random variables.
2. **(Bernoulli):** A similar example of a discrete isotropic distribution is given by a *Bernoulli random vector* X in $\mathbb{R}^n$ whose coordinates are independent symmetric Bernoulli random variables.
3. **(Product distributions):** More generally, consider a random vector X in $\mathbb{R}^n$ whose coordinates are independent random variables with zero mean and unit variance. Then clearly X is an isotropic vector in $\mathbb{R}^n$.
4. **(Coordinate):** Consider a *coordinate random vector* X, which is uniformly distributed in the set $\{\sqrt{n}\,e_i\}_{i=1}^n$ where $\{e_i\}_{i=1}^n$ is the canonical basis of $\mathbb{R}^n$. Clearly X is an isotropic random vector in $\mathbb{R}^n$.[13]

---

[12] This transformation (usually preceded by centering) is a higher-dimensional version of *standardizing* of random variables, which enforces zero mean and unit variance.

[13] The examples of Gaussian and coordinate random vectors are somewhat opposite – one is very continuous and the other is very discrete. They may be used as test cases in our study of random matrices.

5. **(Frame):** This is a more general version of the coordinate random vector. A *frame* is a set of vectors $\{u_i\}_{i=1}^{M}$ in $\mathbb{R}^n$ which obeys an approximate Parseval's identity, i.e. there exist numbers $A, B > 0$ called *frame bounds* such that

$$A\|x\|_2^2 \le \sum_{i=1}^{M} \langle u_i, x \rangle^2 \le B\|x\|_2^2 \quad \text{for all } x \in \mathbb{R}^n.$$

If $A = B$ the set is called a *tight frame*. Thus, tight frames are generalizations of orthogonal bases without linear independence. Given a tight frame $\{u_i\}_{i=1}^{M}$ with bounds $A = B = M$, the random vector X uniformly distributed in the set $\{u_i\}_{i=1}^{M}$ is clearly isotropic in $\mathbb{R}^n$.[14]

6. **(Spherical):** Consider a random vector X uniformly distributed on the unit Euclidean sphere in $\mathbb{R}^n$ with center at the origin and radius $\sqrt{n}$. Then X is isotropic. Indeed, by rotation invariance $\mathbb{E}\langle X, x \rangle^2$ is proportional to $\|x\|_2^2$; the correct normalization $\sqrt{n}$ is derived from Lemma 5.20.

7. **(Uniform on a convex set):** In convex geometry, a convex set K in $\mathbb{R}^n$ is called isotropic if a random vector X chosen uniformly from K according to the volume is isotropic. As we noted, every full-dimensional convex set can be made into an isotropic one by an affine transformation. Isotropic convex sets look "well conditioned," which is advantageous in geometric algorithms (e.g. volume computations).

We generalize the concepts of sub-gaussian random variables to higher dimensions using one-dimensional marginals.

DEFINITION 5.22 (Sub-gaussian random vectors) *We say that a random vector X in $\mathbb{R}^n$ is sub-gaussian if the one-dimensional marginals $\langle X, x \rangle$ are sub-gaussian random variables for all $x \in \mathbb{R}^n$. The sub-gaussian norm of X is defined as*

$$\|X\|_{\psi_2} = \sup_{x \in S^{n-1}} \|\langle X, x \rangle\|_{\psi_2}.$$

Remark 5.23 (Properties of high-dimensional distributions) The definitions of isotropic and sub-gaussian distributions suggest that, more generally, natural properties of high-dimensional distributions may be defined via one-dimensional marginals. This is a natural way to generalize properties of random variables to random vectors. For example, we shall call a random vector sub-exponential if all of its one-dimensional marginals are sub-exponential random variables, etc.

One simple way to create sub-gaussian distributions in $\mathbb{R}^n$ is by taking a product of n sub-gaussian distributions on the line:

---

[14] There is clearly a reverse implication, too, which shows that the class of tight frames can be identified with the class of discrete isotropic random vectors.

LEMMA 5.24 (Product of sub-gaussian distributions) *Let $X_1,\ldots,X_n$ be independent centered sub-gaussian random variables. Then $X = (X_1,\ldots,X_n)$ is a centered sub-gaussian random vector in $\mathbb{R}^n$, and*

$$\|X\|_{\psi_2} \leq C \max_{i \leq n} \|X_i\|_{\psi_2}$$

where C is an absolute constant.

Proof. This is a direct consequence of the rotation invariance principle, Lemma 5.9. Indeed, for every $x = (x_1,\ldots,x_n) \in S^{n-1}$ we have

$$\|\langle X,x\rangle\|_{\psi_2} = \left\|\sum_{i=1}^n x_i X_i\right\|_{\psi_2} \leq C\sum_{i=1}^n x_i^2 \|X_i\|_{\psi_2}^2 \leq C \max_{i \leq n} \|X_i\|_{\psi_2}$$

where we used that $\sum_{i=1}^n x_i^2 = 1$. This completes the proof. □

Example 5.25 Let us analyze the basic examples of random vectors introduced earlier in Example 5.21.

1. **(Gaussian, Bernoulli):** Gaussian and Bernoulli random vectors are sub-gaussian; their sub-gaussian norms are bounded by an absolute constant. These are particular cases of Lemma 5.24.
2. **(Spherical):** A spherical random vector is also sub-gaussian; its sub-gaussian norm is bounded by an absolute constant. Unfortunately, this does not follow from Lemma 5.24 because the coordinates of the spherical vector are not independent. Instead, by rotation invariance, the claim clearly follows from the following geometric fact. For every $\varepsilon \geq 0$, the spherical cap $\{x \in S^{n-1} : x_1 > \varepsilon\}$ makes up at most $\exp(-\varepsilon^2 n/2)$ proportion of the total area on the sphere.[15] This can be proved directly by integration, and also by elementary geometric considerations [9, Lemma 2.2].
3. **(Coordinate):** Although the coordinate random vector X is formally sub-gaussian as its support is finite, its sub-gaussian norm is too big: $\|X\|_{\psi_2} = \sqrt{n} \gg 1$. So we would not think of X as a sub-gaussian random vector.
4. **(Uniform on a convex set):** For many isotropic convex sets K (called ψ_2 bodies), a random vector X uniformly distributed in K is sub-gaussian with $\|X\|_{\psi_2} = O(1)$. For example, the cube $[-1,1]^n$ is a ψ_2 body by Lemma 5.24, while the appropriately normalized cross-polytope $\{x \in \mathbb{R}^n : \|x\|_1 \leq M\}$ is not. Nevertheless, Borell's lemma (which is a consequence of Brunn–Minkowski inequality) implies a weaker property, that X is always *sub-exponential*, and $\|X\|_{\psi_1} = \sup_{x \in S^{n-1}} \|\langle X,x\rangle\|_{\psi_1}$ is bounded by an absolute constant. See [33, Section 2.2.b3] for a proof and discussion of these ideas.

---

[15] This fact about spherical caps may seem counter-intuitive. For example, for $\varepsilon = 0.1$ the cap looks similar to a hemisphere, but the proportion of its area goes to zero very fast as dimension n increases. This is a starting point of the study of the *concentration of measure phenomenon*, see [43].

5.2.6 Sums of independent random matrices

In this section, we mention without proof some results of classical probability theory in which scalars can be replaced by matrices. Such results are useful in particular for problems on random matrices, since we can view a random matrix as a generalization of a random variable. One such remarkable generalization is valid for the Khintchine inequality, Corollary 5.12. The scalars a_i can be replaced by matrices, and the absolute value by the *Schatten norm*. Recall that for $1 \leq p \leq \infty$, the p-Schatten norm of an $n \times n$ matrix A is defined as the ℓ_p norm of the sequence of its singular values:

$$\|A\|_{C_p^n} = \|(s_i(A))_{i=1}^n\|_p = \Big(\sum_{i=1}^n s_i(A)^p\Big)^{1/p}.$$

For $p = \infty$, the Schatten norm equals the spectral norm $\|A\| = \max_{i \leq n} s_i(A)$. Using this one can quickly check that already for $p = \log n$ the Schatten and spectral norms are equivalent: $\|A\|_{C_p^n} \leq \|A\| \leq e\|A\|_{C_p^n}$.

THEOREM 5.26 (Non-commutative Khintchine inequality, see [61] Section 9.8) *Let $A_1, \ldots, A_N$ be self-adjoint $n \times n$ matrices and $\varepsilon_1, \ldots, \varepsilon_N$ be independent symmetric Bernoulli random variables. Then, for every $2 \leq p < \infty$, we have*

$$\mathbb{E}\Big\|\sum_{i=1}^N \varepsilon_i A_i\Big\| \leq C_1 \sqrt{\log n}\,\Big\|\Big(\sum_{i=1}^N A_i^2\Big)^{1/2}\Big\|$$

where C is an absolute constant.

REMARK 5.27 1. The scalar case of this result, for $n = 1$, recovers the classical Khintchine inequality, Corollary 5.12, for $X_i = \varepsilon_i$.

2. By the equivalence of Schatten and spectral norms for $p = \log n$, a version of non-commutative Khintchine inequality holds for the spectral norm:

$$\frac{c_1}{\sqrt{\log n}}\Big\|\Big(\sum_{i=1}^N A_i^2\Big)^{1/2}\Big\| \leq \mathbb{E}\Big\|\sum_{i=1}^N \varepsilon_i A_i\Big\| \leq C_1 \sqrt{\log n}\,\Big\|\Big(\sum_{i=1}^N A_i^2\Big)^{1/2}\Big\| \quad (5.18)$$

where $C_1, c_1 > 0$ are absolute constants. The logarithmic factor is unfortunately essential; its role will be clear when we discuss applications of this result to random matrices in the next sections.

COROLLARY 5.28 (Rudelson's inequality [65]) *Let $x_1, \ldots, x_N$ be vectors in $\mathbb{R}^n$ and $\varepsilon_1, \ldots, \varepsilon_N$ be independent symmetric Bernoulli random variables. Then*

$$\mathbb{E}\Big\|\sum_{i=1}^N \varepsilon_i x_i \otimes x_i\Big\| \leq C\sqrt{\log \min(N,n)} \cdot \max_{i \leq N}\|x_i\|_2 \cdot \Big\|\sum_{i=1}^N x_i \otimes x_i\Big\|^{1/2}$$

where C is an absolute constant.

Proof. One can assume that $n \leq N$ by replacing $\mathbb{R}^n$ with the linear span of $\{x_1, \ldots, x_N\}$ if necessary. The claim then follows from (5.18), since

$$\Big\|\Big(\sum_{i=1}^N (x_i \otimes x_i)^2\Big)^{1/2}\Big\| = \Big\|\sum_{i=1}^N \|x_i\|_2^2 \, x_i \otimes x_i\Big\|^{1/2} \leq \max_{i \leq N} \|x_i\|_2 \Big\|\sum_{i=1}^N x_i \otimes x_i\Big\|^{1/2}. \quad \square$$

Ahlswede and Winter [4] pioneered a different approach to matrix-valued inequalities in probability theory, which was based on trace inequalities like the Golden–Thompson inequality. A development of this idea leads to remarkably sharp results. We quote one such inequality from [77]:

THEOREM 5.29 (Non-commutative Bernstein-type inequality [77]) *Consider a finite sequence X_i of independent centered self-adjoint random $n \times n$ matrices. Assume we have for some numbers K and σ that*

$$\|X_i\| \leq K \text{ almost surely,} \quad \Big\|\sum_i \mathbb{E} X_i^2\Big\| \leq \sigma^2.$$

Then, for every $t \geq 0$ we have

$$\mathbb{P}\Big\{\Big\|\sum_i X_i\Big\| \geq t\Big\} \leq 2n \cdot \exp\Big(\frac{-t^2/2}{\sigma^2 + Kt/3}\Big). \quad (5.19)$$

REMARK 5.30 This is a direct matrix generalization of a classical Bernstein's inequality for bounded random variables. To compare it with our version of Bernstein's inequality for sub-exponentials, Proposition 5.16, note that the probability bound in (5.19) is equivalent to $2n \cdot \exp\big[-c \min\big(\frac{t^2}{\sigma^2}, \frac{t}{K}\big)\big]$ where $c > 0$ is an absolute constant. In both results we see a mixture of Gaussian and exponential tails.

5.3 Random matrices with independent entries

We are ready to study the extreme singular values of random matrices. In this section, we consider the classical model of random matrices whose entries are independent and centered random variables. Later we will study the more difficult models where only the rows or the columns are independent.

The reader may keep in mind some classical examples of $N \times n$ random matrices with independent entries. The most classical example is the *Gaussian random matrix A* whose entries are independent standard normal random variables. In this case, the $n \times n$ symmetric matrix $A^* A$ is called a Wishart matrix; it is a higher-dimensional version of chi-square distributed random variables.

The simplest example of discrete random matrices is the *Bernoulli random matrix A* whose entries are independent symmetric Bernoulli random variables. In other words, Bernoulli random matrices are distributed uniformly in the set of all $N \times n$ matrices with ± 1 entries.

5.3.1 Limit laws and Gaussian matrices

Consider an $N \times n$ random matrix A whose entries are independent centered identically distributed random variables. By now, the *limiting behavior* of the extreme singular values of A, as the dimensions $N, n \to \infty$, is well understood:

THEOREM 5.31 (Bai–Yin's law, see [8]) *Let $A = A_{N,n}$ be an $N \times n$ random matrix whose entries are independent copies of a random variable with zero mean, unit variance, and finite fourth moment. Suppose that the dimensions N and n grow to infinity while the aspect ratio n/N converges to a constant in $[0, 1]$. Then*

$$s_{\min}(A) = \sqrt{N} - \sqrt{n} + o(\sqrt{n}), \quad s_{\max}(A) = \sqrt{N} + \sqrt{n} + o(\sqrt{n}) \quad \text{almost surely.}$$

As we pointed out in the introduction, our program is to find non-asymptotic versions of Bai–Yin's law. There is precisely one model of random matrices, namely Gaussian, where an *exact* non-asymptotic result is known:

THEOREM 5.32 (Gordon's theorem for Gaussian matrices) *Let A be an $N \times n$ matrix whose entries are independent standard normal random variables. Then*

$$\sqrt{N} - \sqrt{n} \leq \mathbb{E} s_{\min}(A) \leq \mathbb{E} s_{\max}(A) \leq \sqrt{N} + \sqrt{n}.$$

The proof of the upper bound, which we borrowed from [21], is based on Slepian's comparison inequality for Gaussian processes.[16]

LEMMA 5.33 (Slepian's inequality, see [45] Section 3.3) *Consider two Gaussian processes $(X_t)_{t \in T}$ and $(Y_t)_{t \in T}$ whose increments satisfy the inequality $\mathbb{E}|X_s - X_t|^2 \leq \mathbb{E}|Y_s - Y_t|^2$ for all $s, t \in T$. Then $\mathbb{E}\sup_{t \in T} X_t \leq \mathbb{E}\sup_{t \in T} Y_t$.*

Proof of Theorem 5.32. We recognize $s_{\max}(A) = \max_{u \in S^{n-1},\, v \in S^{N-1}} \langle Au, v \rangle$ to be the supremum of the Gaussian process $X_{u,v} = \langle Au, v \rangle$ indexed by the pairs of vectors $(u, v) \in S^{n-1} \times S^{N-1}$. We shall compare this process to the following one whose supremum is easier to estimate: $Y_{u,v} = \langle g, u \rangle + \langle h, v \rangle$ where $g \in \mathbb{R}^n$ and $h \in \mathbb{R}^N$ are independent standard Gaussian random vectors. The rotation invariance of Gaussian measures makes it easy to compare the increments of these processes. For every $(u, v), (u', v') \in S^{n-1} \times S^{N-1}$, one can check that

$$\mathbb{E}|X_{u,v} - X_{u',v'}|^2 = \sum_{i=1}^{n}\sum_{j=1}^{N} |u_i v_j - u'_i v'_j|^2 \leq \|u - u'\|_2^2 + \|v - v'\|_2^2 = \mathbb{E}|Y_{u,v} - Y_{u',v'}|^2.$$

Therefore Lemma 5.33 applies, and it yields the required bound

$$\mathbb{E} s_{\max}(A) = \mathbb{E} \max_{(u,v)} X_{u,v} \leq \mathbb{E} \max_{(u,v)} Y_{u,v} = \mathbb{E}\|g\|_2 + \mathbb{E}\|h\|_2 \leq \sqrt{N} + \sqrt{n}.$$

---

[16] Recall that a Gaussian process $(X_t)_{t \in T}$ is a collection of centered normal random variables X_t on the same probability space, indexed by points t in an abstract set T.

Similar ideas are used to estimate $\mathbb{E} s_{\min}(A) = \mathbb{E} \max_{v \in S^{N-1}} \min_{u \in S^{n-1}} \langle Au, v \rangle$, see [21]. One uses in this case Gordon's generalization of Slepian's inequality for minimax of Gaussian processes [35, 36, 37], see [45, Section 3.3]. □

While Theorem 5.32 is about the expectation of singular values, it also yields a large deviation inequality for them. It can be deduced formally by using the *concentration of measure* in the Gauss space.

PROPOSITION 5.34 (Concentration in Gauss space, see [43]) *Let f be a real-valued Lipschitz function on $\mathbb{R}^n$ with Lipschitz constant K, i.e. $|f(x) - f(y)| \leq K \|x - y\|_2$ for all $x, y \in \mathbb{R}^n$ (such functions are also called K-Lipschitz). Let X be the standard normal random vector in $\mathbb{R}^n$. Then for every $t \geq 0$ one has*

$$\mathbb{P}\{f(X) - \mathbb{E}f(X) > t\} \leq \exp(-t^2/2K^2).$$

COROLLARY 5.35 (Gaussian matrices, deviation; see [21]) *Let A be an $N \times n$ matrix whose entries are independent standard normal random variables. Then for every $t \geq 0$, with probability at least $1 - 2\exp(-t^2/2)$ one has*

$$\sqrt{N} - \sqrt{n} - t \leq s_{\min}(A) \leq s_{\max}(A) \leq \sqrt{N} + \sqrt{n} + t.$$

Proof. Note that $s_{\min}(A)$, $s_{\max}(A)$ are 1-Lipschitz functions of matrices A considered as vectors in $\mathbb{R}^{Nn}$. The conclusion now follows from the estimates on the expectation (Theorem 5.32) and Gaussian concentration (Proposition 5.34). □

Later in these notes, we find it more convenient to work with the $n \times n$ positive-definite symmetric matrix A^*A rather than with the original $N \times n$ matrix A. Observe that the normalized matrix $\bar{A} = \frac{1}{\sqrt{N}} A$ is an approximate isometry (which is our goal) if and only if $\bar{A}^* \bar{A}$ is an approximate identity:

LEMMA 5.36 (Approximate isometries) *Consider a matrix B that satisfies*

$$\|B^*B - I\| \leq \max(\delta, \delta^2) \tag{5.20}$$

for some $\delta > 0$. Then

$$1 - \delta \leq s_{\min}(B) \leq s_{\max}(B) \leq 1 + \delta. \tag{5.21}$$

Conversely, if B satisfies (5.21) for some $\delta > 0$ then $\|B^*B - I\| \leq 3\max(\delta, \delta^2)$.

Proof. Inequality (5.20) holds if and only if $\big|\|Bx\|_2^2 - 1\big| \leq \max(\delta, \delta^2)$ for all $x \in S^{n-1}$. Similarly, (5.21) holds if and only if $\big|\|Bx\|_2 - 1\big| \leq \delta$ for all $x \in S^{n-1}$. The conclusion then follows from the elementary inequality

$$\max(|z-1|, |z-1|^2) \leq |z^2 - 1| \leq 3\max(|z-1|, |z-1|^2) \quad \text{for all } z \geq 0. \quad \square$$

Lemma 5.36 reduces our task of proving inequalities (5.2) to showing an equivalent (but often more convenient) bound

$$\left\|\frac{1}{N}A^*A - I\right\| \leq \max(\delta, \delta^2) \quad \text{where } \delta = O(\sqrt{n/N}).$$

5.3.2 General random matrices with independent entries

Now we pass to a more general model of random matrices whose entries are independent centered random variables with some general distribution (not necessarily normal). The largest singular value (the spectral norm) can be estimated by Latala's theorem for general random matrices with non-identically distributed entries:

THEOREM 5.37 (Latala's theorem [42]) *Let A be a random matrix whose entries a_{ij} are independent centered random variables with finite fourth moment. Then*

$$\mathbb{E} s_{\max}(A) \leq C\left[\max_i \Big(\sum_j \mathbb{E} a_{ij}^2\Big)^{1/2} + \max_j \Big(\sum_i \mathbb{E} a_{ij}^2\Big)^{1/2} + \Big(\sum_{i,j} \mathbb{E} a_{ij}^4\Big)^{1/4}\right].$$

If the variance and the fourth moments of the entries are uniformly bounded, then Latala's result yields $s_{\max}(A) = O(\sqrt{N} + \sqrt{n})$. This is slightly weaker than our goal (5.2), which is $s_{\max}(A) = \sqrt{N} + O(\sqrt{n})$ but still satisfactory for most applications. Results of the latter type will appear later in the more general model of random matrices with independent rows or columns.

Similarly, our goal (5.2) for the smallest singular value is $s_{\min}(A) \geq \sqrt{N} - O(\sqrt{n})$. Since the singular values are non-negative anyway, such an inequality would only be useful for sufficiently tall matrices, $N \gg n$. For almost square and square matrices, estimating the smallest singular value (known also as the *hard edge* of spectrum) is considerably more difficult. The progress on estimating the hard edge is summarized in [69]. If A has independent entries, then indeed $s_{\min}(A) \geq c(\sqrt{N} - \sqrt{n})$, and the following is an optimal probability bound:

THEOREM 5.38 (Independent entries, hard edge [68]) *Let A be an $n \times n$ random matrix whose entries are independent identically distributed sub-gaussian random variables with zero mean and unit variance. Then for $\varepsilon \geq 0$,*

$$\mathbb{P}\big(s_{\min}(A) \leq \varepsilon(\sqrt{N} - \sqrt{n-1})\big) \leq (C\varepsilon)^{N-n+1} + c^N$$

where $C > 0$ and $c \in (0,1)$ depend only on the sub-gaussian norm of the entries.

This result gives an optimal bound for square matrices as well ($N = n$).

5.4 Random matrices with independent rows

In this section, we focus on a more general model of random matrices, where we only assume independence of the rows rather than all entries. Such matrices are naturally *generated by high-dimensional distributions.* Indeed, given an arbitrary probability distribution in $\mathbb{R}^n$, one takes a sample of N independent points and arranges them as the rows of an $N \times n$ matrix A. By studying spectral properties of A one should be able to learn something useful about the underlying distribution. For example, as we will see in Section 5.4.3, the extreme singular values of A would tell us whether the covariance matrix of the distribution can be estimated from a sample of size N.

The picture will vary slightly depending on whether the rows of A are sub-gaussian or have arbitrary distribution. For heavy-tailed distributions, an extra logarithmic factor has to appear in our desired inequality (5.2). The analysis of sub-gaussian and heavy-tailed matrices will be completely different.

There is an abundance of examples where the results of this section may be useful. They include all matrices with independent entries, whether sub-gaussian such as Gaussian and Bernoulli, or completely general distributions with mean zero and unit variance. In the latter case one is able to surpass the fourth moment assumption which is necessary in Bai–Yin's law, Theorem 5.31.

Other examples of interest come from non-product distributions, some of which we saw in Example 5.21. Sampling from discrete objects (matrices and frames) fits well in this framework, too. Given a deterministic matrix B, one puts a uniform distribution on the set of the rows of B and creates a random matrix A as before – by sampling some N random rows from B. Applications to sampling will be discussed in Section 5.4.4.

5.4.1 Sub-gaussian rows

The following result goes in the direction of our goal (5.2) for random matrices with independent sub-gaussian rows.

THEOREM 5.39 (Sub-gaussian rows) *Let A be an $N \times n$ matrix whose rows A_i are independent sub-gaussian isotropic random vectors in $\mathbb{R}^n$. Then for every $t \geq 0$, with probability at least $1 - 2\exp(-ct^2)$ one has*

$$\sqrt{N} - C\sqrt{n} - t \leq s_{\min}(A) \leq s_{\max}(A) \leq \sqrt{N} + C\sqrt{n} + t. \qquad (5.22)$$

Here $C = C_K$, $c = c_K > 0$ depend only on the sub-gaussian norm $K = \max_i \|A_i\|_{\psi_2}$ of the rows.

This result is a general version of Corollary 5.35 (up to absolute constants); instead of independent Gaussian entries we allow independent sub-gaussian rows. This of course covers all matrices with independent sub-gaussian entries such as Gaussian and Bernoulli. It also applies to some natural matrices whose entries are not independent. One such example is a matrix whose rows are independent spherical random vectors (Example 5.25).

Proof. The proof is a basic version of a *covering argument*, and it has three steps. We need to control $\|Ax\|_2$ for all vectors x on the unit sphere S^{n-1}. To this end, we discretize the sphere using a net $\mathcal{N}$ (the approximation step), establish a tight control of $\|Ax\|_2$ for every fixed vector $x \in \mathcal{N}$ with high probability (the concentration step), and finish off by taking a union bound over all x in the net. The concentration step will be based on the deviation inequality for sub-exponential random variables, Corollary 5.17.

Step 1: Approximation. Recalling Lemma 5.36 for the matrix $B = A/\sqrt{N}$ we see that the conclusion of the theorem is equivalent to

$$\left\|\frac{1}{N}A^*A - I\right\| \leq \max(\delta, \delta^2) =: \varepsilon \quad \text{where} \quad \delta = C\sqrt{\frac{n}{N}} + \frac{t}{\sqrt{N}}. \tag{5.23}$$

Using Lemma 5.4, we can evaluate the operator norm in (5.23) on a $\frac{1}{4}$-net $\mathcal{N}$ of the unit sphere S^{n-1}:

$$\left\|\frac{1}{N}A^*A - I\right\| \leq 2\max_{x \in \mathcal{N}}\left|\langle(\frac{1}{N}A^*A - I)x, x\rangle\right| = 2\max_{x \in \mathcal{N}}\left|\frac{1}{N}\|Ax\|_2^2 - 1\right|.$$

So to complete the proof it suffices to show that, with the required probability,

$$\max_{x \in \mathcal{N}}\left|\frac{1}{N}\|Ax\|_2^2 - 1\right| \leq \frac{\varepsilon}{2}.$$

By Lemma 5.2, we can choose the net $\mathcal{N}$ so that it has cardinality $|\mathcal{N}| \leq 9^n$.

Step 2: Concentration. Let us fix any vector $x \in S^{n-1}$. We can express $\|Ax\|_2^2$ as a sum of independent random variables

$$\|Ax\|_2^2 = \sum_{i=1}^{N}\langle A_i, x\rangle^2 =: \sum_{i=1}^{N} Z_i^2 \tag{5.24}$$

where A_i denote the rows of the matrix A. By assumption, $Z_i = \langle A_i, x\rangle$ are independent sub-gaussian random variables with $\mathbb{E}Z_i^2 = 1$ and $\|Z_i\|_{\psi_2} \leq K$. Therefore, by Remark 5.18 and Lemma 5.14, $Z_i^2 - 1$ are independent centered sub-exponential random variables with $\|Z_i^2 - 1\|_{\psi_1} \leq 2\|Z_i^2\|_{\psi_1} \leq 4\|Z_i\|_{\psi_2}^2 \leq 4K^2$.

We can therefore use an exponential deviation inequality, Corollary 5.17, to control the sum (5.24). Since $K \geq \|Z_i\|_{\psi_2} \geq \frac{1}{\sqrt{2}}(\mathbb{E}|Z_i|^2)^{1/2} = \frac{1}{\sqrt{2}}$, this gives

$$\mathbb{P}\left\{\left|\frac{1}{N}\|Ax\|_2^2 - 1\right| \geq \frac{\varepsilon}{2}\right\} = \mathbb{P}\left\{\left|\frac{1}{N}\sum_{i=1}^{N}Z_i^2 - 1\right| \geq \frac{\varepsilon}{2}\right\} \leq 2\exp\left[-\frac{c_1}{K^4}\min(\varepsilon^2, \varepsilon)N\right]$$

$$= 2\exp\left[-\frac{c_1}{K^4}\delta^2 N\right] \leq 2\exp\left[-\frac{c_1}{K^4}(C^2 n + t^2)\right]$$

where the last inequality follows by the definition of δ and using the inequality $(a+b)^2 \geq a^2 + b^2$ for $a, b \geq 0$.

Step 3: Union bound. Taking the union bound over all vectors x in the net $\mathcal{N}$ of cardinality $|\mathcal{N}| \leq 9^n$, we obtain

$$\mathbb{P}\left\{\max_{x \in \mathcal{N}}\left|\frac{1}{N}\|Ax\|_2^2 - 1\right| \geq \frac{\varepsilon}{2}\right\} \leq 9^n \cdot 2\exp\left[-\frac{c_1}{K^4}(C^2 n + t^2)\right] \leq 2\exp\left(-\frac{c_1 t^2}{K^4}\right)$$

where the second inequality follows for $C = C_K$ sufficiently large, e.g. $C = K^2\sqrt{\ln 9/c_1}$. As we noted in Step 1, this completes the proof of the theorem. □

REMARK 5.40 (Non-isotropic distributions)

1. A version of Theorem 5.39 holds for general, non-isotropic sub-gaussian distributions. Assume that A is an $N \times n$ matrix whose rows A_i are independent sub-gaussian random vectors in $\mathbb{R}^n$ with second moment matrix Σ. Then for every $t \geq 0$, the following inequality holds with probability at least $1 - 2\exp(-ct^2)$:

$$\left\| \frac{1}{N} A^* A - \Sigma \right\| \leq \max(\delta, \delta^2) \quad \text{where} \quad \delta = C\sqrt{\frac{n}{N}} + \frac{t}{\sqrt{N}}. \tag{5.25}$$

Here as before $C = C_K$, $c = c_K > 0$ depend only on the sub-gaussian norm $K = \max_i \|A_i\|_{\psi_2}$ of the rows. This result is a general version of (5.23). It follows by a straightforward modification of the argument of Theorem 5.39.

2. A more natural, multiplicative form of (5.25) is the following. Assume that $\Sigma^{-1/2} A_i$ are isotropic sub-gaussian random vectors, and let K be the maximum of their sub-gaussian norms. Then for every $t \geq 0$, the following inequality holds with probability at least $1 - 2\exp(-ct^2)$:

$$\left\| \frac{1}{N} A^* A - \Sigma \right\| \leq \max(\delta, \delta^2) \|\Sigma\|, \quad \text{where} \quad \delta = C\sqrt{\frac{n}{N}} + \frac{t}{\sqrt{N}}. \tag{5.26}$$

Here again $C = C_K$, $c = c_K > 0$. This result follows from Theorem 5.39 applied to the isotropic random vectors $\Sigma^{-1/2} A_i$.

5.4.2 Heavy-tailed rows

The class of sub-gaussian random variables in Theorem 5.39 may sometimes be too restrictive in applications. For example, if the rows of A are independent coordinate or frame random vectors (Examples 5.21 and 5.25), they are poorly sub-gaussian and Theorem 5.39 is too weak. In such cases, one would use the following result instead, which operates in remarkable generality.

THEOREM 5.41 (Heavy-tailed rows) *Let A be an $N \times n$ matrix whose rows A_i are independent isotropic random vectors in $\mathbb{R}^n$. Let m be a number such that $\|A_i\|_2 \leq \sqrt{m}$ almost surely for all i. Then for every $t \geq 0$, one has*

$$\sqrt{N} - t\sqrt{m} \leq s_{\min}(A) \leq s_{\max}(A) \leq \sqrt{N} + t\sqrt{m} \tag{5.27}$$

with probability at least $1 - 2n \cdot \exp(-ct^2)$, where $c > 0$ is an absolute constant.

Recall that $(\mathbb{E}\|A_i\|_2^2)^{1/2} = \sqrt{n}$ by Lemma 5.20. This indicates that one would typically use Theorem 5.41 with $m = O(n)$. In this case the result takes the form

$$\sqrt{N} - t\sqrt{n} \leq s_{\min}(A) \leq s_{\max}(A) \leq \sqrt{N} + t\sqrt{n} \tag{5.28}$$

with probability at least $1 - 2n \cdot \exp(-c't^2)$. This is a form of our desired inequality (5.2) for heavy-tailed matrices. We shall discuss this more after the proof.

Proof. We shall use the non-commutative Bernstein's inequality, Theorem 5.29.

Step 1: Reduction to a sum of independent random matrices. We first note that $m \geq n \geq 1$ since by Lemma 5.20 we have $\mathbb{E}\|A_i\|_2^2 = n$. Now we start an argument parallel to Step 1 of Theorem 5.39. Recalling Lemma 5.36 for the matrix $B = A/\sqrt{N}$ we see that the desired inequalities (5.27) are equivalent to

$$\left\|\frac{1}{N}A^*A - I\right\| \leq \max(\delta, \delta^2) =: \varepsilon \quad \text{where} \quad \delta = t\sqrt{\frac{m}{N}}. \tag{5.29}$$

We express this random matrix as a sum of independent random matrices:

$$\frac{1}{N}A^*A - I = \frac{1}{N}\sum_{i=1}^{N} A_i \otimes A_i - I = \sum_{i=1}^{N} X_i, \quad \text{where } X_i := \frac{1}{N}(A_i \otimes A_i - I);$$

note that X_i are independent centered $n \times n$ random matrices.

Step 2: Estimating the mean, range, and variance. We are going to apply the non-commutative Bernstein inequality, Theorem 5.29, for the sum $\sum_i X_i$. Since A_i are isotropic random vectors, we have $\mathbb{E}A_i \otimes A_i = I$ which implies that $\mathbb{E}X_i = 0$ as required in the non-commutative Bernstein inequality.

We estimate the range of X_i using that $\|A_i\|_2 \leq \sqrt{m}$ and $m \geq 1$:

$$\|X_i\| \leq \frac{1}{N}(\|A_i \otimes A_i\| + 1) = \frac{1}{N}(\|A_i\|_2^2 + 1) \leq \frac{1}{N}(m+1) \leq \frac{2m}{N} =: K.$$

To estimate the total variance $\|\sum_i \mathbb{E}X_i^2\|$, we first compute

$$X_i^2 = \frac{1}{N^2}\left[(A_i \otimes A_i)^2 - 2(A_i \otimes A_i) + I\right],$$

so using that the isotropy assumption $\mathbb{E}A_i \otimes A_i = I$ we obtain

$$\mathbb{E}X_i^2 = \frac{1}{N^2}\left[\mathbb{E}(A_i \otimes A_i)^2 - I\right]. \tag{5.30}$$

Since $(A_i \otimes A_i)^2 = \|A_i\|_2^2 A_i \otimes A_i$ is a positive semidefinite matrix and $\|A_i\|_2^2 \leq m$ by assumption, we have $\|\mathbb{E}(A_i \otimes A_i)^2\| \leq m \cdot \|\mathbb{E}A_i \otimes A_i\| = m$. Putting this into (5.30) we obtain

$$\|\mathbb{E}X_i^2\| \leq \frac{1}{N^2}(m+1) \leq \frac{2m}{N^2}$$

where we again used that $m \geq 1$. This yields[17]

$$\left\|\sum_{i=1}^{N}\mathbb{E}X_i^2\right\| \leq N \cdot \max_i \|\mathbb{E}X_i^2\| = \frac{2m}{N} =: \sigma^2.$$

---

[17] Here the seemingly crude application of the triangle inequality is actually not so loose. If the rows A_i are identically distributed, then so are X_i^2, which makes the triangle inequality above into an equality.

Step 3: Application of the non-commutative Bernstein's inequality. Applying Theorem 5.29 (see Remark 5.30) and recalling the definitions of ε and δ in (5.29), we bound the probability in question as

$$\mathbb{P}\left\{\left\|\frac{1}{N}A^*A - I\right\| \geq \varepsilon\right\} = \mathbb{P}\left\{\left\|\sum_{i=1}^{N} X_i\right\| \geq \varepsilon\right\} \leq 2n \cdot \exp\left[-c\min\left(\frac{\varepsilon^2}{\sigma^2}, \frac{\varepsilon}{K}\right)\right]$$

$$\leq 2n \cdot \exp\left[-c\min(\varepsilon^2, \varepsilon) \cdot \frac{N}{2m}\right] = 2n \cdot \exp\left(-\frac{c\delta^2 N}{2m}\right) = 2n \cdot \exp(-ct^2/2).$$

This completes the proof. $\square$

Theorem 5.41 for heavy-tailed rows is different from Theorem 5.39 for sub-gaussian rows in two ways: the boundedness assumption[18] $\|A_i\|_2^2 \leq m$ appears, and the probability bound is weaker. We will now comment on both differences.

REMARK 5.42 (Boundedness assumption) Observe that some boundedness assumption on the distribution is needed in Theorem 5.41. Let us see this on the following example. Choose $\delta \geq 0$ arbitrarily small, and consider a random vector $X = \delta^{-1/2}\xi Y$ in $\mathbb{R}^n$ where ξ is a $\{0,1\}$-valued random variable with $\mathbb{E}\xi = \delta$ (a "selector") and Y is an independent isotropic random vector in $\mathbb{R}^n$ with an arbitrary distribution. Then X is also an isotropic random vector. Consider an $N \times n$ random matrix A whose rows A_i are independent copies of X. However, if $\delta \geq 0$ is suitably small then $A = 0$ with high probability, hence no nontrivial lower bound on $s_{\min}(A)$ is possible.

Inequality (5.28) fits our goal (5.2), but not quite. The reason is that the probability bound is only nontrivial if $t \geq C\sqrt{\log n}$. Therefore, in reality Theorem 5.41 asserts that

$$\sqrt{N} - C\sqrt{n \log n} \leq s_{\min}(A) \leq s_{\max}(A) \leq \sqrt{N} + C\sqrt{n \log n} \qquad (5.31)$$

with probability, say 0.9. This achieves our goal (5.2) up to a logarithmic factor.

REMARK 5.43 (Logarithmic factor) The logarithmic factor cannot be removed from (5.31) for some heavy-tailed distributions. Consider for instance the coordinate distribution introduced in Example 5.21. In order that $s_{\min}(A) > 0$ there must be no zero columns in A. Equivalently, each coordinate vector $e_1, \ldots, e_n$ must be picked at least once in N independent trials (each row of A picks an independent coordinate vector). Recalling the classical coupon collector's problem, one must make at least $N \geq Cn \log n$ trials to make this occur with high probability. Thus the logarithm is necessary in the left-hand side of (5.31).[19]

---

[18] Going a little ahead, we would like to point out that the almost sure boundedness can be relaxed to the bound in expectation $\mathbb{E} \max_i \|A_i\|_2^2 \leq m$, see Theorem 5.45.

[19] This argument moreover shows the optimality of the probability bound in Theorem 5.41. For example, for $t = \sqrt{N}/2\sqrt{n}$ the conclusion (5.28) implies that A is well conditioned (i.e. $\sqrt{N}/2 \leq s_{\min}(A) \leq s_{\max}(A) \leq 2\sqrt{N}$) with probability $1 - n \cdot \exp(-cN/n)$. On the other hand, by the coupon collector's problem we estimate the probability that $s_{\min}(A) > 0$ as $1 - n \cdot (1 - \frac{1}{n})^N \approx 1 - n \cdot \exp(-N/n)$.

A version of Theorem 5.41 holds for general, non-isotropic distributions. It is convenient to state it in terms of the equivalent estimate (5.29):

THEOREM 5.44 (Heavy-tailed rows, non-isotropic) *Let A be an $N \times n$ matrix whose rows A_i are independent random vectors in $\mathbb{R}^n$ with the common second moment matrix $\Sigma = \mathbb{E} A_i \otimes A_i$. Let m be a number such that $\|A_i\|_2 \leq \sqrt{m}$ almost surely for all i. Then for every $t \geq 0$, the following inequality holds with probability at least $1 - n \cdot \exp(-ct^2)$:*

$$\left\|\frac{1}{N} A^* A - \Sigma\right\| \leq \max(\|\Sigma\|^{1/2} \delta, \delta^2) \quad \text{where} \quad \delta = t\sqrt{\frac{m}{N}}. \qquad (5.32)$$

Here $c > 0$ is an absolute constant. In particular, this inequality yields

$$\|A\| \leq \|\Sigma\|^{1/2} \sqrt{N} + t\sqrt{m}. \qquad (5.33)$$

Proof. We note that $m \geq \|\Sigma\|$ because $\|\Sigma\| = \|\mathbb{E} A_i \otimes A_i\| \leq \mathbb{E} \|A_i \otimes A_i\| = \mathbb{E}\|A_i\|_2^2 \leq m$. Then (5.32) follows by a straightforward modification of the argument of Theorem 5.41. Furthermore, if (5.32) holds then by the triangle inequality

$$\frac{1}{N}\|A\|^2 = \left\|\frac{1}{N} A^* A\right\| \leq \|\Sigma\| + \left\|\frac{1}{N} A^* A - \Sigma\right\|$$
$$\leq \|\Sigma\| + \|\Sigma\|^{1/2} \delta + \delta^2 \leq (\|\Sigma\|^{1/2} + \delta)^2.$$

Taking square roots and multiplying both sides by $\sqrt{N}$, we obtain (5.33). □

The *almost sure* boundedness requirement in Theorem 5.41 may sometimes be too restrictive in applications, and it can be relaxed to a bound *in expectation*:

THEOREM 5.45 (Heavy-tailed rows; expected singular values) *Let A be an $N \times n$ matrix whose rows A_i are independent isotropic random vectors in $\mathbb{R}^n$. Let $m := \mathbb{E} \max_{i \leq N} \|A_i\|_2^2$. Then*

$$\mathbb{E} \max_{j \leq n} |s_j(A) - \sqrt{N}| \leq C\sqrt{m \log \min(N, n)}$$

where C is an absolute constant.

The proof of this result is similar to that of Theorem 5.41, except that this time we will use Rudelson's Corollary 5.28 instead of matrix Bernstein's inequality. To this end, we need a link to symmetric Bernoulli random variables. This is provided by a general *symmetrization argument*:

LEMMA 5.46 (Symmetrization) *Let (X_i) be a finite sequence of independent random vectors valued in some Banach space, and (ε_i) be independent symmetric Bernoulli random variables. Then*

$$\mathbb{E}\left\|\sum_i (X_i - \mathbb{E} X_i)\right\| \leq 2\mathbb{E}\left\|\sum_i \varepsilon_i X_i\right\|. \qquad (5.34)$$

Proof. We define random variables $\tilde{X}_i = X_i - X_i'$ where (X_i') is an independent copy of the sequence (X_i). Then $\tilde{X}_i$ are independent symmetric random variables, i.e. the sequence $(\tilde{X}_i)$ is distributed identically with $(-\tilde{X}_i)$ and thus also with $(\varepsilon_i \tilde{X}_i)$. Replacing $\mathbb{E}X_i$ by $\mathbb{E}X_i'$ in (5.34) and using Jensen's inequality, symmetry, and triangle inequality, we obtain the required inequality

$$\mathbb{E}\Big\|\sum_i (X_i - \mathbb{E}X_i)\Big\| \leq \mathbb{E}\Big\|\sum_i \tilde{X}_i\Big\| = \mathbb{E}\Big\|\sum_i \varepsilon_i \tilde{X}_i\Big\|$$

$$\leq \mathbb{E}\Big\|\sum_i \varepsilon_i X_i\Big\| + \mathbb{E}\Big\|\sum_i \varepsilon_i X_i'\Big\| = 2\mathbb{E}\Big\|\sum_i \varepsilon_i X_i\Big\|. \qquad \square$$

We will also need a probabilistic version of Lemma 5.36 on approximate isometries. The proof of that lemma was based on the elementary inequality $|z^2 - 1| \geq \max(|z-1|, |z-1|^2)$ for $z \geq 0$. Here is a probabilistic version:

LEMMA 5.47 *Let Z be a non-negative random variable. Then $\mathbb{E}|Z^2 - 1| \geq \max(\mathbb{E}|Z-1|, (\mathbb{E}|Z-1|)^2)$.*

Proof. Since $|Z-1| \leq |Z^2 - 1|$ pointwise, we have $\mathbb{E}|Z-1| \leq \mathbb{E}|Z^2 - 1|$. Next, since $|Z-1|^2 \leq |Z^2 - 1|$ pointwise, taking square roots and expectations we obtain $\mathbb{E}|Z-1| \leq \mathbb{E}|Z^2 - 1|^{1/2} \leq (\mathbb{E}|Z^2 - 1|)^{1/2}$, where the last bound follows by Jensen's inequality. Squaring both sides completes the proof. $\qquad \square$

Proof of Theorem 5.45. **Step 1: Application of Rudelson's inequality.** As in the proof of Theorem 5.41, we are going to control

$$E := \mathbb{E}\Big\|\frac{1}{N}A^*A - I\Big\| = \mathbb{E}\Big\|\frac{1}{N}\sum_{i=1}^N A_i \otimes A_i - I\Big\| \leq \frac{2}{N}\mathbb{E}\Big\|\sum_{i=1}^N \varepsilon_i A_i \otimes A_i\Big\|$$

where we used Symmetrization Lemma 5.46 with independent symmetric Bernoulli random variables ε_i (which are independent of A as well). The expectation in the right-hand side is taken both with respect to the random matrix A and the signs (ε_i). Taking first the expectation with respect to (ε_i) (conditionally on A) and afterwards the expectation with respect to A, we obtain by Rudelson's inequality (Corollary 5.28) that

$$E \leq \frac{C\sqrt{l}}{N}\mathbb{E}\Big(\max_{i \leq N}\|A_i\|_2 \cdot \Big\|\sum_{i=1}^N A_i \otimes A_i\Big\|^{1/2}\Big)$$

where $l = \log\min(N, n)$. We now apply the Cauchy–Schwarz inequality. Since by the triangle inequality $\mathbb{E}\|\frac{1}{N}\sum_{i=1}^N A_i \otimes A_i\| = \mathbb{E}\|\frac{1}{N}A^*A\| \leq E + 1$, it follows that

$$E \leq C\sqrt{\frac{ml}{N}}(E+1)^{1/2}.$$

This inequality is easy to solve in E. Indeed, considering the cases $E \leq 1$ and $E > 1$ separately, we conclude that

$$E = \mathbb{E}\|\frac{1}{N}A^*A - I\| \leq \max(\delta, \delta^2), \quad \text{where } \delta := C\sqrt{\frac{2ml}{N}}.$$

Step 2: Diagonalization. Diagonalizing the matrix A^*A one checks that

$$\|\frac{1}{N}A^*A - I\| = \max_{j \leq n}|\frac{s_j(A)^2}{N} - 1| = \max\left(|\frac{s_{\min}(A)^2}{N} - 1|, |\frac{s_{\max}(A)^2}{N} - 1|\right).$$

It follows that

$$\max\left(\mathbb{E}|\frac{s_{\min}(A)^2}{N} - 1|, \mathbb{E}|\frac{s_{\max}(A)^2}{N} - 1|\right) \leq \max(\delta, \delta^2)$$

(we replaced the expectation of maximum by the maximum of expectations). Using Lemma 5.47 separately for the two terms on the left-hand side, we obtain

$$\max\left(\mathbb{E}|\frac{s_{\min}(A)}{\sqrt{N}} - 1|, \mathbb{E}|\frac{s_{\max}(A)}{\sqrt{N}} - 1|\right) \leq \delta.$$

Therefore

$$\mathbb{E}\max_{j \leq n}|\frac{s_j(A)}{\sqrt{N}} - 1| = \mathbb{E}\max\left(|\frac{s_{\min}(A)}{\sqrt{N}} - 1|, |\frac{s_{\max}(A)}{\sqrt{N}} - 1|\right)$$
$$\leq \mathbb{E}\left(|\frac{s_{\min}(A)}{\sqrt{N}} - 1| + |\frac{s_{\max}(A)}{\sqrt{N}} - 1|\right) \leq 2\delta.$$

Multiplying both sides by $\sqrt{N}$ completes the proof. $\square$

In a way similar to Theorem 5.44 we note that a version of Theorem 5.45 holds for general, non-isotropic distributions.

THEOREM 5.48 (Heavy-tailed rows, non-isotropic, expectation) *Let A be an $N \times n$ matrix whose rows A_i are independent random vectors in $\mathbb{R}^n$ with the common second moment matrix $\Sigma = \mathbb{E}A_i \otimes A_i$. Let $m := \mathbb{E}\max_{i \leq N}\|A_i\|_2^2$. Then*

$$\mathbb{E}\|\frac{1}{N}A^*A - \Sigma\| \leq \max(\|\Sigma\|^{1/2}\delta, \delta^2) \quad \text{where} \quad \delta = C\sqrt{\frac{m\log\min(N,n)}{N}}.$$

Here C is an absolute constant. In particular, this inequality yields

$$(\mathbb{E}\|A\|^2)^{1/2} \leq \|\Sigma\|^{1/2}\sqrt{N} + C\sqrt{m\log\min(N,n)}.$$

Proof. The first part follows by a simple modification of the proof of Theorem 5.45. The second part follows from the first like in Theorem 5.44. $\square$

REMARK 5.49 (Non-identical second moments) The assumption that the rows A_i have a common second moment matrix Σ is not essential in Theorems 5.44 and 5.48. The reader will be able to formulate more general versions of these results. For example, if A_i have arbitrary second moment matrices $\Sigma_i = \mathbb{E} A_i \otimes A_i$ then the conclusion of Theorem 5.48 holds with $\Sigma = \frac{1}{N}\sum_{i=1}^{N} \Sigma_i$.

5.4.3 Applications to estimating covariance matrices

One immediate application of our analysis of random matrices is in statistics, for the fundamental problem of *estimating covariance matrices*. Let X be a random vector in $\mathbb{R}^n$; for simplicity we assume that X is centered, $\mathbb{E} X = 0$.[20] Recall that the covariance matrix of X is the $n \times n$ matrix $\Sigma = \mathbb{E} X \otimes X$, see Section 5.2.5.

The simplest way to estimate Σ is to take some N independent samples X_i from the distribution and form the *sample covariance matrix* $\Sigma_N = \frac{1}{N}\sum_{i=1}^{N} X_i \otimes X_i$. By the law of large numbers, $\Sigma_N \to \Sigma$ almost surely as $N \to \infty$. So, taking sufficiently many samples we are guaranteed to estimate the covariance matrix as well as we want. This, however, does not address the quantitative aspect: what is the minimal *sample size N* that guarantees approximation with a given accuracy?

The relation of this question to random matrix theory becomes clear when we arrange the samples $X_i =: A_i$ as rows of the $N \times n$ random matrix A. Then the sample covariance matrix is expressed as $\Sigma_N = \frac{1}{N} A^* A$. Note that A is a matrix with independent rows but usually not independent entries (unless we sample from a product distribution). We worked out the analysis of such matrices in Section 5.4, separately for sub-gaussian and general distributions. As an immediate consequence of Theorem 5.39, we obtain:

COROLLARY 5.50 (Covariance estimation for sub-gaussian distributions) *Consider a sub-gaussian distribution in $\mathbb{R}^n$ with covariance matrix Σ, and let $\varepsilon \in (0,1)$, $t \geq 1$. Then with probability at least $1 - 2\exp(-t^2 n)$ one has*

$$\text{If } N \geq C(t/\varepsilon)^2 n \quad \text{then } \|\Sigma_N - \Sigma\| \leq \varepsilon.$$

Here $C = C_K$ depends only on the sub-gaussian norm $K = \|X\|_{\psi_2}$ of a random vector taken from this distribution.

Proof. It follows from (5.25) that for every $s \geq 0$, with probability at least $1 - 2\exp(-cs^2)$ we have $\|\Sigma_N - \Sigma\| \leq \max(\delta, \delta^2)$ where $\delta = C\sqrt{n/N} + s/\sqrt{N}$. The conclusion follows for $s = C't\sqrt{n}$ where $C' = C'_K$ is sufficiently large. □

Summarizing, Corollary 5.50 shows that the size

$$N = O(n)$$

suffices to approximate the covariance matrix of a sub-gaussian distribution in $\mathbb{R}^n$ by the sample covariance matrix.

---

[20] More generally, in this section we estimate the *second moment matrix* $\mathbb{E} X \otimes X$ of an arbitrary random vector X (not necessarily centered).

REMARK 5.51 (Multiplicative estimates, Gaussian distributions) A weak point of Corollary 5.50 is that the sub-gaussian norm K may in turn depend on $\|\Sigma\|$.

To overcome this drawback, instead of using (5.25) in the proof of this result one can use the multiplicative version (5.26). The reader is encouraged to state a general result that follows from this argument. We just give one special example for arbitrary *centered Gaussian distributions* in $\mathbb{R}^n$. For every $\varepsilon \in (0,1)$, $t \geq 1$, the following holds with probability at least $1 - 2\exp(-t^2 n)$:

$$\text{If } N \geq C(t/\varepsilon)^2 n \quad \text{then } \|\Sigma_N - \Sigma\| \leq \varepsilon \|\Sigma\|.$$

Here C is an absolute constant.

Finally, Theorem 5.44 yields a similar estimation result for arbitrary distributions, possibly heavy-tailed:

COROLLARY 5.52 (Covariance estimation for arbitrary distributions) *Consider a distribution in $\mathbb{R}^n$ with covariance matrix Σ and supported in some centered Euclidean ball whose radius we denote $\sqrt{m}$. Let $\varepsilon \in (0,1)$ and $t \geq 1$. Then the following holds with probability at least $1 - n^{-t^2}$:*

$$\text{If } N \geq C(t/\varepsilon)^2 \|\Sigma\|^{-1} m \log n \quad \text{then } \|\Sigma_N - \Sigma\| \leq \varepsilon \|\Sigma\|.$$

Here C is an absolute constant.

Proof. It follows from Theorem 5.44 that for every $s \geq 0$, with probability at least $1 - n \cdot \exp(-cs^2)$ we have $\|\Sigma_N - \Sigma\| \leq \max(\|\Sigma\|^{1/2}\delta, \delta^2)$ where $\delta = s\sqrt{m/N}$. Therefore, if $N \geq (s/\varepsilon)^2 \|\Sigma\|^{-1} m$ then $\|\Sigma_N - \Sigma\| \leq \varepsilon \|\Sigma\|$. The conclusion follows with $s = C't\sqrt{\log n}$ where C' is a sufficiently large absolute constant. □

Corollary 5.52 is typically used with $m = O(\|\Sigma\|n)$. Indeed, if X is a random vector chosen from the distribution in question, then its expected norm is easy to estimate: $\mathbb{E}\|X\|_2^2 = \text{tr}(\Sigma) \leq n\|\Sigma\|$. So, by Markov's inequality, most of the distribution is supported in a centered ball of radius $\sqrt{m}$ where $m = O(n\|\Sigma\|)$. If all distribution is supported there, i.e. if $\|X\| = O(\sqrt{n\|\Sigma\|})$ almost surely, then the conclusion of Corollary 5.52 holds with sample size $N \geq C(t/\varepsilon)^2 n \log n$.

REMARK 5.53 (Low-rank estimation) In certain applications, the distribution in $\mathbb{R}^n$ lies close to a low-dimensional subspace. In this case, a smaller sample suffices for covariance estimation. The intrinsic dimension of the distribution can be measured with the *effective rank* of the matrix Σ, defined as

$$r(\Sigma) = \frac{\text{tr}(\Sigma)}{\|\Sigma\|}.$$

One always has $r(\Sigma) \leq \text{rank}(\Sigma) \leq n$, and this band is sharp.

For example, if X is an isotropic random vector in $\mathbb{R}^n$ then $\Sigma = I$ and $r(\Sigma) = n$. A more interesting example is where X takes values in some r-dimensional subspace E, and the restriction of the distribution of X onto E is isotropic. The latter means that $\Sigma = P_E$, where P_E denotes the orthogonal projection in $\mathbb{R}^n$ onto E. Therefore in this case $r(\Sigma) = r$. The effective rank is a stable quantity compared with the usual rank. For distributions that are approximately low-dimensional, the effective rank is still small.

The effective rank $r = r(\Sigma)$ always controls the typical norm of X, as $\mathbb{E}\|X\|_2^2 = \operatorname{tr}(\Sigma) = r\|\Sigma\|$. It follows by Markov's inequality that most of the distribution is supported in a ball of radius $\sqrt{m}$ where $m = O(r\|\Sigma\|)$. Assume that all of the distribution is supported there, i.e. if $\|X\| = O(\sqrt{r\|\Sigma\|})$ almost surely. Then the conclusion of Corollary 5.52 holds with sample size $N \geq C(t/\varepsilon)^2 r \log n$.

We can summarize this discussion in the following way: the sample size

$$N = O(n \log n)$$

suffices to approximate the covariance matrix of a general distribution in $\mathbb{R}^n$ by the sample covariance matrix. Furthermore, for distributions that are approximately low-dimensional, a smaller sample size is sufficient. Namely, if the effective rank of Σ equals r then a sufficient sample size is

$$N = O(r \log n).$$

REMARK 5.54 (Boundedness assumption) Without the boundedness assumption on the distribution, Corollary 5.52 may fail. The reasoning is the same as in Remark 5.42: for an isotropic distribution which is highly concentrated at the origin, the sample covariance matrix will likely equal 0.

Still, one can weaken the boundedness assumption using Theorem 5.48 instead of Theorem 5.44 in the proof of Corollary 5.52. The weaker requirement is that $\mathbb{E}\max_{i \leq N}\|X_i\|_2^2 \leq m$ where X_i denote the sample points. In this case, the covariance estimation will be guaranteed in expectation rather than with high probability; we leave the details for the interested reader.

A different way to enforce the boundedness assumption is to reject any sample points X_i that fall outside the centered ball of radius $\sqrt{m}$. This is equivalent to sampling from the conditional distribution inside the ball. The conditional distribution satisfies the boundedness requirement, so the results discussed above provide a good covariance estimation for it. In many cases, this estimate works even for the original distribution – namely, if only a small part of the distribution lies outside the ball of radius $\sqrt{m}$. We leave the details for the interested reader; see e.g. [81].

5.4.4 Applications to random sub-matrices and sub-frames

The absence of any moment hypotheses on the distribution in Section 5.4.2 (except finite variance) makes these results especially relevant for discrete distributions. One such situation arises when one wishes to sample entries or rows from a given matrix B, thereby creating a *random sub-matrix* A. It is a big program to understand what we can learn about B by seeing A, see [34, 25, 66]. In other words, we ask – what properties of B pass onto A? Here we shall only scratch the surface of this problem: we notice that random sub-matrices of certain size preserve the property of being an *approximate isometry*.

COROLLARY 5.55 (Random sub-matrices) *Consider an $M \times n$ matrix B such that $s_{\min}(B) = s_{\max}(B) = \sqrt{M}$.[21] Let m be such that all rows B_i of B satisfy $\|B_i\|_2 \le \sqrt{m}$. Let A be an $N \times n$ matrix obtained by sampling N random rows from B uniformly and independently. Then for every $t \ge 0$, with probability at least $1 - 2n \cdot \exp(-ct^2)$ one has*

$$\sqrt{N} - t\sqrt{m} \le s_{\min}(A) \le s_{\max}(A) \le \sqrt{N} + t\sqrt{m}.$$

Here $c > 0$ is an absolute constant.

Proof. By assumption, $I = \frac{1}{M} B^* B = \frac{1}{M} \sum_{i=1}^{M} B_i \otimes B_i$. Therefore, the uniform distribution on the set of the rows $\{B_1, \ldots, B_M\}$ is an isotropic distribution in $\mathbb{R}^n$. The conclusion then follows from Theorem 5.41. □

Note that the conclusion of Corollary 5.55 does not depend on the dimension M of the ambient matrix B. This happens because this result is a specific version of sampling from a discrete isotropic distribution (uniform on the rows of B), where size M of the support of the distribution is irrelevant.

The hypothesis of Corollary 5.55 implies that $\frac{1}{M} \sum_{i=1}^{M} \|B_i\|_2^2 = n$.[22] Hence by Markov's inequality, most of the rows B_i satisfy $\|B_i\|_2 = O(\sqrt{n})$. This indicates that Corollary 5.55 would be often used with $m = O(n)$. Also, to ensure a positive probability of success, the useful magnitude of t would be $t \sim \sqrt{\log n}$. With this in mind, the extremal singular values of A will be close to each other (and to $\sqrt{N}$) if $N \gg t^2 m \sim n \log n$.

Summarizing, Corollary 5.55 states that a random $O(n \log n) \times n$ sub-matrix of an $M \times n$ isometry is an approximate isometry.[23]

Another application of random matrices with heavy-tailed isotropic rows is for *sampling from frames*. Recall that frames are generalizations of bases without linear independence, see Example 5.21. Consider a tight frame $\{u_i\}_{i=1}^{M}$ in $\mathbb{R}^n$, and for the sake of convenient normalization, assume that it has bounds $A = B = M$. We are interested in whether a small random subset of $\{u_i\}_{i=1}^{M}$ is still a nice frame in $\mathbb{R}^n$. Such a question arises naturally because frames are used in signal processing to create *redundant representations* of signals. Indeed, every signal $x \in \mathbb{R}^n$ admits frame expansion $x = \frac{1}{M} \sum_{i=1}^{M} \langle u_i, x \rangle u_i$. Redundancy makes frame representations more robust to errors and losses than basis representations. Indeed, we will show that if one loses all except $N = O(n \log n)$ random coefficients $\langle u_i, x \rangle$ one is still able to reconstruct x from the received coefficients $\langle u_{i_k}, x \rangle$ as $x \approx \frac{1}{N} \sum_{k=1}^{N} \langle u_{i_k}, x \rangle u_{i_k}$. This boils down to showing

---

[21] The first hypothesis says $B^* B = MI$. Equivalently, $\bar{B} := \frac{1}{\sqrt{M}} B$ is an isometry, i.e. $\|\bar{B}x\|_2 = \|x\|_2$ for all x. Equivalently, the columns of $\bar{B}$ are orthonormal.
[22] To recall why this is true, take trace of both sides in the identity $I = \frac{1}{M} \sum_{i=1}^{M} B_i \otimes B_i$.
[23] For the purposes of compressed sensing, we shall study the more difficult *uniform* problem for random sub-matrices in Section 5.6. There B itself will be chosen as a column sub-matrix of a given $M \times M$ matrix (such as DFT), and one will need to control all such B simultaneously, see Example 5.73.

that a random subset of size $N = O(n \log n)$ of a tight frame in $\mathbb{R}^n$ is an approximate tight frame.

COROLLARY 5.56 (Random sub-frames, see [80])   *Consider a tight frame $\{u_i\}_{i=1}^M$ in $\mathbb{R}^n$ with frame bounds $A = B = M$. Let number m be such that all frame elements satisfy $\|u_i\|_2 \leq \sqrt{m}$. Let $\{v_i\}_{i=1}^N$ be a set of vectors obtained by sampling N random elements from the frame $\{u_i\}_{i=1}^M$ uniformly and independently. Let $\varepsilon \in (0,1)$ and $t \geq 1$. Then the following holds with probability at least $1 - 2n^{-t^2}$:*

$$\text{If } N \geq C(t/\varepsilon)^2 m \log n \text{ then } \{v_i\}_{i=1}^N \text{ is a frame in } \mathbb{R}^n$$

with bounds $A = (1-\varepsilon)N$, $B = (1+\varepsilon)N$. Here C is an absolute constant.

In particular, if this event holds, then every $x \in \mathbb{R}^n$ admits an approximate representation using only the sampled frame elements:

$$\left\| \frac{1}{N} \sum_{i=1}^N \langle v_i, x \rangle v_i - x \right\| \leq \varepsilon \|x\|.$$

Proof. The assumption implies that $I = \frac{1}{M} \sum_{i=1}^M u_i \otimes u_i$. Therefore, the uniform distribution on the set $\{u_i\}_{i=1}^M$ is an isotropic distribution in $\mathbb{R}^n$. Applying Corollary 5.52 with $\Sigma = I$ and $\Sigma_N = \frac{1}{N} \sum_{i=1}^N v_i \otimes v_i$ we conclude that $\|\Sigma_N - I\| \leq \varepsilon$ with the required probability. This clearly completes the proof. □

As before, we note that $\frac{1}{M} \sum_{i=1}^M \|u_i\|_2^2 = n$, so Corollary 5.56 would be often used with $m = O(n)$. This shows, liberally speaking, that a random subset of a frame in $\mathbb{R}^n$ of size $N = O(n \log n)$ is again a frame.

REMARK 5.57 (Non-uniform sampling) The boundedness assumption $\|u_i\|_2 \leq \sqrt{m}$, although needed in Corollary 5.56, can be removed by non-uniform sampling. To this end, one would sample from the set of normalized vectors $\bar{u}_i := \sqrt{n} \frac{u_i}{\|u_i\|_2}$ with probabilities proportional to $\|u_i\|_2^2$. This defines an isotropic distribution in $\mathbb{R}^n$, and clearly $\|\bar{u}_i\|_2 = \sqrt{n}$. Therefore, by Theorem 5.56, a random sample of $N = O(n \log n)$ vectors obtained this way forms an almost tight frame in $\mathbb{R}^n$. This result does not require any bound on $\|u_i\|_2$.

5.5 Random matrices with independent columns

In this section we study the extreme singular values of $N \times n$ random matrices A with independent columns A_j. We are guided by our ideal bounds (5.2) as before. The same phenomenon occurs in the column independent model as in the row independent model – sufficiently tall random matrices A are approximate isometries. As before, being tall will mean $N \gg n$ for sub-gaussian distributions and $N \gg n \log n$ for arbitrary distributions.

The problem is equivalent to studying *Gram matrices* $G = A^* A = (\langle A_j, A_k \rangle)_{j,k=1}^n$ of independent isotropic random vectors $A_1, \ldots, A_n$ in $\mathbb{R}^N$. Our results can be interpreted using Lemma 5.36 as showing that the normalized Gram matrix $\frac{1}{N}G$ is an *approximate identity* for N, n as above.

Let us first try to prove this with a heuristic argument. By Lemma 5.20 we know that the diagonal entries of $\frac{1}{N}G$ have mean $\frac{1}{N}\mathbb{E}\|A_j\|_2^2 = 1$ and off-diagonal ones have zero mean and standard deviation $\frac{1}{N}(\mathbb{E}\langle A_j, A_k\rangle^2)^{1/2} = \frac{1}{\sqrt{N}}$. If, hypothetically, the off-diagonal entries were independent, then we could use the results of matrices with independent entries (or even rows) developed in Section 5.4. The off-diagonal part of $\frac{1}{N}G$ would have norm $O(\sqrt{\frac{n}{N}})$ while the diagonal part would approximately equal I. Hence we would have

$$\|\frac{1}{N}G - I\| = O\left(\sqrt{\frac{n}{N}}\right), \tag{5.35}$$

i.e. $\frac{1}{N}G$ is an approximate identity for $N \gg n$. Equivalently, by Lemma 5.36, (5.35) would yield the ideal bounds (5.2) on the extreme singular values of A.

Unfortunately, the entries of the Gram matrix G are obviously not independent. To overcome this obstacle we shall use the *decoupling* technique of probability theory [22]. We observe that there is still enough independence encoded in G. Consider a principal sub-matrix $(A_S)^*(A_T)$ of $G = A^*A$ with disjoint index sets S and T. If we condition on $(A_k)_{k \in T}$ then this sub-matrix has independent rows. Using an elementary decoupling technique, we will indeed seek to replace the full Gram matrix G by one such decoupled $S \times T$ matrix with independent rows, and finish off by applying results of Section 5.4.

By transposition one can try to reduce our problem to studying the $n \times N$ matrix A^*. It has independent rows and the same singular values as A, so one can apply results of Section 5.4. The conclusion would be that, with high probability,

$$\sqrt{n} - C\sqrt{N} \leq s_{\min}(A) \leq s_{\max}(A) \leq \sqrt{n} + C\sqrt{N}.$$

Such an estimate is only good for *flat* matrices ($N \leq n$). For *tall* matrices ($N \geq n$) the lower bound would be trivial because of the (possibly large) constant C. So, from now on we can focus on tall matrices ($N \geq n$) with independent columns.

5.5.1 Sub-gaussian columns

Here we prove a version of Theorem 5.39 for matrices with independent columns.

THEOREM 5.58 (Sub-gaussian columns) *Let A be an $N \times n$ matrix ($N \geq n$) whose columns A_i are independent sub-gaussian isotropic random vectors in $\mathbb{R}^N$ with $\|A_j\|_2 = \sqrt{N}$ a. s. Then for every $t \geq 0$, the inequality holds*

$$\sqrt{N} - C\sqrt{n} - t \leq s_{\min}(A) \leq s_{\max}(A) \leq \sqrt{N} + C\sqrt{n} + t \tag{5.36}$$

with probability at least $1 - 2\exp(-ct^2)$, where $C = C'_K$, $c = c'_K > 0$ depend only on the sub-gaussian norm $K = \max_j \|A_j\|_{\psi_2}$ of the columns.

The only significant difference between Theorem 5.39 for independent rows and Theorem 5.58 for independent columns is that the latter requires *normalization of columns*,

$\|A_j\|_2 = \sqrt{N}$ almost surely. Recall that by isotropy of A_j (see Lemma 5.20) one always has $(\mathbb{E}\|A_j\|_2^2)^{1/2} = \sqrt{N}$, but the normalization is a bit stronger requirement. We will discuss this more after the proof of Theorem 5.58.

REMARK 5.59 (Gram matrices are an approximate identity) By Lemma 5.36, the conclusion of Theorem 5.58 is equivalent to

$$\left\|\frac{1}{N}A^*A - I\right\| \leq C\sqrt{\frac{n}{N}} + \frac{t}{\sqrt{N}}$$

with the same probability $1 - 2\exp(-ct^2)$. This establishes our ideal inequality (5.35). In words, the normalized Gram matrix of n independent sub-gaussian isotropic random vectors in $\mathbb{R}^N$ is an approximate identity whenever $N \gg n$.

The proof of Theorem 5.58 is based on the decoupling technique [22]. What we will need here is an elementary decoupling lemma for double arrays. Its statement involves the notion of a *random subset* of a given finite set. To be specific, we define a random set T of $[n]$ with a given average size $m \in [0, n]$ as follows. Consider independent $\{0,1\}$ valued random variables $\delta_1, \ldots, \delta_n$ with $\mathbb{E}\delta_i = m/n$; these are sometimes called *independent selectors*. Then we define the random subset $T = \{i \in [n] : \delta_i = 1\}$. Its average size equals $\mathbb{E}|T| = \mathbb{E}\sum_{i=1}^n \delta_i = m$.

LEMMA 5.60 (Decoupling) *Consider a double array of real numbers* $(a_{ij})_{i,j=1}^n$ *such that* $a_{ii} = 0$ *for all* i. *Then*

$$\sum_{i,j \in [n]} a_{ij} = 4\mathbb{E} \sum_{i \in T, j \in T^c} a_{ij}$$

where T *is a random subset of* $[n]$ *with average size* $n/2$. *In particular,*

$$4 \min_{T \subseteq [n]} \sum_{i \in T, j \in T^c} a_{ij} \leq \sum_{i,j \in [n]} a_{ij} \leq 4 \max_{T \subseteq [n]} \sum_{i \in T, j \in T^c} a_{ij}$$

where the minimum and maximum are over all subsets T *of* $[n]$.

Proof. Expressing the random subset as $T = \{i \in [n] : \delta_i = 1\}$ where δ_i are independent selectors with $\mathbb{E}\delta_i = 1/2$, we see that

$$\mathbb{E} \sum_{i \in T, j \in T^c} a_{ij} = \mathbb{E} \sum_{i,j \in [n]} \delta_i(1-\delta_j) a_{ij} = \frac{1}{4} \sum_{i,j \in [n]} a_{ij},$$

where we used that $\mathbb{E}\delta_i(1-\delta_j) = 1/4$ for $i \neq j$ and the assumption $a_{ii} = 0$. This proves the first part of the lemma. The second part follows trivially by estimating expectation by maximum and minimum. □

Proof of Theorem 5.58. **Step 1: Reductions.** Without loss of generality we can assume that the columns A_i have zero mean. Indeed, multiplying each column A_i by ± 1 arbitrarily preserves the extreme singular values of A, the isotropy of A_i, and the sub-gaussian norms of A_i. Therefore, by multiplying A_i by independent symmetric Bernoulli random variables we achieve that A_i have zero mean.

For $t = O(\sqrt{N})$ the conclusion of Theorem 5.58 follows from Theorem 5.39 by transposition. Indeed, the $n \times N$ random matrix A^* has independent rows, so for $t \geq 0$ we have

$$s_{\max}(A) = s_{\max}(A^*) \leq \sqrt{n} + C_K \sqrt{N} + t \tag{5.37}$$

with probability at least $1 - 2\exp(-c_K t^2)$. Here $c_K > 0$ and we can obviously assume that $C_K \geq 1$. For $t \geq C_K \sqrt{N}$ it follows that $s_{\max}(A) \leq \sqrt{N} + \sqrt{n} + 2t$, which yields the conclusion of Theorem 5.58 (the left-hand side of (5.36) being trivial). So, it suffices to prove the conclusion for $t \leq C_K \sqrt{N}$. Let us fix such t.

It would be useful to have some a priori control of $s_{\max}(A) = \|A\|$. We thus consider the desired event

$$\mathcal{E} := \{ s_{\max}(A) \leq 3C_K \sqrt{N} \}.$$

Since $3C_K \sqrt{N} \geq \sqrt{n} + C_K \sqrt{N} + t$, by (5.37) we see that $\mathcal{E}$ is likely to occur:

$$\mathbb{P}(\mathcal{E}^c) \leq 2\exp(-c_K t^2). \tag{5.38}$$

Step 2: Approximation. This step is parallel to Step 1 in the proof of Theorem 5.39, except now we shall choose $\varepsilon := \delta$. This way we reduce our task to the following. Let $\mathcal{N}$ be a $\frac{1}{4}$-net of the unit sphere S^{n-1} such that $|\mathcal{N}| \leq 9^n$. It suffices to show that with probability at least $1 - 2\exp(-c'_K t^2)$ one has

$$\max_{x \in \mathcal{N}} \left| \frac{1}{N} \|Ax\|_2^2 - 1 \right| \leq \frac{\delta}{2}, \quad \text{where } \delta = C\sqrt{\frac{n}{N}} + \frac{t}{\sqrt{N}}.$$

By (5.38), it is enough to show that the probability

$$p := \mathbb{P}\left\{ \max_{x \in \mathcal{N}} \left| \frac{1}{N} \|Ax\|_2^2 - 1 \right| > \frac{\delta}{2} \text{ and } \mathcal{E} \right\} \tag{5.39}$$

satisfies $p \leq 2\exp(-c''_K t^2)$, where $c''_K > 0$ may depend only on K.

Step 3: Decoupling. As in the proof of Theorem 5.39, we will obtain the required bound for a fixed $x \in \mathcal{N}$ with high probability, and then take a union bound over x. So let us fix any $x = (x_1, \ldots, x_n) \in S^{n-1}$. We expand

$$\|Ax\|_2^2 = \left\| \sum_{j=1}^n x_j A_j \right\|_2^2 = \sum_{j=1}^n x_j^2 \|A_j\|_2^2 + \sum_{j,k \in [n], j \neq k} x_j x_k \langle A_j, A_k \rangle. \tag{5.40}$$

Since $\|A_j\|_2^2 = N$ by assumption and $\|x\|_2 = 1$, the first sum equals N. Therefore, subtracting N from both sides and dividing by N, we obtain the bound

$$\left| \frac{1}{N} \|Ax\|_2^2 - 1 \right| \leq \left| \frac{1}{N} \sum_{j,k \in [n], j \neq k} x_j x_k \langle A_j, A_k \rangle \right|.$$

The sum in the right-hand side is $\langle G_0 x, x \rangle$ where G_0 is the off-diagonal part of the Gram matrix $G = A^* A$. As we indicated at the beginning of Section 5.5, we are going to replace G_0 by its decoupled version whose rows and columns are indexed by disjoint sets. This is achieved by Decoupling Lemma 5.60: we obtain

$$\left| \frac{1}{N} \|Ax\|_2^2 - 1 \right| \le \frac{4}{N} \max_{T \subseteq [n]} |R_T(x)|, \quad \text{where } R_T(x) = \sum_{j \in T, k \in T^c} x_j x_k \langle A_j, A_k \rangle.$$

We substitute this into (5.39) and take the union bound over all choices of $x \in \mathcal{N}$ and $T \subseteq [n]$. As we know, $|\mathcal{N}| \le 9^n$, and there are 2^n subsets T in $[n]$. This gives

$$p \le \mathbb{P}\left\{ \max_{x \in \mathcal{N}, T \subseteq [n]} |R_T(x)| > \frac{\delta N}{8} \text{ and } \mathcal{E} \right\}$$

$$\le 9^n \cdot 2^n \cdot \max_{x \in \mathcal{N}, T \subseteq [n]} \mathbb{P}\left\{ |R_T(x)| > \frac{\delta N}{8} \text{ and } \mathcal{E} \right\}. \tag{5.41}$$

Step 4: Conditioning and concentration. To estimate the probability in (5.41), we fix a vector $x \in \mathcal{N}$ and a subset $T \subseteq [n]$ and we condition on a realization of random vectors $(A_k)_{k \in T^c}$. We express

$$R_T(x) = \sum_{j \in T} x_j \langle A_j, z \rangle \quad \text{where } z = \sum_{k \in T^c} x_k A_k. \tag{5.42}$$

Under our conditioning z is a fixed vector, so $R_T(x)$ is a sum of independent random variables. Moreover, if event $\mathcal{E}$ holds then z is nicely bounded:

$$\|z\|_2 \le \|A\| \|x\|_2 \le 3 C_K \sqrt{N}. \tag{5.43}$$

If in turn (5.43) holds then the terms $\langle A_j, z \rangle$ in (5.42) are independent centered subgaussian random variables with $\|\langle A_j, z \rangle\|_{\psi_2} \le 3 K C_K \sqrt{N}$. By Lemma 5.9, their linear combination $R_T(x)$ is also a sub-gaussian random variable with

$$\|R_T(x)\|_{\psi_2} \le C_1 \Big(\sum_{j \in T} x_j^2 \|\langle A_j, z \rangle\|_{\psi_2}^2 \Big)^{1/2} \le \widehat{C}_K \sqrt{N} \tag{5.44}$$

where $\widehat{C}_K$ depends only on K.

We can summarize these observations as follows. Denoting the conditional probability by $\mathbb{P}_T = \mathbb{P}\{\cdot \,|\, (A_k)_{k \in T^c}\}$ and the expectation with respect to $(A_k)_{k \in T^c}$ by $\mathbb{E}_{T^c}$, we obtain by (5.43) and (5.44) that

$$\mathbb{P}\left\{ |R_T(x)| > \frac{\delta N}{8} \text{ and } \mathcal{E} \right\} \le \mathbb{E}_{T^c} \mathbb{P}_T \left\{ |R_T(x)| > \frac{\delta N}{8} \text{ and } \|z\|_2 \le 3 C_K \sqrt{N} \right\}$$

$$\le 2 \exp\left[-c_1 \Big(\frac{\delta N/8}{\widehat{C}_K \sqrt{N}} \Big)^2 \right] = 2 \exp\Big(-\frac{c_2 \delta^2 N}{\widehat{C}_K^2} \Big) \le 2 \exp\Big(-\frac{c_2 C^2 n}{\widehat{C}_K^2} - \frac{c_2 t^2}{\widehat{C}_K^2} \Big).$$

The second inequality follows because $R_T(x)$ is a sub-gaussian random variable (5.44) whose tail decay is given by (5.10). Here $c_1, c_2 > 0$ are absolute constants. The last inequality follows from the definition of δ. Substituting this into (5.41) and choosing C sufficiently large (so that $\ln 36 \leq c_2 C^2/\widehat{C}_K^2$), we conclude that

$$p \leq 2\exp\left(-c_2 t^2/\widehat{C}_K^2\right).$$

This proves an estimate that we desired in Step 2. The proof is complete. □

REMARK 5.61 (Normalization assumption) Some a priori control of the norms of the columns $\|A_j\|_2$ is necessary for estimating the extreme singular values, since

$$s_{\min}(A) \leq \min_{i \leq n} \|A_j\|_2 \leq \max_{i \leq n} \|A_j\|_2 \leq s_{\max}(A).$$

With this in mind, it is easy to construct an example showing that a normalization assumption $\|A_i\|_2 = \sqrt{N}$ is essential in Theorem 5.58; it cannot even be replaced by a boundedness assumption $\|A_i\|_2 = O(\sqrt{N})$.

Indeed, consider a random vector $X = \sqrt{2}\xi Y$ in $\mathbb{R}^N$ where ξ is a $\{0,1\}$-valued random variable with $\mathbb{E}\xi = 1/2$ (a "selector") and X is an independent spherical random vector in $\mathbb{R}^n$ (see Example 5.25). Let A be a random matrix whose columns A_j are independent copies of X. Then A_j are independent centered sub-gaussian isotropic random vectors in $\mathbb{R}^n$ with $\|A_j\|_{\psi_2} = O(1)$ and $\|A_j\|_2 \leq \sqrt{2N}$ a.s. So all assumptions of Theorem 5.58 except normalization are satisfied. On the other hand $\mathbb{P}\{X = 0\} = 1/2$, so matrix A has a zero column with overwhelming probability $1 - 2^{-n}$. This implies that $s_{\min}(A) = 0$ with this probability, so the lower estimate in (5.36) is false for all nontrivial N, n, t.

5.5.2 Heavy-tailed columns

Here we prove a version of Theorem 5.45 for independent heavy-tailed columns.

We thus consider $N \times n$ random matrices A with independent columns A_j. In addition to the normalization assumption $\|A_j\|_2 = \sqrt{N}$ already present in Theorem 5.58 for sub-gaussian columns, our new result must also require an a priori control of the off-diagonal part of the Gram matrix $G = A^*A = (\langle A_j, A_k \rangle)_{j,k=1}^n$.

THEOREM 5.62 (Heavy-tailed columns) *Let A be an $N \times n$ matrix ($N \geq n$) whose columns A_j are independent isotropic random vectors in $\mathbb{R}^N$ with $\|A_j\|_2 = \sqrt{N}$ a. s. Consider the incoherence parameter*

$$m := \frac{1}{N}\mathbb{E}\max_{j \leq n} \sum_{k \in [n], k \neq j} \langle A_j, A_k \rangle^2.$$

Then $\mathbb{E}\|\frac{1}{N}A^*A - I\| \leq C_0\sqrt{\frac{m \log n}{N}}$. In particular,

$$\mathbb{E}\max_{j \leq n}|s_j(A) - \sqrt{N}| \leq C\sqrt{m \log n}. \tag{5.45}$$

Let us briefly clarify the role of the incoherence parameter m, which controls the lengths of the rows of the off-diagonal part of G. After the proof we will see that a control of m is essential in Theorem 5.41. But for now, let us get a feel of the typical size of m. We have $\mathbb{E}\langle A_j, A_k\rangle^2 = N$ by Lemma 5.20, so for every row j we see that $\frac{1}{N}\sum_{k\in[n], k\neq j}\langle A_j, A_k\rangle^2 = n-1$. This indicates that Theorem 5.62 would be often used with $m = O(n)$.

In this case, Theorem 5.41 establishes our ideal inequality (5.35) up to a logarithmic factor. In words, the normalized Gram matrix of n independent isotropic random vectors in $\mathbb{R}^N$ is an approximate identity whenever $N \gg n\log n$.

Our proof of Theorem 5.62 will be based on decoupling, symmetrization and an application of Theorem 5.48 for a decoupled Gram matrix with independent rows. The decoupling is done similarly to Theorem 5.58. However, this time we will benefit from formalizing the decoupling inequality for Gram matrices:

LEMMA 5.63 (Matrix decoupling) *Let B be a $N \times n$ random matrix whose columns B_j satisfy $\|B_j\|_2 = 1$. Then*

$$\mathbb{E}\|B^*B - I\| \leq 4 \max_{T \subseteq [n]} \mathbb{E}\|(B_T)^* B_{T^c}\|.$$

Proof. We first note that $\|B^*B - I\| = \sup_{x \in S^{n-1}} \left|\|Bx\|_2^2 - 1\right|$. We fix $x = (x_1, \ldots, x_n) \in S^{n-1}$ and, expanding as in (5.40), observe that

$$\|Bx\|_2^2 = \sum_{j=1}^n x_j^2 \|B_j\|_2^2 + \sum_{j,k\in[n], j\neq k} x_j x_k \langle B_j, B_k\rangle.$$

The first sum equals 1 since $\|B_j\|_2 = \|x\|_2 = 1$. So by Decoupling Lemma 5.60, a random subset T of $[n]$ with average cardinality $n/2$ satisfies

$$\|Bx\|_2^2 - 1 = 4\mathbb{E}_T \sum_{j\in T, k\in T^c} x_j x_k \langle B_j, B_k\rangle.$$

Let us denote by $\mathbb{E}_T$ and $\mathbb{E}_B$ the expectations with respect to the random set T and the random matrix B respectively. Using Jensen's inequality we obtain

$$\mathbb{E}_B\|B^*B - I\| = \mathbb{E}_B \sup_{x\in S^{n-1}} \left|\|Bx\|_2^2 - 1\right|$$

$$\leq 4\mathbb{E}_B\mathbb{E}_T \sup_{x\in S^{n-1}} \left|\sum_{j\in T, k\in T^c} x_j x_k \langle B_j, B_k\rangle\right| = 4\mathbb{E}_T\mathbb{E}_B\|(B_T)^*B_{T^c}\|.$$

The conclusion follows by replacing the expectation by the maximum over T. □

Proof of Theorem 5.62. **Step 1: Reductions and decoupling.** It would be useful to have an a priori bound on $s_{\max}(A) = \|A\|$. We can obtain this by transposing A and applying one of the results of Section 5.4. Indeed, the random $n \times N$ matrix A^* has independent rows A_i^* which by our assumption are normalized as $\|A_i^*\|_2 = \|A_i\|_2 = \sqrt{N}$.

Applying Theorem 5.45 with the roles of n and N switched, we obtain by the triangle inequality that

$$\mathbb{E}\|A\| = \mathbb{E}\|A^*\| = \mathbb{E}s_{\max}(A^*) \leq \sqrt{n} + C\sqrt{N\log n} \leq C\sqrt{N\log n}. \quad (5.46)$$

Observe that $n \leq m$ since by Lemma 5.20 we have $\frac{1}{N}\mathbb{E}\langle A_j, A_k\rangle^2 = 1$ for $j \neq k$. We use Matrix Decoupling Lemma 5.63 for $B = \frac{1}{\sqrt{N}}A$ and obtain

$$E \leq \frac{4}{N}\max_{T \subseteq [n]} \mathbb{E}\|(A_T)^* A_{T^c}\| = \frac{4}{N}\max_{T \subseteq [n]} \mathbb{E}\|\Gamma\| \quad (5.47)$$

where $\Gamma = \Gamma(T)$ denotes the decoupled Gram matrix

$$\Gamma = (A_T)^* A_{T^c} = \big(\langle A_j, A_k\rangle\big)_{j\in T, k\in T^c}.$$

Let us fix T; our problem then reduces to bounding the expected norm of Γ.

Step 2: The rows of the decoupled Gram matrix. For a subset $S \subseteq [n]$, we denote by $\mathbb{E}_{A_S}$ the conditional expectation given A_{S^c}, i.e. with respect to $A_S = (A_j)_{j\in S}$. Hence $\mathbb{E} = \mathbb{E}_{A_{T^c}}\mathbb{E}_{A_T}$.

Let us condition on A_{T^c}. Treating $(A_k)_{k\in T^c}$ as fixed vectors we see that, conditionally, the random matrix Γ has independent rows

$$\Gamma_j = \big(\langle A_j, A_k\rangle\big)_{k\in T^c}, \quad j \in T.$$

So we are going to use Theorem 5.48 to bound the norm of Γ. To do this we need estimates on (a) the norms and (b) the second moment matrices of the rows Γ_j.

(a) Since for $j \in T$, Γ_j is a random vector valued in $\mathbb{R}^{T^c}$, we estimate its second moment matrix by choosing $x \in \mathbb{R}^{T^c}$ and evaluating the scalar second moment

$$\mathbb{E}_{A_T}\langle\Gamma_j, x\rangle^2 = \mathbb{E}_{A_T}\Big(\sum_{k\in T^c}\langle A_j, A_k\rangle x_k\Big)^2 = \mathbb{E}_{A_T}\Big\langle A_j, \sum_{k\in T^c}x_k A_k\Big\rangle^2$$

$$= \Big\|\sum_{k\in T^c}x_k A_k\Big\|^2 = \|A_{T^c} x\|_2^2 \leq \|A_{T^c}\|_2^2 \|x\|_2^2.$$

In the third equality we used isotropy of A_j. Taking the maximum over all $j \in T$ and $x \in \mathbb{R}^{T^c}$, we see that the second moment matrix $\Sigma(\Gamma_j) = \mathbb{E}_{A_T}\Gamma_j \otimes \Gamma_j$ satisfies

$$\max_{j\in T}\|\Sigma(\Gamma_j)\| \leq \|A_{T^c}\|^2. \quad (5.48)$$

(b) To evaluate the norms of Γ_j, $j \in T$, note that $\|\Gamma_j\|_2^2 = \sum_{k\in T^c}\langle A_j, A_k\rangle^2$. This is easy to bound, because the assumption says that the random variable

$$M := \frac{1}{N}\max_{j\in[n]}\sum_{k\in[n], k\neq j}\langle A_j, A_k\rangle^2 \quad \text{satisfies } \mathbb{E}M = m.$$

This produces the bound $\mathbb{E}\max_{j\in T}\|\Gamma_j\|_2^2 \leq N \cdot \mathbb{E}M = Nm$. But at this moment we need to work conditionally on A_{T^c}, so for now we will be satisfied with

$$\mathbb{E}_{A_T} \max_{j\in T}\|\Gamma_j\|_2^2 \leq N \cdot \mathbb{E}_{A_T} M. \tag{5.49}$$

Step 3: The norm of the decoupled Gram matrix. We bound the norm of the random $T \times T^c$ Gram matrix Γ with (conditionally) independent rows using Theorem 5.48 and Remark 5.49. Since by (5.48) we have $\|\frac{1}{|T|}\sum_{j\in T}\Sigma(\Gamma_j)\| \leq \frac{1}{|T|}\sum_{j\in T}\|\Sigma(\Gamma_j)\| \leq \|A_{T^c}\|^2$, we obtain using (5.49) that

$$\mathbb{E}_{A_T}\|\Gamma\| \leq (\mathbb{E}_{A_T}\|\Gamma\|^2)^{1/2} \leq \|A_{T^c}\|\sqrt{|T|} + C\sqrt{N \cdot \mathbb{E}_{A_T}(M)\log|T^c|}$$
$$\leq \|A_{T^c}\|\sqrt{n} + C\sqrt{N \cdot \mathbb{E}_{A_T}(M)\log n}. \tag{5.50}$$

Let us take expectation of both sides with respect to A_{T^c}. The left side becomes the quantity we seek to bound, $\mathbb{E}\|\Gamma\|$. The right side will contain the term which we can estimate by (5.46):

$$\mathbb{E}_{A_{T^c}}\|A_{T^c}\| = \mathbb{E}\|A_{T^c}\| \leq \mathbb{E}\|A\| \leq C\sqrt{N\log n}.$$

The other term that will appear in the expectation of (5.50) is

$$\mathbb{E}_{A_{T^c}}\sqrt{\mathbb{E}_{A_T}(M)} \leq \sqrt{\mathbb{E}_{A_{T^c}}\mathbb{E}_{A_T}(M)} \leq \sqrt{\mathbb{E}M} = \sqrt{m}.$$

So, taking the expectation in (5.50) and using these bounds, we obtain

$$\mathbb{E}\|\Gamma\| = \mathbb{E}_{A_{T^c}}\mathbb{E}_{A_T}\|\Gamma\| \leq C\sqrt{N\log n}\sqrt{n} + C\sqrt{Nm\log n} \leq 2C\sqrt{Nm\log n}$$

where we used that $n \leq m$. Finally, using this estimate in (5.47) we conclude

$$E \leq 8C\sqrt{\frac{m\log n}{N}}.$$

This establishes the first part of Theorem 5.62. The second part follows by the diagonalization argument as in Step 2 of the proof of Theorem 5.45. □

REMARK 5.64 (Incoherence) A priori control on the *incoherence* is essential in Theorem 5.62. Consider for instance an $N \times n$ random matrix A whose columns are independent coordinate random vectors in $\mathbb{R}^N$. Clearly $s_{\max}(A) \geq \max_j \|A_i\|_2 = \sqrt{N}$. On the other hand, if the matrix is not too tall, $n \gg \sqrt{N}$, then A has two identical columns with high probability, which yields $s_{\min}(A) = 0$.

5.6 Restricted isometries

In this section we consider an application of the non-asymptotic random matrix theory in compressed sensing. For a thorough introduction to compressed sensing, see the introductory chapter of this book and [28, 20].

In this area, $m \times n$ matrices A are considered as measurement devices, taking as input a signal $x \in \mathbb{R}^n$ and returning its measurement $y = Ax \in \mathbb{R}^m$. One would like to take measurements economically, thus keeping m as small as possible, and still to be able to recover the signal x from its measurement y.

The interesting regime for compressed sensing is where we take very few measurements, $m \ll n$. Such matrices A are not one-to-one, so recovery of x from y is not possible for all signals x. But in practical applications, the amount of "information" contained in the signal is often small. Mathematically this is expressed as *sparsity* of x. In the simplest case, one assumes that x has few nonzero coordinates, say $|\mathrm{supp}(x)| \leq k \ll n$. In this case, using any non-degenerate matrix A one can check that x can be recovered whenever $m > 2k$ using the optimization problem $\min\{|\mathrm{supp}(x)| : Ax = y\}$.

This optimization problem is highly non-convex and generally NP-complete. So instead one considers a convex relaxation of this problem, $\min\{\|x\|_1 : Ax = y\}$. A basic result in compressed sensing, due to Candès and Tao [17, 16], is that for sparse signals $|\mathrm{supp}(x)| \leq k$, the convex problem recovers the signal x from its measurement y exactly, provided that the measurement matrix A is quantitatively non-degenerate. Precisely, the non-degeneracy of A means that it satisfies the following *restricted isometry property* with $\delta_{2k}(A) \leq 0.1$.

Definition (Restricted isometries) An $m \times n$ matrix A satisfies the *restricted isometry property* of order $k \geq 1$ if there exists $\delta_k \geq 0$ such that the inequality

$$(1 - \delta_k)\|x\|_2^2 \leq \|Ax\|_2^2 \leq (1 + \delta_k)\|x\|_2^2 \tag{5.51}$$

holds for all $x \in \mathbb{R}^n$ with $|\mathrm{supp}(x)| \leq k$. The smallest number $\delta_k = \delta_k(A)$ is called the *restricted isometry constant* of A.

In words, A has a restricted isometry property if A acts as an approximate isometry on all sparse vectors. Clearly,

$$\delta_k(A) = \max_{|T| \leq k} \|A_T^* A_T - I_{\mathbb{R}^T}\| = \max_{|T| = \lfloor k \rfloor} \|A_T^* A_T - I_{\mathbb{R}^T}\| \tag{5.52}$$

where the maximum is over all subsets $T \subseteq [n]$ with $|T| \leq k$ or $|T| = \lfloor k \rfloor$.

The concept of restricted isometry can also be expressed via extreme singular values, which brings us to the topic we studied in the previous sections. A is a restricted isometry if and only if all $m \times k$ sub-matrices A_T of A (obtained by selecting arbitrary k columns from A) are approximate isometries. Indeed, for every $\delta \geq 0$, Lemma 5.36 shows that the following two inequalities are equivalent up to an absolute constant:

$$\delta_k(A) \leq \max(\delta, \delta^2); \tag{5.53}$$

$$1 - \delta \leq s_{\min}(A_T) \leq s_{\max}(A_T) \leq 1 + \delta \quad \text{for all } |T| \leq k. \tag{5.54}$$

More precisely, (5.53) implies (5.54) and (5.54) implies $\delta_k(A) \leq 3\max(\delta, \delta^2)$.

Our goal is thus to find matrices that are good restricted isometries. What good means is clear from the goals of compressed sensing described above. First, we need to keep the restricted isometry constant $\delta_k(A)$ below some small absolute constant, say 0.1. Most importantly, we would like the number of measurements m to be small, ideally proportional to the sparsity $k \ll n$.

This is where non-asymptotic random matrix theory enters. We shall indeed show that, with high probability, $m \times n$ random matrices A are good restricted isometries of order k with $m = O^*(k)$. Here the O^* notation hides some logarithmic factors of n. Specifically, in Theorem 5.65 we will show that

$$m = O(k \log(n/k))$$

for sub-gaussian random matrices A (with independent rows or columns). This is due to the strong concentration properties of such matrices. A general observation of this kind is Proposition 5.66. It says that if for a given x, a random matrix A (taken from any distribution) satisfies inequality (5.51) with high probability, then A is a good restricted isometry.

In Theorem 5.71 we will extend these results to random matrices without concentration properties. Using a uniform extension of Rudelson's inequality, Corollary 5.28, we shall show that

$$m = O(k \log^4 n) \tag{5.55}$$

for heavy-tailed random matrices A (with independent rows). This includes the important example of random Fourier matrices.

5.6.1 Sub-gaussian restricted isometries

In this section we show that $m \times n$ sub-gaussian random matrices A are good restricted isometries. We have in mind either of the following two models, which we analyzed in Sections 5.4.1 and 5.5.1 respectively:

Row-independent model: the rows of A are independent sub-gaussian isotropic random vectors in $\mathbb{R}^n$;

Column-independent model: the columns A_i of A are independent sub-gaussian isotropic random vectors in $\mathbb{R}^m$ with $\|A_i\|_2 = \sqrt{m}$ a.s.

Recall that these models cover many natural examples, including Gaussian and Bernoulli matrices (whose entries are independent standard normal or symmetric Bernoulli random variables), general sub-gaussian random matrices (whose entries are independent sub-gaussian random variables with mean zero and unit variance), "column spherical" matrices whose columns are independent vectors uniformly distributed on the centered Euclidean sphere in $\mathbb{R}^m$ with radius $\sqrt{m}$, "row spherical" matrices whose rows are independent vectors uniformly distributed on the centered Euclidean sphere in $\mathbb{R}^d$ with radius $\sqrt{d}$, etc.

THEOREM 5.65 (Sub-gaussian restricted isometries) *Let A be an $m \times n$ sub-gaussian random matrix with independent rows or columns, which follows either of the two models* above. Then the normalized matrix $\bar{A} = \frac{1}{\sqrt{m}}A$ satisfies the following for every sparsity level $1 \leq k \leq n$ and every number $\delta \in (0,1)$:

$$\text{if } m \geq C\delta^{-2}k\log(en/k) \quad \text{then } \delta_k(\bar{A}) \leq \delta$$

with probability at least $1 - 2\exp(-c\delta^2 m)$. Here $C = C_K$, $c = c_K > 0$ depend only on the sub-gaussian norm $K = \max_i \|A_i\|_{\psi_2}$ of the rows or columns of A.

Proof. Let us check that the conclusion follows from Theorem 5.39 for the row-independent model, and from Theorem 5.58 for the column-independent model. We shall control the restricted isometry constant using its equivalent description (5.52). We can clearly assume that k is a positive integer.

Let us fix a subset $T \subseteq [n]$, $|T| = k$ and consider the $m \times k$ random matrix A_T. If A follows the row-independent model, then the rows of A_T are orthogonal projections of the rows of A onto $\mathbb{R}^T$, so they are still independent sub-gaussian isotropic random vectors in $\mathbb{R}^T$. If alternatively, A follows the column-independent model, then trivially the columns of A_T satisfy the same assumptions as the columns of A. In either case, Theorem 5.39 or Theorem 5.58 applies to A_T. Hence for every $s \geq 0$, with probability at least $1 - 2\exp(-cs^2)$ one has

$$\sqrt{m} - C_0\sqrt{k} - s \leq s_{\min}(A_T) \leq s_{\max}(A_T) \leq \sqrt{m} + C_0\sqrt{k} + s. \tag{5.56}$$

Using Lemma 5.36 for $\bar{A}_T = \frac{1}{\sqrt{m}}A_T$, we see that (5.56) implies that

$$\|\bar{A}_T^* \bar{A}_T - I_{\mathbb{R}^T}\| \leq 3\max(\delta_0, \delta_0^2) \quad \text{where } \delta_0 = C_0\sqrt{\frac{k}{m}} + \frac{s}{\sqrt{m}}.$$

Now we take a union bound over all subsets $T \subset [n]$, $|T| = k$. Since there are $\binom{n}{k} \leq (en/k)^k$ ways to choose T, we conclude that

$$\max_{|T|=k} \|\bar{A}_T^* \bar{A}_T - I_{\mathbb{R}^T}\| \leq 3\max(\delta_0, \delta_0^2)$$

with probability at least $1 - \binom{n}{k} \cdot 2\exp(-cs^2) \geq 1 - 2\exp\left(k\log(en/k) - cs^2\right)$. Then, once we choose $\varepsilon > 0$ arbitrarily and let $s = C_1\sqrt{k\log(en/k)} + \varepsilon\sqrt{m}$, we conclude with probability at least $1 - 2\exp(-c\varepsilon^2 m)$ that

$$\delta_k(\bar{A}) \leq 3\max(\delta_0, \delta_0^2) \quad \text{where } \delta_0 = C_0\sqrt{\frac{k}{m}} + C_1\sqrt{\frac{k\log(en/k)}{m}} + \varepsilon.$$

Finally, we apply this statement for $\varepsilon := \delta/6$. By choosing constant C in the statement of the theorem sufficiently large, we make m large enough so that $\delta_0 \leq \delta/3$, which yields $3\max(\delta_0, \delta_0^2) \leq \delta$. The proof is complete. $\square$

The main reason Theorem 5.65 holds is that the random matrix A has a strong concentration property, i.e. that $\|\bar{A}x\|_2 \approx \|x\|_2$ with high probability for every fixed sparse vector x. This concentration property alone implies the restricted isometry property, regardless of the specific random matrix model:

PROPOSITION 5.66 (Concentration implies restricted isometry, see [10]) *Let A be an $m \times n$ random matrix, and let $k \geq 1$, $\delta \geq 0$, $\varepsilon > 0$. Assume that for every fixed $x \in \mathbb{R}^n$, $|\mathrm{supp}(x)| \leq k$, the inequality*

$$(1-\delta)\|x\|_2^2 \leq \|Ax\|_2^2 \leq (1+\delta)\|x\|_2^2$$

holds with probability at least $1 - \exp(-\varepsilon m)$. Then we have the following:

$$\text{if } m \geq C\varepsilon^{-1} k \log(en/k) \quad \text{then } \delta_k(\bar{A}) \leq 2\delta$$

with probability at least $1 - \exp(-\varepsilon m/2)$. Here C is an absolute constant.

In words, the restricted isometry property can be checked on each individual vector x with high probability.

Proof. We shall use the expression (5.52) to estimate the restricted isometry constant. We can clearly assume that k is an integer, and focus on the sets $T \subseteq [n]$, $|T| = k$. By Lemma 5.2, we can find a net $\mathcal{N}_T$ of the unit sphere $S^{n-1} \cap \mathbb{R}^T$ with cardinality $|\mathcal{N}_T| \leq 9^k$. By Lemma 5.4, we estimate the operator norm as

$$\|A_T^* A_T - I_{\mathbb{R}^T}\| \leq 2 \max_{x \in \mathcal{N}_T} |\langle (A_T^* A_T - I_{\mathbb{R}^T})x, x\rangle| = 2 \max_{x \in \mathcal{N}_T} \big|\|Ax\|_2^2 - 1\big|.$$

Taking the maximum over all subsets $T \subseteq [n]$, $|T| = k$, we conclude that

$$\delta_k(A) \leq 2 \max_{|T|=k} \max_{x \in \mathcal{N}_T} \big|\|Ax\|_2^2 - 1\big|.$$

On the other hand, by assumption we have for every $x \in \mathcal{N}_T$ that

$$\mathbb{P}\big\{\big|\|Ax\|_2^2 - 1\big| > \delta\big\} \leq \exp(-\varepsilon m).$$

Therefore, taking a union bound over $\binom{n}{k} \leq (en/k)^k$ choices of the set T and over 9^k elements $x \in \mathcal{N}_T$, we obtain that

$$\mathbb{P}\{\delta_k(A) > 2\delta\} \leq \binom{n}{k} 9^k \exp(-\varepsilon m) \leq \exp\big(k \ln(en/k) + k \ln 9 - \varepsilon m\big)$$

$$\leq \exp(-\varepsilon m/2)$$

where the last line follows by the assumption on m. The proof is complete. □

5.6.2 Heavy-tailed restricted isometries

In this section we show that $m \times n$ random matrices A with independent heavy-tailed rows (and uniformly bounded coefficients) are good restricted isometries. This result will be established in Theorem 5.71. As before, we will prove this by controlling the extreme singular values of all $m \times k$ sub-matrices A_T. For each individual subset T, this can be achieved using Theorem 5.41: one has

$$\sqrt{m} - t\sqrt{k} \leq s_{\min}(A_T) \leq s_{\max}(A_T) \leq \sqrt{m} + t\sqrt{k} \qquad (5.57)$$

with probability at least $1 - 2k \cdot \exp(-ct^2)$. Although this optimal probability estimate has optimal order, it is too weak to allow for a union bound over all $\binom{n}{k} = (O(1)n/k)^k$ choices of the subset T. Indeed, in order that $1 - \binom{n}{k} 2k \cdot \exp(-ct^2) > 0$ one would need to take $t > \sqrt{k \log(n/k)}$. So in order to achieve a nontrivial lower bound in (5.57), one would be forced to take $m \geq k^2$. This is too many measurements; recall that our hope is $m = O^*(k)$.

This observation suggests that instead of controlling each sub-matrix A_T separately, we should learn how to control all A_T at once. This is indeed possible with the following uniform version of Theorem 5.45:

THEOREM 5.67 (Heavy-tailed rows; uniform) *Let $A = (a_{ij})$ be an $N \times d$ matrix ($1 < N \leq d$) whose rows A_i are independent isotropic random vectors in $\mathbb{R}^d$. Let K be a number such that all entries $|a_{ij}| \leq K$ almost surely. Then for every $1 < n \leq d$, we have*

$$\mathbb{E} \max_{|T| \leq n} \max_{j \leq |T|} |s_j(A_T) - \sqrt{N}| \leq Cl\sqrt{n}$$

where $l = \log(n)\sqrt{\log d}\sqrt{\log N}$ and where $C = C_K$ may depend on K only. The maximum is, as usual, over all subsets $T \subseteq [d]$, $|T| \leq n$.

The nonuniform prototype of this result, Theorem 5.45, was based on Rudelson's inequality, Corollary 5.28. In a very similar way, Theorem 5.67 is based on the following uniform version of Rudelson's inequality.

PROPOSITION 5.68 (Uniform Rudelson's inequality [67]) *Let $x_1, \ldots, x_N$ be vectors in $\mathbb{R}^d$, $1 < N \leq d$, and let K be a number such that all $\|x_i\|_\infty \leq K$. Let $\varepsilon_1, \ldots, \varepsilon_N$ be independent symmetric Bernoulli random variables. Then for every $1 < n \leq d$ one has*

$$\mathbb{E} \max_{|T| \leq n} \left\| \sum_{i=1}^N \varepsilon_i (x_i)_T \otimes (x_i)_T \right\| \leq Cl\sqrt{n} \cdot \max_{|T| \leq n} \left\| \sum_{i=1}^N (x_i)_T \otimes (x_i)_T \right\|^{1/2}$$

where $l = \log(n)\sqrt{\log d}\sqrt{\log N}$ and where $C = C_K$ may depend on K only.

The nonuniform Rudelson's inequality (Corollary 5.28) was a consequence of a non-commutative Khintchine inequality. Unfortunately, there does not seem to exist a way to deduce Proposition 5.68 from any known result. Instead, this proposition is proved using Dudley's integral inequality for Gaussian processes and estimates of covering numbers going back to Carl, see [67]. It is known however that such usage of Dudley's inequality is not optimal (see e.g. [75]). As a result, the logarithmic factors in Proposition 5.68 are probably not optimal.

In contrast to these difficulties with Rudelson's inequality, proving uniform versions of the other two ingredients of Theorem 5.45 – the deviation Lemma 5.47 and Symmetrization Lemma 5.46 – is straightforward.

LEMMA 5.69   *Let $(Z_t)_{t \in \mathcal{T}}$ be a stochastic process[24] such that all $Z_t \geq 0$. Then $\mathbb{E} \sup_{t \in \mathcal{T}} |Z_t^2 - 1| \geq \max(\mathbb{E} \sup_{t \in \mathcal{T}} |Z_t - 1|, (\mathbb{E} \sup_{t \in \mathcal{T}} |Z_t - 1|)^2)$.*

Proof.   The argument is entirely parallel to that of Lemma 5.47.   □

LEMMA 5.70 (Symmetrization for stochastic processes)   *Let X_{it}, $1 \leq i \leq N$, $t \in \mathcal{T}$, be random vectors valued in some Banach space B, where $\mathcal{T}$ is a finite index set. Assume that the random vectors $X_i = (X_{ti})_{t \in \mathcal{T}}$ (valued in the product space $B^{\mathcal{T}}$) are independent. Let $\varepsilon_1, \ldots, \varepsilon_N$ be independent symmetric Bernoulli random variables. Then*

$$\mathbb{E} \sup_{t \in \mathcal{T}} \Big\| \sum_{i=1}^{N} (X_{it} - \mathbb{E} X_{it}) \Big\| \leq 2 \mathbb{E} \sup_{t \in \mathcal{T}} \Big\| \sum_{i=1}^{N} \varepsilon_i X_{it} \Big\|.$$

Proof.   The conclusion follows from Lemma 5.46 applied to random vectors X_i valued in the product Banach space $B^{\mathcal{T}}$ equipped with the norm $|||(Z_t)_{t \in \mathcal{T}}||| = \sup_{t \in \mathcal{T}} \|Z_t\|$. The reader should also be able to prove the result directly, following the proof of Lemma 5.46.   □

Proof of Theorem 5.67.   Since the random vectors A_i are isotropic in $\mathbb{R}^d$, for every fixed subset $T \subseteq [d]$ the random vectors $(A_i)_T$ are also isotropic in $\mathbb{R}^T$, so $\mathbb{E}(A_i)_T \otimes (A_i)_T = I_{\mathbb{R}^T}$. As in the proof of Theorem 5.45, we are going to control

$$E := \mathbb{E} \max_{|T| \leq n} \Big\| \frac{1}{N} A_T^* A_T - I_{\mathbb{R}^T} \Big\| = \mathbb{E} \max_{|T| \leq n} \Big\| \frac{1}{N} \sum_{i=1}^{N} (A_i)_T \otimes (A_i)_T - I_{\mathbb{R}^T} \Big\|$$

$$\leq \frac{2}{N} \mathbb{E} \max_{|T| \leq n} \Big\| \sum_{i=1}^{N} \varepsilon_i (A_i)_T \otimes (A_i)_T \Big\|$$

where we used Symmetrization Lemma 5.70 with independent symmetric Bernoulli random variables $\varepsilon_1, \ldots, \varepsilon_N$. The expectation in the right-hand side is taken both with respect to the random matrix A and the signs (ε_i). First taking the expectation with respect to (ε_i) (conditionally on A) and afterwards the expectation with respect to A, we obtain by Proposition 5.68 that

$$E \leq \frac{C_K l \sqrt{n}}{N} \mathbb{E} \max_{|T| \leq n} \Big\| \sum_{i=1}^{N} (A_i)_T \otimes (A_i)_T \Big\|^{1/2} = \frac{C_K l \sqrt{n}}{\sqrt{N}} \mathbb{E} \max_{|T| \leq n} \Big\| \frac{1}{N} A_T^* A_T \Big\|^{1/2}$$

---

[24] A stochastic process (Z_t) is simply a collection of random variables on a common probability space indexed by elements t of some abstract set $\mathcal{T}$. In our particular application, $\mathcal{T}$ will consist of all subsets $T \subseteq [d]$, $|T| \leq n$.

By the triangle inequality, $\mathbb{E}\max_{|T|\le n}\|\frac{1}{N}A_T^* A_T\| \le E+1$. Hence we obtain

$$E \le C_K l \sqrt{\frac{n}{N}}(E+1)^{1/2}$$

by Hölder's inequality. Solving this inequality in E we conclude that

$$E = \mathbb{E}\max_{|T|\le n}\left\|\frac{1}{N}A_T^* A_T - I_{\mathbb{R}^T}\right\| \le \max(\delta,\delta^2) \quad \text{where } \delta = C_K l\sqrt{\frac{2n}{N}}. \tag{5.58}$$

The proof is completed by a diagonalization argument similar to Step 2 in the proof of Theorem 5.45. One uses there a uniform version of deviation inequality given in Lemma 5.69 for stochastic processes indexed by the sets $|T|\le n$. We leave the details to the reader. $\square$

THEOREM 5.71 (Heavy-tailed restricted isometries) *Let $A=(a_{ij})$ be an $m\times n$ matrix whose rows A_i are independent isotropic random vectors in $\mathbb{R}^n$. Let K be a number such that all entries $|a_{ij}|\le K$ almost surely. Then the normalized matrix $\bar{A}=\frac{1}{\sqrt{m}}A$ satisfies the following for $m\le n$, for every sparsity level $1 < k \le n$ and every number $\delta\in(0,1)$:*

$$\text{if } m \ge C\delta^{-2}k\log n \log^2(k)\log(\delta^{-2}k\log n \log^2 k) \quad \text{then } \mathbb{E}\delta_k(\bar{A})\le\delta. \tag{5.59}$$

Here $C = C_K > 0$ may depend only on K.

Proof. The result follows from Theorem 5.67, more precisely from its equivalent statement (5.58). In our notation, it says that

$$\mathbb{E}\delta_k(\bar{A}) \le \max(\delta,\delta^2) \quad \text{where } \delta = C_K l\sqrt{\frac{k}{m}} = C_K\sqrt{\frac{k\log m}{m}}\log(k)\sqrt{\log n}.$$

The conclusion of the theorem easily follows. $\square$

In the interesting sparsity range $k\ge \log n$ and $k\ge \delta^{-2}$, the condition in Theorem 5.71 clearly reduces to

$$m \ge C\delta^{-2}k\log(n)\log^3 k.$$

REMARK 5.72 (Boundedness requirement) The *boundedness assumption* on the entries of A is essential in Theorem 5.71. Indeed, if the rows of A are independent coordinate vectors in $\mathbb{R}^n$, then A necessarily has a zero column (in fact $n-m$ of them). This clearly contradicts the restricted isometry property.

Example 5.73

1. **(Random Fourier measurements):** An important example for Theorem 5.41 is where A realizes random Fourier measurements. Consider the $n\times n$ Discrete Fourier Transform (DFT) matrix W with entries

$$W_{\omega,t} = \exp\left(-\frac{2\pi i\omega t}{n}\right), \quad \omega,t \in \{0,\ldots,n-1\}.$$

Consider a random vector X in $\mathbb{C}^n$ which picks a random row of W (with uniform distribution). It follows from Parseval's inequality that X is isotropic.[25] Therefore the $m \times n$ random matrix A whose rows are independent copies of X satisfies the assumptions of Theorem 5.41 with $K = 1$. Algebraically, we can view A as a *random row sub-matrix of the DFT matrix*.

In compressed sensing, such a matrix A has a remarkable meaning – it realizes m *random Fourier measurements* of a signal $x \in \mathbb{R}^n$. Indeed, $y = Ax$ is the DFT of x evaluated at m random points; in words, y consists of m random frequencies of x. Recall that in compressed sensing, we would like to guarantee that with high probability every sparse signal $x \in \mathbb{R}^n$ (say, $|\text{supp}(x)| \leq k$) can be effectively recovered from its m random frequencies $y = Ax$. Theorem 5.71 together with Candès–Tao's result (recalled at the beginning of Section 5.6) imply that an exact recovery is given by the convex optimization problem $\min\{\|x\|_1 : Ax = y\}$ provided that we observe *slightly more frequencies than the sparsity of a signal*: $m \gtrsim C\delta^{-2} k \log(n) \log^3 k$.

2. **(Random sub-matrices of orthogonal matrices):** In a similar way, Theorem 5.71 applies to a random row sub-matrix A of an *arbitrary bounded orthogonal matrix* W. Precisely, A may consist of m randomly chosen rows, uniformly and without replacement,[26] from an arbitrary $n \times n$ matrix $W = (w_{ij})$ such that $W^*W = nI$ and with uniformly bounded coefficients, $\max_{ij} |w_{ij}| = O(1)$. The examples of such W include the class of *Hadamard matrices* – orthogonal matrices in which all entries equal ± 1.

5.7 Notes

For Section 5.1

We work with two kinds of moment assumptions for random matrices: sub-gaussian and heavy-tailed. These are the two extremes. By the central limit theorem, the sub-gaussian tail decay is the strongest condition one can demand from an isotropic distribution. In contrast, our heavy-tailed model is completely general – no moment assumptions (except the variance) are required. It would be interesting to analyze random matrices with independent rows or columns in the intermediate regime, *between sub-gaussian and heavy-tailed* moment assumptions. We hope that for distributions with an appropriate finite moment (say, $(2+\varepsilon)$th or 4th), the results should be the same as for sub-gaussian distributions, i.e. no $\log n$ factors should occur. In particular, tall random matrices ($N \gg n$) should still be approximate isometries. This indeed holds for sub-exponential distributions [2]; see [82] for an attempt to go down to finite moment assumptions.

---

[25] For convenience we have developed the theory over $\mathbb{R}$, while this example is over $\mathbb{C}$. As we noted earlier, all our definitions and results can be carried over to the complex numbers. So in this example we use the obvious complex versions of the notion of isotropy and of Theorem 5.71.

[26] Since in the interesting regime very few rows are selected, $m \ll n$, sampling with or without replacement are formally equivalent. For example, see [67] which deals with the model of sampling without replacement.

For Section 5.2

The material presented here is well known. The volume argument presented in Lemma 5.2 is quite flexible. It easily generalizes to covering numbers of more general metric spaces, including convex bodies in Banach spaces. See [60, Lemma 4.16] and other parts of [60] for various methods to control covering numbers.

For Section 5.2.3

The concept of sub-gaussian random variables is due to Kahane [39]. His definition was based on the moment generating function (Property 4 in Lemma 5.5), which automatically required sub-gaussian random variables to be centered. We found it more convenient to use the equivalent Property 3 instead. The characterization of sub-gaussian random variables in terms of tail decay and moment growth in Lemma 5.5 also goes back to [39].

The rotation invariance of sub-gaussian random variables (Lemma 5.9) is an old observation [15]. Its consequence, Proposition 5.10, is a general form of *Hoeffding's inequality*, which is usually stated for bounded random variables. For more on large deviation inequalities, see also notes for Section 5.2.4.

The Khintchine inequality is usually stated for the particular case of symmetric Bernoulli random variables. It can be extended for $0 < p < 2$ using a simple extrapolation argument based on Hölder's inequality, see [45, Lemma 4.1].

For Section 5.2.4

Sub-gaussian and sub-exponential random variables can be studied together in a general framework. For a given exponent $0 < \alpha < \infty$, one defines general ψ_α random variables, those with moment growth $(\mathbb{E}|X|^p)^{1/p} = O(p^{1/\alpha})$. Sub-gaussian random variables correspond to $\alpha = 2$ and sub-exponentials to $\alpha = 1$. The reader is encouraged to extend the results of Sections 5.2.3 and 5.2.4 to this general class.

Proposition 5.16 is a form of *Bernstein's inequality*, which is usually stated for bounded random variables in the literature. These forms of Hoeffding's and Bernstein's inequalities (Propositions 5.10 and 5.16) are partial cases of a large deviation inequality for general ψ_α norms, which can be found in [72, Corollary 2.10] with a similar proof. For a thorough introduction to large deviation inequalities for sums of independent random variables (and more), see the books [59, 45, 24] and the tutorial [11].

For Section 5.2.5

Sub-gaussian distributions in $\mathbb{R}^n$ are well studied in geometric functional analysis; see [53] for a link with compressed sensing. General ψ_α distributions in $\mathbb{R}^n$ are discussed e.g. in [32].

Isotropic distributions on convex bodies, and more generally isotropic log-concave distributions, are central to asymptotic convex geometry (see [31, 57]) and computational geometry [78]. A completely different way in which isotropic distributions appear in convex geometry is from *John's decompositions* for contact points of convex bodies, see [9, 63, 79]. Such distributions are finitely supported and therefore are usually heavy-tailed.

For an introduction to the concept of *frames* (Example 5.21), see [41, 19].

For Section 5.2.6

The non-commutative Khintchine inequality, Theorem 5.26, was first proved by Lust-Piquard [48] with an unspecified constant B_p in place of $C\sqrt{p}$. The optimal value of B_p was computed by Buchholz [13, 14]; see [62, Section 6.5] for a thorough introduction to Buchholz's argument. For the complementary range $1 \leq p \leq 2$, a corresponding version of non-commutative Khintchine inequality was obtained by Lust-Piquard and Pisier [47]. By a duality argument implicitly contained in [47] and independently observed by Marius Junge, this latter inequality also implies the optimal order $B_p = O(\sqrt{p})$, see [65] and [61, Section 9.8].

Rudelson's Corollary 5.28 was initially proved using a majorizing measure technique; our proof follows Pisier's argument from [65] based on the non-commutative Khintchine inequality.

For Section 5.3

The "Bai–Yin law" (Theorem 5.31) was established for $s_{\max}(A)$ by Geman [30] and Yin, Bai, and Krishnaiah [84]. The part for $s_{\min}(A)$ is due to Silverstein [70] for Gaussian random matrices. Bai and Yin [8] gave a unified treatment of both extreme singular values for general distributions. The fourth moment assumption in Bai–Yin's law is known to be necessary [7].

Theorem 5.32 and its argument is due to Gordon [35, 36, 37]. Our exposition of this result and of Corollary 5.35 follows [21].

Proposition 5.34 is just a tip of an iceberg called *concentration of measure phenomenon*. We do not discuss it here because there are many excellent sources, some of which were mentioned in Section 5.1. Instead we give just one example related to Corollary 5.35. For a general random matrix A with independent centered entries bounded by 1, one can use Talagrand's concentration inequality for convex Lipschitz functions on the cube [73, 74]. Since $s_{\max}(A) = \|A\|$ is a convex function of A, Talagrand's concentration inequality implies $\mathbb{P}\{|s_{\max}(A) - \mathrm{Median}(s_{\max}(A))| \geq t\} \leq 2e^{-ct^2}$. Although the precise value of the median may be unknown, integration of this inequality shows that $|\mathbb{E}s_{\max}(A) - \mathrm{Median}(s_{\max}(A))| \leq C$.

For the recent developments related to the *hard edge* problem for almost square and square matrices (including Theorem 5.38) see the survey [69].

For Section 5.4

Theorem 5.39 on random matrices with sub-gaussian rows, as well as its proof by a covering argument, is a folklore in geometric functional analysis. The use of covering arguments in a similar context goes back to Milman's proof of Dvoretzky's theorem [55]; see e.g. [9] and [60, Chapter 4] for an introduction. In the more narrow context of extreme singular values of random matrices, this type of argument appears recently e.g. in [2].

The breakthrough work on heavy-tailed isotropic distributions is due to Rudelson [65]. He used Corollary 5.28 in the way we described in the proof of Theorem 5.45 to show that $\frac{1}{N}A^*A$ is an approximate isometry. Probably Theorem 5.41 can also be deduced by a modification of this argument; however it is simpler to use the non-commutative Bernstein's inequality.

The symmetrization technique is well known. For a slightly more general two-sided inequality than Lemma 5.46, see [45, Lemma 6.3].

The problem of estimating covariance matrices described in Section 5.4.3 is a basic problem in statistics, see e.g. [38]. However, most work in the statistical literature is focused on the normal distribution or general product distributions (up to linear transformations), which corresponds to studying random matrices with independent entries. For non-product distributions, an interesting example is for uniform distributions on convex sets [40]. As we mentioned in Example 5.25, such distributions are sub-exponential but not necessarily sub-gaussian, so Corollary 5.50 does not apply. Still, the sample size $N = O(n)$ suffices to estimate the covariance matrix in this case [2]. It is conjectured that the same should hold for general distributions with finite (e.g. fourth) moment assumption [82].

Corollary 5.55 on random sub-matrices is a variant of Rudelson's result from [64]. The study of random sub-matrices was continued in [66]. Random sub-frames were studied in [80] where a variant of Corollary 5.56 was proved.

For Section 5.5

Theorem 5.58 for sub-gaussian columns seems to be new. However, historically the efforts of geometric functional analysts were immediately focused on the more difficult case of sub-exponential tail decay (given by uniform distributions on convex bodies). An indication to prove results like Theorem 5.58 by decoupling and covering is present in [12] and is followed in [32, 2].

The normalization condition $\|A_j\|_2 = \sqrt{N}$ in Theorem 5.58 cannot be dropped but can be relaxed. Namely, consider the random variable $\delta := \max_{i \leq n} \left| \frac{\|A_j\|_2^2}{N} - 1 \right|$. Then the conclusion of Theorem 5.58 holds with (5.36) replaced by

$$(1-\delta)\sqrt{N} - C\sqrt{n} - t \leq s_{\min}(A) \leq s_{\max}(A) \leq (1+\delta)\sqrt{N} + C\sqrt{n} + t.$$

Theorem 5.62 for heavy-tailed columns also seems to be new. The incoherence parameter m is meant to prevent collisions of the columns of A in a quantitative way. It is not clear whether the *logarithmic factor* is needed in the conclusion of Theorem 5.62, or whether the incoherence parameter alone takes care of the logarithmic factors whenever they appear. The same question can be raised for all other results for heavy-tailed matrices in Section 5.4.2 and their applications – can we replace the logarithmic factors by more sensitive quantities (e.g. the logarithm of the incoherence parameter)?

For Section 5.6

For a mathematical introduction to compressed sensing, see the introductory chapter of this book and [28, 20].

A version of Theorem 5.65 was proved in [54] for the row-independent model; an extension from sub-gaussian to sub-exponential distributions is given in [3]. A general framework of stochastic processes with sub-exponential tails is discussed in [52]. For the column-independent model, Theorem 5.65 seems to be new.

Proposition 5.66 that formalizes a simple approach to restricted isometry property based on concentration is taken from [10]. Like Theorem 5.65, it can also be used to show that Gaussian and Bernoulli random matrices are restricted isometries. Indeed, it is not difficult to check that these matrices satisfy a concentration inequality as required in Proposition 5.66 [1].

Section 5.6.2 on heavy-tailed restricted isometries is an exposition of the results from [67]. Using concentration of measure techniques, one can prove a version of Theorem 5.71 with high probability $1 - n^{-c \log^3 k}$ rather than in expectation [62]. Earlier, Candès and Tao [18] proved a similar result for random Fourier matrices, although with a slightly higher exponent in the logarithm for the number of measurements in (5.55), $m = O(k \log^6 n)$. The survey [62] offers a thorough exposition of the material presented in Section 5.6.2 and more.

References

[1] D. Achlioptas. Database-friendly random projections: Johnson-Lindenstrauss with binary coins, in: Special issue on PODS 2001 (Santa Barbara, CA). *J Comput Syst Sci*, 66:671–687, 2003.

[2] R. Adamczak, A. Litvak, A. Pajor, and N. Tomczak-Jaegermann. Quantitative estimates of the convergence of the empirical covariance matrix in log-concave ensembles. *J Am Math Soc*, 23:535–561, 2010.

[3] R. Adamczak, A. Litvak, A. Pajor, and N. Tomczak-Jaegermann. Restricted isometry property of matrices with independent columns and neighborly polytopes by random sampling. *Const. Approx.*, to appear, 2010.

[4] R. Ahlswede and A. Winter. Strong converse for identification via quantum channels. *IEEE Trans. Inform. Theory*, 48:569–579, 2002.

[5] G. Anderson, A. Guionnet, and O. Zeitouni. *An Introduction to Random Matrices*. Cambridge: Cambridge University Press, 2009.

[6] Z. Bai and J. Silverstein. *Spectral Analysis of Large Dimensional Random Matrices*. Second edition. New York: Springer, 2010.

[7] Z. Bai, J. Silverstein, and Y. Yin. A note on the largest eigenvalue of a large-dimensional sample covariance matrix. *J Multivariate Anal*, 26:166–168, 1988.

[8] Z. Bai and Y. Yin. Limit of the smallest eigenvalue of a large-dimensional sample covariance matrix. *Ann Probab*, 21:1275–1294, 1993.

[9] K. Ball. An elementary introduction to modern convex geometry. *Flavors of Geometry*, pp. 1–58. Math. Sci. Res. Inst. Publ., 31, Cambridge: Cambridge University Press., 1997.

[10] R. Baraniuk, M. Davenport, R. DeVore, and M. Wakin. A simple proof of the restricted isometry property for random matrices. *Constr. Approx.*, 28:253–263, 2008.

[11] S. Boucheron, O. Bousquet, and G. Lugosi. Concentration inequalities. *Advanced Lectures in Machine Learning*, eds. O. Bousquet, U. Luxburg, and G. Rätsch, 208–240. Springer, 2004.

[12] J. Bourgain. Random points in isotropic convex sets, in: *Convex Geometric Analysis, Berkeley, CA, 1996*, pp. 53–58. Math. Sci. Res. Inst. Publ., 34, Cambridge: Cambridge University Press, 1999.

[13] A. Buchholz. Operator Khintchine inequality in non-commutative probability. *Math Ann*, 319:1–16, 2001.

[14] A. Buchholz. Optimal constants in Khintchine type inequalities for fermions, Rademachers and q-Gaussian operators. *Bull Pol Acad Sci Math*, 53:315–321, 2005.

[15] V. V. Buldygin and Ju. V. Kozachenko. Sub-Gaussian random variables. *Ukrainian Math J*, 32:483–489, 1980.

[16] E. Candès. The restricted isometry property and its implications for compressed sensing. *C. R. Acad. Sci. Paris Ser. I*, 346:589–592.

[17] E. Candès and T. Tao. Decoding by linear programming. *IEEE Trans Inform Theory*, 51:4203–4215, 2005.

[18] E. Candès and T. Tao. Near-optimal signal recovery from random projections: universal encoding strategies? *IEEE Trans Inform Theory*, 52:5406–5425, 2006.

[19] O. Christensen. *Frames and Bases. An Introductory Course. Applied and Numerical Harmonic Analysis*. Boston, MA: Birkhäuser Boston, Inc., 2008.

[20] Compressive Sensing Resources, http://dsp.rice.edu/cs

[21] K. R. Davidson and S. J. Szarek. Local operator theory, random matrices and Banach spaces, *Handbook of the Geometry of Banach Spaces*, vol. I, pp. 317–366. Amsterdam: North-Holland, 2001.

[22] V. de la Peña and E. Giné. *Decoupling. From Dependence to Independence. Randomly Stopped Processes. U-statistics and Processes. Martingales and Beyond*. New York: Springer-Verlag, 1999.

[23] P. Deift and D. Gioev. *Random Matrix Theory: Invariant Ensembles and Universality*. Courant Lecture Notes in Mathematics, 18. Courant Institute of Mathematical Sciences, New York; Providence, RI: American Mathematical Society, 2009.

[24] A. Dembo and O. Zeitouni. *Large Deviations Techniques and Applications*. Boston, MA: Jones and Bartlett Publishers, 1993.

[25] P. Drineas, R. Kannan, and M. Mahoney. Fast Monte Carlo algorithms for matrices. I, II III. *SIAM J Comput*, 36:132–206, 2006.

[26] R. Durrett. *Probability: Theory and Examples*. Belmont: Duxbury Press, 2005.

[27] O. Feldheim and S. Sodin. A universality result for the smallest eigenvalues of certain sample covariance matrices. *Geom Funct Anal*, to appear, 2008.

[28] M. Fornasier and H. Rauhut. Compressive sensing, in *Handbook of Mathematical Methods in Imaging*, eds. O. Scherzer, Springer, to appear, 2010.

[29] Z. Füredi and J. Komlós. The eigenvalues of random symmetric matrices. *Combinatorica*, 1:233–241, 1981.

[30] S. Geman. A limit theorem for the norm of random matrices. *Ann. Probab.*, 8:252–261, 1980.

[31] A. Giannopoulos. *Notes on Isotropic Convex Bodies*. Warsaw, 2003.

[32] A. Giannopoulos and V. Milman. Concentration property on probability spaces. *Adv. Math.*, 156:77–106, 2000.

[33] A. Giannopoulos and V. Milman. Euclidean structure in finite dimensional normed spaces, in *Handbook of the Geometry of Banach Spaces*, vol. I, pp. 707–779. Amsterdam: North-Holland, 2001.

[34] G. Golub, M. Mahoney, P. Drineas, and L.-H. Lim. Bridging the gap between numerical linear algebra, theoretical computer science, and data applications. *SIAM News*, 9: Number 8, 2006.

[35] Y. Gordon. On Dvoretzky's theorem and extensions of Slepian's lemma, in *Israel Seminar on geometrical Aspects of Functional Analysis (1983/84), II*. Tel Aviv: Tel Aviv University, 1984.

[36] Y. Gordon. Some inequalities for Gaussian processes and applications. *Israel J Math*, 50:265–289, 1985.

[37] Y. Gordon. Majorization of Gaussian processes and geometric applications. *Probab. Theory Related Fields*, 91:251–267, 1992.

[38] I. Johnstone. On the distribution of the largest eigenvalue in principal components analysis. *Ann Statist*, 29:295–327, 2001.

[39] J.-P. Kahane. Propriétés locales des fonctions à séries de Fourier aléatoires. *Studia Math*, 19:1–25, 1960.

[40] R. Kannan, L. Lovász, and M. Simonovits. Isoperimetric problems for convex bodies and a localization lemma. *Discrete Comput Geom*, 13:541–559, 1995.

[41] J. Kovačević and A. Chebira. *An Introduction to Frames*. Foundations and Trends in Signal Processing. Now Publishers, 2008.

[42] R. Latala. Some estimates of norms of random matrices. *Proc Am Math Soc*, 133, 1273–1282, 2005.

[43] M. Ledoux. *The Concentration of Measure Phenomenon*. Mathematical Surveys and Monographs, 89. Providence: American Mathematical Society, 2005.

[44] M. Ledoux. Deviation inequalities on largest eigenvalues, in *Geometric Aspects of Functional Analysis*, pp. 167–219. Lecture Notes in Math., 1910. Berlin: Springer, 2007.

[45] M. Ledoux and M. Talagrand. *Probability in Banach Spaces*. Berlin: Springer-Verlag, 1991.

[46] A. Litvak, A. Pajor, M. Rudelson, and N. Tomczak-Jaegermann. Smallest singular value of random matrices and geometry of random polytopes. *Adv Math*, 195:491–523, 2005.

[47] F. Lust-Piquard and G. Pisier. Noncommutative Khintchine and Paley inequalities. *Ark Mat*, 29:241–260, 1991.

[48] F. Lust-Piquard. Inégalités de Khintchine dans $C_p(1 < p < \infty)$. *C R Acad Sci Paris Sér I Math*, 303:289–292, 1986.

[49] J. Matoušek. *Lectures on Discrete Geometry*. Graduate Texts in Mathematics, 212. New York: Springer-Verlag, 2002.

[50] M. Meckes. Concentration of norms and eigenvalues of random matrices. *J Funct Anal*, 211:508–524, 2004.

[51] M. L. Mehta. *Random Matrices*. Pure and Applied Mathematics (Amsterdam), 142. Amsterdam: Elsevier/Academic Press, 2004.

[52] S. Mendelson. On weakly bounded empirical processes. *Math. Ann.*, 340:293–314, 2008.

[53] S. Mendelson, A. Pajor, and N. Tomczak-Jaegermann. Reconstruction and subgaussian operators in asymptotic geometric analysis. *Geom Funct Anal*, 17:1248–1282, 2007.

[54] S. Mendelson, A. Pajor, and N. Tomczak-Jaegermann. Uniform uncertainty principle for Bernoulli and subgaussian ensembles. *Constr Approx*, 28:277–289, 2008.

[55] V. D. Milman. A new proof of A. Dvoretzky's theorem on cross-sections of convex bodies. *Funkcional Anal i Prilozhen*, 5:28–37, 1974.

[56] V. Milman and G. Schechtman. *Asymptotic Theory of Finite-Dimensional Normed Spaces. With an appendix by M. Gromov*. Lecture Notes in Mathematics, 1200. Berlin: Springer-Verlag, 1986.

[57] G. Paouris. Concentration of mass on convex bodies. *Geom Funct Anal*, 16:1021–1049, 2006.

[58] S. Péché and A. Soshnikov. On the lower bound of the spectral norm of symmetric random matrices with independent entries. *Electron. Commun. Probab.*, 13:280–290, 2008.

[59] V. V. Petrov. *Sums of Independent Random Variables.* New York-Heidelberg: Springer-Verlag, 1975.

[60] G. Pisier. *The Volume of Convex Bodies and Banach Space Geometry.* Cambridge Tracts in Mathematics, 94. Cambridge: Cambridge University Press, 1989.

[61] G. Pisier. *Introduction to Operator Space Theory.* London Mathematical Society Lecture Note Series, 294. Cambridge: Cambridge University Press, 2003.

[62] H. Rauhut. Compressive sensing and structured random matrices, in *Theoretical Foundations and Numerical Methods for Sparse Recovery*, eds. M. Fornasier, Radon Series Comp Appl Math, vol 9, pp. 1–92. deGruyter, 2010.

[63] M. Rudelson. Contact points of convex bodies. *Israel J. Math.*, 101:93–124, 1997.

[64] M. Rudelson. Almost orthogonal submatrices of an orthogonal matrix. *Israel J Math*, 111:143–155, 1999.

[65] M. Rudelson. Random vectors in the isotropic position. *J. Funct. Anal.*, 164:60–72, 1999.

[66] M. Rudelson and R. Vershynin. Sampling from large matrices: an approach through geometric functional analysis. *J ACM*, 54, Art. 21, 2007.

[67] M. Rudelson and R. Vershynin. On sparse reconstruction from Fourier and Gaussian measurements. *Comm Pure Appl Math*, 61:1025–1045, 2008.

[68] M. Rudelson and R. Vershynin. Smallest singular value of a random rectangular matrix. *Comm. Pure Appl Math*, 62:1707–1739, 2009.

[69] M. Rudelson and R. Vershynin. Non-asymptotic theory of random matrices: extreme singular values. *Proce Int Congr Math*, Hyderabad, India, to appear, 2010.

[70] J. Silverstein. The smallest eigenvalue of a large-dimensional Wishart matrix. *Ann Probab*, 13:1364–1368, 1985.

[71] A. Soshnikov. A note on universality of the distribution of the largest eigenvalues in certain sample covariance matrices. *J Statist Phys*, 108:1033–1056, 2002.

[72] M. Talagrand. The supremum of some canonical processes. *Am Math*, 116:283–325, 1994.

[73] M. Talagrand. Concentration of measure and isoperimetric inequalities in product spaces. *Inst Hautes Études Sci Publ Math*, 81:73–205, 1995.

[74] M. Talagrand. A new look at independence. *Ann. of Probab.*, 24:1–34, 1996.

[75] M. Talagrand. *The Generic Chaining. Upper and Lower Bounds of Stochastic Processes.* Springer Monographs in Mathematics. Berlin: Springer-Verlag, 2005.

[76] T. Tao and V. Vu. From the Littlewood-Offord problem to the circular law: universality of the spectral distribution of random matrices. *Bull Am Math Soc (NS)*, 46:377–396, 2009.

[77] J. Tropp. User-friendly tail bounds for sums of random matrices, submitted, 2010.

[78] S. Vempala. Geometric random walks: a survey, in *Combinatorial and Computational Geometry*, pp. 577–616. Math Sci Res Inst Publ, 52. Cambridge: Cambridge University Press, 2005.

[79] R. Vershynin. John's decompositions: selecting a large part. *Israel J Math*, 122:253–277, 2001.

[80] R. Vershynin. Frame expansions with erasures: an approach through the non-commutative operator theory. *Appl Comput Harmon Anal*, 18:167–176, 2005.

[81] R. Vershynin. Approximating the moments of marginals of high-dimensional distributions. *Ann Probab*, to appear, 2010.

[82] R. Vershynin. How close is the sample covariance matrix to the actual covariance matrix?. *J Theor Probab*, to appear, 2010.

[83] V. Vu. Spectral norm of random matrices. *Combinatorica*, 27:721–736, 2007.

[84] Y. Q. Yin, Z. D. Bai, and P. R. Krishnaiah. On the limit of the largest eigenvalue of the large-dimensional sample covariance matrix. *Probab Theory Related Fields*, 78:509–521, 1998.

6 Adaptive sensing for sparse recovery

Jarvis Haupt and Robert Nowak

In recent years, tremendous progress has been made in high-dimensional inference problems by exploiting intrinsic low-dimensional structure. Sparsity is perhaps the simplest model for low-dimensional structure. It is based on the assumption that the object of interest can be represented as a linear combination of a small number of elementary functions, which are assumed to belong to a larger collection, or dictionary, of possible functions. Sparse recovery is the problem of determining which components are needed in the representation based on measurements of the object. Most theory and methods for sparse recovery are based on an assumption of non-adaptive measurements. This chapter investigates the advantages of sequential measurement schemes that adaptively focus sensing using information gathered throughout the measurement process. In particular, it is shown that adaptive sensing can be significantly more powerful when the measurements are contaminated with additive noise.

6.1 Introduction

High-dimensional inference problems cannot be accurately solved without enormous amounts of data or prior assumptions about the nature of the object to be inferred. Great progress has been made in recent years by exploiting intrinsic low-dimensional structure in high-dimensional objects. *Sparsity* is perhaps the simplest model for taking advantage of reduced dimensionality. It is based on the assumption that the object of interest can be represented as a linear combination of a small number of elementary functions. The specific functions needed in the representation are assumed to belong to a larger collection or dictionary of functions, but are otherwise unknown. The *sparse recovery* problem is to determine which functions are needed in the representation based on measurements of the object. This general problem can usually be cast as a problem of identifying a vector $x \in \mathbb{R}^n$ from measurements. The vector is assumed to have $k \ll n$ non-zero elements, however the locations of the non-zero elements are unknown.

Most of the existing theory and methods for the sparse recovery problem are based on non-adaptive measurements. In this chapter we investigate the advantages of sequential

Compressed Sensing: Theory and Applications, ed. Yonina C. Eldar and Gitta Kutyniok. Published by Cambridge University Press. © Cambridge University Press 2012.

sampling schemes that adapt to x using information gathered throughout the sampling process. The distinction between adaptive and non-adaptive measurement can be made more precise, as follows. Information is obtained from samples or measurements of the form $y_1(x), y_2(x) \ldots$, where y_t are functionals from a space $\mathcal{Y}$ representing all possible measurement forms and $y_t(x)$ are the values the functionals take for x. We distinguish between two types of information:

Non-adaptive information: $y_1, y_2, \ldots \in \mathcal{Y}$ are chosen non-adaptively (deterministically or randomly) and independently of x.

Adaptive information: $y_1, y_2, \ldots \in \mathcal{Y}$ are selected sequentially, and the choice of y_{t+1} may depend on the previously gathered information, $y_1(x), \ldots, y_t(x)$.

In this chapter we will see that adaptive information can be significantly more powerful when the measurements are contaminated with additive noise. In particular, we will discuss a variety of adaptive measurement procedures that gradually focus on the subspace, or sparse support set, where x lives, allowing for increasingly precise measurements to be obtained. We explore adaptive schemes in the context of two common scenarios, which are described in some detail below.

6.1.1 Denoising

The classic denoising problem deals with the following. Suppose we observe x in noise according to the *non-adaptive* measurement model

$$y = x + e, \tag{6.1}$$

where $e \in \mathbb{R}^n$ represents a vector of additive Gaussian white noise; i.e., $e_j \stackrel{i.i.d.}{\sim} \mathcal{N}(0,1)$, $j = 1, \ldots, n$, where i.i.d. stands for *independent and identically distributed* and $\mathcal{N}(0,1)$ denotes the standard Gaussian distribution. It is sufficient to consider unit variance noises in this model, since other values can be accounted for by an appropriate scaling of the entries of x.

Let x be deterministic and sparse, but otherwise unknown. The goal of the denoising problem we consider here is to determine the locations of the non-zero elements in x from the measurement y. Because the noises are assumed to be i.i.d., the usual strategy is to simply threshold the components of y at a certain level τ, and declare those that exceed the threshold as detections. This is challenging for the following simple reason. Consider the probability $\Pr(\max_j e_j > \tau)$ for some $\tau > 0$. Using a simple bound on the Gaussian tail and the union bound, we have

$$\Pr(\max_j e_j > \tau) \leq \frac{n}{2} \exp\left(-\frac{\tau^2}{2}\right) = \exp\left(-\frac{\tau^2}{2} + \log n - \log 2\right). \tag{6.2}$$

This shows that if $\tau > \sqrt{2\log n}$, then the probability of false detections can be controlled. In fact, in the high-dimensional limit [1]

$$\Pr\left(\lim_{n\to\infty} \frac{\max_{j=1,\ldots,n} e_j}{\sqrt{2\log n}} = 1\right) = 1 \qquad (6.3)$$

and therefore, for large n, we see that false detections cannot be avoided with $\tau < \sqrt{2\log n}$. These basic facts imply that this classic denoising problem cannot be reliably solved unless the non-zero components of x exceed $\sqrt{2\log n}$ in magnitude. This dependence on the problem size n can be viewed as a statistical "curse of dimensionality."

The classic model is based on non-adaptive measurements. Suppose instead that the measurements could be performed sequentially, as follows. Assume that each measurement y_j results from integration over time or averaging of repeated independent observations. The classic non-adaptive model allocates an equal portion of the full *measurement budget* to each component of x. In the sequential adaptive model, the budget can be distributed in a more flexible and adaptive manner. For example, a sequential sensing method could first measure all of the components using a third of the total budget, corresponding to observations of each component plus an additive noise distributed as $\mathcal{N}(0,3)$. The measurements are very noisy, but may be sufficiently informative to reliably rule out the presence of non-zero components at a large fraction of the locations. After ruling out many locations, the remaining two thirds of the measurement budget can be directed at the locations still in question. Now, because there are fewer locations to consider, the variance associated with the subsequent measurements can be even smaller than in the classic model. An illustrative example of this process is depicted in Figure 6.1. We will see in later sections that such sequential measurement models can effectively mitigate the curse of dimensionality in high-dimensional sparse inference problems. This permits the recovery of signals having nonzero components whose magnitudes grow much more slowly than $\sqrt{2\log n}$.

6.1.2 Inverse problems

The classic inverse problem deals with the following observation model. Suppose we observe x in noise according to the *non-adaptive* measurement model

$$y = Ax + e, \qquad (6.4)$$

where $A \in \mathbb{R}^{m\times n}$ is a known *measurement matrix*, $e \in \mathbb{R}^n$ again represents a vector of independent Gaussian white noise realizations, and x is assumed to be deterministic and sparse, but otherwise unknown. We will usually assume that the columns of A have unit norm. This normalization is used so that the SNR is not a function of m, the number of rows. Note that in the denoising problem we have $A = I_{n\times n}$, the identity operator, which also has unit norm columns.

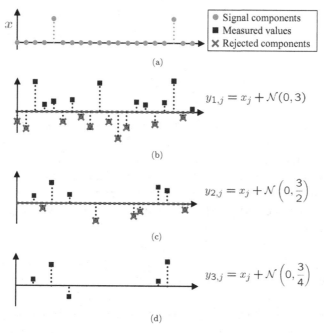

Figure 6.1 Qualitative illustration of a sequential sensing process. A total of three observation steps are utilized, and the measurement budget is allocated uniformly over the steps. The original signal is depicted in panel (a). In the first observation step, shown in panel (b), all components are observed and a simple test identifies two subsets – one corresponding to locations to be measured next, and another set of locations to subsequently ignore. In the second observation step (panel (c)), each observation has twice the precision of the measurements in the previous step, since the same portion of the measurement budget is being used to measure half as many locations. Another refinement step leads to the final set of observations depicted in panel (d). Note that a single-step observation process would yield measurements with variance 1, while the adaptive procedure results in measurements with lower variance at the locations of interest.

The goal of the inverse problem is to recover x from y. A natural approach to this problem is to find a solution to the constrained optimization

$$\min_x \|y - Ax\|_2^2, \text{ subject to } \|x\|_0 \leq k, \tag{6.5}$$

where, as stated in Chapter 1, $\|x\|_0$ is the ℓ_0 (pseudo-)norm which counts the number of non-zero components in x. It is common to refer to an ℓ_0 constraint as a *sparsity constraint*. Note that in the special case where $A = I_{n \times n}$ the solution of the optimization (6.5) corresponds to hard-thresholding of y at the level of the magnitude of the minimum of the k largest (in magnitude) components of y. Therefore the ℓ_0-constrained optimization (6.5) coincides with the denoising problem described above.

For the general inverse problem, A is not proportional to the identity matrix and it may even be non-invertible. Nevertheless, the optimization above can still have a unique solution due to the sparsity constraint. Unfortunately, in this case the optimization (6.5) is combinatorial in nature, generally requiring a brute-force search over all $\binom{n}{k}$ sparsity patterns. A common alternative is to instead solve a convex relaxation of the form

$$\min_x \|y - Ax\|_2^2, \text{ subject to } \|x\|_1 \leq \tau, \tag{6.6}$$

for some $\tau > 0$. This ℓ_1-constrained optimization is relatively easy to solve using convex optimization techniques. It is well known that the solutions of the optimization (6.6) are sparse, and the smaller τ, the sparser the solution.

If the columns of A are not too correlated with one another and τ is chosen appropriately, then the solution to this optimization is close to the solution of the ℓ_0-constrained optimization. In fact in the absence of noise, perfect recovery of the sparse vector x is possible. For example, compressed sensing methods often employ an A comprised of realizations of i.i.d. symmetric random variables. If m (the number of rows) is just slightly larger than k, then every subset of k columns from such an A will be close to orthogonal [2, 3, 4]. This condition suffices to guarantee that any sparse signal with k or fewer non-zero components can be recovered from $\{y, A\}$ – see, for example, [5].

When noise is present in the measurements, reliably determining the locations of the non-zero components in x requires that these components are significantly large relative to the noise level. For example, if the columns of A are scaled to have unit norm, recent work [6] suggests that the optimization in (6.6) will succeed (with high probability) only if the magnitudes of the non-zero components exceed a fixed constant times $\sqrt{\log n}$. In this chapter we will see that this fundamental limitation can again be overcome by sequentially designing the rows of A so that they tend to focus on the relevant components as information is gathered.

6.1.3 A Bayesian perspective

The denoising and inverse problems each have a simple Bayesian interpretation which is a convenient perspective for the development of more general approaches. Recall the ℓ_0-constrained optimization in (6.5). The Lagrangian formulation of this optimization is

$$\min_x \{\|y - Ax\|_2^2 + \lambda \|x\|_0\}, \tag{6.7}$$

where $\lambda > 0$ is the Lagrange multiplier. The optimization can be viewed as a Bayesian procedure, where the term $\|y - Ax\|_2^2$ is the negative Gaussian log-likelihood of x, and $\lambda \|x\|_0$ is the negative log of a prior distribution on the support of x. That is, the mass allocated to x with k non-zero elements is uniformly distributed over the $\binom{n}{k}$ possible sparsity patterns. Minimizing the sum of these two quantities is equivalent to solving for the *Maximum a Posteriori* (MAP) estimate.

The ℓ_1-constrained optimization also has a Bayesian interpretation. The Lagrangian form of that optimization is

$$\min_x \{\|y - Ax\|_2^2 + \lambda \|x\|_1\}. \tag{6.8}$$

In this case the prior is proportional to $\exp(-\lambda \|x\|_1)$, which models components of x independently with a heavy-tailed (double-exponential, or Laplace) distribution. Both the ℓ_0 and ℓ_1 priors, in a sense, reflect a belief that the x we are seeking is sparse (or approximately so) but otherwise unstructured in the sense that all patterns of sparsity are equally probable a priori.

6.1.4 Structured sparsity

The Bayesian perspective also provides a natural framework for more structured models. By modifying the prior (and hence the penalizing term in the optimization), it is possible to encourage solutions having more structured patterns of sparsity. A very general information-theoretic approach to this sort of problem was provided in [7], and we adopt that approach in the following examples. Priors can be constructed by assigning a binary code to each possible x. The prior probability of any given x is proportional to $\exp(-\lambda K(x))$, where $\lambda > 0$ is a constant and $K(x)$ is the bit-length of the code assigned to x. If A is an $m \times n$ matrix with entries drawn independently from a symmetric binary-valued distribution, then the expected mean square error of the estimate

$$\widehat{x} = \arg\min_{x \in \mathcal{X}} \left\{ \|y - Ax\|_2^2 + \lambda K(x) \right\} \qquad (6.9)$$

selected by optimizing over a set $\mathcal{X}$ of candidates (which, for example, could be a discretized subset of the set of all vectors in $\mathbb{R}^n$ with ℓ_2 norm bounded by some specified value), satisfies

$$\mathbb{E}\|\widehat{x} - x^*\|_2^2/n \leq C \min_{x \in \mathcal{X}} \left\{ \|x - x^*\|_2^2/n + cK(x)/m \right\}. \qquad (6.10)$$

Here, x^* is the vector that generated y and $C, c > 0$ are constants depending on the choice of λ. The notation $\mathbb{E}$ denotes expectation, which here is taken with respect to the distribution on A and the additive noise in the observation model (6.4). The $\|x\|_0$ prior/penalty is recovered as a special case in which $\log n$ bits are allocated to encode the location and value of each non-zero element of x (so that $K(x)$ is proportional to $\|x\|_0 \log n$). Then the error satisfies the bound

$$\mathbb{E}\|\widehat{x} - x^*\|_2^2/n \leq C' \|x^*\|_0 \log n/m, \qquad (6.11)$$

for some constant $C' > 0$.

The Bayesian perspective also allows for more structured models. To illustrate, consider a simple sparse binary signal x^* (i.e., all non-zero components take the value 1). If we make no assumptions on the sparsity pattern, then the location of each non-zero component can be encoded using $\log n$ bits, resulting in a bound of the same form as (6.11). Suppose instead that the sparsity pattern of x can be represented by a binary tree whose vertices correspond to the elements of x. This is a common model for the typical sparsity patterns of wavelet coefficients, for example see [8]. The tree-structured restriction means that a node can be non-zero if and only if its "parent" node is also non-zero. Thus, each possible sparsity pattern corresponds to a particular branch of the full binary tree. There exist simple prefix codes for binary trees, and the codelength for a tree with

k vertices is at most $2k+1$ (see, for example, [9]). In other words, we require just over 2 bits per component, rather than $\log n$. Applying the general error bound (6.10) we obtain

$$\mathbb{E}\|\widehat{x} - x^*\|_2^2/n \leq C''\|x^*\|_0/m\,, \tag{6.12}$$

for some constant $C'' > 0$ which, modulo constants, is a factor of $\log n$ better than the bound under the unstructured assumption. Thus, we see that the Bayesian perspective provides a formalism for handling a wider variety of modeling assumptions and deriving performance bounds. Several authors have explored various other approaches to exploiting structure in the patterns of sparsity – see [10, 11, 12, 13], as well as [14] and the exposition in Chapter 4.

Another possibility offered by the Bayesian perspective is to customize the sensing matrix in order to exploit more informative prior information (other than simple unstructured sparsity) that may be known about x. This has been formulated as a Bayesian experimental design problem [15, 16]. Roughly speaking, the idea is to identify a good prior distribution for x and then optimize the choice of the sensing matrix A in order to maximize the expected information of the measurement. In the next section we discuss how this idea can be taken a step further, to sequential Bayesian experimental designs that automatically adapt the sensing to the underlying signal in an online fashion.

6.2 Bayesian adaptive sensing

The Bayesian perspective provides a natural framework for sequential adaptive sensing, wherein information gleaned from previous measurements is used to automatically adjust and focus the sensing. In principle the idea is very simple. Let $\mathcal{Q}_1$ denote a probability measure over all $m \times n$ matrices having unit Frobenius norm in expectation. This normalization generalizes the column normalization discussed earlier. It still implies that the SNR is independent of m, but it also allows for the possibility of distributing the measurement budget more flexibly throughout the columns. This will be crucial for adaptive sensing procedures. For example, in many applications the sensing matrices have entries drawn i.i.d. from a symmetric distribution (see Chapter 5 for a detailed discussion of random matrices). Adaptive sensing procedures, including those discussed in later sections of this chapter, are often also constructed from entries drawn from symmetric, but not identical, distributions. By adaptively adjusting the variance of the distributions used to generate the entries, these sensing matrices can place more or less emphasis on certain components of the signal.

Now consider how we might exploit adaptivity in sparse recovery. Suppose that we begin with a prior probability distribution $p(x)$ for x. Initially collect a set of measurements $y \in \mathbb{R}^m$ according to the sensing model $y = Ax + w$ with $A \sim \mathcal{Q}_1$, where $\mathcal{Q}_1$ is a prior probability distribution on $m \times n$ sensing matrices. For example, $\mathcal{Q}_1$ could correspond to drawing the entries of A independently from a common symmetric distribution. A *posterior* distribution for x can be calculated by combining these data with a

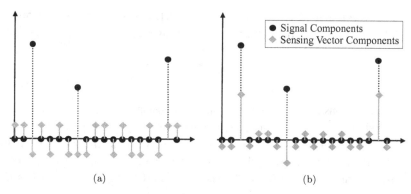

Figure 6.2 Traditional vs. focused sensing. Panel (a) depicts a sensing vector that may be used in a traditional non-adaptive measurement approach. The components of the sensing vector have uniform amplitudes, implying that an equal amount of "sensing energy" is being allocated to all locations regardless of the signal being measured. Panel (b) depicts a focused sensing vector where most of the sensing energy is focused on a small subset of the components corresponding to the relevant entries of the signal.

prior probability model for x, using Bayes' rule. Let $p(x|y)$ denote this posterior distribution. It then becomes natural to ask, which sensing actions will provide the most new information about x? In other words, we are interested in designing $\mathcal{Q}_2$ so that the next measurement using a sensing matrix $A \sim \mathcal{Q}_2$ maximizes our gain in information about x. For example, if certain locations are less likely (or even completely ruled-out) given the observed data y, then $\mathcal{Q}_2$ should be designed to place little (or zero) probability mass on the corresponding columns of the sensing matrix. Our goal will be to develop strategies that utilize information from previous measurements to effectively "focus" the sensing energy of subsequent measurements into subspaces corresponding to the true signal of interest (and away from locations of less interest). An example depicting the notion of focused sensing is shown in Figure 6.2.

More generally, the goal of the next sensing action should be to reduce the uncertainty about x as much as possible. There is a large literature dealing with this problem, usually under the topic of "sequential experiments." The classical Bayesian perspective is nicely summarized in the work of DeGroot [17]. He credits Lindley [18] with first proposing the use of Shannon entropy as a measure of uncertainty to be optimized in the sequential design of experiments. Using the notion of Shannon entropy, the "information-gain" of an experiment can be quantified by the change that the new data produces in the entropy associated with the unknown parameter(s). The optimal design of a number of sequential experiments can be defined recursively and viewed as a dynamic programming problem. Unfortunately, the optimization is intractable in all but the most simple situations. The usual approach, instead, operates in a greedy fashion, maximizing the information-gain at each step in a sequence of experiments. This can be suboptimal, but often is computationally feasible and effective.

An adaptive sensing procedure of this sort can be devised as follows. Let $p(x)$ denote the probability distribution of x after the tth measurement step. Imagine that in the

$(t + 1)$-th step we measure $y = Ax + e$, where $A \sim \mathcal{Q}$ and $\mathcal{Q}$ is a distribution we can design as we like. Let $p(x|y)$ denote the posterior distribution according to Bayes' rule. The "information" provided by this measurement is quantified by the Kullback–Leibler (KL) divergence of $p(x)$ from $p(x|y)$ which is given by

$$\mathbb{E}_X \left[\log \frac{p(x|y)}{p(x)} \right], \qquad (6.13)$$

where the expectation is with respect to the distribution of a random variable $X \sim p(x|y)$. Notice that this expression is a function of y, which is undetermined until the measurement is made. Thus, it is natural to consider the expectation of the KL divergence with respect to the distribution of y, which depends on the prior $p(x)$, the distribution of the noise, and most importantly, on the choice of $\mathcal{Q}$. Let $p(y)$ denote the distribution of the random measurement obtained using the observation matrix $A \sim \mathcal{Q}$. The *expected information gain* from a measurement based on A is defined to be

$$\mathbb{E}_{Y_\mathcal{Q}} \mathbb{E}_X \left[\log \frac{p(x|y)}{p(x)} \right], \qquad (6.14)$$

where the outer expectation is with respect to the distribution of a random variable $Y_\mathcal{Q} \sim p(y)$. This suggests choosing a distribution for the sensing matrix for the next measurement to maximize the expected information gain, that is

$$\mathcal{Q}_{t+1} = \arg\max_\mathcal{Q} \mathbb{E}_{Y_\mathcal{Q}} \mathbb{E}_X \left[\log \frac{p(x|y)}{p(x)} \right], \qquad (6.15)$$

where the optimization is over a space of possible distributions on $m \times n$ matrices.

One useful interpretation of this selection criterion follows by observing that maximizing the expected information gain is equivalent to minimizing the conditional entropy of the posterior distribution [18]. Indeed, simplifying the above expression we obtain

$$\begin{aligned}
\mathcal{Q}_{t+1} &= \arg\max_\mathcal{Q} \mathbb{E}_{Y_\mathcal{Q}} \mathbb{E}_X \left[\log \frac{p(x|y)}{p(x)} \right] \\
&= \arg\min_\mathcal{Q} -\mathbb{E}_{Y_\mathcal{Q}} \mathbb{E}_X \log p(x|y) + \mathbb{E}_{Y_\mathcal{Q}} \mathbb{E}_X \log p(x) \\
&= \arg\min_\mathcal{Q} H(X|Y_\mathcal{Q}) - H(X) \\
&= \arg\min_\mathcal{Q} H(X|Y_\mathcal{Q}), \qquad (6.16)
\end{aligned}$$

where $H(X)$ denotes the Shannon entropy and $H(X|Y_\mathcal{Q})$ the entropy of X conditional on $Y_\mathcal{Q}$. Another intuitive interpretation of the information gain criterion follows from the fact that

$$\mathbb{E}_{Y_\mathcal{Q}} \mathbb{E}_X \left[\log \frac{p(x|y)}{p(x)} \right] = \mathbb{E}_{X,Y_\mathcal{Q}} \left[\log \frac{p(x,y)}{p(x)p(y)} \right] \qquad (6.17)$$

where the right-hand side is just the mutual information between the random variables X and Y_Q. Thus, the information gain criterion equivalently suggests that the next measurements should be constructed in a way that maximizes the mutual information between X and Y_Q.

Now, given this selection of Q_{t+1}, we may draw $A \sim Q_{t+1}$, collect the next measurement $y = Ax + e$, and use Bayes' rule to obtain the new posterior. The rationale is that at each step we are choosing a sensing matrix that maximizes the expected information gain, or equivalently minimizes the expected entropy of the new posterior distribution. Ideally, this adaptive and sequential approach to sensing will tend to focus on x so that sensing energy is allocated to the correct subspace, increasing the SNR of the measurements relative to non-adaptive sensing. The performance could be evaluated, for example, by comparing the result of several adaptive steps to that obtained using a single non-adaptively chosen A.

The approach outlined above suffers from a few inherent limitations. First, while maximizing the expected information gain is a sensible criterion for focusing, the exposition makes no guarantees about the performance of such methods. That is, one cannot immediately conclude that this procedure will lead to an improvement in performance. Second, and perhaps more importantly in practice, selecting the sensing matrix that maximizes the expected information gain can be computationally prohibitive. In the next few sections, we discuss several efforts where approximations or clever choices of the prior are employed to alleviate the computational burden of these procedures.

6.2.1 Bayesian inference using a simple generative model

To illustrate the principles behind the implementation of Bayesian sequential experimental design, we begin with a discussion of the approach proposed in [19]. Their work employed a simple signal model in which the signal vector $x \in \mathbb{R}^n$ was assumed to consist of only a single nonzero entry. Despite the potential model misspecification, this simplification enables the derivation of closed-form expressions for model parameter update rules. It also leads to a simple and intuitive methodology for the shaping of projection vectors in the sequential sampling process.

6.2.1.1 Single component generative model

We begin by constructing a generative model for this class of signals. This model will allow us to define the problem parameters of interest, and to perform inference on them. First, we define L to be a random variable whose range is the set of indices of the signal, $j = \{1, 2, \ldots, n\}$. The entries of the probability mass function of L, denoted by $q_j = \Pr(L = j)$, encapsulate our belief regarding which index corresponds to the true location of the single nonzero component. The amplitude of the single nonzero signal component is a function of its location L, and is denoted by α. Further, conditional on the outcome $L = j$, we model the amplitude of the nonzero component as a Gaussian random variable with location-dependent mean and variance, μ_j and ν_j, respectively.

That is, the distribution of α given $L = j$ is given by

$$p(\alpha|L=j) \sim \mathcal{N}(\mu_j, \nu_j). \tag{6.18}$$

Thus, our prior on the signal x is given by $p(\alpha, L)$, and is described by the hyperparameters $\{q_j, \mu_j, \nu_j\}_{j=1}^n$.

We will perform inference on the hyperparameters, updating our knowledge of them using scalar observations collected according to the standard observation model,

$$y_t = A_t x + e_t, \tag{6.19}$$

where A_t is a $1 \times n$ vector and the noises $\{e_t\}$ are assumed to be i.i.d. $\mathcal{N}(0, \sigma^2)$ for some known $\sigma > 0$. We initialize the hyperparameters of the prior to $q_j(0) = 1/n$, $\mu_j(0) = 0$, and $\nu_j(0) = \sigma_0^2$ for some specified σ_0, for all $j = 1, 2, \ldots, n$. Now, at time step $t \geq 1$, the posterior distribution for the unknown parameters at a particular location j can be written as

$$p(\alpha, L = j|y_t, A_t) = p(\alpha|y_t, A_t, L=j) \cdot q_j(t-1). \tag{6.20}$$

Employing Bayes' rule, we can rewrite the first term on the right-hand side to obtain

$$p(\alpha|y_t, A_t, L=j) \propto p(y_t|A_t, \alpha, L=j) \cdot p(\alpha|L=j), \tag{6.21}$$

and thus the posterior distribution for the unknown parameters satisfies

$$p(\alpha, L=j|y_t, A_t) \propto p(y_t|A_t, \alpha, L=j) \cdot p(\alpha|L=j) \cdot q_j(t-1). \tag{6.22}$$

The proportionality notation has been used to suppress the explicit specification of the normalizing factor. Notice that, by construction, the likelihood function $p(y_t|A_t, \alpha, L=j)$ is conjugate to the prior $p(\alpha|L=j)$, since each is Gaussian. Substituting in the corresponding density functions, and following some straightforward algebraic manipulation, we obtain the following update rules for the hyperparameters:

$$\mu_j(t) = \frac{A_{t,j}\nu_j(t-1)y_t + \mu_j(t-1)\sigma^2}{A_{t,j}^2\nu_j(t-1) + \sigma^2}, \tag{6.23}$$

$$\nu_j(t) = \frac{\nu_j(t-1)\sigma^2}{A_{t,j}^2\nu_j(t-1) + \sigma^2}, \tag{6.24}$$

$$q_j(t) \propto \frac{q_j(t-1)}{\sqrt{A_{t,j}^2\nu_j(t-1) + \sigma^2}} \exp\left(-\frac{1}{2}\frac{(y_t - A_{t,j}\mu_j(t-1))^2}{A_{t,j}^2\nu_j(t-1) + \sigma^2}\right). \tag{6.25}$$

6.2.1.2 Measurement adaptation

Now, as mentioned above, our goal here is twofold. On one hand, we want to *estimate* the parameters corresponding to the location and amplitude of the unknown signal component. On the other hand, we want to devise a strategy for *focusing* subsequent measurements onto the features of interest to boost the performance of our inference

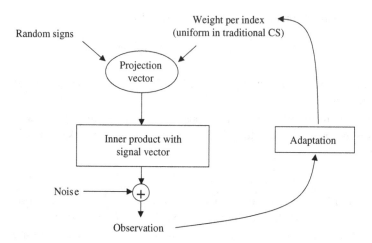

Figure 6.3 Block diagram of the adaptive focusing procedure. Previous observations are utilized to "shape" the weights associated with each location of the random vectors which will be used in the sensing process.

methods. This can be accomplished by employing the information gain criterion; that is, selecting our next measurement to be the most informative measurement that can be made given our current state of knowledge of the unknown quantities. This knowledge is encapsulated by our current estimates of the problem parameters.

We adopt the criterion in (6.16), as follows. Suppose that the next measurement vector A_{t+1} is drawn from some distribution $\mathcal{Q}$ over $1 \times n$ vectors. Let $Y_\mathcal{Q}$ denote the random measurement obtained using this choice of A_{t+1}. Our goal is to select $\mathcal{Q}$ to minimize the conditional entropy of a random variable X distributed according to our generative model with parameters that reflect information obtained up to time t, given $Y_\mathcal{Q}$. In other words, the information gain criterion suggests that we choose the distribution from which the next sensing vector will be drawn according to (6.16).

To facilitate the optimization, we will consider a simple construction for the space from which $\mathcal{Q}$ is to be chosen. Namely, we will assume that the next projection vector is given by the element-wise product between a random sign vector $\xi \in \{-1, 1\}^n$ and a non-negative, unit-norm weight vector $\psi \in \mathbb{R}^n$, so that $A_{t+1,j} = \xi_j \psi_j$. Further, we assume that the entries of the sign vector are equally likely and independent. In other words, we will assume that the overall observation process is as depicted in Figure 6.3, and our goal will be to determine the weight vector ψ.

Recall that the optimization (6.16) is equivalent to maximizing the mutual information between X and $Y_\mathcal{Q}$. Thus the optimization (6.16) is equivalent to

$$\mathcal{Q}_{t+1} = \arg\max_{\mathcal{Q}} H(Y_\mathcal{Q}) - H(Y_\mathcal{Q}|X).$$

This is the formulation we will use here, but rather than solving this optimization directly we will instead employ a bound optimization approach. Namely, we consider maximizing a lower bound of this objective function, which we obtain as follows.

First, we establish a lower bound on the first term in the objective using the fact the conditional differential entropy is a lower bound for the differential entropy [20]. Conditioned on $L = j$ and the corresponding sign vector entry ϵ_j, the distribution of Y_Q is $\mathcal{N}(\psi_j \xi_j \mu_j(t), \psi_j^2 \xi_j^2 \nu_j(t) + \sigma^2)$, which is equivalent to $\mathcal{N}(\psi_j \xi_j \mu_j(t), \psi_j^2 \nu_j(t) + \sigma^2)$ since $\xi_j^2 = 1$. It thus follows that

$$H(Y_Q) \geq H(Y_Q | L, \xi) = \frac{1}{2} \sum_{j=1}^{n} q_j(t) \log(2\pi e (\psi_j^2 \nu_j(t) + \sigma^2)). \quad (6.26)$$

Second, note that conditioned on $X = x$ (or, equivalently, the realizations $L = j$ and $X_j = x_j$), Y_Q is distributed according to a two-component Gaussian mixture, where each component has variance σ^2. Applying the definition of differential entropy directly to this distribution, it is straightforward to establish the bound

$$H(Y_Q | X) \leq \log(2) + \frac{1}{2} \log(2\pi e \sigma^2). \quad (6.27)$$

Now, note that of the bounds (6.26) and (6.27), only (6.26) exhibits any dependence on the choice of Q, and this dependence is only through the projection weights, ψ_j. Thus, our criteria for selecting the projection weights simplifies to

$$\psi = \arg \max_{z \in \mathbb{R}^n : \|z\|_2 = 1} \sum_{j=1}^{n} q_j(t) \log \left(z_j^2 \nu_j(t) + \sigma^2 \right).$$

This constrained optimization can be solved by a simple application of Lagrange multipliers, but it is perhaps more illustrative to consider one further simplification that is appropriate in low-noise settings. In particular, let us assume that $\sigma^2 \approx 0$, then the optimization becomes

$$\psi = \arg \max_{z \in \mathbb{R}^n : \|z\|_2 = 1} \sum_{j=1}^{n} q_j(t) \log \left(z_j^2 \right). \quad (6.28)$$

It is easy to show that the objective in this formulation is maximized by selecting $z_j = \sqrt{q_j(t)}$.

The focusing criterion obtained here is generally consistent with our intuition for this problem. It suggests that the amount of "sensing energy" that should be allocated to a given location j be proportional to our current belief that the nonzero signal component is indeed at location j. Initially, when we assume that the location of the nonzero component is uniformly distributed among the set of indices, this criterion instructs us to allocate our sensing energy uniformly, as is the case in traditional "non-adaptive" CS methods. On the other hand, as we become more confident in our belief that we have identified a set of promising locations at which the nonzero component could be present, the criterion suggests that we focus our energy on those locations to reduce the measurement uncertainty (i.e., to obtain the highest SNR measurements possible).

The procedure outlined here can be extended, in a straightforward way, to settings where the unknown vector x has multiple nonzero entries. The basic idea is to identify the nonzero entries of the signal one-at-a-time, using a sequence of iterations of the proposed procedure. For each iteration, the procedure is executed as described above until one entry of the posterior distribution for the location parameter exceeds a specified threshold $\tau \in (0,1)$. That is, the current iteration of the sequential sensing procedure terminates when the posterior likelihood of a true nonzero component at any of the locations becomes large, which corresponds to the event that $q_j(t) > \tau$ for any $j \in \{1,2,\ldots,n\}$, for a specified τ that we choose to be close to 1. At that point, we conclude that a nonzero signal component is present at the corresponding location. The sequential sensing procedure is then restarted and the parameters $\{q_j, \mu_j, \nu_j\}_{j=1}^n$ are reinitialized, except that the initial values of $\{q_j\}_{j=1}^n$ are set to zero at locations identified as signal components in previous iterations of the procedure, and uniformly distributed over the remaining locations. The resulting multi-step procedure is akin to an "onion peeling" process.

6.2.2 Bayesian inference using multi-component models

The simple single-component model for the unknown signal x described above is but one of many possible generative models that might be employed in a Bayesian treatment of the sparse inference problem. Another, perhaps more natural, option is to employ a more sophisticated model that explicitly allows for the signal to have multiple nonzero components.

6.2.2.1 Multi-component generative model

As discussed above, a widely used sparsity promoting prior is the Laplace distribution,

$$p(x|\lambda) = \left(\frac{\lambda}{2}\right)^n \cdot \exp\left(-\lambda \sum_{j=1}^n |x_j|\right). \tag{6.29}$$

From an analytical perspective in Bayesian inference, however, this particular choice of prior on x can lead to difficulties. In particular, under a Gaussian noise assumption, the resulting likelihood function for the observations (which is conditionally Gaussian given x and the projection vectors) is not conjugate to the Laplace prior, and so closed-form update rules cannot be easily obtained.

Instead, here we discuss the method that was examined in [21], which utilizes a *hierarchical* prior on the signal x, similar to a construction proposed in the context of sparse Bayesian learning in [22]. As before, we begin by constructing a generative model for the signal x. To each x_j, $j = 1, 2, \ldots, n$, we associate a parameter $\rho_j > 0$. The joint distribution of the entries of x, conditioned on the parameter vector $\rho = (\rho_1, \rho_2, \ldots, \rho_n)$, is given in the form of a product distribution,

$$p(x|\rho) = \prod_{j=1}^n p(x_j|\rho_j), \tag{6.30}$$

and we let $p(x_j|\rho_j) \sim \mathcal{N}(0, \rho_j^{-1})$. Thus, we may interpret the ρ_j as precision or "inverse variance" parameters. In addition, we impose a prior on the entries of ρ, as follows. For global parameters $\alpha, \beta > 0$, we set

$$p(\rho|\alpha,\beta) = \prod_{j=1}^{n} p(\rho_j|\alpha,\beta), \tag{6.31}$$

where $p(\rho_j|\alpha,\beta) \sim \text{Gamma}(\alpha,\beta)$ is distributed according to a Gamma distribution with parameters α and β. That is,

$$p(\rho_j|\alpha,\beta) = \frac{\rho_j^{\alpha-1} \beta^{\alpha} \exp(-\beta\rho_j)}{\Gamma(\alpha)}, \tag{6.32}$$

where

$$\Gamma(\alpha) = \int_0^{\infty} z^{\alpha-1} \exp(-z) dz \tag{6.33}$$

is the Gamma function. We model the noise as zero-mean Gaussian with unknown variance, and impose a Gamma prior on the distribution of the noise precision. This results in a hierarchial prior similar to that utilized for the signal vector. Formally, we model our observations using the standard matrix-vector formulation,

$$y = Ax + e, \tag{6.34}$$

where $y \in \mathbb{R}^m$ and $A \in \mathbb{R}^{m \times n}$, and we let $p(e|\rho_0) \sim \mathcal{N}(0, \rho_0 I_{m \times m})$, and $p(\rho_0|\gamma,\delta) \sim \text{Gamma}(\gamma, \delta)$. A graphical summary of the generative signal and observation models is depicted in Figure 6.4.

Now, the hierarchical model was chosen primarily to facilitate analysis, since the Gaussian prior on the signal components is conjugate to the Gaussian (conditional) likelihood of the observations. Generally speaking, a Gaussian prior itself will not promote sparsity; however, incorporating the effect of the Gamma hyperprior lends some additional insight into the situation here. By marginalizing over the parameters ρ, we can

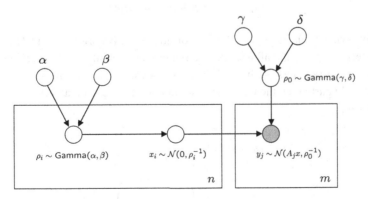

Figure 6.4 Graphical model associated with the multi-component Bayesian CS model.

obtain an expression for the overall prior distribution of the signal components in terms of the parameters α and β,

$$p(x|\alpha,\beta) = \prod_{j=1}^{n} \int_0^\infty p(x_j|\rho_j) \cdot p(\rho_j|\alpha,\beta) d\rho_j. \tag{6.35}$$

The integral(s) can be evaluated directly, giving

$$p(x_j|\alpha,\beta) = \int_0^\infty p(x_j|\rho_j) \cdot p(\rho_j|\alpha,\beta) d\rho_j$$

$$= \frac{\beta^\alpha \Gamma(\alpha+1/2)}{(2\pi)^{1/2}\Gamma(\alpha)} \left(\beta + \frac{x_j^2}{2}\right)^{-(\alpha+1/2)}. \tag{6.36}$$

In other words, the net effect of the prescribed hierarchical prior on the signal coefficients is that of imposing a Student's t prior distribution on each signal component. The upshot is that, for certain choices of the parameters α and β, the product distribution can be strongly peaked about zero, similar (in spirit) to the Laplace distribution – see [22] for further discussion.

Given the hyperparameters ρ and ρ_0, as well as the observation vector y and corresponding measurement matrix A, the posterior for x is conditionally a multivariate Gaussian distribution with mean μ and covariance matrix Σ. Letting $R = \text{diag}(\rho)$, and assuming that the matrix $(\rho_0 A^T A + R)$ is full-rank, we have

$$\Sigma = (\rho_0 A^T A + R)^{-1}, \tag{6.37}$$

and

$$\mu = \rho_0 \Sigma A^T y. \tag{6.38}$$

The goal of the inference procedure, then, is to estimate the hyperparameters ρ and ρ_0 from the observed data y. From Bayes' rule, we have that

$$p(\rho,\rho_0|y) \propto p(y|\rho,\rho_0)p(\rho)p(\rho_0). \tag{6.39}$$

Now, following the derivation in [22], we consider improper priors obtained by setting the parameters α,β,γ, and δ all to zero, and rather than seeking a fully specified posterior for the hyperparameters we instead obtain point estimates via a maximum likelihood procedure. In particular, the maximum likelihood estimates of ρ and ρ_0 are obtained by maximizing

$$p(y|\rho,\rho_0) = (2\pi)^{-m/2} \left|\frac{1}{\rho_0} I_{m\times m} + AR^{-1}A^T\right|^{-1/2}$$

$$\exp\left\{-\frac{1}{2}y^T \left(\frac{1}{\rho_0} I_{m\times m} + AR^{-1}A^T\right)^{-1} y\right\}. \tag{6.40}$$

This yields the following update rules:

$$\rho_j^{\text{new}} = \frac{1 - \rho_j \Sigma_{j,j}}{\mu_j^2}, \tag{6.41}$$

$$\rho_0^{\text{new}} = \frac{m - \sum_{j=1}^{n}(1 - \rho_j \Sigma_{j,j})}{\|y - A\mu\|_2^2}. \tag{6.42}$$

Overall, the inference procedure alternates between solving for ρ_0 and ρ as functions of μ and Σ using (6.41) and (6.42), and solving for μ and Σ as functions of ρ_0 and ρ using (6.37) and (6.38).

6.2.2.2 Measurement adaptation

As in the previous section, we may devise a sequential sensing procedure by first formulating a criterion under which the next projection vector can be chosen to be the most informative. Let us denote the distribution of x given the first t measurements by $p(x)$. Suppose that the $(t+1)$-th measurement is obtained by projecting onto a vector $A_{t+1} \sim \mathcal{Q}$, and let $p(x|y)$ denote the posterior. Now, the criterion for selecting the distribution $\mathcal{Q}_{t+1}$ from which the next measurement vector should be drawn is given by (6.16). As in the previous example, we will simplify the criterion by first restricting the space of distributions over which the objective is to be optimized. In this case, we will consider a space of degenerate distributions. We assume that each $\mathcal{Q}$ corresponds to a distribution that takes a deterministic value $Q \in \mathbb{R}^n$ with probability one, where $\|Q\|_2 = 1$. The goal of the optimization, then, is to determine the "direction" vector Q.

Recall that by construction, given the hyperparameters ρ_0 and ρ the signal x is multivariate Gaussian with mean vector μ and covariance matrix Σ as given in (6.38) and (6.37), respectively. The hierarchical prior(s) imposed on the hyperparameters ρ_0 and ρ make it difficult to evaluate $H(X|Y_\mathcal{Q})$ directly. Instead, we simplify the problem further by assuming that x is unconditionally Gaussian (i.e., ρ_0 and ρ are deterministic and known). In this case the objective function of the information gain criterion can be evaluated directly, and the criterion for selecting the next measurement vector becomes

$$A_{t+1} = \arg\min_{Q \in \mathbb{R}^n, \|Q\|_2=1} -\frac{1}{2}\log\left(1 + \rho_0 Q \Sigma Q^T\right), \tag{6.43}$$

where Σ and ρ_0 reflect the knowledge of the parameters up to time t. From this it is immediately obvious that A_{t+1} should be in the direction of the eigenvector corresponding to the largest eigenvalue of the covariance matrix Σ.

As with the simple single-component signal model case described in the previous section, the focusing rule obtained here also lends itself to some intuitive explanations. Recall that at a given step of the sequential sensing procedure, Σ encapsulates our knowledge of both our level of uncertainty about which entries of the unknown signal are relevant as well as our current level of uncertainty about the actual component value. In particular, note that under the zero-mean Gaussian prior assumption on the signal amplitudes, large values of the diagonal entries of R can be understood to imply the existence of a true nonzero signal component at the corresponding location. Thus, the focusing criterion described above suggests that we focus our sensing energy onto locations at which we are both fairly certain that a signal component is present (as quantified by large entries of the diagonal matrix R), and fairly uncertain about its actual value because of the measurement noise (as quantified by the $\rho_0 A^T A$ term in (6.37)). Further, the relative contribution of each is determined by the level of the additive noise or, more precisely, our current estimate of it.

6.2.3 Quantifying performance

The adaptive procedures discussed in the previous sections can indeed provide realizable performance improvements relative to non-adaptive CS methods. It has been shown, via simulation, that these adaptive sensing procedures can outperform traditional CS in noisy settings. For example, adaptive methods can provide a reduction in mean square reconstruction error, relative to non-adaptive CS, in situations where each utilizes the same total number of observations. Similarly, it has been shown that in some settings adaptive methods can achieve the same error performance as non-adaptive methods using a smaller number of measurements. We refer the reader to [19, 21], as well as [23, 24] for extensive empirical results and more detailed performance comparisons of these procedures.

A complete analysis of these adaptive sensing procedures would ideally also include an analytical performance evaluation. Unfortunately, it appears to be very difficult to devise quantitative error bounds, like those known for non-adaptive sensing, for Bayesian sequential methods. Because each sensing matrix depends on the data collected in the previous steps, the overall process is riddled with complicated dependencies that prevent the use of the usual approaches to obtain error bounds based, for example, on concentration of measure and other tools.

In the next section, we present a recently developed alternative to Bayesian sequential design called *distilled sensing* (DS). In essence, the DS framework encapsulates the spirit of sequential Bayesian methods, but uses a much simpler strategy for exploiting the information obtained from one sensing step to the next. The result is a powerful, computationally efficient procedure that is also amenable to analysis, allowing us to quantify the dramatic performance improvements that can be achieved through adaptivity.

6.3 Quasi-Bayesian adaptive sensing

In the previous section, the Bayesian approach to adaptive sensing was discussed, and several examples were reviewed to show how this approach might be implemented in practice. The salient aspect of these techniques, in essence, was the use of information from prior measurements to guide the acquisition of subsequent measurements in an effort to obtain samples that are most informative. This results in sensing actions that focus sensing resources toward locations that are more likely to contain signal components, and away from locations that likely do not. While this notion is intuitively pleasing, its implementation introduces statistical dependencies that make an analytical treatment of the performance of such methods quite difficult.

In this section we discuss a recently developed adaptive sensing procedure called *distilled sensing* (DS) [25] which is motivated by Bayesian adaptive sensing techniques, but also has the added benefit of being amenable to theoretical performance analysis. The DS procedure is quite simple, consisting of a number of iterations, each of which is comprised of an observation stage followed by a refinement stage. In each observation stage, measurements are obtained at a set of locations which could potentially correspond to nonzero components. In the corresponding refinement stage, the set of locations at which observations were collected in the measurement stage is partitioned into two disjoint sets – one corresponding to locations at which additional measurements are to be obtained in the next iteration, and a second corresponding to locations to subsequently ignore. This type of adaptive procedure was the basis for the example in Figure 6.1. The refinement strategy utilized in DS is a sort of "poor-man's Bayesian" methodology intended to approximate the focusing behavior achieved by methods that employ the information gain criterion. The upshot here is that this simple refinement is still quite effective at focusing sensing resources toward locations of interest. In this section we examine the performance guarantees that can be attained using the DS procedure.

For the purposes of comparison, we begin with a brief discussion of the performance limits for non-adaptive sampling procedures, expanding on the discussion of the denoising problem in Section 6.1.1. We then present and discuss the DS procedure in some detail, and we provide theoretical guarantees on its performance which quantify the gains that can be achieved via adaptivity. In the last subsection we discuss extensions of DS to underdetermined compressed sensing observation models, and we provide some preliminary results on that front.

6.3.1 Denoising using non-adaptive measurements

Consider the general problem of recovering a sparse vector $x \in \mathbb{R}^n$ from its samples. Let us assume that the observations of x are described by the simple model

$$y = x + e, \qquad (6.44)$$

where $e \in \mathbb{R}^n$ represents a vector of additive perturbations, or "noise." The signal x is assumed to be sparse, and for the purposes of analysis in this section we will assume that all the non-zero components of x take the same value $\mu > 0$. Even with this restriction on the form of x, we will see that non-adaptive sensing methods cannot reliably recover signals unless the amplitude μ is considerably larger than the noise level. Recall that the support of x, denoted by $\mathcal{S} = \mathcal{S}(x) = \text{supp}(x)$, is defined to be the set of all indices at which the vector x has a nonzero component. The sparsity level $\|x\|_0$ is simply the cardinality of this set, $\|x\|_0 = |\mathcal{S}|$. To quantify the effect of the additive noise, we will suppose that the entries of e are i.i.d. $\mathcal{N}(0,1)$. Our goal will be to perform *support recovery* (also called model selection), or to obtain an accurate estimate of the support set of x, using the noisy data y. We denote our support estimate by $\hat{\mathcal{S}} = \hat{\mathcal{S}}(y)$.

Any estimation procedure based on noisy data is, of course, subject to error. To assess the quality of a given support estimate $\hat{S}$, we define two metrics to quantify the two different types of errors that can occur in this setting. The first type of error corresponds to the case where we declare that nonzero signal components are present at some locations where they are not, and we refer to such mistakes as *false discoveries*. We quantify the number of these errors using the false discovery proportion (FDP), defined here as

$$\text{FDP}(\hat{S}) := \frac{|\hat{S} \setminus S|}{|\hat{S}|}, \qquad (6.45)$$

where the notation $\hat{S} \setminus S$ denotes the set difference. In words, the FDP of $\hat{S}$ is the ratio of the number of components falsely declared as non-zero to the total number of components declared non-zero. The second type of error occurs when we decide that a particular location does not contain a true nonzero signal component when it actually does. We refer to these errors as *non-discoveries*, and we quantify them using the non-discovery proportion (NDP), defined as

$$\text{NDP}(\hat{S}) := \frac{|S \setminus \hat{S}|}{|S|}. \qquad (6.46)$$

In words, the NDP of $\hat{S}$ is the ratio of the number of non-zero components missed to the number of actual non-zero components. For our purposes, we will consider a testing procedure to be effective if its errors in these two metrics are suitably small.

In contrast to the Bayesian treatments discussed above, here we will assume that x is fixed, but it is otherwise unknown. Recall that by assumption the nonzero components of x are assumed to be non-negative. In this case it is natural to focus on a specific type of estimator for S, which is obtained by applying a simple, coordinate-wise, one-sided thresholding test to the outcome of each of the observations. In particular, the support estimate we will consider here is

$$\hat{S} = \hat{S}(y, \tau) = \{j : y_j > \tau\}, \qquad (6.47)$$

where $\tau > 0$ is a specified threshold.

To quantify the error performance of this estimator, we examine the behavior of the resulting FDP and NDP for a sequence of estimation problems indexed by the dimension parameter n. Namely, for each value of n, we consider the estimation procedure applied to a signal $x \in \mathbb{R}^n$ having $k = k(n)$ nonzero entries of amplitude $\mu = \mu(n)$, observed according to (6.44). Analyzing the procedure for increasing values of n is a common approach to quantify performance in high-dimensional settings, as a function of the corresponding problem parameters. To that end we consider letting n tend to infinity to identify a critical value of the signal amplitude μ below which the estimation procedure fails, and above which it succeeds. The result is stated here as a theorem [26, 25].

THEOREM 6.1 *Assume x has $n^{1-\beta}$ non-zero components of amplitude $\mu = \sqrt{2r \log n}$ for some $\beta \in (0,1)$ and $r > 0$. If $r > \beta$, there exists a coordinate-wise thresholding* procedure with corresponding threshold value $\tau(n)$ that yields an estimator $\hat{S}$ for which

$$\text{FDP}(\hat{S}) \xrightarrow{P} 0, \quad \text{NDP}(\hat{S}) \xrightarrow{P} 0, \qquad (6.48)$$

as $n \to \infty$, where $\xrightarrow{P}$ denotes convergence in probability. Moreover, if $r < \beta$, then there does not exist a coordinate-wise thresholding procedure that can guarantee that both the FDP and NDP tend to 0 as $n \to \infty$.

This result can be easily extended to settings where the nonzero entries of x are both positive and negative, and may also have unequal amplitudes. In those cases, an analogous support estimation procedure can be devised which applies the threshold test to the magnitudes of the observations. Thus, Theorem 6.1 can be understood as a formalization of the general statement made in Section 6.1.1 regarding the denoising problem. There it was argued, based on simple Gaussian tail bounds, that the condition $\mu \approx \sqrt{2 \log n}$ was required in order to reliably identify the locations of the relevant signal components from noisy entry-wise measurements. The above result was obtained using a more sophisticated analysis, though the behavior with respect to the problem dimension n is the same. In addition, and perhaps more interestingly, Theorem 6.1 also establishes a *converse* result – that reliable recovery from non-adaptive measurements is *impossible* unless μ increases in proportion to $\sqrt{\log n}$ as n gets large. This result gives us a baseline with which to compare the performance of adaptive sensing, which is discussed in the following section.

6.3.2 Distilled sensing

We begin our discussion of the distilled sensing procedure by introducing a slight generalization of the sampling model (6.44). This will facilitate explanation of the procedure and allow for direct comparison with non-adaptive methods. Suppose that we are able to collect measurements of the components of x in a sequence of T observation steps, according to the model

$$y_{t,j} = x_j + \rho_{t,j}^{-1/2} e_{t,j}, \; j = 1, 2, \ldots, n, \; t = 1, 2, \ldots, T, \qquad (6.49)$$

where $e_{t,j}$ are i.i.d. $\mathcal{N}(0,1)$ noises, t indexes the observation step, and the $\rho_{t,j}$ are non-negative "precision" parameters that can be chosen to modify the noise variance associated with a given observation. In other words, the variance of additive noise associated with observation $y_{t,j}$ is $\rho_{t,j}^{-1}$, so larger values of $\rho_{t,j}$ correspond to more precise observations. Here, we adopt the convention that setting $\rho_{t,j} = 0$ for some pair (t,j) means that component j is not observed at step t.

This multi-step observation model has natural practical realizations. For example, suppose that observations are obtained by measuring at each location one or more times and averaging the measurements. Then $\sum_{t,j} \rho_{t,j}$ expresses a constraint on the total number of measurements that can be made. This measurement budget can be distributed uniformly over the locations (as in non-adaptive sensing), or nonuniformly and adaptively. Alternatively, suppose that each observation is based on a sensing mechanism that integrates over time to reduce noise. The quantity $\sum_{t,j} \rho_{t,j}$, in this case, corresponds to a constraint on the total observation time. In any case, the model encapsulates an inherent flexibility in the sampling process, in which sensing resources may be preferentially allocated to locations of interest. Note that, by dividing through by $\rho_{t,j} > 0$, we arrive at an equivalent observation model, $\tilde{y}_{t,j} = \rho_{t,j}^{1/2} x_j + e_{t,j}$, which fits the general linear observation model utilized in the previous sections. Our analysis would proceed similarly in either case; we choose to proceed here using the model as stated in (6.49) because of its natural interpretation.

To fix the parameters of the problem, and to facilitate comparison with non-adaptive methods, we will impose a constraint on the overall measurement budget. In particular, we assume that $\sum_{t=1}^{T} \sum_{j=1}^{n} \rho_{t,j} \leq B(n)$. In the case $T = 1$ and $\rho_{1,j} = 1$ for $j = 1, 2, \ldots, n$, which corresponds to the choice $B(n) = n$, the model (6.49) reduces to the canonical non-adaptive observation model (6.44). For our purposes here we will adopt the same measurement budget constraint, $B(n) = n$.

With this framework in place, we now turn to the description of the DS procedure. To begin, we initialize by selecting the number of observation steps T that are to be performed. The total measurement budget $B(n)$ is then divided among the T steps so that a portion B_t is allocated to the tth step, for $t = 1, 2, \ldots, T$, and $\sum_{t=1}^{T} B_t \leq B(n)$. The set of indices to be measured in the first step is initialized to be the set of all indices, $\mathcal{I}_1 = \{1, 2, \ldots, n\}$. Now, the portion of the measurement budget B_1 designated for the first step is allocated uniformly over the indices to be measured, resulting in the precision allocation $\rho_{1,j} = B_1/|\mathcal{I}_1|$ for $j \in \mathcal{I}_1$. Noisy observations are collected, with the given precision, for each entry $j \in \mathcal{I}_1$. The set of observations to be measured in the next step, $\mathcal{I}_2$, is obtained by applying a simple threshold test to each of the observed values. Specifically, we identify the locations to be measured in the next step as those corresponding to observations that are strictly greater than zero, giving $\mathcal{I}_2 = \{j \in \mathcal{I}_1 : y_j > 0\}$. This procedure is repeated for each of the T measurement steps, where (as stated above) the convention $\rho_{t,j} = 0$ implies that the signal component at location j is not observed in measurement step t. The output of the procedure consists of the final set of locations measured, $\mathcal{I}_T$, and the observations collected at those locations $y_{T,j}$, $j \in \mathcal{I}_T$. The entire process is summarized as Algorithm 6.1.

A few aspects of the DS procedure are worth further explanation. First, we comment on the apparent simplicity of the refinement step, which identifies the set of locations to be measured in the subsequent observation step. This simple criterion encapsulates the notion that, given that the nonzero signal components are assumed to have positive amplitude, we expect that their corresponding noisy observation should be non-negative as well. Interpreting this from a Bayesian perspective, the hard-thresholding selection operation encapsulates the idea that the probability of $y_{t,j} > 0$ given $x_j = \mu$ and $\rho_{t,j} > 0$ is approximately equal to one. In reality, using a standard bound on the tail of the Gaussian distribution, we have that

$$\Pr(y_{t,j} > 0 | \rho_{t,j} > 0, x_j = \mu) \geq 1 - \exp\left(-\frac{\rho_{t,j}\mu^2}{2}\right), \qquad (6.50)$$

Algorithm 6.1 (Distilled sensing)

Input:
> Number of observation steps T
> Resource allocation sequence $\{B_t\}_{t=1}^T$ satisfying $\sum_{t=1}^T B_t \leq B(n)$

Initialize:
> Initial index set $\mathcal{I}_1 = \{1, 2, \ldots, n\}$

Distillation:
> For $t = 1$ to T
> Allocate resources: $\rho_{t,j} = \begin{cases} B_t/|\mathcal{I}_t| & j \in \mathcal{I}_t \\ 0 & j \notin \mathcal{I}_t \end{cases}$
> Observe: $y_{t,j} = x_j + \rho_{t,j}^{-1/2} e_{t,j}, j \in \mathcal{I}_t$
> Refine: $\mathcal{I}_{t+1} = \{j \in \mathcal{I}_t : y_{t,j} > 0\}$
> End for

Output:
> Final index set $\mathcal{I}_T$
> Distilled observations $y_T = \{y_{T,j} : j \in \mathcal{I}_T\}$

suggesting that the quality of this approximation may be very good, depending on the particular values of the signal amplitude μ and the precision parameter for the given observation, $\rho_{t,j}$.

Second, as in the simple testing problem described in Section 6.3.1, the DS procedure can also be extended in a straightforward way to account for signals with both positive and negative entries. One possible approach would be to further divide the measurement budget allocation for each step B_t in half, and then perform the whole DS procedure twice. For the first pass, the procedure is performed as stated in Algorithm 6.1 with the goal of identifying positive signal components. For the second pass, replacing the refinement criterion by $\mathcal{I}_{t+1} = \{i \in \mathcal{I}_t : y_{t,j} < 0\}$ would enable the procedure to identify the locations corresponding to negative signal components.

6.3.2.1 Analysis of distilled sensing

The simple adaptive behavior of DS, relative to a fully Bayesian treatment of the problem, renders the procedure amenable to analysis. As in Section 6.3.1, our objects of interest here will be sparse vectors $x \in \mathbb{R}^n$ having $n^{1-\beta}$ nonzero entries, where $\beta \in (0, 1)$ is a fixed (and typically unknown) parameter. Recall that our goal is to obtain an estimate $\hat{S}$ of the signal support S, for which the errors as quantified by the False Discovery Proportion (6.45) and Non-Discovery Proportion (6.46) are simultaneously controlled. The following theorem shows that the DS procedure results in significant improvements over the comparable non-adaptive testing procedure using the same measurement budget [25]. This is achieved by carefully calibrating the problem parameters, i.e., the number of observation steps T and the measurement budget allocation $\{B_t\}_{t=1}^T$.

THEOREM 6.2 *Assume x has $n^{1-\beta}$ non-zero components, where $\beta \in (0,1)$ is fixed, and that each nonzero entry has amplitude exceeding $\mu(n)$. Sample x using the distilled sensing procedure with*

- *$T = T(n) = \max\{\lceil \log_2 \log n \rceil, 0\} + 2$ measurement steps,*
- *measurement budget allocation $\{B_t\}_{t=1}^T$ satisfying $\sum_{t=1}^T B_t \leq n$, and for which*
- *$B_{t+1}/B_t \geq \delta > 1/2$, and*
- *$B_1 = c_1 n$ and $B_T = c_T n$ for some $c_1, c_T \in (0,1)$.*

If $\mu(n) \to \infty$ as a function of n, then the support set estimator constructed using the output of the DS algorithm

$$\hat{\mathcal{S}}_{\mathrm{DS}} := \{j \in \mathcal{I}_T : y_{T,j} > \sqrt{2/c_T}\} \qquad (6.51)$$

satisfies

$$\mathrm{FDP}(\hat{\mathcal{S}}_{\mathrm{DS}}) \xrightarrow{P} 0, \quad \mathrm{NDP}(\hat{\mathcal{S}}_{\mathrm{DS}}) \xrightarrow{P} 0, \qquad (6.52)$$

as $n \to \infty$.

This result can be compared directly to the result of Theorem 6.1, where it was shown that the errors associated with the estimator obtained from non-adaptive observations would converge to zero in probability only in the case $\mu > \sqrt{2\beta \log n}$. In contrast, the result of Theorem 6.2 states that the same performance metrics can be met for an estimator obtained from adaptive samples, under the much weaker constraint $\mu(n) \to \infty$. This includes signals whose nonzero components have amplitude on the order of $\mu \sim \sqrt{\log \log n}$, or $\mu \sim \sqrt{\log \log \log \cdots \log n}$, in fact, the result holds if $\mu(n)$ is *any arbitrarily slowly growing function of n*. If we interpret the ratio between the squared amplitude of the nonzero signal components and the noise variance as the SNR, the result in Theorem 6.2 establishes that adaptivity can provide an improvement in *effective* SNR of up to a factor of $\log n$ over comparable non-adaptive methods. This improvement can be very significant in high-dimensional testing problems where n can be in the hundreds or thousands, or more.

Interpreted another way, the result of Theorem 6.2 suggests that adaptivity can dramatically mitigate the "curse of dimensionality," in the sense that the error performance for DS exhibits much less dependence on the ambient signal dimension than does the error performance for non-adaptive procedures. This effect is demonstrated in finite-sample regimes by the simulation results in Figure 6.5. Each panel of the figure depicts a scatter plot of the FDP and NDP values resulting from 1000 trials of both the adaptive DS procedure, and the non-adaptive procedure whose performance was quantified in Theorem 6.1. Each trial used a different (randomly selected) threshold value to form the support estimate. Panels (a)–(d) correspond to four different values of n: $n = 2^{10}, 2^{13}$,

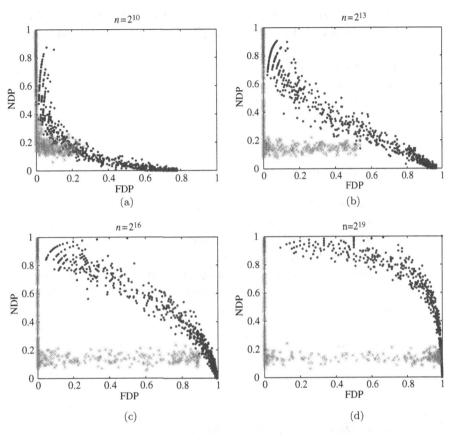

Figure 6.5 The curse of dimensionality and the virtue of adaptivity. Each panel depicts a scatter plot of FDP and NDP values resulting for non-adaptive sensing (•) and the adaptive DS procedure (∗). Not only does DS outperform the non-adaptive method, it exhibits much less dependence on the ambient dimension.

2^{16}, and 2^{19}, respectively. In all cases, the signals being estimated have 128 nonzero entries of amplitude μ, and the SNR is fixed by the selection $\mu^2 = 8$. For each value of n, the measurement budget allocation parameters B_t were chosen so that $B_{t+1} = 0.75 B_t$ for $t = 1, \ldots, T-2$, $B_1 = B_T$, and $\sum_{t=1}^{T} B_t = n$. Comparing the results across panels, we see that the error performance of the non-adaptive procedure degrades significantly as a function of the ambient dimension, while the error performance of DS is largely unchanged across 9 orders of magnitude. This demonstrates the effectiveness of DS for acquiring high-precision observations primarily at the signal locations of interest.

The analysis of the DS procedure relies inherently upon two key ideas pertaining to the action of the refinement step(s) at each iteration. First, for any iteration of the procedure, observations collected at locations where no signal component is present will be independent samples of a zero-mean Gaussian noise process. Despite the fact that the variance of the measured noise will depend on the allocation of sensing resources, the symmetry of the Gaussian distribution ensures that the value obtained for each such observation will be (independently) positive with probability $1/2$. This notion can be made formal by a straightforward application of Hoeffding's inequality.

LEMMA 6.1 *Let* $\{y_j\}_{j=1}^m \overset{iid}{\sim} \mathcal{N}(0,\sigma^2)$. *For any* $0 < \epsilon < 1/2$, *the number of* y_j *exceeding zero satisfies*

$$\left(\frac{1}{2} - \epsilon\right) m \leq \left|\left\{j \in \{1,2,\ldots,m\} : y_j > 0\right\}\right| \leq \left(\frac{1}{2} + \epsilon\right) m, \qquad (6.53)$$

with probability at least $1 - 2\exp\left(-2m\epsilon^2\right)$.

In other words, each refinement step will eliminate about half of the (remaining) locations at which no signal component is present with high probability.

The second key idea is that the simple refinement step will not incorrectly eliminate too many of the locations corresponding to nonzero signal components from future consideration. A formal statement of this result, which is fundamentally a statement about the tails of the Binomial distribution, is given in the following lemma [25]. The proof is repeated here for completeness.

LEMMA 6.2 *Let* $\{y_j\}_{j=1}^m \overset{iid}{\sim} \mathcal{N}(\mu,\sigma^2)$ *with* $\sigma > 0$ *and* $\mu > 2\sigma$. *Let*

$$\delta = \frac{\sigma}{\mu\sqrt{2\pi}}, \qquad (6.54)$$

and note that $\delta < 0.2$, *by assumption. Then,*

$$(1-\delta)m \leq \left|\left\{j \in \{1,2,\ldots,m\} : y_j > 0\right\}\right| \leq m, \qquad (6.55)$$

with probability at least

$$1 - \exp\left(-\frac{\mu m}{4\sigma\sqrt{2\pi}}\right). \qquad (6.56)$$

Proof. Let $q = \Pr(y_j > 0)$. Using a standard bound on the tail of the Gaussian distribution, we have

$$1 - q \leq \frac{\sigma}{\mu\sqrt{2\pi}} \exp\left(-\frac{\mu^2}{2\sigma^2}\right). \qquad (6.57)$$

Next, we employ the Binomial tail bound from [27]: for any $0 < b < \mathbb{E}[\sum_{j=1}^m \mathbf{1}_{\{y_j > 0\}}] = mq$,

$$\Pr\left(\sum_{j=1}^m \mathbf{1}_{\{y_j > 0\}} \leq b\right) \leq \left(\frac{m - mq}{m - b}\right)^{m-b} \left(\frac{mq}{b}\right)^b. \qquad (6.58)$$

Note that $\delta > 1 - q$ (or equivalently, $1 - \delta < q$), so we can apply the Binomial tail bound to the sum $\sum_{j=1}^{m} \mathbf{1}_{\{y_j > 0\}}$ with $b = (1 - \delta)m$ to obtain

$$\Pr\left(\sum_{j=1}^{m} \mathbf{1}_{\{y_j > 0\}} \leq (1 - \delta)m\right) \leq \left(\frac{1-q}{\delta}\right)^{\delta m} \left(\frac{q}{1-\delta}\right)^{(1-\delta)m} \quad (6.59)$$

$$\leq \exp\left(-\frac{\mu^2 \delta m}{2\sigma^2}\right) \left(\frac{1}{1-\delta}\right)^{(1-\delta)m}. \quad (6.60)$$

Now, to establish the stated result, it suffices to show that

$$\exp\left(-\frac{\mu^2 \delta m}{2\sigma^2}\right) \left(\frac{1}{1-\delta}\right)^{(1-\delta)m} \leq \exp\left(-\frac{\mu m}{4\sigma\sqrt{2\pi}}\right). \quad (6.61)$$

Taking logarithms and dividing through by δm, the condition to establish becomes

$$-\frac{\mu^2}{2\sigma^2} + \left(\frac{1-\delta}{\delta}\right) \log\left(\frac{1}{1-\delta}\right) \leq -\frac{\mu}{4\delta\sigma\sqrt{2\pi}}$$

$$= -\frac{\mu^2}{4\sigma^2}, \quad (6.62)$$

where the last equality follows from the definition of δ. The bound holds provided $\mu \geq 2\sigma$, since $0 < \delta < 1$ and

$$\left(\frac{1-\delta}{\delta}\right) \log\left(\frac{1}{1-\delta}\right) \leq 1 \quad (6.63)$$

for $\delta \in (0,1)$. □

Overall, the analysis of the DS procedure entails the repeated application of these two lemmas across iterations of the procedure. Note that the result in Lemma 6.1 is independent of the noise power, while the parameter δ in Lemma 6.2 is a function of both the signal amplitude and the observation noise variance. The latter is a function of how the sensing resources are allocated to each iteration and how many locations are being measured in that step. In other words, statistical dependencies are present across iterations with this procedure, as in the case of the Bayesian methods described above. However, unlike in the Bayesian methods, here the dependencies can be tolerated in a straightforward manner by conditioning on the output of the previous iterations of the procedure.

Rather than presenting the full details of the proof here, we instead provide a short sketch of the general idea. To clarify the exposition, we will find it useful to fix some additional notation. First, we let $S_t = |\mathcal{S} \cap \mathcal{I}_t|$ be the number of locations corresponding to nonzero signal components that are to be observed in step t. Similarly, let $N_t = |\mathcal{S}^c \cap \mathcal{I}_t| = |\mathcal{I}_t| - S_t$ denote the number of remaining locations that are to be measured in the tth iteration. Let $\sigma_1 = \sqrt{|\mathcal{I}_1|/B_1}$ denote the standard deviation of the observation noise in the first iteration, and let δ_1 be the corresponding quantity from Lemma 6.2

described in terms of the quantity σ_1. Notice that since the quantity $|\mathcal{I}_1|$ is fixed and known, the quantities σ_1 and δ_1 are deterministic.

Employing Lemmas 6.1 and 6.2, we determine that the result of the refinement step in the first iteration is that for any $0 < \epsilon < 1/2$, the bounds $(1 - \delta_1)S_1 \leq S_2 \leq S_1$ and $(1/2 - \epsilon) N_1 \leq N_2 \leq (1/2 + \epsilon) N_1$ hold simultaneously, except in an event of probability no greater than

$$2\exp(-2N_1\epsilon^2) + \exp\left(-\frac{\mu S_1}{4\sigma_1\sqrt{2\pi}}\right). \quad (6.64)$$

To evaluate the outcome of the second iteration, we condition on the event that the bounds on S_2 and N_2 stated above hold. In this case, we can obtain bounds on the quantity $\mathcal{I}_2 = S_2 + N_2$, which in turn imply an upper bound on the variance of the observation noise in the second iteration. Let σ_2 denote such a bound, and δ_2 its corresponding quantity from Lemma 6.2. Following the second iteration step, we have that the bounds $(1 - \delta_1)(1 - \delta_2)S_1 \leq S_3 \leq S_1$ and $(1/2 - \epsilon)^2 N_1 \leq N_3 \leq (1/2 + \epsilon)^2 N_1$ hold simultaneously, except in an event of probability no greater than

$$2\exp(-2N_1\epsilon^2) + \exp\left(-\frac{\mu S_1}{4\sigma_1\sqrt{2\pi}}\right) + \quad (6.65)$$

$$2\exp(-2(1-\epsilon)N_1\epsilon^2) + \exp\left(-\frac{\mu(1-\delta_1)S_1}{4\sigma_2\sqrt{2\pi}}\right). \quad (6.66)$$

The analysis proceeds in this fashion, by iterated applications of Lemmas 6.1 and 6.2 conditioned on the outcome of all previous refinement steps. The end result is a statement quantifying the probability that the bounds $\prod_{t=1}^{T-1}(1-\delta_t)S_1 \leq S_T \leq S_1$ and $(1/2 - \epsilon)^{T-1} N_1 \leq N_s \leq (1/2 + \epsilon)^{T-1} N_1$ hold simultaneously following the refinement step in the $(T-1)$-th iteration, prior to the Tth observation step. It follows that the final testing problem is equivalent in structure to a general testing problem of the form considered in Section 6.1.1, but with a different *effective* observation noise variance. The final portion of the proof of Theorem 6.2 entails a careful balancing between the design of the resource allocation strategy, the number of observation steps T, and the specification of the parameter ϵ. The goal is to ensure that as $n \to \infty$ the stated bounds on S_T and N_T are valid with probability tending to one, the fraction of signal components missed throughout the refinement process tends to zero, and the effective variance of the observation noise for the final set of observations is small enough to enable the successful testing of signals with very weak features. The full details can be found in [25].

6.3.3 Distillation in compressed sensing

While the results above demonstrate that adaptivity in sampling can provide a tremendous improvement in effective measurement SNR in certain sparse recovery problems, the benefits of adaptivity are somewhat less clear with respect to the other problem parameters. In particular, the comparison outlined above was made on the basis that each procedure was afforded the same measurement budget, as quantified by a global quantity having a natural interpretation in the context of a total sample budget or a total time constraint. Another basis for comparison would be the total number of measurements collected with each procedure. In the non-adaptive method in Section 6.3.1, a total of n measurements were collected (one per signal component). In contrast, the number of measurements obtained via the DS procedure is necessarily larger, since each component is directly measured at least once, and some components may be measured up to a total of T times – once for each iteration of the procedure. Strictly speaking, the total number of measurements collected during the DS procedure is a random quantity which depends implicitly on the outcome of the refinements at each step, which in turn are functions of the noisy measurements. However, our high-level intuition regarding the behavior of the procedure allows us to make some illustrative approximations. Recall that each refinement step eliminates (on average) about half of the locations at which no signal component is present. Further, under the sparsity level assumed in our analysis, the signals being observed are vanishingly sparse – that is, the fraction of locations of x corresponding to non-zero components tends to zero as $n \to \infty$. Thus, for large n, the number of measurements collected in the tth step of the DS procedure is *approximately* given by $n \cdot 2^{-(t-1)}$, which implies (upon summing over t) that the DS procedure requires on the order of $2n$ total measurements.

By this analysis, the SNR benefits of adaptivity come at the expense of a (modest) relative increase in the number of measurements collected. Motivated by this comparison, it is natural to ask whether the distilled sensing approach might also be extended to the so-called underdetermined observation settings, such as those found in standard compressed sensing (CS) problems. In addition, and perhaps more importantly, can an analysis framework similar to that employed for DS be used to obtain performance guarantees for adaptive CS procedures? We will address these questions here, beginning with a discussion of how the DS procedure might be applied in CS settings.

At a high level, the primary implementation differences relative to the original DS procedure result from the change in observation model. Recall that, for $\rho_{t,j} > 0$, the observation model (6.49) from the previous section could alternatively be written as

$$y_{t,j} = \rho_{t,j}^{1/2} x_j + e_{t,j}, \; j = 1, 2, \ldots, n, \;\; t = 1, 2, \ldots, T, \tag{6.67}$$

subject to a global constraint on $\sum_{t,j} \rho_{t,j}$. Under this alternative formulation, the overall sampling process can be effectively described using the matrix-vector formulation $y = Ax + e$ where A is a matrix whose entries are either zero (at times and locations where no measurements were obtained) or equal to some particular $\rho_{t,j}^{1/2}$. The first point we address relates to the specification of the sampling or measurement budget. In this setting, we can interpret our budget of measurement resources in terms of the matrix A, in a natural way. Recall that in our original formulation, the constraint was imposed on the quantity $\sum_{t,j} \rho_{t,j}$. Under the matrix-vector formulation, this translates directly to a constraint on the sum of the squared entries of A. Thus, we can generalize the measurement budget constraint to the current setting by imposing a condition on the Frobenius norm of A. To account for the possibly random nature of the sensing matrix (as in traditional CS applications), we impose the constraint in expectation:

$$\mathbb{E}\left[\|A\|_F^2\right] = \mathbb{E}\left[\sum_{t,j} A_{t,j}^2\right] \leq B(n). \tag{6.68}$$

Note that, since the random matrices utilized in standard CS settings typically are constructed to have unit-norm columns, they satisfy this constraint when $B(n) = n$.

The second point results from the fact that each observation step will now comprise a number of noisy projection samples of x. This gives rise to another set of algorithmic parameters to specify how many measurements are obtained in each step, and these will inherently depend on the sparsity of the signal being acquired. In general, we will denote by m_t the number of rows in the measurement matrix utilized in step t.

The final point to address in this setting pertains to the refinement step. In the original DS formulation, because the measurement process obtained direct samples of the signal components plus independent Gaussian noises, the simple one-sided threshold test was a natural choice. Here the problem is slightly more complicated. Fundamentally the goal is the same – to process the current observations in order to accurately determine promising locations to measure in subsequent steps. However in the current setting, the decisions must be made using (on average) much less than one measurement per location. In this context, each refinement decision can itself be thought of as a coarse-grained model selection task.

We will discuss one instance of this *Compressive Distilled Sensing* (CDS) procedure, corresponding to particular choices of the algorithm parameters and refinement strategy. Namely, for each step, indexed by $t = 1, 2, \ldots, T$, we will obtain measurements using an $m_t \times n$ sampling matrix A_t constructed as follows. For $u = 1, 2, \ldots, m_t$ and $v \in \mathcal{I}_t$, the (u, v)-th entry of A_t is drawn independently from the distribution $\mathcal{N}(0, \tau_t/m_t)$ where $\tau_t = B_t/|\mathcal{I}_t|$. The entries of A_t are zero otherwise. Notice that this choice automatically guarantees that the overall measurement budget constraint $\mathbb{E}\left[\|A\|_F^2\right] \leq B(n)$ is satisfied. The refinement at each step is performed by coordinate-wise thresholding of the crude estimate $\widehat{x}_t = A_t^T y_t$. Specifically, the set $\mathcal{I}_{t+1}$ of locations to subsequently consider is obtained as the subset of $\mathcal{I}_t$ corresponding to locations at which $\widehat{x}_t$ is positive. This approach is outlined in Algorithm 6.2.

The final support estimate is obtained by applying the Least Absolute Shrinkage and Selection Operator (LASSO) to the distilled observations. Namely, for some $\lambda > 0$, we obtain the estimate

$$\widetilde{x} = \arg\min_{z \in \mathbb{R}^n} \|y_T - A_T z\|_2^2 + \lambda \|z\|_1, \tag{6.69}$$

and from this, the support estimate $\widehat{\mathcal{S}}_{\text{DS}} = \{j \in \mathcal{I}_T : \widetilde{x}_j > 0\}$ is constructed. The following theorem describes the error performance of this support estimator obtained using the CDS adaptive compressive sampling procedure. The result follows from iterated application of Lemmas 1 and 2 in [28], which are analogous to Lemmas 6.1 and 6.2 here, as well as the results in [29] which describe the model selection performance of the LASSO.

THEOREM 6.3 *Let $x \in \mathbb{R}^n$ be a vector having at most $k(n) = n^{1-\beta}$ nonzero entries for some fixed $\beta \in (0,1)$, and suppose that every nonzero entry of x has the same value $\mu = \mu(n) > 0$. Sample x using the compressive distilled sensing procedure described above with*

- *$T = T(n) = \max\{\lceil \log_2 \log n \rceil, 0\} + 2$ measurement steps,*
- *measurement budget allocation $\{B_t\}_{t=1}^T$ satisfying $\sum_{t=1}^T B_t \leq n$, and for which*
- *$B_{t+1}/B_t \geq \delta > 1/2$, and*
- *$B_1 = c_1 n$ and $B_T = c_T n$ for some $c_1, c_T \in (0,1)$.*

There exist constants $c, c', c'' > 0$ and $\lambda = O(1)$ such that if $\mu \geq c\sqrt{\log \log \log n}$ and the number of measurements collected satisfies $m_t = c' \cdot k \cdot \log \log \log n$ for $t = 1, \ldots, T-1$ and $m_T = c'' \cdot k \cdot \log n$, then the support estimate $\hat{S}_{\text{DS}}$ obtained as described above satisfies

$$\text{FDP}(\hat{S}_{\text{DS}}) \xrightarrow{P} 0, \quad \text{NDP}(\hat{S}_{\text{DS}}) \xrightarrow{P} 0, \tag{6.70}$$

as $n \to \infty$.

A few comments are in order regarding the results of Theorem 6.2 and Theorem 6.3. First, while Theorem 6.2 guaranteed recovery provided only that $\mu(n)$ be a growing function of n, the result in Theorem 6.3 is slightly more restrictive, requiring that $\mu(n)$ grow like $\sqrt{\log \log \log n}$. Even so, this still represents a significant improvement relative to the non-adaptive testing case in Section 6.3.1. Second, we note that Theorem 6.3 actually *requires* that the signal components have *the same amplitudes* (or, more precisely, that their amplitudes be within a constant multiple of each other), whereas the result in Theorem 6.2 placed no restrictions on the values of the signal amplitudes relative to each other. In essence these two points arise from the choice of refinement procedure. Here, the threshold tests are no longer statistically independent as they were in the original DS formulation, and the methods employed to tolerate this dependence give rise to these differences.

The effectiveness of CDS can also be observed in finite sample regimes. Here, we examine (by experiment) the performance of CDS relative to a non-adaptive compressed sensing that utilizes a random measurement matrix with i.i.d. zero-mean Gaussian entries. For both cases, the support estimators we consider are constructed as the positive components of the LASSO estimate that is obtained using the corresponding adaptive or non-adaptive measurements. Our application of the CDS recovery procedure differs slightly from the conditions of Theorem 6.3, in that we apply the LASSO to *all* of the adaptively collected measurements.

The results of the comparison are depicted in Figure 6.6. Each panel of the figure shows a scatter plot of the FDP and NDP values resulting from 1000 trials of both the CDS procedure and the non-adaptive sensing approach, each using a different randomly selected LASSO regularization parameter. For each trial, the unknown signals $x \in \mathbb{R}^n$ were constructed to have 128 nonzero entries of uniform (positive) amplitude μ, and the SNR is fixed by the selection $\mu^2 = 12$. Panels (a)–(d) correspond to

---
Algorithm 6.2 (Compressive distilled sensing)

Input:
  Number of observation steps T
  Measurement allocation sequence $\{m_t\}_{t=1}^T$
  Resource allocation sequence $\{B_t\}_{t=1}^T$ satisfying $\sum_{t=1}^T B_t \leq B(n)$

Initialize:
  Initial index set: $\mathcal{I}_1 = \{1, 2, \ldots, n\}$

Distillation:
  For $t = 1$ to T
    Construct $m_t \times n$ measurement matrix:
    $A_t(u,v) \sim \mathcal{N}\left(0, \frac{B_t}{m_t |\mathcal{I}_t|}\right),\ u = 1, 2, \ldots, m_t,\ v \in \mathcal{I}_t$
    $A_t(u,v) = 0,\ u = 1, 2, \ldots, m_t,\ v \in \mathcal{I}_t^c$
    Observe: $y_t = A_t x + e_t$
    Compute: $\widehat{x}_{t,i} = A_t^T y_t$
    Refine: $\mathcal{I}_{t+1} = \{i \in \mathcal{I}_t : \widehat{x}_{t,i} > 0\}$
  End for

Output:
  Index sets $\{\mathcal{I}_t\}_{t=1}^T$
  Distilled observations $\{y_t, A_t\}_{t=1}^T$

$n = 2^{13}$, 2^{14}, 2^{15}, and 2^{16} respectively, and the number of measurements in all cases was $m = 2^{12}$. The measurement budget allocation parameters for CDS, B_t, were chosen so that $B_{t+1} = 0.75 B_t$ for $j = 1, \ldots, T-2$, $B_1 = B_T$, and $\sum_{t=1}^T B_t = n$, where n is the ambient signal dimension in each case. Measurement allocation parameters m_t were chosen so that $\lfloor m/3 \rfloor = 1365$ measurements were utilized for the last step of the procedure, and the remaining $\lfloor 2m/3 \rfloor$ measurements were equally allocated to the first $T - 1$ observation steps. Simulations were performed using the *Gradient Projection for Sparse Reconstruction* (GPSR) software [30].

Comparing the results across all panels of Figure 6.6, we see that CDS exhibits much less dependence on the ambient dimension than does the non-adaptive procedure. In particular, note that the performance of the CDS procedure remains relatively unchanged across four orders of magnitude of the ambient dimension while the performance of the non-adaptive procedure degrades markedly with increasing dimension. As with the examples for DS above, we see that CDS is an effective approach to mitigate the "curse of dimensionality" here as well.

In conclusion, we note that the result of Theorem 6.3 has successfully addressed our initial question, at least in part. We have shown that in some special settings, the CDS

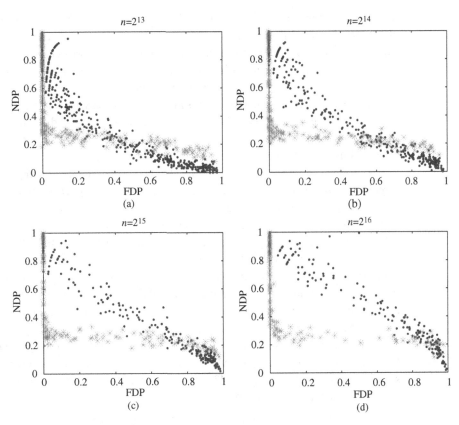

Figure 6.6 Adaptivity in compressed sensing. Each panel depicts a scatter plot of FDP and NDP values resulting for non-adaptive CS (•) and the adaptive CDS procedure (∗).

procedure can achieve similar performance to the DS procedure but using many fewer total measurements. In particular, the total number of measurements required to obtain the result in Theorem 6.3 is $m = O(k \cdot \log\log\log n \cdot \log\log n + k \log n) = O(k \log n)$, while the result of Theorem 6.2 required $O(n)$ total measurements. The discussion in this section demonstrates that it is possible to obtain the benefits of both adaptive sampling *and* compressed sensing. This is a significant step toward a full understanding of the benefits of adaptivity in CS.

6.4 Related work and suggestions for further reading

Adaptive sensing methods for high-dimensional inference problems are becoming increasingly common in many modern applications of interest, primarily due to the continuing tremendous growth of our acquisition, storage, and computational abilities. For instance, multiple testing and denoising procedures are an integral component of many modern bioinformatics applications (see [31] and the references therein), and sequential acquisition techniques similar in spirit to those discussed here are becoming quite popular in this domain. In particular, two-stage testing approaches in gene association and expression studies were examined in [32, 33, 34]. Those works described procedures where a large number of genes is initially tested to identify a promising subset, which is then examined more closely in a second stage. Extensions to multi-stage approaches were discussed in [35]. Two-stage sequential sampling techniques have also been examined recently in the signal processing literature. In [36], two-stage target detection procedures were examined, and a follow-on work examined a Bayesian approach for incorporating prior information into such two-step detection procedures [37].

The problem of target detection and localization from sequential compressive measurements was recently examined in [38]. That work examined a multi-step binary bisection procedure to identify signal components from noisy projection measurements, and provided bounds for its sample complexity. Similar adaptive compressive sensing techniques based on binary bisection were examined in [39]. In [40], an adaptive compressive sampling method for acquiring wavelet-sparse signals was proposed. Leveraging the inherent tree structure often present in the wavelet decompositions of natural images, that work discussed a procedure where the sensing action is guided by the presence (or absence) of significant features at a given scale to determine which coefficients to acquire at finer scales.

Finally, we note that sequential experimental design continues to be popular in other fields as well, such as in computer vision and machine learning. We refer the reader to the survey article [41] as well as [42, 43] and the references therein for further information on active vision and active learning.

References

[1] M. R. Leadbetter, G. Lindgren, and H. Rootzen, *Extremes and Related Properties of Random Sequences and Processes.* Springer-Verlag; 1983.

[2] D. Donoho. Compressed sensing. *IEEE Trans Inform Theory*, 52(4):1289–1306, 2006.

[3] E. Candés and T. Tao. Near optimal signal recovery from random projections: Universal encoding strategies? *IEEE Trans Information Theory*, 52(12):5406–5425, 2006.

[4] R. Baraniuk, M. Davenport, R. DeVore, and M. Wakin. A simple proof of the restricted isometry property for random matrices. *Construc Approx*, 28(3):253–263, 2008.

[5] E. Candés. The restricted isometry property and its implications for compressed sensing. *C Re l'Acad Scie Paris*, 346:589–592, 2008.

[6] E. Candés and Y. Plan. Near-ideal model selection by ℓ_1 minimization. *Ann Stat*, 37(5A):2145–2177, 2009.

[7] J. Haupt and R. Nowak. Signal reconstruction from noisy random projections. *IEEE Trans Information Theory*, 52(9):4036–4048, 2006.

[8] M. Crouse, R. Nowak, and R. Baraniuk. Wavelet-based statistical signal processing using hidden Markov models. *IEEE Trans Sig Proc*, 46(4):886–902, 1998.

[9] G. Jacobson. Space-efficient static trees and graphs. *Proc 30th Ann Symp Found Comp. Sci.* 549–554, 1989.

[10] J. Huang, T. Zhang, and D. Metaxas. Learning with structured sparsity. In: *ICML '09: Proc 26th Ann Int Conf Machine Learning*. New York: ACM, 417–424, 2009.

[11] Y. C. Eldar and M. Mishali. Robust recovery of signals from a structured union of subspaces. *IEEE Trans Information Theory*, 55(11):5302–5316, 2009.

[12] R. Baraniuk, V. Cevher, M. Duarte, and C. Hedge. Model-based compressive sensing. *IEEE Trans Information Theory*, 56(4):1982–2001, 2010.

[13] Y. C. Eldar, P. Kuppinger, and H. Bolcskei. Block-sparse signals: Uncertainty relations and efficient recovery. *IEEE Trans Sig Proc*, 58(6):3042–3054, 2010.

[14] M. Mishali and Y. C. Eldar. From theory to practice: Sub-Nyquist sampling of sparse wideband analog signals. *IEEE J Sel Topics Sig Proc* 4(2):375–391, 2010.

[15] M. Seeger. Bayesian inference and optimal design in the sparse linear model. *J Machine Learning Res* 9:759–813, 2008.

[16] M. Seeger, H. Nickisch, R. Pohmann, and B. Schölkopf. Optimization of k-Space trajectories for compressed sensing by Bayesian experimental design. *Magn Res Med*, 63:116–126, 2009.

[17] DeGroot, M. Uncertainty, information, and sequential experiments. *Ann Math Stat*, 33(2):404–419, 1962.

[18] D. Lindley. On the measure of the information provided by an experiment. *Ann Math Stat*, 27(4):986–1005, 1956.

[19] R. Castro, J. Haupt, R. Nowak, and G. Raz. Finding needles in noisy haystacks. *Proc IEEE Int Conf Acoust, Speech, Sig Proc*, Las Vegas, NV, pp. 5133–5136, 2008.

[20] T. Cover and J. Thomas. *Elements of Information Theory*. 2nd edn. Wiley, 2006.

[21] S. Ji, Y. Xue, and L. Carin. Bayesian compressive sensing. *IEEE Trans Signal Proc* 56(6):2346–2356, 2008.

[22] M. Tipping. Sparse Bayesian learning and the relevance vector machine. *J Machine Learning Res*, 1:211–244, 2001.

[23] M. Seeger. Bayesian inference and optimal design for the sparse linear model. *J Machine Learning Res*, 9:759–813, 2008.

[24] M. Seeger and H. Nickisch. Compressed sensing and Bayesian experimental design. In *Proc Int Conf Machine Learning*. Helsinki, Finland, 2008.

[25] J. Haupt, R. Castro, and R. Nowak. Distilled sensing: Adaptive sampling for sparse detection and estimation. *IEEF Trans Inform Theory*, 57(9):6222–6235, 2011.

[26] D. Donoho and J. Jin. Higher criticism for detecting sparse heterogeneous mixtures. *Ann Stat* 32(3):962–994, 2004.

[27] H. Chernoff. A measure of asymptotic efficiency for tests of a hypothesis based on the sum of observations. *Ann Stat*, 23(4):493–507, 1952.

[28] J. Haupt, R. Baraniuk, R. Castro, and R. Nowak. Compressive distilled sensing: Sparse recovery using adaptivity in compressive measurements. *Proc 43rd Asilomar Conf Sig, Syst, Computers*. Pacific Grove, CA, pp. 1551–1555, 2009.

[29] M. Wainwright. Sharp thresholds for high-dimensional and noisy sparsity pattern recovery using ℓ_1 constrained quadratic programming (Lasso). *IEEE Trans Inform Theory*, 55(5):2183–2202, 2009.

[30] M. A. T. Figueiredo, R. D. Nowak, and S. J. Wright. Gradient projection for sparse reconstruction. *IEEE J Sel Topics Sig Proc*, 1(4):586–597, 2007.

[31] S. Dudoit and M. van der Laan. *Multiple Testing Procedures with Applications to Genomics*. Springer Series in Statistics. Springer, 2008.

[32] H. H. Muller, R. Pahl, and H. Schafer. Including sampling and phenotyping costs into the optimization of two stage designs for genomewide association studies. *Genet Epidemiol*, 31(8):844–852, 2007.

[33] S. Zehetmayer, P. Bauer, and M. Posch. Two-stage designs for experiments with large number of hypotheses. *Bioinformatics*, 21(19):3771–3777, 2005.

[34] J. Satagopan and R. Elston. Optimal two-stage genotyping in population-based association studies. *Genet Epidemiol*, 25(2):149–157, 2003.

[35] S. Zehetmayer, P. Bauer, and M. Posch. Optimized multi-stage designs controlling the false discovery or the family-wise error rate. *Stat Med*, 27(21):4145–4160, 2008.

[36] E. Bashan, R. Raich, and A. Hero. Optimal two-stage search for sparse targets using convex criteria. *IEEE Trans Sig Proc*, 56(11):5389–5402, 2008.

[37] G. Newstadt, E. Bashan, and A. Hero. Adaptive search for sparse targets with informative priors. In: *Proc. Int Conf Acoust, Speech, Sig Proc*. Dallas, TX, 3542–3545, 2010.

[38] M. Iwen and A. Tewfik. Adaptive group testing strategies for target detection and localization in noisy environments. Submitted. 2010 June.

[39] A. Aldroubi, H. Wang, and K. Zarringhalam. Sequential adaptive compressed sampling via Huffman codes. Preprint, http://arxiv.org/abs/0810.4916v2, 2009.

[40] S. Deutsch, A. Averbuch, and S. Dekel. Adaptive compressed image sensing based on wavelet modeling and direct sampling. *Proc. 8th Int Conf Sampling Theory Appli*, Marseille, France, 2009.

[41] Various authors. Promising directions in active vision. *Int J Computer Vision*, 11(2):109–126, 1991.

[42] D. Cohn. Neural network exploration using optimal experiment design. *Adv Neural Information Proc Syst (NIPS)*, 679–686, 1994.

[43] D. Cohn, Z. Ghahramani, and M. Jordan. Active learning with statistical models. *J Artif Intelligence Res*, 4:129–145, 1996.

7 Fundamental thresholds in compressed sensing: a high-dimensional geometry approach

Weiyu Xu and Babak Hassibi

In this chapter, we introduce a unified high-dimensional geometric framework for analyzing the phase transition phenomenon of ℓ_1 minimization in compressive sensing. This framework connects studying the phase transitions of ℓ_1 minimization with computing the Grassmann angles in high-dimensional convex geometry. We demonstrate the broad applications of this Grassmann angle framework by giving sharp phase transitions for ℓ_1 minimization recovery robustness, weighted ℓ_1 minimization algorithms, and iterative reweighted ℓ_1 minimization algorithms.

7.1 Introduction

Compressive sensing is an area of signal processing which has attracted a lot of recent attention for its broad applications and rich mathematical background [7] [19] and Chapter 1. In compressive sensing, we would like to recover an $n \times 1$ real-numbered signal vector x, but we can only get $m < n$ measurement samples through a linear mixing of x. Namely

$$y = Ax, \qquad (7.1)$$

where A is an $m \times n$ measurement matrix and y is an $m \times 1$ measurement result. In an ideal model for compressive sensing, x is an $n \times 1$ unknown k-sparse signal vector, which is defined as a vector having only k nonzero elements. This special structure of x makes recovering x from the compressed measurement y possible.

A naive way to decode or solve for the k-sparse x from y is to enumerate the $\binom{n}{k}$ possible supports of x and then try to see whether there exists such an x satisfying $y = Ax$. But this is of exponential complexity if k is proportionally growing with n and is not computationally feasible. What enables practical compressive sensing is the existence of efficient decoding algorithms to recover the sparse signal x from the compressed measurements y. Arguably the most prominent and powerful decoding approach is the Basis Pursuit programming, namely the ℓ_1 minimization method [13, 15]. The ℓ_1 minimization method solves the following problem

$$\min \|z\|_1$$
$$\text{subject to } y = Az, \qquad (7.2)$$

where $\|z\|_1$ denotes the ℓ_1 norm of z, namely the sum of the absolute values of all the elements in z. This is a convex program which is easy to solve. It has been empirically observed to work very well in producing sparse solutions. Breakthroughs, for example, [11][15][20], in understanding why ℓ_1 minimization successfully promotes sparsity have emerged in recent years, and have triggered an explosive growth of research in compressive sensing, see Chapter 1.

We should remark that in this chapter we are particularly interested in the parameter regime where k and m grow proportionally with n, as n grows large. In other words, the number of measurements is $m = \delta n$, and the number of nonzero elements of x is $k = \rho \delta n = \zeta n$, where $0 < \rho < 1$ and $0 < \delta < 1$ are constants independent of n, and $\delta > \zeta$.

It has been empirically observed and theoretically shown [15, 20] that the ℓ_1 minimization method often exhibits a "phase transition" phenomenon: when the signal support size is below a certain threshold, ℓ_1 minimization will recover the signal vector with overwhelming probability; while when the signal support size is above this threshold, ℓ_1 minimization will fail to recover the signal vector with high probability. Studying this phase transition phenomenon and characterizing the threshold for the support size k has been a very important and active research branch in the development of compressive sensing theories [15, 20] [4] [29, 43, 44] [28, 40, 45, 46][21, 38, 39] [34, 36] [16, 17] and Chapter 9. This branch of research gives precise prediction of sparse recovery algorithms, brings theoretical rigor to compressive sensing theories, and inspires new powerful sparse recovery algorithms.

The first work in the literature that precisely and rigorously characterized the phase transition was [15, 20], through beautifully connecting the projection of high-dimensional convex polytopes and the success of ℓ_1 minimization. In [15, 20], Donoho and Tanner formulated a k-neighborly polytope condition on the measurement matrix A for ℓ_1 minimization to generate the original sparse signal. As shown in [15], this k-neighborly polytope A is in fact a necessary and sufficient condition for (7.2) to produce the sparse solution x satisfying (7.1). This geometric insight, together with known results on the neighborliness of projected polytopes in the literature of convex geometry [1, 41], has led to sharp bounds on the performance of ℓ_1 minimization. In [15], it was shown that if the matrix A has i.i.d. zero-mean Gaussian entries, then the k-neighborly polytope condition holds with overwhelming probability if k is sufficiently small. In the linear scaling setting for m, n, and k discussed in this chapter, the relation between m, n, and k in order for the k-neighborly polytope condition to hold is precisely characterized and calculated in [15]. In fact, the computed values of ζ for the so-called "weak" threshold, obtained for different values of δ through the neighborly polytope condition in [15], match exactly with the phase transitions obtained by simulation when n is large.

However, the neighborly polytope approach in [15] only addressed the phase transitions for ideally sparse signal vectors whose residual elements are exactly zero excluding the k nonzero components. By comparison, the popular restricted isometry property (RIP) [11][4] and Chapter 1 can also be used to analyze the robustness of ℓ_1 minimization [10], even though the RIP analysis generally produces much looser phase transition results than the neighborly polytope condition [4]. Then the question is whether we can have a unified method of determining precise phase transitions for ℓ_1 minimization in broader applications. More specifically, this method should give us tighter phase transitions for ℓ_1 minimization than the RIP condition; but it should also work in deriving phase transitions in more general settings such as:

- phase transitions for recovering approximately sparse signals, instead of only perfectly sparse ones [43];
- phase transitions when the compressed observations are corrupted with noises [42];
- phase transitions for weighted ℓ_1 minimization, instead of regular ℓ_1 minimization [29];
- phase transitions for iterative reweighted ℓ_1 algorithms [12][44].

In this chapter, we are interested in presenting a unified high-dimensional geometric framework to analyze the phase transition phenomenon of ℓ_1 minimization. As we will see, in many applications, it turns out that the performance of ℓ_1 minimization and its variants often depends on the null space "balancedness" properties of the measurement matrix A, see Chapter 1. This unified high-dimensional geometric analysis framework investigates the phase transitions for the null space "balancedness" conditions using the notion of a Grassmann angle. This framework generalizes the neighborly polytope approach in [15, 20] for deriving phase transitions of recovering perfectly sparse signals; however, this Grassmann angle framework can be further used in analyzing the performance thresholds of ℓ_1 minimization for *approximately sparse* signals, weighted ℓ_1 minimization algorithms, and iterative reweighted ℓ_1 minimization algorithms. In this chapter, we will present the Grassmann angle framework for analyzing the null space "balancedness" properties in detail by focusing on the example of characterizing the threshold bounds for ℓ_1 minimization robustness in recovering approximately sparse signals. Then we will briefly illustrate the application of this Grassmann angle framework in characterizing the phase transitions for weighted ℓ_1 minimization and iterative reweighted ℓ_1 minimization algorithms. This framework and results of this chapter have earlier appeared in [43, 29, 44].

Before demonstrating how the Grassmann angle geometric framework can be used to analyze the null space "balancedness" properties, we will give an overview of the main results, comparisons of this Grassmann angle approach with other approaches in the literature, the geometrical concepts to be used frequently in this chapter, and also the organization of this chapter.

7.1.1 Threshold bounds for ℓ_1 minimization robustness

Instead of assuming that x is an exactly k-sparse signal, we now assume that k components of x have large magnitudes and that the vector comprised of the remaining $(n-k)$ components has an ℓ_1-norm less than some value, say, $\sigma_k(x)_1$. We will refer to this type of signal as an approximately k-sparse signal, or for brevity only an approximately sparse signal. It is also possible that the y can be further corrupted with measurement noise. In this case exact recovery of the unknown vector x from a reduced number of measurements is generally not possible. Instead, we focus on obtaining a reconstruction of the signal that is "close" to the true one. More precisely, if we denote the unknown signal as x and denote $\hat{x}$ as one solution to (7.2), we prove that for any given constant $0 < \delta < 1$ and any given constant $C > 1$ (representing how close in ℓ_1 norm the recovered vector $\hat{x}$ should be to x), there exists a constant $\zeta > 0$ and a sequence of measurement matrices $A \in \mathbb{R}^{m \times n}$ as $n \to \infty$ such that

$$||\hat{x} - x||_1 \leq \frac{2(C+1)\sigma_k(x)_1}{C-1}, \quad (7.3)$$

holds for *all* $x \in \mathbb{R}^n$, where $\sigma_k(x)_1$ is the minimum possible ℓ_1 norm value for any $(n-k)$ elements of x (recall $k = \zeta n$). Here ζ will be a function of C and δ, but independent of the problem dimension n. In particular, we have the following theorem.

THEOREM 7.1 *Let n, m, k, x, $\hat{x}$ and $\sigma_k(x)_1$ be defined as above. Let K denote a subset of $\{1, 2, \ldots, n\}$ such that $|K| = k$, where $|K|$ is the cardinality of K, and let K_i denote the ith element of K and $\overline{K} = \{1, 2, \ldots, n\} \setminus K$.*

Then the solution $\hat{x}$ produced by (7.2) will satisfy

$$||\hat{x} - x||_1 \leq \frac{2(C+1)\sigma_k(x)_1}{C-1}. \quad (7.4)$$

for all $x \in \mathbb{R}^n$, if for all vectors $w \in \mathbb{R}^n$ in the null space of A, and for all K such that $|K| = k$, we have

$$C\|w_K\|_1 \leq \|w_{\overline{K}}\|_1, \quad (7.5)$$

where w_K denotes the part of w over the subset K.

Furthermore, if $A \in \mathbb{R}^{m \times n}$ a random matrix with i.i.d. standard Gaussian $\mathcal{N}(0,1)$ entries, then as $n \to \infty$, for any constant $C > 1$ and any $\delta = m/n > 0$, there exists a $\zeta(\delta, C) = k/n > 0$, so that both (7.4) and (7.5) hold with overwhelming probability.

As we said, the generalized Grassmann angle geometric framework can be used to analyze such null space "balancedness" conditions in (7.5), thus establishing a sharp relationship between δ, ζ, and C. For example, when $\delta = m/n$ varies, we have Figure 7.1 showing the tradeoff between the signal sparsity ζ and the parameter C, which determines the robustness[1] of ℓ_1 minimization. We remark that the above theorem clearly subsumes

---
[1] The "robustness" concept in this sense is often called "stability" in other papers, for example, [7].

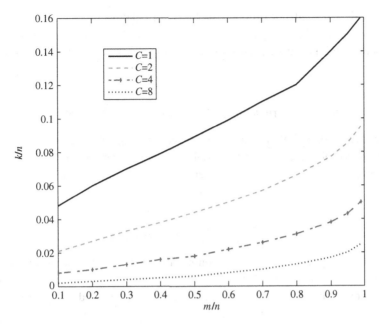

Figure 7.1 Tradeoff between signal sparsity and ℓ_1 recovery robustness as a function of C (allowable imperfection of the recovered signal is $\frac{2(C+1)\sigma_k(x)_1}{C-1}$).

perfect recovery in the perfectly k-sparse setting. For the perfectly k-sparse signal, $\sigma_k(x)_1 = 0$, and so from (7.4) we have $\|\hat{x} - x\|_1 = 0$ and therefore $\hat{x} = x$, when we allow $C \to 1$. This null space characterization is in fact equivalent to the neighborly polytope characterization from [15] in the perfectly k-sparse case when $C = 1$.

The Grassmann angle framework can be applied to characterize sparse recovery performance when there are observation noises involved and $y = Ax + e$, where e is an $m \times 1$ additive noise vector with $\|e\|_2 \leq \epsilon$. We can still use (7.2) to solve for $\hat{x}$. As long as A has rank m with its smallest nonzero singular value $\sigma_m > 0$, there always exists an $n \times 1$ vector Δx such that $\|\Delta x\|_2 \leq \frac{1}{\sigma_r}\epsilon$ and $y = A(x + \Delta x)$. Namely y can be seen as generated by a perturbation of x. This added perturbation Δx can then be treated in the same way as x_K in analyzing $\|x - \hat{x}\|_1$. To further bound $\|x - \hat{x}\|_2$, one can use the almost Euclidean property described in Subsection 7.1.3. For more details on dealing with observation noises based on the Grassmann angle analysis, please refer to [42].

Besides this concept of robustness in ℓ_1 norm discussed in Theorem 7.2, there are also discussions of robustness in general ℓ_p norms, which will involve other types of null space properties. The readers are encouraged to refer to Chapter 1 for an overview of this topic. A similar formulation of ℓ_1 norm robustness recovering an approximately sparse problem was considered in [14], where the null space characterization for recovering approximately sparse signals was analyzed using the RIP condition; however, no explicit values of ζ were given. Through message passing analysis, [16] recently deals with a related but different problem formulation of characterizing the trade-off between signal sparsity and noise sensitivity of the LASSO recovery method. An overview of this new message passing analysis approach is presented in Chapter 9 of this book.

7.1.2 Weighted and iterative reweighted ℓ_1 minimization thresholds

When there is statistical prior information about the signal x, a better decoding method would be to solve the weighted ℓ_1 minimization programming [29]. The weighted ℓ_1 minimization solves the following general version of ℓ_1 minimization

$$\min_{Az=y} \|z\|_{w,1} = \min_{Az=y} \sum_{i=1}^{n} w_i |z_i|, \qquad (7.6)$$

where the weights w_i can be any non-negative real numbers to accommodate prior information. For example, if the prior information shows that x_i is more likely to be zero or small, then a larger corresponding weight can be applied to suppress its decoding result to be zero or a small number.

Again, the successful decoding condition for weighted ℓ_1 minimization is a weighted version of the null space "balancedness" condition. It turns out that the null space Grassmann angle framework can also be readily applied to give sharp sparse recovery threshold analysis for weighted ℓ_1 minimization algorithms [29], where better phase transitions are shown when prior information is available. When no prior information about the signal is available, this null space Grassmann angle framework can also be used to analyze iterative reweighted ℓ_1 minimization [44], where we can rigorously show for the first time that iterative reweighted ℓ_1 algorithms, compared with the plain ℓ_1 minimization algorithms, can increase the phase transition thresholds for interesting types of signals.

7.1.3 Comparisons with other threshold bounds

In this section, we will review other approaches to establishing sparse recovery threshold bounds for ℓ_1 minimization, and compare their strengths and limitations.

Restricted isometry property

In [9][11], it was shown that if the matrix A satisfies the now well-known restricted isometry property (RIP), then any unknown vector x with no more than $k = \zeta n$ nonzero elements can be recovered by solving (7.2), where ζ is an absolute constant as a function of δ, but independent of n, and explicitly bounded in [11]. Please see Chapter 1 for the definition of the RIP conditions and its applications. However, it should be noted that the RIP condition is only a sufficient condition for ℓ_1 minimization to produce a sparse solution to (7.2). Partially because of this fact, the threshold bounds on ζ obtained by the RIP condition are not very sharp so far and are often a very small fraction of the bounds on ζ obtained by the neighborly polytope approach or its generalization to the Grassmann angle approach in this chapter.

One strength of the RIP condition lies in its applicability to a large range of measurement matrices. It turns out that for measurement matrices with i.i.d. zero-mean Gaussian entries, measurement matrices with i.i.d. Bernoulli entries, or matrices for random Fourier measurements, the RIP condition holds with overwhelming probability [11, 2, 33]. In contrast, the neighborly polytope approach and the Grassmann angle approach so far only rigorously work for the measurement matrices with i.i.d. Gaussian entries, even though the universality of the predicted phase transitions by these approaches beyond the Gaussian matrices has been observed [18]. The RIP analysis is also convenient for bounding the reconstruction error in ℓ_2 norm when the observation noises are present.

Neighborly polytope approach

As discussed earlier, by relating a k-neighborly polytope condition to the success of ℓ_1 minimization for decoding ideally k-sparse signals, Donoho and Tanner gave the precise phase transitions for decoding ideally sparse signals in [15, 20]. The Grassmann angle approach in this chapter is a generalization of the neighborly polytope approach in [15, 20]. Compared with the neighborly polytope condition which only works for analyzing the ideally sparse signal vectors, the generalized Grassmann approach is intended to give sharp phase transitions for the null space "balancedness" conditions, which are useful in a more general setting, for example, in analyzing the robustness of ℓ_1 minimization, weighted ℓ_1 minimization, and iterative reweighted ℓ_1 minimization. Mathematically, in this chapter we need to derive new formulas for the various geometric angles in the Grassmann angle approach. This chapter uses the same computational techniques in estimating the asymptotics of the Grassmann angle as in estimating the asymptotic face counts in [15].

Spherical section property approach

The threshold bounds on ζ for the null space condition to hold was also analyzed in [28, 40, 45, 46], using the *spherical section property* of linear subspaces derived from the Kashin–Garnaev–Gluskin Inequality [23, 27, 46]. The Kashin–Garnaev–Gluskin Inequality claims that for a uniformly distributed $(n-m)$-dimensional subspace, with overwhelming probability, all vectors w from this subspace will satisfy the spherical section property,

$$\|w\|_1 \geq \frac{c_1 \sqrt{m}}{\sqrt{1+\log(n/m)}} \|w\|_2, \qquad (7.7)$$

where c_1 is a constant independent of the problem dimension. Note that $\|w\|_1 \leq \sqrt{n}\|w\|_2$ always holds, taking equality only if w is a perfectly balanced vector of constant magnitude, so it is natural to see that this spherical section property can be used to investigate the subspace "balancedness" property [45]. This approach extends to general matrices such as random Gaussian matrices, random Bernoulli matrices, and random Fourier mapping matrices [40], and is applicable to the analysis of sparse recovery robustness [46]. The threshold bounds on ζ given by this approach are sometimes better than those obtained from the RIP condition [45], but are generally worse than those obtained by the neighborly polytope approach, partially because of the coarsely estimated c_1 used in the literature.

Sphere covering approach

The null space condition has also been analyzed by a sphere covering approach in [21, 38, 39]. The subspace property (7.5) is supposed to hold for every vector w in the null space of A, and we can restrict our attention to all the points w in the form of $w = Bv$, where $B \in \mathbb{R}^{n \times (n-m)}$ is a fixed basis for the null space of A, and v is any point from the unit Euclidean sphere in $\mathbb{R}^{n-m}$. The sphere covering approach proposed to cover the unit Euclidean sphere densely with discrete points such that any point on this unit Euclidean sphere is close enough to a discrete point. If the null space condition (7.5) holds for the vectors generated by these discrete points, it would be possible to infer that the null space condition will also hold for all the points generated by the unit Euclidean sphere and for all the points in the null space of A.

Following this methodology, various threshold bounds have been established in [21, 38, 39]. These bounds are generally better than the threshold bounds from the RIP condition, but weaker than the bounds from the Grassmann angle approach. But in the limiting case when m is very close to n, the threshold bounds from the sphere covering approach can match or are better than the ones obtained from the neighborly polytope approach.

"Escape-through-mesh" approach

More recently, an alternative framework for establishing sharp ℓ_1 minimization thresholds has been proposed in [36] by craftily using the "escape-through-mesh" theorem [24]. The "escape-through-mesh" theorem quantifies the probability that a uniformly distributed $(n-m)$-dimensional subspace in $\mathbb{R}^n$ misses a set of points on the unit Euclidean sphere in $\mathbb{R}^n$. The "escape-through-mesh" theorem was first used in analyzing sparse reconstructions in [34]. Based on this theorem, a careful calculation was devised in [36] to evaluate the probability that a uniformly distributed $(n-m)$-dimensional subspace in $\mathbb{R}^n$ escapes the set of points that violate the null space "balancedness" condition (7.5) for $C = 1$.

The method of [36] yields almost the same threshold bounds for weak recovery that the neighborly polytope approach does; however, for sectional and strong recoveries, it gives different threshold bounds (in some regimes the neighborly polytope approach gives a better bound, and in other regimes the "escape-through-mesh" approach does better). Fully understanding the relation between this "escape-through-mesh" approach and the neighborly polytope approach should be of great interest.

Message passing analysis approach

More recent works [16][17] give threshold bounds for large-scale ℓ_1 minimization and ℓ_1 regularized regression problems through graphical models and novel message passing analysis. For this very interesting approach, the readers are encouraged to refer to Chapter 9 for more details. By comparison, the Grassmann angle approach can provide ℓ_1 norm bounded robustness results in the "weak," "sectional," and "strong" senses (see Section 7.8), while the message passing analysis is more powerful in providing average-case robustness results in terms of mean squared error [17].

7.1.4 Some concepts in high-dimensional geometry

In this part, we will give the explanations of several geometric terminologies often used in this chapter for the purpose of quick reference.

Grassmann manifold

The Grassmann manifold $\mathrm{Gr}_i(j)$ refers to the set of i-dimensional subspaces in the j-dimensional Euclidean space $\mathbb{R}^j$. It is known that there exists a unique invariant measure μ' on $\mathrm{Gr}_i(j)$ such that $\mu'(\mathrm{Gr}_i(j)) = 1$.

For more facts on the Grassmann manifold, please see [5].

Polytope, face, vertex

A polytope in this chapter refers to the convex hull of a finite number of points in the Euclidean space. Any extreme point of a polytope is a vertex of this polytope. A face of a polytope is defined as the convex hull of a set of its vertices such that no point in this convex hull is an interior point of the polytope. The dimension of a face refers to the dimension of the affine hull of that face. The book [26] offers a nice reference on convex polytopes.

Cross-polytope

The n-dimensional cross-polytope is the polytope of the unit ℓ_1 ball, namely it is the set

$$\{x \in \mathbb{R}^n \mid \|x\|_1 = 1\}.$$

The n-dimensional cross-polytope has $2n$ vertices, namely $\pm e_1, \pm e_2, \ldots, \pm e_n$, where e_i, $1 \leq i \leq n$, is the unit vector with its ith coordinate element being 1. Any k extreme points without opposite pairs at the same coordinate will constitute a $(k-1)$-dimensional face of the cross-polytope. So the cross-polytope will have $2^k \binom{n}{k}$ faces of dimension $(k-1)$.

Grassmann angle

The Grassmann angle for an n-dimensional cone $\mathcal{C}$ under the Grassmann manifold $\mathrm{Gr}_i(n)$, is the measure of the set of i-dimensional subspaces (over $\mathrm{Gr}_i(n)$) which intersect the cone $\mathcal{C}$ nontrivially (namely at some other point besides the origin). For more details on the Grassmann angle, internal angle, and external angle, please refer to [25][26][31].

Cone obtained by observing a set B from a set A

In this chapter, when we say "cone obtained by observing B from A," we mean the conic hull of all the vectors in the form of $x_1 - x_2$, where $x_1 \in B$ and $x_2 \in A$.

Internal angle

An internal angle $\beta(F_1, F_2)$, between two faces F_1 and F_2 of a polytope or a polyhedral cone, is the fraction of the hypersphere S covered by the cone obtained by observing the face F_2 from the face F_1. The internal angle $\beta(F_1, F_2)$ is defined to be zero when

$F_1 \not\subseteq F_2$ and is defined to be one if $F_1 = F_2$. Note the dimension of the hypersphere S here matches the dimension of the corresponding cone discussed. Also, the center of the hypersphere is the apex of the corresponding cone. All these defaults also apply to the definition of the external angles.

External angle

An external angle $\gamma(F_3, F_4)$, between two faces F_3 and F_4 of a polytope or a polyhedral cone, is the fraction of the hypersphere S covered by the cone of outward normals to the hyperplanes supporting the face F_4 at the face F_3. The external angle $\gamma(F_3, F_4)$ is defined to be zero when $F_3 \not\subseteq F_4$ and is defined to be one if $F_3 = F_4$.

7.1.5 Organization

The rest of the chapter is organized as follows. In Section 7.2, we introduce a null space characterization for guaranteeing robust signal recovery using ℓ_1 minimization. Section 7.3 presents a Grassmann angle-based high-dimensional geometrical framework for analyzing the null space characterization. In Sections 7.4, 7.5, 7.6, and 7.7, analytical performance bounds are given for the null space characterization for matrices that are rotationally invariant, such as those constructed from i.i.d. Gaussian entries. Section 7.8 shows how the Grassmann angle analytical framework can be extended to analyzing the "weak," "sectional," and "strong" notations of robust signal recovery.

In Section 7.9, numerical evaluations of the performance bounds for robust signal recovery are given. Section 7.10 and 7.11 will introduce the applications of the Grassmann angle approach to analyzing weighted ℓ_1 minimization and iterative reweighted ℓ_1 minimization algorithms.

Section 7.12 concludes the chapter. In the Appendix, we provide the proofs of related lemmas and theorems.

7.2 The null space characterization

In this section we introduce a useful characterization of the matrix A. The characterization will establish a necessary and sufficient condition on the matrix A so that the solution of (7.2) approximates the solution of (7.1) such that (7.3) holds. (See [22, 30, 45, 14, 38, 39, 28] etc. for variations of this result).

THEOREM 7.2 *Assume that A is a general $m \times n$ measurement matrix. Let $C > 1$ be a positive number. Further, assume that $y = Ax$ and that w is an $n \times 1$ vector. Let K be a subset of $\{1, 2, \ldots, n\}$ such that $|K| = k$, where $|K|$ is the cardinality of K and let K_i denote the ith element of K. Further, let $\overline{K} = \{1, 2, \ldots, n\} \setminus K$. Then for any $x \in \mathbb{R}^n$, for any K such that $|K| = k$, any solution $\hat{x}$ produced by (7.2) will satisfy*

$$\|x - \hat{x}\|_1 \leq \frac{2(C+1)}{C-1} \|x_{\overline{K}}\|_1, \tag{7.8}$$

if $\forall w \in \mathbb{R}^n$ *such that*

$$Aw = 0$$

and $\forall K$ *such that* $|K| = k$, *we have*

$$C\|w_K\|_1 \leq \|w_{\overline{K}}\|_1. \tag{7.9}$$

Conversely, there exists a measurement matrix A, an x, and corresponding $\hat{x}$ ($\hat{x}$ is a minimizer to the programming (7.2)) such that (7.9) is violated for some set K with cardinality k and some vector w from the null space of A, and

$$\|x - \hat{x}\|_1 > 2\frac{(C+1)}{C-1}\|x_{\overline{K}}\|_1.$$

Proof. First, suppose the matrix A has the claimed null space property as in (7.9) and we want to prove that any solution $\hat{x}$ satisfies (7.8). Note that the solution $\hat{x}$ of (7.2) satisfies

$$\|\hat{x}\|_1 \leq \|x\|_1,$$

where x is the original signal. Since $A\hat{x} = y$, it easily follows that $w = \hat{x} - x$ is in the null space of A. Therefore we can further write $\|x\|_1 \geq \|x + w\|_1$. Using the triangular inequality for the ℓ_1 norm we obtain

$$\begin{aligned}\|x_K\|_1 + \|x_{\overline{K}}\|_1 &= \|x\|_1 \\ &\geq \|\hat{x}\|_1 = \|x + w\|_1 \\ &\geq \|x_K\|_1 - \|w_K\|_1 + \|w_{\overline{K}}\|_1 - \|x_{\overline{K}}\|_1 \\ &\geq \|x_K\|_1 - \|x_{\overline{K}}\|_1 + \frac{C-1}{C+1}\|w\|_1\end{aligned}$$

where the last inequality is from the claimed null space property. Relating the head and tail of the inequality chain above,

$$2\|x_{\overline{K}}\|_1 \geq \frac{(C-1)}{C+1}\|w\|_1.$$

Now we prove the second part of the theorem, namely when (7.9) is violated, there exist scenarios where the error performance bound (7.8) fails.

Consider a generic $m \times n$ matrix A'. For each integer $1 \leq k \leq n$, let us define the quantity h_k as the supremum of $\frac{\|w_K\|_1}{\|w_{\overline{K}}\|_1}$ over all such sets K of size $|K| \leq k$ and over all nonzero vectors w in the null space of A'. Let k^* be the biggest k such that $h_k \leq 1$. Then there must be a nonzero vector w' in the null space of A and a set K^* of size k^*, such that

$$\|w'_{K^*}\|_1 = h_{k^*}\|w'_{\overline{K^*}}\|_1.$$

Now we generate a new measurement matrix A by multiplying the portion A'_{K^*} of the matrix A' by h_{k^*}. Then we will have a vector w in the null space of A satisfying

$$\|w_{K^*}\|_1 = \|w_{\overline{K^*}}\|_1.$$

Now we take a signal vector $x = (-w_{K^*}, 0_{\overline{K^*}})$ and claim that $\hat{x} = (0, w_{\overline{K^*}})$ is a minimizer to the programming (7.2). In fact, recognizing the definition of h_{k^*}, we know all the vectors w'' in the null space of the measurement matrix A will satisfy $\|x + w''\|_1 \geq \|x\|_1$. Let us assume that $k^* \geq 2$ and take $K'' \subseteq K^*$ as the index set corresponding to the largest $(k^* - i)$ elements of x_{K^*} in amplitude, where $1 \leq i \leq (k^* - 1)$. From the definition of k^*, it is apparent that $C' = \frac{\|w_{\overline{K''}}\|_1}{\|w_{K''}\|_1} > 1$ since w is nonzero for any index in the set K^*. Let us now take $C = \frac{\|w_{\overline{K''}}\|_1}{\|w_{K''}\|_1} + \epsilon$, where $\epsilon > 0$ is any arbitrarily small positive number. Thus the condition (7.9) is violated for the vector w, the set K'', and the defined constant C.

Now by inspection, the decoding error is

$$\|x - \hat{x}\|_1 = \frac{2(C'+1)}{C'-1}\|x_{K''}\|_1 > \frac{2(C+1)}{C-1}\|x_{K''}\|_1,$$

violating the error bound (7.8) (for the set K''). $\square$

Discussion

It should be noted that if the condition (7.9) is true for all the sets K of cardinality k, then

$$2\|x_{\overline{K}}\|_1 \geq \frac{(C-1)}{C+1}\|\hat{x} - x\|_1$$

is also true for the set K which corresponds to the k largest (in amplitude) components of the vector x. So

$$2\sigma_k(x)_1 \geq \frac{(C-1)}{C+1}\|\hat{x} - x\|_1$$

which exactly corresponds to (7.3). It is an interesting result that, for a particular *fixed* measurement matrix A, the violation of (7.9) for some $C > 1$ does not necessarily imply that the existence of a vector x and a minimizer solution $\hat{x}$ to (7.2) such that the performance guarantee (7.8) is violated. For example, assume $n = 2$ and the null space of the measurement matrix A is a one-dimensional subspace and has the vector $(1, 100)$ as its basis. Then the null space of the matrix A violates (7.9) with $C = 101$ and the set $K = \{1\}$. But a careful examination shows that the biggest possible $\frac{\|x - \hat{x}\|_1}{\|x_{\overline{K}}\|_1}$ ($\|x_{\overline{K}}\|_1 \neq 0$) is equal to $\frac{100+1}{100} = \frac{101}{100}$, achieved by such an x as $(-1, -1)$. In fact, all those vectors $x = (a, b)$ with $b \neq 0$ will achieve $\frac{\|x - \hat{x}\|_1}{\|x_{\overline{K}}\|_1} = \frac{101}{100}$. However, (7.8) has $\frac{2(C+1)}{C-1} = \frac{204}{100}$. This suggests that for a specific measurement matrix A, the tightest error bound for $\frac{\|x - \hat{x}\|_1}{\|x_{\overline{K}}\|_1}$ should involve the detailed structure of the null space of A. But for general measurement matrices A, as suggested by Theorem 7.2, the condition (7.9) is a necessary and sufficient condition to offer the performance guarantee (7.8).

Analyzing the null space condition: the Gaussian ensemble

In the remaining part of this chapter, for a given value $\delta = m/n$ and any value $C \geq 1$, we will devote our efforts to determining the value of feasible $\zeta = \rho\delta = k/n$ for which there exists a sequence of A such that the null space condition (7.9) is satisfied for all the sets K of size k when n goes to infinity and $m/n = \delta$. For a specific A, it is very hard to check whether the condition (7.9) is satisfied or not. Instead, we consider randomly choosing A from the Gaussian ensemble, namely A has i.i.d. $\mathcal{N}(0,1)$ entries, and analyze for what ζ, the condition (7.9) for its null space is satisfied with overwhelming probability as n goes to infinity. This Gaussian matrix ensemble is widely used in compressive sensing research, see for example, Chapters 1 and 9.

The following lemma gives a characterization of the resulting null space of A, which is a fairly well-known result [8][32].

LEMMA 7.1 *Let $A \in \mathbb{R}^{m \times n}$ be a random matrix with i.i.d. $\mathcal{N}(0,1)$ entries. Then the following statements hold:*

- *The distribution of A is right-rotationally invariant: for any Θ satisfying $\Theta\Theta^* = \Theta^*\Theta = I$, $P_A(A) = P_A(A\Theta)$.*
- *There exists a basis Z of the null space of A, such that the distribution of Z is left-rotationally invariant: for any Θ satisfying $\Theta\Theta^* = \Theta^*\Theta = I$, $P_Z(Z) = P_Z(\Theta^*Z)$.*
- *It is always possible to choose a basis Z for the null space such that Z has i.i.d. $\mathcal{N}(0,1)$ entries.*

In view of Theorem 7.2 and Lemma 7.1 what matters is that the null space of A be rotationally invariant. Sampling from this rotationally invariant distribution is equivalent to uniformly sampling a random $(n-m)$-dimensional subspace from the Grassmann manifold $\mathrm{Gr}_{(n-m)}(n)$. For any such A and ideally sparse signals, the sharp bounds of [15] apply. However, we shall see that the neighborly polytope condition for ideally sparse signals does not readily apply to the proposed null space condition analysis for approximately sparse signals, since the null space condition cannot be transformed to the k-neighborly property in a *single* high-dimensional polytope [15]. Instead, in this chapter, we shall give a unified Grassmann angle framework to directly analyze the proposed null space property.

7.3 The Grassmann angle framework for the null space characterization

In this section we detail the Grassmann angle-based framework for analyzing the bounds on $\zeta = k/n$ such that (7.9) holds for every vector in the null space, which we denote by Z. Put more precisely, given a certain constant $C > 1$ (or $C \geq 1$), which corresponds to a certain level of recovery accuracy for the approximately sparse signals, we are interested in what scaling k/n we can achieve while satisfying the following condition on Z ($|K| = k$):

$$\forall w \in Z, \forall K \subseteq \{1, 2, \ldots, n\}, C\|w_K\|_1 \leq \|w_{\overline{K}}\|_1. \tag{7.10}$$

From the definition of the condition (7.10), there is a tradeoff between the largest sparsity level k and the parameter C. As C grows, clearly the largest k satisfying (7.10) will likely decrease, and, at the same time, ℓ_1 minimization will be more robust in terms of the residual norm $\|x_{\overline{K}}\|_1$. The key in our derivation is the following lemma:

LEMMA 7.2 *For a certain subset $K \subseteq \{1,2,\ldots,n\}$ with $|K| = k$, the event that the null space Z satisfies*

$$C\|w_K\|_1 \leq \|w_{\overline{K}}\|_1, \forall w \in Z$$

is equivalent to the event that $\forall x$ supported on the k-set K (or supported on a subset of K):

$$\|x_K + w_K\|_1 + \|\frac{w_{\overline{K}}}{C}\|_1 \geq \|x_K\|_1, \forall w \in Z. \quad (7.11)$$

Proof. First, let us assume that $C\|w_K\|_1 \leq \|w_{\overline{K}}\|_1, \forall w \in Z$. Using the triangular inequality, we obtain

$$\|x_K + w_K\|_1 + \|\frac{w_{\overline{K}}}{C}\|_1$$
$$\geq \|x_K\|_1 - \|w_K\|_1 + \|\frac{w_{\overline{K}}}{C}\|_1$$
$$\geq \|x_K\|_1$$

thus proving the forward part of this lemma. Now let us assume instead that $\exists w \in Z$, such that $C\|w_K\|_1 > \|w_{\overline{K}}\|_1$. Then we can construct a vector x supported on the set K (or a subset of K), with $x_K = -w_K$. Then we have

$$\|x_K + w_K\|_1 + \|\frac{w_{\overline{K}}}{C}\|_1$$
$$= 0 + \|\frac{w_{\overline{K}}}{C}\|_1$$
$$< \|x_K\|_1,$$

proving the converse part of this lemma. □

Now let us consider the probability that condition (7.10) holds for the sparsity $|K| = k$ if we uniformly sample a random $(n-m)$-dimensional subspace Z from the Grassmann manifold $\mathrm{Gr}_{(n-m)}(n)$. Based on Lemma 7.2, we can equivalently consider the complementary probability P that there exists a subset $K \subseteq \{1,2,\ldots,n\}$ with $|K| = k$, and a vector $x \in \mathbb{R}^n$ supported on the set K (or a subset of K) failing the condition (7.11). With the linearity of the subspace Z in mind, to obtain P, we can restrict our attention to those vectors x from the cross-polytope (the unit ℓ_1 ball) $\{x \in \mathbb{R}^n \mid \|x\|_1 = 1\}$ that are only supported on the set K (or a subset of K).

First, we upper bound the probability P by a union bound over all the possible support sets $K \subseteq \{1,2,\ldots,n\}$ and all the sign patterns of the k-sparse vector x. Since the k-sparse vector x has $\binom{n}{k}$ possible support sets of cardinality k and 2^k possible sign patterns (non-negative or non-positive), we have

$$P \leq \binom{n}{k} \times 2^k \times P_{K,-}, \tag{7.12}$$

where $P_{K,-}$ is the probability that for a specific *support set* K, there exists a k-sparse vector x of a specific *sign pattern* which fails the condition (7.11). By symmetry, without loss of generality, we assume the signs of the elements of x to be nonpositive.

So now let us focus on deriving the probability $P_{K,-}$. Since x is a nonpositive k-sparse vector supported on the set K (or a subset of K) and can be restricted to the cross-polytope $\{x \in \mathbb{R}^n \mid \|x\|_1 = 1\}$, x is also on a $(k-1)$-dimensional face, denoted by F, of the skewed cross-polytope (weighted ℓ_1 ball) SP:

$$\text{SP} = \{y \in \mathbb{R}^n \mid \|y_{\overline{K}}\|_1 + \|\frac{y_K}{C}\|_1 \leq 1\}. \tag{7.13}$$

Then $P_{K,-}$ is the probability that there exists an $x \in F$, and there exists a $w \in Z$ ($w \neq 0$) such that

$$\|x_{\overline{K}} + w_{\overline{K}}\|_1 + \|\frac{w_K}{C}\|_1 \leq \|x_K\|_1 = 1. \tag{7.14}$$

We first focus on studying a specific *single* point $x \in F$, without loss of generality, assumed to be in the relative interior of this $(k-1)$-dimensional face F. For this single particular x on F, the probability, denoted by P'_x, that $\exists w \in Z$ ($w \neq 0$) such that (7.14) holds is essentially the probability that a uniformly chosen $(n-m)$-dimensional subspace Z shifted by the point x, namely $(Z+x)$, intersects the skewed cross-polytope

$$\text{SP} = \{y \in \mathbb{R}^n \mid \|y_{\overline{K}}\|_1 + \|\frac{y_K}{C}\|_1 \leq 1\} \tag{7.15}$$

nontrivially, namely, at some other point besides x.

From the linear property of the subspace Z, the event that $(Z+x)$ intersects the skewed cross-polytope SP is equivalent to the event that Z intersects nontrivially with the cone SP-Cone(x) obtained by observing the skewed polytope SP from the point x. (Namely, SP-Cone(x) is the conic hull of the point set (SP $-x$) and SP-Cone(x) has the origin of the coordinate system as its apex.) However, as noticed in the geometry of convex polytopes [25][26], the SP-Cone(x) is identical for any x lying in the relative interior of the face F. This means that the probability $P_{K,-}$ is equal to P'_x, regardless of the fact that x is only a single point in the relative interior of the face F. (The acute reader may have noticed some singularities here because $x \in F$ may not be in the relative interior of F, but it turns out that the SP-Cone(x) is then only a subset of the cone we get when x is in the relative interior of F. So we do not lose anything if we restrict x to be in the relative interior of the face F.) In summary, we have

$$P_{K,-} = P'_x.$$

Now we only need to determine P'_x. From its definition, P'_x is exactly the *complementary Grassmann angle* [25] for the face F with respect to the polytope SP under the Grassmann manifold $\text{Gr}_{(n-m)}(n)$: the probability of a uniformly distributed $(n-m)$-dimensional subspace Z from the Grassmann manifold $\text{Gr}_{(n-m)}(n)$ intersecting nontrivially with the cone SP-Cone(x) formed by observing the skewed cross-polytope SP from the relative interior point $x \in F$.

Building on the works by L. A. Santaló [35] and P. McMullen [31] etc. in high-dimensional integral geometry and convex polytopes, the complementary Grassmann angle for the $(k-1)$-dimensional face F can be explicitly expressed as the sum of products of internal angles and external angles [26]:

$$2 \times \sum_{s \geq 0} \sum_{G \in \Im_{m+1+2s}(\text{SP})} \beta(F,G)\gamma(G,\text{SP}), \qquad (7.16)$$

where s is any non-negative integer, G is any $(m+1+2s)$-dimensional face of the skewed cross-polytope ($\Im_{m+1+2s}(\text{SP})$ is the set of all such faces), $\beta(\cdot,\cdot)$ stands for the internal angle, and $\gamma(\cdot,\cdot)$ stands for the external angle.

The internal angles and external angles are basically defined as follows [26][31]:

- An internal angle $\beta(F_1, F_2)$ is the fraction of the hypersphere S covered by the cone obtained by observing the face F_2 from the face F_1.[2] The internal angle $\beta(F_1, F_2)$ is defined to be zero when $F_1 \not\subseteq F_2$ and is defined to be one if $F_1 = F_2$.
- An external angle $\gamma(F_3, F_4)$ is the fraction of the hypersphere S covered by the cone of outward normals to the hyperplanes supporting the face F_4 at the face F_3. The external angle $\gamma(F_3, F_4)$ is defined to be zero when $F_3 \not\subseteq F_4$ and is defined to be one if $F_3 = F_4$.

Let us take for example the 2-dimensional skewed cross-polytope

$$\text{SP} = \{(y_1, y_2) \in \mathbb{R}^2 | \ \|y_2\|_1 + \|\frac{y_1}{C}\|_1 \leq 1\}$$

(namely the diamond) in Figure 7.2, where $n = 2$, $(n-m) = 1$ and $k = 1$. Then the point $x = (0, -1)$ is a 0-dimensional face (namely a vertex) of the skewed polytope SP. Now from their definitions, the internal angle $\beta(x, \text{SP}) = \beta/2\pi$ and the external angle $\gamma(x, \text{SP}) = \gamma/2\pi$, $\gamma(\text{SP}, \text{SP}) = 1$. The complementary Grassmann angle for the vertex x with respect to the polytope SP is the probability that a uniformly sampled 1-dimensional subspace (namely a line, we denote it by Z) shifted by x intersects nontrivially with SP $= \{(y_1, y_2) \in \mathbb{R}^2 | \ \|y_2\|_1 + \|\frac{y_1}{C}\|_1 \leq 1\}$ (or equivalently the probability that Z intersects nontrivially with the cone obtained by observing SP from the point x). It is obvious that this probability is β/π. The readers can also verify the correctness of the formula (7.16) very easily for this toy example.

---

[2] Note the dimension of the hypersphere S here matches the dimension of the corresponding cone discussed. Also, the center of the hypersphere is the apex of the corresponding cone. All these defaults also apply to the definition of the external angles.

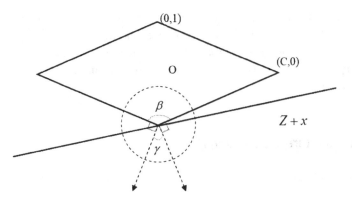

Figure 7.2 The Grassmann angle for a skewed cross-polytope. [42] © 2008 IEEE.

Generally, it might be hard to give explicit formulae for the external and internal angles involved, but fortunately in the skewed cross-polytope case, both the internal angles and the external angles can be explicitly derived. The derivations of these quantities involve the computations of the volumes of cones in high-dimensional geometry and will be presented in the appendix. Here we only present the final results.

Firstly, let us look at the internal angle $\beta(F,G)$ between the $(k-1)$-dimensional face F and a $(l-1)$-dimensional face G. Notice that the only interesting case is when $F \subseteq G$ since $\beta(F,G) \neq 0$ only if $F \subseteq G$. We will see if $F \subseteq G$, the cone formed by observing G from F is the direct sum of a $(k-1)$-dimensional linear subspace and a convex polyhedral cone formed by $(l-k)$ unit vectors with inner product $\frac{1}{1+C^2k}$ between each other. So the internal angle derived in the appendix is given by

$$\beta(F,G) = \frac{V_{l-k-1}(\frac{1}{1+C^2k}, l-k-1)}{V_{l-k-1}(S^{l-k-1})}, \quad (7.17)$$

where $V_i(S^i)$ denotes the ith-dimensional surface measure on the unit sphere S^i, while $V_i(\alpha', i)$ denotes the surface measure for a regular spherical simplex with $(i+1)$ vertices on the unit sphere S^i and with inner product as α' between these $(i+1)$ vertices. Thus in the appendix, (7.17) is shown to be equal to $B(\frac{1}{1+C^2k}, l-k)$, where

$$B(\alpha', m') = \theta^{\frac{m'-1}{2}} \sqrt{(m'-1)\alpha' + 1} \pi^{-m'/2} \alpha'^{-1/2} J(m', \theta), \quad (7.18)$$

with $\theta = (1 - \alpha')/\alpha'$ and

$$J(m', \theta) = \frac{1}{\sqrt{\pi}} \int_{-\infty}^{\infty} (\int_0^\infty e^{-\theta v^2 + 2iv\lambda} dv)^{m'} e^{-\lambda^2} d\lambda. \quad (7.19)$$

Secondly, we derive in the appendix the external angle $\gamma(G, \mathrm{SP})$ between the $(l-1)$-dimensional face G and the skewed cross-polytope SP as:

$$\gamma(G, \mathrm{SP}) = \frac{2^{n-l}}{\sqrt{\pi}^{n-l+1}} \int_0^\infty e^{-x^2} (\int_0^{\frac{x}{C\sqrt{k+\frac{l-k}{C^2}}}} e^{-y^2} dy)^{n-l} dx. \quad (7.20)$$

In summary, combining (7.12), (7.16), (7.17), and (7.20), we get an upper bound on the probability P. If we can show that for a certain $\zeta = k/n$, P goes to zero exponentially in n as $n \to \infty$, then we know that for such ζ, the null space condition (7.10) holds with overwhelming probability. This is the guideline for computing the bound on ζ in the following sections.

7.4 Evaluating the threshold bound ζ

In summary, we have

$$P \leq \binom{n}{k} \times 2^k \times 2 \times \sum_{s \geq 0} \sum_{G \in \mathfrak{S}_{m+1+2s}(\text{SP})} \beta(F,G)\gamma(G,\text{SP}). \tag{7.21}$$

This upper bound on the failure probability is similar to the upper bound on the expected number of faces lost in the random projection of the standard ℓ_1 ball through the random projection A, which was originally derived in [1] and used in [15]. However, there are two differences between these two upper bounds. Firstly, different from [15], in (7.21), there do not exist terms dealing with faces F whose dimension is smaller than $(k-1)$. This is because we do not lose anything by only considering the Grassmann angle for a point in the relative interior of a $(k-1)$-dimensional face F, as explained in the previous section. Secondly, the internal angles and external angles expressions in (7.21) will change as a function of $C \geq 1$, while the corresponding angles in (7.21) are for the neighborly polytopes, where $C = 1$.

In the next few sections, we will build on the techniques developed in [15, 41] to evaluate the bounds on ζ from (7.21) such that P asymptotically goes to 0 as n grows, taking into account the variable $C > 1$. To illustrate the effect of C on the bound ζ, also for the sake of completeness, we will keep the detailed derivations. In the meanwhile, to make the steps easier for the readers to follow, we adopt the same set of notations as in [15] for corresponding quantities.

For simplicity of analysis, we define $l = (m+1+2s)+1$ and $\nu = l/n$. In the skewed cross-polytope SP, we notice that there are in total $\binom{n-k}{l-k}2^{l-k}$ faces G of dimension $(l-1)$ such that $F \subseteq G$ and $\beta(F,G) \neq 0$. Because of symmetry, it follows from (7.21) that

$$P \leq \sum_{s \geq 0} 2\underbrace{\binom{n}{k}2^l \times \binom{n-k}{l-k}}_{COM_s}\beta(F,G)\gamma(G,\text{SP}), \tag{7.22}$$

$$\underbrace{\phantom{\sum_{s \geq 0} 2\binom{n}{k}2^l \times \binom{n-k}{l-k}\beta(F,G)\gamma(G,\text{SP})}}_{D_s}$$

where $l = (m+1+2s)+1$ and $G \subseteq \text{SP}$ is any single face of dimension $(l-1)$ such that $F \subseteq G$. We also define each sum term and its coefficient as D_s and COM_s, as illustrated in (7.22).

In order for the upper bound on P in (7.22) to decrease to 0 as $n \to \infty$, one sufficient condition is that every sum term D_s in (7.22) goes to 0 exponentially fast in n. Since

$$n^{-1}\log(D_s) = n^{-1}\log(COM_s) + n^{-1}\log(\gamma(G,\mathrm{SP})) + n^{-1}\log(\beta(F,G)),$$

if we want the natural logarithm $n^{-1}\log(D_s)$ to be negative, $n^{-1}\log(COM_s)$, which is non-negative, needs to be overwhelmed by the sum of the logarithms, which are non-positive, for internal angles and external angles.

For fixed ρ, δ, and C, it turns out that there exists a decay exponent $\psi_{ext}(\nu;\rho,\delta,C)$, as a function of $\nu = l/n$, at which rate $\gamma(G,\mathrm{SP})$ decays exponentially. Namely for each $\epsilon > 0$, we have

$$n^{-1}\log(\gamma(G,\mathrm{SP})) \le -\psi_{ext}(\nu;\rho,\delta,C) + \epsilon,$$

uniformly in $l \ge \delta n$, $n \ge n_0(\rho,\delta,\epsilon,C)$, where $n_0(\rho,\delta,\epsilon,C)$ is a large enough natural number depending only on ρ, δ, ϵ, and C. This exponent $\psi_{ext}(\nu;\rho,\delta,C)$ is explicitly specified in Section 7.6.

Similarly for fixed ρ, δ, and C, the internal angle $\beta(F,G)$ decays at a rate $\psi_{int}(\nu;\rho,\delta,C)$, which is defined in Section 7.5. Namely, for any $\epsilon > 0$, we will have the scaling

$$n^{-1}\log(\beta(F,G)) = -\psi_{int}(\nu;\rho,\delta,C) + \epsilon,$$

uniformly over $l \ge \delta n$ and $n \ge n_0(\rho,\delta,\epsilon,C)$, where $n_0(\rho,\delta,\epsilon,C)$ is a large enough natural number.

For the coefficient term COM_s in (7.21), after some algebra, we know that for any $\epsilon > 0$,

$$n^{-1}\log(COM_s) = \underbrace{\nu\log(2) + H(\rho\delta) + H\left(\frac{\nu - \rho\delta}{1 - \rho\delta}\right)(1 - \rho\delta)}_{\text{combinatorial growth exponent } \psi_{com}(\nu;\rho,\delta)} + \epsilon, \quad (7.23)$$

uniformly when $l \ge \delta n$ and $n > n_0(\rho,\delta,\epsilon)$ (where $n_0(\rho,\delta,\epsilon)$ is some big enough natural number), and $H(p) = p\log(1/p) + (1-p)\log(1/(1-p))$. In getting (7.23), we used the well-known fact that $\frac{1}{n}\log\left(\binom{n}{\lfloor pn \rfloor}\right)$ approaches $H(p)$ arbitrarily close as n grows to infinity [15].

In summary, if we define the net exponent $\psi_{net}(\nu;\rho,\delta,C) = \psi_{com}(\nu;\rho,\delta) - \psi_{int}(\nu;\rho,\delta,C) - \psi_{ext}(\nu;\rho,\delta,C)$, then for an arbitrary $C \ge 1$, for any fixed choice of ρ, δ, $\epsilon > 0$, and for large enough n,

$$n^{-1}\log(D_s) \le \psi_{net}(\nu;\rho,\delta,C) + 3\epsilon, \quad (7.24)$$

holds uniformly over the sum parameter s in (7.16).

Now we are ready to define the threshold bound $\rho_N(\delta,C)$ such that whenever $\rho < \rho_N(\delta,C)$, the probability P in (7.21) will be decaying to 0 exponentially fast as n grows.

DEFINITION 7.1 *For any $\delta \in (0,1]$, and any $C \ge 1$, we define the critical threshold $\rho_N(\delta,C)$ as the supremum of $\rho \in [0,1]$ such that for any $\nu \in [\delta,1]$,*

$$\psi_{net}(\nu;\rho,\delta,C) < 0.$$

Now it is time to describe how to calculate the exponents ψ_{int} and ψ_{ext} for the internal angles and external angles respectively. When the parameters ρ, δ, and C are clear from the context, we will omit them from the notations for the combinatorial, internal, and external exponents.

7.5 Computing the internal angle exponent

In this section, we first state how the internal angle exponent $\psi_{int}(\nu; \rho, \delta, C)$ is computed and then justify this computation.

For each ν, we take the function

$$\xi_{\gamma'}(y) = \frac{1-\gamma'}{\gamma'} y^2/2 + \Lambda^*(y), \tag{7.25}$$

where

$$\gamma' = \frac{\rho\delta}{\frac{C^2-1}{C^2}\rho\delta + \frac{\nu}{C^2}},$$

and $\Lambda^*(\cdot)$ is the dual large deviation rate function given by

$$\Lambda^*(y) = \max_s sy - \Lambda(s).$$

Here $\Lambda(s)$ is the cumulant generating function

$$\Lambda(s) = \log(E(\exp(sY))) = \frac{s^2}{2} + \log(2\Phi(s)),$$

for a standard half-normal random variable Y, where Φ is the cumulative distribution function of a standard Gaussian random variable $\mathcal{N}(0,1)$. Note that a standard half-normal random variable $Y \sim HN(0,1)$ is the absolute value $|X|$ of a standard Gaussian random variable $X \sim \mathcal{N}(0,1)$.

Since the dual large deviation rate function $\Lambda^*(\cdot)$ is a convex function that takes its minimum at $E(Y) = \sqrt{2/\pi}$, so $\xi_{\gamma'}(y)$ is also a convex function that takes its minimum in the interval $(0, \sqrt{2/\pi})$. Let us denote the minimizer of $\xi_{\gamma'}(y)$ as $y_{\gamma'}$. Then for any $C \geq 1$, the internal angle exponent can be computed as

$$\psi_{int}(\nu; \rho, \delta, C) = (\xi_{\gamma'}(y_{\gamma'}) + \log(2))(\nu - \rho\delta). \tag{7.26}$$

Next, we will show (7.26) is indeed the internal angle decay exponent for an arbitrary $C \geq 1$; namely we will prove the following lemma in the same spirit as Lemma 6.1 from [15]:

LEMMA 7.3 *For $k = \rho\delta n$, any $\epsilon > 0$, and any $C \geq 1$, when $n > n_0(\rho, \delta, \epsilon, C)$, where $n_0(\rho, \delta, \epsilon, C)$ is a large enough number,*

$$n^{-1} \log(\beta(F,G)) \leq -\psi_{int}(l/n; \rho, \delta, C) + \epsilon,$$

uniformly for any $l \geq \delta n$.

In fact, using the formula for the internal angle derived in the appendix, we know that

$$-n^{-1}\log(\beta(F,G)) = -n^{-1}\log(B(\frac{1}{1+C^2k}, l-k)), \quad (7.27)$$

where

$$B(\alpha', m') = \theta^{\frac{m'-1}{2}}\sqrt{(m'-1)\alpha' + 1}\pi^{-m'/2}\alpha'^{-1/2}J(m', \theta), \quad (7.28)$$

with $\theta = (1-\alpha')/\alpha'$ and

$$J(m', \theta) = \frac{1}{\sqrt{\pi}}\int_{-\infty}^{\infty}(\int_0^{\infty} e^{-\theta v^2 + 2iv\lambda}\, dv)^{m'} e^{-\lambda^2}\, d\lambda. \quad (7.29)$$

To evaluate (7.27), we need to evaluate the complex integral in $J(m', \theta')$. A saddle point method based on contour integration was sketched for similar integral expressions in [41]. A probabilistic method using large deviation theory for evaluating similar integrals was developed in [15]. Both of these two methods can be applied in our case and of course they will produce the same final results. We will follow the probabilistic method in this chapter. The basic idea is to see the integral in $J(m', \theta')$ as the convolution of $(m'+1)$ probability densities being expressed in the Fourier domain.

LEMMA 7.4 [15] *Let $\theta = (1-\alpha')/\alpha'$. Let T be a random variable with the $N(0, \frac{1}{2})$ distribution, and let $W_{m'}$ be a sum of m' i.i.d. half normals $U_i \sim HN(0, \frac{1}{2\theta})$. Let T and $W_{m'}$ be stochastically independent, and let $g_{T+W_{m'}}$ denote the probability density function of the random variable $T + W_{m'}$. Then*[3]

$$B(\alpha', m') = \sqrt{\frac{\alpha'(m'-1)+1}{1-\alpha'}} \cdot 2^{-m'} \cdot \sqrt{\pi} \cdot g_{T+W_{m'}}(0). \quad (7.30)$$

Here we apply this lemma to $\alpha' = \frac{1}{C^2k+1}$ for general $C \geq 1$. Applying this probabilistic interpretation and large deviation techniques, it is evaluated as in [15] that

$$g_{T+W_{m'}}(0) \leq \frac{2}{\sqrt{\pi}} \cdot \left(\int_0^{\mu_{m'}} v e^{-v^2 - m'\Lambda^*(\frac{\sqrt{2\theta}}{m'}v)}\, dv + e^{-\mu_{m'}^2}\right), \quad (7.31)$$

where Λ^* is the rate function for the standard half-normal random variable $HN(0, 1)$ and $\mu_{m'}$ is the expectation of $W_{m'}$. In fact, the second term in the sum is negligible because it decays at a greater speed than the first term as the dimension m' grows (to see this, note that $-v^2 - m'\Lambda^*(\frac{\sqrt{2\theta}}{m'}v)$ is a concave function achieving its maximum when $v < \mu_{m'}$; and $-v^2 - m'\Lambda^*(\frac{\sqrt{2\theta}}{m'}v)$ is equal to $-\mu_{m'}^2$ when $v = \mu_{m'}$. Laplace's methods discussed below then show the integral in the first term indeed decays at a slower speed than $e^{-\mu_{m'}^2}$). And after taking $y = \frac{\sqrt{2\theta}}{m'}v$, we have an upper bound for the first term:

$$\frac{2}{\sqrt{\pi}} \cdot \frac{m'^2}{2\theta} \cdot \int_0^{\sqrt{2/\pi}} y e^{-m'(\frac{m'}{2\theta})y^2 - m'\Lambda^*(y)}\, dy. \quad (7.32)$$

---

[3] In [15], the term $2^{-m'}$ was $2^{1-m'}$, but we believe that $2^{-m'}$ is the right term.

As we know, m' in the exponent of (7.32) is defined as $(l-k)$. Now we notice that the function $\xi_{\gamma'}$ in (7.25) appears in the exponent of (7.32), with $\gamma' = \frac{\theta}{m'+\theta}$. Since $\theta = \frac{1-\alpha'}{\alpha'} = C^2 k$, we have

$$\gamma' = \frac{\theta}{m'+\theta} = \frac{C^2 k}{(C^2-1)k+l}.$$

Since k scales as $\rho \delta n$ and l scales as νn, we further have

$$\gamma' = \frac{k}{\frac{l}{C^2} + \frac{C^2-1}{C^2}k} = \frac{\rho\delta}{\frac{C^2-1}{C^2}\rho\delta + \frac{\nu}{C^2}},$$

which is apparently consistent with our previous definition of γ' in (7.25).

Recall $\xi_{\gamma'}(y)$ defined in (7.25), then standard Laplace's method will give the upper bound

$$g_{T+W_{m'}}(0) \leq e^{-m'\xi_{\gamma'}(y_{\gamma'})} R_{m'}(\gamma'),$$

where $m'^{-1}\sup_{\gamma' \in [\eta, 1]} \log(R_{m'}(\gamma')) = o(1)$ as $m' \to \infty$ for any $\eta > 0$.

Plugging this upper bound back into (7.30) and recalling $m' = (\nu - \rho\delta)n$, for any $\epsilon > 0$, with large enough n,

$$n^{-1}\log(\beta(F,G)) \leq (-\xi_{\gamma'}(y_{\gamma'}) - \log(2))(\nu - \rho\delta) + \epsilon,$$

holds uniformly over $l \geq \nu n$, generalizing the $C=1$ case in [15].

For any $C \geq 1$, as shown in [15], $\xi_{\gamma'}(y_{\gamma'})$ scales like

$$\frac{1}{2}\log(\frac{1-\gamma'}{\gamma'}), \text{ as } \gamma' \to 0. \tag{7.33}$$

Because $\gamma' = \frac{\rho\delta}{\frac{C^2-1}{C^2}\rho\delta + \frac{\nu}{C^2}}$, for any $\nu \in [\delta, 1]$, if we take ρ small enough, γ' can become arbitrarily small. The asymptotic (7.33) means that as $\rho \to 0$,

$$\psi_{int}(\nu; \rho, \delta, C) \geq (\frac{1}{2} \cdot \log(\frac{1-\gamma'}{\gamma'})(1-\eta) + \log(2))(\nu - \rho\delta). \tag{7.34}$$

This generalizes the $C=1$ case in [15]. Notice as C increases, the internal angle exponent asymptotic (7.34) decreases.

7.6 Computing the external angle exponent

Closely following [15], let X be a half-normal $HN(0, 1/2)$ random variable, namely a random variable $X = |Z|$ where $Z \sim \mathcal{N}(0, 1/2)$. For $\nu \in (0, 1]$, define x_ν as the solution of

$$\frac{2xG(x)}{g(x)} = \frac{1-\nu}{\nu'}, \tag{7.35}$$

where $\nu' = (C^2 - 1)\rho\delta + \nu$, $G(x)$ is the cumulative distribution function of X and thus $G(x)$ is the error function

$$G(x) = \frac{2}{\sqrt{\pi}} \int_0^x e^{-y^2} dy, \qquad (7.36)$$

and $g(x) = \frac{2}{\sqrt{\pi}} \exp(-x^2)$ for $x \geq 0$ is the density function for X.

Keeping in mind the dependence of x_ν on $C \geq 1$, we define

$$\psi_{ext}(\nu; \rho, \delta, C) = -(1-\nu)\log(G(x_\nu)) + \nu' x_\nu^2.$$

When $C = 1$, we have the asymptotic from [15]

$$\psi_{ext}(\nu; \rho, \delta, 1) \sim \nu \log(\frac{1}{\nu}) - \frac{1}{2}\nu \log(\log(\frac{1}{\nu})) + o(\nu), \nu \to 0. \qquad (7.37)$$

We now set out to prove that the defined external angle exponent is indeed the right exponent. We first give the explicit formula for the external angle formula as a function of the parameter $C \geq 1$ in the appendix. Extracting the exponent from the external angle formula follows [15] and includes the necessary changes to take into account the parameter $C \geq 1$. The justification is summarized in this following lemma:

LEMMA 7.5   For any $C \geq 1$, $\rho = k/n$, and $\delta = m/n$, then for any fixed $\epsilon_1 > 0$,

$$n^{-1} \log(\gamma(G, SP)) < -\psi_{ext}(\frac{l}{n}; \rho, \delta, C) + \epsilon_1, \qquad (7.38)$$

uniformly in $l \geq \delta n$, *when n is large enough.*

Proof.   In the appendix, we derived the explicit integral formula for the external angle:

$$\gamma(G, SP) = \frac{2^{n-l}}{\sqrt{\pi}^{n-l+1}} \int_0^\infty e^{-x^2} (\int_0^{\frac{x}{C\sqrt{k + \frac{l-k}{C^2}}}} e^{-y^2} dy)^{n-l} dx. \qquad (7.39)$$

After changing integral variables, we have

$$\gamma(G, SP) = \sqrt{\frac{(C^2 - 1)k + l}{\pi}} \qquad (7.40)$$

$$\int_0^\infty e^{-((C^2-1)k+l)x^2} (\frac{2}{\sqrt{\pi}} \int_0^x e^{-y^2} dy)^{n-l} dx.$$

Let $\nu = l/n$, $\nu' = (C^2 - 1)\rho\delta + \nu$ then the integral formula can be written as

$$\sqrt{\frac{n\nu'}{\pi}} \int_0^\infty e^{-n\nu' x^2 + n(1-\nu)\log(G(x))} dx, \qquad (7.41)$$

where G is the error function from (7.36). To look at the asymptotic behavior of (7.41), following the same methodology as in [15], we first define

$$f_{\rho,\delta,\nu,n}(y) = e^{-n\psi_{\rho,\delta,\nu}(y)} \cdot \sqrt{\frac{n\nu'}{\pi}} \qquad (7.42)$$

with

$$\psi_{\rho,\delta,\nu}(y) = \nu' y^2 - (1-\nu)\log(G(y)).$$

Applying Laplace's method to $\psi_{\rho,\delta,\nu}$ gives Lemma 7.6, which is in the spirit of Lemma 5.2 in [15], and we omit its proof in this chapter.

LEMMA 7.6 *For $C \geq 1$ and $\nu \in (0,1)$, let x_ν denote the minimizer of $\psi_{\rho,\delta,\nu}$. Then*

$$\int_0^\infty f_{\rho,\delta,\nu,n}(x)\,dx \leq e^{-n\psi_{\rho,\delta,\nu}(x_\nu)(1+R_n(\nu))},$$

where for $\delta, \eta > 0$,

$$\sup_{\nu \in [\delta, 1-\eta]} R_n(\nu) = o(1) \text{ as } n \to \infty,$$

and x_ν is exactly the same x_ν defined earlier in (7.35).

Recall that the defined exponent ψ_{ext} is given by

$$\psi_{ext}(\nu; \rho, \delta, C) = \psi_{\rho,\delta,\nu}(x_\nu). \qquad (7.43)$$

From the definition of $\psi_{\rho,\delta,\nu}(x_\nu)$ and (7.43), it is not hard to see that as $\nu \to 1$, $x_\nu \to 0$ and $\psi_{ext}(\nu; \rho, \delta, C) \to 0$. So from (7.43) and Lemma 7.6,

$$n^{-1}\log(\gamma(G, \text{SP})) < -\psi_{ext}(l/n; \rho, \delta, C) + \epsilon_1,$$

uniformly in $l \geq \delta n$, when n is large enough. □

7.7 Existence and scaling of $\rho_N(\delta, C)$

Recall that in determining $\rho_N(\delta, C)$, ψ_{com} is the exponent which must be overwhelmed by the other two exponents $\psi_{int} + \psi_{net}$. The asymptotic relations (7.37) and (7.33) allow us to see the following key facts about $\rho_N(\delta, C)$, the proofs of which are given in the appendix.

LEMMA 7.7 *For any $\delta > 0$ and any $C > 1$, we have*

$$\rho_N(\delta, C) > 0, \delta \in (0,1). \qquad (7.44)$$

This generalizes the nontriviality of $\rho_N(\delta,C)$ to arbitrary $C \geq 1$. Finally, we have the lower and upper bounds for $\rho_N(\delta,C)$, which shows the scaling bounds for $\rho_N(\delta,C)$ as a function of C.

LEMMA 7.8 *When $C \geq 1$, for any fixed $\delta > 0$,*

$$\Omega(\frac{1}{C^2}) \leq \rho_N(\delta,C) \leq \frac{1}{C+1}, \qquad (7.45)$$

where $\Omega(\frac{1}{C^2}) \leq \rho_N(\delta,C)$ means that there exists a constant $\iota(\delta)$,

$$\frac{\iota(\delta)}{C^2} \leq \rho_N(\delta,C), \quad \text{as } C \to \infty,$$

where we can take $\iota(\delta) = \rho_N(\delta,1)$.

7.8 "Weak," "sectional," and "strong" robustness

So far, we have discussed the robustness of ℓ_1 minimization for sparse signal recovery in the "strong" case, namely we required robust signal recovery for all the approximately k-sparse signal vectors x. But in applications or performance analysis, we are also often interested in the signal recovery robustness in weaker senses. As we shall see, the framework given in the previous sections can be naturally extended to the analysis of other notions of robustness for sparse signal recovery, resulting in a coherent analysis scheme. For example, we hope to get a tighter performance bound for a particular signal vector instead of a more general, but looser, performance bound for all the possible signal vectors. In this section, we will present our null space conditions on the matrix A to guarantee the performance of the programming (7.2) in the "weak," "sectional," and "strong" senses. Here the robustness in the "strong" sense is exactly the robustness we discussed in the previous sections.

THEOREM 7.3 *Let A be a general $m \times n$ measurement matrix, x be an n-element vector, and $y = Ax$. Denote K as a subset of $\{1,2,\ldots,n\}$ such that its cardinality $|K| = k$ and further denote $\overline{K} = \{1,2,\ldots,n\} \setminus K$. Let w denote an $n \times 1$ vector. Let $C > 1$ be a fixed number.*

- *(Weak Robustness) Given a specific set K and suppose that the part of x on K, namely x_K is fixed. $\forall x_{\overline{K}}$, any solution $\hat{x}$ produced by (7.2) satisfies*

$$\|x_K\|_1 - \|\hat{x}_K\|_1 \leq \frac{2}{C-1}\|x_{\overline{K}}\|_1$$

and

$$\|(x-\hat{x})_{\overline{K}}\|_1 \leq \frac{2C}{C-1}\|x_{\overline{K}}\|_1,$$

if and only if $\forall w \in \mathbb{R}^n$ such that $Aw = 0$, we have

$$\|x_K + w_K\|_1 + \|\frac{w_{\overline{K}}}{C}\|_1 \geq \|x_K\|_1; \tag{7.46}$$

- *(Sectional Robustness)* Given a specific set $K \subseteq \{1, 2, \ldots, n\}$. Then $\forall x \in \mathbb{R}^n$, any solution $\hat{x}$ produced by (7.2) will satisfy

$$\|x - \hat{x}\|_1 \leq \frac{2(C+1)}{C-1}\|x_{\overline{K}}\|_1,$$

if and only if $\forall x' \in \mathbb{R}^n$, $\forall w \in \mathbb{R}^n$ such that $Aw = 0$,

$$\|x'_K + w_K\|_1 + \|\frac{w_{\overline{K}}}{C}\|_1 \geq \|x'_K\|_1; \tag{7.47}$$

- *(Strong Robustness)* If for all possible $K \subseteq \{1, 2, \ldots, n\}$, and for all $x \in \mathbb{R}^n$, any solution $\hat{x}$ produced by (7.2) satisfies

$$\|x - \hat{x}\|_1 \leq \frac{2(C+1)}{C-1}\|x_{\overline{K}}\|_1,$$

if and only if $\forall K \subseteq \{1, 2, \ldots, n\}, \forall x' \in \mathbb{R}^n$, $\forall w \in \mathbb{R}^n$ such that $Aw = 0$,

$$\|x'_K + w_K\|_1 + \|\frac{w_{\overline{K}}}{C}\|_1 \geq \|x'_K\|_1. \tag{7.48}$$

Proof. We will first show the sufficiency of the null space conditions for the various definitions of robustness. Let us begin with the "weak" robustness part. Let $w = \hat{x} - x$ and we must have $Aw = A(\hat{x} - x) = 0$. From the triangular inequality for ℓ_1 norm and the fact that $\|x\|_1 \geq \|x + w\|_1$, we have

$$\|x_K\|_1 - \|x_K + w_K\|_1$$
$$\geq \|w_{\overline{K}} + x_{\overline{K}}\|_1 - \|x_{\overline{K}}\|_1$$
$$\geq \|w_{\overline{K}}\|_1 - 2\|x_{\overline{K}}\|_1.$$

But the condition (7.46) guarantees that

$$\|w_{\overline{K}}\|_1 \geq C(\|x_K\|_1 - \|x_K + w_K\|_1),$$

so we have

$$\|w_{\overline{K}}\|_1 \leq \frac{2C}{C-1}\|x_{\overline{K}}\|_1,$$

and

$$\|x_K\|_1 - \|\hat{x}_K\|_1 \leq \frac{2}{C-1}\|x_{\overline{K}}\|_1.$$

For the "sectional" robustness, again, we let $w = \hat{x} - x$. Then there must exist an $x' \in \mathbb{R}^n$ such that
$$\|x'_K + w_K\|_1 = \|x'_K\|_1 - \|w_K\|_1.$$

Following the condition (7.47), we have
$$\|w_K\|_1 \leq \|\frac{w_{\overline{K}}}{C}\|_1.$$

Since
$$\|x\|_1 \geq \|x + w\|_1,$$

following the proof of Theorem 7.2, we have
$$\|x - \hat{x}\|_1 \leq \frac{2(C+1)}{C-1}\|x_{\overline{K}}\|_1.$$

The sufficiency of the condition (7.48) for strong robustness also follows.

Necessity: Since in the proof of the sufficiency, equalities can be achieved in the triangular equalities, the conditions (7.46), (7.47), and (7.48) are also necessary conditions for the respective robustness to hold for every x (otherwise, for certain x's, there will be $x' = x + w$ with $\|x'\|_1 < \|x\|_1$ which violates the respective robustness definitions. Also, such x' can be the solution to (7.2)). The detailed arguments will similarly follow the proof of the second part of Theorem 7.2. □

The conditions for "weak," "sectional," and "strong" robustness seem to be very similar, and yet there are key differences. The "weak" robustness condition is for x with a specific x_K on a specific subset K, the "sectional" robustness condition is for x with arbitrary value x_K on a specific subset K, and the "strong" robustness condition is for x's with arbitrary value x_K on all possible subsets. Basically, the "weak" robustness condition (7.46) guarantees that the ℓ_1 norm of $\hat{x}_K$ is not too far away from the ℓ_1 norm of x_K and the error vector $w_{\overline{K}}$ scales linearly in ℓ_1 norm as a function of $\|x_{\overline{K}}\|_1$. Notice that if we define
$$\kappa = \max_{Aw=0, w \neq 0} \frac{\|w_K\|_1}{\|w_{\overline{K}}\|_1},$$
then
$$\|x - \hat{x}\|_1 \leq \frac{2C(1+\kappa)}{C-1}\|x_{\overline{K}}\|_1.$$

That means, if κ is not ∞ for a measurement matrix A, $\|x - \hat{x}\|_1$ also approaches 0 when $\|x_{\overline{K}}\|_1$ approaches 0. Indeed, it is not hard to see that, for a given matrix A, $\kappa < \infty$ as long as the rank of matrix A_K is equal to $|K| = k$, which is generally satisfied for $k < m$.

While the "weak" robustness condition is only for one specific signal x, the "sectional" robustness condition instead guarantees that given *any* approximately k-sparse signal mainly supported on the subset K, the ℓ_1 minimization gives a solution $\hat{x}$ close to the original signal by satisfying (7.3). When we measure an approximately k-sparse signal x (the support of the k largest-magnitude components is fixed though unknown to the decoder) using a randomly generated measurement matrix A, the "sectional" robustness conditions characterize the probability that the ℓ_1 minimization solution satisfies (7.3) for *any* signals for the set K. If that probability goes to 1 as $n \to \infty$ for any subset K, we know that there exist measurement matrices A's that guarantee (7.3) on "almost all" support sets (namely, (7.3) is "almost always" satisfied). The "strong" robustness condition instead guarantees the recovery for approximately sparse signals mainly supported on *any* subset K. The "strong" robustness condition is useful in guaranteeing the decoding bound *simultaneously* for *all* approximately k-sparse signals under a single measurement matrix A.

REMARK: We should mention that from a practical point of view weak robustness is the most meaningful and is what can be observed in simulations (since it is impossible to check all x_K and all subsets K to check for sectional and strong robustness).

As expected, after we take $C = 1$ and let (7.46), (7.47), and (7.48) take strict inequality for all $w \neq 0$ in the null space of A, the conditions (7.46), (7.47), and (7.48) are also sufficient and necessary conditions for unique exact recovery of ideally k-sparse signals in "weak," "sectional," and "strong" senses [15].

For a given value $\delta = m/n$ and any value $C \geq 1$, we will determine the value of feasible $\zeta = k/n$ for which there exist a sequence of A's such that these three conditions are satisfied when $n \to \infty$ and $m/n = \delta$. As manifested by the statements of the three conditions (7.46), (7.47), and (7.48) and the previous discussions in Section 7.3, we can naturally extend the Grassmann angle approach to analyze the bounds for the probabilities that (7.46), (7.47), and (7.48) fail. Here we will denote these probabilities as P_1, P_2, and P_3, respectively. Note that there are $\binom{n}{k}$ possible support sets K and there are 2^k possible sign patterns for signal x_K. From previous discussions, we know that the event that the condition (7.46) fails is the same for all x_K's of a specific support set and a specific sign pattern. Then following the same line of reasoning as in Section 7.3, we have $P_1 = P_{K,-}$, $P_2 \leq 2^k \times P_1$, and $P_3 \leq \binom{n}{k} \times 2^k \times P_1$, where $P_{K,-}$ is the probability as in (7.12).

It is worthwhile mentioning that the formula for P_1 is exact since there is no union bound involved and so the threshold bound for the "weak" robustness is tight. In summary, the results in this section suggest that even if k is very close to the weak threshold for ideally sparse signals, we can still have robustness results for approximately sparse signals while the results using restricted isometry conditions [10] suggest smaller sparsity level for recovery robustness. This is the first such result.

7.9 Numerical computations on the bounds of ζ

In this section, we will numerically evaluate the performance bounds on $\zeta = k/n$ such that the conditions (7.9), (7.46), (7.47), and (7.48) are satisfied with overwhelming probability as $n \to \infty$.

First, we know that the condition (7.9) fails with probability

$$P \leq \binom{n}{k} \times 2^k \times 2 \times \sum_{s \geq 0} \sum_{G \in \Im_{m+1+2s}(\text{SP})} \beta(F, G) \gamma(G, \text{SP}). \quad (7.49)$$

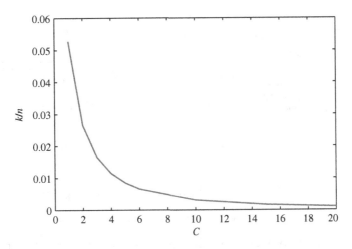

Figure 7.3 Allowable sparsity as a function of C (allowable imperfection of the recovered signal is $\frac{2(C+1)\sigma_k(x)_1}{C-1}$).

Recall that we assume $m/n = \delta$, $l = (m+1+2s)+1$ and $\nu = l/n$. In order to make P overwhelmingly converge to zero as $n \to \infty$, following the discussions in Section 7.4, one sufficient condition is to make sure that the exponent for the combinatorial factors

$$\psi_{com} = \lim_{n \to \infty} \frac{\log\left(\binom{n}{k}2^k 2\binom{n-k}{l-k}2^{l-k}\right)}{n} \qquad (7.50)$$

and the negative exponent for the angle factors

$$\psi_{angle} = -\lim_{n \to \infty} \frac{\log\left(\beta(F,G)\gamma(G,\mathrm{SP})\right)}{n} \qquad (7.51)$$

satisfy $\psi_{com} - \psi_{angle} < 0$ uniformly over $\nu \in [\delta, 1)$.

Following [15] we take $m = 0.5555n$. By analyzing the decaying exponents of the external angles and internal angles through the Laplace methods as in Section 7.6, and 7.5, we can compute the numerical results as shown in Figure 7.3, Figure 7.5, and Figure 7.6. In Figure 7.3, we show the largest sparsity level $\zeta = k/n$ (as a function of C) which makes the failure probability of the condition (7.11) approach zero asymptotically as $n \to \infty$. As we can see, when $C = 1$, we get the same bound $\zeta = 0.095 \times 0.5555 \approx 0.0528$ as obtained for the "weak" threshold for the ideally sparse signals in [15]. As expected, as C grows, the ℓ_1 minimization requires a smaller sparsity level ζ to achieve higher signal recovery accuracy.

In Figure 7.4(a), we show the exponents ψ_{com}, ψ_{int}, ψ_{ext} under the parameters $C = 2$, $\delta = 0.5555$ and $\zeta = 0.0265$. For the same set of parameters, in Figure 7.4(b), we compare the exponents ψ_{com} and ψ_{angle}: the solid curve denotes ψ_{angle} and the dashed curve denotes ψ_{com}. It shows that, under $\zeta = 0.0265$, $\psi_{com} - \psi_{angle} < 0$ uniformly over $\delta \leq \nu \leq 1$. Indeed, $\zeta = 0.0265$ is the bound shown in Figure 7.3 for $C = 2$. In Figure 7.5, for the parameter $\delta = 0.5555$, we give the bounds ζ as a function of C for satisfying

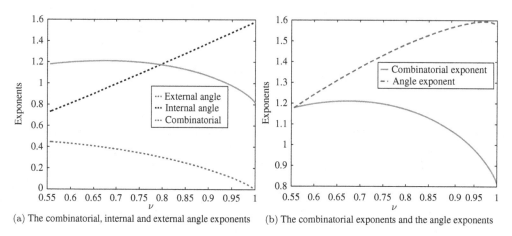

(a) The combinatorial, internal and external angle exponents

(b) The combinatorial exponents and the angle exponents

Figure 7.4 The combinatorial, external and internal angle exponent. (a) The combinational interref and Externel angle exponents. (b) The combinational exponents and angle exponents.

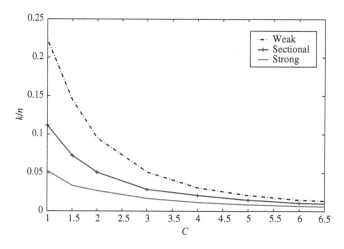

Figure 7.5 The "weak," "sectional," and "strong" robustness bounds.

the signal recovery robustness conditions (7.46), (7.47), and (7.48), respectively in the "weak," "sectional," and "strong" senses. In Figure 7.6, fixing $C = 2$, we plot how large $\rho = \zeta/\delta$ can be for different δ's while satisfying the signal recovery robustness conditions (7.46), (7.47), and (7.48), respectively in "weak," "sectional," and "strong" senses.

7.10 Recovery thresholds for weighted ℓ_1 minimization

So far, we have used a null space Grassmann angle geometric approach to give sharp characterizations for the sparsity and ℓ_1 recovery stability tradeoff in compressive sensing. It turns out that the null space Grassmann angle approach is a general framework

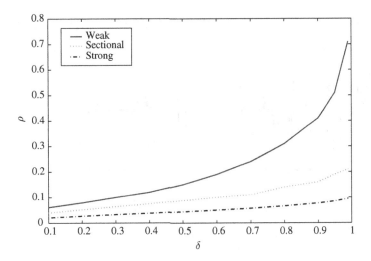

Figure 7.6 The "weak," "sectional," and "strong" robustness bounds.

which can be used to give sharp performance bounds for other sparse recovery algorithms, for example, weighted ℓ_1 minimization algorithms and iterative reweighted ℓ_1 algorithms. In these applications, the success of these algorithms can also be reduced to the event that the null space of the measurement matrix intersects trivially with different polyhedral cones. So similarly for these applications, we will be able to characterize the sharp sparsity transition thresholds and in turn, these threshold results will help us optimize the configurations of the weighted algorithms.

The conventional approach to compressive sensing assumes no prior information on the unknown signal vector other than the fact that it is sufficiently sparse over a particular basis. In many applications, however, additional prior information is available, such as in natural images, medical imaging, and in DNA microarrays. How to exploit the structure information in the sparse signals has led to the development of structured sparse models in recent years, see Chapter 1. In the DNA microarrays applications, for instance, signals are often *block sparse*, i.e., the signal is more likely to be nonzero in certain blocks rather than in others [37]. Even when no prior information is available, the preprocessing phases of some sparse recovery algorithms feed "prior" information on the sparse signal (e.g., its sparsity pattern) to the inner-loops of these sparse recovery algorithms [12, 29].

In [29] we consider a particular model for the sparse signal where the entries of the unknown vector fall into a number u of classes, with each class having a specific fraction of nonzero entries. The standard compressed sensing model is therefore a special case where there is only one class. We will focus on the case where the entries of the unknown signal fall into a fixed number u of categories; in the ith set K_i with cardinality n_i, the fraction of nonzero entries is p_i. This model is rich enough to capture many of the salient features regarding prior information. We refer to the signals generated based on this model as *nonuniform sparse* signals. For completeness, we present a general definition.

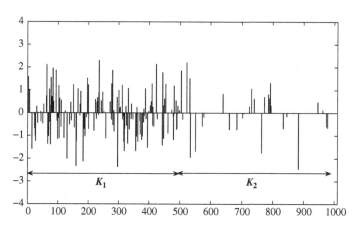

Figure 7.7 Illustration of a nonuniform sparse signal. [29] © 2009 IEEE.

DEFINITION 7.2 *Let $\mathcal{K} = \{K_1, K_2, \ldots, K_u\}$ be a partition of $\{1, 2, \ldots, n\}$, i.e. ($K_i \cap K_j = \emptyset$ for $i \neq j$, and $\bigcup_{i=1}^{u} K_i = \{1, 2, \ldots, n\}$), and $P = \{p_1, p_2, \ldots, p_u\}$ be a set of positive numbers in $[0, 1]$. An $n \times 1$ vector $x = (x_1, x_2, \cdots, x_n)^T$ is said to be a random* **nonuniform sparse** *vector with sparsity fraction p_i over the set K_i for $1 \leq i \leq u$, if x is generated from the following random procedure:*

- *Over each set K_i, $1 \leq i \leq u$, the set of nonzero entries of x is a random subset of size $p_i|K_i|$. In other words, a fraction p_i of the entries are nonzero in K_i. p_i is called the sparsity fraction over K_i. The values of the nonzero entries of x can be arbitrary nonzero real numbers.*

In Figure 7.7, a sample nonuniform sparse signal with Gaussian distribution for nonzero entries is plotted. The number of sets is considered to be $u = 2$ and both classes have the same size $n/2$, with $n = 1000$. The sparsity fraction for the first class K_1 is $p_1 = 0.3$, and for the second class K_2 is $p_2 = 0.05$.

To accommodate the prior information, one can simply think of modifying ℓ_1 minimization to a weighted ℓ_1 minimization as follows:

$$\min_{Az=y} \|z\|_{w,1} = \min_{Az=y} \sum_{i=1}^{n} w_i |z_i|. \tag{7.52}$$

The index w on the norm is an indication of the $n \times 1$ non-negative weight vector. Naturally, if we want to suppress the ith entry to be zero in the decoding result, we would like to assign a bigger value to w_i. To boost the performance of sparse recovery, it may benefit to give bigger weights to the blocks where there are more zero elements. For example, in Figure 7.7, we can assign weight $W_1 = 1$ to the first block K_1 and assign another weight $W_2 > 1$ to the sparser block K_2.

Now the question is, what are the optimal sets of weights for weighted ℓ_1 minimization (7.52) to minimize the number of measurements (or the threshold on the number of measurements) ensuring a signal vector of the nonuniform sparse model is recovered with overwhelming probability?

This seems to be a very different problem from the ℓ_1 minimization robustness problem we have considered earlier in this chapter. However, these two problems are connected through the null space conditions for the measurement matrix A, and so the Grassmann angle approach in the earlier work can also be applied to this problem. More explicitly, suppose K is the support of a signal vector from the nonuniform sparse model and $\overline{K}$ is the complement of the support set, then the weighted ℓ_1 minimization succeeds in recovering all the vectors supported on K if and only if

$$\|v_K\|_{w_K,1} < \|v_{\overline{K}}\|_{w_{\overline{K}},1} \qquad (7.53)$$

holds for every nonzero vector v from the null space of A. The proof of this weighted null space condition is relatively obvious following the same reasoning as in the proof of Theorem 7.2.

In studying this weighted null space condition, one can then extend the Grassmann angle framework to analyze the "failure" probability that the null space of a random A intersects nontrivially with the "weighted" cone of vectors v satisfying

$$\|v_K\|_{w_K,1} \geq \|v_{\overline{K}}\|_{w_{\overline{K}},1}. \qquad (7.54)$$

As in the analysis for ℓ_1 minimization robustness, this "failure" probability can be reduced to studying the event that the null space of a random A intersects with a union of "weighted" polyhedral cones. This of course reduces to the computation and evaluation of Grassmann angles for individual cones, only this time for "weighted" polyhedral cones. In fact, for any set of specialized block and sparsity parameters, and for any particular set of weights, one can compute via the Grassmann angle approach the threshold for $\delta_c = m/n$ (the number of measurements needed) such that a sparse signal vector from the nonuniform sparse signal model is recovered with high probability. The derivations and calculations follow the same steps as in previous sections for ℓ_1 minimization robustness, and we will omit them here for the sake of space. For the technical details, the reader is encouraged to read [29]. The main result is stated in the following theorem and its proof can be found in [29].

THEOREM 7.4 *Let $\delta = m/n$ for the Gaussian measurement matrix $A \in \mathbb{R}^{m \times n}$, $\gamma_1 = n_1/n$ and $\gamma_2 = n_1/n$. For fixed values of γ_1, γ_2, p_1, p_2, $\omega = w_{K_2}/w_{K_1}$, define E to be the event that a random nonuniform sparse vector x_0 (Definition 7.2) with sparsity fractions p_1 and p_2 over the sets K_1 and K_2 respectively with $|K_1| = \gamma_1 n$ and $|K_2| = \gamma_2 n$ is recovered via the weighted ℓ_1 minimization. There exists a computable critical threshold $\delta_c = \delta_c(\gamma_1, \gamma_2, p_1, p_2, \omega)$ such that if $\delta = m/n \geq \delta_c$, then E happens with overwhelming probability as $n \to \infty$.*

Let us again look at the sample nonuniform sparse signal model in Figure 7.7. For $u = 2$, $\gamma_1 = |K_1|/n = 0.5$, $\gamma_2 = |K_2|/n = 0.5$, $p_1 = 0.3$, and $p_2 = 0.05$, we have numerically computed $\delta_c(\gamma_1, \gamma_2, p_1, p_2, w_{K_2}/w_{K_1})$ as a function of w_{K_2}/w_{K_1} and depicted the

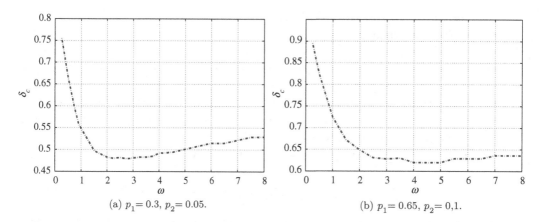

Figure 7.8 δ_c as a function of $\omega = w_{K_2}/w_{K_1}$ for $\gamma_1 = \gamma_2 = 0.5$. [29] © 2009 IEEE.

resulting curve in Figure 7.8(a). This suggests that $w_{K_2}/w_{K_1} \approx 2.5$ is the optimal ratio that one can choose. The value of δ_c for another choice of p_1, p_2 is shown in Figure 7.8(b).

7.11 Approximate support recovery and iterative reweighted ℓ_1

Despite its simplicity and extensive research on other polynomial-complexity sparse recovery algorithms, when no prior information is available, regular ℓ_1 minimization still has the best theoretically established sparse recovery threshold performance for decoding general sparse signal vectors. However, using the Grassmann angle analysis, even when no prior information is available, we are able to show for the first time that a class of (iterative) reweighted ℓ_1 minimization algorithms have strictly higher recovery thresholds on recoverable sparsity levels than regular ℓ_1 minimization, for certain classes of signal vectors whose nonzero elements have fast decaying amplitudes. The technical details of this claim are not presented here due to space limitations, and a more comprehensive study on this can be found in [44].

The reweighted ℓ_1 recovery algorithm proposed in [44] is composed of two steps. In the first step a standard ℓ_1 minimization is used to decode the signal. Note that when the number of nonzero elements is above the recovery threshold, ℓ_1 minimization generally will not give the original sparse signal. Based on ℓ_1 minimization output, a set of entries where the nonzero elements are more likely to reside (the so-called approximate support) are identified. The elements of the unknown signal are thus divided into two classes: one is the approximate support with a relatively higher density of nonzero entries, and the other one is the complement of the approximate support, which has a smaller density of nonzero entries. This corresponds to a nonuniform sparse model in the previous section. The second step of reweighted ℓ_1 recovery is then to perform a weighted ℓ_1 minimization (see the previous section) where elements outside the approximate support set are penalized with a weight larger than 1.

The algorithm is then given as follows, where k is the number of nonzero elements and ω is a weighting parameter which can be adjusted.

Algorithm 7.1

1. Solve the ℓ_1 minimization problem:

$$\hat{x} = \arg\min \|z\|_1 \text{ subject to } Az = Ax. \tag{7.55}$$

2. Obtain an approximation for the support set of x: find the index set $L \subset \{1, 2, \ldots, n\}$ which corresponds to the largest k elements of $\hat{x}$ in magnitude.
3. Solve the following weighted ℓ_1 minimization problem and declare the solution as output:

$$x^* = \arg\min \|z_L\|_1 + \omega \|z_{\overline{L}}\|_1 \text{ subject to } Az = Ax. \tag{7.56}$$

For a given number of measurements, if the support size of x, namely $k = |K|$, is slightly larger than the sparsity threshold of ℓ_1 minimization, then the robustness of ℓ_1 minimization, as analyzed via the Grassmann angle approach in this chapter, helps find a lower bound for $\frac{|L \cap K|}{|L|}$, i.e. the density of nonzero elements of x over the set L. With the help of this type of "prior" information about the support of x, the weighted ℓ_1 algorithm, as analyzed via the Grassmann angle approach in the previous section, can be shown to guarantee a full recovery of the original sparse vector even though the number of its nonzero elements is beyond the ℓ_1 minimization recovery threshold, and, at the beginning, we do not have prior support information.

It should be noted that, at the cost of not having any prior information at the beginning of this algorithm, the sparse recovery threshold improvement is not universal over all types of signals. For example, if the nonzero elements of the signal are of a constant amplitude, the support estimate in the first step can be very misleading [12] and leads to bad recovery performance in the second step.

Other variations of reweighted ℓ_1 minimization are given in the literature. For example the algorithm in [12] assigns a different weight to each single entry based on the inverse of the absolute value of its decoding result in regular ℓ_1 minimization, $\hat{x}$. In some sense, the theoretical results in [44], via the Grassmann angle analysis, explain the threshold improvements observed empirically in [12].

7.12 Conclusion

In this chapter we analyzed a null space characterization of the measurement matrix to guarantee a specific performance for ℓ_1-norm optimization for approximately sparse signals. Using high-dimensional geometry tools, we give a unified *null space Grassmann angle*-based analytical framework for compressive sensing. This new framework gives sharp quantitative tradeoffs between the signal sparsity parameter and the recovery accuracy of the ℓ_1 optimization for general signals or approximately sparse signals. As expected, the neighborly polytopes result of [15] for ideally sparse signals can be viewed as a special case on this tradeoff curve. It can therefore be of practical use in applications where the underlying signal is not ideally sparse and where we are interested in the quality of the recovered signal. For example, using the results and their extensions in this chapter and [15], we are able to give a precise sparsity threshold analysis for weighted ℓ_1 minimization when prior information about the signal vector is available [29]. In [44], using the robustness result from this chapter, we are able to show that a two-step weighted ℓ_1 minimization algorithm can provably improve over the sparsity threshold of ℓ_1 minimization for interesting classes of signals, even when prior information is not available.

In essence, this work investigates the fundamental "balancedness" property of linear subspaces, and may be of independent mathematical interest. In future work, it will be interesting to obtain more accurate analysis for compressive sensing under noisy measurements than presented in the current chapter.

7.13 Appendix

7.13.1 Derivation of the internal angles

LEMMA 7.9 *Suppose that F is a $(k-1)$-dimensional face of the skewed cross-polytope*

$$SP = \{y \in \mathbb{R}^n \mid \|y_K\|_1 + \|\frac{y_{\overline{K}}}{C}\|_1 \leq 1\}$$

supported on the subset K with $|K| = k$. Then the internal angle $\beta(F,G)$ between the $(k-1)$-dimensional face F and a $(l-1)$-dimensional face G ($F \subseteq G$, $G \neq SP$) is given by

$$\beta(F,G) = \frac{V_{l-k-1}(\frac{1}{1+C^2k}, l-k-1)}{V_{l-k-1}(S^{l-k-1})}, \qquad (7.57)$$

where $V_i(S^i)$ denotes the ith dimensional surface measure on the unit sphere S^i, while $V_i(\alpha', i)$ denotes the surface measure for a regular spherical simplex with $(i+1)$ vertices on the unit sphere S^i and with inner product as α' between these $(i+1)$ vertices. Equation (7.57) is equal to $B(\frac{1}{1+C^2k}, l-k)$, where

$$B(\alpha', m') = \theta^{\frac{m'-1}{2}} \sqrt{(m'-1)\alpha'+1} \pi^{-m'/2} \alpha'^{-1/2} J(m', \theta) \qquad (7.58)$$

with $\theta = (1-\alpha')/\alpha'$ and

$$J(m', \theta) = \frac{1}{\sqrt{\pi}} \int_{-\infty}^{\infty} (\int_0^{\infty} e^{-\theta v^2 + 2iv\lambda} dv)^{m'} e^{-\lambda^2} d\lambda. \qquad (7.59)$$

Proof. Without loss of generality, assume that F is a $(k-1)$-dimensional face with k vertices as $e_p, 1 \leq p \leq k$, where e_p is the n-dimensional standard unit vector with the pth element as "1"; and also assume that the $(l-1)$-dimensional face G be the convex hull of the l vertices: $e_p, 1 \leq p \leq k$ and $Ce_p, (k+1) \leq p \leq l$. Then the cone $\text{Con}_{F,G}$ formed by observing the $(l-1)$-dimensional face G of the skewed cross-polytope SP from an interior point x^F of the face F is the positive cone of the vectors:

$$Ce_j - e_i, \text{ for all } j \in J \backslash K, i \in K, \tag{7.60}$$

and also the vectors

$$e_{i_1} - e_{i_2}, \text{ for all } i_1 \in K, i_2 \in K, \tag{7.61}$$

where $J = \{1, 2, \ldots, l\}$ is the support set for the face G.

So the cone $\text{Con}_{F,G}$ is the direct sum of the linear hull $L_F = \text{lin}\{F - x^F\}$ formed by the vectors in (7.61) and the cone $\text{Con}_{F^\perp, G} = \text{Con}_{F,G} \cap L_F^\perp$, where $L_F^\perp$ is the orthogonal complement to the linear subspace L_F. Then $\text{Con}_{F^\perp, G}$ has the same spherical volume as $\text{Con}_{F,G}$.

Now let us analyze the structure of $\text{Con}_{F^\perp, G}$. We notice that the vector

$$e_0 = \sum_{p=1}^{k} e_p$$

is in the linear space $L_F^\perp$ and is also the only such vector (up to linear scaling) supported on K. Thus a vector x in the positive cone $\text{Con}_{F^\perp, G}$ must take the form

$$-\sum_{i=1}^{k} b_i \times e_i + \sum_{i=k+1}^{l} b_i \times e_i, \tag{7.62}$$

where $b_i, 1 \leq i \leq l$ are non-negative real numbers and

$$C \sum_{i=1}^{k} b_i = \sum_{i=k+1}^{l} b_i,$$

$$b_1 = b_2 = \cdots = b_k.$$

That is to say, the $(l-k)$-dimensional $\text{Con}_{F^\perp, G}$ is the positive cone of $(l-k)$ vectors $a^1, a^2, \ldots, a^{l-k}$, where

$$a^i = C \times e_{k+i} - \sum_{p=1}^{k} e_p / k, \quad 1 \leq i \leq (l-k).$$

The normalized inner products between any two of these $(l-k)$ vectors is

$$\frac{<a^i, a^j>}{\|a^i\| \|a^j\|} = \frac{k \times \frac{1}{k^2}}{C^2 + k \times \frac{1}{k^2}} = \frac{1}{1 + kC^2}.$$

(In fact, a^i's are also the vectors obtained by observing the vertices $e_{k+1}, \ldots, e_l$ from $Ec = \sum_{p=1}^{k} e_p/k$, the epicenter of the face F.)

We have so far reduced the computation of the internal angle to evaluating (7.57), the relative spherical volume of the cone $\mathrm{Con}_{F^\perp, G}$ with respect to the sphere surface S^{l-k-1}. This was computed as given in this lemma [41, 6] for the positive cones of vectors with equal inner products by using a transformation of variables and the well-known formula

$$V_{i-1}(S^{i-1}) = \frac{i\pi^{\frac{i}{2}}}{\Gamma(\frac{i}{2}+1)},$$

where $\Gamma(\cdot)$ is the usual Gamma function. □

7.13.2 Derivation of the external angles

LEMMA 7.10 *Suppose that F is a $(k-1)$-dimensional face of the skewed cross-polytope*

$$SP = \{y \in \mathbb{R}^n \mid \|y_K\|_1 + \|\frac{y_{\overline{K}}}{C}\|_1 \leq 1\}$$

supported on a subset K with $|K| = k$. Then the external angle $\gamma(G, SP)$ between a $(l-1)$-dimensional face G ($F \subseteq G$) and the skewed cross-polytope SP is given by

$$\gamma(G, SP) = \frac{2^{n-l}}{\sqrt{\pi}^{n-l+1}} \int_0^\infty e^{-x^2} \left(\int_0^{\frac{x}{C\sqrt{k + \frac{l-k}{C^2}}}} e^{-y^2} \, dy \right)^{n-l} dx. \quad (7.63)$$

Proof. We take the same proof technique of transforming external angle calculation into the integral of Gaussian distribution over the outward normal cone [3]. Without loss of generality, we assume $K = \{1, \ldots, k\}$. Since the $(l-1)$-dimensional face G is the convex hull of k regular vertices with length 1 and $(l-k)$ vertices of length C, without of loss of generality, again we have the $(l-1)$-dimensional face

$$G = \mathrm{conv}\{e^1, \ldots, e^k, C \times e^{k+1}, \ldots, C \times e^l\}$$

of the skewed cross-polytope SP. Since there are 2^{n-l} facets containing the face G, the 2^{n-l} outward normal vectors of the supporting hyperplanes of the facets containing G are given by

$$\{\sum_{p=1}^{k} e_p + \sum_{p=l+1}^{n} j_p e_p / C + \sum_{p=k+1}^{l} e_p / C, j_p \in \{-1, 1\}\}.$$

The outward normal cone $c(G, SP)$ at the face G is the positive hull of these normal vectors. We also have

$$\int_{c(G,SP)} e^{-\|x\|^2} \, dx = \gamma(G, SP) V_{n-l}(S^{n-l})$$

$$\times \int_0^\infty e^{-r^2} r^{n-l} \, dr = \gamma(G, SP) \cdot \pi^{(n-l+1)/2}, \quad (7.64)$$

where $V_{n-l}(S^{n-l})$ is the spherical volume of the $(n-l)$-dimensional sphere S^{n-l}.

Suppose a vector in the cone $c(G,\text{SP})$ takes value t at index $i=1$, then that vector can take any value in the interval $[-t/C, t/C]$ at those indices $(l+1) \leq i \leq n$ (due to the negative and positive signs of the outward normal vectors at those indices) and that vector must take the value t/C at the indices $(k+1) \leq i \leq l$. So we only need $(n-l+1)$ free variables in describing the outward normal cone $c(G,\text{SP})$, and we define a set U as

$$\{x \in \mathbb{R}^{n-l+1} \mid x_{n-l+1} \geq 0, |x_p| \leq \frac{x_{n-l+1}}{C}, 1 \leq p \leq (n-l)\}.$$

So we further define a one-one mapping from the describing variables to the cone $c(G,\text{SP})$ $f(x_1, \ldots, x_{n-l+1}) : U \to c(G,\text{SP})$

$$f(x_1, \ldots, x_{n-l+1}) = \sum_{p=l+1}^{n} x_{p-l} e_p + \sum_{p=k+1}^{l} \frac{x_{n-l+1}}{C} e_p + \sum_{p=1}^{k} x_{n-l+1} \times e_p.$$

Then we can evaluate

$$\int_{c(G,\text{SP})} e^{-\|x'\|^2} \, dx'$$

$$= \sqrt{k + \frac{l-k}{C^2}} \int_{U} e^{-\|f(x)\|^2} \, dx$$

$$= \sqrt{k + \frac{l-k}{C^2}} \int_{0}^{\infty} \int_{-\frac{x_{n-l+1}}{C}}^{\frac{x_{n-l+1}}{C}} \cdots \int_{-\frac{x_{n-l+1}}{C}}^{\frac{x_{n-l+1}}{C}} e^{-x_1^2 - \cdots - x_{n-l}^2 - (k + \frac{l-k}{C^2}) x_{n-l+1}^2} \, dx_1 \cdots dx_{n-l+1}$$

$$= \sqrt{k + \frac{l-k}{C^2}} \int_{0}^{\infty} e^{-(k + \frac{l-k}{C^2}) x^2} \times \left(\int_{-\frac{x}{C}}^{\frac{x}{C}} e^{-y^2} \, dy \right)^{n-l} dx$$

$$= 2^{n-l} \int_{0}^{\infty} e^{-x^2} \left(\int_{0}^{\frac{x}{C \sqrt{k + \frac{l-k}{C^2}}}} e^{-y^2} \, dy \right)^{n-l} dx,$$

where $\sqrt{k + \frac{l-k}{C^2}}$ is due to the change of integral variables. We obtain the conclusion of this lemma by combining this integral result with (7.64). $\square$

7.13.3 Proof of Lemma 7.7

Proof. Consider any fixed $\delta > 0$. First, we consider the internal angle exponent ψ_{int}, where we define $\gamma' \doteq \frac{\rho \delta}{\frac{C^2-1}{C^2} \rho \delta + \frac{\nu}{C^2}}$. Then for this fixed δ,

$$\frac{1 - \gamma'}{\gamma'} \geq \frac{\frac{C^2-1}{C^2} \rho \delta + \frac{\delta}{C^2}}{\rho \delta} - 1$$

uniformly over $\nu \in [\delta, 1]$.

Now if we take ρ small enough, $\frac{\frac{C^2-1}{C^2}\rho\delta + \frac{\delta}{C^2}}{\rho\delta}$ can be arbitrarily large. By the asymptotic expression (7.34), this leads to large enough internal decay exponent ψ_{int}. At the same time, the external angle exponent ψ_{ext} is lower-bounded by zero and the combinatorial exponent is upper-bounded by some finite number. Then if ρ is small enough, we will get the net exponent ψ_{net} to be negative uniformly over the range $\nu \in [\delta, 1]$. □

7.13.4 Proof of Lemma 7.8

Proof. Suppose instead that $\rho_N(\delta, C) > \frac{1}{C+1}$. Then for every vector w from the null space of the measurement matrix A, any $\rho_N(\delta, C)$ fraction of the n components in w take no more than $\frac{1}{C+1}$ fraction of $\|w\|_1$. But this cannot be true if we consider the $\rho_N(\delta, C)$ fraction of w with the largest magnitudes.

Now we only need to prove the lower bound for $\rho_N(\delta, C)$; in fact, we argue that

$$\rho_N(\delta, C) \geq \frac{\rho_N(\delta, C=1)}{C^2}.$$

We know from Lemma 7.7 that $\rho_N(\delta, C) > 0$ for any $C \geq 1$. Denote $\psi_{net}(C)$, $\psi_{com}(\nu; \rho, \delta, C)$, $\psi_{int}(\nu; \rho, \delta, C)$, and $\psi_{ext}(\nu; \rho, \delta, C)$ as the respective exponents for a certain C. Because $\rho_N(\delta, C=1) > 0$, for any $\rho = \rho_N(\delta, C=1) - \epsilon$, where $\epsilon > 0$ is an arbitrarily small number, the net exponent $\psi_{net}(C=1)$ is negative uniformly over $\nu \in [\delta, 1]$.

By examining the formula (7.20) for the external angle $\gamma(G, SP)$, where G is a $(l-1)$-dimensional face of the skewed cross-polytope SP, we have $\gamma(G, SP)$ is a decreasing function in both k and C for a fixed l. So $\gamma(G, SP)$ is upper-bounded by

$$\frac{2^{n-l}}{\sqrt{\pi}^{n-l+1}} \int_0^\infty e^{-x^2} \left(\int_0^{\frac{x}{\sqrt{l}}} e^{-y^2}\, dy \right)^{n-l} dx, \tag{7.65}$$

namely the expression for the external angle when $C = 1$. Then for any $C > 1$ and any k, $\psi_{ext}(\nu; \rho, \delta, C)$ is lower-bounded by $\psi_{ext}(\nu; \rho, \delta, C=1)$.

Now let us check $\psi_{int}(\nu; \rho, \delta, C)$ by using the formula (7.26). With

$$\gamma' = \frac{\rho\delta}{\frac{C^2-1}{C^2}\rho\delta + \frac{\nu}{C^2}},$$

we have

$$\frac{1-\gamma'}{\gamma'} = -\frac{1}{C^2} + \frac{\nu}{C^2\rho\delta}. \tag{7.66}$$

Then for any fixed $\delta > 0$, if we take $\rho = \frac{\rho_N(\delta, C=1) - \epsilon}{C^2}$, where ϵ is an arbitrarily small positive number, then for any $\nu \geq \delta$, $\frac{1-\gamma'}{\gamma'}$ is an increasing function in C. So, following easily from its definition, $\xi_{\gamma'}(y_{\gamma'})$ is an increasing function in C. This further implies that $\psi_{int}(\nu; \rho, \delta)$ is an increasing function in C if we take $\rho = \frac{\rho_N(\delta, C=1) - \epsilon}{C^2}$, for any $\nu \geq \delta$.

Also, for any fixed ν and δ, it is not hard to show that $\psi_{com}(\nu; \rho, \delta, C)$ is a decreasing function in C if $\rho = \frac{\rho_N(\delta, C=1)}{C^2}$. This is because in (7.16),

$$\binom{n}{k}\binom{n-k}{l-k} = \binom{n}{l}\binom{l}{k}.$$

Thus for any $C > 1$, if $\rho = \frac{\rho_N(\delta, C=1) - \epsilon}{C^2}$, the net exponent $\psi_{net}(C)$ is also negative uniformly over $\nu \in [\delta, 1]$. Since the parameter ϵ can be arbitrarily small, our claim and Lemma 7.8 then follow. $\square$

Acknowledgement

This work was supported in part by the National Science Foundation under grant no. CCF-0729203, by the David and Lucille Packard Foundation, and by Caltech's Lee Center for Advanced Networking.

References

[1] F. Affentranger and R. Schneider. Random projections of regular simplices. *Discrete Comput Geom*, 7(3):219–226, 1992.
[2] R. Baraniuk, M. Davenport, R. DeVore, and M. Wakin. A simple proof of the restricted isometry property for random matrices. *Construc Approx*, 28(3):253–263, 2008.
[3] U. Betke and M. Henk. Intrinsic volumes and lattice points of crosspolytopes. *Monat für Math*, 115(1-2):27–33, 1993.
[4] J. Blanchard, C. Cartis, and J. Tanner. Compressed sensing: How sharp is the restricted isometry property? 2009. www.maths.ed.ac.uk/ tanner/.
[5] W. M. Boothby. *An Introduction to Differential Manifolds and Riemannian Geometry*. Springer-Verlag, 1986. 2nd edn. San Diego, CA: Academic Press.
[6] K. Böröczky and M. Henk. Random projections of regular polytopes. *Arch Math (Basel)*, 73(6):465–473, 1999.
[7] E. J. Candès. Compressive sampling. In *Int Congr Math. Vol. III*: 1433–1452. Eur. Math. Soc., Zürich, 2006.
[8] E. J. Candès and P. Randall. Highly robust error correction by convex programming. *IEEE Trans Inform Theory*, 54:2829–2840, 2008.
[9] E. J. Candès, J. Romberg, and T. Tao. Robust uncertainty principles: exact signal reconstruction from highly incomplete frequency information. *IEEE Trans Inform Theory*, 52(2):489–509, 2006.
[10] E. J. Candès, J. Romberg, and T. Tao. Stable signal recovery from incomplete and inaccurate measurements. *Commun Pure Appl Math*, 59:1208–1223, 2006.
[11] E. J. Candès and T. Tao. Decoding by linear programming. *IEEE Trans Inform Theory*, 51(12):4203–4215, 2005.
[12] E. J. Candès, M. P. Wakin, and S. P. Boyd. Enhancing sparsity by reweighted ℓ_1 minimization. *J Fourier Anal Appl*, 14(5):877–905, 2008.
[13] J. F. Claerbout and F. Muir. Robust modeling with erratic data. *Geophysics*, 38(5):826–844, 1973.

[14] A. Cohen, W. Dahmen, and R. DeVore. Compressed sensing and best k-term approximation. *J Am Math Soc*, 22:211–231, 2008.

[15] D. Donoho. High-dimensional centrally symmetric polytopes with neighborliness proportional to dimension. *Discrete Comput Geom*, 35(4):617–652, 2006.

[16] D. Donoho, A. Maleki, and A. Montanari. The noise-sensitivity phase transition in compressed sensing. *arXiv:1004.1218*, 2010.

[17] D. Donoho, A. Maleki, and A. Montanari. Message passing algorithms for compressed sensing. In *Proc Nat Acad Sci (PNAS)*, November 2009.

[18] D. Donoho and J. Tanner. Observed universality of phase transitions in high-dimensional geometry. *Phil Trans Roy Soc A*, 367:4273–4293, 2009.

[19] D. L. Donoho. Compressed sensing. *IEEE Trans Inform Theory*, 52(4): 1289–1306, 2006.

[20] D. L. Donoho and J. Tanner. Neighborliness of randomly projected simplices in high dimensions. *Proc Nat Acad Sci USA*, 102(27):9452–9457, 2005.

[21] C. Dwork, F. McSherry, and K. Talwar. The price of privacy and the limits of lp decoding. *Proc 39th Ann ACM Symp Theory of Comput (STOC)*, 2007.

[22] A. Feuer and A. Nemirovski. On sparse representation in pairs of bases. *IEEE Trans Information Theory*, 49(6):1579–1581, 2003.

[23] A. Garnaev and E. Gluskin. The widths of a Euclidean ball. *Dokl Acad Nauk USSR*, 1048–1052, 1984.

[24] Y. Gordon. On Milman's inequality and random subspaces which escape through a mesh in R^n. *Geome Asp Funct Anal*, 84–106, 1987.

[25] B. Grünbaum. Grassmann angles of convex polytopes. *Acta Math*, 121:293–302, 1968.

[26] B. Grünbaum. Convex polytopes, *Graduate Texts in Mathematics*, vol 221. Springer-Verlag, New York, 2nd edn., 2003. Prepared and with a preface by Volker Kaibel, Victor Klee, and Günter M. Ziegler.

[27] B. Kashin. The widths of certain finite dimensional sets and classes of smooth functions. *Izvestia*, 41:334–351, 1977.

[28] B. S. Kashin and V. N. Temlyakov. A remark on compressed sensing. *Math Notes*, 82(5):748–755, November 2007.

[29] M. A. Khajehnejad, W. Xu, A. S. Avestimehr, and B. Hassibi. Weighted ℓ_1 minimization for sparse recovery with prior information. In *Proc Int Symp Inform Theory*, 2009.

[30] N. Linial and I. Novik. How neighborly can a centrally symmetric polytope be? *Discrete Comput Geom*, 36(6):273–281, 2006.

[31] P. McMullen. Non-linear angle-sum relations for polyhedral cones and polytopes. *Math Proc Camb Phil Soc*, 78(2):247–261, 1975.

[32] M. L. Mehta. *Random Matrices*. Amsterdam: Academic Press, 2004.

[33] M. Rudelson and R. Vershynin. Geometric approach to error correcting codes and reconstruction of signals. *Int Math Res Notices*, 64:4019–4041, 2005.

[34] M. Rudelson and R. Vershynin. On sparse reconstruction from Fourier and Gaussian measurements. *Commun Pure and Appl Math*, 61, 2007.

[35] L. A. Santaló. Geometría integral enespacios de curvatura constante. *Rep. Argentina Publ. Com. Nac. Energí Atómica, Ser. Mat 1, No.1*, 1952.

[36] M. Stojnic. Various thresholds for ℓ_1-optimization in compressed sensing. 2009. Preprint available at http://arxiv.org/abs/0907.3666.

[37] M. Stojnic, F. Parvaresh, and B. Hassibi. On the reconstruction of block-sparse signals with an optimal number of measurements. *IEEE Trans Signal Proc*, 57(8):3075–3085, 2009.

[38] M. Stojnic, W. Xu, and B. Hassibi. Compressed sensing – probabilistic analysis of a null-space characterization. *Proc IEEE Int Conf Acoust, Speech, Signal Proc (ICASSP)*, 2008.

[39] M. Stojnic, W. Xu, and B. Hassibi. Compressed sensing of approximately sparse signals. *IEEE Int Symp Inform Theory*, 2008.

[40] S. Vavasis. Derivation of compressive sensing theorems for the spherical section property. *University of Waterloo, CO 769 lecture notes*, 2009.

[41] A. M. Vershik and P. V. Sporyshev. Asymptotic behavior of the number of faces of random polyhedra and the neighborliness problem. *Sel Math Soviet*, 11(2):181–201, 1992.

[42] W. Xu and B. Hassibi. Compressed sensing over the Grassmann manifold: A unified geometric framework. Accepted to *IEEE Trans Inform Theory*.

[43] W. Xu and B. Hassibi. Compressed sensing over the Grassmann manifold: A unified analytical framework. *Proc 46th Ann Allerton Conf Commun, Control Comput*, 2008.

[44] W. Xu, A. Khajehnejad, S. Avestimehr, and B. Hassibi. Breaking through the thresholds: an analysis for iterative reweighted ℓ_1 minimization via the Grassmann angle framework. *Proc Int Conf Acoust, Speech, Signal Proc (ICASSP)*, 2010.

[45] Y. Zhang. When is missing data recoverable? 2006. Available online at www.caam.rice.edu/~zhang/reports/index.html.

[46] Y. Zhang. Theory of compressive sensing via ℓ_1-minimization: a non-RIP analysis and extensions. 2008. Available online at www.caam.rice.edu/~zhang/reports/index.html.

8 Greedy algorithms for compressed sensing

Thomas Blumensath, Michael E. Davies, and Gabriel Rilling

Compressed sensing (CS) is often synonymous with ℓ_1-based optimization. However, when choosing an algorithm for a particular application, there is a range of different properties that have to be considered and weighed against each other. Important algorithm properties, such as speed and storage requirements, ease of implementation, flexibility, and recovery performance have to be compared. In this chapter we will therefore present a range of alternative algorithms that can be used to solve the CS recovery problem and which outperform convex optimization-based methods in some of these areas. These methods therefore add important versatility to any CS recovery toolbox.

8.1 Greed, a flexible alternative to convexification

The thread that binds all of the approaches of this chapter is their "greediness." In this context, the moniker "*greedy*" implies strategies that, at each step, make a "hard" decision usually based upon some locally optimal optimization criterion. Recall the noisy CS recovery problem,[1]

$$y = Ax + e, \qquad (8.1)$$

where, for a given $y \in \mathbb{R}^m$, we want to recover an approximately k-sparse vector $x \in \mathbb{R}^n$ under the assumption that the error $e \in \mathbb{R}^m$ is bounded and that the measurement matrix A satisfies the restricted isometry property (RIP)

$$(1 - \delta_{2k})\|x\|_2^2 \leq \|Ax\|_2^2 \leq (1 + \delta_{2k})\|x\|_2^2, \qquad (8.2)$$

for all $2k$-sparse vectors x and for some $0 \leq \delta_{2k} < 1$.

In this chapter, we consider two broad categories of greedy methods to recover x. The first set of strategies, which we jointly call "greedy pursuits" and discuss in Section 8.2,

---

[1] We restrict our discussion here primarily to real vectors, though all ideas discussed here extend trivially also to vectors with elements in $\mathbb{C}$.

Compressed Sensing: Theory and Applications, ed. Yonina C. Eldar and Gitta Kutyniok. Published by Cambridge University Press. © Cambridge University Press 2012.

can be defined as a set of methods that iteratively build up an estimate x. Starting with a zero vector, these methods estimate a set of nonzero components of x by iteratively adding new components that are deemed nonzero. This greedy selection is alternated with an estimation step in which the values for the nonzero components are optimized. These methods can often lead to very fast algorithms that are applicable to very large data-sets, however, theoretical performance guarantees are typically weaker than those of some other methods.

The second set of routines alternates both element selection as well as element pruning steps. Due to their ability to remove nonzero elements, these will be called "Thresholding" algorithms and will be discussed in Section 8.3. These methods are often very easy to implement and can be relatively fast. They have theoretical performance guarantees that rival those guarantees derived for convex optimization-based approaches as discussed in Chapter 1. Furthermore, as discussed in more detail in Section 8.4, not only can they be used to recover sparse signals, but are easily adapted to take account of many additional signal structures and are even applicable to recovery problems where non-sparse signal models are used.

8.2 Greedy pursuits

In this section a class of algorithms collectively called greedy pursuits will be discussed. The term "pursuits" dates back to 1974 [1] when the concept of projection pursuit was introduced. The technique projected the data in a given direction and tested for deviation from Gaussianity and it was this idea of projecting data onto different directions that was taken up by Mallat and Zhang [2] and used for signal approximation.

There is now a large and growing family of greedy pursuit techniques for signal approximation with a long and multi-rooted history. Similar ideas have appeared independently in different disciplines. For example, the notions of: *forward stepwise regression* in statistics [3]; the *pure greedy algorithm* in nonlinear approximation [4]; *Matching Pursuit* in signal processing [2]; and the *CLEAN* algorithm in radio astronomy [5] are all closely related yet discovered independently.

Here we will concentrate on a handful of greedy pursuit algorithms, briefly highlighting interesting variations.

8.2.1 General framework

Let us introduce greedy pursuits as a family of algorithms that share the following two fundamental steps: element selection and coefficient update. These methods are usually initialized with a zero estimate, $\widehat{x}^{[0]} = \mathbf{0}$. With this initialization, the initial residual error is $r^{[0]} = y - A\widehat{x}^{[0]} = y$ and the support set (i.e. the indices of the nonzero elements) of the first estimate $\widehat{x}^{[0]}$ is $T = \emptyset$. Each iteration then updates these quantities by adding additional elements (columns from A) to the support set T and by updating the signal estimate $\widehat{x}$, thereby decreasing the residual observation error r. This is done, with minor variations, as outlined in Algorithm 8.1.

Algorithm 8.1 General greedy pursuit framework

Input: y, A, and k
for $i = 1; i := i + 1$ till stopping criterion is met **do**
  Calculate $g^{[i]} = A^T r^{[i]}$ and select elements (columns from A) based on the magnitude of the elements of $g^{[i]}$
  Calculate a revised estimate for $\widehat{x}^{[i]}$ (and hence $\widehat{y}^{[i]}$) by decreasing the cost function
$$F(\widehat{x}^{[i]}) = \|y - A\widehat{x}^{[i]}\|_2^2 \qquad (8.3)$$

end for
Output: $r^{[i]}$ and $\widehat{x}^{[i]}$

Algorithm 8.2 Matching Pursuit (MP)

Input: y, A, and k
$r^{[0]} = y$, $\widehat{x}^{[0]} = \mathbf{0}$
for $i = 1; i := i + 1$ till stopping criterion is met **do**
  $g^{[i]} = A^T r^{[i-1]}$
  $j^{[i]} = \mathrm{argmax}_j |g_j^{[i]}|/\|A_j\|_2$
  $\widehat{x}_{j^{[i]}}^{[i]} = \widehat{x}_{j^{[i]}}^{[i-1]} + g_{j^{[i]}}^{[i]}/\|A_{j^{[i]}}\|_2^2$
  $r^{[i]} = r^{[i-1]} - A_{j^{[i]}} g_{j^{[i]}}^{[i]}/\|A_{j^{[i]}}\|_2^2$
end for
Output: $r^{[i]}$ and $\widehat{x}^{[i]}$

Matching Pursuit

One of the simplest pursuit algorithms is Matching Pursuit (MP) [2] (known as the Pure Greedy Algorithm in approximation theory [4]) summarized in Algorithm 8.2. The approximation is incremental, selecting one column from A at a time and, at each iteration, only the coefficient associated with the selected column is updated.

At each iteration, the update $\widehat{x}_{j^{[i]}}^{[i]} = \widehat{x}_{j^{[i]}}^{[i-1]} + g_{j^{[i]}}^{[i]}/\|A_{j^{[i]}}\|_2^2$ minimizes the approximation cost $\|y - A\widehat{x}^{[i]}\|_2^2$ with respect to the selected coefficient. Here and throughout the chapter A_j is the jth column of the matrix A. Note that MP will generally repeatedly select the same columns from A in order to further refine the approximation. However, it is known that $\|r^{[i]}\|$ converges linearly to zero whenever the columns of A span $\mathbb{R}^m$ [2]. MP will therefore stop in a finite number of iterations if the norm of $r^{[i]}$ is used to define a stopping criterion for the algorithm.

MP requires repeated evaluation of matrix multiplications involving A^T which dominate the computational complexity. Therefore MP is generally proposed for use with matrices A that admit a fast implementation, often based on the fast Fourier transform (FFT). Extremely fast implementations of MP are now available for problems where A has columns with restricted support [6].

Algorithm 8.3 Orthogonal Matching Pursuit (OMP)
Input: y, A, and k
Initialize: $r^{[0]} = y, \widehat{x}^{[0]} = \mathbf{0}, T^{[0]} = \emptyset$
for $i = 1; i := i+1$ till stopping criterion is met **do**
$\quad g^{[i]} = A^T r^{[i-1]}$
$\quad j^{[i]} = \operatorname{argmax}_j |g_j^{[i]}|/\|A_j\|_2$
$\quad T^{[i]} = T^{[i-1]} \cup j^{[i]}$
$\quad \widehat{x}_{T^{[i]}}^{[i]} = A_{T^{[i]}}^\dagger y$
$\quad r^{[i]} = y - A\widehat{x}^{[i]}$
end for
Output: $r^{[i]}$ and $\widehat{x}^{[i]}$

Orthogonal Matching Pursuit

A more sophisticated strategy is implemented in Orthogonal Matching Pursuit (OMP) [7, 8] (known as the Orthogonal Greedy Algorithm in approximation theory [4]). In OMP the approximation for x is updated in each iteration by projecting y orthogonally onto the columns of A associated with the current support set $T^{[i]}$. OMP therefore minimizes $\|y - A\widehat{x}\|_2$ over all $\widehat{x}$ with support $T^{[i]}$. The full algorithm is listed in Algorithm 8.3 where † in step 7 represents the pseudo-inverse operator. Note that in contrast to MP the minimization is performed with respect to all of the currently selected coefficients:

$$\widehat{x}_{T^{[i]}}^{[i]} = \underset{\widetilde{x}_{T^{[i]}}}{\operatorname{argmin}} \|y - A_{T^{[i]}} \widetilde{x}_{T^{[i]}}\|_2^2. \tag{8.4}$$

Unlike MP, OMP never re-selects an element and the residual at any iteration is always orthogonal to all currently selected elements.

Although for general dictionaries, as with MP, the computational cost of OMP is dominated by the matrix-vector products, when fast transforms are used the orthogonalization step is usually the bottleneck. Various techniques for solving the least-squares problem have been proposed including QR factorization [8], Cholesky factorization [9], or iterative techniques such as conjugate gradient methods. While OMP is more computationally complex than MP it generally enjoys superior performance, particularly in the context of CS. A detailed comparison of computation and storage costs will be given in Section 8.2.4.

There are two main problems with applying OMP to large-scale data. First, the computation and storage costs of a single iteration of OMP are quite high for large-scale problems and second, the selection of one atom at a time means that exactly k iterations are needed to approximate y with k atoms of A. When k is large this can be impractically slow. Some of the variations discussed below were proposed to specifically address these issues.

8.2.2 Variations in coefficient updates

Although both MP and OMP have identical selection strategies and update their coefficients by minimizing the squared error criterion, $\|y - A\widehat{x}^{[i]}\|_2^2$, the form of the update is substantially different. OMP minimizes over the coefficients for all selected elements at iteration i while in MP the minimization only involves the coefficient of the most recently selected element. These are, however, only two possibilities and it is interesting to consider what other updates can be used. For example, a relaxed form of MP has been considered where a damping factor is included [10].[2]

A framework for pursuits with different directional updates was presented in [11]. Consider at the ith iteration updating the selected coefficients $\widehat{x}_{T^{[i]}}^{[i-1]}$ along some other, yet to be defined, direction $d_{T^{[i]}}^{[i]}$.

$$\widehat{x}_{T^{[i]}}^{[i]} = x_{T^{[i]}}^{[i-1]} + a^{[i]} d_{T^{[i]}}^{[i]}. \tag{8.5}$$

The step size $a^{[i]}$ can be explicitly chosen to minimize the same quadratic cost as before.

$$a^{[i]} = \frac{\langle r^{[i]}, c^{[i]} \rangle}{\|c^{[i]}\|_2^2}, \tag{8.6}$$

where $c^{[i]} = A_{T^{[i]}} d_{T^{[i]}}^{[i]}$. If we use such an update along with the standard MP/OMP selection criterion then this directional pursuit is a member of the family of *general Matching Pursuit* algorithms as defined in [12] and shares the same necessary and sufficient conditions for (worst case) exact recovery as OMP [13].

Note also that both MP and OMP naturally fit in this framework with update directions: $\delta_{j^{[i]}}$ and $A_{T^{[i]}}^\dagger g_{T^{[i]}}$ respectively.

The directional pursuit family of algorithms is summarized in Algorithm 8.4. The aim of introducing directional updates is to produce an approximation to the orthogonal projection with a reduced computation cost. Here we will focus on gradient-based update strategies [11].[3]

Gradient Pursuits

A natural choice for the update direction is the negative gradient of the cost function $\|y - A_{T^{[i]}} \widetilde{x}_{T^{[i]}}\|_2^2$, i.e.

$$d_{T^{[i]}}^{[i]} := g_{T^{[i]}}^{[i]} = A_{T^{[i]}}^T (y - A_{T^{[i]}} \widehat{x}_{T^{[i]}}^{[i-1]}). \tag{8.7}$$

Fortunately we already have this as a by-product of the selection process. We simply restrict the vector $g^{[i]}$ (which has already been calculated) to the elements $T^{[i]}$. Using (8.7) as the directional update results in the most basic form of directional pursuit which we call *Gradient Pursuit* (GP) [11].

---

[2] A similar idea is used in the CLEAN algorithm to remove the effects of the point spread function in astronomical imaging [5].

[3] Another instance of directional pursuit is presented in [14] and uses local directional updates to exploit the localized structure present in certain dictionaries.

Algorithm 8.4 Directional Pursuit

Input: y, A, and k
Initialize: $r^{[0]} = y$, $\widehat{x}^{[0]} = \mathbf{0}$, $T^{[0]} = \emptyset$
for $i = 1; i := i+1$ till stopping criterion is met **do**
  $g^{[i]} = A^T r^{[i-1]}$
  $j^{[i]} = \mathrm{argmax}_j |g_j^{[i]}|/\|A_j\|_2$
  $T^{[i]} = T^{[i-1]} \cup j^{[i]}$
  calculate update direction $d_{T^{[i]}}^{[i]}$; $c^{[i]} = A_{T^{[i]}} d_{T^{[i]}}^{[i]}$ and $a^{[i]} = \frac{\langle r^{[i]}, c^{[i]} \rangle}{\|c^{[i]}\|_2^2}$
  $x_{T^{[i]}}^{[i]} := x_{T^{[i]}}^{[i-1]} + a^{[i]} d_{T^{[i]}}^{[i]}$
  $r^{[i]} = r^{[i-1]} - a^{[i]} c^{[i]}$
end for
Output: $r^{[i]}$ and $\widehat{x}^{[i]}$

The increase in computational complexity over MP is small and when the submatrices of A are well conditioned (i.e. A has a good restricted isometry property) then minimizing along the gradient direction can provide a good approximation to solving the full least-squares problem.

Specifically, suppose that A has a small restricted isometry constant δ_k then we can bound the condition number κ of the Gram matrix restricted to $T^{[i]}$, $G^{[i]} = A_{T^{[i]}}^T A_{T^{[i]}}$ by

$$\kappa(G^{[i]}) \leq \left(\frac{1+\delta_k}{1-\delta_k}\right) \tag{8.8}$$

for all $T^{[i]}$, $|T^{[i]}| \leq k$. A worst case analysis of the gradient line search [15] then shows that for small δ_k the gradient update achieves most of the minimization:

$$\frac{F(\widehat{x}_{T^{[i]}}^{[i]}) - F(\widehat{x}_{T^{[i]}}^*)}{F(\widehat{x}_{T^{[i]}}^{[i-1]}) - F(\widehat{x}_{T^{[i]}}^*)} \leq \left(\frac{\kappa-1}{\kappa+1}\right)^2 \tag{8.9}$$

$$\leq \delta_k^2$$

where $\widehat{x}_{T^{[i]}}^*$ denotes the least-squares solution of $F(\widetilde{x}_{T^{[i]}}) = \|y - A\widetilde{x}_{T^{[i]}}\|_2^2$. Hence for small δ_k the convergence, even of a single gradient iteration, is good.

Conjugate gradients

An alternative gradient-based update, popular in solving quadratic optimization problems is the *conjugate gradient*. The conjugate gradient method uses successive line minimizations along directions that are G-conjugate where a set of vectors $\{d^{[1]}, d^{[2]}, \ldots, d^{[i]}\}$ is defined as G-conjugate if $\langle d^{[i]}, G d^{[j]} \rangle = 0$ for all $i \neq j$, see [16, Section 10.2]. The magic of the conjugate gradient method is that if we start with $d^{[1]} = -g^{[1]}$, the current negative gradient, then after line minimization a new conjugate direction can be calculated by:

$$d^{[i+1]} = g^{[i+1]} + \beta^{[i]} d^{[i]} \tag{8.10}$$

where $\beta^{[i]} = \langle g^{[i+1]}, Gd^{[i]} \rangle / \langle d^{[i]}, Gd^{[i]} \rangle$ to ensure that $\langle d^{[i+1]}, Gd^{[i]} \rangle = 0$. Note that this guarantees that $d^{[i+1]}$ is conjugate to *all* previous search directions even though $\beta^{[i]}$ explicitly forces conjugacy only with respect to $d^{[i]}$.

The same principle can also be applied to the pursuit optimization at any given iteration with $G^{[i]} = A_{T^{[i]}}^T A_{T^{[i]}}$ and this has indeed been advocated as an efficient method of orthogonalization [17]. However this ignores the work done in the previous iterations. In [11] a conjugate gradient method that spans across the pursuit iterations was investigated. This enables the pursuit cost function (8.3) to be fully minimized using a single update direction and without the need for a direct calculation of the pseudo-inverse $A_{T^{[i]}}^\dagger$. However, unlike the classical conjugate gradient method each new direction must be selected to be explicitly conjugate to all previous directions. This is a consequence of the change in dimensionality of the solution space at each iteration. The resulting algorithm has a similar structure to the QR factorization implementation of OMP and a similar computational cost.

In principle one can implement a conjugate direction pursuit that enforces conjugacy with as many previous update directions as one wants, thus providing a family of pursuit algorithms that interpolate between the conjugate gradient implementation of OMP (conjugate to all directions) and GP (conjugate to no previous directions).

For the remainder of this section we will concentrate on the specific case where we enforce conjugacy with only the last preceding direction. We call this *Conjugate Gradient Pursuit* (CGP) [11].

Following (8.10) at the ith iteration we can select the update direction

$$d_{T^{[i]}}^{[i]} = g_{T^{[i]}}^{[i]} + \beta^{[i]} d_{T^{[i]}}^{[i-1]}, \tag{8.11}$$

where we can calculate $\beta^{[i]}$ as follows.

$$\beta^{[i]} = \frac{\left\langle (A_{T^{[i-1]}} d_{T^{[i-1]}}^{[i-1]}), (A_{T^{[i]}} g_{T^{[i]}}^{[i]}) \right\rangle}{\|A_{T^{[i-1]}} d_{T^{[i-1]}}^{[i-1]}\|_2^2} \tag{8.12}$$

$$= \frac{\left\langle c^{[i]}, A_{T^{[i]}} g_{T^{[i]}}^{[i]} \right\rangle}{\|c^{[i]}\|_2^2} \tag{8.13}$$

where $c^{[i]} = A_{T^{[i-1]}} d_{T^{[i-1]}}^{[i-1]}$ as before. This suggests that CGP will require the additional matrix-vector product $A_{T^{[i]}} g_{T^{[i]}}^{[i]}$ in comparison to GP. However, calculating $A_{T^{[i]}} g_{T^{[i]}}^{[i]}$ allows us to evaluate $c^{[i+1]}$ without additional matrix vector products through the following recursion [18]

$$\begin{aligned} c^{[i+1]} &= A_{T^{[i]}} d_{T^{[i]}}^{[i]} \\ &= A_{T^{[i]}} \left(g_{T^{[i]}}^{[i]} + \beta^{[i]} d_{T^{[i]}}^{[i-1]} \right) \\ &= A_{T^{[i]}} g_{T^{[i]}}^{[i]} + \beta^{[i]} c^{[i]}. \end{aligned} \tag{8.14}$$

Therefore CGP has exactly the same order of computational cost as GP. Furthermore, using arguments from [19] it is possible to show that the CGP update will always be at least as good as the GP update in terms of reducing the cost function (8.3) [18], making CGP the preferred option.

8.2.3 Variations in element selection

The second problem with MP/OMP type strategies is the need to perform at least as many iterations as there are atoms to be selected. This does not scale for large dimensions: i.e. where the number of elements to be selected is large (but still small with respect to the size of A). In order to speed up pursuit algorithms it is thus necessary to select multiple elements at a time. This idea, first proposed in [17] is termed *stagewise* selection.

In MP/OMP, the selection step chooses the element that is maximally correlated with the residual: $j^{[i]} = \mathrm{argmax}_j |g_j^{[i]}|/\|A_j\|_2$. A very natural stagewise strategy is to replace the maximum by a threshold criterion.

Let $\lambda^{[i]}$ define the threshold at iteration i. Then the stagewise selection becomes:

$$T^{[i]} = T^{[i-1]} \cup \{j : |g_j^{[i]}|/\|A_j\|_2 \geq \lambda^{[i]}\}. \tag{8.15}$$

Various choices for $\lambda^{[i]}$ are possible. For example (8.15) includes simple (non-iterative) thresholding [20]. While this is by far the most computationally simple procedure it has limited recovery guarantees – see Schnass and Vandergheynst [20]. We therefore concentrate on iterative thresholding strategies.[4] In particular we will focus on two proposed schemes: *Stagewise Orthogonal Matching Pursuit* (StOMP) in which $\lambda^{[i]}$ is a function of the residual $r^{[i-1]}$ [17] and *Stagewise Weak Gradient Pursuit* (StWGP) where $\lambda^{[i]}$ is a function of the correlations with the residual $g^{[i]}$ [18].

StOMP

In [17] StOMP was proposed with the aim of providing good reconstruction performance for CS applications while keeping computational costs low enough for application to large-scale problems. The threshold strategy is:

$$\lambda_{stomp}^{[i]} = t^{[i]} \|r^{[i-1]}\|_2/\sqrt{m}, \tag{8.16}$$

where the authors give the guidance that a good choice of $t^{[i]}$ will usually take a value: $2 \leq t^{[i]} \leq 3$. Specific formulae for $t^{[i]}$ are derived in the appendix of [17] for the case of Bernoulli distributed sparse coefficients and A generated from a uniform spherical ensemble.[5] Theoretical performance guarantees for this method when applied to more

---

[4] Note that the thresholding methods discussed here differ from methods such as Iterative Hard Thresholding [21], CoSaMP [22], and Subspace Pursuit [23]. Such algorithms do not simply use thresholding to augment the support set. They also use it to prune out previously selected elements. These algorithms and their impressive theoretical guarantees will be discussed in detail in Section 8.3.

[5] Two thresholds are derived based upon classical detection criteria: constant false alarm rates and constant false discovery rates.

general matrices A and more general coefficient values are not available. Furthermore from a practical point of view, the selection of the parameter t appears critical for good performance.

A specific problem that can occur is that the algorithm terminates prematurely when all inner products fall below the threshold. Indeed in the range of experiments presented in [18] StOMP gave mixed results.

Stagewise weak element selection

The selection strategy in StOMP is difficult to generalize beyond specific scenarios. Blumensath and Davies [18] therefore proposed an alternative selection strategy that can be more tightly linked to general MP/OMP recovery results based upon a *weak* selection strategy.

Weak selection was originally introduced in [2] to deal with the issue of infinite-dimensional dictionaries where only a finite number of inner products can be evaluated. Weak selection allows the selection of a *single* element $A_{j^{[i]}}$ whose correlation with the residual is close to maximum:

$$\frac{|g_{j^{[i]}}^{[i]}|}{\|A_{j^{[i]}}\|_2} \geq \alpha \max_j \frac{|g_j^{[i]}|}{\|A_j\|_2}. \tag{8.17}$$

A nice property of weak orthogonal matching pursuit (WOMP) is that it inherits a weakened version of the recovery properties of MP/OMP [13].

Instead of selecting a *single* element the *stagewise weak* selection chooses *all* elements whose correlation is close to the maximum. That is we set the threshold in (8.16) as:

$$\lambda_{weak}^{[i]} = \alpha \max_j \frac{|g_j^{[i]}|}{\|A_j\|_2}.$$

In practice, variations in selection strategy are complementary to variations in directional updates. In [18] the combination of CGP and the stagewise weak selection is advocated as it has good theoretical properties as well as good empirical performance. The combination is called *Stagewise Weak Conjugate Gradient Pursuit* (StWGP).

ROMP

Another alternative multi-element selection strategy that has been proposed is the Regularized OMP (ROMP) [24], [22] which groups the inner products g_i into sets J_k such that the elements in each set have a similar magnitude, i.e. they satisfy

$$\frac{|g_i|}{\|A_i\|_2} \leq \frac{1}{r} \frac{|g_j|}{\|A_j\|_2}, \text{ for all } i,j \in J_k.$$

ROMP then selects the set J_k for which $\sum_{j \in J_k}(|g_j|/\|A_j\|_2)^2$ is largest.

For the ROMP selection strategy proposed in [24] and [22], r was assumed to be 0.5. In this case, the algorithm was shown to have uniform performance guarantees closer to those of ℓ_1-based methods than exist for OMP and its derivatives.

ROMP has played an important historical role in the research on greedy algorithms, being the first to enjoy such "good" uniform recovery guarantees. However the constants in the theoretical guarantees are significantly larger than those for ℓ_1 minimization and ROMP has been quickly superseded by the thresholding techniques that we will discuss in Section 8.3. This, combined with the fact that empirically ROMP is not competitive with other pursuit algorithms [18] means that it is generally not considered as a good practical algorithm for CS.

ORMP

A different variant in element selection that is worth mentioning occurs in Order Recursive Matching Pursuit (ORMP). Unlike the stagewise selection, where the aim is to reduce the computational complexity, the aim of ORMP is to improve the approximation performance over OMP.

ORMP has gone by many names such as Stepwise Projection in the approximation theory literature [10] and Orthogonal Least Squares in neural networks [25], amongst others. Furthermore, historically there has also been much confusion between ORMP and OMP – for further discussion on the history see Blumensath [26].

Although OMP selects the element most correlated to the current residual at each iteration this does not guarantee the largest reduction in error after orthogonalization. This is because it does not account for the correlation between the elements under consideration with those we have already selected.

ORMP rectifies this deficiency. The ORMP selection step chooses the element that will most decrease the residual in the subsequent coefficient update (using orthogonal projection). This can be written as a joint selection and update step as

$$j^{[i]} = \operatorname*{argmin}_{j} \min_{\{\widetilde{x}_{T^{[i-1]}}, \widetilde{x}_j\}} \|y - A_{T^{[i-1]}}\widetilde{x}_{T^{[i-1]}} - A_j \widetilde{x}_j\|_2^2. \tag{8.18}$$

To calculate this it is possible to exploit the orthogonality between $r^{[i-1]}$ and the selected dictionary elements $A_{T^{[i-1]}}$. Let us define the orthogonal projection operator $\mathbf{P}_{T^{[i]}}^{\perp} := (\mathbf{I} - A_{T^{[i]}} A_{T^{[i]}}^{\dagger})$. The ORMP selection can then be written as:

$$j^{[i]} = \operatorname*{argmax}_{j} \frac{|A_j^T r^{[i-1]}|}{\|\mathbf{P}_{T^{[i-1]}}^{\perp} A_j\|_2}. \tag{8.19}$$

That is, ORMP selects the element whose *normalized* projection orthogonal to $\operatorname{span}(A_{T^{[i-1]}})$ is best correlated to $r^{[i-1]}$.

While ORMP has a theoretical advantage over OMP it comes at a computational cost. An efficient implementation of ORMP can use a QR factorization similar to that of OMP [25], but there is the additional burden of projecting the elements $A_j, j \notin T^{[i-1]}$, orthogonal to $A_{T^{[i-1]}}$.

Moreover, for CS applications, if the dictionary has been judiciously designed to a small RIP constant, the benefits of the orthogonal projection may be small. Indeed empirical studies suggest that ORMP does not perform significantly better than OMP.

Table 8.1. Comparison of pursuit methods in terms of computational cost (flops) per iteration and storage requirements (number of floating point numbers) where k refers to the size of the support set in the current iteration i and A is the computational cost of applying or storing the transform A or A^T. For StOMP (CG), ν is the number of conjugate gradient steps used per iteration, which in the worst case is equal to the number of elements selected.

Algorithm	Computation cost	Storage cost
MP	$m + A + n$	$A + m + 2k + n$
OMP (QR)	$2mk + m + A + n$	$2(m+1)k + 0.5k(k+1) + A + n$
OMP (Chol)	$3A + 3k^2 + 2m + n$	$0.5k(i+1) + A + m + 2k + n$
GP	$2A + k + 3m + n$	$2m + A + 2k + n$
CGP	$2A + k + 3m + n$	$2m + A + 2k + n$
StWGP	$2A + k + 3m + 2n$	$2m + A + 2k + n$
StOMP (CG)	$(\nu + 2)A + k + 3m + n$	$2m + A + 2k + n$
ORMP	$2m(n-k) + 3m + A + n$	$2(m+1)k + 0.5k(k+1) + nm + n$

8.2.4 Computational considerations

The computational requirements of each of the pursuit algorithms depend on the specific implementation details, the structure of the sensing matrix A, and the number of iterations used. Due to the variability of the computation and storage costs of A (from $\mathcal{O}(n\log(m))$ for an FFT-type operation to $\mathcal{O}(nm)$ for unstructured matrices) we explicitly count the number of matrix-vector products Ax and $A^T y$ without specifying an associated number of flops. Note that even with fast transforms the matrix-vector product generally dominates the computational cost, so this quantity is very important. A summary of the overall computational and storage costs *per iteration* for each of the pursuit algorithms is presented in Table 8.1.

8.2.5 Performance guarantees

One of the cornerstones of CS is the theoretical recovery guarantee that it provides for certain algorithms. Unfortunately, although empirically greedy pursuit algorithms often appear competitive with ℓ_1 minimization, a key weakness in these algorithms lies in the strength of their theoretical recovery results. Indeed it was shown in [27, 28] that for certain random matrices A, when $m \sim k\log(n)$ then with high probability there exists a k-sparse vector x for which OMP will fail to select a correct element at the first iteration. That is: if we use OMP to recover k elements, then such matrices do not provide uniform recovery of all k-sparse vectors in this regime. As all the pursuit algorithms discussed in this section will also select the same element as OMP in the first step, this weakness is universal to all greedy pursuit algorithms.

A possible criticism of such an analysis lies in the definition of recovery used: namely OMP must *only* select correct elements from A. This precludes the possibility that the algorithm makes some mistakes but that these can be rectified by selecting more than k

elements[6] (if the columns of A are in general position any exact solution $A\hat{x} = y$ found that contains less than m elements will provide a correct recovery [27]). Whilst it has been suspected that uniform recovery is also not possible in this scenario [27, 28], this remains an active topic of current research.

So when do greedy pursuit algorithms provide uniform sparse recovery? The original worst case analysis of OMP was presented in [13] and most of these results carry over to general weak MP algorithms (i.e. MP, GP, CGP, StWGP) with minor modifications. In the context of CS the following approximation result on $\|\hat{x}^{[i]} - x\|_2$ was derived for StWGP or any weak gradient pursuit [18] in terms of the Restricted Isometry Property for A (a similar result was derived independently for OMP in [29]).

THEOREM 8.1 (Uniform recovery for (stagewise) weak gradient pursuits) *For any x, let $y = Ax + e$ and stop the algorithm before it selects more than k nonzero elements. Let the last iteration be iteration i^* and let $\hat{x}^{[i^*]}$ be the estimation of x calculated at this iteration. If*

$$\delta_{k+1} < \frac{\alpha}{\sqrt{k} + \alpha}, \tag{8.20}$$

then there exists a constant C (depending on α and δ_{2k}), such that

$$\|\hat{x}^{[i^*]} - x\|_2 \leq C\left(\|(x - x_k)\|_2 + \frac{\|(x - x_k)\|_1}{\sqrt{k}} + \|e\|_2\right), \tag{8.21}$$

where x_k is the best k-term approximation to x.

The bound on the error ϵ is optimal up to a constant and is of the same form as that for ℓ_1 minimization discussed in Chapter 1 [30]. However here we require that $k^{0.5}\delta_{2k}$ is small, which translates into the requirement $m \geq \mathcal{O}(k^2 \log(n/k))$, which is similar to other OMP recovery results [13].

Although pursuit algorithms do not enjoy a uniform recovery property when $m = \mathcal{O}(k \log(n/k))$, we typically observe recovery behavior for greedy algorithms similar to ℓ_1 minimization. This suggests that the worst case bound in Theorem 8.1 is not necessarily indicative of typical algorithm performance. In order to understand the typical behavior of OMP and its relatives we can examine the typical (nonuniform) recovery behavior for random dictionaries. Specifically suppose we are given an arbitrary k-sparse vector x and we then draw A at random from a suitable random set. Under what conditions will OMP recover x with high probability? Note that A is only being asked to recover a specific x not all k-sparse vectors. This question was first investigated in [31] where it was shown that successful recovery of sparse vectors using OMP can be done with $m = \mathcal{O}(k \log(n))$ measurements. Specifically, Tropp and Gilbert [31] give the following result:

THEOREM 8.2 ([31] OMP with random measurements) *Suppose that x is an arbitrary k-sparse signal in $\mathbb{R}^n$ and draw a random $m \times n$ matrix A with i.i.d. Gaussian or*

---

[6] Allowing a small number of mistakes in support selection appears to significantly improve the sparse recovery performance for OMP in practice.

Bernoulli entries. Given the data $y = Ax$ and choosing $m \geq Ck\log(n/\sqrt{\delta})$ where C is a constant depending on the random variables used for A, then OMP can reconstruct the signal with probability at least $1 - \delta$.

Theorem 8.2 shows that if we select A independently from the signal to be recovered then OMP should also exhibit good performance even when m scales linearly with k. When A is Gaussian then asymptotically for $n \to \infty$ the constant can be calculated to be $C = 2$ [32]. Moreover, it is possible to consider noisy measurements so long as asymptotically the signal-to-noise ratio tends to infinity.

It is also very simple to adapt the results of Tropp and Gilbert [31] to deal with general weak MP algorithms. The only property specific to OMP used in the proof of Theorem 8.2 is that if successful OMP terminates in k iterations. For general weak MP algorithms we can state the following [18]

THEOREM 8.3 (General weak MP with random measurements) *Suppose that x is an arbitrary k-sparse signal in $\mathbb{R}^n$ and draw a random $m \times n$ matrix A with i.i.d. Gaussian or Bernoulli entries. Given the data $y = Ax$ any general weak MP algorithm with weakness parameter α will only select correct elements within the first L iterations with probability at least*

$$1 - (4L(n-k))e^{-\alpha^2 m/(Ck)},$$

where C is a constant dependent on the random variables used for A.

This gives a "result dependent" guarantee that can be stated as:

COROLLARY 8.1 *Suppose a given general weak MP algorithm selects k elements in the first $L \leq k$ iterations. Then if $m \geq 2C\alpha^{-2}L\log(n/\sqrt{\delta})$ the correct support of x has been found with probability at least $1 - \delta$.*

Note that in the noiseless case and for exact-sparse x, if we use exact orthogonalization as in StOMP, then the condition $L \leq k$ is automatically satisfied.

8.2.6 Empirical comparisons

We use a simple toy problem: 10 000 dictionaries of size 128×256 were generated with columns A_i drawn uniformly from the unit sphere. From each dictionary, and at a number of different degrees of sparsity, elements were selected at random and multiplied with unit variance, zero mean Gaussian coefficients[7] to generate 10 000 different signals per sparsity level. We first analyze the average performance of various greedy pursuit methods in terms of exact recovery of the elements used to generate the signal.

The results are shown in Figure 8.1. We here show the results for MP, GP, StWGP, ACGP, and OMP. All algorithms were stopped after they had selected exactly the number of elements used to generate the signal. It is clear that weakening the selection criterion

---

[7] It is worth pointing out that the observed average performance of many pursuit algorithms varies with the distribution of the nonzero coefficients and is often worse than shown here if the nonzero elements are set to 1 or -1 with equal probability – see for example [18].

Greedy algorithms for compressed sensing

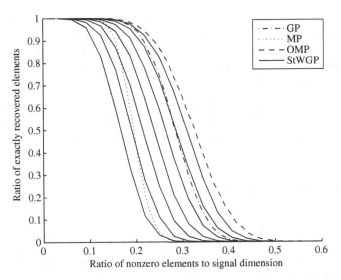

Figure 8.1 Comparison between MP (dotted), OMP (dashed), GP (dash-dotted), and StWGP (solid) in terms of exactly recovering the original coefficients. The ordinate shows the fraction of runs in which the algorithms exactly recovered the index set T used to generate the data while the abscissa shows the ratio of the size of T to the dimension of x. Results averaged over 10 000 runs. The solid lines correspond to (from left to right): $\alpha = 0.7, 0.75, 0.8, 0.85, 0.9, 0.95$, and 1.0 (CGP). Adapted from [33], ©[2008] IEEE.

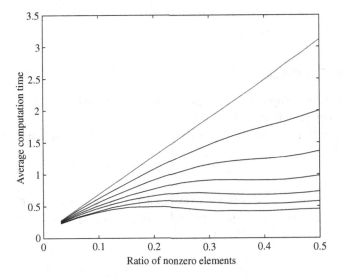

Figure 8.2 Comparison of the computation time for StWGP with the different values of α as in Figure 8.1. The curves correspond to (going from top to bottom): $\alpha = 1.0, 0.95, 0.9, 0.85, 0.8, 0.75$, and 0.7. Adapted from [33], ©[2008] IEEE.

reduces (in a controlled manner) the recovery performance. The advantage of this is a reduction in computational cost. This is shown in Figure 8.2. Here the curves correspond to (going from top to bottom): $\alpha = 1.0, 0.95, 0.9, 0.85, 0.8, 0.75$, and 0.7. The top curve indicates that the computational cost for ACGP (StWGP with $\alpha = 1.0$) grows linearly with the number of nonzero coefficients. In contrast for $\alpha < 1.0$ the computational cost grows much more slowly. It should be noted here that these figures do not fully capture the performance of StWGP since the dictionaries used do not have a fast implementation. However they do provide a fair relative comparison between different values of α.

8.3   Thresholding type algorithms

As shown above, greedy pursuits are easy to implement and use and can be extremely fast. However, they do not have recovery guarantees as strong as methods based on convex relaxation. The methods discussed in this section bridge this gap. They are fairly easy to implement and can be extremely fast but also show the strong performance guarantees available with methods based on convex relaxation. These methods therefore offer a powerful set of tools for CS applications. Furthermore, as discussed in detail in Section 8.4, these methods can easily be adapted to more general low-dimensional signal models.

We here concentrate on three algorithms, the Iterative Hard Thresholding (IHT) algorithm [21], the Compressive Sampling Matching Pursuit (CoSaMP) [34], and the Subspace Pursuit (SP) [23], though methods based on similar ideas have been suggested in [35], [36], [37], and [38].

8.3.1   Iterative Hard Thresholding

Motivated by the developments of Kingsbury and Reeves [35], the IHT algorithm was introduced independently in [21] and [36].

IHT is a greedy algorithm that iteratively solves a local approximation to the CS recovery problem

$$\min_{\widetilde{x}} \|y - A\widetilde{x}\|_2^2 \quad \text{subject to} \quad \|\widetilde{x}\|_0 \leq k. \quad (8.22)$$

The local approximation to this non-convex problem can be derived based on the optimization transfer framework of [39]. Instead of optimizing equation (8.22) directly, a surrogate objective function is introduced

$$C_k^S(\widetilde{x}, z) = \mu \|y - A\widetilde{x}\|_2^2 - \mu \|A\widetilde{x} - Az\|_2^2 + \|\widetilde{x} - z\|_2^2. \quad (8.23)$$

The advantage of this cost function is that we can write (8.23) as

$$C_k^S(\widetilde{x}, z) \propto \sum_j [\widetilde{x}_j^2 - 2\widetilde{x}_j(z_j + \mu A_j^T y - A_j^T Az)], \quad (8.24)$$

where A_j is again the jth column of A.

Equation (8.24) can be optimized for each $\widetilde{x}_j$ independently. If we were to ignore the constraint $\|\widetilde{x}\|_0 \leq k$, then (8.23) has a minimizer

$$x^\star = z + \mu A^T(y - Az). \tag{8.25}$$

At this minimum, the cost function (8.23) has a value proportional to

$$C_k^S(x^\star, z) \propto \|x^\star\|_2^2 - 2\langle x^\star, (z + \mu A^T(y - Az))\rangle = -\|x^\star\|_2^2. \tag{8.26}$$

The constraint can therefore be enforced by simply choosing the k largest (in magnitude) coefficients of $x^\star$, and by setting all other coefficients to zero.

The minimum of (8.23) subject to the constraint that $\|\widetilde{x}\|_0 \leq k$ is thus attained at

$$\widehat{x} = H_k(z + \mu A^T(y - Az)), \tag{8.27}$$

where H_k is the nonlinear *projection* that sets all but the largest k elements of its argument to zero. In cases where the k largest coefficients are not uniquely defined, we assume that the algorithm selects from the offending coefficients using a predefined order.

This local optimization approach can be turned into an iterative algorithm by setting $z = \widehat{x}^{[i]}$, in which case we get the Iterative Hard Thresholding (IHT) algorithm 8.5. Based on the current estimate $\widehat{x}^{[i]}$, this algorithm greedily finds a global minimum of the constrained surrogate objective.

Algorithm 8.5 Iterative Hard Thresholding algorithm (IHT)

Input: y, A, k, and μ
Initialize: $\widehat{x}^{[0]} = \mathbf{0}$
for $i = 0$, $i := i + 1$, until stopping criterion is met **do**
$\quad \widehat{x}^{[i+1]} = H_k(\widehat{x}^{[i]} + \mu A^T(y - A\widehat{x}^{[i]}))$
end for
Output: $\widehat{x}^{[i]}$

The IHT algorithm is easy to implement and is computationally efficient. Apart from vector additions, the main computational steps are the multiplication of vectors by A and its transpose, as well as the partial sorting required for the thresholding step. Storage requirements are therefore small and, if structured measurement matrices are used, multiplication by A and A^T can also often be done efficiently.

Performance guarantees

One fundamental property of any iterative algorithm is its convergence. For the IHT algorithm, convergence can be shown conditionally on the step size μ. This condition will be based on a property of A, namely the quantity β_{2k}, defined as the smallest quantity, such that

$$\|A(x_1 - x_2)\|_2^2 \leq \beta_{2k}\|x_1 - x_2\|_2^2 \tag{8.28}$$

holds for all k-sparse vectors x_1 and x_2. It is clear that $\beta_{2k} \leq (1 + \delta_{2k}) \leq \|A\|_2^2$, where δ_{2k} is the symmetric RIP constant as defined in (8.2).

The main convergence result, derived by Blumensath and Davies [21], can be stated as follows.

THEOREM 8.4 *If $\beta_{2k} < \mu^{-1}$, $k \leq m$ and assume A is of full rank, then the sequence $\{\widehat{x}^{[i]}\}_i$ defined by the IHT algorithm converges to a local minimum of (8.22).*

Whilst convergence is clearly desirable, in CS an even more important property of any algorithm is its ability to recover sparse or nearly sparse signals, that is, the ability of a method to calculate and estimate $\widehat{x}$ of x that is close to x. One of the fundamental properties of the IHT algorithm is that it possesses this property with a near optimal error guarantee whenever the RIP holds. Indeed, its performance in this respect is similar to the performance of convex optimization-based approaches discussed in Chapter 1.

As the restricted isometry property introduced in (8.2) is sensitive to a re-scaling of the matrix A, we will here state the main result based on the non-symmetric RIP

$$\alpha_{2k}\|x_1 - x_2\|_2^2 \leq \|A(x_1 - x_2)\|_2^2 \leq \beta_{2k}\|x_1 - x_2\|_2^2, \quad (8.29)$$

for all k-sparse x_1 and x_2. Note that we are primarily interested in how A acts on the difference between two k-sparse vectors, hence the particular form of the RIP used here.[8]

The first recovery result of the IHT algorithm was derived in [40], with later refinements based on the work by Meka *et al.* [41] and Garg and Khandekar [42] as first reported in [43]. A somewhat more careful argument which can be found in Subsection 8.3.4 gives the following performance guarantee.

THEOREM 8.5 *For arbitrary x, given $y = Ax + e$ where A satisfies the non-symmetric RIP with $\beta_{2k} \leq \mu^{-1} < 1.5\alpha_{2k}$, after*

$$i^\star = \left\lceil 2\frac{\log(\|\widetilde{e}\|_2 / \|x_k\|_2)}{\log(2/(\mu\alpha_{2k}) - 2)} \right\rceil \quad (8.30)$$

iterations, the IHT algorithm calculates a solution $\widehat{x}^{[i^\star]}$ satisfying

$$\|x - \widehat{x}^{[i^\star]}\|_2 \leq (1 + c\sqrt{\beta_{2k}})\|x - x_k\|_2 + c\sqrt{\beta_{2k}}\frac{\|x - x_k\|_1}{\sqrt{k}} + c\|e\|_2, \quad (8.31)$$

where $c \leq \sqrt{\frac{4}{3\alpha_{2k} - 2\mu}} + 1$.

Similarly, if $s\alpha_{2k} \leq \mu^{-1} \leq \beta_{2k}$ holds for some $s < \frac{6}{5}$ and if $\frac{\beta_{2k}}{\alpha_{2k}} < \frac{3 + \frac{6}{s}}{8}$, then the same bound holds with $c \leq \sqrt{\frac{4 + 16s - 16\frac{\beta_{2k}}{\alpha_{2k}}}{(3 + 6s)\alpha_{2k} - 8\beta_{2k}}} + 1$ and $i^\star = \left\lceil 2\frac{\log(\|\widetilde{e}\|_2 / \|x_k\|_2)}{\log\left(\frac{4\frac{\beta_{2k}}{\alpha_{2k}} - 2(1+s)}{1 + 4s - 4\frac{\beta_{2k}}{\alpha_{2k}}}\right)} \right\rceil$.

---

[8] In this form, the RIP can also be understood as a bi-Lipschitz condition, where $\sqrt{\beta_{2k}}$ is the Lipschitz constant of the map A acting on the set of all k-sparse signals, whilst $1/\sqrt{\alpha_{2k}}$ is the Lipschitz constant of its inverse. Our version of the RIP can be easily translated into the symmetric RIP using $\alpha_{2k} \geq (1 - \delta_{2k})$ and $\beta_{2k} \leq (1 + \delta_{2k})$.

Step size determination

In order to make use of the recovery result of Theorem 8.5, we need to choose $1/\mu$ appropriately. However, as discussed at length in Chapter 1, for any given measurement matrix A, we do not normally know if the RIP holds and, even if it holds, what the exact RIP constants are. So how do we choose μ if we do not know α_{2k} and β_{2k}? If we have designed A using a random construction and we have good bounds on β_{2k} and α_{2k}, such that $\beta_{2k} \leq 1.5\alpha_{2k}$ holds with high probability, then we could set $\beta_{2k} \leq \mu^{-1} \leq 1.5\alpha_{2k}$, which would guarantee both convergence of the method as well as give near optimal recovery as shown in Theorem 8.5. However, in many practical situations, physical and computational constraints prevent us from using any of the well-understood random constructions of A so that the RIP constants are typically unknown. Furthermore, it is often desirable to also use the method in a regime in which the RIP conditions do not hold. Empirical results achievable using a fixed step size μ have shown mixed results in this regime [44] and, as suggested by Blumensath and Davies [44], it is instead advisable to choose μ adaptively in each iteration.

Let $T^{[i]}$ be the support set of $\widehat{x}^{[i]}$ and let $g^{[i]} = A^T(y - A\widehat{x}^{[i]})$ be the negative gradient of $\|y - A\widetilde{x}\|_2^2$ evaluated at the current estimate $\widehat{x}^{[i]}$. (Should $\widehat{x}^{[i]}$ be zero, which typically only happens in the first iteration when we initialize the algorithm with the zero vector, we use the index set of the largest k (in magnitude) elements of $A^T y$ as the set $T^{[i]}$.) Assume that we have identified the correct support, that is, $T^{[i]}$ is the support of the best k-term approximation to x. In this case we would want to minimize $\|y - A_{T^{[i]}}\widetilde{x}_{T^{[i]}}\|_2^2$. Using a gradient descent algorithm, this would be done using the iteration $\widehat{x}_{T^{[i]}}^{[i+1]} = \widehat{x}_{T^{[i]}}^{[i]} + \mu A_{T^{[i]}}^T(y - A_{T^{[i]}}x_{T^{[i]}}^{[i]})$. Importantly, in the case in which the support is fixed, we can calculate an optimal step size, that is, a step size that maximally reduces the error in each iteration. It is easy to see that this step size is [16]

$$\mu = \frac{\|g_{T^{[i]}}^{[i]}\|_2^2}{\|A_{T^{[i]}}g_{T^{[i]}}^{[i]}\|_2^2}, \qquad (8.32)$$

where $g_{T^{[i]}}^{[i]}$ is the sub-vector of $g^{[i]}$ obtained by discarding all elements apart from those in $T^{[i]}$ and where $A_{T^{[i]}}$ is defined similarly by discarding columns of A. Evaluation of this quantity requires the calculation of $A_{T^{[i]}}g_{T^{[i]}}^{[i]}$, the computational cost of which is equivalent to the evaluation of $A\widehat{x}^{[i]}$ and $A^T(y - A\widehat{x}^{[i]})$, so that the complexity of the IHT algorithm only increases by a constant fraction. Importantly, using (8.32), we do not require exact knowledge of the RIP constants, because if A has non-symmetric RIP with α_{2k} and β_{2k}, then as $g_{T^{[i]}}^{[i]}$ has only k nonzero elements, we have the bound

$$\alpha_{2k} \leq \frac{\|A_{T^{[i]}}g_{T^{[i]}}^{[i]}\|_2^2}{\|g_{T^{[i]}}^{[i]}\|_2^2} \leq \beta_{2k}, \qquad (8.33)$$

so that, if $\beta_{2k}/\alpha_{2k} < 9/8$, then Theorem 8.5 holds. We thus have the corollary.

COROLLARY 8.2 *Given $y = Ax + e$. If A satisfies the non-symmetric RIP with $\beta_{2k}/\alpha_{2k} < 9/8$, then, after*

$$i^\star = \left\lceil 2 \frac{\log(\|\widetilde{e}\|_2/\|x_k\|_2)}{\log\left(\frac{4\frac{\beta_{2k}}{\alpha_{2k}}-4}{5-4\frac{\beta_{2k}}{\alpha_{2k}}}\right)} \right\rceil \quad (8.34)$$

iterations, the algorithm defined by the iteration

$$\widehat{x}^{[i+1]} = H_k\left(\widehat{x}^{[i]} + \frac{\|g^{[i]}_{T^{[i]}}\|_2^2}{\|A_{T^{[i]}}g^{[i]}_{T^{[i]}}\|_2^2} A^T(y - A\widehat{x}^{[i]})\right) \quad (8.35)$$

calculates a solution $\widehat{x}^{[i^\star]}$ satisfying

$$\|x - \widehat{x}^{[i^\star]}\|_2 \leq (1 + c\sqrt{\beta_{2k}})\|y - x_k\|_2 + c\sqrt{\beta_{2k}}\frac{\|x - x_k\|_1}{\sqrt{k}} + c\|e\|_2, \quad (8.36)$$

where $c \leq \sqrt{\frac{20 - 16\frac{\beta_{2k}}{\alpha_{2k}}}{9\alpha_{2k} - 8\beta_{2k}}} + 1$.

There remains however a further problem. We cannot typically guarantee that our measurement system satisfies the required bound on the RIP constants. We thus need to ensure that the algorithm will be robust also in the case in which the condition of the theorem fails. Whilst we can no longer guarantee that we will be able to recover x with high accuracy, we can at least ensure that the algorithm will converge if the RIP condition fails. In order to do this, we need to monitor convergence and will use a line search approach to guarantee stability of the method.

In the case in which the support of $\widetilde{x}^{[i+1]}$ is the same as that of $\widehat{x}^{[i]}$, our choice of the step size ensures that we are guaranteed to have a maximal reduction in the cost function, which in turn ensures stability of the method. However, if the support of $\widetilde{x}^{[i+1]}$ differs from the support of $\widehat{x}^{[i]}$, the optimality of μ is no longer guaranteed. In this case, a sufficient condition that guarantees convergence is [44]

$$\mu \leq (1-c)\frac{\|\widetilde{x}^{[i+1]} - \widehat{x}^{[i]}\|_2^2}{\|A(\widetilde{x}^{[i+1]} - \widehat{x}^{[i]})\|_2^2} \quad (8.37)$$

for any small fixed constant c.

Hence, if our first proposal $\widetilde{x}^{[i+1]}$ has a different support, we need to check whether μ satisfies the above convergence condition. If the condition holds, then we keep the new update and set $\widehat{x}^{[i+1]} = \widetilde{x}^{[i+1]}$. Otherwise we need to shrink the step size μ. The simplest way to do this is to set $\mu \leftarrow \mu/(\kappa(1-c))$, for some constant $\kappa > 1/(1-c)$. With this new step size, a new proposal $\widetilde{x}^{[i+1]}$ is calculated. This procedure is terminated when condition (8.37) is satisfied or if the support of $\widetilde{x}^{[i+1]}$ is the same as that of $\widehat{x}^{[i]}$, in which case we accept the latest proposal and continue with the next iteration.[9]

---

[9] Note that it can be shown that the algorithm will accept a new step size after a finite number of such tries.

Using this automatic step-size reduction, it can be shown that the algorithm, which is summarized as Algorithm 8.6, will converge to a fixed point [44].

THEOREM 8.6 *If* $\text{rank}(A) = m$ *and* $\text{rank}(A_T) = k$ *for all T such that* $|T| = k$, *then the normalized IHT algorithm converges to a local minimum of the optimization problem (8.22).*

Algorithm 8.6 Normalized Iterative Hard Thresholding algorithm (IHT)

Input: y, A, and k
Initialize: $\widehat{x}^{[0]} = 0$, $T^{[0]} = \text{supp}(H_k(A^T y))$
for $i = 0$, $i := i + 1$, until stopping criterion is met **do**
  $g^{[i]} = A^T(y - A\widehat{x}^{[i]})$
  $\mu^{[i]} = \frac{\|g^{[i]}_{T^{[i]}}\|_2^2}{\|A_{T^{[i]}} g^{[i]}_{T^{[i]}}\|_2^2}$
  $\widetilde{x}^{[i+1]} = H_k(\widehat{x}^{[i]} + \mu^{[i]} g^{[i]})$
  $T^{[i+1]} = \text{supp}(\widetilde{x}^{[i+1]})$
  if $T^{[i+1]} = T^{[i]}$ **then**
    $\widehat{x}^{[i+1]} = \widetilde{x}^{[i+1]}$
  else if $T^{[i+1]} \neq T^{[i]}$ **then**
    if $\mu^{[i]} \leq (1-c)\frac{\|\widetilde{x}^{[i+1]} - \widehat{x}^{[i]}\|_2^2}{\|A(\widetilde{x}^{[i+1]} - \widehat{x}^{[i]})\|_2^2}$ **then**
      $\widehat{x}^{[i+1]} = \widetilde{x}^{[i+1]}$
    else if $\mu^{[i]} > (1-c)\frac{\|\widetilde{x}^{[i+1]} - \widehat{x}^{[i]}\|_2^2}{\|A(\widetilde{x}^{[i+1]} - \widehat{x}^{[i]})\|_2^2}$ **then**
      repeat
        $\mu^{[i]} \leftarrow \mu^{[i]}/(\kappa(1-c))$
        $\widetilde{x}^{[i+1]} = H_k(\widehat{x}^{[i]} + \mu^{[i]} g^{[i]})$
      until $\mu^{[i]} \leq (1-c)\frac{\|\widetilde{x}^{[i+1]} - \widehat{x}^{[i]}\|_2^2}{\|A(\widetilde{x}^{[i+1]} - \widehat{x}^{[i]})\|_2^2}$
      $T^{[i+1]} = \text{supp}(\widetilde{x}^{[i+1]})$
      $\widehat{x}^{[i+1]} = \widetilde{x}^{[i+1]}$
    end if
  end if
end for
Output: $r^{[i]}$ and $\widehat{x}^{[i]}$

8.3.2 Compressive Sampling Matching Pursuit and Subspace Pursuit

The Compressive Sampling Matching Pursuit (CoSaMP) algorithm by Needell and Tropp [34] and the Subspace Pursuit (SP) algorithm by Dai and Milenkovic [23] are very similar and share many of their properties. We will therefore treat both methods jointly.

General framework

Both CoSaMP and SP keep track of an active set T of nonzero elements and both add as well as remove elements in each iteration. At the beginning of each iteration, a k-sparse estimate $\widehat{x}^{[i]}$ is used to calculate a residual error $y - A\widehat{x}^{[i]}$, whose inner products with the column vectors of A are calculated. The indexes of those columns of A with the k (or $2k$) largest inner products are then selected and added to the support set of $\widehat{x}^{[i]}$ to get a larger set $T^{[i+0.5]}$. An intermediate estimate $\widehat{x}^{[i+0.5]}$ is then calculated as the least-squares solution $\operatorname{argmin}_{\widetilde{x}_{T^{[i+0.5]}}} \|y - A\widetilde{x}_{T^{[i+0.5]}}^{[i+0.5]}\|_2$. The largest k elements of this intermediate estimate are now found and used as the new support set $T^{[i+1]}$. The methods differ in the last step, the CoSaMP algorithm takes as a new estimate the intermediate estimate $\widehat{x}^{[i+0.5]}$ restricted to the new smaller support set $T^{[i+1]}$, whilst SP solves a second least-squares problem restricted to this reduced support.

CoSaMP

The CoSaMP algorithm, which was introduced and analyzed by Needell and Tropp [34], is summarized in Algorithm 8.7.

Algorithm 8.7 Compressive Sampling Matching Pursuit (CoSaMP)

Input: y, A, and k
Initialize: $T^{[0]} = \operatorname{supp}(H_k(A^T y))$, $\widehat{x}^{[0]} = 0$
for $i = 0$, $i := i+1$, until stopping criterion is met **do**
$\quad g^{[i]} = A^T(y - A\widehat{x}^{[i]})$
$\quad T^{[i+0.5]} = T^{[i]} \bigcup \operatorname{supp}(g_{2k}^{[i]})$
$\quad \widehat{x}_{T^{[i+0.5]}}^{[i+0.5]} = A_{T^{[i+0.5]}}^\dagger y,\ \widehat{x}_{\overline{T^{[i+0.5]}}}^{[i+0.5]} = 0$
$\quad T^{[i+1]} = \operatorname{supp}(\widehat{x}_k^{[i+0.5]})$
$\quad \widehat{x}_{T^{[i+1]}}^{[i+1]} = \widehat{x}_{T^{[i+1]}}^{[i+0.5]},\ \widehat{x}_{\overline{T^{[i+1]}}}^{[i+1]} = 0$
end for
Output: $r^{[i]}$ and $\widehat{x}^{[i]}$

Instead of calculating $A_{T^{[i+0.5]}}^\dagger y$ exactly in each iteration, which can be computationally demanding, a faster approximate implementation of the CoSaMP algorithm was also proposed in [34]. This fast version replaces the exact least-squares estimate $\widehat{x}_{T^{[i+0.5]}}^{[i+0.5]} = A_{T^{[i+0.5]}}^\dagger y$ with three iterations of a gradient descent or a conjugate gradient solver. Importantly, this modified algorithm still retains the same theoretical guarantee given in Theorem 8.7 below, though the empirical performance tends to deteriorate as shown in the simulations in Subsection 8.3.3.

Needell and Tropp [34] suggested different strategies to stop the CoSaMP algorithm. If the RIP holds, then the size of the error $y - A\widehat{x}^{[i]}$ can be used to bound the error $x - \widehat{x}^{[i]}$, which in turn can be used to stop the algorithm. However, in practice, if it is unknown if the RIP holds, then this relationship is not guaranteed. In this case, an alternative would be to stop the iterations as soon as $\|\widehat{x}^{[i]} - \widehat{x}^{[i+1]}\|_2$ is small or whenever the approximation error $\|y - A\widehat{x}^{[i]}\|_2 < \|y - A\widehat{x}^{[i+1]}\|_2$ increases. Whilst the first of these methods does not guarantee the convergence of the method, the second approach is guaranteed to prevent instability, however, it is also somewhat too strict in the case in which the RIP holds.

SP

The SP algorithm, developed and analyzed by Dai and Milenkovic [23], is very similar to CoSaMP as shown in Algorithm 8.8.

Algorithm 8.8 Subspace Pursuit (SP)

Input: y, A, and k
Initialize: $T^{[0]} = \text{supp}(H_k(A^T y))$, $\widehat{x}^{[0]} = A^\dagger_{T^{[0]}} y$
for $i = 0$, $i := i+1$, until $\|y - A\widehat{x}^{[i+1]}\|_2 \geq \|y - A\widehat{x}^{[i]}\|_2$ **do**
$\quad g^{[i]} = A^T(y - A\widehat{x}^{[i]})$
$\quad T^{[i+0.5]} = T^{[i]} \bigcup \text{supp}(g_k^{[i]})$
$\quad \widehat{x}^{[i+0.5]}_{T^{[i+0.5]}} = A^\dagger_{T^{[i+0.5]}} y$, $\widehat{x}^{[i+0.5]}_{\overline{T^{[i+0.5]}}} = 0$
$\quad T^{[i+1]} = \text{supp}(\widehat{x}_k^{[i+0.5]})$
$\quad \widehat{x}^{[i+1]} = A^\dagger_{T^{+1}} y$
end for
Output: $r^{[i]}$ and $\widehat{x}^{[i]}$

Here the same stopping rule based on the difference $\|y - A\widehat{x}^{[i]}\|_2 - \|y - A\widehat{x}^{[i+1]}\|_2$ has been proposed by Dai and Milenkovic [23]. This guarantees that the method remains stable, even in a regime in which the RIP condition fails.

The main difference between the two approaches is the size of the set added to $T^{[i]}$ in each iteration as well as the additional least-squares solution required in the SP. Furthermore, the possibility of replacing the least-squares solution in CoSaMP with three gradient-based updates means that it can be implemented much more efficiently than SP.

Performance guarantees

Just as the IHT algorithm, both CoSaMP as well as SP offer near optimal performance guarantees under conditions on the RIP. The CoSaMP algorithm was the first greedy method to be shown to possess similar performance guarantees to ℓ_1-based methods [34]. The proof of the result uses similar ideas to those used in Subsection 8.3.4 to prove IHT performance, though as the CoSaMP algorithm involves significantly more steps than the IHT algorithm, the performance proof is also significantly longer and is therefore omitted here.

THEOREM 8.7 ([34]) *For any x, given $y = Ax + e$ where A satisfies the RIP with $0.9 \leq \frac{\|Ax_1 - Ax_2\|_2^2}{\|x_1 - x_2\|_2^2} \leq 1.1$ for all $2k$-sparse vectors x_1 and x_2, after*

$$i^\star = \lceil \log(\|x_k\|_2 / \|\widetilde{e}\|_2) \log 2 \rceil \tag{8.38}$$

iterations, the CoSaMP algorithm calculates a solution $\widehat{x}^{[i^*]}$ satisfying

$$\|x - \widehat{x}^{[i^*]}\|_2 \leq 21 \left(\|y - x_k\|_2 + \frac{\|x - x_k\|_1}{\sqrt{k}} + \|e\|_2 \right). \tag{8.39}$$

For SP, Dai and Milenkovic [23], using a similar approach, derived analogous results.

THEOREM 8.8 ([23])   *For any x, given $y = Ax + e$ where A satisfies, for all k-sparse vectors x_1 and all $2k$-sparse vectors x_2, the RIP with $0.927 \leq \frac{\|Ax_1 - Ax_2\|_2^2}{\|x_1 - x_2\|_2^2} \leq 1.083$, then SP calculates a solution $\widehat{x}$ satisfying*

$$\|x - \widehat{x}\|_2 \leq 1.18 \left(\|y - x_k\|_2 + \frac{\|x - x_k\|_1}{\sqrt{k}} + \|e\|_2 \right). \tag{8.40}$$

8.3.3 Empirical comparison

To highlight the performance of the different approaches discussed in this section, we repeated the experiment of Subsection 8.2.6, comparing the normalized IHT algorithm, the CoSaMP algorithm, and an ℓ_1-based approach (minimize $\|x\|_1$ such that $y = Ax$). Here, the two different implementations of CoSaMP have been used (CoSaMP performance with exact least-squares solver is shown in Figure 8.3 with dash-dotted lines and CoSaMP performance with a fixed number of conjugate gradient steps is shown with dotted lines, from left to right, using 3, 6, and 9 conjugate gradient steps within each iteration).

The normalized IHT algorithm performs well in this average case analysis, even in a regime when simple bounds on the RIP constants indicate that the RIP condition of the theorem is violated. Here, the normalized IHT algorithm outperforms CoSaMP with exact least-squares solver and performs nearly as well as the ℓ_1-based approach. Though it should be noted that, as reported by Blumensath and Davies [44], if the nonzero entries of x all have equal magnitude, then CoSaMP with exact least-squares solver slightly outperforms the normalized IHT algorithm.

Figure 8.4 shows the computation times for the experiment reported in Figure 8.3 in the region where the algorithms perform well.[10] The speed advantage of the normalized IHT algorithm is clearly evident.

8.3.4 Recovery proof

Proof of Theorem 8.5.   To prove the recovery bound for IHT, the estimation error norm is split into two parts using the triangle inequality

$$\|x - \widehat{x}^{[i+1]}\|_2 \leq \|x_k - \widehat{x}^{[i+1]}\|_2 + \|x_k - x\|_2. \tag{8.41}$$

---

[10] To solve the ℓ_1 optimization problem, we here used the spgl1 [45] algorithm (available on www.cs.ubc.ca/labs/scl/spgl1/).

Greedy algorithms for compressed sensing

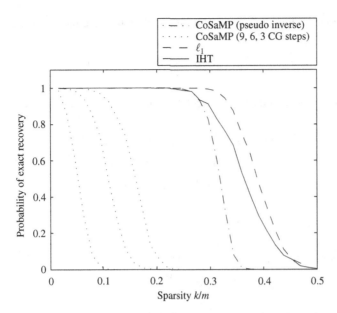

Figure 8.3 Percentage of instances in which the algorithms could identify exactly which of the elements in x were non-zero. The non-zero elements in x were i.i.d. normal distributed. Results are shown for normalized IHT (solid), ℓ_1 solution (dashed), CoSaMP implemented using the pseudo-inverse (dash-dotted), and CoSaMP using (from left to right) 3, 6, or 9 conjugate gradient iterations (dotted). Adapted from [44] ©[2010] IEEE.

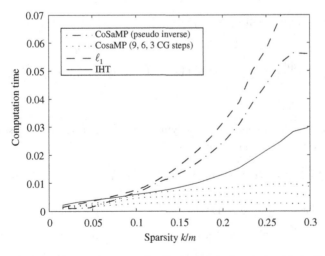

Figure 8.4 Computation time for normalized IHT (solid), ℓ_1 solution (dashed), CoSaMP using the pseudo-inverse (dash-dotted), and CoSaMP using (from top to bottom) 9, 6, or 3 conjugate gradient iterations (dotted). Adapted from [44] ©[2010] IEEE.

We now continue by bounding the first term on the right using the non-symmetric RIP

$$\|x_k - \widehat{x}^{[i+1]}\|_2^2 \leq \frac{1}{\alpha_{2k}} \|A(x_k - \widehat{x}^{[i+1]})\|_2^2, \tag{8.42}$$

which in turn can be bounded (using the definition $\widetilde{e} = A(x - x_k) + e$) by

$$\begin{aligned}\|A(x_k - \widehat{x}^{[i+1]})\|_2^2 &= \|y - A\widehat{x}^{[i+1]}\|_2^2 + \|\widetilde{e}\|_2^2 - 2\langle \widetilde{e}, (y - A\widehat{x}^{[i+1]})\rangle \\ &\leq \|y - A\widehat{x}^{[i+1]}\|_2^2 + \|\widetilde{e}\|_2^2 + \|\widetilde{e}\|_2^2 + \|y - A\widehat{x}^{[i+1]}\|_2^2 \\ &= 2\|y - A\widehat{x}^{[i+1]}\|_2^2 + 2\|\widetilde{e}\|_2^2, \end{aligned} \tag{8.43}$$

where the last inequality follows from

$$\begin{aligned}-2\langle \widetilde{e}, (y - A\widehat{x}^{[i+1]})\rangle &= -\|\widetilde{e} + (y - A\widehat{x}^{[i+1]})\|_2^2 + \|\widetilde{e}\|_2^2 + \|(y - A\widehat{x}^{[i+1]})\|_2^2 \\ &\leq \|\widetilde{e}\|_2^2 + \|(y - A\widehat{x}^{[i+1]})\|_2^2. \end{aligned} \tag{8.44}$$

To bound the first term in (8.43), we use the bound

$$\|y - A\widehat{x}^{[i+1]}\|_2^2 \leq (\mu^{-1} - \alpha_{2k})\|(x_k - \widehat{x}^{[i]})\|_2^2 + \|\widetilde{e}\|_2^2 + (\beta_{2k} - \mu^{-1})\|\widehat{x}^{[i+1]} - \widehat{x}^{[i]}\|_2^2, \tag{8.45}$$

which follows from (using $g^{[i]} = 2A^T(y - A\widehat{x}^{[i]})$)

$$\begin{aligned}&\|y - A\widehat{x}^{[i+1]}\|_2^2 - \|y - A\widehat{x}^{[i]}\|_2^2 \\ &\leq -\langle (x_k - \widehat{x}^{[i]}), g^{[i]}\rangle + \mu^{-1}\|x_k - \widehat{x}^{[i]}\|_2^2 + (\beta_{2k} - \mu^{-1})\|\widehat{x}^{[i+1]} - \widehat{x}^{[i]}\|_2^2 \\ &\leq -\langle (x_k - \widehat{x}^{[i]}), g^{[i]}\rangle + \|A(x_k - \widehat{x}^{[i]})\|_2^2 \\ &\quad + (\mu^{-1} - \alpha_{2k})\|x_k - \widehat{x}^{[i]}\|_2^2 + (\beta_{2k} - \mu^{-1})\|\widehat{x}^{[i+1]} - \widehat{x}^{[i]}\|_2^2 \\ &= \|\widetilde{e}\|_2^2 - \|y - A\widehat{x}^{[i]}\|_2^2 + (\mu^{-1} - \alpha_{2k})\|(x_k - \widehat{x}^{[i]})\|_2^2 + (\beta_{2k} - \mu^{-1})\|\widehat{x}^{[i+1]} - \widehat{x}^{[i]}\|_2^2\end{aligned} \tag{8.46}$$

where the second inequality uses the non-symmetric RIP and where the first inequality is due to the following lemma proven in [43]

LEMMA 8.1 *If* $\widehat{x}^{[i+1]} = H_k(\widehat{x}^{[i]} + \mu A^T(y - A\widehat{x}^{[i]}))$, *then*

$$\begin{aligned}&\|y - A\widehat{x}^{[i+1]}\|_2^2 - \|y - A\widehat{x}^{[i]}\|_2^2 \\ &\leq -\langle (x_k - \widehat{x}^{[i]}), g^{[i]}\rangle + \mu^{-1}\|x_k - \widehat{x}^{[i]}\|_2^2 + (\beta_{2k} - \mu^{-1})\|\widehat{x}^{[i+1]} - \widehat{x}^{[i]}\|_2^2. \end{aligned} \tag{8.47}$$

Combining inequalities (8.41), (8.42), and (8.45) thus shows that if $\beta_{2k} \leq \mu^{-1}$, then

$$\|x_k - \widehat{x}^{[i+1]}\|_2^2 \leq 2\left(\frac{1}{\mu\alpha_{2k}} - 1\right)\|(x_k - \widehat{x}^{[i]})\|_2^2 + \frac{4}{\alpha_{2k}}\|\widetilde{e}\|_2^2. \tag{8.48}$$

Therefore, the condition $2(\frac{1}{\mu\alpha_{2k}} - 1) < 1$ implies that

$$\|x_k - x^{[i]}\|_2^2 \leq \left(2\left(\frac{1}{\mu\alpha_{2k}} - 1\right)\right)^i \|x_k\|_2^2 + c\|\widetilde{e}\|_2^2, \quad (8.49)$$

where $c \leq \frac{4}{3\alpha_{2k} - 2\mu^{-1}}$. The theorem then follows from the bound

$$\|x - x^{[i]}\|_2 \leq \sqrt{\left(\frac{2}{\mu\alpha_{2k}} - 2\right)^i \|x_k\|_2^2 + c\|\widetilde{e}\|_2^2} + \|x_k - x\|_2$$

$$\leq \left(\frac{2}{\mu\alpha_{2k}} - 2\right)^{i/2} \|x_k\|_2 + c^{0.5}\|\widetilde{e}\|_2 + \|x_k - x\|_2. \quad (8.50)$$

Thus after $i^\star = \left\lceil 2\frac{\log(\|\widetilde{e}\|_2/\|x_k\|_2)}{\log(2/(\mu\alpha_{2k})-2)}\right\rceil$ iterations we have

$$\|x - x^{[i^\star]}\|_2 \leq (c^{0.5} + 1)\|\widetilde{e}\|_2 + \|x_k - x\|_2. \quad (8.51)$$

Alternatively, if $\beta_{2k} > \mu^{-1}$, then

$$\|x_k - \widehat{x}^{[i+1]}\|_2^2 \leq 2\left(\frac{1}{\mu\alpha_{2k}} - 1\right)\|(x_k - \widehat{x}^{[i]})\|_2^2 + \frac{2}{\alpha_{2k}}(\beta_{2k} - \mu^{-1})\|\widehat{x}^{[i+1]} - \widehat{x}^{[i]}\|_2^2$$
$$+ \frac{4}{\alpha_{2k}}\|\widetilde{e}\|_2^2$$
$$\leq 2\left(\frac{1}{\mu\alpha_{2k}} - 1\right)\|(x_k - \widehat{x}^{[i]})\|_2^2 + \frac{4}{\alpha_{2k}}(\beta_{2k} - \mu^{-1})\|x_k - \widehat{x}^{[i]}\|_2^2$$
$$+ \frac{4}{\alpha_{2k}}(\beta_{2k} - \mu^{-1})\|x_k - \widehat{x}^{[i+1]}\|_2^2 + \frac{4}{\alpha_{2k}}\|\widetilde{e}\|_2^2, \quad (8.52)$$

so that

$$\|x_k - \widehat{x}^{[i+1]}\|_2^2 \leq \frac{4\frac{\beta_{2k}}{\alpha_{2k}} - \frac{2}{\mu\alpha_{2k}} - 2}{1 - 4\frac{\beta_{2k}}{\alpha_{2k}} + \frac{4}{\mu\alpha_{2k}}}\|(x_k - \widehat{x}^{[i]})\|_2^2 + \frac{4}{\alpha_{2k}}\|\widetilde{e}\|_2^2. \quad (8.53)$$

The second part of the theorem follows from this using the same reasoning used to derive the first part from (8.48). Finally, $\|\widetilde{e}\|_2$ is bounded by

LEMMA 8.2 (Needell and Tropp, lemma 6.1 in [34]) *If A satisfies the RIP of order 2k, then*

$$\|\widetilde{e}\|_2 \leq \sqrt{\beta_{2k}}\|y - x_k\|_2 + \sqrt{\beta_{2k}}\frac{\|x - x_k\|_1}{\sqrt{k}} + \|e\|_2. \quad (8.54)$$

□

8.4 Generalizations of greedy algorithms to structured models

Apart from their ease of implementation and computational advantages, greedy algorithms have another very important feature that distinguishes them from methods based on convex relaxation. They are easily adapted to more complex signal models that go beyond sparsity. For example, wavelet representations of natural images are known to exhibit tree-structures, so that it is natural in CS imaging, to not only look for sparse wavelet representations, but to further restrict the search to sparse *and tree-structured* representations. The set of all tree-structured sparse signals is much smaller than the set of all sparse signals, thus this additional information should allow us to reconstruct images using far fewer measurements.

For many structured sparse problems such as the tree sparse recovery problem, convex optimization approaches are difficult to design. It is indeed not clear what the correct convex cost function should be or even if there exist convex cost functions that offer better performance than standard ℓ_1-methods not exploiting the structure. Greedy algorithms, on the other hand, can often be adapted much more easily to incorporate and exploit additional structures. How this is done and what the benefits of such an approach can be will be the topic of this section.

8.4.1 The union of subspaces model

Most structures that have been used in applications of CS can be based on *union of subspaces* (UoS) models and we here concentrate on this important class. However, some of the ideas developed here for UoS models can also be derived for other low-dimensional signal models (including for example manifold models as discussed in Chapter 1) [43]. UoS models were first studied by Lu and Do [46] with additional theoretical developments presented by Blumensath and Davies [47]. Explicit recovery results and practical algorithms for block-sparse signal models were first derived in [48] whilst results for *structured sparse models*, were derived by Baraniuk et al. [49]. Blumensath [43] derived recovery results for an IHT-type algorithm adapted to the most general UoS setting.

The model

In the sparse signal model, the signal is assumed to live in, or close to, the union of several k-dimensional subspaces, where each of these subspaces is one of the k-dimensional canonical subspaces generated by k canonical vectors of the signal space $\mathbb{R}^n$. The UoS model generalizes this notion.

Chapter 1 already stated several examples of UoS models. We here give a definition in a quite general setting. Assume x lives in a Hilbert space $\mathcal{H}$ (with norm $\|\cdot\|_{\mathcal{H}}$) and, furthermore, assume that x lies in, or close to, a union of linear subspaces $\mathcal{U} \subset \mathcal{H}$, defined as

$$\mathcal{U} = \bigcup_{j \in I} \mathcal{S}_j, \tag{8.55}$$

where the $\mathcal{S}_j \subset \mathcal{H}$ are arbitrary closed subspaces of $\mathcal{H}$. In the most general setting, $\mathcal{H}$ can be an infinite-dimensional Hilbert space and $\mathcal{U}$ might possibly be an uncountably infinite union of infinite-dimensional subspaces. In many practical situations, however, the subspaces $\mathcal{S}_j$ have finite dimension and their number is also finite (in which case we write their number as $L := |I|$). In UoS models with finite-dimensional subspaces, the largest dimension of the subspaces plays the same role as the sparsity in standard CS. We therefore use a similar notation and define $k := \sup \dim \mathcal{S}_j$.

An important subset of UoS models are structured sparse signals. These are signals that are k-sparse, but whose support is further restricted. These models were introduced by Baraniuk et al. [49], where they were called "model sparse." The key difference between sparse and structured sparse signal models is that the number of included sparse subspaces in the structured case can be (and should be to see any benefit) much smaller than the total number of k-sparse subspaces. Intuitively this means that it should be easier to identify the subspace where the signal lives. This in turn suggests that the signal can be recovered from fewer measurements. Important examples of structured sparse models include tree-sparse models and block-sparse models which will be discussed in more detail below.

Examples of UoS models

We now describe a selection of important UoS models that have been considered for CS. As pointed out above, the standard sparse model, where x is assumed to have no more than k nonzero elements, is a canonical UoS model.

Another UoS model is defined by signals that are sparse in a dictionary Φ. A dictionary in $\mathbb{R}^n$ is a matrix $\Phi \in \mathbb{R}^{n \times d}$, whose column vectors $\{\phi_j, j \in \{1,\ldots,d\}\}$ span $\mathbb{R}^n$. Given a dictionary Φ (which may be a basis when $d = n$), the subspaces spanned by all combinations of k columns of Φ constitute a UoS model. In other words, for each x in such a model, there is a k-sparse vector c, such that $x = \Phi c$. If Φ is a basis, then the number of such subspaces is $\binom{n}{k}$, whilst if Φ is an overcomplete dictionary (i.e. $d > n$), then the number of k-sparse subspaces is at most $\binom{d}{k}$.

Block-sparse signals form another UoS model in $\mathbb{R}^n$ and are an important example of structured sparse models. In the block-sparse model, x is partitioned into blocks. Let $B_j \subset \{1,\ldots,n\}, j \in \{1,\ldots,J\}$ be the set of indices in block j. A signal that is k-block sparse is then defined as any x whose support is contained in no more than k different sets B_j, that is

$$\mathrm{supp}(x) \subset \bigcup_{j \in \mathcal{J}: \mathcal{J} \subset \{1,2,\ldots,J\}, |\mathcal{J}| \leq k} B_j. \tag{8.56}$$

Typically, it is assumed that the B_j's are non-overlapping ($B_j \cap B_l = \emptyset$ if $j \neq l$).

Other structured sparse models are so called tree-sparse models, in which the nonzero coefficients of x have an underlying tree structure. This type of model typically arises in wavelet decompositions of images [50, 51, 52] and is exploited in state-of-the-art image compression methods [53, 54]. An important special case of tree-sparse models, which are typically used to model wavelet coefficients, are rooted-tree-sparse models, where the sparse subtrees have to contain the root of the wavelet tree.

Alternatively, instead of assuming that x is structured sparse, we can also define a UoS model in which $x = \Phi c$, where Φ is again a dictionary and where c is block- or tree-sparse. Note that, as noted by Eldar and Mishali [55], all finite-dimensional UoS models can be written in the form $x = \Phi c$, where c is block-sparse for some Φ. Unfortunately, reformulating a UoS problem as a block-sparse problem is not always optimal, as the distance between points $x_1 = \Phi c_1$ and $x_2 = \Phi c_2$ is not generally the same as the distance between c_1 and c_2. Due to this, stable recovery in the reformulated block-sparse model might become impossible even though the UoS recovery problem itself is stable. A similar example where the reformulation of a model into block-sparse form is suboptimal is the MMV recovery problem introduced in Chapter 1, where the performance of block-sparse recovery methods can be suboptimal [56] as it ignores the fact that all vectors are sampled by the same operator. We discuss this in more detail in Section 8.4.6.

Other examples of UoS models (see Chapter 1) include low rank matrix models, which consist of matrices X of a given size with rank-k or less, continuous multiband models (also called analog shift-invariant subspace models), and finite rate of innovation signals, which are all UoS models where the number of subspaces is infinite.

Projecting onto UoS models

As for the k-sparse model, real-world signals rarely follow a UoS model exactly. In this case, CS reconstruction algorithms for structured models will only be able to recover an approximation of the actual signal in the UoS model. In the best case, that approximation would be the UoS signal that is closest to the actual signal. Using the norm of $\mathcal{H}$ $\|\cdot\|_{\mathcal{H}}$ to measure this distance, this approximation is the *projection* of the signal onto the UoS model.

This projection can be defined more generally for a union $\mathcal{U}$ of closed subspaces as any map $P_{\mathcal{U}}$ from $\mathcal{H}$ to $\mathcal{U}$ satisfying

$$\forall x \in \mathcal{H}, P_{\mathcal{U}}(x) \in \mathcal{U}, \text{ and } \|x - P_{\mathcal{U}}(x)\|_{\mathcal{H}} = \inf_{x' \in \mathcal{U}} \|x - x'\|_{\mathcal{H}}. \tag{8.57}$$

Note that, when dealing with infinite unions, the existence of the projection defined in (8.57) is not necessarily guaranteed for certain $\mathcal{U}$, even when the subspaces are closed. We will here assume that such a projection exists, though this is not a very stringent requirement, firstly, because for finite unions this is guaranteed whenever the subspaces are closed and secondly, because for infinite unions this restriction can be removed by using a slight modification of the definition of the projection [43].

8.4.2 Sampling and reconstructing union of subspaces signals

As in standard CS, in the UoS setting, sampling can be done using a linear sampling system.

$$y = Ax + e, \tag{8.58}$$

where $x \in \mathcal{H}$ and where y and e are elements of a Hilbert space $\mathcal{L}$ (with norm $\|\cdot\|_{\mathcal{L}}$); e models observation errors and A is a linear map from $\mathcal{H}$ to $\mathcal{L}$. As for sparse signal models, when x is assumed to lie close to one of the subspaces of a UoS model, efficient recovery methods are required that can exploit the UoS structure. Ideally we would pose an optimization problem of the form

$$x_{opt} = \operatorname*{argmin}_{\widetilde{x}:\, \widetilde{x} \in \mathcal{U}} \|y - A\widetilde{x}\|_{\mathcal{L}}. \tag{8.59}$$

However, even if the above minimizer exists, as in standard sparse CS, this problem can only be solved explicitly for the simplest of UoS models. Instead, one has to turn to approximations to this problem and again two approaches can be distinguished, methods based on convex optimization and greedy algorithms. For certain structured sparse models, such as block-sparse models, convex optimization methods have been derived (see for example [55]). Our focus is however on greedy algorithms which have the advantage that they are applicable to all UoS models, as long as we can efficiently calculate the projection (8.57).

Extensions of greedy pursuit-type algorithms to tree models have been proposed by La and Do [57] and Duarte *et al.* [58]. While these extensions seem to perform well in practice, there are unfortunately no strong theoretical guarantees supporting them. Thresholding-type algorithms on the other hand are more widely applicable and have been easier to approach from a theoretical point of view. In particular, the greedy thresholding algorithms presented in Section 8.3 can be easily modified to take into account more general models. The key observation is that these algorithms handle the sparse signal model by *projecting* at each iteration on the set of k-sparse (or $2k$-sparse) signals so that for UoS models, all one has to do is to replace this thresholding operation with the projection (8.57). This idea was first introduced by Baraniuk *et al.* [49] for the use with structured sparse models and later by Blumensath [43] for general UoS models. This extension of the IHT algorithm to UoS models was called the *Projected Landweber Algorithm* (PLA) by Blumensath [43] and is summarized in Algorithm 8.9. The only difference to Algorithm 8.5 is that we have replaced the hard thresholding operator H_k by the UoS model projector $P_\mathcal{U}$. Note that the PLA could use a similar strategy to determine μ adaptively as has been suggested for the IHT method.

Algorithm 8.9 Projected Landweber Algorithm (PLA)

Input: y, A, and μ
Initialize: $x^{[0]} = 0$
for $i = 0$, $i := i + 1$, until stopping criterion is met **do**
$\quad \widehat{x}^{[i+1]} = P_\mathcal{U}(\widehat{x}^{[i]} + \mu A^*(y - A\widehat{x}^{[i]}))$, where A^* is the adjoint of A
end for
Output: $\widehat{x}^{[i]}$

The PLA extends the IHT algorithm to general UoS methods. For structured sparse models, a similar approach can also be used to extend the CoSaMP (or Subspace Pursuit) algorithms. Algorithm 8.10 describes a modified version of the standard CoSaMP algorithm applicable to structured sparse models. For these models, the only required modification is the replacement of the hard thresholding step by the UoS projection (8.57).

Examples of UoS projection operators

Unlike the standard IHT and CoSaMP algorithms, the complexity of algorithms based on the projection onto more complicated UoS models is often not dominated by the

Algorithm 8.10 CoSaMP algorithm for structured sparse models

Input: y, A, and k
Initialize: $T^{[0]} = \text{supp}(P_{\mathcal{U}}(A^T y))$, $x^{[0]} = 0$
for $i = 0$, $i := i+1$, until stopping criterion is met **do**
$\quad \mathbf{g} = A^T(y - Ax^n)$
$\quad T^{[i+0.5]} = T^{[i]} \cup \text{supp}(P_{\mathcal{U}^2}(\mathbf{g}))$
$\quad x^{[i+0.5]}_{T^{[i+0.5]}} = A^{\dagger}_{T^{[i+0.5]}} y, x^{[i+0.5]}_{\overline{T^{[i+0.5]}}} = 0$
$\quad T^{[i+1]} \text{supp}(P_{\mathcal{U}}(x^{[i+0.5]}))$
$\quad x^{[i+1]}_{T^{[i+1]}} = x^{[i+0.5]}_{T^{[i+1]}}, x^{[i+1]}_{\overline{T^{[i+1]}}} = 0$
end for
Output: $r^{[i]}$ and $\widehat{x}^{[i]}$

---

matrix vector multiplications (possibly implemented via a fast transform) but by the complexity of the projection itself. While the projection on the set of k-sparse signals is a simple sort and threshold operation, the projection on an arbitrary UoS is generally a nontrivial operation which may in the worst case require an exhaustive search through the subspaces. In practice, unless the number of subspaces is small, such an exhaustive search is prohibitive. It is thus very important for applications that there exists an efficient algorithm to calculate the projection (8.57) for the UoS model used. Fortunately, for many UoS models of interest, such efficient projection methods exist.

In particular, for block-sparse models, tree-sparse models, low rank matrix models, and continuous multiband signal models, computationally efficient methods are available.

For example, assume $\mathcal{U}$ is a block-sparse model, which is defined by clustering the indices $\{1, 2, \ldots, n\}$ into disjoint sets B_j. The projection on $\mathcal{U}$ is then straightforward and directly extends the sort and threshold principle for k-sparse vectors. Given a signal $x \in \mathbb{R}^n$, we can compute the energy of the signal in block B_j as $\sum_{i \in B_j} |x_i|^2$. The projection then sets all elements in x to zero apart from those elements whose indices are in the k-blocks B_j with the highest energy values, which are kept unaltered.

Another important example are rooted tree-sparse models, where the projection onto $\mathcal{U}$ can be done efficiently using an algorithm known as *condensing sort and select algorithm* (CSSA). This method provides a solution with a computational complexity of $\mathcal{O}(n \log n)$ [59].

The projection onto the closest rank-k matrix is simply calculated by computing the singular value decomposition and by setting to zero all but the k largest singular values [60].

The final example, where this projection can be calculated at least conceptually, are continuous multiband signals as defined in Section 8.2.3. A continuous multiband signal is a signal whose Fourier transform is concentrated in a few narrow frequency bands. To define the projection onto these models, it is easier to consider a simplified model where the Fourier transform is partitioned into small blocks and where the Fourier transform of the signal is assumed to lie in a few of these blocks [61, 62]. This differs from the original model in which the location of the nonzero frequency bands is not constrained to lie in pre-defined blocks but can be arbitrary.[11] Importantly, for this simplified model, the projection operator is essentially the same as that for a block-sparse model: compute the energy within each block (by integrating the squared Fourier transform over each frequency band) and only keep the k-blocks with the largest energies.

For other UoS models such as signal models where the signal is sparse in a dictionary, calculating the projection (8.57) is much more difficult. For example, if the UoS contains all $x \in \mathbb{R}^n$ which can be written as $x = \Phi c$, where $c \in \mathbb{R}^d$ is a k-sparse vector and where $\Phi \in \mathbb{R}^{n \times d}$ with $d > n$, then solving the projection onto UoS is a computational problem that is as difficult as solving a standard sparse CS problem, which is computationally demanding even if the dictionary Φ itself satisfies the RIP condition. However, in this case, one might also be able to show that the product $A\Phi$ satisfies the RIP condition [63], in which case it might be better to directly try to solve the problem

$$\min_{\widetilde{c}:\ \|\widetilde{c}\|_0 \leq k} \|y - A\Phi\widetilde{c}\|_2^2. \tag{8.60}$$

8.4.3 Performance guarantees

Replacing the hard thresholding steps by more general projections is conceptually trivial, however, it is not clear if the modified algorithms are able to recover UoS signals and under which conditions this might be possible. As for sparse signals, the first requirement we need for a sampling operator A to be able to sample signals that exactly follow a UoS model, is that A should be *one to one* so that the samples uniquely define a UoS signal

$$\forall x_1, x_2 \in \mathcal{U}, Ax_1 = Ax_2 \implies x_1 = x_2.$$

When $\mathcal{L} = \mathbb{R}^m$ and if $\mathcal{U}$ is the union of countably many subspaces of dimension $k < \infty$ or less, then the set of linear maps A that are one to one for signals in $\mathcal{U}$ is *dense* in the set of all linear maps from $\mathcal{H}$ to $\mathbb{R}^m$ [46] as soon as $m \geq k_2$, with $k_2 = k + k_1$, where k_1 is the dimension of the second largest subspace in $\mathcal{U}$. Moreover, if $\mathcal{L} = \mathbb{R}^m$ and $\mathcal{H} = \mathbb{R}^n$ and if $\mathcal{U}$ is a finite union, then almost all linear maps are *one to one* if $m \geq k_2$, whilst if we additionally assume a smooth measure on each of the subspaces of $\mathcal{U}$, then *almost all* linear maps are one to one for *almost all* elements in $\mathcal{U}$ whenever $m \geq k$ [47].

More generally, a good sampling strategy also needs to be robust to measurement inaccuracies. As for the sparse signal model, this can be quantified by a property similar to the RIP, which we refer to as $\mathcal{U}$-*RIP* [47] and which we here defined quite generally for arbitrary subsets $\mathcal{U} \subset \mathcal{H}$. This property is referred to as "model-based RIP" in [49] and is very closely related to the stable sampling condition of Lu and Do [46].

---

[11] In practice, many recovery algorithms consider such a simplified model as a signal following the general continuous multiband signal model with arbitrary band locations which can always be described with the simplified model at the cost of a doubling of the number of active bands.

DEFINITION 8.1 *For any matrix A and any subset $\mathcal{U} \subset \mathcal{H}$, we can define the $\mathcal{U}$-restricted isometry constants $\alpha_{\mathcal{U}}(A)$ and $\beta_{\mathcal{U}}(A)$ as the tightest constants such that*

$$\alpha_{\mathcal{U}}(A) \leq \frac{\|Ax\|_{\mathcal{L}}^2}{\|x\|_{\mathcal{H}}^2} \leq \beta_{\mathcal{U}}(A) \tag{8.61}$$

holds for all $x \in \mathcal{U}$.

If we define $\mathcal{U}^2 = \{x = x_1 + x_2 : x_1, x_2 \in \mathcal{U}\}$, then $\alpha_{\mathcal{U}^2}(A)$ and $\beta_{\mathcal{U}^2}(A)$ characterize the sampling strategy defined by A for signals in $\mathcal{U}$. In particular, the sampling operation A is one to one for signals $x \in \mathcal{U}$ if and only if $\alpha_{\mathcal{U}^2}(A) > 0$. Also, the interpretation of $\alpha_{\mathcal{U}^2}(A)$ and $\beta_{\mathcal{U}^2}(A)$ in terms of Lipschitz constants of the map $x \mapsto Ax$ for $x \in \mathcal{U}$ (and its inverse) mentioned in Section 8.3.1 is still valid in this more general setting.

Similarly to the k-sparse case for which the recovery guarantees require $2k$-RIP, $3k$-RIP, etc., some theoretical guarantees will require the $\mathcal{U}^2$-RIP, $\mathcal{U}^3$-RIP, etc. The enlarged UoSs $\mathcal{U}^p$ can be defined as Minkowski sums of the UoS defining the signal model and generalize the previous definition of $\mathcal{U}^2$. Given a UoS model $\mathcal{U}$, we thus define for $p > 0$

$$\mathcal{U}^p = \left\{ x = \sum_{j=1}^{p} x^{(j)}, x^{(j)} \in \mathcal{U} \right\}. \tag{8.62}$$

PLA recovery result

The PLA enjoys the following theoretical result which generalizes the result of Theorem 8.5 [43].

THEOREM 8.9 *For arbitrary $x \in \mathcal{H}$, given $y = Ax + e$ where A satisfies the $\mathcal{U}^2$-RIP with $\beta_{\mathcal{U}^2}(A)/\alpha_{\mathcal{U}^2}(A) < 1.5$, then given some arbitrary $\delta > 0$ and a step size μ satisfying $\beta_{\mathcal{U}^2}(A) \leq 1/\mu < 1.5\alpha_{\mathcal{U}^2}(A)$, after*

$$i^\star = \left\lceil 2 \frac{\log(\delta \frac{\|\tilde{e}\|_{\mathcal{L}}}{\|P_{\mathcal{U}}(x)\|_{\mathcal{H}}})}{\log(2/(\mu\alpha) - 2)} \right\rceil \tag{8.63}$$

iterations, the PLA calculates a solution $\hat{x}$ satisfying

$$\|x - \hat{x}\|_{\mathcal{H}} \leq (\sqrt{c} + \delta) \|\tilde{e}\|_{\mathcal{L}} + \|P_{\mathcal{U}}(x) - x\|_{\mathcal{H}} \tag{8.64}$$

where $c \leq \frac{4}{3\alpha_{\mathcal{U}^2}(A) - 2\mu}$ and $\tilde{e} = A(x - P_{\mathcal{U}}(x)) + e$.

It is important to stress that, for UoS models, the above result is near optimal, that is, even if we were able to solve the problem (8.59) exactly, the estimate x_{opt} would have a worst case error bound that also depends linearly on $\|\tilde{e}\|_{\mathcal{L}} + \|P_{\mathcal{U}}(x) - x\|_{\mathcal{H}}$ [43].

Improved bounds for structured sparse models

In the k-sparse signal case in $\mathbb{R}^n$, if A satisfies the condition on the RIP of order $2k$, then the size of the error $\|\widetilde{e}\|_2 = \|e + A(x - x_k)\|_2$ can be bounded by (see Lemma 8.2)

$$\|\widetilde{e}\|_2 \leq \sqrt{\beta_{2k}} \|y - x_k\|_2 + \sqrt{\beta_{2k}} \frac{\|x - x_k\|_1}{\sqrt{k}} + \|e\|_2. \tag{8.65}$$

Whilst in the worst case, $\|e + A(x - x_k)\|_2^2 = \|e\|_2^2 + \|A\|_{2,2}^2 \|(x - x_k)\|_2^2$, if we assume that the ordered coefficients of x decay rapidly, i.e. that $\|x - x_k\|_1$ is small, and if A satisfies the RIP condition, then Lemma 8.2 gives a much less pessimistic error bound. In order to derive a similar improved bound for UoS models, we similarly need to impose additional constraints. Unfortunately, no general conditions and bounds are available for the most general UoS model, however, for structured sparse models, conditions have been proposed by Baraniuk et al. [49] which allow similar tighter bounds to be derived.

The first condition to be imposed is the *nested approximation property* (NAP). This property requires structured sparse models to have a nesting property, which requires that the structured sparse model has to be defined *for arbitrary* $k \in \{1, \ldots, n\}$, so that we have a family of UoS models $\mathcal{U}_k$, with one model for each k. When talking about these models, we will write $\mathcal{U}$ (without subscript) to refer to the whole family of models. To have the nesting property, the models $\mathcal{U}_k$ of a family $\mathcal{U}$ have to generate *nested approximations*.

DEFINITION 8.2 *A family of models* $\mathcal{U}_k$, $k \in \{1, 2, \ldots, n\}$ *has the* nested approximation property *if*

$$\forall k < k', \forall x \in \mathbb{R}^n, \operatorname{supp}(P_{\mathcal{U}_k}(x)) \subset \operatorname{supp}(P_{\mathcal{U}_{k'}}(x)). \tag{8.66}$$

In terms of UoS, this means that one of the k-sparse subspaces containing $P_{\mathcal{U}_k}(x)$ must be included in at least one of the k'-sparse subspaces containing $P_{\mathcal{U}_{k'}}(x)$.

Given a family of models with the NAP, we will also need to consider the *residual subspaces* of size k.

DEFINITION 8.3 *For a given family* $\mathcal{U}$ *of UoS models, the residual subspaces of size* k *are defined as*

$$\mathcal{R}_{j,k} = \{x \in \mathbb{R}^n : \exists x' \in \mathbb{R}^n, x = P_{\mathcal{U}_{jk}}(x') - P_{\mathcal{U}_{(j-1)k}}(x')\}, \tag{8.67}$$

for $j \in \{1, \ldots, \lceil n/k \rceil\}$.

The residual subspaces allow us to partition the support of any signal $x \in \mathbb{R}^n$ into sets no larger than k. Indeed, x can be written as $x = \sum_{j=1}^{\lceil n/k \rceil} x_{T_j}$ with $x_{T_j} \in \mathcal{R}_{j,k}$.

This machinery allows us to define a notion similar to that of compressible signals discussed in Chapter 1. In analogy with ℓ_p-compressible signals which are characterized by the decay of $\sigma_k(x)_p = \|x - x_k\|_p$ as k increases, these so-called $\mathcal{U}$ compressible signals are characterized by a fast decay of $\|P_{\mathcal{U}_k}(x) - x\|_2$, which implies a fast decay of the energies of the x_{T_j} with increasing j.

DEFINITION 8.4 *Given a family of structured sparse UoS models $\{\mathcal{U}_k\}_{k\in\{1,\ldots,n\}}$, a signal $x \in \mathbb{R}^n$ is s-model-compressible if*

$$\forall k \in \{1,\ldots,n\}, \|x - P_{\mathcal{U}_k} x\|_2 \leq c k^{-1/s}, \tag{8.68}$$

for some $c < \infty$. Moreover, the smallest value of c for which (8.68) holds for x and s will be referred to as s-model-compressibility constant $c_s(x)$.

An s-compressible signal in a structured sparse model that satisfies the NAP is characterized by a fast decay of its coefficients in the successive residual subspaces $\mathcal{R}_{j,k}$ for increasing j. When sampling such a signal, the sensing matrix A may amplify some of the residual components $x_{T_j} \in \mathcal{R}_{j,k}$ of x more than others. Intuitively, as long as this amplification is compensated by the decay of the energies of the x_{T_j}, the signal can still be efficiently sampled by A. This behavior is controlled by a second property called the *restricted amplification property* (RAmP), which controls how much the sensing matrix can amplify the x_{T_j}.

DEFINITION 8.5 *Given a scalable model $\mathcal{U}$ satisfying the nested approximation property, a matrix A has the (ϵ_k, r)-restricted amplification property for the residual subspaces $\mathcal{R}_{j,k}$ of $\mathcal{U}$ if*

$$\forall j \in \{1,\ldots,\lceil n/k \rceil\}, \forall x \in \mathcal{R}_{j,k} \|Ax\|_2^2 \leq (1+\epsilon_k)j^{2r}\|x\|_2^2). \tag{8.69}$$

Note that for standard sparse models, the RIP automatically gives a RAmP for the nested sparse subspaces, so that no additional property is required for CS with the standard sparse signal model. Importantly, s-compressible NAP sparse models sampled with a map A that satisfy the $(\epsilon_k, s-1)$-RAmP, satisfy a lemma that is analogous to Lemma 8.2 of the standard sparse model [49].

LEMMA 8.3 *Any s-model-compressible signal x sampled with a matrix A with $(\epsilon_k, s-1)$-RAmP satisfies*

$$\|A(x - P_{\mathcal{U}_k}(x))\|_2 \leq \sqrt{1+\epsilon_k} k^{-s} c_s(x) \log \lceil \frac{n}{k} \rceil. \tag{8.70}$$

Through control provided by the additional assumptions, Lemma 8.3 provides a tighter bound on $\|\widetilde{e}\|_2$ in Theorem 8.9. This leads directly to an improved theoretical guarantee for PLA recovery of s-model-compressible signals measured with a system that satisfies the RAmP of the lemma. With the tighter bound on $\|A(x - P_{\mathcal{U}_k}(x))\|_2$, the PLA is thus guaranteed to achieve a better reconstruction in this more constrained setting.

A similar recovery guarantee has also been proved for the model-based CoSaMP algorithm [49].

THEOREM 8.10 *Given $y = Ax + e$, where A has $\mathcal{U}_k^4$-RIP constants satisfying*[12] *$\beta_{\mathcal{U}_k^4}/\alpha_{\mathcal{U}_k^4} \leq 1.22$, and verifies the $(\epsilon_k, s-1)$-RAmP, then after i iterations the signal*

---

[12] The result in [49] is actually based on a symmetric $\mathcal{U}_k^4$-RIP while an asymmetric version is considered here. In the symmetric $\mathcal{U}_k^4$-RIP, the constants $\alpha_{\mathcal{U}_k^4}$ and $\beta_{\mathcal{U}_k^4}$ are replaced by $1 - \delta_{\mathcal{U}_k^4}$ and $1 + \delta_{\mathcal{U}_k^4}$

estimate $\widehat{x}$ satisfies

$$\|x - \widehat{x}\|_2 \leq 2^{-i}\|x\|_2 + 35\left(\|e\|_2 + c_s(x)k^{-s}\left(1 + \log\left\lceil\frac{n}{k}\right\rceil\right)\right), \quad (8.71)$$

where $c_s(x)$ is the s-model-compressibility constant of x defined in Definition 8.4.

8.4.4 When do the recovery conditions hold?

The strong recovery bounds for UoS as well as for s-model-compressible signals stated in the previous section require A to satisfy the $\mathcal{U}^2$-RIP as well as the $\mathcal{U}_k^4$-RIP and $(\epsilon_k, s-1)$-RAmP conditions respectively. What sampling systems satisfy these conditions? How do we design these systems and how do we check if a system satisfies these conditions? As with standard RIP, checking a linear map A for these conditions is computationally infeasible, however, it is again possible to show that certain random constructions of sampling systems will satisfy the appropriate conditions with high probability.

A sufficient condition for the existence of matrices with given $\mathcal{U}$-RIP

Considering a *finite* union of L *finite*-dimensional subspaces $\mathcal{U}$, each of dimension less than k, it has been shown [47] that i.i.d. sub-gaussian $m \times n$ matrices will satisfy the $\mathcal{U}$-RIP with overwhelming probability as soon as the number of measurements is large enough.

More precisely, for any $t > 0$ and $0 < \delta < 1$, then i.i.d. sub-gaussian random matrices $A \in \mathbb{R}^{m \times n}$ with

$$m \geq \frac{1}{c(\delta/6)}\left(2\log(L) + 2k\log\left(\frac{36}{\delta}\right) + t\right) \quad (8.72)$$

have $\mathcal{U}$-RIP constants $\alpha_{\mathcal{U}}(A) \geq 1 - \delta$ and $\beta_{\mathcal{U}}(A) \leq 1 + \delta$ with probability at least $1 - e^{-t}$. The value of $c(\delta)$ in (8.72) depends on the specific sub-gaussian distribution of the values in A. For i.i.d. Gaussian and Bernoulli $\pm 1/\sqrt{n}$ matrices, $c(\delta)$ is given by $c(\delta) = \delta^2/4 - \delta^3/6$.

In the traditional CS setting, the number of k-sparse subspaces is given by $L = \binom{n}{k} \approx (ne/k)^k$ and the above sufficient condition boils down to the standard CS requirement of $\mathcal{O}(k\log(n/k))$ measurements. In structured sparse models $\mathcal{U}$ with sparsity k, the number of subspaces L can be significantly smaller than in the traditional CS case. This suggests that acceptable $\mathcal{U}$-RIP constants can be obtained for matrices with fewer rows, which in turn implies that signals may be recovered from fewer measurements. It is also worth noticing that the required number of measurements (8.72) does not directly depend on the dimension of the ambient space n unlike in the traditional CS setting.

For the theoretical guarantees presented above, the sensing matrix A will need to verify the $\mathcal{U}^2$-RIP, $\mathcal{U}^3$-RIP, or $\mathcal{U}^4$-RIP. These unions contain more subspaces than $\mathcal{U}$, $\binom{L}{p}$ for $\mathcal{U}^p$ which scales as L^p for small p. Therefore, for a UoS model $\mathcal{U}$ with $L < \binom{n}{k}$ subspaces, the difference in the number of subspaces in $\mathcal{U}^p$ and the related number of subspaces in

respectively. In terms of the symmetric $\mathcal{U}_k^4$-RIP, the condition in [49] is $\delta_{\mathcal{U}_k^4} \leq 0.1$ (for some appropriately rescaled A), which is guaranteed to hold true when $\beta_{\mathcal{U}_k^4}/\alpha_{\mathcal{U}_k^4} \leq 1.22$.

a pk-sparse model ($\approx L^p - \binom{n}{k}^p$) is thus significantly larger than the difference between the original models $L - \binom{n}{k}$.

Whilst the above result covers all finite unions of finite-dimensional subspaces, similar results can also be derived for other UoS models where infinitely many and/or infinite-dimensional subspaces are considered [43]. For example, Recht *et al.* [64] have shown that randomly constructed linear maps satisfy $\mathcal{U}^2$-RIP on all rank-k matrices.

THEOREM 8.11 *Let $A \in \mathbb{R}^{m \times n_1 n_2}$ be a matrix with appropriately scaled i.i.d. Gaussian entries. Define the linear map $PX_1 = Ax_1$ where x_1 is the vectorized version of matrix X_1, then with probability $1 - e^{-c_1 m}$,*

$$(1-\delta)\|X_1 - X_2\|_F \leq \|P(X_1 - X_2)\|_2 \leq (1+\delta)\|X_1 - X_2\|_F \quad (8.73)$$

for all rank-k matrices $X_1 \in \mathbb{R}^{n_1 \times n_2}$ and $X_2 \in \mathbb{R}^{n_1 \times n_2}$, whenever $m \geq c_0 k(n_1 + n_2)\log(n_1 n_2)$, where c_1 and c_0 are constants depending on δ only.

As shown by Mishali and Eldar [62] and Blumensath [43], similar results also hold for the sampling strategy proposed in [62] for the analog shift-invariant subspace model. Here a real-valued time series $x(t)$ whose Fourier transform $\mathcal{X}(f)$ has limited bandwidth B_n is mapped into a real-valued time series $y(t)$ with bandwidth B_m and Fourier transform $\mathcal{Y}(f)$. This mapping is done by partitioning the maximal support of $\mathcal{X}(f)$ into n equal size blocks $\mathcal{X}(S_j)$ and the maximal support of $\mathcal{Y}(f)$ into m blocks of the same size $\mathcal{Y}(\widetilde{S}_i)$. The map is then defined by an $m \times n$ matrix A as

$$\mathcal{Y}(\widetilde{S}_i) = \sum_{j=1}^{n} [A]_{i,j} \mathcal{X}(S_j). \quad (8.74)$$

THEOREM 8.12 *Let $\mathcal{U} \subset L_{\mathbb{C}}^2([0, B_n])$ be the subset of the set of square integrable real-valued functions whose Fourier transform has positive support $S \subset [0, B_n]$, where S is the union of no more than k intervals of width no more than B_k. If the matrix $A \in \mathbb{R}^{m \times n}$ satisfies the RIP as a map from the set of all $2k$-sparse vectors in $\mathbb{R}^n$ into $\mathbb{R}^m$, with constants α_{2k} and β_{2k}, then the map defined by Equation (8.74) satisfies the $\mathcal{U}^2$-RIP with constants α_{2k} and β_{2k}.*

A sufficient condition for the existence of matrices with given RAmP constants

Similarly to the sufficient condition for matrices to have a given $\mathcal{U}$-RIP, one can show that sub-gaussian matrices with a large enough number of rows also satisfy the RAmP with overwhelming probability.

Given a family of models $\mathcal{U}$ satisfying the nested approximation property, a matrix $A \in \mathbb{R}^{m \times n}$ with i.i.d. sub-gaussian entries has the (ϵ_k, r)-RAmP with probability $1 - e^{-t}$ as soon as

$$m \geq \max_{j \in \{1, \ldots, \lceil n/k \rceil\}} \frac{1}{\left(j^r \sqrt{1+\epsilon_k} - 1\right)^2} \left(2k + 4\log\frac{R_j n}{k} + 2t\right), \quad (8.75)$$

where R_j is the number of k-sparse subspaces contained in $\mathcal{R}_{j,k}$. As for the $\mathcal{U}$-RIP, one can show that for the k-sparse model, $R_j = \binom{n}{k}$ and the requirement on m to verify the RAmP is of the order of $k\log(n/k)$ as for the RIP. For a structured sparse model satisfying the NAP, R_j is smaller than $\binom{n}{k}$ which allows one to relax the constraint on the number of measurements m.

To demonstrate the importance of this result, it is worth considering rooted-tree-sparse models. Importantly, there are significantly fewer subspaces in these models than are found in unconstrained sparse models. For these models, it is therefore possible to show [49] that, for certain randomly constructed matrices, both the $\mathcal{U}$-RIP as well as the (ϵ_k, r)-RAmP hold with high probability whenever

$$m = \mathcal{O}(k). \tag{8.76}$$

The number of observations required to sample rooted-tree-compressible signals is thus significantly lower than $m = \mathcal{O}(k\log(n/k))$, which would be the number of samples required if we were to assume sparsity alone.

8.4.5 Empirical comparison

To demonstrate how exploiting additional structure can improve the recovery performance of greedy algorithms, we consider the application of the IHT algorithm and the PLA to signals with a rooted-sparse-tree structure. Tree supports were generated according to a uniform distribution on a k-sparse tree and the nonzero coefficients were i.i.d. normal distributed. As in the simulations presented in Section 8.2.6, $m = 128$, $n = 256$ and for each value of k/m, 1000 measurement matrices A were generated with columns drawn independently and uniformly from the unit sphere. As can be observed in Figure 8.5, the tree-based PLA significantly outperforms the standard IHT algorithm and achieves perfect reconstruction up to $k/m \approx 0.35$.

8.4.6 Rank structure in the MMV problem

We end this section by briefly discussing an additional rank structure that can be exploited in the Multiple Measurement Vector (MMV) sparse recovery problem (here we will only discuss the noiseless MMV case).

Recall from Chapter 1, that in the MMV problem we observe multiple sparse signals, $\{x_i\}_{i=1:l}$, $x_i \in \mathbb{R}^n$, through a common sensing matrix A giving multiple observation vectors $\{y_i\}_{i=1:l}$, $y_i \in \mathbb{R}^m$. In addition we assume that the sparse signals, x_i, share the same underlying support. As always the goal is to recover the unknown signals x_i from the observed data. To this end we can write the equivalent ℓ_0-based optimization as:

$$\hat{X} = \underset{X}{\mathrm{argmin}}\, |\mathrm{supp}(X)| \text{ s.t. } AX = Y, \tag{8.77}$$

where $X = \{x_i\}_{i=1:l}$, $Y = \{y_i\}_{i=1:l}$, and $\mathrm{supp}(X) := \bigcup_i \mathrm{supp}(x_i)$ is the row support of X. This problem is one of the earliest forms of structured sparse approximation to be considered and as we will see has strong links with the field of array signal processing.

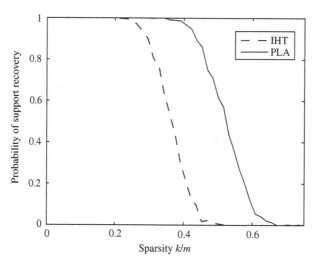

Figure 8.5 Probability of finding the correct tree-support using IHT (dashed) and tree-based PLA (solid) on signals with a tree structured support. The nonzero elements in x are i.i.d. normal distributed and the tree-supports are generated according to a uniform distribution on the trees of size k.

Exact recovery

When $\text{rank}(X) = 1$ the MMV problem reduces to the single measurement vector (SMV) problem as no additional information can be leveraged. However, when $\text{rank}(X) > 1$ the multiple signals do provide additional information that can aid recovery. Geometrically, we are now observing subspaces of data rather than simply vectors.

For the SMV problem it is well known (see Chapter 1) that a necessary and sufficient condition for the measurements $y = Ax$ to uniquely determine each k-sparse vector x is given by $k < \text{spark}(A)/2$ where the *spark* denotes the smallest number of columns of A that are linearly dependent. In contrast, in the MMV problem it can be shown that the k-joint sparse solution to $AX = Y$ is unique if and only if [65, 56]:

$$|\text{supp}(X)| < \frac{\text{spark}(A) - 1 + \text{rank}(Y)}{2} \tag{8.78}$$

(or equivalently replacing $\text{rank}(Y)$ with $\text{rank}(X)$ [56]). This indicates that, at least when tackling (8.77), the problem becomes easier as the rank of X increases. Indeed, in the best-case scenario, when X and Y have maximal rank, $\text{rank}(X) = \text{rank}(Y) = k$, and $\text{spark}(A) > k+1$, only $k+1$ measurements per signal are needed to ensure uniqueness.

Popular approaches to MMV

Despite the fact that the rank of the observation matrix Y can be exploited to improve recovery performance, to date most popular techniques have ignored this fact. Indeed most of the approaches advocated for solving the MMV problem are simple extensions of single measurement vector algorithms that typically replace a vector norm $\|x\|_p$ in the SMV problem with a matrix norm of the form:

$$\|X\|_{p,q} := \left(\sum_i \|X_{i,:}\|_p^q\right)^{1/q}. \tag{8.79}$$

For example, convex optimization approaches [65, 66] would minimize $\|X\|_{1,q}$ for some $1 \leq q \leq \infty$. While MMV greedy pursuit algorithms, such as Simultaneous Orthogonal Matching Pursuit (SOMP) [67, 65, 68], would replace the maximum correlation selection by a selection involving $\|A^T X\|_{q,\infty}$ for some $q \geq 1$. Unfortunately norms of type (8.79) do not 'see' the rank information of the matrix and it can be shown that the worst case performance of MMV algorithms based upon these mixed norms are essentially equivalent to the worst case bounds of the associated SMV problem [56]. In [56] such algorithms are called *rank blind*.

MUSIC

Another surprise in the MMV scenario is that although most popular algorithms are rank blind, when $\mathrm{rank}(Y) = k$ the computational complexity of the recovery problem ceases to be NP-hard and the problem can be solved using an exhaustive search procedure with complexity that scales polynomially with the dimension of the coefficient space. The approach is to apply the MUSIC (MUltiple SIgnal Classification) algorithm [69] popular in array signal processing. Its first appearance as an algorithm for solving a more abstract MMV problem was by Feng and Bresler [70, 71] in the context of multiband (compressive) sampling.

Let us briefly describe the procedure. Since $\mathrm{rank}(Y) = k$ it follows that $\mathrm{range}(Y) = \mathrm{range}(A_\Omega)$, $\Omega = \mathrm{supp}(X)$. We can therefore consider each column of A individually. Each A_j, $j \in \Omega$, must lie within the range of Y. Furthermore, assuming uniqueness, these will be the only columns contained in $\mathrm{range}(Y)$. We therefore form an orthnormal basis for the range of Y, $U = \mathrm{orth}(Y)$ using for example an eigenvalue or singular value decomposition and identify the support of X by:

$$\frac{\|A_j^T U\|_2}{\|A_j\|} = 1, \text{ if and only if } j \in \Omega. \tag{8.80}$$

Then X can be recovered through $X = A_\Omega^\dagger Y$. Note that MUSIC can also efficiently deal with noise through appropriate eigenvalue thresholding [69].

Rank aware pursuits

While MUSIC provides guaranteed recovery for the MMV problem in the maximal rank case there are no performance guarantees if $\mathrm{rank}(X) < k$ and empirically MUSIC does not perform well in this scenario. This motivated a number of researchers [56, 72, 73] to investigate the possibility of an algorithm that in some way interpolates between a classical greedy algorithm for the SMV problem and MUSIC when $\mathrm{rank}(X) = k$.

One possible solution that was proposed in [72, 73] for various greedy algorithms is to adopt a hybrid approach and apply one of a variety of greedy algorithms to select the first $i = k - \mathrm{rank}(Y)$ elements. The remaining components can then be found by applying

MUSIC to an augmented data matrix $[Y, A_T^{[i]}]$ which, under identifiability assumptions, will span the range of A_Ω. A drawback of such techniques is that they require prior knowledge of the signal's sparsity, k in order to know when to apply MUSIC.

A more direct solution can be achieved by careful modification of the ORMP algorithm [56] to make it rank aware, the key ingredient of which is a modified selection step.

Suppose that at the start of the ith iteration we have a selected support set $T^{[i-1]}$. A new column A_j is then chosen based upon the following selection rule:

$$j^{[i]} = \underset{j}{\operatorname{argmax}} \frac{\|A_j^T U^{[i-1]}\|_2}{\|\mathbf{P}_{T^{[i-1]}}^\perp A_j\|_2}. \tag{8.81}$$

In comparison with Equation (8.19) we have simply replaced the inner product $A_j^T r^{[i-1]}$ by the inner product $A_j^T U^{[i-1]}$. Therefore the right-hand side of (8.81) measures the distance of the normalized vector $\mathbf{P}_{T^{[i-1]}}^\perp A_j / \|\mathbf{P}_{T^{[i-1]}}^\perp A_j\|$ from the subspace spanned by $U^{[i-1]}$ thereby accounting for the subspace geometry of $U^{[i-1]}$. The full description of the Rank Aware ORMP algorithm is summarized in Algorithm 8.11. As with the original ORMP efficient implementations are possible based around QR factorization.

Algorithm 8.11 Rank Aware Order Recursive Matching Pursuit (RA-ORMP)

Input: Y, A
Initialize $R^{[0]} = Y, \hat{X}^{[0]} = \mathbf{0}, T^{[0]} = \emptyset$
for $i = 1; i := i + 1$ until stopping criterion is met **do**
  Calculate orthonormal basis for residual: $U^{[i-1]} = \operatorname{orth}(R^{[i-1]})$
  $j^{[i]} = \operatorname{argmax}_{j \notin T^{[i-1]}} \|A_j^T U^{[i-1]}\|_2 / \|\mathbf{P}_{T^{[i-1]}}^\perp A_j\|_2$
  $T^{[i]} = T^{[i-1]} \cup j^{[i]}$
  $\hat{X}^{[i]}_{\{T^{[i]},:\}} = A_{T^{[i]}}^\dagger Y$
  $R^{[i]} = Y - A\hat{X}^{[i]}$
end for

Note that it is crucial to use the ORMP normalized column vectors in order to guarantee repeated correct selection in the maximal rank case. When a similar selection strategy is employed in an MMV OMP algorithm there is a rank degeneration effect – the rank of the residual matrix decreases with each correct selection while the sparsity level typically remains at k. This means the algorithm can and does result in incorrect selections. For further details see [56]. In contrast, like MUSIC, RA-ORMP is guaranteed to correctly identify X when $\operatorname{rank}(X) = k$ and the identifiability conditions are satisfied [56].

The improvements in performance obtainable by the correct use of rank information are illustrated in Figure 8.6. In these simulations dictionaries of size 32×256 were generated with columns A_j drawn uniformly from a unit sphere. The number of measurement vectors was varied between 1 and 32 while the nonzero elements of X were also independently drawn from a unit variance Gaussian distribution. Note that this implies that $\operatorname{rank}(X) = \operatorname{rank}(Y) = \min\{l, k\}$ with probability one.

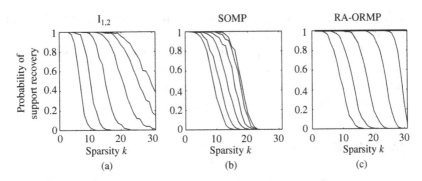

Figure 8.6 The empirical probability of recovery for mixed $\ell_{1,2}$ minimization (a), SOMP (b), and RA-ORMP (c) as a function of sparsity level k. The curves in each plot relate to $l = 1, 2, 4, 16,$ and 32 (from left to right). Note that the right-most curve ($l = 32$) for RA-ORMP is not visible as perfect recovery is achieved for all values of k shown.

The plots indicate that recovery with $\ell_{1,2}$ minimization and SOMP exhibits improved performance when multiple measurement vectors are present, but they do however not approach MUSIC-like performance as the rank of X increases. In contrast, RA-ORMP is able to achieve OMP-like performance in the SMV scenario ($l = 1$) and exhibits improved recovery with increasing rank, with guaranteed recovery in the maximal rank case.

8.5 Conclusions

As stressed repeatedly throughout this chapter, greedy methods offer powerful approaches for CS signal recovery. Greedy pursuit algorithms, which were discussed in Section 8.2, can be much faster than ℓ_1-based methods and are often applicable to very large sparse recovery problems. Iterative thresholding approaches, which were discussed in Section 8.3, offer another fast alternative, which in addition, share the near optimal recovery guarantees offered by ℓ_1-based approaches. Furthermore, as discussed at length in Section 8.4, these methods are easily adapted to much more general signal models.

Whilst we have here looked particularly at the class of UoS models, it is again worth pointing out that similar results hold for much broader classes of low-dimensional signal models [43]. Furthermore, these methods can also be adapted to a setting where the observation model is nonlinear [74], that is, where $y = A(x) + e$, with A a nonlinear map. The Projected Landweber Algorithm then becomes

$$\widehat{x}^{[i+1]} = P_\mathcal{U} \left(\widehat{x}^{[i]} - \frac{\mu}{2} \nabla(\widehat{x}^{[i]}) \right), \qquad (8.82)$$

where $\nabla(\widehat{x}^{[i]})$ is the (generalized) gradient of the error $\|y - A\widetilde{x}\|_\mathcal{L}^2$ evaluated at $\widehat{x}^{[i]}$. Note that this formulation allows more general cost functions $\|y - A\widetilde{x}\|_\mathcal{L}^2$ (not necessarily only

Hilbert space norms). Importantly, Blumensath [74] recently derived the first theoretical recovery guarantees that generalize CS theory also to this nonlinear setting.

Greedy methods are thus powerful tools for CS recovery problems. They are fast and versatile and applicable far beyond the standard sparse setting traditionally considered in CS. These methods therefore remain at the forefront of active CS research and provide important tools for a wide range of CS applications.

Acknowledgement

This work was supported in part by the UK's Engineering and Physical Science Research Council grant EP/F039697/1, by the European Commission through the SMALL project under FET-Open, grant number 225913 and a Research Fellowship from the School of Mathematics at the University of Southampton.

References

[1] J. H. Friedman and J. W. Tukey. A projection pursuit algorithm for exploratory data analysis. *IEEE Trans Comput*, 23(9):881–890, 1974.

[2] S. Mallat and Z. Zhang. Matching pursuits with time-frequency dictionaries. *IEEE Trans Signal Proc*, 41(12):3397–3415, 1993. Available from: citeseer.nj.nec.com/mallat93matching.html.

[3] A. Miller. *Subset Selection in Regression.* 2nd edn. Chapman and Hall; 2002.

[4] R. A. DeVore and V. N. Temlyakov. Some remarks on greedy algorithms. *Adv Comput Math*, 5:173–187, 1996.

[5] J. Hogbom. Aperture synthesis with a non-regular distribution of interferometer baselines. *Astrophys J Suppl Ser*, 15:417–426, 1974.

[6] S. Krstulovic and R. Gribonval. MPTK: Matching pursuit made tractable. *Proc of Int Conf Acoust, Speech, Signal Proc*, Vol. 3. Toulouse, France; 2006. pp. III–496–III–499.

[7] Y. C. Pati, R. Rezaifar, and P. S. Krishnaprasad. Orthogonal matching pursuit: recursive function approximation with applications to wavelet decomposition. *Rec 27th Asilomar Conf Sign, Syst Comput*, 1993.

[8] S. Mallat, G. Davis, and Z. Zhang. Adaptive time-frequency decompositions. *SPIE J Opt Eng*, 33(7):2183–2191, 1994.

[9] S. F. Cotter, J. Adler, B. D. Rao, and K. Kreutz-Delgado. Forward sequential algorithms for best basis selection. In: *IEE Proc Vision, Image Signal Proc*, 235–244, 1999.

[10] A. R. Barron, A. Cohen, R. A. DeVore, and W. Dahmen. Approximation and learning by greedy algorithms. *Ann Stati*, 2008;36(1):64–94.

[11] T. Blumensath and M. Davies. Gradient pursuits. *IEEE Trans Sign Proc*, 56(6):2370–2382, 2008.

[12] R. Gribonval and P. Vandergheynst. On the exponential convergence of Matching Pursuits in Quasi-Incoherent Dictionaries. *IEEE Trans Inform Theory*, 52(1):255–261, 2006.

[13] J. A. Tropp. Greed is good: algorithmic results for Sparse Approximation. *IEEE Trans Inform Theory*, 2004;50(10):2231–2242, 2004.

[14] B. Mailhe, R. Gribonval, F. Bimbot, and P. Vandergheynst. A low complexity Orthogonal Matching Pursuit for sparse signal approximation with shift-invariant dictionaries.

Proc IEEE Int Conf Acoust, Speech, Signal Proc, Washington, DC, USA, pp. 3445–3448, 2009.

[15] J. R. Shewchuk. An introduction to the conjugate gradient method without the agonizing pain. School of Computer Science, Carnegie Mellon University, 1994.

[16] G. H. Golub and F. Van Loan. *Matrix Computations*. 3rd edn. Johns Hopkins University Press, 1996.

[17] D. L. Donoho, I. Drori, Y. Tsaig, and J. L. Starck. Sparse solution of underdetermined linear equations by stagewise orthogonal matching pursuit. Stanford University, 2006.

[18] T. Blumensath and M. Davies. Stagewise weak gradient pursuits. *IEEE Trans Sig Proc*, 57(11):4333–4346, 2009.

[19] H. Crowder and P. Wolfe. Linear convergence of the Conjugate Gradient Method. *Numer Comput*, 16(4):431–433, 1972.

[20] K. Schnass and P. Vandergheynst. Average performance analysis for thresholding. *IEEE Sig Proc Letters*, 14(11):431–433, 2007.

[21] T. Blumensath and M. Davies. Iterative thresholding for sparse approximations. *J Fourier Anal Appl*, 14(5):629–654, 2008.

[22] D. Needell and R. Vershynin. Signal recovery from incomplete and inaccurate measurements via regularized orthogonal matching pursuit. *IEEE J Sel Topics Sig Proc*, 4(2):310–316, 2010.

[23] W. Dai and O. Milenkovic. Subspace pursuit for compressive sensing signal reconstruction. *IEEE Trans Inform Theory*, 55(5):2230–2249, 2009.

[24] D. Needell and R. Vershynin. Uniform uncertainty principle and signal recovery via regularized Orthogonal Matching Pursuit. *Found Comput Math*, 9(3):317–334, 2008.

[25] S. Chen, S. A. Billings, and W. Luo. Orthogonal least-squares methods and their application to non-linear system identification. *Int J Control*, 50(5):1873–1896, 1989.

[26] T. Blumensath and M. Davies. On the difference between Orthogonal Matching Pursuit and Orthogonal Least Squares; 2007. Unpublished manuscript, available at: http://eprints.soton.ac.uk/142469/. Available from: www.see.ed.ac.uk/~tblumens/publications.html.

[27] D. L. Donoho. For most large underdetermined systems of linear equations the minimal 1-norm solution is also the sparsest solution. *Commun Pure Appl Math*, 59(6):797–829, 2006.

[28] H. Rauhut. On the impossibility of uniform sparse reconstruction using greedy methods. *Sampling Theory Signal Image Proc*, 7(2):197–215, 2008.

[29] M. A. Davenport and M. B. Wakin. Analysis of orthogonal matching pursuit using the restricted isometry property. *IEEE Trans Inform Theory*, 56(9):4395–4401, 2009.

[30] E. Candès. The restricted isometry property and its implications for compressed sensing. *C R Acad Sci, Paris, Serie I*, 346:589–592, 2008.

[31] J. A. Tropp and A. C. Gilbert. Signal recovery from partial information via Orthogonal Matching Pursuit. *IEEE Trans Inform Theory*, 53(12):4655–4666, 2006.

[32] A. K. Fletcher, S. Rangan, and V. K. Goyal. Necessary and sufficient conditions for sparsity pattern recovery. *IEEE Trans Inform Theory*, 55(12):5758–5772, 2009.

[33] M. E. Davies and T. Blumensath. Faster and greedier: algorithms for sparse reconstruction of large datasets. In: *3rd Int Symp Commun, Control Signal Proc*, pp. 774–779, 2008.

[34] D. Needell and J. A. Tropp. CoSaMP: iterative signal recovery from incomplete and inaccurate samples. *Appl Comput Harmonic Anal*, 26(3):301–321, 2008.

[35] N. G. Kingsbury and T. H. Reeves. Iterative image coding with overcomplete complex wavelet transforms. *Proc Conf Visual Commun Image Proc*, 2003.

[36] J. Portilla. Image restoration through l0 analysis-based sparse optimization in tight frames. *Proc IEEE Int Conf Image Proc*, 2009.

[37] A. Cohen, W. Dahmen, and R. DeVore. Instance optimal decoding by thresholding in compressed sensing. *Proc 8th Int Conf Harmonic Anal Partial Differential Equations*. Madrid, Spain, pp. 1–28, 2008.

[38] R. Berind and P. Indyk. Sequential Sparse Matching Pursuit. In: *Allerton'09 Proc 47th Ann Allerton Conf Commun, Control, Comput*, 2009.

[39] K. Lange, D. R. Hunter, and I. Yang. Optimization transfer using surrogate objective functions. *J Comput Graphical Stat*, 9:1–20, 2006.

[40] T. Blumensath and M. Davies. Iterative hard thresholding for compressed sensing. *Appl Comput Harmonic Anal*, 27(3):265–274, 2009.

[41] R. Meka, P. Jain, and I. S. Dhillon. Guaranteed rank minimization via singular value projection. arXiv:09095457v3, 2009.

[42] R. Garg and R. Khandekar. Gradient Descent with Sparsification: an iterative algorithm for sparse recovery with restricted isometry property. *Proc Int Conf Machine Learning*, Montreal, Canada; 2009.

[43] T. Blumensath. Sampling and reconstructing signals from a union of linear subspaces. Submitted to *IEEE Trans Inform Theory*, 2010.

[44] T. Blumensath and M. Davies. Normalised iterative hard thresholding; guaranteed stability and performance. *IEEE J Sel Topics Sig Proc*, 4(2):298–309, 2010.

[45] E. van den Berg and M. P. Friedlander. Probing the Pareto frontier for basis pursuit solutions. *SIAM J Sci Comput*, 31(2):890–912, 2008.

[46] Y. Lu and M. Do. A theory for sampling signals from a union of subspaces. *IEEE Trans Signal Proc*, 56(6):2334–2345, 2008.

[47] T. Blumensath and M. E. Davies. Sampling theorems for signals from the union of finite-dimensional linear subspaces. *IEEE Trans Inform Theory*, 55(4):1872–1882, 2009.

[48] Y. C. Eldar, P. Kuppinger, and H. Bolcskei. Block-sparse signals: uncertainty relations and efficient recovery. *IEEE Trans Signal Proc*, 58(6):3042–3054, 2010.

[49] R. G. Baraniuk, V. Cevher, M. F. Duarte, and C. Hegde. Model-based compressive sensing. *IEEE Trans Inform Theory*, 56(4):1982–2001, 2010.

[50] M. Crouse, R. Nowak, and R. Baraniuk. Wavelet-based statistical signal processing using hidden Markov models. *IEEE Trans Sig Proc*, 46(4):886–902, 1997.

[51] J. K. Romberg, H. Choi, and R. G. Baraniuk. Bayesian tree-structured image modeling using wavelet-domain hidden Markov models. *IEEE Trans Image Proc*, 10(7):1056–1068, 2001.

[52] J. Portilla, V. Strela, M. Wainwright, and E. P. Simoncelli. Image denoising using scale mixtures of Gaussians in the wavelet domain. *IEEE Trans Image Proc*, 12(11):1338–1351, 2003.

[53] J. M. Shapiro. Embedded image coding using zerotrees of wavelet coefficients. *IEEE Trans Signal Proc*, 41(12):3445–3462, 1993.

[54] A. Cohen, W. Dahmen, I. Daubechies, and R. Devore. Tree approximation and optimal encoding. *J Appl Comp Harmonic Anal*, 11:192–226, 2000.

[55] Y. Eldar and M. Mishali. Robust recovery of signals from a structured union of subspaces. *IEEE Trans Inform Theory*, 55(11):5302–5316, 2009.

[56] M. E. Davies and Y. C. Eldar. Rank awareness in joint sparse recovery. arXiv:10044529v1. 2010.

[57] C. La and M. Do. Signal reconstruction using sparse tree representations. *Proc SPIE Conf Wavelet Appl Signal Image Proc XI*. San Diego, California, 2005.

[58] M. F. Duarte, M. B. Wakin, and R. G. Baraniuk. Fast reconstruction of piecewise smooth signals from random projections. *Proc Workshop Sig Proc Adapt Sparse Struct Repres*, Rennes, France; 2005.

[59] R. G. Baraniuk. Optimal tree approximation with wavelets. *Wavelet Applications in Signal and Image Processing VII*, vol 3813, pp. 196–207, 1999.

[60] D. Goldfarb and S. Ma. Convergence of fixed point continuation algorithms for matrix rank minimization. arXiv:09063499v3. 2010.

[61] M. Mishali and Y. C. Eldar. From theory to practice: sub-Nyquist sampling of sparse wideband analog signals. *IEEE J Sel Topics Sig Proc*, 4(2):375–391, 2010.

[62] M. Mishali and Y. C. Eldar. Blind multi-band signal reconstruction: compressed sensing for analog signals. *IEEE Trans Sig Proc*, 57(3):993–1009, 2009.

[63] H. Rauhut, K. Schnass, and P. Vandergheynst. Compressed sensing and redundant dictionaries. *IEEE Trans Inform Theory*, 54(5):2210–2219, 2008.

[64] B. Recht, M. Fazel, and P. A. Parrilo. Guaranteed minimum-rank solution of linear matrix equations via nuclear norm minimization. To appear in *SIAM Rev*. 2010.

[65] J. Chen and X. Huo. Theoretical results on sparse representations of multiple-measurement vectors. *IEEE Trans Sig Proc*, 54(12):4634–4643, 2006.

[66] J. A. Tropp. Algorithms for simultaneous sparse approximation. Part II: Convex relaxation. *Sig Proc*, 86(3):589–602, 2006.

[67] S. F. Cotter, B. D. Rao, K. Engan, and K. Delgado. Sparse solutions to linear inverse problems with multiple measurement vectors. *IEEE Trans Sig Proc*, 53(7):2477–2488, 2005.

[68] J. A. Tropp. Algorithms for simultaneous sparse approximation. Part II: Convex relaxation. *Sig Proc*, 86:589–602, 2006.

[69] R. O. Schmidt. Multiple emitter location and signal parameter estimation. *Proc RADC Spectral Estimation Workshop*, 243–258, 1979.

[70] P. Feng. Universal minimum-rate sampling and spectrum-blind reconstruction for multiband signals. University of Illinois; 1998.

[71] P. Feng and Y. Bresler. Spectrum-blind minimum-rate sampling and reconstruction of multiband signals. *Proc IEEE Int Conf Acoust, Speech, Signal Proc*, 1688–1691, 1996.

[72] K. Lee and Y. Bresler. iMUSIC: iterative MUSIC Algorithm for joint sparse recovery with any rank. 2010;Arxiv preprint: arXiv:1004.3071v1.

[73] J. M. Kim, O. K. Lee, and J. C. Ye. Compressive MUSIC: a missing link between compressive sensing and array signal processing;Arxiv preprint, arXiv:1004.4398v1.

[74] T. Blumensath. Compressed sensing with nonlinear observations. Preprint, available at: http://eprintssotonacuk/164753. 2010.

9 Graphical models concepts in compressed sensing

Andrea Montanari

This chapter surveys recent work in applying ideas from graphical models and message passing algorithms to solve large-scale regularized regression problems. In particular, the focus is on compressed sensing reconstruction via ℓ_1 penalized least-squares (known as LASSO or BPDN). We discuss how to derive fast approximate message passing algorithms to solve this problem. Surprisingly, the analysis of such algorithms allows one to prove exact high-dimensional limit results for the LASSO risk.

9.1 Introduction

The problem of reconstructing a high-dimensional vector $x \in \mathbb{R}^n$ from a collection of observations $y \in \mathbb{R}^m$ arises in a number of contexts, ranging from statistical learning to signal processing. It is often assumed that the measurement process is approximately linear, i.e. that

$$y = Ax + w, \qquad (9.1)$$

where $A \in \mathbb{R}^{m \times n}$ is a known measurement matrix, and w is a noise vector.

The graphical models approach to such a reconstruction problem postulates a joint probability distribution on (x, y) which takes, without loss of generality, the form

$$p(\mathrm{d}x, \mathrm{d}y) = p(\mathrm{d}y|x)\, p(\mathrm{d}x). \qquad (9.2)$$

The conditional distribution $p(\mathrm{d}y|x)$ models the noise process, while the prior $p(\mathrm{d}x)$ encodes information on the vector x. In particular, within compressed sensing, it can describe its sparsity properties. Within a *graphical models* approach, either of these distributions (or both) factorizes according to a specific graph structure. The resulting posterior distribution $p(\mathrm{d}x|y)$ is used for inferring x given y.

Compressed Sensing: Theory and Applications, ed. Yonina C. Eldar and Gitta Kutyniok. Published by Cambridge University Press. © Cambridge University Press 2012.

There are many reasons to be skeptical about the idea that the joint probability distribution $p(\mathrm{d}x, \mathrm{d}y)$ can be determined, and used for reconstructing x. To name one such reason for skepticism, any finite sample will allow one to determine the prior distribution of x, $p(\mathrm{d}x)$ only within limited accuracy. A reconstruction algorithm based on the posterior distribution $p(\mathrm{d}x|y)$ might be sensitive with respect to changes in the prior thus leading to systematic errors.

One might be tempted to drop the whole approach as a consequence. We argue that sticking to this point of view is instead fruitful for several reasons:

1. *Algorithmic.* Several existing reconstruction methods are in fact M-estimators, i.e. they are defined by minimizing an appropriate cost function $\mathcal{C}_{A,y}(x)$ over $x \in \mathbb{R}^n$ [80]. Such estimators can be derived as Bayesian estimators (e.g. maximum a posteriori probability) for specific forms of $p(\mathrm{d}x)$ and $p(\mathrm{d}y|x)$ (for instance by letting $p(\mathrm{d}x|y) \propto \exp\{-\mathcal{C}_{A,y}(x)\}\,\mathrm{d}x$). The connection is useful both in interpreting/comparing different methods, and in adapting known algorithms for Bayes estimation. A classical example of this cross-fertilization is the paper [33]. This review discusses several other examples that build on graphical models inference algorithms.

2. *Minimax.* When the prior $p(\mathrm{d}x)$ or the noise distributions, and therefore the conditional distribution $p(\mathrm{d}y|x)$, "exist" but are unknown, it is reasonable to assume that they belong to specific structure classes. By this term we refer generically to a class of probability distributions characterized by a specific property. For instance, within compressed sensing one often assumes that x has at most k nonzero entries. One can then take $p(\mathrm{d}x)$ to be a distribution supported on k-sparse vectors $x \in \mathbb{R}^n$. If $\mathcal{F}_{n,k}$ denotes the class of such distributions, the minimax approach strives to achieve the best uniform guarantee over $\mathcal{F}_{n,k}$. In other words, the minimax estimator achieves the *smallest* expected error (e.g. mean square error) for the "worst" distribution in $\mathcal{F}_{n,k}$.

   It is a remarkable fact in statistical decision theory [52] (which follows from a generalization of Von Neumann minimax theorem) that the minimax estimator coincides with the Bayes estimator for a specific (worst case) prior $p \in \mathcal{F}_{n,k}$. In one dimension considerable information is available about the worst case distribution and asymptotically optimal estimators (see Section 9.3). The methods developed here allow one to develop similar insights in high-dimension.

3. *Modeling.* In some applications it is possible to construct fairly accurate models both of the prior distribution $p(\mathrm{d}x)$ and of the measurement process $p(\mathrm{d}y|x)$. This is the case for instance in some communications problems, whereby x is the signal produced by a transmitter (and generated uniformly at random according to a known codebook), and w is the noise produced by a well-defined physical process (e.g. thermal noise in the receiver circuitry). A discussion of some families of practically interesting priors $p(\mathrm{d}x)$ can be found in [11].

Further, the question of modeling the prior in compressed sensing is discussed from the point of view of Bayesian theory in [43].

The rest of this chapter is organized as follows. Section 9.2 describes a graphical model naturally associated to the compressed sensing reconstruction problem. Section 9.3

provides important background on the one-dimensional case. Section 9.4 describes a standard message passing algorithm – the min-sum algorithm – and how it can be simplified to solve the LASSO optimization problem. The algorithm is further simplified in Section 9.5 yielding the AMP algorithm. The analysis of this algorithm is outlined in Section 9.6. As a consequence of this analysis, it is possible to compute exact high-dimensional limits for the behavior of the LASSO estimator. Finally in Section 9.7 we discuss a few examples of how the approach developed here can be used to address reconstruction problems in which a richer structural information is available.

9.1.1 Some useful notation

Throughout this review, probability measures over the real line $\mathbb{R}$ or the Euclidean space $\mathbb{R}^K$ play a special role. It is therefore useful to be careful about the probability-theory notation. The less careful reader who prefers to pass directly to the "action" is invited to skip these remarks at a first reading.

We will use the notation p or $p(\mathrm{d}x)$ to indicate probability measures (eventually with subscripts). Notice that, in the last form, the $\mathrm{d}x$ is only a reminder of which variable is distributed with measure p. (Of course one is tempted to think of $\mathrm{d}x$ as an infinitesimal interval but this intuition is accurate only if p admits a density.)

A special measure (positive, but not normalized and hence not a probability measure) is the Lebesgue measure for which we reserve the special notation $\mathrm{d}x$ (something like $\mu(\mathrm{d}x)$ would be more consistent but, in our opinion, less readable). This convention is particularly convenient for expressing in formulae statements of the form "*p admits a density f with respect to Lebesgue measure* $\mathrm{d}x$*, with* $f : x \mapsto f(x) \equiv \exp(-x^2/(2a))/\sqrt{2\pi a}$ *a Borel function,*" which we write simply

$$p(\mathrm{d}x) = \frac{1}{\sqrt{2\pi a}} e^{-x^2/2a} \, \mathrm{d}x. \tag{9.3}$$

It is well known that expectations are defined as integrals with respect to the probability measure which we denote as

$$\mathbb{E}_p\{f\} = \int_{\mathbb{R}} f(x) p(\mathrm{d}x), \tag{9.4}$$

sometimes omitting the subscript p in $\mathbb{E}_p$ and $\mathbb{R}$ in $\int_{\mathbb{R}}$. Unless specified otherwise, we do not assume such probability measures to have a density with respect to Lebesgue measure. The probability measure p is a set function defined on the Borel σ-algebra, see e.g. [7, 82]. Hence it makes sense to write $p((-1, 3])$ (the probability of the interval $(-1, 3]$ under measure p) or $p(\{0\})$ (the probability of the point 0). Equally valid would be expressions such as $\mathrm{d}x((-1, 3])$ (the Lebesgue measure of $(-1, 3]$) or $p(\mathrm{d}x)((-1, 3])$ (the probability of the interval $(-1, 3]$ under measure p) but we avoid them as somewhat clumsy.

A (joint) probability measure over $x \in \mathbb{R}^K$ and $y \in \mathbb{R}^L$ will be denoted by $p(\mathrm{d}x, \mathrm{d}y)$ (this is just a probability measure over $\mathbb{R}^K \times \mathbb{R}^L = \mathbb{R}^{K+L}$). The corresponding conditional probability measure of y given x is denoted by $p(\mathrm{d}x|y)$ (for a rigorous definition we refer to [7, 82]).

Finally, we will not make use of cumulative distribution functions – commonly called distribution functions in probability theory – and instead use "probability distribution" interchangeably with "probability measure."

Some fairly standard discrete mathematics notation will also be useful. The set of first K integers is to be denoted by $[K] = \{1, \ldots, K\}$. Order of growth of various functions will be characterized by the standard "big-O" notation. Recall in particular that, for $M \to \infty$, one writes $f(M) = O(g(M))$ if $f(M) \le C g(M)$ for some finite constant C, $f(M) = \Omega(g(M))$ if $f(M) \ge g(M)/C$, and $f(M) = \Theta(g(M))$ if $g(M)/C \le f(M) \le C g(M)$. Further $f(M) = o(g(M))$ if $f(M)/g(M) \to 0$. Analogous notations are used when the argument of f and g go to 0.

9.2 The basic model and its graph structure

Specifying the conditional distribution of y given x is equivalent to specifying the distribution of the noise vector w. In most of this chapter we shall take $p(w)$ to be a Gaussian distribution of mean 0 and variance $\beta^{-1}\mathbf{I}$, whence

$$p_\beta(\mathrm{d}y|x) = \left(\frac{\beta}{2\pi}\right)^{n/2} \exp\left\{-\frac{\beta}{2}\|y - Ax\|_2^2\right\} \mathrm{d}y. \tag{9.5}$$

The simplest choice for the prior consists in taking $p(\mathrm{d}x)$ to be a product distribution with identical factors $p(\mathrm{d}x) = p(\mathrm{d}x_1) \times \cdots \times p(\mathrm{d}x_n)$. We thus obtain the joint distribution

$$p_\beta(\mathrm{d}x, \mathrm{d}y) = \left(\frac{\beta}{2\pi}\right)^{n/2} \exp\left\{-\frac{\beta}{2}\|y - Ax\|_2^2\right\} \mathrm{d}y \prod_{i=1}^n p(\mathrm{d}x_i). \tag{9.6}$$

It is clear at the outset that generalizations of this basic model can be easily defined, in such a way to incorporate further information on the vector x or on the measurement process. As an example, consider the case of block-sparse signals: The index set $[n]$ is partitioned into blocks $B(1), B(2), \ldots B(\ell)$ of equal length n/ℓ, and only a small fraction of the blocks is non-vanishing. This situation can be captured by assuming that the prior $p(\mathrm{d}x)$ factors over blocks. One thus obtains the joint distribution

$$p_\beta(\mathrm{d}x, \mathrm{d}y) = \left(\frac{\beta}{2\pi}\right)^{n/2} \exp\left\{-\frac{\beta}{2}\|y - Ax\|_2^2\right\} \mathrm{d}y \prod_{j=1}^\ell p(\mathrm{d}x_{B(j)}), \tag{9.7}$$

where $x_{B(j)} \equiv (x_i : i \in B(j)) \in \mathbb{R}^{n/\ell}$. Other examples of structured priors will be discussed in Section 9.7.

The posterior distribution of x given observations y admits an explicit expression, that can be derived from Eq. (9.6):

$$p_\beta(\mathrm{d}x|y) = \frac{1}{Z(y)} \exp\left\{-\frac{\beta}{2}\|y - Ax\|_2^2\right\} \prod_{i=1}^n p(\mathrm{d}x_i), \qquad (9.8)$$

where $Z(y) = (2\pi/\beta)^{n/2} p(y)$ ensures the normalization $\int p(\mathrm{d}x|y) = 1$. Let us stress that while this expression is explicit, computing expectations or marginals of this distribution is a hard computational task.

Finally, the square residuals $\|y - Ax\|_2^2$ decompose in a sum of m terms yielding

$$p_\beta(\mathrm{d}x|y) = \frac{1}{Z(y)} \prod_{a=1}^m \exp\left\{-\frac{\beta}{2}(y_a - A_a^T x)^2\right\} \prod_{i=1}^n p(\mathrm{d}x_i), \qquad (9.9)$$

where A_a is the ath row of the matrix a. This factorized structure is conveniently described by a *factor graph*, i.e. a bipartite graph including a "variable node" $i \in [n]$ for each variable x_i, and a "factor node" $a \in [m]$ for each term $\psi_a(x) = \exp\{-\beta(y_a - A_a^T x)^2/2\}$. Variable i and factor a are connected by an edge if and only if $\psi_a(x)$ depends nontrivially on x_i, i.e. if $A_{ai} \neq 0$. One such factor graph is reproduced in Figure 9.1.

An estimate of the signal can be extracted from the posterior distribution (9.9) in various ways. One possibility is to use conditional expectation

$$\widehat{x}_\beta(y; p) \equiv \int_{\mathbb{R}^n} x \, p_\beta(\mathrm{d}x|y). \qquad (9.10)$$

Classically, this estimator is justified by the fact that it achieves the minimal mean square error provided the $p_\beta(\mathrm{d}x, \mathrm{d}y)$ is the *actual* joint distribution of (x, y). In the present context we will not assume that the "postulated" prior $p_\beta(\mathrm{d}x)$ coincides with the actual distribution of x, and hence $\widehat{x}_\beta(y; p)$ is not necessarily optimal (with respect to

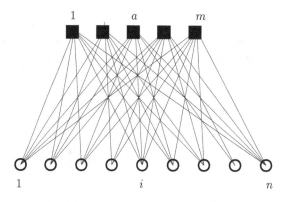

Figure 9.1 Factor graph associated to the probability distribution (9.9). Empty circles correspond to variables x_i, $i \in [n]$ and squares correspond to measurements y_a, $a \in [m]$.

mean square error). The best justification for $\widehat{x}_\beta(y;p)$ is that a broad class of estimators can be written in the form (9.10).

An important problem with the estimator (9.10) is that it is in general hard to compute. In order to obtain a tractable proxy, we assume that $p(\mathrm{d}x_i) = p_{\beta,h}(\mathrm{d}x_i) = c f_{\beta,h}(x_i)\,\mathrm{d}x_i$ for $f_{\beta,h}(x_i) = e^{-\beta h(x_i)}$ an un-normalized probability density function. As β get large, the integral in Eq. (9.10) becomes dominated by the vector x with the highest posterior probability p_β. One can then replace the integral in $\mathrm{d}x$ with a maximization over x and define

$$\widehat{x}(y;h) \equiv \mathrm{argmin}_{z \in \mathbb{R}^n} \mathcal{C}_{A,y}(z;h), \tag{9.11}$$

$$\mathcal{C}_{A,y}(z;h) \equiv \frac{1}{2}\|y - Az\|_2^2 + \sum_{i=1}^n h(z_i),$$

where we assumed for simplicity that $\mathcal{C}_{A,y}(z;h)$ has a unique minimum.

According to the above discussion, the estimator $\widehat{x}(y;h)$ can be thought of as the $\beta \to \infty$ limit of the general estimator (9.10). Indeed, it is easy to check that, provided $x_i \mapsto h(x_i)$ is upper semicontinuous, we have

$$\lim_{\beta \to \infty} \widehat{x}_\beta(y; p_{\beta,h}) = \widehat{x}(y;h).$$

In other words, the posterior mean converges to the mode of the posterior in this limit. Further, $\widehat{x}(y;h)$ takes the familiar form of a regression estimator with separable regularization. If $h(\cdot)$ is convex, the computation of $\widehat{x}$ is tractable. Important special cases include $h(x_i) = \lambda x_i^2$, which corresponds to ridge regression [39], and $h(x_i) = \lambda|x_i|$ which corresponds to the LASSO [77] or basis pursuit denoising (BPDN) [10]. Due to the special role it plays in compressed sensing, we will devote special attention to the latter case, that we rewrite explicitly below with a slight abuse of notation

$$\widehat{x}(y) \equiv \mathrm{argmin}_{z \in \mathbb{R}^n} \mathcal{C}_{A,y}(z), \tag{9.12}$$

$$\mathcal{C}_{A,y}(z) \equiv \frac{1}{2}\|y - Az\|_2^2 + \lambda\|z\|_1.$$

9.3 Revisiting the scalar case

Before proceeding further, it is convenient to pause for a moment and consider the special case of a single measurement of a scalar quantity, i.e. the case $m = n = 1$. We therefore have

$$y = x + w, \tag{9.13}$$

and want to estimate x from y. Despite the apparent simplicity, there exists a copious literature on this problem with many open problems [24, 23, 22, 41]. Here we only want to clarify a few points that will come up again in what follows.

In order to compare various estimators we will assume that (x,y) are indeed random variables with some underlying probability distribution $p_0(\mathrm{d}x, \mathrm{d}y) = p_0(\mathrm{d}x) p_0(\mathrm{d}y|x)$. It is important to stress that this distribution is conceptually distinct from the one used in inference, cf. Eq. (9.10). In particular we cannot assume to know the actual prior distribution of x, at least not exactly, and hence $p(\mathrm{d}x)$ and $p_0(\mathrm{d}x)$ do not coincide. The "actual" prior p_0 is the distribution of the vector to be inferred, while the "postulated" prior p is a device used for designing inference algorithms.

For the sake of simplicity we also consider Gaussian noise $w \sim \mathsf{N}(0, \sigma^2)$ with known noise level σ^2. Various estimators will be compared with respect to the resulting mean square error

$$\mathrm{MSE} = \mathbb{E}\{|\widehat{x}(y) - x|^2\} = \int_{\mathbb{R} \times \mathbb{R}} |\widehat{x}(y) - x|^2 \, p_0(\mathrm{d}x, \mathrm{d}y).$$

We can distinguish two cases:

I. The signal distribution $p_0(x)$ is known as well. This can be regarded as an "oracle" setting. To make contact with compressed sensing, we consider distributions that generate sparse signals, i.e. that put mass at least $1-\varepsilon$ on $x=0$. In formulae $p_0(\{0\}) \geq 1-\varepsilon$.

II. The signal distribution is unknown but it is known that it is "sparse," namely that it belongs to the class

$$\mathcal{F}_\varepsilon \equiv \{p_0 : p_0(\{0\}) \geq 1-\varepsilon\}. \tag{9.14}$$

The *minimum mean square error*, is the minimum MSE achievable by any estimator $\widehat{x}: \mathbb{R} \to \mathbb{R}$:

$$\mathrm{MMSE}(\sigma^2; p_0) = \inf_{\widehat{x}:\mathbb{R} \to \mathbb{R}} \mathbb{E}\{|\widehat{x}(y) - x|^2\}.$$

It is well known that the infimum is achieved by the conditional expectation

$$\widehat{x}^{\,\mathrm{MMSE}}(y) = \int_{\mathbb{R}} x \, p_0(\mathrm{d}x|y).$$

However, this estimator assumes that we are in situation I above, i.e. that the prior p_0 is known.

In Figure 9.2 we plot the resulting MSE for a three-point distribution,

$$p_0 = \frac{\varepsilon}{2} \delta_{+1} + (1-\varepsilon) \delta_0 + \frac{\varepsilon}{2} \delta_{-1}. \tag{9.15}$$

The MMSE is non-decreasing in σ^2 by construction, converges to 0 in the noiseless limit $\sigma \to 0$ (indeed the simple rule $\widehat{x}(y) = y$ achieves MSE equal to σ^2) and to ε in the large noise limit $\sigma \to \infty$ (MSE equal to ε is achieved by $\widehat{x} = 0$).

In the more realistic situation II, we do not know the prior p_0. A principled way to deal with this ignorance would be to minimize the MSE for the worst case distribution

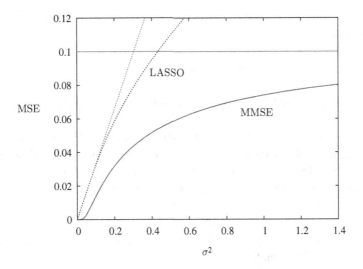

Figure 9.2 Mean square error for estimating a three points random variable, with probability of nonzero $\varepsilon = 0.1$, in Gaussian noise. Lower curves: Minimal mean square error achieved by conditional expectation (thick) and its large noise asymptote (thin). Upper curves: Mean square error for LASSO or equivalently for soft thresholding (thick) and its small noise asymptote (thin).

in the class $\mathcal{F}_\varepsilon$, i.e. to replace the minimization in Eq. (9.15) with the following minimax problem

$$\inf_{\widehat{x}:\mathbb{R}\to\mathbb{R}} \sup_{p_0 \in \mathcal{F}_\varepsilon} \mathbb{E}\{|\widehat{x}(y) - x|^2\}. \tag{9.16}$$

A lot is known about this problem [22–24, 41]. In particular general statistical decision theory [52, 41] implies that the optimum estimator is just the posterior expectation for a specific worst case prior. Unfortunately, even a superficial discussion of this literature goes beyond the scope of the present review.

Nevertheless, an interesting exercise (indeed not a trivial one) is to consider the LASSO estimator (9.12), which in this case reduces to

$$\widehat{x}(y;\lambda) = \mathrm{argmin}_{z\in\mathbb{R}}\left\{\frac{1}{2}(y-z)^2 + \lambda|z|\right\}. \tag{9.17}$$

Notice that this estimator is insensitive to the details of the prior p_0. Instead of the full minimax problem (9.16), one can then simply optimize the MSE over λ.

The one-dimensional optimization problem (9.17) admits an explicit solution in terms of the *soft thresholding function* $\eta : \mathbb{R} \times \mathbb{R}_+ \to \mathbb{R}$ defined as follows

$$\eta(y;\theta) = \begin{cases} y - \theta & \text{if } y > \theta, \\ 0 & \text{if } -\theta \leq y \leq \theta, \\ y + \theta & \text{if } y < -\theta. \end{cases} \tag{9.18}$$

The *threshold* value θ has to be chosen equal to the regularization parameter λ yielding the simple solution

$$\widehat{x}(y;\lambda) = \eta(y;\theta), \quad \text{for } \lambda = \theta. \tag{9.19}$$

(We emphasize the identity of λ and θ in the scalar case, because it breaks down in the vector case.)

How should the parameter θ (or equivalently λ) be fixed? The rule is conceptually simple: θ should minimize the maximal mean square error for the class $\mathcal{F}_\varepsilon$. Remarkably this complex saddle point problem can be solved rather explicitly. The key remark is that the worst case distribution over the class $\mathcal{F}_\varepsilon$ can be identified and takes the form $p^{\#} = (\varepsilon/2)\delta_{+\infty} + (1-\varepsilon)\delta_0 + (\varepsilon/2)\delta_{-\infty}$ [23, 22, 41].

Let us outline how the solution follows from this key fact. First of all, it makes sense to scale λ as the noise standard deviation, because the estimator is supposed to filter out the noise. We then let $\theta = \alpha\sigma$. In Figure 9.2 we plot the resulting MSE when $\theta = \alpha\sigma$, with $\alpha \approx 1.1402$. We denote the LASSO/soft thresholding mean square error by $\mathrm{mse}(\sigma^2;p_0,\alpha)$ when the noise variance is σ^2, $x \sim p_0$, and the regularization parameter is $\lambda = \theta = \alpha\sigma$. The worst case mean square error is given by $\sup_{p_0 \in \mathcal{F}_\varepsilon} \mathrm{mse}(\sigma^2;p_0,\alpha)$. Since the class $\mathcal{F}_\varepsilon$ is invariant by rescaling, this worst case MSE must be proportional to the only scale in the problem, i.e. σ^2. We get

$$\sup_{p_0 \in \mathcal{F}_\varepsilon} \mathrm{mse}(\sigma^2;p_0,\alpha) = M(\varepsilon,\alpha)\sigma^2. \tag{9.20}$$

The function M can be computed explicitly by evaluating the mean square error on the worst case distribution $p^{\#}$ [23, 22, 41]. A straightforward calculation (see also [26, Supplementary Information], and [28]) yields

$$M(\varepsilon,\alpha) = \varepsilon(1+\alpha^2) + (1-\varepsilon)[2(1+\alpha^2)\Phi(-\alpha) - 2\alpha\phi(\alpha)] \tag{9.21}$$

where $\phi(z) = e^{-z^2/2}/\sqrt{2\pi}$ is the Gaussian density and $\Phi(z) = \int_{-\infty}^{z} \phi(u)\,du$ is the Gaussian distribution. It is also not hard to show that $M(\varepsilon,\alpha)$ is the slope of the soft thresholding MSE at $\sigma^2 = 0$ in a plot like the one in Figure 9.2.

Minimizing the above expression over α, we obtain the soft thresholding minimax risk, and the corresponding optimal threshold value

$$M^{\#}(\varepsilon) \equiv \min_{\alpha \in \mathbb{R}_+} M(\varepsilon,\alpha), \quad \alpha^{\#}(\varepsilon) \equiv \arg\min_{\alpha \in \mathbb{R}_+} M(\varepsilon,\alpha). \tag{9.22}$$

The functions $M^{\#}(\varepsilon)$ and $\alpha^{\#}(\varepsilon)$ are plotted in Figure 9.3. For comparison we also plot the analogous functions when the class $\mathcal{F}_\varepsilon$ is replaced by $\mathcal{F}_\varepsilon(a) = \{p_0 \in \mathcal{F}_\varepsilon : \int x^2 p_0(\mathrm{d}x) \leq a^2\}$ of sparse random variables with bounded second moment. Of particular interest is the behavior of these curves in the very sparse limit $\varepsilon \to 0$,

$$\alpha^{\#}(\varepsilon) = \sqrt{2\log(1/\varepsilon)}\{1 + o(1)\}. \tag{9.23}$$

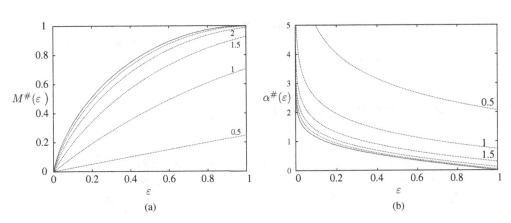

Figure 9.3 (a) (Continuous line): minimax mean square error under soft thresholding for estimation of ε-sparse random variable in Gaussian noise. Dotted lines correspond to signals of bounded second moment (labels on the curves refer to the maximum allowed value of $[\int x^2 \, p_0(\mathrm{d}x)]^{1/2}$). (b) (Continuous line): optimal threshold level for the same estimation problem. Dotted lines again refer to the case of bounded second moment.

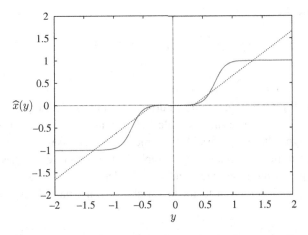

Figure 9.4 Continuous line: The MMSE estimator for the three-point distribution (9.15) with $\varepsilon = 0.1$, when the noise has standard deviation $\sigma = 0.3$. Dotted line: the minimax soft threshold estimator for the same setting. The corresponding mean square errors are plotted in Figure 9.2.

Getting back to Figure 9.2, the reader will notice that there is a significant gap between the minimal MSE and the MSE achieved by soft thresholding. This is the price paid by using an estimator that is *uniformly good* over the class $\mathcal{F}_\varepsilon$ instead of one that is tailored for the distribution p_0 at hand. Figure 9.4 compares the two estimators for $\sigma = 0.3$. One might wonder whether *all* this price has to be paid, i.e. whether we can reduce the gap by using a more complex function instead of the soft threshold $\eta(y; \theta)$. The answer is both yes and no. On one hand, there exist provably superior – in minimax sense – estimators over $\mathcal{F}_\varepsilon$. Such estimators are of course more complex than simple soft thresholding. On the other hand, better estimators have the same minimax risk $M^{\#}(\varepsilon) = (2\log(1/\varepsilon))^{-1}\{1+o(1)\}$ in the very sparse limit, i.e. they improve only the $o(1)$ term as $\varepsilon \to 0$ [23, 22, 41].

9.4 Inference via message passing

The task of extending the theory of the previous section to the vector case (9.1) might appear daunting. It turns out that such extension is instead possible in specific high-dimensional limits. The key step consists in introducing an appropriate message passing algorithm to solve the optimization problem (9.12) and then analyzing its behavior.

9.4.1 The min-sum algorithm

We start by considering the min-sum algorithm. Min-sum is a popular optimization algorithm for graph-structured cost functions (see for instance [65, 42, 58, 61] and references therein). In order to introduce the algorithm, we consider a general cost function over $x = (x_1, \ldots, x_n)$, that decomposes according to a factor graph such as the one shown in Figure 9.1:

$$\mathcal{C}(x) = \sum_{a \in F} \mathcal{C}_a(x_{\partial a}) + \sum_{i \in V} \mathcal{C}_i(x_i). \tag{9.24}$$

Here F is the set of m *factor nodes* (squares in Figure 9.1) and V is the set of n *variable nodes* (circles in the same figure). Further ∂a is the set of neighbors of node a and $x_{\partial a} = (x_i : i \in \partial a)$. The min-sum algorithm is an iterative algorithm of the belief-propagation type. Its basic variables are messages: a message is associated to each directed edge in the underlying factor graph. In the present case, messages are functions on the optimization variables, and we will denote them as $J_{i \to a}^t(x_i)$ (from variable to factor), $\widehat{J}_{a \to i}^t(x_i)$ (from factor to variable), with t indicating the iteration

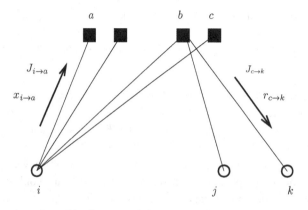

Figure 9.5 A portion of the factor graph from Figure 9.1 with notation for messages.

number. Figure 9.5 describes the association of messages to directed edges in the factor graph. Messages are meaningful up to an additive constant, and therefore we will use the special symbol $\cong$ to denote identity up to an additive constant independent of the argument x_i. At the t-th iteration they are updated as follows[1]

$$J_{i\to a}^{t+1}(x_i) \cong \mathcal{C}_i(x_i) + \sum_{b\in\partial i\setminus a} \widehat{J}_{b\to i}^t(x_i), \qquad (9.25)$$

$$\widehat{J}_{a\to i}^t(x_i) \cong \min_{x_{\partial a\setminus i}} \left\{ \mathcal{C}_a(x_{\partial a}) + \sum_{j\in\partial a\setminus i} J_{j\to a}^t(x_j) \right\}. \qquad (9.26)$$

Eventually, the optimum is approximated by

$$\widehat{x}_i^{t+1} = \arg\min_{x_i\in\mathbb{R}} J_i^{t+1}(x_i), \qquad (9.27)$$

$$J_i^{t+1}(x_i) \cong \mathcal{C}_i(x_i) + \sum_{b\in\partial i} \widehat{J}_{b\to i}^t(x_i). \qquad (9.28)$$

There exists a vast literature justifying the use of algorithms of this type, applying them on concrete problems, and developing modifications of the basic iteration with better properties [65, 42, 58, 61, 83, 44]. Here we limit ourselves to recalling that the iteration (9.25), (9.26) can be regarded as a dynamic programming iteration that computes the minimum cost when the underlying graph is a tree. Its application to loopy graphs (i.e. graphs with closed loops) is not generally guaranteed to converge.

At this point we notice that the LASSO cost function Eq. (9.12) can be decomposed as in Eq. (9.24),

$$\mathcal{C}_{A,y}(x) \equiv \frac{1}{2}\sum_{a\in F}(y_a - A_a^T x)^2 + \lambda\sum_{i\in V}|x_i|. \qquad (9.29)$$

The min-sum updates read

$$J_{i\to a}^{t+1}(x_i) \cong \lambda|x_i| + \sum_{b\in\partial i\setminus a} \widehat{J}_{b\to i}^t(x_i), \qquad (9.30)$$

$$\widehat{J}_{a\to i}^t(x_i) \cong \min_{x_{\partial a\setminus i}} \left\{ \frac{1}{2}(y_a - A_a^T x)^2 + \sum_{j\in\partial a\setminus i} J_{j\to a}^t(x_j) \right\}. \qquad (9.31)$$

9.4.2 Simplifying min-sum by quadratic approximation

Unfortunately, an exact implementation of the min-sum iteration appears extremely difficult because it requires to keep track of $2mn$ messages, each being a function on the

---

[1] The reader will notice that for a dense matrix A, $\partial i = [n]$ and $\partial a = [m]$. We will nevertheless stick to the more general notation, since it is somewhat more transparent.

real axis. A possible approach consists in developing *numerical* approximations to the messages. This line of research was initiated in [69].

Here we will overview an alternative approach that consists in deriving *analytical* approximations [26, 27, 28]. Its advantage is that it leads to a remarkably simple algorithm, which will be discussed in the next section. In order to justify this algorithm we will first derive a simplified message passing algorithm, whose messages are simple real numbers (instead of functions), and then (in the next section) reduce the number of messages from $2mn$ to $m+n$.

Throughout the derivation we shall assume that the matrix A is normalized in such a way that its columns have zero mean and unit ℓ_2 norm. Explicitly, we have $\sum_{a=1}^{m} A_{ai} = 0$ and $\sum_{a=1}^{m} A_{ai}^2 = 1$. In fact it is only sufficient that these conditions are satisfied asymptotically for large system sizes. Since however we are only presenting a heuristic argument, we defer a precise formulation of this assumption until Section 9.6.2. We also assume that its entries have roughly the same magnitude $O(1/\sqrt{m})$. Finally, we assume that m scales linearly with n. These assumptions are verified by many examples of sensing matrices in compressed sensing, e.g. random matrices with i.i.d. entries or random Fourier sections. Modifications of the basic algorithm that cope with strong violations of these assumptions are discussed in [8].

It is easy to see by induction that the messages $J_{i \to a}^t(x_i)$, $\widehat{J}_{a \to i}^t(x_i)$ remain, for any t, convex functions, provided they are initialized as convex functions at $t=0$. In order to simplify the min-sum equations, we will approximate them by quadratic functions. Our first step consists in noticing that, as a consequence of Eq. (9.31), the function $\widehat{J}_{a \to i}^t(x_i)$ depends on its argument only through the combination $A_{ai} x_i$. Since $A_{ai} \ll 1$, we can approximate this dependence through a Taylor expansion (without loss of generality setting $\widehat{J}_{a \to i}^t(0) = 0$):

$$\widehat{J}_{a \to i}^t(x_i) \cong -\alpha_{a \to i}^t(A_{ai} x_i) + \frac{1}{2} \beta_{a \to i}^t (A_{ai} x_i)^2 + O(A_{ai}^3 x_i^3). \tag{9.32}$$

The reason for stopping this expansion at third order should become clear in a moment. Indeed substituting in Eq. (9.30) we get

$$J_{i \to a}^{t+1}(x_i) \cong \lambda |x_i| - \bigg(\sum_{b \in \partial i \setminus a} A_{bi} \alpha_{b \to i}^t \bigg) x_i + \frac{1}{2} \bigg(\sum_{b \in \partial i \setminus a} A_{bi}^2 \beta_{a \to i}^t \bigg) x_i^2 + O(n A_{\cdot i}^3 x_i^3). \tag{9.33}$$

Since $A_{ai} = O(1/\sqrt{n})$, the last term is negligible. At this point we want to approximate $J_{i \to a}^t$ by its second-order Taylor expansion around its minimum. The reason for this is that only this order of the expansion matters when plugging these messages in Eq. (9.31) to compute $\alpha_{a \to i}^t$, $\beta_{a \to i}^t$. We thus define the quantities $x_{i \to a}^t$, $\gamma_{i \to a}^t$ as parameters of this Taylor expansion:

$$J_{i \to a}^t(x_i) \cong \frac{1}{2\gamma_{i \to a}^t}(x_i - x_{i \to a}^t)^2 + O((x_i - x_{i \to a}^t)^3). \tag{9.34}$$

Here we include also the case in which the minimum of $J^t_{i \to a}(x_i)$ is achieved at $x_i = 0$ (and hence the function is not differentiable at its minimum) by letting $\gamma^t_{i \to a} = 0$ in that case. Comparing Eqs. (9.33) and (9.34), and recalling the definition of $\eta(\cdot\,;\cdot)$, cf. Eq. (9.18), we get

$$x^{t+1}_{i \to a} = \eta(\mathsf{a}_1;\mathsf{a}_2), \qquad \gamma^{t+1}_{i \to a} = \eta'(\mathsf{a}_1;\mathsf{a}_2), \qquad (9.35)$$

where $\eta'(\cdot\,;\cdot)$ denotes the derivative of η with respect to its first argument and we defined

$$\mathsf{a}_1 \equiv \frac{\sum_{b \in \partial i \setminus a} A_{bi} \alpha^t_{b \to i}}{\sum_{b \in \partial i \setminus a} A^2_{bi} \beta^t_{b \to i}}, \qquad \mathsf{a}_2 \equiv \frac{\lambda}{\sum_{b \in \partial i \setminus a} A^2_{bi} \beta^t_{b \to i}}. \qquad (9.36)$$

Finally, by plugging the parameterization (9.34) in Eq. (9.31) and comparing with Eq. (9.32), we can compute the parameters $\alpha^t_{a \to i}$, $\beta^t_{a \to i}$. A long but straightforward calculation yields

$$\alpha^t_{a \to i} = \frac{1}{1 + \sum_{j \in \partial a \setminus i} A^2_{aj} \gamma^t_{j \to a}} \left\{ y_a - \sum_{j \in \partial a \setminus i} A_{aj} x^t_{j \to a} \right\}, \qquad (9.37)$$

$$\beta^t_{a \to i} = \frac{1}{1 + \sum_{j \in \partial a \setminus i} A^2_{aj} \gamma^t_{j \to a}}. \qquad (9.38)$$

Equations (9.35) to (9.38) define a message passing algorithm that is considerably simpler than the original min-sum algorithm: each message consists of a pair of real numbers, namely $(x^t_{i \to a}, \gamma^t_{i \to a})$ for variable-to-factor messages and $(\alpha_{a \to i}, \beta_{a \to i})$ for factor-to-variable messages. In the next section we will simplify it further and construct an algorithm (AMP) with several interesting properties. Let us pause a moment to make two observations:

1. The soft-thresholding operator that played an important role in the scalar case, cf. Eq. (9.3), reappeared in Eq. (9.35). Notice however that the threshold value that follows as a consequence of our derivation is not the naive one, namely equal to the regularization parameter λ, but rather a rescaled one.
2. Our derivation leveraged on the assumption that the matrix entries A_{ai} are all of the same order, namely $O(1/\sqrt{m})$. It would be interesting to repeat the above derivation under different assumptions on the sensing matrix.

9.5 Approximate message passing

The algorithm derived above is still complex in that its memory requirements scale proportionally to the *product* of the number of dimensions of the signal and the number of measurements. Further, its computational complexity per iteration scales quadratically as well. In this section we will introduce a simpler algorithm, and subsequently discuss its derivation from the one in the previous section.

9.5.1 The AMP algorithm, some of its properties, ...

The AMP (for approximate message passing) algorithm is parameterized by two sequences of scalars: the thresholds $\{\theta_t\}_{t\geq 0}$ and the "reaction terms" $\{\mathsf{b}_t\}_{t\geq 0}$. Starting with initial condition $x^0 = 0$, it constructs a sequence of estimates $x^t \in \mathbb{R}^n$, and residuals $r^t \in \mathbb{R}^m$, according to the following iteration

$$x^{t+1} = \eta(x^t + A^T r^t; \theta_t), \tag{9.39}$$

$$r^t = y - Ax^t + \mathsf{b}_t r^{t-1}, \tag{9.40}$$

for all $t \geq 0$ (with convention $r^{-1} = 0$). Here and below, given a scalar function $f: \mathbb{R} \to \mathbb{R}$, and a vector $u \in \mathbb{R}^\ell$, we adopt the convention of denoting by $f(u)$ the vector $(f(u_1), \ldots, f(u_\ell))$.

The choice of parameters $\{\theta_t\}_{t\geq 0}$ and $\{\mathsf{b}_t\}_{t\geq 0}$ is tightly constrained by the connection with the min-sum algorithm, as will be discussed below, but the connection with the LASSO is more general. Indeed, as formalized by the proposition below, general sequences $\{\theta_t\}_{t\geq 0}$ and $\{\mathsf{b}_t\}_{t\geq 0}$ can be used as long as (x^t, z^t) converges.

PROPOSITION 9.1 *Let (x^*, r^*) be a fixed point of the iteration (9.39), (9.40) for $\theta_t = \theta$, $\mathsf{b}_t = \mathsf{b}$ fixed. Then x^* is a minimum of the LASSO cost function (9.12) for*

$$\lambda = \theta(1 - \mathsf{b}). \tag{9.41}$$

Proof. From Eq. (9.39) we get the fixed point condition

$$x^* + \theta v = x^* + A^T r^*, \tag{9.42}$$

for $v \in \mathbb{R}^n$ such that $v_i = \text{sign}(x_i^*)$ if $x_i^* \neq 0$ and $v_i \in [-1, +1]$ otherwise. In other words, v is a subgradient of the ℓ_1-norm at x^*, $v \in \partial \|x^*\|_1$. Further from Eq. (9.40) we get $(1 - \mathsf{b})r^* = y - Ax^*$. Substituting in the above equation, we get

$$\theta(1 - \mathsf{b})v^* = A^T(y - Ax^*),$$

which is just the stationarity condition for the LASSO cost function if $\lambda = \theta(1 - \mathsf{b})$. □

As a consequence of this proposition, if we find sequences $\{\theta_t\}_{t\geq 0}$, $\{\mathsf{b}_t\}_{t\geq 0}$ that converge, and such that the estimates x^t converge as well, then we are guaranteed that the limit is a LASSO optimum. The connection with the message passing min-sum algorithm (see Section 9.5.2) implies an unambiguous prescription for b_t:

$$\mathsf{b}_t = \frac{1}{m} \|x^t\|_0, \tag{9.43}$$

where $\|u\|_0$ denotes the 0 pseudo-norm of vector u, i.e. the number of its nonzero components. The choice of the sequence of thresholds $\{\theta_t\}_{t\geq 0}$ is somewhat more flexible. Recalling the discussion of the scalar case, it appears to be a good choice to use $\theta_t = \alpha \tau_t$ where $\alpha > 0$ and τ_t is the root mean square error of the un-thresholded estimate

$(x^t + A^T r^t)$. It can be shown that the latter is (in a high-dimensional setting) well approximated by $(\|r^t\|_2^2/m)^{1/2}$. We thus obtain the prescription

$$\theta_t = \alpha \widehat{\tau}_t, \qquad \widehat{\tau}_t^2 = \frac{1}{m}\|r^t\|_2^2. \tag{9.44}$$

Alternative estimators can be used instead of $\widehat{\tau}_t$ as defined above. For instance, the median of $\{|r_i^t|\}_{i\in[m]}$, can be used to define the alternative estimator:

$$\widehat{\tau}_t^2 = \frac{1}{\Phi^{-1}(3/4)}|r^t|_{(m/2)}, \tag{9.45}$$

where $|u|_{(\ell)}$ is the ℓth largest magnitude among the entries of a vector u, and $\Phi^{-1}(3/4) \approx 0.6745$ denotes the median of the absolute values of a Gaussian random variable.

By Proposition 9.1, if the iteration converges to $(\widehat{x}, \widehat{r})$, then this is a minimum of the LASSO cost function, with regularization parameter

$$\lambda = \alpha \frac{\|\widehat{r}\|_2}{\sqrt{m}}\left(1 - \frac{\|\widehat{x}\|_0}{m}\right) \tag{9.46}$$

(in case the threshold is chosen as per Eq. (9.44)). While the relation between α and λ is not fully explicit (it requires to find the optimum $\widehat{x}$), in practice α is as useful as λ: both play the role of knobs that adjust the level of sparsity of the solution we are seeking.

We conclude by noting that the AMP algorithm (9.39), (9.40) is quite close to iterative soft thresholding (IST), a well-known algorithm for the same problem that proceeds by

$$x^{t+1} = \eta(x^t + A^T r^t; \theta_t), \tag{9.47}$$

$$r^t = y - Ax^t. \tag{9.48}$$

The only (but important) difference lies in the introduction of the term $\mathsf{b}_t r^{t-1}$ in the second equation, cf. Eq. (9.40). This can be regarded as a momentum term with a very specific prescription on its size, cf. Eq. (9.43). A similar term – with motivations analogous to the one presented below – is popular under the name of 'Onsager term' in statistical physics [64, 76, 60].

9.5.2 ... and its derivation

In this section we present a heuristic derivation of the AMP iteration in Eqs. (9.39), (9.40) starting from the standard message passing formulation given by Eqs. (9.35) to (9.38). Our objective is to develop an intuitive understanding of the AMP iteration, as well as of the prescription (9.43). Throughout our argument, we treat m as scaling linearly with n. A full justification of the derivation presented here is beyond the scope of this review: the actual rigorous analysis of the AMP algorithm goes through an indirect and very technical mathematical proof [9].

We start by noticing that the sums $\sum_{j\in\partial a\setminus i} A_{aj}^2 \gamma_{j\to a}^t$ and $\sum_{b\in\partial i\setminus a} A_{bi}^2 \beta_{b\to i}^t$ are sums of $\Theta(n)$ terms, each of order $1/n$ (because $A_{ai}^2 = O(1/n)$). Notice that the terms in these sums are not independent: nevertheless by analogy to what happens in the case of sparse graphs [62, 59, 68, 3], one can hope that dependencies are weak. It is then reasonable to think that a law of large numbers applies and that therefore these sums can be replaced by quantities that do not depend on the instance or on the row/column index.

We then let $r_{a \to i}^t = \alpha_{a \to i}^t / \beta_{a \to i}^t$ and rewrite the message passing iteration, cf. Eqs. (9.35) to (9.38), as

$$r_{a \to i}^t = y_a - \sum_{j \in [n] \setminus i} A_{aj} x_{j \to a}^t, \qquad (9.49)$$

$$x_{i \to a}^{t+1} = \eta \Big(\sum_{b \in [m] \setminus a} A_{bi} r_{b \to i}^t; \theta_t \Big), \qquad (9.50)$$

where $\theta_t \approx \lambda / \sum_{b \in \partial i \setminus a} A_{bi}^2 \beta_{b \to i}^t$ is – as mentioned – treated as independent of b.

Notice that on the right-hand side of both equations above, the messages appear in sums over $\Theta(n)$ terms. Consider for instance the messages $\{r_{a \to i}^t\}_{i \in [n]}$ for a fixed node $a \in [m]$. These depend on $i \in [n]$ only because the term excluded from the sum on the right-hand side of Eq. (9.49) changes. It is therefore natural to guess that $r_{a \to i}^t = r_a^t + O(n^{-1/2})$ and $x_{i \to a}^t = x_i^t + O(m^{-1/2})$, where r_a^t only depends on the index a (and not on i), and x_i^t only depends on i (and not on a).

A naive approximation would consist in neglecting the $O(n^{-1/2})$ correction but this approximation turns out to be inaccurate even in the large-n limit. We instead set

$$r_{a \to i}^t = r_a^t + \delta r_{a \to i}^t, \qquad x_{i \to a}^t = x_i^t + \delta x_{i \to a}^t.$$

Substituting in Eqs. (9.49) and (9.50), we get

$$r_a^t + \delta r_{a \to i}^t = y_a - \sum_{j \in [n]} A_{aj}(x_j^t + \delta x_{j \to a}^t) + A_{ai}(x_i^t + \delta x_{i \to a}^t),$$

$$x_i^{t+1} + \delta x_{i \to a}^{t+1} = \eta \Big(\sum_{b \in [m]} A_{bi}(r_b^t + \delta r_{b \to i}^t) - A_{ai}(r_a^t + \delta r_{a \to i}^t); \theta_t \Big).$$

We will now drop the terms that are negligible without writing explicitly the error terms. First of all notice that single terms of the type $A_{ai} \delta r_{a \to i}^t$ are of order $1/n$ and can be safely neglected. Indeed $\delta r_{a \to i} = O(n^{-1/2})$ by our ansatz, and $A_{ai} = O(n^{-1/2})$ by definition. We get

$$r_a^t + \delta r_{a \to i}^t = y_a - \sum_{j \in [n]} A_{aj}(x_j^t + \delta x_{j \to a}^t) + A_{ai} x_i^t,$$

$$x_i^{t+1} + \delta x_{i \to a}^{t+1} = \eta \Big(\sum_{b \in [m]} A_{bi}(r_b^t + \delta r_{b \to i}^t) - A_{ai} r_a^t; \theta_t \Big).$$

We next expand the second equation to linear order in $\delta x^t_{i \to a}$ and $\delta r^t_{a \to i}$:

$$r^t_a + \delta r^t_{a \to i} = y_a - \sum_{j \in [n]} A_{aj}(x^t_j + \delta x^t_{j \to a}) + A_{ai}x^t_i,$$

$$x^{t+1}_i + \delta x^{t+1}_{i \to a} = \eta\Big(\sum_{b \in [m]} A_{bi}(r^t_b + \delta r^t_{b \to i}); \theta_t\Big) - \eta'\Big(\sum_{b \in [m]} A_{bi}(r^t_b + \delta r^t_{b \to i}); \theta_t\Big) A_{ai} r^t_a.$$

The careful reader might be puzzled by the fact that the soft thresholding function $u \mapsto \eta(u; \theta)$ is non-differentiable at $u \in \{+\theta, -\theta\}$. However, the rigorous analysis carried out in [9] through a different (and more technical) method reveals that almost-everywhere differentiability is sufficient here.

Notice that the last term on the right-hand side of the first equation above is the only one dependent on i, and we can therefore identify this term with $\delta r^t_{a \to i}$. We obtain the decomposition

$$r^t_a = y_a - \sum_{j \in [n]} A_{aj}(x^t_j + \delta x^t_{j \to a}), \tag{9.51}$$

$$\delta r^t_{a \to i} = A_{ai} x^t_i. \tag{9.52}$$

Analogously for the second equation we get

$$x^{t+1}_i = \eta\Big(\sum_{b \in [m]} A_{bi}(r^t_b + \delta r^t_{b \to i}); \theta_t\Big), \tag{9.53}$$

$$\delta x^{t+1}_{i \to a} = -\eta'\Big(\sum_{b \in [m]} A_{bi}(r^t_b + \delta r^t_{b \to i}); \theta_t\Big) A_{ai} r^t_a. \tag{9.54}$$

Substituting Eq. (9.52) in Eq. (9.53) to eliminate $\delta r^t_{b \to i}$ we get

$$x^{t+1}_i = \eta\Big(\sum_{b \in [m]} A_{bi} r^t_b + \sum_{b \in [m]} A^2_{bi} x^t_i; \theta_t\Big), \tag{9.55}$$

and using the normalization of A, we get $\sum_{b \in [m]} A^2_{bi} \to 1$, whence

$$x^{t+1} = \eta(x^t + A^T r^t; \theta_t). \tag{9.56}$$

Analogously substituting Eq. (9.54) in (9.51), we get

$$z^t_a = y_a - \sum_{j \in [n]} A_{aj} x^t_j + \sum_{j \in [n]} A^2_{aj} \eta'(x^{t-1}_j + (A^T r^{t-1})_j; \theta_{t-1}) r^{t-1}_a. \tag{9.57}$$

Again, using the law of large numbers and the normalization of A, we get

$$\sum_{j \in [n]} A^2_{aj} \eta'(x^{t-1}_j + (A^T r^{t-1})_j; \theta_{t-1}) \approx \frac{1}{m} \sum_{j \in [n]} \eta'(x^{t-1}_j + (A^T r^{t-1})_j; \theta_{t-1}) = \frac{1}{m} \|x^t\|_0,$$

$$\tag{9.58}$$

whence substituting in (9.57), we obtain Eq. (9.40), with the prescription (9.43) for the Onsager term. This finishes our derivation.

9.6 High-dimensional analysis

The AMP algorithm enjoys several unique properties. In particular it admits an *asymptotically exact* analysis along sequences of instances of diverging size. This is quite remarkable, since all analyses available for other algorithms that solve the LASSO hold only "up to undetermined constants."

In particular in the large system limit (and with the exception of a "phase transition" line), AMP can be shown to converge exponentially fast to the LASSO optimum. Hence the analysis of AMP yields asymptotically exact predictions on the behavior of the LASSO, including in particular the asymptotic mean square error per variable.

9.6.1 Some numerical experiments with AMP

How is it possible that an *asymptotically exact* analysis of AMP can be carried out? Figure 9.6 illustrates the key point. It shows the distribution of un-thresholded estimates $(x^t + A^T r^t)_i$ for coordinates i such that the original signal had value $x_i = +1$. These estimates were obtained using the AMP algorithm (9.39), (9.40) with choice (9.43) of b_t (a) and the iterative soft thresholding algorithm (9.47), (9.48) (b). The same instances (i.e. the same matrices A and measurement vectors y) were used in the two cases, but the resulting distributions are dramatically different. In the case of AMP, the distribution is

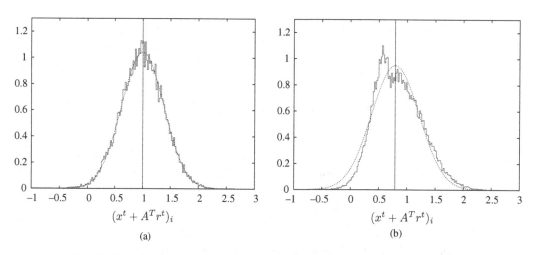

Figure 9.6 Distributions of un-thresholded estimates for AMP (a) and IST (b), after $t = 10$ iterations. These data were obtained using sensing matrices with $m = 2000$, $n = 4000$ and i.i.d. entries uniform in $\{+1/\sqrt{m}, -1/\sqrt{m}\}$. The signal x contained 500 nonzero entries uniform in $\{+1, -1\}$. A total of 40 instances was used to build the histograms. Dotted lines are Gaussian fits and vertical lines represent the fitted mean.

close to Gaussian, with mean on the correct value, $x_i = +1$. For iterative soft thresholding the estimates do not have the correct mean and are not Gaussian.

This phenomenon appears here as an empirical observation, valid for a specific iteration number t, and specific dimensions m, n. In the next section we will explain that it can be proved rigorously in the limit of a large number of dimensions, for all values of iteration number t. Namely, as $m, n \to \infty$ at t fixed, the empirical distribution of $\{(x^t + A^T r^t)_i - x_i\}_{i \in [n]}$ converges to a Gaussian distribution, when x^t and r^t are computed using AMP. The variance of this distribution depends on t, and its evolution with t can be computed exactly. Vice versa, for iterative soft thresholding, the distribution of the same quantities remains non-gaussian.

This dramatic difference remains true for any t, even when AMP and IST converge to the same minimum. Indeed even at the fixed point, the resulting residual r^t is different in the two algorithms, as a consequence of the introduction of the Onsager term.

More importantly, the two algorithms differ dramatically in the rate of convergence. One can interpret the vector $(x^t + A^T r^t) - x$ as "effective noise" after t iterations. Both AMP and IST "denoise" the vector $(x^t + A^T r^t)$ using the soft thresholding operator. As discussed in Section 9.3, the soft thresholding operator is essentially optimal for denoising in Gaussian noise. This suggests that AMP should have superior performances (in the sense of faster convergence to the LASSO minimum) with respect to simple IST.

Figure 9.7 presents the results of a small experiment confirming this expectation. Measurement matrices A with dimensions $m = 1600$, $n = 8000$, were generated randomly with i.i.d. entries $A_{ai} \in \{+1/\sqrt{m}, -1/\sqrt{m}\}$ uniformly at random. We consider here the problem of reconstructing a signal x with entries $x_i \in \{+1, 0, -1\}$ from noiseless measurements $y = Ax$, for different levels of sparsity. Thresholds were set according to

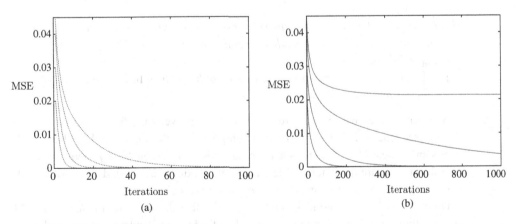

Figure 9.7 Evolution of the mean square error as a function of the number of iterations for AMP (a) and iterative soft thresholding (b), for random measurement matrices A, with i.i.d. entries $A_{ai} \in \{+1/\sqrt{m}, -1/\sqrt{m}\}$ uniformly. Notice the different scales used for the horizontal axis! Here $n = 8000$, $m = 1600$. Different curves depend on different levels of sparsity. The number of nonzero entries of the signal x is, for the various curves, $\|x\|_0 = 160, 240, 320, 360$ (from bottom to top).

the prescription (9.44) with $\alpha = 1.41$ for AMP (the asymptotic theory of [26] yields the prescription $\alpha \approx 1.40814$) and $\alpha = 1.8$ for IST (optimized empirically). For the latter algorithm, the matrix A was rescaled in order to get an operator norm $\|A\|_2 = 0.95$.

Convergence to the original signal x is slower and slower as this becomes less and less sparse.[2] Overall, AMP appears to be at least 10 times faster even on the sparsest vectors (lowest curves in the figure).

9.6.2 State evolution

State evolution describes the asymptotic limit of the AMP estimates as $m, n \to \infty$, for any fixed t. The word "evolution" refers to the fact that one obtains an "effective" evolution with t. The word "state" refers to the fact that the algorithm behavior is captured in this limit by a single parameter (a state) $\tau_t \in \mathbb{R}$.

We will consider sequences of instances of increasing sizes, along which the AMP algorithm behavior admits a nontrivial limit. An instance is completely determined by the measurement matrix A, the signal x, and the noise vector w, the vector of measurements y being given by $y = Ax + w$, cf. Eq. (9.1). While rigorous results have been proved so far only in the case in which the sensing matrices A have i.i.d. Gaussian entries, it is nevertheless useful to collect a few basic properties that the sequence needs to satisfy in order for state evolution to hold.

DEFINITION 9.1 *The sequence of instances* $\{x(n), w(n), A(n)\}_{n \in \mathbb{N}}$ *indexed by* n *is said to be a* converging sequence *if* $x(n) \in \mathbb{R}^n$, $w(n) \in \mathbb{R}^m$, $A(n) \in \mathbb{R}^{m \times n}$ *with* $m = m(n)$ *is such that* $m/n \to \delta \in (0, \infty)$, *and in addition the following conditions hold:*

(a) *The empirical distribution of the entries of* $x(n)$ *converges weakly to a probability measure* p_0 *on* $\mathbb{R}$ *with bounded second moment. Further* $n^{-1} \sum_{i=1}^n x_i(n)^2 \to \mathbb{E}_{p_0}\{X_0^2\}$.
(b) *The empirical distribution of the entries of* $w(n)$ *converges weakly to a probability measure* p_W *on* $\mathbb{R}$ *with bounded second moment. Further* $m^{-1} \sum_{i=1}^m w_i(n)^2 \to \mathbb{E}_{p_W}\{W^2\} \equiv \sigma^2$.
(c) *If* $\{e_i\}_{1 \leq i \leq n}$, $e_i \in \mathbb{R}^n$ *denotes the canonical basis, then* $\lim_{n \to \infty} \max_{i \in [n]} \|A(n) e_i\|_2 = 1$ *and* $\lim_{n \to \infty} \min_{i \in [n]} \|A(n) e_i\|_2 = 1$.

As mentioned above, rigorous results have been proved only for a subclass of converging sequences, namely under the assumption that the matrices $A(n)$ have i.i.d. Gaussian entries. Notice that such matrices satisfy condition (c) by elementary tail bounds on χ-square random variables. The same condition is satisfied by matrices with i.i.d. sub-gaussian entries thanks to concentration inequalities [53].

On the other hand, numerical simulations show that the same limit behavior should apply within a much broader domain, including for instance random matrices with i.i.d. entries under an appropriate moment condition. This *universality* phenomenon is well-known in random matrix theory whereby asymptotic results initially established for

---

[2] Indeed basis pursuit (i.e. reconstruction via ℓ_1 minimization) fails with high probability if $\|x\|_0 / m \gtrsim 0.243574$, see [29] and Section 9.6.6.

Gaussian matrices were subsequently proved for broader classes of matrices. Rigorous evidence in this direction is presented in [46]. This paper shows that the normalized cost $\min_{x \in \mathbb{R}^n} \mathcal{C}_{A(n),y(n)}(x)/n$ has a limit for $n \to \infty$, which is universal with respect to random matrices A with i.i.d. entries. (More precisely, it is universal provided $\mathbb{E}\{A_{ij}\} = 0$, $\mathbb{E}\{A_{ij}^2\} = 1/m$ and $\mathbb{E}\{A_{ij}^6\} \leq C/m^3$ for some n-independent constant C.)

For a converging sequence of instances $\{x(n), w(n), A(n)\}_{n \in \mathbb{N}}$, and an arbitrary sequence of thresholds $\{\theta_t\}_{t \geq 0}$ (independent of n), the AMP iteration (9.39), (9.40) admits a high-dimensional limit which can be characterized exactly, provided Eq. (9.43) is used for fixing the Onsager term. This limit is given in terms of the trajectory of a simple one-dimensional iteration termed *state evolution* which we will describe next.

Define the sequence $\{\tau_t^2\}_{t \geq 0}$ by setting $\tau_0^2 = \sigma^2 + \mathbb{E}\{X_0^2\}/\delta$ (for $X_0 \sim p_0$ and $\sigma^2 \equiv \mathbb{E}\{W^2\}$, $W \sim p_W$) and letting, for all $t \geq 0$:

$$\tau_{t+1}^2 = \mathsf{F}(\tau_t^2, \theta_t), \tag{9.59}$$

$$\mathsf{F}(\tau^2, \theta) \equiv \sigma^2 + \frac{1}{\delta} \mathbb{E}\{[\eta(X_0 + \tau Z; \theta) - X_0]^2\}, \tag{9.60}$$

where $Z \sim \mathsf{N}(0,1)$ is independent of $X_0 \sim p_0$. Notice that the function F depends implicitly on the law p_0. Further, the state evolution $\{\tau_t^2\}_{t \geq 0}$ depends on the specific converging sequence through the law p_0, and the second moment of the noise $\mathbb{E}_{p_W}\{W^2\}$, cf. Definition 9.1.

We say a function $\psi : \mathbb{R}^k \to \mathbb{R}$ is *pseudo-Lipschitz* if there exists a constant $L > 0$ such that for all $x, y \in \mathbb{R}^k$: $|\psi(x) - \psi(y)| \leq L(1 + \|x\|_2 + \|y\|_2)\|x - y\|_2$. (This is a special case of the definition used in [9] where such a function is called pseudo-Lipschitz *of order 2*.)

The following theorem was conjectured in [26], and proved in [9]. It shows that the behavior of AMP can be tracked by the above state evolution recursion.

THEOREM 9.1 ([9]) *Let $\{x(n), w(n), A(n)\}_{n \in \mathbb{N}}$ be a converging sequence of instances with the entries of $A(n)$ i.i.d. normal with mean 0 and variance $1/m$, while the signals $x(n)$ and noise vectors $w(n)$ satisfy the hypotheses of Definition 9.1. Let $\psi_1 : \mathbb{R} \to \mathbb{R}$, $\psi_2 : \mathbb{R} \times \mathbb{R} \to \mathbb{R}$ be pseudo-Lipschitz functions. Finally, let $\{x^t\}_{t \geq 0}$, $\{r^t\}_{t \geq 0}$ be the sequences of estimates and residuals produced by AMP, cf. Eqs. (9.39), (9.40). Then, almost surely*

$$\lim_{n \to \infty} \frac{1}{m} \sum_{a=1}^{m} \psi_1(r_a^t) = \mathbb{E}\{\psi_1(\tau_t Z)\}, \tag{9.61}$$

$$\lim_{n \to \infty} \frac{1}{n} \sum_{i=1}^{n} \psi_2(x_i^{t+1}, x_i) = \mathbb{E}\{\psi_2(\eta(X_0 + \tau_t Z; \theta_t), X_0)\}, \tag{9.62}$$

where $Z \sim \mathsf{N}(0,1)$ is independent of $X_0 \sim p_0$.

It is worth pausing for a few remarks.

REMARK 9.1 Theorem 9.1 holds for any choice of the sequence of thresholds $\{\theta_t\}_{t\geq 0}$. It does not require – for instance – that the latter converge. Indeed [9] proves a more general result that holds for a broad class of approximate message passing algorithms. The more general theorem establishes the validity of state evolution *in this broad context*.

For instance, the soft thresholding functions $\eta(\,\cdot\,;\theta_t)$ can be replaced by a generic sequence of Lipschitz continuous functions, provided the coefficients b_t in Eq. (9.40) are suitably modified.

REMARK 9.2 This theorem does not require the vectors $x(n)$ to be sparse. The use of other functions instead of the soft thresholding functions $\eta(\,\cdot\,;\theta_t)$ in the algorithm can be useful for estimating such non-sparse vectors.

Alternative nonlinearities, can also be useful when additional information on the entries of $x(n)$ is available.

REMARK 9.3 While the theorem requires the matrices $A(n)$ to be random, neither the signal $x(n)$ nor the noise vectors $w(n)$ need to be random. They are generic deterministic sequences of vectors under the conditions of Definition 9.1.

The fundamental reason for this universality is that the matrix A is both row and column exchangeable. Row exchangeability guarantees universality with respect to the signals $x(n)$, while column exchangeability guarantees universality with respect to the noise $w(n)$. To see why, observe that, by row exchangeability (for instance), $x(n)$ can be replaced by the random vector obtained by randomly permuting its entries. Now, the distribution of such a random vector is very close (in appropriate sense) to the one of a random vector with i.i.d. entries whose distribution matches the empirical distribution of $x(n)$.

Theorem 9.1 strongly supports both the use of soft thresholding, and the choice of the threshold level in Eq. (9.44) or (9.45). Indeed Eq. (9.61) states that the components of r^t are approximately i.i.d. $\mathsf{N}(0,\tau_t^2)$, and hence both definitions of $\widehat{\tau}_t$ in Eq. (9.44) or (9.45) provide consistent estimators of τ_t. Further, Eq. (9.61) implies that the components of the deviation $(x^t + A^T r^t - x)$ are also approximately i.i.d. $\mathsf{N}(0,\tau_t^2)$. In other words, the estimate $(x^t + A^T r^t)$ is equal to the actual signal plus noise of variance τ_t^2, as illustrated in Figure 9.6. According to our discussion of scalar estimation in Section 9.3, the correct way of reducing the noise is to apply soft thresholding with threshold level $\alpha\tau_t$.

The choice $\theta_t = \alpha\tau_t$ with α fixed has another important advantage. In this case, the sequence $\{\tau_t\}_{t\geq 0}$ is determined by the one-dimensional recursion

$$\tau_{t+1}^2 = \mathsf{F}(\tau_t^2, \alpha\tau_t). \tag{9.63}$$

The function $\tau^2 \mapsto \mathsf{F}(\tau^2, \alpha\tau)$ depends on the distribution of X_0 as well as on the other parameters of the problem. An example is plotted in Figure 9.8. It turns out that the behavior shown here is generic: the function is always non-decreasing and concave. This remark allows one to easily prove the following.

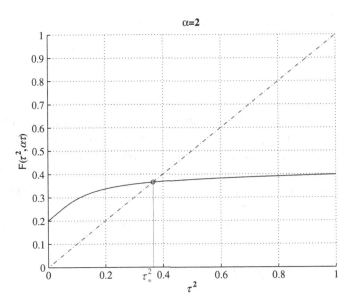

Figure 9.8 Mapping $\tau^2 \mapsto \mathsf{F}(\tau^2, \alpha\tau)$ for $\alpha = 2$, $\delta = 0.64$, $\sigma^2 = 0.2$, $p_0(\{+1\}) = p_0(\{-1\}) = 0.064$ and $p_0(\{0\}) = 0.872$.

PROPOSITION 9.2 (28) *Let $\alpha_{\min} = \alpha_{\min}(\delta)$ be the unique non-negative solution of the equation*

$$(1+\alpha^2)\Phi(-\alpha) - \alpha\phi(\alpha) = \frac{\delta}{2}, \qquad (9.64)$$

with $\phi(z) \equiv e^{-z^2/2}/\sqrt{2\pi}$ the standard Gaussian density and $\Phi(z) \equiv \int_{-\infty}^{z} \phi(x)\,dx$.
For any $\sigma^2 > 0$, $\alpha > \alpha_{\min}(\delta)$, the fixed point equation $\tau^2 = \mathsf{F}(\tau^2, \alpha\tau)$ admits a unique solution. Denoting by $\tau_* = \tau_*(\alpha)$ this solution, we have $\lim_{t\to\infty} \tau_t = \tau_*(\alpha)$.

It can also be shown that, under the choice $\theta_t = \alpha\tau_t$, convergence is exponentially fast unless the problem parameters take some "exceptional" values (namely on the phase transition boundary discussed below).

9.6.3 The risk of the LASSO

State evolution provides a scaling limit of the AMP dynamics in the high-dimensional setting. By showing that AMP converges to the LASSO estimator, one can transfer this information to a scaling limit result of the LASSO estimator itself.

Before stating the limit, we have to describe a *calibration* mapping between the AMP parameter α (that defines the sequence of thresholds $\{\theta_t\}_{t\geq 0}$) and the LASSO regularization parameter λ. The connection was first introduced in [28].

We define the function $\alpha \mapsto \lambda(\alpha)$ on $(\alpha_{\min}(\delta), \infty)$, by

$$\lambda(\alpha) \equiv \alpha \tau_* \left[1 - \frac{1}{\delta}\mathbb{P}\{|X_0 + \tau_* Z| \geq \alpha \tau_*\}\right], \qquad (9.65)$$

where $\tau_* = \tau_*(\alpha)$ is the state evolution fixed point defined as per Proposition 9.2. Notice that this relation corresponds to the scaling limit of the general relation (9.41), provided we assume that the solution of the LASSO optimization problem (9.12) is indeed described by the fixed point of state evolution (equivalently, by its $t \to \infty$ limit). This follows by noting that $\theta_t \to \alpha \tau_*$ and that $\|x\|_0/n \to \mathbb{E}\{\eta'(X_0 + \tau_* Z; \alpha \tau_*)\}$. While this is just an interpretation of the definition (9.65), the result presented next implies that the interpretation is indeed correct.

In the following we will need to invert the function $\alpha \mapsto \lambda(\alpha)$. We thus define $\alpha : (0, \infty) \to (\alpha_{\min}, \infty)$ in such a way that

$$\alpha(\lambda) \in \{a \in (\alpha_{\min}, \infty) : \lambda(a) = \lambda\}.$$

The fact that the right-hand side is non-empty, and therefore the function $\lambda \mapsto \alpha(\lambda)$ is well defined, is part of the main result of this section.

THEOREM 9.2 *Let $\{x(n), w(n), A(n)\}_{n \in \mathbb{N}}$ be a converging sequence of instances with the entries of $A(n)$ i.i.d. normal with mean 0 and variance $1/m$. Denote by $\widehat{x}(\lambda)$ the LASSO estimator for instance $(x(n), w(n), A(n))$, with $\sigma^2, \lambda > 0$, and let $\psi : \mathbb{R} \times \mathbb{R} \to \mathbb{R}$ be a pseudo-Lipschitz function. Then, almost surely*

$$\lim_{n \to \infty} \frac{1}{n} \sum_{i=1}^{n} \psi(\widehat{x}_i, x_i) = \mathbb{E}\{\psi(\eta(X_0 + \tau_* Z; \theta_*), X_0)\}, \qquad (9.66)$$

where $Z \sim \mathsf{N}(0,1)$ is independent of $X_0 \sim p_0$, $\tau_* = \tau_*(\alpha(\lambda))$ and $\theta_* = \alpha(\lambda) \tau_*(\alpha(\lambda))$. Further, the function $\lambda \mapsto \alpha(\lambda)$ is well defined and unique on $(0, \infty)$.

The assumption of a converging problem sequence is important for the result to hold, while the hypothesis of Gaussian measurement matrices $A(n)$ is necessary for the proof technique to be applicable. On the other hand, the restrictions $\lambda, \sigma^2 > 0$, and $\mathbb{P}\{X_0 \neq 0\} > 0$ (whence $\tau_* \neq 0$ using Eq. (9.65)) are made in order to avoid technical complications due to degenerate cases. Such cases can be resolved by continuity arguments.

Let us emphasize that some of the remarks made in the case of state evolution, cf. Theorem 9.1, hold for the last theorem as well. More precisely:

REMARK 9.4 Theorem 9.2 does not require either the signal $x(n)$ or the noise vectors $w(n)$ to be random. They are generic deterministic sequences of vectors under the conditions of Definition 9.1.

In particular, it does not require the vectors $x(n)$ to be sparse. Lack of sparsity will reflect in a large risk as computed through the mean square error computed through Eq. (9.66).

On the other hand, when restricting $x(n)$ to be k-sparse for $k = n\varepsilon$ (i.e. to be in the class $\mathcal{F}_{n,k}$), one can derive asymptotically exact estimates for the minimax risk over this class. This will be further discussed in Section 9.6.6.

REMARK 9.5 As a special case, for noiseless measurements $\sigma = 0$, and as $\lambda \to 0$, the above formulae describe the asymptotic risk (e.g. mean square error) for the basis pursuit estimator, minimize $\|x\|$ subject to $y = Ax$. For sparse signals $x(n) \in \mathcal{F}_{n,k}$, $k = n\rho\delta$, the risk vanishes below a certain phase transition line $\rho < \rho_c(\delta)$: this point is further discussed in Section 9.6.6.

Let us now discuss some limitations of this result. Theorem 9.2 assumes that the entries of matrix A are i.i.d. Gaussians. Further, our result is asymptotic, and one might wonder how accurate it is for instances of moderate dimensions.

Numerical simulations were carried out in [28, 4] and suggest that the result is universal over a broader class of matrices and that it is relevant already for n of the order of a few hundreds. As an illustration, we present in Figures 9.9 and 9.10 the outcome of such simulations for two types of random matrices. Simulations with real data can be found in [4]. We generated the signal vector randomly with entries in $\{+1, 0, -1\}$ and $\mathbb{P}(x_{0,i} = +1) = \mathbb{P}(x_{0,i} = -1) = 0.064$. The noise vector w was generated by using i.i.d. $N(0, 0.2)$ entries.

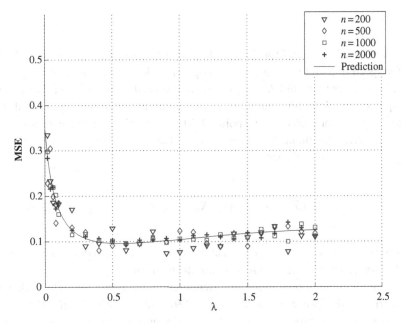

Figure 9.9 MSE as a function of the regularization parameter λ compared to the asymptotic prediction for $\delta = 0.64$ and $\sigma^2 = 0.2$. Here the measurement matrix A has i.i.d. $N(0, 1/m)$ entries. Each point in this plot is generated by finding the LASSO predictor $\widehat{x}$ using a measurement vector $y = Ax + w$ for an independent signal vector x, an independent noise vector w, and an independent matrix A.

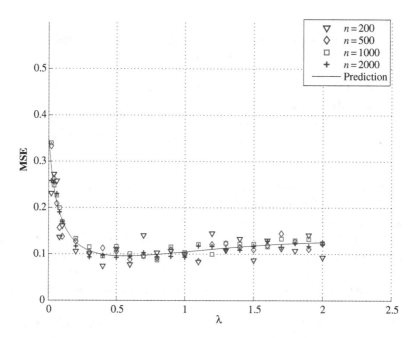

Figure 9.10 As in Figure 9.9, but the measurement matrix A has i.i.d. entries that are equal to $\pm 1/\sqrt{m}$ with equal probabilities.

We solved the LASSO problem (9.12) and computed estimator $\widehat{x}$ using CVX, a package for specifying and solving convex programs [34] and OWLQN, a package for solving large-scale versions of LASSO [1]. We used several values of λ between 0 and 2 and n equal to 200, 500, 1000, and 2000. The aspect ratio of matrices was fixed in all cases to $\delta = 0.64$. For each case, the point (λ, MSE) was plotted and the results are shown in the figures. Continuous lines correspond to the asymptotic prediction by Theorem 9.2 for $\psi(a,b) = (a-b)^2$, namely

$$\lim_{n \to \infty} \frac{1}{n} \|\widehat{x} - x\|_2^2 = \mathbb{E}\big\{ [\eta(X_0 + \tau_* Z; \theta_*) - X_0]^2 \big\} = \delta(\tau_*^2 - \sigma^2).$$

The agreement is remarkably good already for n, m of the order of a few hundreds, and deviations are consistent with statistical fluctuations.

The two figures correspond to different entries distributions: (i) Random Gaussian matrices with aspect ratio δ and i.i.d. $N(0, 1/m)$ entries (as in Theorem 9.2); (ii) Random ± 1 matrices with aspect ratio δ. Each entry is independently equal to $+1/\sqrt{m}$ or $-1/\sqrt{m}$ with equal probability. The resulting MSE curves are hardly distinguishable. Further evidence towards universality will be discussed in Section 9.6.7.

Notice that the asymptotic prediction has a minimum as a function of λ. The location of this minimum can be used to select the regularization parameter.

9.6.4 A decoupling principle

There exists a suggestive interpretation of the state evolution result in Theorem 9.1, as well as of the scaling limit of the LASSO established in Theorem 9.2: *The estimation problem in the vector model $y = Ax + w$ reduces – asymptotically – to n uncoupled scalar estimation problems $\widetilde{y}_i = x_i + \widetilde{w}_i$.* However the noise variance is increased from σ^2 to τ_t^2 (or τ_*^2 in the case of the LASSO), due to "interference" between the original coordinates:

$$y = Ax + w \quad \Leftrightarrow \quad \begin{cases} \widetilde{y}_1 = x_1 + \widetilde{w}_1 \\ \widetilde{y}_2 = x_2 + \widetilde{w}_2 \\ \vdots \\ \widetilde{y}_n = x_n + \widetilde{w}_n \end{cases}. \tag{9.67}$$

An analogous phenomenon is well known in statistical physics and probability theory and takes sometimes the name of "correlation decay" [81, 36, 58]. In the context of CDMA system analysis via replica method, the same phenomenon was also called "decoupling principle" [75, 38].

Notice that the AMP algorithm gives a precise realization of this decoupling principle, since for each $i \in [n]$, and for each number of iterations t, it produces an estimate, namely $(x^t + A^T r^t)_i$ that can be considered a realization of the observation $\widetilde{y}_i$ above. Indeed Theorem 9.1 (see also discussion below the theorem) states that $(x^t + A^T r^t)_i = x_i + \widetilde{w}_i$ with $\widetilde{w}_i$ asymptotically Gaussian with mean 0 and variance τ_t^2.

The fact that observations of distinct coordinates are asymptotically decoupled is stated precisely below.

COROLLARY 9.1 (Decoupling principle, [9]) *Under the assumption of Theorem 9.1, fix $\ell \geq 2$, let $\psi : \mathbb{R}^{2\ell} \to \mathbb{R}$ be any Lipschitz function, and denote by E expectation with respect to a uniformly random subset of distinct indices $J(1), \ldots, J(\ell) \in [n]$.*

Further, for some fixed $t > 0$, let $\widetilde{y}^t = x^t + A^T r^t \in \mathbb{R}^n$. Then, almost surely

$$\lim_{n \to \infty} \mathsf{E}\psi(\widetilde{y}_{J(1)}^t, \ldots, \widetilde{y}_{J(\ell)}^t, x_{J(1)}, \ldots, x_{J(\ell)})$$
$$= \mathbb{E}\{\psi(X_{0,1} + \tau_t Z_1, \ldots, X_{0,\ell} + \tau_t Z_\ell, X_{0,1}, \ldots, X_{0,\ell})\},$$

for $X_{0,i} \sim p_0$ and $Z_i \sim \mathsf{N}(0,1)$, $i = 1, \ldots, \ell$ mutually independent.

9.6.5 A heuristic derivation of state evolution

The state evolution recursion has a simple heuristic description that is useful to present here since it clarifies the difficulties involved in the proof. In particular, this description brings up the key role played by the "Onsager term" appearing in Eq. (9.40) [26].

Consider again the recursion (9.39), (9.40) but introduce the following three modifications: (i) Replace the random matrix A with a new independent copy $A(t)$ at each iteration t; (ii) correspondingly replace the observation vector y with $y^t = A(t)x + w$;

(*iii*) eliminate the last term in the update equation for r^t. We thus get the following dynamics:

$$x^{t+1} = \eta(A(t)^T r^t + x^t; \theta_t), \qquad (9.68)$$
$$r^t = y^t - A(t)x^t, \qquad (9.69)$$

where $A(0), A(1), A(2), \ldots$ are i.i.d. matrices of dimensions $m \times n$ with i.i.d. entries $A_{ij}(t) \sim \mathsf{N}(0, 1/m)$. (Notice that, unlike in the rest of the chapter, we use here the argument of A to denote the iteration number, and not the matrix dimensions.)

This recursion is most conveniently written by eliminating r^t:

$$\begin{aligned} x^{t+1} &= \eta\big(A(t)^T y^t + (\mathbf{I} - A(t)^T A(t))x^t; \theta_t\big), \\ &= \eta\big(x + A(t)^T w + B(t)(x^t - x); \theta_t\big), \end{aligned} \qquad (9.70)$$

where we defined $B(t) = \mathbf{I} - A(t)^T A(t) \in \mathbb{R}^{n \times n}$. Let us stress that this recursion does not correspond to any concrete algorithm, since the matrix A changes from iteration to iteration. It is nevertheless useful for developing intuition.

Using the central limit theorem, it is easy to show that each entry of $B(t)$ is approximately normal, with zero mean and variance $1/m$. Further, distinct entries are approximately pairwise independent. Therefore, if we let $\widetilde{\tau}_t^2 = \lim_{n \to \infty} \|x^t - x\|_2^2 / n$, we obtain that $B(t)(x^t - x)$ converges to a vector with i.i.d. normal entries with 0 mean and variance $n \widetilde{\tau}_t^2 / m = \widetilde{\tau}_t^2 / \delta$. Notice that this is true because $A(t)$ is independent of $\{A(s)\}_{1 \le s \le t-1}$ and, in particular, of $(x^t - x)$.

Conditional on w, $A(t)^T w$ is a vector of i.i.d. normal entries with mean 0 and variance $(1/m)\|w\|_2^2$ which converges by assumption to σ^2. A slightly longer exercise shows that these entries are approximately independent from the ones of $B(t)(x^t - x_0)$. Summarizing, each entry of the vector in the argument of η in Eq. (9.70) converges to $X_0 + \tau_t Z$ with $Z \sim \mathsf{N}(0, 1)$ independent of X_0, and

$$\tau_t^2 = \sigma^2 + \frac{1}{\delta}\widetilde{\tau}_t^2, \qquad (9.71)$$
$$\widetilde{\tau}_t^2 = \lim_{n \to \infty} \frac{1}{n}\|x^t - x\|_2^2.$$

On the other hand, by Eq. (9.70), each entry of $x^{t+1} - x$ converges to $\eta(X_0 + \tau_t Z; \theta_t) - X_0$, and therefore

$$\widetilde{\tau}_{t+1}^2 = \lim_{n \to \infty} \frac{1}{n}\|x^{t+1} - x\|_2^2 = \mathbb{E}\big\{[\eta(X_0 + \tau_t Z; \theta_t) - X_0]^2\big\}. \qquad (9.72)$$

Using together Eq. (9.71) and (9.72) we finally obtain the state evolution recursion, Eq. (9.59).

We conclude that state evolution would hold if the matrix A was drawn independently from the same Gaussian distribution at each iteration. In the case of interest, A does not change across iterations, and the above argument falls apart because x^t and A are dependent. This dependency is non-negligible even in the large system limit $n \to \infty$. This point can be clarified by considering the IST algorithm given by Eqs. (9.47), (9.48). Numerical studies of iterative soft thresholding [57, 26] show that its behavior is dramatically different from that of AMP and in particular *state evolution does not hold for IST*, even in the large system limit.

This is not a surprise: the correlations between A and x^t simply cannot be neglected. On the other hand, adding the Onsager term leads to an asymptotic cancelation of these correlations. As a consequence, state evolution holds for the AMP iteration.

9.6.6 The noise sensitivity phase transition

The formalism developed so far allows one to extend the minimax analysis carried out in the scalar case in Section 9.3 to the vector estimation problem [28]. We define the LASSO mean square error per coordinate when the empirical distribution of the signal converges to p_0, as

$$\mathrm{MSE}(\sigma^2; p_0, \lambda) = \lim_{n \to \infty} \frac{1}{n} \mathbb{E}\{\|\widehat{x}(\lambda) - x\|_2^2\}, \qquad (9.73)$$

where the limit is taken along a converging sequence. This quantity can be computed using Theorem 9.2 for any specific distribution p_0.

We consider again the sparsity class $\mathcal{F}_\varepsilon$ with $\varepsilon = \rho\delta$. Hence $\rho = \|x\|_0/m$ measures the number of nonzero coordinates per measurement. Taking the worst case MSE over this class, and then the minimum over the regularization parameter λ, we get a result that depends on ρ, δ, as well as on the noise level σ^2. The dependence on σ^2 must be linear because the class $\mathcal{F}_{\rho\delta}$ is scale invariant, and we obtain therefore

$$\inf_{\lambda} \sup_{p_0 \in \mathcal{F}_{\rho\delta}} \mathrm{MSE}(\sigma^2; p_0, \lambda) = M^*(\delta, \rho)\sigma^2, \qquad (9.74)$$

for some function $(\delta, \rho) \mapsto M^*(\delta, \rho)$. We call this the LASSO minimax risk. It can be interpreted as the sensitivity (in terms of mean square error) of the LASSO estimator to noise in the measurements.

It is clear that the prediction for $\mathrm{MSE}(\sigma^2; p_0, \lambda)$ provided by Theorem 9.2 can be used to characterize the LASSO minimax risk. What is remarkable is that the resulting formula is so simple.

THEOREM 9.3 [28] *Assume the hypotheses of Theorem 9.2, and recall that $M^\#(\varepsilon)$ denotes the soft thresholding minimax risk over the class $\mathcal{F}_\varepsilon$ cf. Eqs. (9.20), (9.22). Further let $\rho_c(\delta)$ be the unique solution of $\rho = M^\#(\rho\delta)$.*

Then for any $\rho < \rho_c(\delta)$ the LASSO minimax risk is bounded and given by

$$M^*(\delta, \rho) = \frac{M^\#(\rho\delta)}{1 - M^\#(\rho\delta)/\delta}. \qquad (9.75)$$

Vice versa, for any $\rho \geq \rho_c(\delta)$, we have $M^*(\delta, \rho) = \infty$.

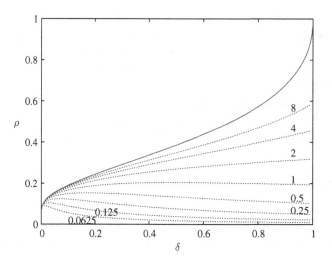

Figure 9.11 Noise sensitivity phase transition in the plane (δ, ρ) (here $\delta = m/n$ is the undersampling ratio and $\rho = \|x\|_0/m$ is the number of nonzero coefficients per measurement). Continuous line: The phase transition boundary $\rho = \rho_c(\delta)$. Dotted lines: Level curves for the LASSO minimax $M^*(\delta, \rho)$. Notice that $M^*(\delta, \rho) \uparrow \infty$ as $\rho \uparrow \rho_c(\delta)$.

Figure 9.11 shows the location of the noise sensitivity boundary $\rho_c(\delta)$ as well as the level lines of $M^*(\delta, \rho)$ for $\rho < \rho_c(\delta)$. Above $\rho_c(\delta)$ the LASSO MSE is not uniformly bounded in terms of the measurement noise σ^2. Other estimators (for instance one step of soft thresholding) can offer better stability guarantees in this region.

One remarkable fact is that the phase boundary $\rho = \rho_c(\delta)$ coincides with the phase transition for $\ell_0 - \ell_1$ equivalence derived earlier by Donoho [29] on the basis of random polytope geometry results by Affentranger–Schneider [2]. The same phase transition was further studied in a series of papers by Donoho, Tanner, and coworkers [30, 31], in connection with the noiseless estimation problem. For $\rho < \rho_c$ estimating x by ℓ_1-norm minimization returns the correct signal with high probability (over the choice of the random matrix A). For $\rho > \rho_c(\delta)$, ℓ_1 minimization fails.

Here this phase transition is derived from a completely different perspective as a special case of a stronger result. We indeed use a new method – the state evolution analysis of the AMP algorithm – which offers quantitative information about the noisy case as well, namely it allows one to compute the value of $M^*(\delta, \rho)$ for $\rho < \rho_c(\delta)$. Within the present approach, the line $\rho_c(\delta)$ admits a very simple expression. In parametric form, it is given by

$$\delta = \frac{2\phi(\alpha)}{\alpha + 2(\phi(\alpha) - \alpha\Phi(-\alpha))}, \tag{9.76}$$

$$\rho = 1 - \frac{\alpha\Phi(-\alpha)}{\phi(\alpha)}, \tag{9.77}$$

where ϕ and Φ are the Gaussian density and Gaussian distribution function, and $\alpha \in [0, \infty)$ is the parameter. Indeed α has a simple and practically important interpretation as well. Recall that the AMP algorithm uses a sequence of thresholds $\theta_t = \alpha \hat{\tau}_t$, cf. Eqs. (9.44) and (9.45). How should the parameter α be fixed? A very simple prescription is obtained in the noiseless case. In order to achieve exact reconstruction for all $\rho < \rho_c(\delta)$ for a given undersampling ratio δ, α should be such that $(\delta, \rho_c(\delta)) = (\delta(\alpha), \rho(\alpha))$ with functions $\alpha \mapsto \delta(\alpha)$, $\alpha \mapsto \rho(\alpha)$ defined as in Eqs. (9.76), (9.77). In other words, this parametric expression yields each point of the phase boundary as a function of the threshold parameter used to achieve it via AMP.

9.6.7 On universality

The main results presented in this section, namely Theorems 9.1, 9.2, and 9.3, are proved for measurement matrices with i.i.d. Gaussian entries. As stressed above, it is expected that the same results hold for a much broader class of matrices. In particular, they should extend to matrices with i.i.d. or weakly correlated entries. For the sake of clarity, it is useful to put forward a formal conjecture, that generalizes Theorem 9.2.

CONJECTURE 9.1 *Let $\{x(n), w(n), A(n)\}_{n \in \mathbb{N}}$ be a converging sequence of instances with the entries of $A(n)$ i.i.d. with mean $\mathbb{E}\{A_{ij}\} = 0$, variance $\mathbb{E}\{A_{ij}^2\} = 1/m$ and such that $\mathbb{E}\{A_{ij}^6\} \leq C/m$ for some fixed constant C. Denote by $\hat{x}(\lambda)$ the LASSO estimator for instance $(x(n), w(n), A(n))$, with $\sigma^2, \lambda > 0$, and let $\psi : \mathbb{R} \times \mathbb{R} \to \mathbb{R}$ be a pseudo-Lipschitz function. Then, almost surely*

$$\lim_{n \to \infty} \frac{1}{n} \sum_{i=1}^{n} \psi(\hat{x}_i, x_i) = \mathbb{E}\left\{\psi\big(\eta(X_0 + \tau_* Z; \theta_*), X_0\big)\right\}, \qquad (9.78)$$

where $Z \sim \mathsf{N}(0,1)$ is independent of $X_0 \sim p_0$, $\tau_* = \tau_*(\alpha(\lambda))$ and $\theta_* = \alpha(\lambda)\tau_*(\alpha(\lambda))$ are given by the same formulae holding for Gaussian matrices, cf. Section 9.6.3.

The conditions formulated in this conjecture are motivated by the universality result in [46], that provides partial evidence towards this claim. Simulations (see for instance Figure 9.10 and [4]) strongly support this claim.

While proving Conjecture 9.1 is an outstanding mathematical challenge, many measurement models of interest do not fit the i.i.d. model. Does the theory developed in this section say anything about such measurements? Systematic numerical simulations [28, 4] reveal that, even for highly structured matrices, the same formula (9.78) is surprisingly close to the actual empirical performances.

As an example, Figure 9.12 presents the empirical mean square error for a partial Fourier measurement matrix A, as a function of the regularization parameter λ. The matrix is obtained by subsampling the rows of the $N \times N$ Fourier matrix F, with entries $F_{ij} = e^{2\pi i j \sqrt{-1}}$. More precisely we sample $n/2$ rows of F with replacement, construct two rows of A by taking real and imaginary parts, and normalize the columns of the resulting matrix.

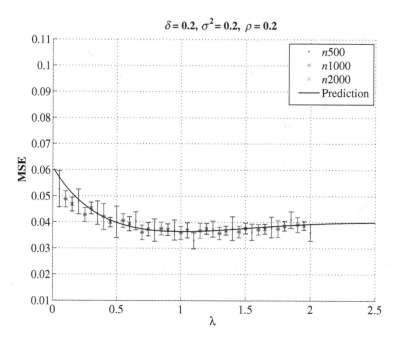

Figure 9.12 Mean square error as a function of the regularization parameter λ for a partial Fourier matrix (see text). The noise variance is $\sigma^2 = 0.2$, the undersampling factor $\delta = 0.2$, and the sparsity ratio $\rho = 0.2$. Data points are obtained by averaging over 20 realizations, and error bars are 95% confidence intervals. The continuous line is the prediction of Theorem 9.2.

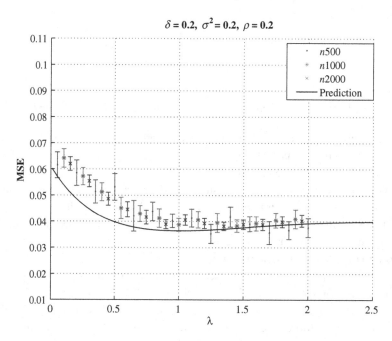

Figure 9.13 As in Figure 9.12, but for a measurement matrix A which models the analog-to-digital converter of [78].

Figure 9.13 presents analogous results for the random demodulator matrix which is at the core of the analog-to-digital converter (ADC) of [78]. Schematically, this is obtained by normalizing the columns of $\widetilde{A} = HDF$, with F a Fourier matrix, D a random diagonal matrix with $D_{ii} \in \{+1, -1\}$ uniformly at random, and H an "accumulator":

$$H = \begin{bmatrix} 1111 & & & \\ & 1111 & & \\ & & \cdots & \\ & & & 1111 \end{bmatrix}.$$

Both these examples show good agreement between the asymptotic prediction provided by Theorem 9.2 and the empirical mean square error. Such an agreement is surprising given that in both cases the measurement matrix is generated with a small amount of randomness, compared to a Gaussian matrix. For instance, the ADC matrix only requires n random bits. Although statistically significant discrepancies can be observed (cf. for instance Figure 9.13), the present approach provides *quantitative* predictions of great interest for design purposes. For a more systematic investigation, we refer to [4].

9.6.8 Comparison with other analysis approaches

The analysis presented here is significantly different from more standard approaches. We derived an *exact* characterization for the high-dimensional limit of the LASSO estimation problem under the assumption of converging sequences of random sensing matrices.

Alternative approaches assume an appropriate "isometry," or "incoherence" condition to hold for A. Under this condition upper bounds are proved for the mean square error. For instance Candès, Romberg, and Tao [19] prove that the mean square error is bounded by $C\sigma^2$ for some constant C. Work by Candès and Tao [21] on the analogous *Dantzig selector*, upper bounds the mean square error by $C\sigma^2(k/n)\log n$, with k the number of nonzero entries of the signal x.

These types of results are very robust but present two limitations: (i) they do not allow one to distinguish reconstruction methods that differ by a constant factor (e.g. two different values of λ); (ii) the restricted isometry condition (or analogous ones) is quite restrictive. For instance, it holds for random matrices only under very strong sparsity assumptions. These restrictions are intrinsic to the worst-case point of view developed in [19, 21].

Guarantees have been proved for correct support recovery in [85], under an incoherence assumption on A. While support recovery is an interesting conceptualization for some applications (e.g. model selection), the metric considered in the present chapter (mean square error) provides complementary information and is quite standard in many different fields.

Close to the spirit of the treatment presented here, [67] derived expressions for the mean square error under the same model considered here. Similar results were presented recently in [50, 35]. These papers argue that a sharp asymptotic characterization of the

LASSO risk can provide valuable guidance in practical applications. Unfortunately, these results were non-rigorous and were obtained through the famously powerful "replica method" from statistical physics [58]. The approach discussed here offers two advantages over these recent developments: (i) it is completely *rigorous*, thus putting on a firmer basis this line of research; (ii) it is *algorithmic* in that the LASSO mean square error is shown to be equivalent to the one achieved by a low-complexity message passing algorithm.

Finally, recently random models for the measurement matrix have been studied in [15, 17]. The approach developed in these papers allows one to treat matrices that do not necessarily satisfy the restricted isometry property or similar conditions, and applies to a general class of random matrices A with i.i.d. rows. On the other hand, the resulting bounds are not asymptotically sharp.

9.7 Generalizations

The single most important advantage of the point of view based on graphical models is that it offers a unified disciplined approach to exploit structural information on the signal x. The use of such information can dramatically reduce the number of required compressed sensing measurements.

"Model-based" compressed sensing [5] provides a general framework for specifying such information. However, it focuses on "hard" combinatorial information about the signal. Graphical models are instead a rich language for specifying "soft" dependencies or constraints, and more complex models. These might include combinatorial constraints, but vastly generalize them. Also, graphical models come with an algorithmic arsenal that can be applied to leverage the potential of such more complex signal models.

Exploring such potential generalizations is – to a large extent – a future research program which is still in its infancy. Here we will only discuss a few examples.

9.7.1 Structured priors…

Block-sparsity [74, 32] is a simple example of combinatorial signal structure. We decompose the signal as $x = (x_{B(1)}, x_{B(2)}, \ldots, x_{B(\ell)})$ where $x_{B(i)} \in \mathbb{R}^{n/\ell}$ is a block for $\ell \in \{1, \ldots, \ell\}$. Only a fraction $\varepsilon \in (0, 1)$ of the blocks is non-vanishing. This type of model naturally arises in many applications: for instance the case $\ell = n/2$ (blocks of size 2) can model signals with complex-valued entries. Larger blocks can correspond to shared sparsity patterns among many vectors, or to clustered sparsity.

It is customary in this setting to replace the LASSO cost function with the following

$$\mathcal{C}_{A,y}^{\text{Block}}(z) \equiv \frac{1}{2}\|y - Az\|_2^2 + \lambda \sum_{i=1}^{\ell} \|z_{B(i)}\|_2. \tag{9.79}$$

The block-ℓ_2 regularization promotes block-sparsity. Of course, the new regularization can be interpreted in terms of a new assumed prior that factorizes over blocks.

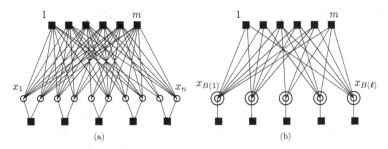

Figure 9.14 Two possible graphical representation of the block-sparse compressed sensing model (and corresponding cost function (9.9)). Upper squares correspond to measurements y_a, $a \in [m]$, and lower squares to the block-sparsity constraint (in this case blocks have size 2). In (a) circles correspond to variables $x_i \in \mathbb{R}$, $i \in [n]$. In (b) double circles correspond to blocks $x_{B(i)} \in \mathbb{R}^{n/\ell}$, $i \in [\ell]$.

Figure 9.14 reproduces two possible graphical structures that encode the block-sparsity constraint. In the first case, this is modeled explicitly as a constraint over blocks of variable nodes, each block comprising n/ℓ variables. In the second case, blocks correspond explicitly to variables taking values in $\mathbb{R}^{n/\ell}$. Each of these graphs dictates a somewhat different message passing algorithm.

An approximate message passing algorithm suitable for this case is developed in [25]. Its analysis allows one to generalize $\ell_0 - \ell_1$ phase transition curves reviewed in Section 9.6.6 to the block-sparse case. This quantifies precisely the benefit of minimizing (9.79) over simple ℓ_1 penalization.

As mentioned above, for a large class of signals sparsity is not uniform: some subsets of entries are sparser than others. Tanaka and Raymond [79], and Som, Potter, and Schniter and [73] studied the case of signals with multiple levels of sparsity. The simplest example consists of a signal $x = (x_{B(1)}, x_{B(2)})$, where $x_{B(1)} \in \mathbb{R}^{n_1}$, $x_{B(2)} \in \mathbb{R}^{n_2}$, $n_1 + n_2 = n$. Block $i \in \{1, 2\}$ has a fraction ε_i of nonzero entries, with $\varepsilon_1 \neq \varepsilon_2$. In the most complex case, one can consider a general factorized prior

$$p(\mathrm{d}x) = \prod_{i=1}^{n} p_i(\mathrm{d}x_i),$$

where each $i \in [n]$ has a different sparsity parameter $\varepsilon_i \in (0, 1)$, and $p_i \in \mathcal{F}_{\varepsilon_i}$. In this case it is natural to use a weighted-ℓ_1 regularization, i.e. to minimize

$$\mathcal{C}_{A,y}^{\text{weight}}(z) \equiv \frac{1}{2}\|y - Az\|_2^2 + \lambda \sum_{i=1}^{n} w_i |z_i|, \qquad (9.80)$$

for a suitable choice of the weights $w_1, \ldots, w_n \geq 0$. The paper [79] studies the case $\lambda \to 0$ (equivalent to minimizing $\sum_i w_i |z_i|$ subject to $y = Az$), using non-rigorous statistical mechanics techniques that are equivalent to the state evolution approach presented here. Within a high-dimensional limit, it determines optimal tuning of the parameters w_i, for given sparsities ε_i. The paper [73] follows instead the state evolution approach explained in the present chapter. The authors develop a suitable AMP iteration and compute the optimal thresholds to be used by the algorithm. These are in correspondence with the optimal weights w_i mentioned above, and can be also interpreted within the minimax framework developed in the previous pages.

The graphical model framework is particularly convenient for exploiting prior information that is probabilistic in nature, see in particular [12, 13]. A prototypical example was studied by Schniter [71] who considered the case in which the signal x is generated by a Hidden Markov Model (HMM). As for the block-sparse model, this can be used to model signals in which the nonzero coefficients are clustered, although in this case one can accommodate greater stochastic variability of the cluster sizes.

In the simple case studied in detail in [71], the underlying Markov chain has two states indexed by $s_i \in \{0, 1\}$, and

$$p(\mathrm{d}x) = \sum_{s_1, \ldots, s_n} \left\{ \prod_{i=1}^{n} p(\mathrm{d}x_i|s_i) \cdot \prod_{i=1}^{n-1} p(s_{i+1}|s_i) \cdot p_1(s_1) \right\}, \tag{9.81}$$

where $p(\,\cdot\,|0)$ and $p(\,\cdot\,|1)$ belong to two different sparsity classes $\mathcal{F}_{\varepsilon_0}$, $\mathcal{F}_{\varepsilon_1}$. For instance one can consider the case in which $\varepsilon_0 = 0$ and $\varepsilon_1 = 1$, i.e. the support of x coincides with the subset of coordinates such that $s_i = 1$.

Figure 9.15 reproduces the graphical structure associated with this type of model. This can be partitioned in two components: a bipartite graph corresponding to the compressed sensing measurements (upper part in Figure 9.15) and a chain graph corresponding to the Hidden Markov Model structure of the prior (lower part in Figure 9.15).

Reconstruction was performed in [71] using a suitable generalization of AMP. Roughly speaking, inference is performed in the upper half of the graph using AMP and in the lower

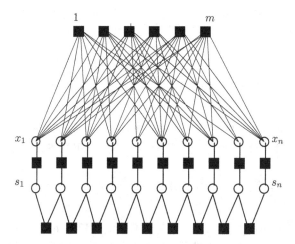

Figure 9.15 Graphical model for compressed sensing of signals with clustered support. The support structure is described by a Hidden Markov Model comprising the lower factor nodes (filled squares) and variable nodes (empty circles). Upper variable nodes correspond to the signal entries x_i, $i \in [n]$, and upper factor nodes to the measurements y_a, $a \in [m]$.

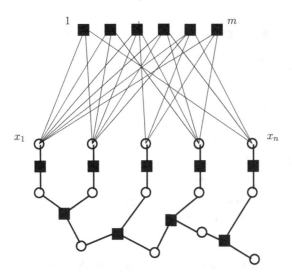

Figure 9.16 Graphical model for compressed sensing of signals with tree-structured prior. The support structure is a tree graphical model, comprising factor nodes and variable nodes in the lower part of the graph. Upper variable nodes correspond to the signal entries x_i, $i \in [n]$, and upper factor nodes to the measurements y_a, $a \in [m]$.

part using the standard forward-backward algorithm. Information is exchanged across the two components in a way that is very similar to what happens in turbo codes [68].

The example of HMM priors clarifies the usefulness of the graphical model structure in eliciting tractable substructures in the probabilistic model and hence leading to natural iterative algorithms. For an HMM prior, inference can be performed efficiently because the underlying graph is a simple chain.

A broader class of priors for which inference is tractable is provided by Markov-tree distributions [72]. These are graphical models that factor according to a tree graph (i.e. a graph without loops). A cartoon of the resulting compressed sensing model is reproduced in Figure 9.16.

The case of tree-structured priors is particularly relevant in imaging applications. Wavelet coefficients of natural images are sparse (an important motivating remark for compressed sensing) and nonzero entries tend to be localized along edges in the image. As a consequence, they cluster in subtrees of the tree of wavelet coefficients. A Markov-tree prior can capture well this structure.

Again, reconstruction is performed exactly on the tree-structured prior (this can be done efficiently using belief propagation), while AMP is used to do inference over the compressed sensing measurements (the upper part of Figure 9.16).

9.7.2 Sparse sensing matrices

Throughout this review we focused for simplicity on dense measurement matrices A. Several of the mathematical results presented in the previous sections do indeed hold

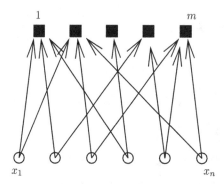

Figure 9.17 Sparse sensing graph arising in a networking application. Each network flow (empty circles below) hashes into $k = 2$ counters (filled squares).

for dense matrices with i.i.d. components. Graphical models ideas are on the other hand particularly useful for sparse measurements.

Sparse sensing matrices present several advantages, most remarkably lower measurement and reconstruction complexities [6]. While sparse constructions are not suitable for all applications, they appear a promising solution for networking applications, most notably in network traffic monitoring [14, 56].

In an over-simplified example, one would like to monitor the sizes of n packet flows at a router. It is a recurring empirical observation that most of the flows consist of a few packets, while most of the traffic is accounted for by a few flows. Denoting by $x_1, x_2, \ldots, x_n$ the flow sizes (as measured, for instance, by the number of packets belonging to the flow), it is desirable to maintain a small sketch of the vector $x = (x_1, \ldots, x_n)$.

Figure 9.17 describes a simple approach: flow i hashes into a small number – say k – of memory spaces, $\partial i = \{a_1(i), \ldots, a_k(i)\} \subseteq [m]$. Each time a new packet arrives for flow i, the counters in ∂i are incremented. If we let $y = (y_1, \ldots, y_m)$ be the contents of the counters, we have

$$y = Ax, \qquad (9.82)$$

where $x \geq 0$ and A is a matrix with i.i.d. columns with k entries per column equal to 1 and all the other entries equal to 0. While this simple scheme requires unpractically deep counters (the entries of y can be large), [56] showed how to overcome this problem by using a multi-layer graph.

Numerous algorithms have been developed for compressed sensing reconstruction with sparse measurement matrices [14, 84, 6, 40]. Most of these algorithms are based on greedy methods, which are essentially of message passing type. Graphical models ideas can be used to construct such algorithms in a very natural way. For instance, the algorithm of [56] (see also [54, 55, 20] for further analysis of the same algorithm) is closely related to the ideas presented in the rest of this chapter. It uses messages $x_{i \to a}^t$ (from variable nodes to function nodes) and $r_{a \to i}^t$ (from function nodes to variable nodes). These are updated according to

$$r_{a\to i}^t = y_a - \sum_{j \in \partial a \setminus i} x_{j\to a}^t, \qquad (9.83)$$

$$x_{i\to a}^{t+1} = \begin{cases} \min\{r_{b\to i}^t : b \in \partial i \setminus a\} & \text{at even iterations } t, \\ \max\{r_{b\to i}^t : b \in \partial i \setminus a\} & \text{at odd iterations } t, \end{cases} \qquad (9.84)$$

where ∂a denotes the set of neighbors of node a in the factor graph. These updates are very similar to Eqs. (9.49), (9.50) introduced earlier in our derivation of AMP.

9.7.3 Matrix completion

"Matrix completion" is the task of inferring an (approximately) low rank matrix from observations on a small subset of its entries. This problem has attracted considerable interest over the last two years due to its relevance in a number of applied domains (collaborative filtering, positioning, computer vision, etc.).

Significant progress has been achieved on the theoretical side. The reconstruction question has been addressed in analogy with compressed sensing in [18, 16, 37, 63], while an alternative approach based on greedy methods was developed in [48, 49, 45]. While the present chapter does not treat matrix completion in any detail, it is interesting to mention that graphical models ideas can be useful in this context as well.

Let $M \in \mathbb{R}^{m \times n}$ be the matrix to be reconstructed, and assume that a subset $E \subseteq [m] \times [n]$ of its entries is observed. It is natural to try to accomplish this task by minimizing the ℓ_2 distance on observed entries. For $X \in \mathbb{R}^{m \times r}, Y \in \mathbb{R}^{n \times r}$, we introduce therefore the cost function

$$\mathcal{C}(X,Y) = \frac{1}{2} \|\mathcal{P}_E(M - XY^T)\|_F^2 \qquad (9.85)$$

where $\mathcal{P}_E$ is the projector that sets to zero the entries outside E (i.e. $\mathcal{P}_E(L)_{ij} = L_{ij}$ if $(i,j) \in E$ and $\mathcal{P}_E(L)_{ij} = 0$ otherwise). If we denote the rows of X as $x_1, \ldots, x_m \in \mathbb{R}^r$ and the rows in Y as $y_1, \ldots, y_n \in \mathbb{R}^r$, the above cost function can be rewritten as

$$\mathcal{C}(X,Y) = \frac{1}{2} \sum_{(i,j) \in E} (M_{ij} - \langle x_i, y_j \rangle)^2, \qquad (9.86)$$

with $\langle \cdot, \cdot \rangle$ the standard scalar product on $\mathbb{R}^r$. This cost function factors accordingly the bipartite graph G with vertex sets $V_1 = [m]$ and $V_2 = [n]$ and edge set E. The cost decomposes as a sum of pairwise terms associated with the edges of G.

Figure 9.18 reproduces the graph G that is associated to the cost function $\mathcal{C}(X,Y)$. It is remarkable that some properties of the reconstruction problem can be "read" from the graph. For instance, in the simple case $r = 1$, the matrix M can be reconstructed if and only if G is connected (banning for degenerate cases) [47]. For higher values of the rank r, rigidity of the graph is related to uniqueness of the solution of the reconstruction problem [70]. Finally, message passing algorithms for this problem were studied in [51, 45].

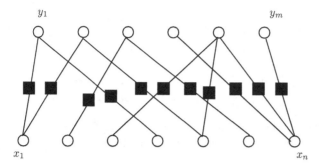

Figure 9.18 Factor graph describing the cost function (9.86) for the matrix completion problem. Variables $x_i, y_j \in \mathbb{R}^r$ are to be optimized over. The cost is a sum of pairwise terms (filled squares) corresponding to the observed entries in M.

9.7.4 General regressions

The basic reconstruction method discussed in this review is the regularized least-squares regression defined in Eq. (9.12), also known as the LASSO. While this is by far the most interesting setting for signal processing applications, for a number of statistical learning problems, the linear model (9.1) is not appropriate. Generalized linear models provide a flexible framework to extend the ideas discussed here.

An important example is logistic regression, which is particularly suited for the case in which the measurements $y_1, \ldots y_m$ are 0–1 valued. Within logistic regression, these are modeled as independent Bernoulli random variables with

$$p(y_a = 1 | x) = \frac{e^{A_a^T x}}{1 + e^{A_a^T x}}, \qquad (9.87)$$

with A_a a vector of "features" that characterizes the ath experiment. The objective is to learn the vector x of coefficients that encodes the relevance of each feature. A possible approach consists in minimizing the regularized (negative) log-likelihood, that is

$$\mathcal{C}_{A,y}^{\text{LogReg}}(z) \equiv -\sum_{a=1}^{m} y_a (A_a^T z) + \sum_{a=1}^{m} \log\left(1 + e^{A_a^T z}\right) + \lambda \|z\|_1. \qquad (9.88)$$

The papers [66, 8] develop approximate message passing algorithms for solving optimization problems of this type.

Acknowledgements

It is a pleasure to thank Mohsen Bayati, Jose Bento, David Donoho, and Arian Maleki, with whom this research has been developed. This work was partially supported by a Terman fellowship, the NSF CAREER award CCF-0743978, and the NSF grant DMS-0806211.

References

[1] G. Andrew and G. Jianfeng, Scalable training of l^1-regularized log-linear models. *Proc 24th Inte Conf Mach Learning*, 2007, pp. 33–40.

[2] R. Affentranger and R. Schneider, Random projections of regular simplices. *Discr Comput Geom*, 7:219–226, 1992.

[3] D. Aldous and J. M. Steele. The objective method: probabilistic combinatorial optimization and local weak convergence. *Probability on Discrete Structures* H. Kesten, ed., Springer Verlag, pp. 1–72, 2003.

[4] M. Bayati, J. Bento, and A. Montanari. Universality in sparse reconstruction: A comparison between theories and empirical results, in preparation, 2011.

[5] R. G. Baraniuk, V. Cevher, M. F. Duarte, and C. Hegde. Model-based compressive sensing, *IEEE Trans Inform Theory* 56:1982–2001, 2010.

[6] R. Berinde, A. C. Gilbert, P. Indyk, H. Karloff, and M. J. Strauss. Combining geometry and combinatorics: a unified approach to sparse signal recovery. *46th Ann Allerton Conf* (Monticello, IL), September 2008.

[7] P. Billingsley. *Probability and Measure*, Wiley, 1995.

[8] M. Bayati and A. Montanari. Approximate message passing algorithms for generalized linear models, in preparation, 2010.

[9] M. Bayati and A. Montanari. The dynamics of message passing on dense graphs, with applications to compressed sensing. *IEEE Trans Inform Theory*, accepted, http://arxiv.org/pdf/1001.3448, 2011.

[10] S. S. Chen and D. L. Donoho. Examples of basis pursuit. *Proc Wavelet Appl Sig Image Proc III*, San Diego, CA, 1995.

[11] V. Cevher, Learning with compressible priors. *Neur Inform Proc Syst*, Vancouver, 2008.

[12] V. Cevher, C. Hegde, M. F. Duarte, and R. G. Baraniuk, Sparse signal recovery using Markov random fields. *Neur Inform Proc Syst*, Vancouver, 2008.

[13] V. Cevher, P. Indyk, L. Carin, and R. G. Baraniuk. Sparse signal recovery and acquisition with graphical models. *IEEE Sig Process Mag* 27:92–103, 2010.

[14] G. Cormode and S. Muthukrishnan. Improved data streams summaries: The count-min sketch and its applications. LATIN, Buenos Aires, pp. 29–38, 2004.

[15] E. J. Candès and Y. Plan. Near-ideal model selection by ℓ_1 minimization. *Ann Statist*, 37:2145–2177, 2009.

[16] E. J. Candès and Y. Plan, Matrix completion with noise. *Proc IEEE* 98:925–936, 2010.

[17] E. J. Candès and Y. Plan. A probabilistic and ripless theory of compressed sensing. arXiv:1011.3854, November 2010.

[18] E. J. Candès and B. Recht. Exact matrix completion via convex optimization. *Found Comput Math* 9:717–772, 2009.

[19] E. Candès, J. K. Romberg, and T. Tao. Stable signal recovery from incomplete and inaccurate measurements. *Commun Pure Appl Math*, 59:1207–1223, 2006.

[20] V. Chandar, D. Shah, and G. W. Wornell, A simple message-passing algorithm for compressed sensing, *Proc IEEE Int Symp Inform Theory (ISIT)* (Austin), 2010.

[21] E. Candès and T. Tao. The Dantzig selector: statistical estimation when p is much larger than n. *Ann Stat*, 35:2313–2351, 2007.

[22] D. L. Donoho and I. M. Johnstone. Ideal spatial adaptation via wavelet shrinkage. *Biometrika*, 81:425–455, 1994.

[23] D. L. Donoho and I. M. Johnstone. Minimax risk over l_p balls. *Prob Th Rel Fields*, 99:277–303, 1994.

[24] D. L. Donoho, I. M. Johnstone, J. C. Hoch, and A. S. Stern. Maximum entropy and the nearly black object. *J Roy Statist Soc, Ser. B (Methodological)*, 54(1):41–81, 1992.

[25] D. Donoho and A. Montanari. Approximate message passing for reconstruction of block-sparse signals, in preparation, 2010.

[26] D. L. Donoho, A. Maleki, and A. Montanari. Message passing algorithms for compressed sensing. *Proc Nat Acad Sci* 106:18914–18919, 2009.

[27] D. L. Donoho, A. Maleki, and A. Montanari. Message passing algorithms for compressed sensing: I. Motivation and construction. *Proc IEEE Inform Theory Workshop*, Cairo, 2010.

[28] D. L. Donoho, A. Maleki, and A. Montanari. The noise sensitivity phase transition in compressed sensing, http://arxiv.org/abs/1004.1218, 2010.

[29] D. Donoho. High-dimensional centrally symmetric polytopes with neighborliness proportional to dimension. *Discr Comput Geom*, 35:617–652, 2006.

[30] D. L. Donoho and J. Tanner. Neighborliness of randomly-projected simplices in high dimensions. *Proc Nat Acad Sci*, 102(27):9452–9457, 2005.

[31] D. L. Donoho and J. Tanner. Counting faces of randomly projected polytopes when the projection radically lowers dimension, *J Am Math Soc*, 22:1–53, 2009.

[32] Y. C. Eldar, P. Kuppinger, and H. Bolcskei. Block-sparse signals: uncertainty relations and efficient recovery. *IEEE Trans Sig Proc*, 58:3042–3054, 2010.

[33] M. A. T. Figueiredo and R. D. Nowak. An EM algorithm for wavelet-based image restoration. *IEEE Trans Image Proc*, 12:906–916, 2003.

[34] M. Grant and S. Boyd. *CVX: Matlab software for disciplined convex programming, version 1.21*, http://cvxr.com/cvx, May 2010.

[35] D. Guo, D. Baron, and S. Shamai. A single-letter characterization of optimal noisy compressed sensing. *47th Ann Allerton Conf* (Monticello, IL), September 2009.

[36] D. Gamarnik and D. Katz. Correlation decay and deterministic FPTAS for counting list-colorings of a graph. *18th Ann ACM-SIAM Symp Discrete Algorithms*, New Orleans, 1245–1254, 2007.

[37] D. Gross. Recovering low-rank matrices from few coefficients in any basis. arXiv:0910.1879v2, 2009.

[38] D. Guo and S. Verdu. Randomly spread CDMA: asymptotics via statistical physics. *IEEE Trans Inform Theory*, 51:1982–2010, 2005.

[39] T. Hastie, R. Tibshirani, and J. Friedman. *The Elements of Statistical Learning*, Springer-Verlag, 2003.

[40] P. Indyk. Explicit constructions for compressed sensing of sparse signals, *19th Ann ACM-SIAM Symp Discrete Algorithm*, San Francisco, 2008.

[41] I. Johnstone, *Function Estimation and Gaussian Sequence Models*. Draft of a book, available at www-stat.stanford.edu/~imj/based.pdf, 2002.

[42] M. Jordan (ed.). *Learning in Graphical Models*, MIT Press, 1998.

[43] S. Ji, Y. Xue, and L. Carin. Bayesian compressive sensing. *IEEE Trans Sig Proc*, 56:2346–2356, 2008.

[44] D. Koller and N. Friedman. *Probabilistic Graphical Models*, MIT Press, 2009.

[45] R. H. Keshavan and A. Montanari. Fast algorithms for matrix completion, in preparation, 2010.

[46] S. Korada and A. Montanari. Applications of Lindeberg Principle in communications and statistical learning. http://arxiv.org/abs/1004.0557, 2010.

[47] R. H. Keshavan, A. Montanari, and S. Oh. Learning low rank matrices from $O(n)$ entries. *Proc Allerton Conf Commun, Control Comput*, September 2008, arXiv:0812.2599.

[48] R. H. Keshavan, A. Montanari, and S. Oh, Matrix completion from a few entries. *IEEE Trans Inform Theory*, 56:2980–2998, 2010.

[49] R. H. Keshavan, A. Montanari, and S. Oh. Matrix completion from noisy entries. *J Mach Learn Res* 11:2057–2078, 2010.

[50] Y. Kabashima, T. Wadayama, and T. Tanaka, A typical reconstruction limit for compressed sensing based on lp-norm minimization. *J Stat Mech*, L09003, 2009.

[51] B.-H. Kim, A. Yedla, and H. D. Pfister. Imp: A message-passing algorithm for matrix completion. *Proc Stand 6th Inte Symp Turbo Codes*, September 2010, arXiv:1007.0481.

[52] E.L. Lehmann and G. Casella. *Theory of Point Estimation*, Springer-Verlag, 1998.

[53] M. Ledoux, *The Concentration of Measure Phenomenon*, American Mathematical Society, 2001.

[54] Y. Lu, A. Montanari, and B. Prabhakar, Detailed network measurements using sparse graph counters: The Theory, *45th Ann Allerton Confe* Monticello, IL, 2007.

[55] Y. Lu, A. Montanari, and B. Prabhakar. Counter braids: asymptotic optimality of the message passing decoding algorithm. *46th Ann Allerton Confe*, Monticello, IL, 2008.

[56] Y. Lu, A. Montanari, B. Prabhakar, S. Dharmapurikar, and A. Kabbani. Counter braids: a novel counter architecture for per-flow measurement. *SIGMETRICS 2010*, 2008.

[57] A. Maleki and D. L. Donoho, Optimally tuned iterative thresholding algorithm for compressed sensing. *IEEE J Sel Topics Sig Process*, 4:330–341, 2010.

[58] M. Mézard and A. Montanari. *Information, Physics and Computation*. Oxford University Press, 2009.

[59] A. Montanari. Estimating random variables from random sparse observations. *Eur Trans Telecom*, 19:385–403, 2008.

[60] M. Mézard, G. Parisi, and M. A. Virasoro. *Spin Glass Theory and Beyond*. World Scientific, 1987.

[61] C. Moallemi and B. Van Roy. Convergence of the min-sum algorithm for convex optimization. *45th Ann Allerton Confe* (Monticello, IL), September 2007.

[62] A. Montanari and D. Tse. Analysis of belief propagation for non-linear problems: the example of CDMA (or: how to prove Tanaka's formula). *Proc IEEE Inform Theory Workshop* (Punta de l'Este, Uruguay), 2006.

[63] S. Negahban and M. J. Wainwright. Estimation of (near) low-rank matrices with noise and high-dimensional scaling. arXiv:0912.5100, 2009.

[64] L. Onsager. Electric moments of molecules in liquids. *J Am Chem Soc*, 58:1486–1493, 1936.

[65] J. Pearl. *Probabilistic Reasoning in Intelligent Systems: Networks of Plausible Inference*. Morgan Kaufmann, 1988.

[66] S. Rangan. Generalized approximate message passing for estimation with random linear mixing. arXiv:1010.5141, 2010.

[67] S. Rangan, A. K. Fletcher, and V. K. Goyal. *Asymptotic Analysis of Map Estimation via the Replica Method and Applications to Compressed Sensing*, 2009.

[68] T. J. Richardson and R. Urbanke. *Modern Coding Theory*, Cambridge University Press, 2008.

[69] S. Sarvotham, D. Baron, and R. Baraniuk. Bayesian compressive sensing via belief propagation. *IEEE Trans Sig Proc*, 58:269–280, 2010.

[70] A. Singer and M. Cucuringu. Uniqueness of low-rank matrix completion by rigidity theory. arXiv:0902.3846, 2009.

[71] P. Schniter. Turbo reconstruction of structured sparse signals. *Proc Conf Inform Sci Syst*, Princeton, NJ, 2010.

[72] S. Som, L. C. Potter, and P. Schniter. Compressive imaging using approximate message passing and a Markov-Tree Prior. *Proc Asilomar Conf Sig Syst Comput*, 2010.

[73] L. C. Potter, S. Som, and P. Schniter. On approximate message passing for reconstruction of non-uniformly sparse signals, *Proc Nati Aereospace Electron Confe*, Dayton, OH, 2010.

[74] M. Stojnic, F. Pavaresh, and B. Hassibi. On the reconstruction of block-sparse signals with an optimal number of measurements. *IEEE Trans Sig Proc*, 57:3075–3085, 2009.

[75] T. Tanaka. A statistical-mechanics approach to large-system analysis of CDMA multiuser detectors. *IEEE Trans Inform Theory*, 48:2888–2910, 2002.

[76] D. J. Thouless, P. W. Anderson, and R. G. Palmer. Solution of "Solvable model of a spin glass." *Phil Mag* 35:593–601, 1977.

[77] R. Tibshirani. Regression shrinkage and selection with the lasso. *J Roy Stat Soc*, B 58:267–288, 1996.

[78] J. A. Tropp, J. N. Laska, M. F. Duarte, J. K. Romberg, and R. G. Baraniuk. Beyond Nyquist: efficient sampling of sparse bandlimited signals. *IEEE Trans Inform. Theory*, 56:520–544, 2010.

[79] T. Tanaka and J. Raymond. Optimal incorporation of sparsity information by weighted L_1 optimization. *Proc IEEE Int Symp Inform Theory (ISIT)*, Austin, 2010.

[80] A. W. van der Vaart. *Asymptotic Statistics*. Cambridge University Press, 2000.

[81] D. Weitz. Combinatorial criteria for uniqueness of Gibbs measures. *Rand Struct Alg*, 27:445–475, 2005.

[82] D. Williams. *Probability with Martingales*. Cambridge University Press, 1991.

[83] M. J. Wainwright and M. I. Jordan. Graphical models, exponential families, and variational inference. *Found Trends Mach Learn*, 1, 2008.

[84] W. Xu and B. Hassibi. Efficient compressive sensing with deterministic guarantees using expander graphs. *Proc IEEE Inform Theory Workshop*, Tahoe City, CA, 2007.

[85] P. Zhao and B. Yu. On model selection consistency of Lasso. *J Mach Learn Res*, 7:2541–2563, 2006.

10 Finding needles in compressed haystacks

Robert Calderbank and Sina Jafarpour

Abstract

In this chapter, we show that compressed learning, learning directly in the compressed domain, is possible. In particular, we provide tight bounds demonstrating that the linear kernel SVM's classifier in the measurement domain, with high probability, has true accuracy close to the accuracy of the best linear threshold classifier in the data domain. We show that this is beneficial both from the compressed sensing and the machine learning points of view. Furthermore, we indicate that for a family of well-known compressed sensing matrices, compressed learning is provided on the fly. Finally, we support our claims with experimental results in the texture analysis application.

10.1 Introduction

In many applications, the data has a sparse representation in some basis in a much higher dimensional space. Examples are the sparse representation of images in the wavelet domain, the bag of words model of text, and the routing tables in data monitoring systems.

Compressed sensing combines measurement to reduce the dimensionality of the underlying data with reconstruction to recover sparse data from the projection in the measurement domain. However, there are many sensing applications where the objective is not full reconstruction but is instead classification with respect to some signature. Examples include radar, detection of trace chemicals, face detection [7, 8], and video streaming [9] where we might be interested in anomalies corresponding to changes in wavelet coefficients in the data domain. In all these cases our objective is pattern recognition in the measurement domain.

Classification in the measurement domain offers a way to resolve this challenge and we show that it is possible to design measurements for which there are performance guarantees. Similar to compressed sensing, *linear* measurements are used to remove the costs of pointwise sampling and compression. However, the ultimate goal of compressed learning is not reconstruction of the sparse data from their linear measurements. In contrast, here we are provided with compressively sampled training data, and the goal is to design a classifier directly in the measurement domain with almost the same accuracy as the best classifier in the data domain.

Being able to learn in the compressed domain is beneficial both from the compressed sensing and the machine learning points of view. From the compressed sensing viewpoint, it eliminates the significant cost of recovering irrelevant data; in other words, classification in the measurement domain is like a sieve and makes it possible to only recover the desired signals, or even remove the recovery phase totally, if we are only interested in the results of classification. This is like finding a needle in a compressively sampled haystack without recovering all the hay. In addition, compressed learning has the potential of working successfully in situations where one cannot observe the data domain or where measurements are difficult or expensive.

Dimensionality reduction is a fundamental step in applications as diverse as the nearest-neighbor approximation [1, 2], data-streaming [3], machine learning [4, 5], graph approximation [6], etc. In compressed learning, the sensing procedure can be also considered as a linear dimensionality reduction step. In this chapter we will show that most compressed sensing matrices also provide the desired properties of good linear dimensionality reduction matrices.

In terms of geometry, the difference between compressed sensing and compressed learning is that the former is concerned with separating pairs of points in the measurement domain to enable unambiguous recovery of sparse data while the latter is concerned with consistent separation of clouds of points in the data and measurement domains. In this chapter we demonstrate feasibility of pattern recognition in the measurement domain. We provide PAC-style bounds guaranteeing that if the data is measured directly in the compressed domain, a soft margin SVM's classifier that is trained based on the compressed data performs almost as well as the best possible SVM's classifier in the data domain. The results are robust against the noise in the measurement.

The compressed learning framework is applicable to any sparse high-dimensional dataset. For instance, in texture analysis [10] the goal is to predict the direction of an image by looking only at its wavelet coefficients. A weighted voting among horizontal and vertical wavelet coefficients of each image can accurately predict whether the image is vertical, horizontal, or neither. However, in compressive imaging, the high-dimensional wavelet representation of the image is not provided. In contrast, a non-adaptive low-rank sensing matrix is used to project the wavelet vector into some low-dimensional space. Here we show that a weighted voting among the entries of these measurement vectors has approximately the same accuracy as the original weighted voting among the wavelet entries in the texture analysis task.

Section 10.2 clarifies the notation that is used in this chapter. Support Vector Machines (SVM's) are introduced in Section 10.3. In Sections 10.4 and 10.5 we introduce the Distance-Preserving Property, and prove that if a sensing matrix is Distance-Preserving with respect to a feature space, then the SVM's classifier in the measurement domain is close to optimal. Section 10.6 shows that the celebrated Johnson-Lindenstrauss property is a sufficient condition for distance-preserving. Then in Sections 10.7 and 10.8 we provide two examples of sensing matrices which are widely used in compressed sensing, and satisfy the Johnson-Lindenstrauss property with respect to sparse feature spaces. Section 10.9 proves the main results about the average-case compressed learning, and Section 10.10 concludes the chapter.

10.2 Background and notation

10.2.1 Notation

Throughout this chapter, in order to distinguish vectors from scalar values, vectors are always represented by bold symbols. Let n be a positive integer, and let k be a positive integer less than n. We sometimes denote $\{1, \cdots, n\}$ by $[n]$. We assume that all data are represented as vectors in $\mathbb{R}^n$. The feature space $\mathcal{X}$ (which we also call the data-domain), is a subset of the whole n-dimensional space, and every data point is a vector in the feature space. Choosing appropriate feature space $\mathcal{X}$ highly depends on the prior information available about the data. For instance, if there is no prior knowledge available, then the feature space is the whole space $\mathbb{R}^n$. However, if it is known *a priori* that all data points are k-sparse, then the feature space can be restricted to the union of all k-dimensional subspaces of $\mathbb{R}^n$.

Let $\mathcal{X}$ denote an n-dimensional feature space. We associate two parameters with $\mathcal{X}$ which we later use in our analysis. The radius R, and the distortion $\ell_{2 \to 1}$ of the feature space $\mathcal{X}$ are defined as

$$R \doteq \max_{x \in \mathcal{X}} \|x\|_2 \quad \text{and} \quad \ell_{2 \to 1}(\mathcal{X}) \doteq \max_{x \in \mathcal{X}} \frac{\|x\|_1}{\|x\|_2}. \tag{10.1}$$

It follows from the Cauchy-Schwarz inequality that on one hand, if every element of $\mathcal{X}$ is k-sparse, then $\ell_{2 \to 1}(\mathcal{X}) = \sqrt{k}$. On the other hand, if the feature space covers the whole space $\mathbb{R}^n$, then $\ell_{2 \to 1}(\mathcal{X}) = \sqrt{n}$.

The *Hamming weight* of a vector $x \in \mathbb{R}^n$, denoted by $\|x\|_0$, is defined as the number of nonzero entries of x. The support of the k-sparse vector x, denoted by $\text{supp}(x)$, contains the indices of the nonzero entries of x. Throughout this chapter we let $\Pi_1^k = \{\Pi_1, \cdots, \Pi_k\}$ be a uniformly random k-subset of $[n]$, and $\pi_1^k = \{\pi_1, \cdots, \pi_k\}$ denote a fixed k-subset of $[n]$. We use $v_1^\top v_2$ to denote the inner product between two vectors v_1 and v_2. Also whenever it is clear from context, we remove the subscript from the ℓ_2 norm of a vector.

Let A be an $m \times n$ matrix. We use the notation A_j for the jth column of the sensing matrix A; its entries will be denoted by a_{ij}, with the row label i varying from 0 to $m-1$. The matrix A is a tight-frame with redundancy $\frac{n}{m}$ if and only if $AA^\dagger = \frac{n}{m} I_{m \times m}$. Note that if A is a tight-frame with redundancy $\frac{n}{m}$, then $\|A\|^2 = \frac{n}{m}$.

10.2.2 Concentration inequalities

Here we provide the main concentration inequalities which are used in the chapter.

PROPOSITION 10.1 (Azuma's inequality [11]) *Let $\langle Z_0, Z_1, \cdots, Z_k \rangle$ be a set of complex random variables such that, for each i, $\mathbb{E}[Z_i] = Z_{i-1}$, and $|Z_i - Z_{i-1}| \leq c_i$. Then for all $\epsilon > 0$,*

$$\Pr\left[|Z_k - Z_0| \geq \epsilon\right] \leq 4\exp\left\{\frac{-\epsilon^2}{8\sum_{i=1}^k c_i^2}\right\}.$$

PROPOSITION 10.2 (Extension to Azuma's inequality when differences are bounded with high probability [12]) *Suppose $\langle Z_0, Z_1, \cdots, Z_k \rangle$ is a complex martingale sequence, that is for each i, $\mathbb{E}[Z_i] = Z_{i-1}$. Moreover, suppose that with probability $1 - \delta$ for all i: $|Z_i - Z_{i-1}| \leq c_i$, and always $|Z_i - Z_{i-1}| \leq b_i$. Then for all $\epsilon > 0$,*

$$\Pr\left[|Z_k - Z_0| \geq \epsilon\right] \leq 4\left(\exp\left\{\frac{-\epsilon^2}{32\sum_{i=1}^k c_i^2}\right\} + \delta\sum_{i=1}^k \frac{b_i}{c_i}\right).$$

10.2.3 Group theory

In this chapter, we will analyze deterministic sensing matrices for which the columns form a group $\mathcal{G}$ under pointwise multiplication. The multiplicative identity is the column **1** with every entry equal to 1. The following property is fundamental.

LEMMA 10.1 *If every row contains some entry not equal to 1, then the column group $\mathcal{G}$ satisfies $\sum_{g \in \mathcal{G}} g = 0$.*

Proof. Given a row i and an element $f_i \neq 1$, we have

$$f_i \left(\sum_g g_i\right) = \sum_g (f_i g_i) = \sum_g g_i.$$

$\square$

10.3 Support Vector Machines

Support vector machines (SVM's) [13] is a linear threshold classifier in some feature space, with maximum margin and consistent with the training examples. Any linear threshold classifier $w(x)$ corresponds to a vector $w \in \mathbb{R}^n$ such that $w(x) = \text{sign}(w^\top x)$; as a result, we identify the linear threshold classifiers with their corresponding vectors. Also for simplicity we only focus on classifiers passing through the origin. The results can be simply extended to the general case.

Whenever the training examples are not linearly separable soft margin SVM's are used. The idea is to simultaneously maximize the margin and minimize the empirical hinge loss. More precisely let

$$H(x) \doteq (1 + x)_+ = \max\{0, 1 + x\},$$

and let $S \doteq \langle (x_1, l_1), \ldots, (x_M, l_M) \rangle$ be a set of M labeled training data sampled i.i.d. from some distribution $\mathcal{D}$. For any linear classifier $w \in \mathbb{R}^n$ we define its true hinge loss as

$$H_\mathcal{D}(w) \doteq \mathbb{E}_{(x,l) \sim \mathcal{D}} \left[\left(1 - l w^\top x\right)_+ \right],$$

and its empirical hinge loss

$$\hat{H}_S(w) \doteq \mathbb{E}_{(x_i, l_i) \sim S} \left[\left(1 - l_i w^\top x_i\right)_+ \right].$$

We also define the true regularization loss of a classifier w as

$$L(w) \doteq H_\mathcal{D}(w) + \frac{1}{2C} \|w\|^2, \tag{10.2}$$

where C is the regularization constant. The empirical regularization loss is similarly defined to be

$$\hat{L}(w) \doteq \hat{H}_S(w) + \frac{1}{2C} \|w\|^2. \tag{10.3}$$

Soft margin SVM's then minimizes the empirical regularization loss which is a convex optimization program.

The following theorem is a direct consequence of the convex duality, and we provide a proof for completeness.

THEOREM 10.1 *Let $\langle (x_1, l_1), \ldots, (x_M, l_M) \rangle$ be a set of M labeled training examples chosen i.i.d. from some distribution $\mathcal{D}$, and let w be the SVM's classifier obtained by minimizing Equation (10.3). Then the SVM's classifier can be represented as a linear combination of the training examples, i.e., $w = \sum_{i=1}^M s_i l_i x_i$, with $\|w\|^2 \leq C$. Moreover,*

$$\text{for all } i \in [M]: \quad 0 \leq s_i \leq \frac{C}{M}.$$

Proof. The optimization problem of Equation (10.3) can be written as

$$\text{minimize } \frac{1}{2}\|w\|^2 + \frac{C}{M} \sum_{i=1}^M \xi_i \text{ such that:} \tag{10.4}$$

$$\forall i: \xi_i \geq 0$$

$$\forall i: \xi_i \geq 1 - l_i w^\top x_i.$$

We form the Lagrangian of the above optimization problem by linearly combining the objective function and the constraints

$$\mathcal{L}(w, \xi, s, \eta) = \frac{1}{2}\|w\|^2 + \frac{C}{M} \sum_{i=1}^M \xi_i + \sum_{i=1}^M s_i (1 - l_i w^\top x_i - \xi_i) - \sum_{i=1}^M \eta_i \xi_i. \tag{10.5}$$

By writing the KKT conditions, we obtain the following conditions for the saddle point of the Lagrangian function:

- The optimal classifier is a linear combination of the training examples:

$$w - \sum_{i=1}^{M} s_i l_i x_i = 0.$$

- The optimal dual variables s_i and η_i are all non-negative.
- For every index i we have $\frac{C}{M} - s_i - \eta_i = 0$, which implies that $s_i \leq \frac{C}{M}$.

Therefore, the dual program can be written as

$$\text{maximize} \sum_{i=1}^{M} s_i - \frac{1}{2} \sum_{i,j=1}^{M} s_i s_j l_i l_j x_i^\top x_j$$

$$\text{s.t. } \forall\, i \in [M]: 0 \leq s_i \leq \frac{C}{M}.$$

Finally we show that the SVM's classifier has bounded ℓ_2 norm. We have

$$\frac{1}{2}\|w\|^2 \leq \frac{1}{2}\|w\|^2 + \frac{C}{M}\sum_{i=1}^{M}\xi_i = \sum_{i=1}^{M} s_i - \frac{1}{2}\|w\|^2 \leq C - \frac{1}{2}\|w\|^2,$$

where the first inequality follows from the positivity of the hinge loss, the next equality follows from the equality of the optimal primal value and the optimal dual value for convex programs and the fact that the optimal classifier is a linear combination of the training examples, and the last inequality follows from the fact that for every index i, $s_i \leq \frac{C}{M}$. □

In order to analyze the performance of the SVM's classifiers, we will use the following Corollary of Theorem 10.1:

COROLLARY 10.1 *Let w be the soft margin SVM's optimizer of Equation* (10.3). *Then w is in the C-convex hull of the feature space, which is defined as:*

$$\mathcal{CH}_C(\mathcal{X}) \doteq \left\{ \sum_{i=1}^{n} t_i x_i : n \in \mathbb{N}, x_i \in \mathcal{X}, \sum_{i=1}^{n} |t_i| \leq C \right\}.$$

Throughout this chapter, we will use the following oracle model to analyze the performance of the SVM's classifiers. As before, let $\mathcal{X}$ denote the feature space, and let $\mathcal{D}$ be some distribution over $\mathcal{X}$. We assume that there exists a linear threshold classifier $\boldsymbol{w}_0 \in \mathbb{R}^n$ that has large margin and hence small norm $\|\boldsymbol{w}_0\|$, and whose true regularization loss achieves a low generalization error (expected hinge loss). For each regularization parameter C, we also define $\boldsymbol{w}^* \doteq \arg\min L_\mathcal{D}(w)$. It follows from the definition that $L_\mathcal{D}(\boldsymbol{w}^*) \leq L_\mathcal{D}(\boldsymbol{w}_0)$. To compare the true hinge loss of the SVM's classifiers with the true hinge loss of the oracle classifier $\boldsymbol{w}_0$, we use the following theorem by Sridharan *et al.* [14].

THEOREM 10.2 *Let $\mathcal{X}$ be the feature space, and suppose $\mathcal{D}$ is some distribution over $\mathcal{X}$. Let*
$$S \doteq \langle (\boldsymbol{x}_1, l_1), \ldots, (\boldsymbol{x}_M, l_M) \rangle$$
denote M labeled training data sampled i.i.d. from $\mathcal{D}$. Then, for every linear threshold classifier w with $\|w\|^2 \leq 2C$, with probability at least $1 - \delta$ over training set S the following holds:
$$L_\mathcal{D}(w) - L_\mathcal{D}(\boldsymbol{w}^*) \leq 2\left[\hat{L}_S(w) - \hat{L}_S(\boldsymbol{w}^*)\right]_+ + \frac{8R^2 C \log(\frac{1}{\delta})}{M}. \tag{10.6}$$

10.4 Near-isometric projections

Given a distribution $\mathcal{D}$ over the n-dimensional feature space $\mathcal{X}$, the goal of the SVM's learner is to approximate the best large margin linear threshold classifier from i.i.d. training examples. However, as mentioned earlier, in many applications including compressed sensing and data streaming, the data is not directly represented in the feature space $\mathcal{X}$, and in contrast an $m \times n$ projection matrix A is used to transform the data to some m-dimensional space. In this section, we provide sufficient conditions (with respect to $\mathcal{X}$) that a projection matrix needs to satisfy to guarantee that the m-dimensional SVM's classifier has near-optimal generalization accuracy.

Here we allow A to be generated through a random process. However, following the standard batch-learning paradigms, we shall assume that the process of generating the projection matrix A is non-adaptive (i.e., independent of the choice of $\mathcal{D}$). In this section we assume no prior sparsity structure for the feature space $\mathcal{X}$. The results provided here are general and can be applied to any projection matrix that approximately preserves the inner products between instances of $\mathcal{X}$. Later, in Section 10.6, we will show examples of projection matrices satisfying the required conditions.

We also assume that the projection is noisy. In our analysis, for simplicity we assume that every projection $x \to y$ is contaminated with a noise vector with bounded ℓ_2 norm. That is $y \doteq Ax + e$, with $\|e\|_2 \leq \sigma R$, where σ is a small number. We define the measurement domain as

$$A\mathcal{X} \doteq \{y = Ax + e : x \in \mathcal{X}, \|e\|_2 \leq \sigma R\}.$$

Intuitively, a projection matrix A is appropriate for compressed learning if it approximately preserves the inner products (margins) between the training examples. First we provide a formal definition of the above intuitive argument, and then we prove that if a projection matrix satisfied this proposed property, then the SVM's classifier learned directly in the measurement domain has the desired near-optimal performance.

Here we define the Distance-Preserving Property for a projection matrix, which we will later use to prove that the SVM's classifier in the measurement domain is close to optimal.

DEFINITION 10.1 ((M, L, ϵ, δ)-Distance Preserving Property) *Let L be a positive number greater than 1, let M be a positive integer, and suppose that ϵ and δ are small positive numbers. Let $S \doteq \langle(\boldsymbol{x}_1, l_1), \ldots, (\boldsymbol{x}_M, l_M)\rangle$ denote a set of M arbitrary training examples. Then the projection matrix A is (M, L, ϵ, δ)-Distance-Preserving $((M, L, \epsilon, \delta_M) - DP)$, if with probability at least $1 - \delta$ (over the space of the projection matrices A) the following conditions are all satisfied:*

- (C1). *For every* $x \in \mathcal{X}$, $\|Ax\|_2 \leq L \|x\|_2$.
- (C2). *For every index i in* $[M]$: $(1-\epsilon)\|x_i\|^2 \leq \|y_i\|^2 \leq (1+\epsilon)\|x_i\|^2$.
- (C3). *For every distinct indices i and j in* $[M]$: $\left|x_i^\top x_j - y_i^\top y_j\right| \leq \epsilon R^2$.

Next we show that the Distance-Preserving Property is a sufficient condition to guarantee that the SVM's classifier trained directly in the measurement domain performs almost as well as the best linear threshold classifier in the data domain. To this end, we need to clarify more the notation used in the rest of this chapter.

We denote any classifier in the high-dimensional space by bold letter w, and any classifier in the low-dimensional space by bold letter z. Let $\boldsymbol{w}^*$ be the classifier in the data domain that minimizes the true regularization loss in the data domain, i.e.,

$$\boldsymbol{w}^* \doteq \arg_w \min L(w), \tag{10.7}$$

and let $\boldsymbol{z}^*$ be the classifier in the measurement domain that minimizes the true regularization loss in the measurement domain, i.e.,

$$\boldsymbol{z}^* \doteq \arg_z \min L(z). \tag{10.8}$$

Also let $\hat{\boldsymbol{w}}_S$ be the soft margin classifier trained with a training set S in the data domain, i.e.,

$$\hat{\boldsymbol{w}}_S \doteq \arg_w \min \hat{L}_{\langle(\boldsymbol{x}_1, l_1), \ldots, (\boldsymbol{x}_M, l_M)\rangle}(w),$$

and similarly $\hat{z}_{AS}$ is the soft margin classifier trained with the compressed training set AS in the measurement domain:

$$\hat{z}_{AS} \doteq \arg_z \min \hat{L}_{\langle(\boldsymbol{y}_1, l_1), \ldots, (\boldsymbol{y}_M, l_M)\rangle}(z).$$

Finally, let $A\hat{\boldsymbol{w}}_S$ be the classifier in the measurement domain obtained by projecting $\hat{\boldsymbol{w}}_S$ from the data domain to the measurement domain. Note that we will only use $A\hat{\boldsymbol{w}}_S$ in our analysis to prove the generalization error bound for the $\hat{z}_{AS}$ classifier.

THEOREM 10.3 *Let L, M, ϵ, and δ be as defined in Definition 10.1. Suppose that the projection matrix A is $(M+1, L, \epsilon, \delta)$-Distance-Preserving with respect to the feature space $\mathcal{X}$. Let $\boldsymbol{w}_0$ denote the data-domain oracle classifier, and fix*

$$C \doteq \sqrt{\frac{\|\boldsymbol{w}_0\|^2 \left(\frac{3R^2\epsilon}{2} + \frac{8(1+(L+\sigma)^2)\log(\frac{1}{\delta})}{M}\right)}{2}},$$

as the SVM's regularization parameter. Let $S \doteq \langle(\boldsymbol{x}_1, l_1), \ldots, (\boldsymbol{x}_M, l_M)\rangle$ represent M training examples in the data domain, and let $AS \doteq \langle(\boldsymbol{y}_1, l_1), \ldots, (\boldsymbol{y}_M, l_M)\rangle$ denote the representation of the training examples in the measurement domain. Finally let $\hat{z}_{AS}$ denote the measurement domain SVM's classifier trained on AS. Then with probability at least $1 - 3\delta$,

$$H_{\mathcal{D}}(\hat{z}_{AS}) \leq H_{\mathcal{D}}(\boldsymbol{w}_0) + O\left(R\|\boldsymbol{w}_0\|\sqrt{\left(\epsilon + \frac{(L+\sigma)^2 \log\left(\frac{1}{\delta}\right)}{M}\right)}\right). \quad (10.9)$$

10.5 Proof of Theorem 10.3

In this section we prove that the Distance-Preserving Property provides fidelity for the SVM's learning in the measurement domain. We use a hybrid argument to show that the SVM's classifier in the measurement domain has approximately the same accuracy as the best SVM's classifier in the data domain. Theorem 10.1 together with the theory of structural risk minimization implies that if the data are represented in high-dimensional space, the high-dimensional SVM's classifier $\hat{\boldsymbol{w}}_S$ has almost the same performance as the best classifier $\boldsymbol{w}_0$. Then we show that if we project the SVM's classifier $\hat{\boldsymbol{w}}_S$ to the measurement domain, the true regularization loss of the classifier $A\hat{\boldsymbol{w}}_S$ is almost the same as the true regularization loss of the SVM's classifier $\hat{\boldsymbol{w}}_S$ in the high-dimensional domain. Again, we emphasize that we only use the projected classifier $A\hat{\boldsymbol{w}}_S$ in the analysis. The compressed learning algorithm only uses its own *learned* SVM's classifier from the measurement domain training examples.

To prove that the measurement domain SVM's classifier is near-optimal, we first show that if a matrix A is Distance-Preserving with respect to a training set S, then it also approximately preserves the distances between any two points in the convex hull of the training examples. This property is then used to show that the distance (margin) between the data-domain SVM's classifier, and any training example is approximately preserved by the sensing matrix A.

LEMMA 10.2 *Let M be a positive integer, and let D_1 and D_2 be two positive numbers. Let A be an $m \times n$ sensing matrix which is $(M+1, L, \epsilon, \delta)$-Distance-Preserving with respect to $\mathcal{X}$. Let $\langle(\boldsymbol{x}_1, l_1), \ldots, (\boldsymbol{x}_M, l_M)\rangle$ denote M arbitrary elements of $\mathcal{X}$, and let $\langle(\boldsymbol{y}_1, l_1), \ldots, (\boldsymbol{y}_M, l_M)\rangle$ denote the corresponding labeled projected vectors. Then with probability $1 - \delta$ the following argument is valid:*

For every non-negative number $s_1, \cdots, s_M$, and $t_1, \cdots, t_M$, with $\sum_{i=1}^{M} s_i \leq D_1$, and $\sum_{j=1}^{M} t_j \leq D_2$, let

$$w_1 \doteq \sum_{i=1}^{M} s_i l_i x_i, \text{ and } w_2 = \sum_{i=1}^{M} t_i l_i x_i, \quad (10.10)$$

and similarly define

$$z_1 \doteq \sum_{i=1}^{M} s_i l_i y_i \text{ and } z_2 \doteq \sum_{i=1}^{M} t_i l_i y_i. \quad (10.11)$$

Then

$$|w_1^\top w_2 - z_1^\top z_2| \leq D_1 D_2 R^2 \epsilon.$$

Proof. It follows from the definition of z_1 and z_2, and from the triangle inequality that

$$\left|z_1^\top z_2 - w_1^\top w_2\right| = \left|\sum_{i=1}^{M}\sum_{j=1}^{M} s_i t_j l_i l_j \left(y_i^\top y_j - x_i^\top x_j\right)\right|$$

$$\leq \sum_{i=1}^{M}\sum_{j=1}^{M} s_i t_j \left|y_i^\top y_j - x_i^\top x_j\right|$$

$$\leq \epsilon R^2 \left(\sum_{i=1}^{M} s_i\right)\left(\sum_{j=1}^{M} t_j\right) \leq D_1 D_2 R^2 \epsilon. \qquad (10.12)$$

$\square$

Next we use Lemma 10.2 and the fact that the SVM's classifier is in the C-convex hull of the training example to show that the margin and the true hinge loss of the SVM's classifier are not significantly distorted by the Distance-Preserving matrix A.

LEMMA 10.3 *Let $\mathcal{D}$ be some distribution over the feature space $\mathcal{X}$, and let $S = \langle (x_1, l_1), \ldots, (x_M, l_M) \rangle$ denote M i.i.d. labeled samples from $\mathcal{D}$. Also let $\langle (y_1, l_1), \ldots, (y_M, l_M) \rangle$ denote the corresponding labeled vectors in the measurement domain. Let $\hat{w}_S \doteq \sum_{i=1}^{M} s_i l_i x_i$ denote the data domain soft-margin SVM's classifier trained on S, and let $A\hat{w}_S = \sum_{i=1}^{M} s_i l_i y_i$ denote the corresponding linear threshold classifier in the measurement domain. Suppose that the sensing matrix is $(M+1, L, \epsilon, \delta)$-Distance-Preserving with respect to $\mathcal{X}$. Then with probability $1-\delta$,*

$$L_{\mathcal{D}}(A\hat{w}_S) \leq L_{\mathcal{D}}(\hat{w}_S) + \frac{3CR^2\epsilon}{2}.$$

Proof. Since the definition of the Distance-Preserving Property is independent of the choice of $\mathcal{D}$, first we assume that $\langle (x_1, l_1), \ldots, (x_M, l_M) \rangle$ are M fixed vectors in the data domain. Let (x_{M+1}, l_{M+1}) be a fresh sample from $\mathcal{D}$ (which we will use in analyzing the true hinge loss of the measurement domain SVM's classifier. Since A is $(M+1, L, \epsilon, \delta)$-Distance-Preserving, with probability $1-\delta$, it preserves the distances between any pair of points x_i and x_j (with $i, j \in [M+1]$). By Theorem 10.1, the soft margin SVM's classifier is a linear combination of the support vectors, where each s_i is a positive number and $\sum_{i=1}^{M} s_i \leq C$. As a result, using Theorem 10.2 with $D_1 = D_2 = C$ yields

$$\frac{1}{2C}\|A\hat{w}_S\|^2 \leq \frac{1}{2C}\|\hat{w}_S\|^2 + \frac{CR^2\epsilon}{2}. \qquad (10.13)$$

Now we show that dimensionality reduction using A does not distort the true hinge loss of $\hat{w}_S$ significantly. Again, since A is $(M+1, L, \epsilon, \delta)$-Distance-Preserving, Lemma 10.2

with $D_1 = C$ and $D_2 = 1$ guarantees that

$$1 - 1_{M+1}(A\hat{\boldsymbol{w}}_S)^\top(y_{M+1}) \leq 1 - 1_{M+1}\hat{\boldsymbol{w}}_S^\top x_{M+1} + CR^2\epsilon. \tag{10.14}$$

Now as
$$1 - 1_{M+1}\hat{\boldsymbol{w}}_S^\top x_{M+1} \leq \left[1 - 1_{M+1}\hat{\boldsymbol{w}}_S^\top x_{M+1}\right]_+,$$

and since
$$\left[1 - 1_{M+1}\hat{\boldsymbol{w}}_S^\top x_{M+1}\right]_+ + CR^2\epsilon$$

is always non-negative, we have

$$\left[1 - 1_{M+1}(A\hat{\boldsymbol{w}}_S)^\top(y_{M+1})\right]_+ \leq \left[1 - 1_{M+1}\hat{\boldsymbol{w}}_S^\top x_{M+1}\right]_+ + CR^2\epsilon. \tag{10.15}$$

Equation 10.15 is valid for *each* sample $(x_{M+1}, 1_{M+1})$ with probability $1 - \delta$ over the distribution of A. Now since the distribution of A is independent from $\mathcal{D}$, we can take the expectation of Equation (10.15) with respect to $\mathcal{D}$, and we obtain the bound

$$H_\mathcal{D}(A\hat{\boldsymbol{w}}_S) \leq H_\mathcal{D}(\hat{\boldsymbol{w}}_S) + CR^2\epsilon.$$

□

So far we have bounded the change in the regularization loss of the SVM's classifier $\hat{\boldsymbol{w}}_S$ after projection via a Distance-Preserving matrix A. Next we show that the regularization loss of $\hat{z}_{AS}$, which is the classifier we can directly learn from the measurement domain training examples, is close to the regularization loss of the linear threshold classifier $A\hat{\boldsymbol{w}}_S$. We use this to conclude that the regularization loss of $\hat{z}_{AS}$ is close to the regularization loss of the oracle best classifier $\boldsymbol{w}^*$ in the data domain. This completes our hybrid argument.

Proof of Theorem 10.3. From the definition of the regularization loss we have:

$$H(\hat{z}_{AS}) \leq H(\hat{z}_{AS}) + \frac{1}{2C}\|\hat{z}_{AS}\|^2 = L(\hat{z}_{AS}).$$

Since A is Distance-Preserving and every noise vector has bounded ℓ_2 norm σR, it follows from the triangle inequality that

$$\max_y \|f\|_2 = \max_{x,e} \|Ax + e\|_2 \leq \max_x \|Ax\|_2 + \max_e \|e\|_2$$
$$\leq L\max_x \|x\|_2 + \max_e \|e\|_2 \leq LR + \sigma R.$$

Theorem 10.2 states that with probability $1 - \delta$, the regularization loss of the SVM's classifier in the measurement domain $\hat{z}_{AS}$, is close to the regularization loss of the best classifier in the measurement domain z^*:

$$L_\mathcal{D}(\hat{z}_{AS}) \leq L_\mathcal{D}(z^*) + \frac{8(L+\sigma)^2 R^2 C \log(\frac{1}{\delta})}{M}. \tag{10.16}$$

By the definition of z^* in Equation (10.8), z^* is the best classifier in the measurement domain, so

$$L_\mathcal{D}(z^*) \leq L_\mathcal{D}(A\hat{w}_S). \tag{10.17}$$

Lemma 10.3 connects the regularization loss of the SVM's classifier in the data domain, $\hat{w}_S$, to the regularization loss of its projected vector $A\hat{w}_S$. That is, with probability $1-\delta$,

$$L_\mathcal{D}(A\hat{w}_S) \leq L_\mathcal{D}(\hat{w}_S) + \frac{3CR^2\epsilon}{2}. \tag{10.18}$$

Theorem 10.2 now applied in the data domain, connects the regularization loss of the SVM's classifier $\hat{w}_S$ to the regularization loss of the best classifier w^*:

$$\Pr\left[L_\mathcal{D}(\hat{w}_S) \geq L_\mathcal{D}(w^*) + \frac{8R^2 C \log(\frac{1}{\delta_{M+1}})}{M}\right] \leq \delta. \tag{10.19}$$

From combining the inequalities of Equations (10.16) to (10.19), it follows that with probability $1 - 3\delta$

$$H_\mathcal{D}(\hat{z}_{AS}) \leq H_\mathcal{D}(w^*) + \frac{1}{2C}\|w^*\|^2 + \frac{3CR^2\epsilon}{2} + \frac{8R^2 C \log(\frac{1}{\delta})}{M} + \frac{8(L+\sigma)^2 R^2 C \log(\frac{1}{\delta})}{M}.$$

Observe that for every choice of C, since w^* is the minimizer of the true regularization loss (with respect to that choice of C), we have $L_\mathcal{D}(w^*) \leq L_\mathcal{D}(w_0)$; therefore, for every choice of C with probability at least $1 - 3\delta$

$$H_\mathcal{D}(\hat{z}_{AS}) \leq H_\mathcal{D}(w_0) + \frac{1}{2C}\|w_0\|^2 + \frac{3CR^2\epsilon}{2} + \frac{8R^2 C \log(\frac{1}{\delta})}{M} + \frac{8(L+\sigma)^2 R^2 C \log(\frac{1}{\delta})}{M}. \tag{10.20}$$

In particular, by choosing

$$C \doteq \sqrt{\frac{\|w_0\|^2 \left(\frac{3R^2\epsilon}{2} + \frac{8(1+(L+\sigma)^2)\log(\frac{1}{\delta})}{M}\right)}{2}},$$

which minimizes the right-hand side of Equation (10.6), we get:

$$H_\mathcal{D}(\hat{z}_{AS}) \leq H_\mathcal{D}(w_0) + O\left(R\|w_0\|\sqrt{\left(\epsilon + \frac{(L+\sigma)^2 \log\left(\frac{1}{\delta}\right)}{M}\right)}\right). \tag{10.21}$$

□

10.6 Distance-Preserving via Johnson–Lindenstrauss Property

In this section, we introduce the Johnson–Lindenstrauss Property [15]. We also show how this property implies the Distance-Preserving Property and also the Restricted Isometry Property which is a sufficient property for compressed sensing. Throughout this section let ε and ρ be two positive numbers with $\varepsilon < 0.2$, and as before, let k be an integer less than n.

DEFINITION 10.2 *Let x_1 and x_2 denote two arbitrary vectors in the feature space $\mathcal{X}$. Then A satisfies the (ε, ρ)-Johnson-Lindenstrauss Property $((\varepsilon, \rho)$-JLP) if with probability at least $1 - \rho$, A approximates the distances between x_1 and x_2 up to a multiplicative distortion ε:*

$$(1-\varepsilon)\|x_1 - x_2\|^2 \leq \|A(x_1 - x_2)\|^2 \leq (1+\varepsilon)\|x_1 - x_2\|^2. \tag{10.22}$$

The Johnson–Lindenstrauss Property is now widely used in different applications including nearest-neighbor approximation [1, 2], data-streaming [3], machine learning [4, 5], graph approximation [6], and compressed sensing.

In this section, we show that matrices satisfying the JL Property are appropriate for compressed learning as well. We start by showing that the JL Property provides sufficient conditions for *worst-case* compressed sensing:

DEFINITION 10.3 *An $m \times n$ matrix A satisfies the (k, ε)-Restricted Isometry Property $((k, \varepsilon)$-RIP) if it acts as a near-isometry on every k-sparse vector x, i.e.,*

$$(1-\varepsilon)\|x\|^2 \leq \|Ax\|^2 \leq (1+\varepsilon)\|x\|^2.$$

The following theorem is proved by Baraniuk et al. [16], and relates the JL Property to the RIP:

THEOREM 10.4 *Let A be an $m \times n$ sensing matrix satisfying (ε, ρ)-JLP. Then for every positive integer k, with probability at least $1 - \binom{n}{k}\left(\frac{6}{\sqrt{\varepsilon}}\right)^k \rho$, A is $(k, 5\varepsilon)$-RIP.*

Proof. First note that if a vector x is k-sparse, then the normalized vector $\frac{x}{\|x\|_2}$ is also k-sparse. Furthermore, since the sensing operator A is linear, we have

$$A\left(\frac{x}{\|x\|_2}\right) = \frac{1}{\|x\|_2}Ax.$$

Therefore, the matrix approximately preserves the norm of x if and only if it approximately preserves the norm of $\frac{x}{\|x\|_2}$. Hence, to prove that a matrix satisfies the RIP, it is sufficient to show that A acts as a near-isometry on every k-sparse vector that has unit Euclidean norm.

Now fix a k-dimensional subspace. Here we denote the set of unit-norm vectors that lie on that subspace by $\mathcal{X}_k$. We calculate the probability that A approximately preserves the norm of every vector in $\mathcal{X}_k$. Define

$$\mathcal{Q} \doteq \left\{ q \in \mathcal{X}_k \text{ s.t. for all } x \in \mathcal{X}_k, \min_{q \in \mathcal{Q}} \|x - q\|_2 \leq \frac{\sqrt{\varepsilon}}{2} \right\}.$$

Also let $\Delta \doteq \max_{x \in \mathcal{X}_k} \|Ax\|$. A combinatorial argument can be used to show that one can find a set $\mathcal{Q}$ of $\left(\frac{6}{\sqrt{\varepsilon}}\right)^k$ vectors in $\mathcal{X}_k$, such that the minimum distance of any vector x in $\mathcal{X}_k$ to $\mathcal{Q}$ is at most $\frac{\sqrt{\varepsilon}}{2}$. The JL Property guarantees that with probability at least $1 - |\mathcal{Q}|\rho$, A approximately preserves the norm of every point in $\mathcal{Q}$. That is, for every $q \in \mathcal{Q}$

$$(1-\varepsilon)\|q\|^2 \leq \|Ax\|^2 \leq (1+\varepsilon)\|x\|^2,$$

and therefore,

$$(1-\sqrt{\varepsilon})\|q\|_2 \leq \|Ax\|_2 \leq (1+\sqrt{\varepsilon})\|x\|_2.$$

Let x be a vector in $\mathcal{X}_k$, and let $q^* \doteq \arg\min_{q \in \mathcal{Q}} \|x - q\|_2$. It follows from the triangle inequality, and from the definition of q^* that

$$\|Ax\|_2 \leq \|A(x - q^*)\|_2 + \|Aq^*\|_2 \leq \Delta \frac{\sqrt{\varepsilon}}{2} + (1 + \sqrt{\epsilon}). \quad (10.23)$$

Equation (10.23) is valid for every $x \in \mathcal{X}_k$. Therefore, we must have

$$\Delta = \max_{x \in \mathcal{X}_k} \|Ax\|_2 \leq \Delta \frac{\sqrt{\varepsilon}}{2} + (1 + \sqrt{\varepsilon}),$$

and as a result,

$$\Delta \leq 1 + \frac{\frac{3}{2}\sqrt{\varepsilon}}{1 - \frac{\sqrt{\varepsilon}}{2}} < 1 + 2\sqrt{\varepsilon}.$$

Also from using the triangle inequality it follows that for every $x \in \mathcal{X}_k$

$$\|Ax\|_2 \geq \|Aq^*\|_2 - \|A(x - q^*)\|_2 \geq (1 - \sqrt{\varepsilon}) - (1 + 2\sqrt{\varepsilon})\frac{\sqrt{\varepsilon}}{2} > (1 - 2\sqrt{\varepsilon}).$$

Consequently, with probability $1 - |\mathcal{Q}|\rho$, for every vector x in the k-dimensional unit sphere $\mathcal{X}_k$ we have

$$(1 - 2\sqrt{\varepsilon})\|x\|_2 \leq \|Ax\|_2 \leq (1 + 2\sqrt{\varepsilon})\|x\|_2,$$

which implies that

$$(1 - 5\varepsilon)\|x\|^2 \leq \|Ax\|^2 \leq (1 + 5\varepsilon)\|x\|^2. \quad (10.24)$$

Equation (10.24) proves that the probability that the sensing matrix A does not act as a near-isometry on each *fixed* subspace is at most $\left(\frac{6}{\sqrt{\varepsilon}}\right)^k \rho$. Since there are $\binom{n}{k}$ total

k-dimensional subspaces, it follows from taking the union bound over all $\binom{n}{k}$ subspaces that with probability at least $1 - \binom{n}{k}\left(\frac{6}{\sqrt{\varepsilon}}\right)^k \rho$, A is $(k, 5\varepsilon)$-RIP. □

The following proposition is a consequence of the RIP property and is proved by Needell and Tropp [17]. It provides an upper bound for the Lipschitz constant of a matrix satisfying the RIP. We will later use this Proposition to prove that matrices that satisfy the JL Property are Distance-Preserving.

PROPOSITION 10.3 *Let A be a $(k, 5\varepsilon)$-RIP sensing matrix. Then for every vector $x \in \mathbb{R}^n$ the following bound holds:*

$$\|Ax\|_2 \leq \sqrt{1+5\varepsilon}\|x\|_2 + \sqrt{\frac{1+5\varepsilon}{k}}\|x\|_1.$$

The RIP property is a *sufficient* condition for sparse recovery. It has been shown by Candès, Romberg, and Tao [18, 19] that if a matrix satisfies the RIP property with sufficiently small distortion parameter ε, then one can use convex optimization methods to estimate the best k-term approximation of *any* vector $x \in \mathbb{R}^n$. Now we show that a matrix that satisfies the JL Property is Distance-Preserving. Therefore, it follows from Theorem 10.3 that if a matrix satisfies the JL Property then the SVM's classifier in the measurement domain is almost optimal.

In the rest of this section, we assume that ε_1 and ε_2 are two positive numbers less than 0.2, and ρ_1 and ρ_2 are two positive numbers less than one. We will assume that the sensing matrix satisfies both the (ε_1, ρ_1)-JL Property (which is used to show that A approximately preserves the regularization loss of the data domain SVM's classifier), and the (ε_2, ρ_2)-JL Property (which is used to show that A satisfies RIP). Potentially, ε_1 can be significantly smaller than ε_2. The reason is that ε_2 only appears in the Lipschitz bound of the sensing matrix A, whereas ε_1 controls the distortion of the regularization loss of the SVM's classifier.

LEMMA 10.4 *Let M be a positive integer, and let A be the $m \times n$ sensing matrix. Then for every integer k, A is $(M+1, \mathrm{L}, \epsilon, \delta)$-Distance-Preserving matrix with respect to $\mathcal{X}$ with*

$$\mathrm{L} = \sqrt{1+5\varepsilon_2}\left(1 + \frac{\ell_{2\to 1}(\mathcal{X})}{\sqrt{k}}\right),$$

$$\epsilon = \left(3\varepsilon_1 + 4\sigma + \sigma^2\right),$$

$$\delta = \binom{M+2}{2}\rho_1 + \binom{n}{k}\left(\frac{6}{\sqrt{\varepsilon_1}}\right)^k \rho_2.$$

Proof. We show that the matrix A satisfies all three conditions of Definition 10.1.
Proof of Condition (C1): (For every $x \in \mathcal{X}$, $\|Ax\|_2 \leq \mathrm{L}\|x\|_2$.)
Theorem 10.4 states that for every integer k, with probability at least $1 - \binom{n}{k}\left(\frac{6}{\sqrt{\varepsilon_1}}\right)^k \rho_2$, A is $(k, 5\varepsilon_2)$-RIP. Therefore, it follows from Proposition 10.3, and from the definition

of $\ell_{2\to 1}$ (Equation (10.1)), that with the same probability, for *every* vector x in $\mathcal{X}$

$$\|Ax\|_2 \leq \sqrt{1+5\varepsilon_2}\left(1+\frac{\ell_{2\to 1}(\mathcal{X})}{\sqrt{k}}\right)\|x\|_2.$$

Therefore A is L-Lipschitz with respect to $\mathcal{X}$ with $L = \sqrt{1+5\varepsilon_2}\left(1+\frac{\ell_{2\to 1}(\mathcal{X})}{\sqrt{k}}\right)$.

Proof of Conditions (C2) and (C3):

- (C2). For every index i in $[M+1]$: $(1-\epsilon)\|x_i\|^2 \leq \|y_i\|^2 \leq (1+\epsilon)\|x_i\|^2$.
- (C3). For every distinct indices i and j in $[M+1]$: $\left|x_i^\top x_j - y_i^\top y_j\right| \leq \epsilon R^2$.

Let 0 denote the n-dimensional all-zero vector. Since A satisfies the (ε_1, ρ_1)-JL Property, with probability $1-\rho_1$ it preserves the distances between two fixed vectors. Here we apply the union bound to the set of all $\binom{M+2}{2}$ pairs of vectors selected from the set $\langle 0, x_1, \ldots, x_{M+1}\rangle$. It follows from the union bound that with probability at least $1 - \binom{M+2}{2}\rho_1$ the following two statements hold simultaneously:

1. For every index i ranging in $[M+1]$:

$$(1-\varepsilon_1)\|x_i\|^2 \leq \|Ax_i\|^2 \leq (1+\varepsilon_1)\|x_i\|^2.$$

2. For every pair of indices i and j in $[M+1]$:

$$(1-\varepsilon_1)\|x_i-x_j\|^2 \leq \|A(x_i-x_j)\|^2 \leq (1+\varepsilon_1)\|x_i-x_j\|^2.$$

Now let i and j be any two fixed indices in $[M+1]$. We have

$$\|A(x_i-x_j)\|^2 \leq (1+\varepsilon_1)\|x_i-x_j\|^2 \tag{10.25}$$
$$= (1+\varepsilon_1)\left(\|x_i\|^2 + \|x_j\|^2 - 2x_i^\top x_j\right).$$

Also

$$(1-\varepsilon_1)\left(\|x_i\|^2+\|x_j\|^2\right) - 2(Ax_i)^\top(Ax_j) \leq \|Ax_i\|^2 + \|Ax_j\|^2 - 2(Ax_i)^\top(Ax_j)$$
$$= \|A(x_i-x_j)\|^2. \tag{10.26}$$

Combining Equations (10.25) and (10.26), and noting that $\|x_i\| \leq R$ and $\|x_j\| \leq R$, we get

$$(1+\varepsilon_1)x_i^\top x_j \leq (Ax_i)^\top(Ax_j) + 2R^2\varepsilon_1, \tag{10.27}$$

which implies that

$$x_i^\top x_j - (Ax_i)^\top(Ax_j) \leq \varepsilon_1\left(2R^2 + x_i^\top x_j\right) \leq 3R^2\varepsilon_1.$$

Similarly, we can show that
$$(Ax_i)^\top (Ax_j) - x_i^\top x_j \leq 3R^2 \varepsilon_1.$$

Therefore
$$\left| x_i^\top x_j - (Ax_i)^\top (Ax_j) \right| \leq 3R^2 \varepsilon_1. \tag{10.28}$$

Having identified the difference between $x_i^\top x_j$ and $(Ax_i)^\top (Ax_j)$ we now use the triangle inequality to calculate the distance between $x_i^\top x_j$ and $y_i^\top y_j$. Recall that $y_i \doteq Ax_i + e_i$ and $y_j \doteq Ax_j + e_j$, with $\|e_i\|_2 \leq \sigma R$ and $\|e_j\|_2 \leq \sigma R$. From the triangle inequality we get

$$\begin{aligned}
\left| x_i^\top x_j - y_i^\top y_j \right| &= \left| x_i^\top x_j - (Ax_i + e_i)^\top (Ax_j + e_j) \right| \\
&\leq \left| x_i^\top x_j - (Ax_i)^\top (Ax_j) \right| + |e_i^\top (Ax_j)| + |(Ax_i)^\top e_j| + |e_i^\top e_j| \\
&\leq \left| x_i^\top x_j - (Ax_i)^\top (Ax_j) \right| + 2\|x_j\|\|e_i\| + 2\|x_i\|\|e_j\| + \|e_i\|\|e_j\| \\
&\leq 3R^2 \varepsilon_1 + 4R^2 \sigma + R^2 \sigma^2 = R^2 \left(3\varepsilon_1 + 4\sigma + \sigma^2 \right).
\end{aligned}$$

$\square$

Next we prove that if a sensing matrix is Distance-Preserving then the performance of the SVM's classifier in the measurement domain is close to the performance of the data domain SVM's.

THEOREM 10.5 *Let w_0 denote the data-domain oracle classifier. Let $S \doteq \langle (x_1, l_1), \ldots, (x_M, l_M) \rangle$ represent M training examples in the data domain, and let $AS \doteq \langle (y_1, l_1), \ldots, (y_M, l_M) \rangle$ denote the representation of the training examples in the measurement domain. Finally let $\hat{z}_{AS}$ denote the measurement domain SVM's classifier trained on AS. Then for every integer k, if*

$$\rho_1 \leq \frac{1}{6} \binom{M+2}{2}^{-1} \quad \text{and} \quad \rho_2 \leq \frac{1}{6} \left[\binom{n}{k} \left(\frac{6}{\sqrt{\varepsilon_1}} \right)^k \right]^{-1},$$

then with probability at least $1 - 3 \left[\binom{n}{k} \left(\frac{6}{\sqrt{\varepsilon_1}} \right)^k \rho_2 + \binom{M+2}{2} \rho_1 \right]$,

$$H_\mathcal{D}(\hat{z}_{AS}) \leq H_\mathcal{D}(w_0) \tag{10.29}$$

$$+ O \left(R\|w_0\| \sqrt{\varepsilon_1 + \sigma + \frac{(1+\varepsilon_2)\left(1 + \frac{\ell_{2 \to 1}(\mathcal{X})}{\sqrt{k}}\right)^2 \log\left(\binom{n}{k} \left(\frac{6}{\sqrt{\varepsilon_1}} \right)^k \rho_2 + \binom{M+2}{2} \rho_1 \right)^{-1}}{M}} \right).$$

Proof. The proof of Theorem 10.5 follows from inserting the values of L, ε, and ρ from Lemma 10.4 into Theorem 10.3. $\square$

10.7 Worst-case JL Property via random projection matrices

10.7.1 Johnson–Lindenstrauss and random sensing

So far we have shown that if a projection matrix satisfies the Johnson–Lindenstrauss Property, then that matrix can be used for compressed learning. Now we provide examples of projection matrices satisfying the Johnson–Lindenstrauss Property. The following theorem is proved by Dasgupta and Gupta [20], and by Indyk and Motwani [21], and guarantees that random Gaussian (or sub-Gaussian) projection matrices satisfy the desired JL Property.

PROPOSITION 10.4 *Let A be an $m \times n$ projection matrix whose entries are sampled i.i.d. from a $\mathcal{N}\left(0, \frac{1}{m}\right)$ Gaussian distribution. Let ε be a positive real number, and let x_1 and x_2 be two fixed vectors in $\mathbb{R}^n$. Then with probability at least $1 - 2\exp\left\{-\frac{m\varepsilon^2}{4}\right\}$, A approximately preserves the distance between x_1 and x_2.*

$$(1-\varepsilon)\|x_1 - x_2\|^2 \leq \|A(x_1 - x_2)\|^2 \leq (1+\varepsilon)\|x_1 - x_2\|^2.$$

By combining Proposition 10.4 and Theorem 10.5 we obtain the following corollaries for Gaussian projection matrices.

COROLLARY 10.2 *Let $\mathcal{X}$ represent the sphere of radius R in $\mathbb{R}^n$, and let A be the $m \times n$-dimensional Gaussian projection matrix. Let $\mathcal{D}$ be some distribution over $\mathcal{X}$, and let w_0 denote the data-domain oracle classifier. Suppose $S \doteq \langle (x_1, l_1), \ldots, (x_M, l_M) \rangle$ represent M training examples in the data domain, and $AS \doteq \langle (y_1, l_1), \ldots, (y_M, l_M) \rangle$ denote the representation of the training examples in the measurement domain. Let $\hat{z}_{AS}$ denote the measurement domain SVM's classifier trained on AS. Then there exist universal constants κ_1 and κ_2, such that for every integer k, if*

$$m \geq \kappa_1 \left(\frac{\log M}{\varepsilon_1^2} + \frac{k \log \frac{n}{k}}{\varepsilon_2^2} \right),$$

then with probability at least $1 - 6 \left[\exp\{-\kappa_2 m \varepsilon_1^2\} + \exp\{-\kappa_2 m \varepsilon_2^2\} \right]$,

$$H_{\mathcal{D}}(\hat{z}_{AS}) \leq H_{\mathcal{D}}(w_0) + O\left(R\|w_0\| \sqrt{\varepsilon_1 + \sigma + \frac{(1+\varepsilon_2) n m \varepsilon_1^2}{kM}} \right). \quad (10.30)$$

Proof. Here the feature space $\mathcal{X}$ contains every point inside the sphere with radius R. Therefore, it follows from the Cauchy–Schwarz inequality that $\ell_{2 \to 1}(\mathcal{X}) \leq \sqrt{n}$. Moreover, from applying Proposition 10.4 it follows that the projection matrix satisfies the both the $\left(\varepsilon_1, 2\exp\left\{-\frac{m\varepsilon_1^2}{4}\right\}\right)$-JL Property, and the $\left(\varepsilon_2, 2\exp\left\{-\frac{m\varepsilon_2^2}{4}\right\}\right)$-JL Property. Now observe that since

$$\log\left[\binom{n}{k} \left(\frac{6}{\sqrt{\varepsilon_1}}\right)^k \right] = O\left(k \log \frac{n}{k}\right) \quad \text{and} \quad \log\left[\binom{M+2}{2} \right] = O(\log M),$$

there exist sufficiently large universal constants κ_1 and κ_2, such that if $m\varepsilon_1^2 \geq \kappa_1 \log M$, and $m\varepsilon_2^2 \geq \kappa_1 \left(k \log \frac{n}{k}\right)$, then

$$2 \binom{n}{k} \left(\frac{6}{\sqrt{\varepsilon_1}}\right)^k \exp\left\{-\frac{m\varepsilon_2^2}{4}\right\} + \binom{M+2}{2} \exp\left\{-\frac{m\varepsilon_1^2}{4}\right\}$$
$$= 2\left[\exp\left\{-\kappa_2 m\varepsilon_1^2\right\} + \exp\left\{-\kappa_2 m\varepsilon_2^2\right\}\right].$$

The result then follows directly from Theorem 10.5. □

COROLLARY 10.3 *Let $\mathcal{X}$ represent the union of all k-sparse vectors x in $\mathbb{R}^n$ restricted to the sphere of radius R. Let $\mathcal{D}$ be some distribution over $\mathcal{X}$, and let $\mathbf{w}_0$ denote the data-domain oracle classifier. Suppose $S \doteq \langle(\mathbf{x}_1, l_1), \ldots, (\mathbf{x}_M, l_M)\rangle$ represent M training examples in the data domain, and $AS \doteq \langle(\mathbf{y}_1, l_1), \ldots, (\mathbf{y}_M, l_M)\rangle$ denote the representation of the training examples in the measurement domain. Finally let $\hat{\mathbf{z}}_{AS}$ denote the measurement domain SVM's classifier trained on AS. Then there exist universal constants κ_1' and κ_2' (independent of k), such that if*

$$m \geq \kappa_1 \left(\frac{\log M}{\varepsilon_1^2} + \frac{k \log \frac{n}{k}}{\varepsilon_2^2}\right),$$

then with probability at least $1 - 6\left[\exp\left\{-\kappa_2' m\varepsilon_2^2\right\} + \exp\left\{-\kappa_2' m\varepsilon_1^2\right\}\right]$,

$$H_\mathcal{D}(\hat{\mathbf{z}}_{AS}) \leq H_\mathcal{D}(\mathbf{w}_0) + O\left(R\|\mathbf{w}_0\|\sqrt{\varepsilon_1 + \sigma + \frac{(1+\varepsilon_2)m\varepsilon_1^2}{M}}\right). \quad (10.31)$$

Proof. The proof of Corollary 10.3 is aligned with the proof of Corollary 10.2. The only difference is that now every vector in the feature space $\mathcal{X}$ is exactly k-sparse. The Cauchy-Schwarz inequality implies that for every k-sparse vector x in $\mathcal{X}$, $\|x\|_1 \leq \sqrt{k}\|x\|_2$. Therefore, $\ell_{2\to 1}(\mathcal{X}) \leq \sqrt{k}$. □

REMARK 10.1 Corollary 10.3 states that if it is known *a priori* that every vector in the feature space is exactly k-sparse, then the number of training examples needed to obtain a fixed hinge loss can be decreased significantly. In other words, prior knowledge of sparsity can facilitate the learning task and can be helpful not only from a compressed sensing point of view, but also from a machine learning perspective.

REMARK 10.2 Note that a larger value for m leads to lower distortion parameter ε_1, and provides lower classification error. In particular, by setting $\varepsilon_1 = \sqrt{\frac{\log M}{m}}$ we get

$$H_\mathcal{D}(\hat{\mathbf{z}}_{AS}) \leq H_\mathcal{D}(\mathbf{w}_0) + \tilde{O}\left(\left(\frac{\log M}{m}\right)^{\frac{1}{4}} + \left(\frac{\log M}{M}\right)^{\frac{1}{2}} + \sigma^{\frac{1}{2}}\right).$$

Finally note that here we only focused on random Gaussian matrices as examples of matrices satisfying the JL Property. However, there exist other families of random projection matrices satisfying the desired JL Property. Examples of such matrices are Sparse

Johnson–Lindenstrauss Transform [22, 23], and Fast Johnson–Lindenstrauss Transform [24] and its variations based on dual BCH-codes [25], Lean Walsh Transform [26], and Fast Fourier Transform [27]. (See [28] for the explanation of the other matrices with JL Property.)

REMARK 10.3 Note that the use of random projection in dimensionality reduction was discovered even before the start of the compressed sensing field. The results of this section were based on a series of past work [29, 30, 31, 32] (see the references therein), and tightens and generalizes them based on the new results in learning theory and stochastic optimization. Interestingly, the JL Property unifies the compressed sensing and compressed sensing tasks. If a matrix satisfies the JL Property, then not only the measurement domain SVM's classification is close to optimal, but it is also possible to successfully reconstruct any desired data point efficiently.

10.7.2 Experimental results

Here we provide experimental results supporting the results of the previous section. Here we fixed $n = 1024$, and $m = 100$, and analyzed the impact of changing the sparsity level k, and the number of training examples M on the average test error of the measurement domain classifier. At each experiment, we repeated the following process 100 times independently. First, we generated an i.i.d. 100×1024 Gaussian matrix. Then we selected a random k-subset of the columns of the matrix and generated M training examples supported on the random k-subset and with independent random signs. Finally we used the inverse wavelet transform and transformed the training examples to the inverse wavelet domain.

We also generated a uniformly random vector $\boldsymbol{w}_0 \in \mathbb{R}^n$, and used the sign of the inner-product between each training example and $\boldsymbol{w}_0$ as the label of the corresponding training example. The training examples were projected to the measurement domain using the random Gaussian matrix, and then an SVM's classifier was trained directly in the measurement domain. We used the 3-fold cross-validation to evaluate the accuracy of the trained classifier.

Figure 10.1(a) shows the dependency between the sparsity level (in the wavelet domain) of the training examples and the average cross-validation error of the SVM's classifier in the measurement domain. The average cross-validation of the data domain SVM's classifier is also provided for comparison. As shown in Figure 10.1(a), the error rate of the measurement domain SVM's classifier is very close to the error rate of the data-domain SVM's. Also note that as k increases the cross-validation error also increases. This is not surprising recalling the curse of dimensionality. Figure 10.1(b) demonstrates the impact of increasing the number of training examples on the cross-validation error of the SVM's classifiers. Again it turns out that the error of the measurement domain SVMs is close to the error of the data-domain SVM's. Also as predicted in theory, by increasing the number of training examples M, the cross-validation error rate consistently decreases.

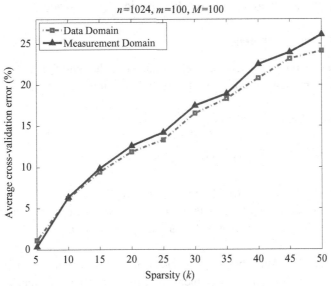

(a) The impact of the sparsity level on the average cross-validation error.

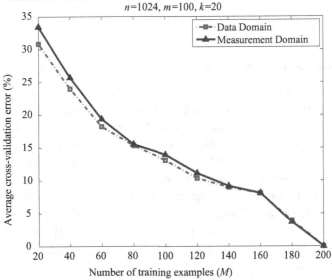

(b) The impact of the number of training examples on the average cross-validation error.

Figure 10.1 A comparison between the cross-validation errors of data-domain and measurement-domain SVM's classifiers. Here a 100×1024 Gaussian projection matrix is generated and the 3-fold cross-validation is used to measure the cross-validation error.

10.8 Average-case JL Property via explicit projection matrices

10.8.1 Global measures of coherence

Let A be an $m \times n$ matrix such that every column of A has unit ℓ_2 norm. The following two quantities measure the coherence between the columns of A [33]:

- Worst-case coherence $\mu \doteq \max_{\substack{i,j \in [n] \\ i \neq j}} \left| A_i^\top A_j \right|$.
- Average coherence $\nu \doteq \frac{1}{n-1} \max_{i \in [n]} \left| \sum_{\substack{j \in [n] \\ j \neq i}} A_i^\top A_j \right|$.

Roughly speaking, we can consider the columns of A as n distinct points on the unit sphere in $\mathbb{R}^n$. Worst-case coherence then measures how close two distinct points can be, whereas the average coherence is a measure of the spread of these points.

REMARK 10.4 The Welch bound [34] states that if all the columns of a matrix A have unit ℓ_2 norm, then unless $m = \Omega(n)$, the worst-case coherence of A cannot be too small, i.e., $\mu = \Omega\left(\frac{1}{\sqrt{m}}\right)$.

10.8.2 Average-case compressed learning

In Section 10.7, we showed how *random* sensing matrices can be used in *worst-case* compressed learning in which the data lie on *arbitrary* k-dimensional subspaces. In this section, we focus on *average-case* compressed learning. Our goal is to show that there exist explicit sensing matrices A which are suitable for average-case compressed sensing, i.e., when the signals are supported on a random k-dimensional subspace. Therefore, the results of this section work for *most* (in contrast to *all*) k-sparse vectors.

Here we assume that the feature space is restricted to a *random* k-dimensional subspace. Let Π_1^k denote a random k-subset of $[n]$. The feature space, denoted as $\mathcal{X}_{\Pi_1^k}$ is then the set of all k-sparse vectors x in $\mathbb{R}^n$, with $\operatorname{supp}(x) = \Pi_1^k$, and with $\|x\|_2 \leq R$. We will show that there exist *deterministic* sensing matrices which act as near-isometry on the feature space $\mathcal{X}_{\Pi_1^k}$.

We start by using the following proposition which is proved by Tropp [34] to show that a large family of explicit sensing matrices are Lipschitz with respect to the feature space $\mathcal{X}_{\Pi_1^k}$.

PROPOSITION 10.5 *There exists a universal constant C such that for every $\eta > 1$ and every positive ε, the following argument holds. Suppose that the $m \times n$ sensing matrix A satisfies the conditions*

$$\|A\|^2 = \frac{n}{m} \quad \text{and} \quad \mu \leq \frac{\varepsilon}{2C\eta \log n},$$

let Π_1^k denote a random k-subset of $[n]$ with $k \leq \frac{m\varepsilon^2}{(2C\eta)^2 \log n}$, and let $A_{\Pi_1^k}$ denote the submatrix of A whose columns are restricted to Π_1^k. Then

$$\Pr\left[\left\|A_{\Pi_1^k}^\dagger A_{\Pi_1^k} - I\right\|_2 \geq \varepsilon\right] \leq \frac{1}{n^\eta}.$$

We can use Proposition 10.5 to argue that if the sensing matrix is a tight-frame with redundancy $\frac{n}{m}$, and if the columns of the sensing matrix are sufficiently low coherent, then with overwhelming probability A is Lipschitz with respect to the feature space $\mathcal{X}_{\Pi_1^k}$.

Now we show that if the sensing matrix has sufficiently low μ and ν, then with overwhelming probability A approximately preserves the distances between any two vectors in $\mathcal{X}_{\Pi_1^k}$ which have arbitrary (but fixed) values.

10.8.3 Two fundamental measures of coherence and their role in compressed learning

We first show that a random k-subset of the columns of A is highly uncorrelated with any remaining column:

LEMMA 10.5 *Let ϵ_1 be a positive number, and let k be a positive integer such that $k \leq \min\left\{\frac{n-1}{2}, \epsilon_1^2 \nu^{-2}\right\}$. Let Π_1^k be a random k-subset of $[n]$. Let x be a k-sparse vector with fixed values such that $\mathrm{supp}(x) = \Pi_1^k$. Then*

$$\Pr_\Pi\left[\exists W \in [n] - \Pi_1^k \text{ s.t } \left|\sum_{i=1}^k x_i A_W^\top A_{\Pi_i}\right| \geq 2\epsilon_1 \|x\|_2\right] \leq 4n \exp\left\{\frac{-\epsilon_1^2}{128\mu^2}\right\}, \quad (10.32)$$

and

$$\Pr_\Pi\left[\exists j \in [k] \text{ s.t } \left|\sum_{\substack{i \in [k] \\ i \neq j}} x_i A_{\Pi_j}^\top A_{\Pi_i}\right| \geq 2\epsilon_1 \|x\|_2\right] \leq 4k \exp\left\{\frac{-\epsilon_1^2}{128\mu^2}\right\}. \quad (10.33)$$

Proof. See Section 10.9.1.

□

Throughout the rest of this section, assume that the projection matrix A is a tight-frame with redundancy $\frac{n}{m}$. Also let M be a positive integer (which we use to denote the number of training examples), and define, $\theta_M \doteq 128\sqrt{2\log(64n^4 M^2)}$. Furthermore, assume that ϵ_2 and ϵ_3 are two positive numbers with $\epsilon_2 \leq \frac{\mu \theta_M}{16}$.

Here we show that if the sparsity level k is $O\left(\min\left\{n^{\frac{2}{3}}, \frac{\mu}{\nu}\right\}\right)$, then with overwhelming probability, the Euclidean norm of a vector x in $\mathcal{X}_{\Pi_1^k}$ is distorted by at most $(1 \pm \epsilon)$, where $\epsilon = O(\mu(\log n + \log M))$.

THEOREM 10.6 *Let*

$$k \leq \min\left\{\left(\frac{n\theta_M}{96}\right)^{\frac{2}{3}}, \epsilon_2^2 \nu^{-2}, \left(\frac{\mu\theta_M}{96\nu}\right)^2, \epsilon_3 \nu^{-1}\right\}, \quad (10.34)$$

and let x be a k-sparse vector in $\mathcal{X}_{\Pi_1^k}$. Then

$$\Pr_{\Pi_1^k}\left[\left|\|Ax\|^2 - \|x\|^2\right| \geq 2\epsilon_3\|x\|^2\right] \leq 4\exp\left\{\frac{-\epsilon_3^2}{32\mu^2\theta_M^2}\right\} + 32k^2n\exp\left\{\frac{-\epsilon_2^2}{128\mu^2}\right\}.$$

Proof. See Section 10.9.2. □

The following Corollary is a direct consequence of Theorem 10.6.

COROLLARY 10.4 *There exists a constant κ such that if $k \leq \min\left\{n^{\frac{2}{3}}, \frac{\mu}{\nu}\right\}$, then for any $x \in \mathcal{X}_{\Pi_1^k}$, the following holds*

$$\Pr_{\Pi_1^k}\left[\left|\|Ax\|^2 - \|x\|^2\right| \geq \kappa\mu\log n\|x\|^2\right] \leq \frac{1}{n}.$$

Proof. The proof follows from Theorem 10.6. Set $\epsilon_2 = \frac{\mu\theta_M}{16}$ and $\epsilon_3 = 4\sqrt{2\log(8n)}\mu\theta_M$. Since $\theta_M > 128$ we have $\epsilon_2 > \mu$, and $\epsilon_3 > \mu$. Therefore, to satisfy Equation (10.34), it is sufficient to have $k \leq \min\left\{n^{\frac{2}{3}}, \left(\frac{\mu}{\nu}\right)^2, \frac{\mu}{\nu}\right\}$. Now observe that from the definition of μ and ν, $\frac{\mu}{\nu}$ is always less than or equal to $\left(\frac{\mu}{\nu}\right)^2$. Therefore

$$k \leq \min\left\{n^{\frac{2}{3}}, \frac{\mu}{\nu}\right\}$$

is a sufficient condition for Equation 10.34. Theorem 10.6 now guarantees that

$$\Pr_{\Pi_1^k}\left[\left|\|Ax\|^2 - \|x\|^2\right| \geq 8\sqrt{2\log(8n)}\mu\theta_M\|x\|^2\right] \leq \frac{4}{8n} + \frac{32k^2n}{64n^4} \leq \frac{1}{n}.$$

□

REMARK 10.5 Corollary 10.4 can be applied to any matrix and in particular to matrices with optimal coherence properties including matrices based on binary linear codes (see Section 10.8.4).

Proposition 10.5 states that if the projection matrix is a low coherent tight-frame, then with overwhelming probability A is Lipschitz with respect to a random k-subspace $\mathcal{X}_{\Pi_1^k}$. Theorem 10.6 guarantees that if the projection matrix also has sufficiently small average-coherence, then it satisfies the Johnson–Lindenstrauss Property with respect to the same feature space. Therefore, we can use Lemma 10.4 to guarantee that with overwhelming probability A is Distance-Preserving with respect to $\mathcal{X}_{\Pi_1^k}$.

LEMMA 10.6 *There exists a universal constant C such that for every $\eta \geq 1$, if the conditions*

$$\mu \leq \frac{\varepsilon_1}{2C\eta\log n}, \text{ and } k \leq \min\left\{\frac{m\varepsilon_1^2}{(2C\eta)^2\log n}, \left(\frac{n\theta_M}{96}\right)^{\frac{2}{3}}, \epsilon_2^2\nu^{-2}, \left(\frac{\mu\theta_M}{96\nu}\right)^2, \epsilon_3\nu^{-1}\right\}$$

hold simultaneously, then A is $(M+1, L, \epsilon, \delta)$-Distance-Preserving with respect to the feature space $\mathcal{X}_{\Pi_1^k}$, with

$$L = \sqrt{1+\varepsilon_1},$$
$$\epsilon = \left(6\epsilon_3 + 4\sigma + \sigma^2\right),$$
$$\delta = 4\binom{M+2}{2}\left[\exp\left\{\frac{-\epsilon_3^2}{32\mu^2\theta_M^2}\right\} + 32\,k^2 n \exp\left\{\frac{-\epsilon_2^2}{128\mu^2}\right\}\right] + \frac{1}{n^\eta}.$$

Proof. Since $\mathcal{X}_{\Pi_1^k}$ is k-dimensional, $\ell_{2\to 1}(\mathcal{X}_{\Pi_1^k}) \leq \sqrt{k}$. The proof now follows from substituting the value of L from Proposition 10.5, and the values of ϵ and δ from Theorem 10.6 into Lemma 10.4. □

We now use Theorem 10.3 to bound the difference in the true hinge loss of the SVM's classifier in the measurement domain and the SVM's classifier in the data domain.

THEOREM 10.7 *Let $\boldsymbol{w}_0$ denote the data-domain oracle classifier. Also let $S \doteq \langle(\boldsymbol{x}_1, l_1), \ldots, (\boldsymbol{x}_M, l_M)\rangle$ represent M training examples in the data domain, and let $AS \doteq \langle(\boldsymbol{y}_1, l_1), \ldots, (\boldsymbol{y}_M, l_M)\rangle$ denote the representation of the training examples in the measurement domain. Finally let $\hat{\boldsymbol{z}}_{AS}$ denote the measurement domain SVM's classifier trained on AS. Then there exists a universal constant C such that for every $\eta \geq 1$ if the conditions*

$$\mu \leq \frac{\varepsilon_1}{2C\eta \log n}, \text{ and } k \leq \min\left\{\frac{m\varepsilon_1^2}{(2C\eta)^2 \log n}, \left(\frac{n\theta_M}{96}\right)^{\frac{2}{3}}, \epsilon_2^2 \nu^{-2}, \left(\frac{\mu\theta_M}{96\nu}\right)^2, \epsilon_3 \nu^{-1}\right\}$$

hold simultaneously, then with probability $1 - 3\delta$, where

$$\delta \doteq \binom{M+2}{2}\left[4\exp\left\{\frac{-\epsilon_3^2}{32\mu^2\theta_M^2}\right\} + 128\,k^2 n \exp\left\{\frac{-\epsilon_2^2}{128\mu^2}\right\}\right] + \frac{1}{n^\eta},$$

the following holds

$$H_\mathcal{D}(\hat{\boldsymbol{z}}_{AS}) \leq H_\mathcal{D}(\boldsymbol{w}_0) + O\left(R\|\boldsymbol{w}_0\|\sqrt{\left(\epsilon_3 + \sigma + \frac{(1+\varepsilon_1)\log\left(\frac{1}{\delta}\right)}{M}\right)}\right). \quad (10.35)$$

Proof. The proof follows from substituting the values of L, ϵ, and δ from Lemma 10.6 into Theorem 10.3. □

COROLLARY 10.5 *Under the conditions of Theorem 10.7, if $k \leq \min\left\{\frac{m\varepsilon_1^2}{(2C\eta)^2 \log n}, n^{\frac{2}{3}}, \frac{\mu}{\nu}\right\}$, then with probability at least $1 - \frac{6}{n}$ the following holds*

$$H_\mathcal{D}(\hat{\boldsymbol{z}}_{AS}) \leq H_\mathcal{D}(\boldsymbol{w}_0) + O\left(R\|\boldsymbol{w}_0\|\sqrt{\left(\mu(\log M + \log n) + \sigma + \frac{(1+\varepsilon_1)\log n}{M}\right)}\right).$$
$$(10.36)$$

Proof. Corollary 10.4 proves that $k \leq \min\left\{n^{\frac{2}{3}}, \frac{\nu}{\mu}\right\}$ is a sufficient condition for $k \leq \min\left\{\left(\frac{n\theta_M}{96}\right)^{\frac{2}{3}}, \epsilon_2^2 \nu^{-2}, \left(\frac{\mu\theta_M}{96\nu}\right)^2, \epsilon_3 \nu^{-1}\right\}$. Now it follows from Theorem 10.7 (with $\eta = 1$) that it is sufficient to guarantee that

$$\binom{M+2}{2}\left[4\exp\left\{\frac{-\epsilon_3^2}{32\mu^2\theta_M^2}\right\} + 128\,k^2 n \exp\left\{\frac{-\epsilon_2^2}{128\mu^2}\right\}\right] + \frac{1}{n} \leq \frac{2}{n}. \tag{10.37}$$

In order to satisfy the requirement of Equation (10.37), it is sufficient to ensure that the following two equalities hold simultaneously:

$$\frac{\epsilon_3^2}{32\mu^2\theta_M^2} = \log\left(4n(M+2)(M+1)\right), \text{ and } \quad \frac{\epsilon_2^2}{128\mu^2} = \log\left(128n^2(M+2)(M+1)\right). \tag{10.38}$$

Consequently, since

$$\sqrt{\theta_M^2 \log[4n(M+2)(M+1)]} = O\left(\log M + \log n\right),$$

there exists a universal constant κ such that if $\epsilon_3 \geq \kappa\mu\left(\log M + \log n\right)$ then $\delta \leq \frac{2}{n}$. □

10.8.4 Average-case Distance-Preserving using Delsarte–Goethals frames

10.8.4.1 Construction of the Delsarte–Goethals frames

In the previous section, we introduced two fundamental measures of coherence between the columns of a tight-frame, and showed how these parameters can be related to the performance of the SVM's classifier in the measurement domain. In this section we construct an explicit sensing matrix (*Delsarte–Goethals frame* [35, 36]) with sufficiently small average coherence ν, and worst-case coherence μ. We start by picking an odd number o. The 2° rows of the Delsarte–Goethals frame A are indexed by the binary o-tuples t, and the $2^{(r+2)o}$ columns are indexed by the pairs (P, b), where P is an $o \times o$ binary symmetric matrix in the Delsarte–Goethals set $DG(o, r)$, and b is a binary o-tuple. The entry $a_{(P,b),t}$ is given by

$$a_{(P,b),t} = \frac{1}{\sqrt{m}} i^{wt(d_P) + 2wt(b)} i^{tPt^\top + 2bt^\top} \tag{10.39}$$

where d_p denotes the main diagonal of P, and wt denotes the *Hamming weight* (the number of 1s in the binary vector). Note that all arithmetic in the expressions $tPt^\top + 2bt^\top$ and $wt(d_P) + 2wt(b)$ takes place in the ring of integers modulo 4, since they appear only as exponents for i. Given P and b, the vector $tPt^\top + 2bt^\top$ is a codeword in the Delsarte–Goethals code. For a fixed matrix P, the 2° columns $A_{(P,b)}$ ($b \in \mathbb{F}_2^o$) form an orthonormal basis Γ_P that can also be obtained by postmultiplying the Walsh–Hadamard basis by the unitary transformation diag $\left[i^{tPt^\top}\right]$.

The Delsarte–Goethals set $DG(o, r)$ is a binary vector space containing $2^{(r+1)o}$ binary symmetric matrices with the property that the difference of any two distinct matrices has rank at least $o - 2r$ (see [37]). The Delsarte–Goethals sets are nested

$$DG(o, 0) \subset DG(o, 1) \subset \cdots \subset DG\left(o, \frac{(o-1)}{2}\right).$$

The first set $DG(o, 0)$ is the classical Kerdock set, and the last set $DG(o, \frac{(o-1)}{2})$ is the set of all binary symmetric matrices. The rth Delsarte–Goethals sensing matrix is determined by $DG(o, r)$ and has $m = 2^o$ rows and $n = 2^{(r+2)o}$ columns.

Throughout the rest of this section let 1 denote the all-one vector. Also let Φ denote the *unnormalized* DG frame, i.e., $A = \frac{1}{\sqrt{m}}\Phi$. We use the following lemmas to show that the Delsarte–Goethals frames are low-coherence tight-frames. First we prove that the columns of the rth Delsarte–Goethals sensing matrix form a group under pointwise multiplication.

LEMMA 10.7 *Let $\mathcal{G} = \mathcal{G}(o, r)$ be the set of unnormalized columns $\Phi_{(P,b)}$ where*

$$\phi_{(P,b),t} = i^{wt(d_P)+2wt(b)} i^{tPt^\top + 2bt^\top}, \text{ where } t \in \mathbb{F}_2^o$$

where $b \in \mathbb{F}_2^o$ and where the binary symmetric matrix P varies over the Delsarte–Goethals set $DG(o, r)$. Then $\mathcal{G}$ is a group of order $2^{(r+2)o}$ under pointwise multiplication.

Proof. The proof of Lemma 10.7 is based on the construction of the DG frames, and is provided in [38]. □

Next we bound the worst-case coherence of the Delsarte–Goethals frames.

THEOREM 10.8 *Let Q be a binary symmetric $o \times o$ matrix from the $DG(o, r)$ set, and let $b \in \mathbb{F}_2^o$. If $S \doteq \sum_t i^{tQt^\top + 2bt^\top}$, then either $S = 0$, or*

$$S^2 = 2^{o+2r} i^{v_1 Q v_1^\top + 2b v_1^\top}, \text{ where } v_1 Q = d_Q.$$

Proof. We have

$$S^2 = \sum_{t,u} i^{tQt^\top + uQu^\top + 2b(t+u)^\top} = \sum_{t,u} i^{(t \oplus u)Q(t \oplus u)^\top + 2tQu^\top + 2b(t \oplus u)^\top}.$$

Changing variables to $v = t \oplus u$ and u gives

$$S^2 = \sum_v i^{vQv^\top + 2bv^\top} \sum_u (-1)^{(d_Q + vQ)u^\top}.$$

Since the diagonal d_Q of a binary symmetric matrix Q is contained in the row space of Q there exists a solution for the Equation $vQ = d_Q$. Moreover, since Q has rank at least $o - 2r$, the solutions to the Equation $vQ = 0$ form a vector space E of dimension at most $2r$, and for all $e, f \in E$

$$eQe^\top + fQf^\top = (e+f)Q(e+f)^\top \pmod{4}.$$

Hence

$$S^2 = 2^o \sum_{e \in E} i^{(v_1+e)Q(v_1+e)^\top + 2(v_1+e)b^\top} = 2^o i^{v_1 Q z_1^\top + 2v_1 b^\top} \sum_{e \in E} i^{eQe^\top + 2eb^\top}.$$

The map $e \to eQe^\top$ is a linear map from E to $\mathbb{Z}_2$, so the numerator $eQe^\top + 2eb^\top$ also determines a linear map from E to $\mathbb{Z}_2$ (here we identify $\mathbb{Z}_2$ and $2\mathbb{Z}_4$). If this linear map is the zero map then

$$S^2 = 2^{o+2r} i^{v_1 Q v_1^\top + 2b v_1^\top},$$

and if it is not zero then $S = 0$. $\square$

COROLLARY 10.6 *Let A be an $m \times n$ $DG(o,r)$ frame whose column entries are defined by (10.39). Then $\mu \leq \frac{2^r}{\sqrt{m}}$.*

Proof. Lemma 10.7 states that the columns of the unnormalized DG frame form a group under pointwise multiplication. Therefore, the inner product between any two columns of this matrix is another column sum. Consequently, we have

$$\mu = \max_{i \neq j} \left| A_i^\dagger A_j \right| = \frac{1}{m} \leq \max_{i \neq j} \left| \Phi_i^\dagger \Phi_j \right| \leq \frac{1}{m} = \max_{i \neq 1} \left| \Phi_i^\dagger 1 \right| \leq \frac{\sqrt{2^{o+2r}}}{m} = \frac{2^r}{\sqrt{m}}.$$

$\square$

LEMMA 10.8 *Let A be a $DG(o,r)$ frame with $m = 2^o$, and $n = 2^{(r+2)o}$. Then $\nu = \frac{1}{n-1}$.*

Proof. We have

$$\nu \doteq \max_i \frac{1}{n-1} \left| \sum_{j \neq i} A_i^\top A_j \right| = \frac{1}{m(n-1)} \left| \sum_{j \neq i} \Phi_i^\top \Phi_j \right| = \frac{1}{m(n-1)} \left| \sum_{i \neq 1} 1^\top \Phi_i \right|.$$

Now it follows from Lemma 10.1 that since every row of Φ has at least one non-identity element, every row sum vanishes. Therefore, $\sum_{i \in [n]} 1^\top \phi_i = 0$, and since $\Phi_1 = 1$ we have

$$\frac{1}{m(n-1)} \left| \sum_{i \neq 0} 1^\top \phi_i \right| = \frac{1}{m(n-1)} \left| -1^\top 1 \right| = \frac{1}{n-1}.$$

$\square$

LEMMA 10.9 *Let A be a $DG(o,r)$ frame. Then A is a tight-frame with redundancy $\frac{n}{m}$.*

Proof. Let t and t' be two indices in $[m]$. We calculate the inner-product between the rows indexed by t and t'. It follows from Equation (10.39) that the inner-product can be written as

$$\sum_{P,b} a_{(P,b),t}\overline{a_{(P,b),t'}} = \frac{1}{m}\sum_{P,b} i^{tPt^\top - t'Pt'^\top + 2bt^\top - 2bt'^\top}$$

$$= \frac{1}{m}\left(\sum_P i^{tPt^\top - t'Pt'^\top}\right)\left(\sum_b (-1)^{b(t\oplus t')^\top}\right).$$

Therefore, it follows from Lemma 10.1 that if $t \neq t'$ then the inner-product is zero, and is $\frac{n}{m}$ otherwise. □

10.8.4.2 Compressed sensing via the Delsarte–Goethals frames

So far we have proved that Delsarte–Goethals frames are tight-frames with optimal coherence values. Designing dictionaries with small spectral norms (tight-frames in the ideal case), and with small coherence ($\mu = O\left(\frac{1}{\sqrt{m}}\right)$) in the ideal case) is useful in compressed sensing for the following reasons.

Uniqueness of sparse representation (ℓ_0 minimization). The following theorem is due to Tropp [34] and shows that with overwhelming probability the ℓ_0 minimization program successfully recovers the original k-sparse signal.

THEOREM 10.9 *Assume the dictionary A satisfies $\mu \leq \frac{C}{\log n}$, where C is an absolute constant. Further assume $k \leq \frac{Cn}{\|A\|^2 \log n}$. Let x be a k-sparse vector, such that the support of x is selected uniformly at random, and the distribution of the k nonzero entries of x is absolutely continuous with respect to the Lebesgue measure on $\mathbb{R}^k$. Then with probability $1 - \frac{1}{n}$, x is the unique k-sparse vector mapped to $y = Ax$ by the measurement matrix A.*

Sparse recovery via lasso (ℓ_1 minimization). Uniqueness of sparse representation is of limited utility given that ℓ_0 minimization is computationally intractable. However, given modest restrictions on the class of sparse signals, Candès and Plan [39] have shown that with overwhelming probability the solution to the ℓ_0 minimization problem coincides with the solution to a convex lasso program.

THEOREM 10.10 *Assume the dictionary A satisfies $\mu \leq \frac{C_1}{\log n}$, where C is an absolute constant. Further assume $k \leq \frac{C_1 n}{\|A\|^2 \log n}$. Let x be a k-sparse vector, such that*

1. *The support of the k nonzero coefficients of x is selected uniformly at random.*
2. *Conditional on the support, the signs of the nonzero entries of x are independent and equally likely to be -1 or 1.*

Let $y = Ax + e$, where e contains m i.i.d. $\mathcal{N}(0,\sigma^2)$ Gaussian elements. Then if $\|x\|_{\min} \geq 8\sigma\sqrt{2\log n}$, with probability $1 - O(n^{-1})$ the lasso estimate

$$x^* \doteq \arg\min_{x^+ \in \mathbb{R}^n} \frac{1}{2}\|y - Ax^+\|^2 + 2\sqrt{2\log n}\,\sigma^2\,\|x^+\|_1$$

has the same support and sign as x, and $\|Ax - Ax^*\|^2 \leq C_2 k \sigma^2$, where C_2 is a constant independent of x.

Stochastic noise in the data domain. The tight-frame property of the sensing matrix makes it possible to map i.i.d. Gaussian noise in the data domain to i.i.d. Gaussian noise in the measurement domain:

LEMMA 10.10 *Let ς be a vector with n i.i.d. $\mathcal{N}(0, \sigma_d^2)$ entries and e be a vector with m i.i.d. $\mathcal{N}(0, \sigma_m^2)$ entries. Let $\hbar = A\varsigma$ and $v = \hbar + e$. Then v contains m entries, sampled i.i.d. from $\mathcal{N}(0, \sigma^2)$, where $\sigma^2 = \frac{n}{m}\sigma_d^2 + \sigma_m^2$.*

Proof. The tight-frame property implies

$$\mathbb{E}\left[\hbar\hbar^\dagger\right] = E[A\varsigma\varsigma^\dagger A^\dagger] = \sigma_d^2 A A^\dagger = \frac{n}{m}\sigma_d^2 I.$$

Therefore, $v = \hbar + e$ contains i.i.d. Gaussian elements with zero mean and variance σ^2. □

Here we present numerical experiments to evaluate the performance of the lasso program with DG frames. The performance of DG frames is compared with that of random Gaussian sensing matrices of the same size. The SpaRSA algorithm [40] with ℓ_1 regularization parameter $\lambda = 10^{-9}$ is used for signal reconstruction in the noiseless case, and the parameter is adjusted according to Theorem 10.10 in the noisy case.

These experiments relate accuracy of sparse recovery to the sparsity level and the Signal to Noise Ratio (SNR). Accuracy is measured in terms of the statistical $0-1$ loss metric which captures the fraction of signal support that is successfully recovered. The reconstruction algorithm outputs a k-sparse approximation $\hat{x}$ to the k-sparse signal x, and the statistical $0-1$ loss is the fraction of the support of x that is not recovered in $\hat{x}$. Each experiment was repeated 2000 times and Figure 10.2 records the average loss.

Figure 10.2 plots statistical $0-1$ loss and complexity (average reconstruction time) as a function of the sparsity level k. We select k-sparse signals with uniformly random support, with random signs, and with the amplitude of nonzero entries set equal to 1. Three different sensing matrices are compared; a Gaussian matrix, a $DG(7,0)$ frame and a $DG(7,1)$ sieve.[1] After compressive sampling the signal, its support is recovered using the SpaRSA algorithm with $\lambda = 10^{-9}$.

Figure 10.3(a) plots statistical $0-1$ loss as a function of noise in the measurement domain and Figure 10.3(b) does the same for noise in the data domain. In the measurement noise study, a $\mathcal{N}(0, \sigma_m^2)$ i.i.d. measurement noise vector is added to the sensed vector to obtain the m-dimensional vector y. The original k-sparse signal x is then approximated by solving the lasso program with $\lambda = 2\sqrt{2\log n}\sigma^2$. Following Lemma 10.10, we used a similar method to study noise in the data domain. Figure 10.3 shows that DG frames and sieves outperform random Gaussian matrices in terms of noisy signal recovery using the lasso.

---

[1] A DG sieve is a submatrix of a DG frame which has no Walsh tones.

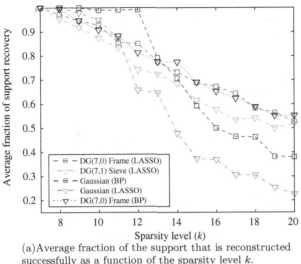

(a) Average fraction of the support that is reconstructed successfully as a function of the sparsity level k.

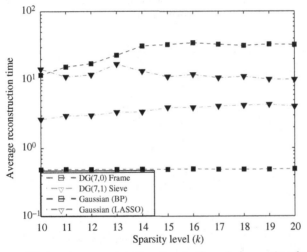

(b) Average reconstruction time in the noiseless regime for different sensing matrices.

Figure 10.2 Comparison between $DG(7,0)$ frame, $DG(7,1)$ sieve, and Gaussian matrices of the same size in the noiseless regime. The regularization parameter for lasso is set to 10^{-9}.

We also performed a Monte Carlo simulation to calculate the probability of *exact* signal recovery. We fixed the number of measurements to $m = 512$ and swept across the sparsity level k, and the data dimension n^2. For each (k,n)-pair, we repeated the following 100 times: (*i*) generate a random sparse vector with unit norm (*ii*) generate

---

[2] To vary n, we selected the first n columns of a $DG(9,0)$ frame (which is still an incoherent tight-frame as long as $\frac{n}{m}$ is an integer).

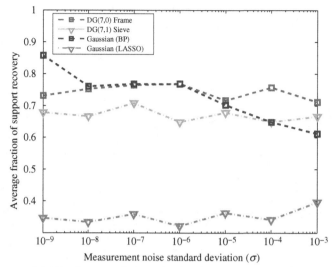

(a) The impact of the noise in the measurement domain on the accuracy of the sparse approximation for different sensing matrices.

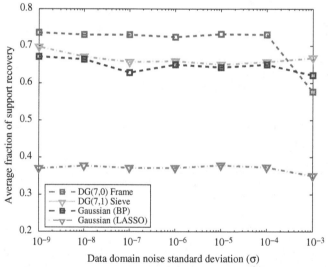

(b) The impact of the noise in the data domain on the accuracy of the sparse approximation for different sensing matrices.

Figure 10.3 Average fraction of the support that is reconstructed successfully as a function of the noise level in the measurement domain (left), and in the data domain (right). Here the sparsity level is 14. The regularization parameter for lasso is determined as a function of the noise variance according to Theorem 10.10.

compressive measurements (no noise) using the DG frame, and (*iii*) recover the signals using lasso. Figure 10.4(a) reports the probability of exact recovery over the 100 trials.

We also performed a similar experiment in the noisy regime. Here we independently changed the standard deviations of the data-domain noise (σ_d) and the measurement noise (σ_m) from 10^{-6} to 10^{-1}. We then used the lasso program to obtain a sparse approximation $\hat{x}$ to the k-sparse vector x. Figure 10.4(b) plots the average reconstruction error ($-10\log_{10}(\|\hat{x} - x\|_2)$) as a function of σ_m and σ_d.

Finally we used the DG frames to compressively sample an image of size $n = 2048 \times 2048$. We used the Daubechies-8 discrete wavelet transform to obtain compressible coefficients for the image. We compared the quality and computational cost of two different DG sensing matrices, a $DG(19,0)$ frame (providing a compression ratio of 25%), and a $DG(17,0)$ frame (providing a compression ratio of 6.25%). The results are shown in Figure 10.5 (see [41] for more detailed comparison with other methods).

10.8.4.3 Compressed learning via the Delsarte–Goethals frames

Since Delsarte–Goethals frames have optimal worst-case and average-case coherence values, we can use Corollary 10.5 to guarantee that the measurement domain SVM's classifier is near-optimal.

COROLLARY 10.7 *Let o be an odd integer, and let $r \leq \frac{o-1}{2}$. Let A be an $m \times n$ DG frame with $m = 2^o$, and $n = 2^{(r+2)o}$. Let w_0 denote the data-domain oracle classifier. Also let $S \doteq \langle(x_1, l_1), \ldots, (x_M, l_M)\rangle$ represent M training examples in the data domain, and let $AS \doteq \langle(y_1, l_1), \ldots, (y_M, l_M)\rangle$ denote the representation of the training examples in the measurement domain. Finally let $\hat{z}_{AS}$ denote the measurement domain SVM's classifier trained on AS. Then there exists a universal constant C such that if*

$$m \geq \left(\frac{2^{r+1} C \log n}{\varepsilon_1}\right)^2, \quad \text{and} \quad k \leq \min\left\{\frac{m\varepsilon_1^2}{(2C)^2 \log n}, n^{\frac{2}{3}}\right\}$$

then with probability at least $1 - \frac{6}{n}$ the following holds

$$H_{\mathcal{D}}(\hat{z}_{AS}) \leq H_{\mathcal{D}}(w_0) + O\left(R\|w_0\|\sqrt{\left(\frac{2^r(\log M + \log n)}{\sqrt{m}} + \sigma + \frac{(1+\varepsilon_1)\log n}{M}\right)}\right). \tag{10.40}$$

Proof. The proof follows from Corollary 10.5 after substituting the values $\mu = \frac{2^r}{\sqrt{m}}$ and $\nu = \frac{1}{n-1}$. □

REMARK 10.6 Corollary 10.7 guarantees that larger measurement domain dimension m leads to lower measurement domain classification loss. In other words

$$H_{\mathcal{D}}(\hat{z}_{AS}) \leq H_{\mathcal{D}}(w_0) + \tilde{O}\left(\left(\frac{(\log M + \log n)^2}{m}\right)^{\frac{1}{4}} + \left(\frac{\log n}{M}\right)^{\frac{1}{2}} + \sigma^{\frac{1}{2}}\right).$$

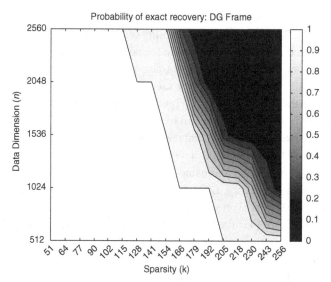

(a) Probability of exact signal recovery as a function of the sparsity level k, and the data domain dimension n using a $DG(9,0)$ frame.

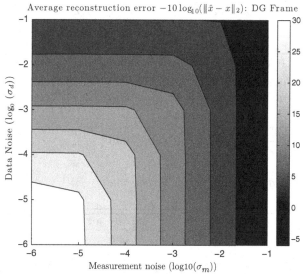

(b) Average reconstruction error as a function of the data domain noise (σ_d), and the measurement domain noise (σ_m) using a $DG(9,0)$ frame.

Figure 10.4 The impact of the sparsity level and noise on the performance of the lasso program using a $DG(9,0)$ frame.

Application: texture classification. Finally we demonstrate an application of compressed learning in texture classification. In texture classification, the goal is to classify images into one of the "horizontal," "vertical," or "other" classes. The information about the direction of an image is stored in the horizontal and vertical wavelet coefficients of

Table 10.1. Comparison between the classification results of the SVM's classifier in the data domain and the SVM's classifier in the measurement domain.

SVM's	# of "horizontals"	# of "verticals"	# of "others"
Data Domain	14	18	23
Measurement Domain	12	15	28

(a) A $DG(19.0)$ frame is used to provide a compression ratio of 25%. The reconstruction SNR is 16.58, and the reconstruction time is 8369 seconds.

(b) A $DG(17.0)$ frame is used to provide a compression ratio of 6.25%. The reconstruction SNR is 12.52, and the reconstruction time is 8606 seconds.

Figure 10.5 Comparison of CS recovery performance of an $n = 2048 \times 2048$ image using the Basis pursuit algorithm from 2^{19} measurements (left), and 2^{17} measurements (right). The reconstruction SNR is $-20\log_{10}\left(\frac{\|\hat{x}-x\|_2}{\|x\|_2}\right)$.

that image. Therefore, an SVM's classifier in the data (pixel or wavelet) domain would provide high texture classification accuracy. Here we show that an SVM's classifier trained directly over the compressively sampled images also has high performance.

We used the Brodatz texture database [42] which contains 111, 128×128 images. First we divided the dataset into 56 training images and 55 test images, and trained an SVM's classifier from the 128×128 images. The images were then projected to a 2^{11} dimensional space using a $DG(11,0)$ matrix. We then used the same procedure to train the measurement domain SVM's classifier and classified the images accordingly. Table 10.1 compares the classification results of the SVM's classifier in the data domain and the SVM's classifier in the measurement domain. Figure 10.6 demonstrates examples of images in each class. The measurement domain classifier misclassifies 3 "horizontal" images, 3 "vertical" images and 1 "others" image. Therefore, the relative classification error of the measurement domain SVM's classifier is $\frac{|14-11|+|18-15|+|23-22|}{55} \approx 12.7\%$.

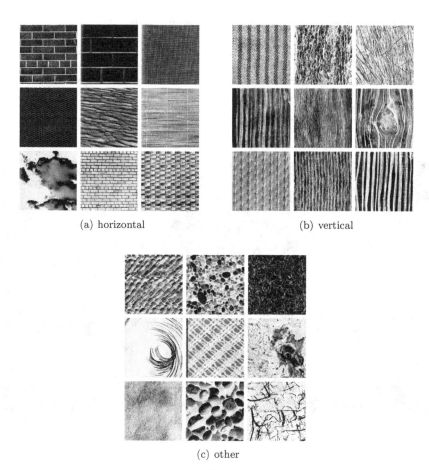

Figure 10.6 Examples of images classified as horizontal (a), vertical (b), and other (c) using the measurement domain SVM's classifier with a $DG(11,0)$ sensing matrix.

REMARK 10.7 In this section we discussed DG frames as a low-coherence tight-frame; however, there exist other families of such matrices satisfying the average-case JL Property. Those matrices are based on dual-BCH codes [43], Binary Chirps [44], Gabor frames [45], and partial Fourier ensembles [38] (see [43] for more discussion about their construction).

10.9 Proof of the main average-case compressed learning results

10.9.1 Proof of Lemma 10.5

Proof. Without loss of generality we can assume that the first k-entries of x are nonzero; and in contrast, the columns of A are permuted by Π. First we prove Equation (10.32).

We apply the union bound over all possible values of W to get

$$\Pr_{\Pi_1^k}\left[\exists W \in [n] - \Pi_1^k \text{ s.t } \left|\sum_{i=1}^k x_i A_W{}^\top A_{\Pi_i}\right| \geq 2\epsilon_1 \|x\|_2\right] \qquad (10.41)$$

$$\leq \sum_{w=1}^n \Pr_{\Pi_1^k}\left[\exists W \in [n] - \Pi_1^k \text{ s.t } \left|\sum_{i=1}^k x_i A_W{}^\top A_{\Pi_i}\right| \geq 2\epsilon_1 \|x\|_2 \,\Big|\, W = w\right].$$

Now fix the value w, and define $\hbar_w(\Pi_1^k) \doteq \sum_{i=1}^k x_i A_w{}^\top A_{\Pi_i}$. Note that conditioned on $w \notin \Pi_1^k$, Π_1^k is a random k-subset of $[n] - \{w\}$. We first prove that the expected value of $\hbar_w(\Pi_1^k)$ is sufficiently small. Then we use Azuma's inequality to show that with overwhelming probability $\hbar_w(\Pi_1^k)$ is concentrated about its expectation.

It follows from the linearity of expectation that

$$\left|\mathbb{E}_{\Pi_1^k}[\hbar_w(\Pi_1^k)]\right| = \left|\sum_{i=1}^k x_i \mathbb{E}_{\Pi_i \neq w}\left[A_w{}^\top A_{\Pi_i}\right]\right| \leq \|x\|_1 \nu \leq \sqrt{k}\|x\|_2 \nu. \qquad (10.42)$$

To prove the concentration, let π_1^k be a fixed k-subset of $[n] - \{w\}$, and define the martingale sequence

$$Z_t \doteq \mathbb{E}_{\Pi_1^k}\left[\hbar_w(\Pi_1^k) \,|\, \Pi_1^t = \pi_1^t\right] = \sum_{i=1}^t x_i A_w{}^\top A_{\pi_i} + \sum_{i=t+1}^k x_i \mathbb{E}_{\Pi_i \notin \{\pi_1,\cdots,\pi_t,w\}}\left[A_w{}^\top A_{\Pi_i}\right]. \qquad (10.43)$$

Here the goal is to bound the difference $|Z_t - Z_{t-1}|$. Note that

$$Z_{t-1} = \sum_{i=1}^{t-1} x_i A_w{}^\top A_{\pi_i} + \sum_{i=t}^k x_i \mathbb{E}_{\Pi_i \notin \{\pi_1,\cdots,\pi_{t-1},w\}}\left[A_w{}^\top A_{\Pi_i}\right].$$

Therefore, it follows from the triangle inequality that

$$|Z_t - Z_{t-1}| \leq |x_t|\left|A_w{}^\top A_{\pi_t} - \mathbb{E}_{\Pi_t \notin \{\pi_1,\cdots,\pi_{t-1},w\}}\left[A_w{}^\top A_{\Pi_t}\right]\right| \qquad (10.44)$$

$$+ \sum_{i=t+1}^k |x_i|\left|\mathbb{E}_{\Pi_i \notin \{\pi_1,\cdots,\pi_t,w\}}\left[A_w{}^\top A_{\Pi_i}\right] - \mathbb{E}_{\Pi_i \notin \{\pi_1,\cdots,\pi_{t-1},w\}}\left[A_w{}^\top A_{\Pi_i}\right]\right|.$$

From marginalizing the expectation, we have

$$\mathbb{E}_{\Pi_i \notin \{\pi_1,\cdots,\pi_{t-1},w\}}\left[A_w{}^\top A_{\Pi_i}\right] = \qquad (10.45)$$

$$\Pr[\Pi_i = \pi_t] A_w{}^\top A_{\pi_t} + \Pr[\Pi_i \neq \pi_t] \mathbb{E}_{\Pi_i \notin \{\pi_1,\cdots,\pi_t,w\}}\left[A_w{}^\top A_{\Pi_i}\right].$$

As a result

$$\left| \mathbb{E}_{\Pi_i \notin \{\pi_1, \cdots, \pi_t, w\}} \left[A_w{}^\top A_{\Pi_i} \right] - \mathbb{E}_{\Pi_i \notin \{\pi_1, \cdots, \pi_{t-1}, w\}} \left[A_w{}^\top A_{\Pi_i} \right] \right| \quad (10.46)$$

$$= \Pr[\Pi_i = \pi_t] \left| \mathbb{E}_{\Pi_i \notin \{\pi_1, \cdots, \pi_t, w\}} \left[A_w{}^\top A_{\Pi_i} \right] - A_w{}^\top A_{\pi_t} \right| \leq \frac{2\mu}{n-1-t}.$$

By substituting (10.46) into (10.44) we get the difference bound

$$c_t \doteq |Z_t - Z_{t-1}| \leq 2|x_t|\mu + \frac{2\mu}{n-(t+1)} \|x\|_1 \leq 2|x_t|\mu + \frac{2\mu}{n-(k+1)} \|x\|_1. \quad (10.47)$$

Now in order to use Azuma's inequality, we need to bound $\sum_{t=1}^{k} c_t^2$:

$$\sum_{t=1}^{k} c_t^2 = 4\mu^2 \sum_{t=1}^{k} \left(|x_t| + \frac{\|x\|_1}{n-(k+1)} \right)^2 \quad (10.48)$$

$$\leq 4\mu^2 \left(\|x\|_2^2 + \frac{k^2}{(n-(k+1))^2} \|x\|_2^2 + 2\frac{k}{n-(k+1)} \|x\|_2^2 \right).$$

If $k \leq \frac{n-1}{2}$, then $\sum_{t=1}^{k} c_t^2 \leq 16\mu^2 \|x\|_2^2$, and it follows from Azuma's inequality (Proposition 10.1) that for every $\epsilon_1 \geq \sqrt{k}\nu$,

$$\Pr_{\Pi_1^k} \left[|\hbar_w(\Pi_1^k)| \geq 2\epsilon_1 \|x\|_2 \right] \leq \Pr \left[\left| \hbar_w(\Pi_1^k) - \mathbb{E}_{\Pi_1^k} \left[\hbar_w(\Pi_1^k) \right] \right| \geq \epsilon_1 \|x\|_2 \right] \quad (10.49)$$

$$\leq 4 \exp \left\{ \frac{-\epsilon_1^2}{128\mu^2} \right\}.$$

Now by taking the union bound over all n choices for w we get

$$\Pr_{\Pi_1^k} \left[\exists W \in [n] - \Pi_1^k \text{ s.t } \left| \sum_{i=1}^{k} x_i A_W{}^\top A_{\Pi_i} \right| \geq 2\epsilon_1 \|x\|_2 \right] \leq 4n \exp \left\{ \frac{-\epsilon_1^2}{128\mu^2} \right\}.$$

The proof of Equation (10.33) is similar: First observe that using the union bound

$$\Pr_{\Pi_1^k} \left[\exists j \in \{1, \cdots, k\} \text{ s.t } \left| \sum_{\substack{i \in \{1, \cdots k\} \\ i \neq j}} x_i A_{\Pi_j}{}^\top A_{\Pi_i} \right| \geq 2\epsilon_2 \|x\|_2 \right] \leq \quad (10.50)$$

$$\sum_{j=1}^{k} \Pr_{\Pi_1^k} \left[\left| \sum_{\substack{i \in \{1, \cdots k\} \\ i \neq j}} x_i A_{\Pi_j}{}^\top A_{\Pi_i} \right| \geq 2\epsilon_2 \|x\|_2 \right].$$

Furthermore, by marginalizing the left-hand side of (10.50) we get

$$\Pr_{\Pi_1^k}\left[\left|\sum_{\substack{i\in\{1,\cdots k\}\\i\neq j}} x_i A_{\Pi_j}{}^\top A_{\Pi_i}\right| \geq 2\epsilon_2\|x\|_2\right] =$$

$$\sum_{w=1}^n \Pr[\Pi_j = w]\Pr_{\Pi_1^k}\left[\left|\sum_{\substack{i\in\{1,\cdots k\}\\i\neq j}} x_i A_w{}^\top A_{\Pi_i}\right| \geq 2\epsilon_2\|x\|_2 \mid \Pi_j = w\right].$$

An argument similar to the proof of Lemma 10.5 can be used to show that for every $\epsilon_2 \geq \sqrt{k}\nu$

$$\Pr_{\Pi_1^k}\left[\left|\sum_{\substack{i\in\{1,\cdots k\}\\i\neq j}} x_i A_{\Pi_i}{}^\top A_{\Pi_j}\right| \geq 2\epsilon_2\|x\|_2 \mid \Pi_j = w\right] \leq 4\exp\left\{\frac{-\epsilon_2^2}{128\mu^2}\right\}.$$

Hence

$$\Pr_{\Pi_1^k}\left[\exists j \text{ s.t } \left|\sum_{\substack{i\in\{1,\cdots k\}\\i\neq j}} x_i A_{\Pi_j}{}^\top A_{\Pi_i}\right| \geq 2\epsilon_2\|x\|_2\right] \leq \sum_{j=1}^k \sum_{w=1}^n \Pr[\Pi_j = w] 4\exp\left\{\frac{-\epsilon_2^2}{128\mu^2}\right\}$$

$$= 4k\exp\left\{\frac{-\epsilon_2^2}{128\mu^2}\right\}.$$

$\square$

10.9.2 Proof of Theorem 10.6

The proof of Theorem 10.6 follows from constructing bounded-difference martingale sequences and applying the Extended Azuma Inequality (Proposition 10.2). Let Π be a random permutation of $[n]$, also let π be a fixed permutation of $[n]$. For every index t, let $\Pi_1^t \doteq \{\Pi_1,\cdots,\Pi_t\}$, and $\pi_1^t \doteq \{\pi_1,\cdots,\pi_t\}$. Let x be k-sparse vector with fixed values, supported on Π_1^k. Without loss of generality we can assume that the first k-entries of x are nonzero; and in contrast, the columns of A are permuted by Π. For $t = 0, 1, \cdots, k$, we define the following martingale sequence

$$Z_t = \mathbb{E}_{\Pi_1^k}\left[\sum_{i=1}^k x_i \left(\sum_{j=1}^k x_j A_{\Pi_i}{}^\top A_{\Pi_j}\right) \mid \Pi_1 = \pi_1,\cdots,\Pi_t = \pi_t\right]. \quad (10.51)$$

First we bound the difference $|Z_t - Z_{t-1}|$. We need the following Lemmas:

LEMMA 10.11 *Let Π_1^k be a random k-subset of $[n]$, and let π_1^k be a fixed k-subset of $[n]$. Then for every $t \leq k$ the following inequalities simultaneously hold:*

- (I1): *for every pair of indices i, j with $1 \leq i < t$ and $t < j$:*

$$\left| \mathbb{E}_{\Pi_j \notin \{\pi_1^t\}} \left[A_{\pi_i}{}^\top A_{\Pi_j} \right] - \mathbb{E}_{\Pi_j \notin \{\pi_1^{t-1}\}} \left[A_{\pi_i}{}^\top A_{\Pi_j} \right] \right| \leq \frac{2\mu}{n-t+1}.$$

- (I2) *for every pair of distinct indices i, j greater than t*

$$\left| \mathbb{E}_{\Pi_i \notin \pi_1^t} \left[\mathbb{E}_{\Pi_j \notin \{\pi_1^t, \Pi_i\}} \left[A_{\Pi_i}{}^\top A_{\Pi_j} \right] \right] - \mathbb{E}_{\Pi_i \notin \pi_1^{t-1}} \left[\mathbb{E}_{\Pi_j \notin \{\pi_1^{t-1}, \Pi_i\}} \left[A_{\Pi_i}{}^\top A_{\Pi_j} \right] \right] \right|$$
$$\leq \frac{2\mu}{n-t+1}.$$

Proof. The first inequality is proved by marginalizing the expectation. The proof of the second inequality is similar and we omit the details.

$$\left| \mathbb{E}_{\Pi_j \notin \{\pi_1^t\}} \left[A_{\pi_i}{}^\top A_{\Pi_j} \right] - \mathbb{E}_{\Pi_j \notin \{\pi_1^{t-1}\}} \left[A_{\pi_i}{}^\top A_{\Pi_j} \right] \right|$$
$$= \left| \mathbb{E}_{\Pi_j \notin \{\pi_1^t\}} \left[A_{\pi_i}{}^\top A_{\Pi_j} \right] - \Pr[\Pi_j = \pi_t] A_{\pi_i}{}^\top A_{\pi_t} - \Pr[\Pi_j \neq \pi_t] \mathbb{E}_{\Pi_j \notin \{\pi_1^t\}} \left[A_{\pi_i}{}^\top A_{\Pi_j} \right] \right|$$
$$= \Pr[\Pi_j = \pi_t] \left| \mathbb{E}_{\Pi_j \notin \{\pi_1^t\}} \left[A_{\pi_i}{}^\top A_{\Pi_j} \right] - A_{\pi_i}{}^\top A_{\pi_t} \right| \leq \frac{2\mu}{n-(t-1)}.$$

□

LEMMA 10.12 *Let Π_1^k and π_1^k be defined as in Lemma 10.11. Then for every $t \leq k$ the following inequalities simultaneously hold:*

- (Q1): *for every $j < t$:*

$$\left| \mathbb{E}_{\Pi_t \notin \{\pi_1^{t-1}\}} \left[A_{\Pi_t}{}^\top A_{\pi_j} \right] \right| \leq \frac{k}{n-k}\mu + \frac{n-1}{n-k}\nu.$$

- (Q2): *for every $i > t$:*

$$\left| \mathbb{E}_{\Pi_i \notin \{\pi_1^t\}} \left[A_{\Pi_i}{}^\top A_{\pi_t} \right] - \mathbb{E}_{\Pi_t \notin \{\pi_1^{t-1}\}} \left[\mathbb{E}_{\Pi_i \notin \{\pi_1, \cdots, \pi_{t-1}, \Pi_t\}} \left[A_{\Pi_i}{}^\top A_{\Pi_t} \right] \right] \right|$$
$$\leq \frac{2k}{n-k}\mu + \frac{2(n-1)}{n-k}\nu.$$

Proof. The proof of Lemma 10.12 is similar to the proof of Lemma 10.11 and is provided in [46]. □

Now we are ready to bound the difference $|Z_t - Z_{t-1}|$.

LEMMA 10.13 *Let Z_t defined by (10.51). Then*

$$|Z_t - Z_{t-1}| \leq 2 \left| \sum_{j=1}^{t-1} x_t x_j A_{\pi_t}{}^\top A_{\pi_j} \right| + 6 \left(\frac{k}{n-k}\mu + \frac{n-1}{n-k}\nu \right) |x_t| \|x\|_1 + \frac{4\mu}{n-k} \|x\|_1^2. \quad (10.52)$$

Proof. The proof of Lemma 10.13 is similar to the proof of Lemma 10.12 after some simplifications, and is provided in [46]. □

Now, in order to use Extended Azuma Inequality (Proposition 10.2), we need to analyze the average-case and worst-case behaviors of $|Z_t - Z_{t-1}|$.

LEMMA 10.14 *Let Z_t defined by (10.51). Then*

$$|Z_t - Z_{t-1}| \leq 2\sqrt{k} \left(\mu + 3 \left(\frac{k}{n-k}\mu + \frac{n-1}{n-k}\nu \right) \right) |x_t| \|x\|_2 + \frac{4k\mu}{n-k} \|x\|_2^2. \quad (10.53)$$

Moreover, for every positive ϵ_2, if $k \leq \min\{\frac{n-1}{2}, \epsilon_2^2 \nu^{-2}\}$, then with probability $1 - 4k\exp\left\{\frac{-\epsilon_2^2}{128\mu^2}\right\}$, for every index t between 1 and k

$$|Z_t - Z_{t-1}| \leq 2 \left(2\epsilon_2 + 3\sqrt{k} \left(\frac{k}{n-k}\mu + \frac{n-1}{n-k}\nu \right) \right) |x_t| \|x\|_2 + \frac{4k\mu}{n-k} \|x\|_2^2. \quad (10.54)$$

Proof. Equation (10.53) follows from Equation (10.52) after applying the Cauchy-Schwarz inequality. Now if $k \leq \min\{\frac{n-1}{2}, \epsilon_2^2 \nu^{-2}\}$ then it follows from Lemma 10.5 that

$$\Pr_{\Pi_1^k} \left[\exists t \in [k] \text{ s.t } \left| \sum_{i=1}^{t-1} x_i A_{\Pi_i}{}^\top A_{\Pi_t} \right| \geq 2\epsilon_2 \|x\|_2 \right] \leq 4k \exp\left\{\frac{-\epsilon_3^2}{128\mu^2}\right\}.$$

□

We shall find an upper bound for the martingale difference $|Z_t - Z_{t-1}|$.

LEMMA 10.15 *Let ϵ_2 be a positive number less than $\frac{\mu\theta_M}{16}$. Suppose*

$$k \leq \min\left\{ \left(\frac{n\theta_M}{96}\right)^{\frac{2}{3}}, \epsilon_2^2 \nu^{-2}, \left(\frac{\theta_M \mu}{96\nu}\right)^2 \right\},$$

and define

$$c_t \doteq 2 \left(2\epsilon_2 + 3\sqrt{k} \left(\frac{k}{n-k}\mu + \frac{n-1}{n-k}\nu \right) \right) |x_t| \|x\|_2 + \frac{4k\mu}{n-k} \|x\|_2^2.$$

Then

$$\sum_{t=1}^{k} c_t^2 \leq \mu^2 \theta_M^2 \|x\|_2^4. \quad (10.55)$$

Proof. The proof of Lemma 10.15 requires some algebraic calculations and is provided in [46]. □

Having bounded the difference $|Z_t - Z_{t-1}|$, we can use the Extended Azuma Inequality to show that with overwhelming probability $\sum_{i=1}^{k} x_i \left(\sum_{j=1}^{k} \boldsymbol{x}_j A_{\Pi_i}{}^\top A_{\Pi_j} \right)$ is concentrated around $\|x\|^2$.

Now we are ready to finish the proof of Theorem 10.6.

Proof of Theorem 10.6. First we show that the expected value of $\|Ax\|^2$ is close to $\|x\|^2$. It follows from the linearity of expectation that

$$\mathbb{E}_{\Pi_1^k}\left[\|Ax\|^2\right] = \sum_{i=1}^{k} x_i \sum_{j=1}^{k} \boldsymbol{x}_j \mathbb{E}_{\Pi_1^k}\left[A_{\Pi_i}{}^\top A_{\Pi_j}\right].$$

Hence

$$\left|\mathbb{E}_{\Pi_1^k}\left[\|Ax\|^2\right] - \|Ax\|^2\right| = \left|\sum_{i=1}^{k} x_i \left(\sum_{\substack{j\in[k]\\j\neq i}} \boldsymbol{x}_j \mathbb{E}_{\Pi}\left[A_{\Pi_i}{}^\top A_{\Pi_j}\right] \right)\right| \leq \nu\|x\|_1^2 \leq k\nu\|x\|^2.$$

Let Z_t defined by Equation (10.51). Then since $\epsilon_3 \geq k\nu$, we have

$$\Pr_{\Pi_1^k}\left[\left|\|Ax\|^2 - \|x\|^2\right| \geq 2\epsilon_3 \|x\|^2\right] \leq \Pr_{\Pi}\left[|Z_k - Z_0| \geq \epsilon_3 \|x\|^2\right].$$

Lemma 10.14 states that always

$$|Z_t - Z_{t-1}| \leq b_t$$

and with probability at least $1 - 4k\exp\left\{\frac{-\epsilon_3^2}{128\mu^2}\right\}$, for every index t between 1 and k

$$|Z_t - Z_{t-1}| \leq c_t,$$

where b_t and c_t are the right-hand sides of Equations (10.53) and (10.54). Lemma 10.15 proves that $\sum_{t=1}^{k} c_t^2 \leq \mu^2 \theta_M^2 \|x\|_2^4$. Moreover, it is easy to verify that for every t, $c_t \geq \frac{4k\mu}{n}\|x\|^2$, and $b_t < 8k\mu\|x\|^2$. Therefore

$$\sum_{t=1}^{k} \frac{b_t}{c_t} \leq \sum_{t=1}^{k} 2n \leq 2kn.$$

The result then follows from the Extended Azuma Inequality. □

10.10 Conclusion

In this chapter we introduced compressed learning, a *linear* dimensionality reduction technique for measurement-domain pattern recognition in compressed sensing applications. We formulated the conditions that a sensing matrix should satisfy to guarantee the near-optimality of the measurement-domain SVM's classifier. We then showed that a large family of compressed sensing matrices satisfy the required properties.

We again emphasize that the dimensionality reduction has been studied for a long time in many different communities. In particular, the development of theory and methods that cope with the curse of dimensionality has been the focus of the machine learning community for at least 20 years (e.g., SVM's, complexity-regularization, model selection, boosting, aggregation, etc.). Besides, there has been a huge amount of research on designing robust dimensionality reducing techniques, e.g., manifold learning [48, 49], locality sensitive hashing [50], etc. (see also [47]). The compressed learning approach of this section is most beneficial in compressed sensing applications. The reason is that compressed sensing already projects the data to some low-dimensional space, and therefore the dimensionality reduction can be done as fast and efficiently as the state-of-the art sensing methods are.

Moreover, even though the linear dimensionality reduction technique is not as complicated as other dimensionality reduction methods, it is sufficient for many applications. It has been confirmed empirically that in many applications including information retrieval [30] and face classification [7], the data already has a good enough structure that this linear technique can perform almost as well as the best classifier in the high-dimensional space.

Acknowledgement

It is a pleasure to thank Waheed Bajwa, Ingrid Daubechies, Marco Duarte, Stephen Howard, Robert Schapire, and Karthik Sridharan for sharing many valuable insights. The work of R. Calderbank and S. Jafarpour is supported in part by NSF under grant DMS 0701226, by ONR under grant N00173-06-1-G006, and by AFOSR under grant FA9550-05-1-0443.

References

[1] E. Kushilevitz, R. Ostrovsky, and Y. Rabani. Efficient search for approximate nearest neighbor in high dimensional spaces. *SIAM J Comput*, 30(2):457–474, 2000.

[2] P. Indyk. On approximation nearest neighbors in non-Euclidean spaces. *Proc 39th Ann IEEE Symp Found Computer Scie (FOCS)*: 148–155, 1998.

[3] N. Alon, Y. Matias, and M. Szegedy. The space complexity of approximating the frequency moments. *Proc 28th Ann ACM Symp Theory Comput (STOC)*: 20–29, 1996.

[4] M. F. Balcan, A. Blum, and S. Vempala. Kernels as features: On kernels, margins, and low-dimensional mappings. *Mach Learning*, 65(1), 79–94, 2006.

[5] K. Weinberger, A. Dasgupta, J. Attenberg, J. Langford, and A. Smola. Feature hashing for large scale multitask learning. *Proc 26th Ann Int Conf Machine Learning (ICML)*, 1113–1120, 2009.

[6] D. A. Spielman and N. Srivastava. Graph sparsification by effective resistances. *Proc 40th Ann ACM Symp Theory Comp (STOC)*, 563–568, 2008.

[7] J. Wright, A. Yang, A. Ganesh, S. Shastry, and Y. Ma. Robust face recognition via sparse representation. *IEEE Trans Pattern Machine Intell*, 32(2), 210–227, 2009.

[8] J. Mairal, F. Bach, J. Ponce, and G. Sapiro. Online learning for matrix factorization and sparse coding. *J Machine Learning Res*, 11, 19–60, 2010.

[9] T. Do, Y. Chen, D. Nguyen, N. Nguyen, L. Gan, and T. Tran. Distributed compressed video sensing. *Proc 16th IEEE Int Conf Image Proc*, 1381–1384, 2009.

[10] J. Han, S. McKenna, and R. Wang. Regular texture analysis as statistical model selection. In *ECCV (4)*, vol 5305 of *Lecture Notes in Comput Sci*, 242–255, 2008.

[11] C. McDiarmid. On the method of bounded differences. *Surv Combinatorics*, 148–188, Cambridge University Press, Cambridge, 1989.

[12] S. Kutin. Extensions to McDiarmid's inequality when differences are bounded with high probability. *Tech Rep TR-2002-045*, University of Chicago, April, 2002.

[13] C. J.C. Burgess. A tutorial on support vector machines for pattern recognition. *Data Mining Knowledge Discovery*, 2:121–167, 1998.

[14] K. Sridharan, N. Srebro, and S. Shalev-Shwartz. Fast rates for regularized objectives. In *Neural Information Processing Systems*, 2008.

[15] W. B. Johnson and J. Lindenstrauss. Extensions of Lipschitz mappings into a Hilbert space. *Contemp Math* 26:189–206, 1984.

[16] R. Baraniuk, M. Davenport, R. DeVore, and M. Wakin. A simple proof of the restricted isometry property for random matrices. *Construc Approx*, 28(3):253–263, 2008.

[17] D. Needell and J. A. Tropp. CoSaMP: Iterative signal recovery from incomplete and inaccurate samples. *Appl Comput Harmonic Anal*, 26(3):301–321, 2009.

[18] E. Candès, J. Romberg, and T. Tao. Stable signal recovery from incomplete and inaccurate measurements. *Commun Pure Appl Math*, 59(8):1207–1223, 2006.

[19] E. Candès, J. Romberg, and T. Tao. Robust uncertainty principles: Exact signal reconstruction from highly incomplete frequency information. *IEEE Trans Inform Theory*, 52(2):489–509, 2006.

[20] S. Dasgupta and A. Gupta. An elementary proof of the Johnson–Lindenstrauss lemma. *Technical Report 99-006*, UC Berkeley, 1999.

[21] P. Indyk and R. Motwani. Approximate nearest neighbors: Towards removing the curse of dimensionality. *Proc 30th Ann ACM Symp Theory of Comput (STOC)*: 604–613, 1998.

[22] A. Dasgupta, R. Kumar, and T. Sarlos. A Sparse Johnson–Lindenstrauss Transform. *42nd ACM Symp Theory of Comput (STOC)*, 2010.

[23] D. Achiloptas. Database-friendly random projections: Johnson–Lindenstrauss with binary coins. *J Comput Syst Sci*, 66:671–687, 2003.

[24] N. Ailon and B. Chazelle. The fast Johnson–Lindenstrauss transform and approximate nearest neighbors. *SIAM J Comput*, 39(1):302–322, 2009.

[25] N. Ailon and E. Liberty. Fast dimension reduction using Rademacher series on dual BCH codes. *Discrete Comput Geom*, 42(4):615–630, 2009.

[26] E. Liberty, N. Ailon, and A. Singer. Dense fast random projections and lean Walsh transforms. *12th Int Workshop Randomization Approx Techniques Comput Sci*: 512–522, 2008.

[27] N. Ailon and E. Liberty. Almost Optimal Unrestricted Fast Johnson–Lindenstrauss Transform. *Preprint*, 2010.

[28] J. Matousek. On variants of Johnson–Lindenstrauss lemma. *Private Communication*, 2006.

[29] D. Achlioptas, F. McSherry, and B. Scholkopf. Sampling techniques for kernel methods. *Adv Neural Inform Proc Syst (NIPS)*, 2001.

[30] D. Fradkin. Experiments with random projections for machine learning. *ACM SIGKDD Int Conf Knowledge Discovery Data Mining*: 517–522, 2003.

[31] A. Blum. Random projection, margins, kernels, and feature-selection. *Lecture Notes Comput Sci*, 3940, 52–68, 2006.

[32] A. Rahimi and B. Recht. Random features for large-scale kernel machines. *Adv Neural Inform Proc Syst (NIPS)*, 2007.

[33] W. Bajwa, R. Calderbank, and S. Jafarpour. Model selection: Two fundamental measures of coherence and their algorithmic significance. *Proc IEEE Symp Inform Theory (ISIT)*, 2010.

[34] J. Tropp. The sparsity gap: Uncertainty principles proportional to dimension. *To appear, Proc. 44th Ann. IEEE Conf. Inform Sci Syst (CISS)*, 2010.

[35] R. Calderbank, S. Howard, and S. Jafarpour. Sparse reconstruction via the Reed-Muller Sieve. *Proc IEEE Symp Inform Theory (ISIT)*, 2010.

[36] R. Calderbank and S. Jafarpour. Reed Muller Sensing Matrices and the LASSO. *Int Conf Sequences Applic (SETA)*: 442–463, 2010.

[37] A. R. Hammons, P. V. Kumar, A. R. Calderbank, N. J. A. Sloane, and P. Sole. The $\mathbb{Z}_4$-linearity of Kerdock Codes, Preparata, Goethals, and related codes. *IEEE Trans Inform Theory*, 40(2):301–319, 1994.

[38] R. Calderbank, S. Howard, and S. Jafarpour. Construction of a large class of matrices satisfying a Statistical Isometry Property. *IEEE J Sel Topics Sign Proc, Special Issue on Compressive Sensing*, 4(2):358–374, 2010.

[39] E. Candès and J. Plan. Near-ideal model selection by ℓ_1 minimization. *Anna Stat*, 37: 2145–2177, 2009.

[40] S. Wright, R. Nowak, and M. Figueiredo. Sparse reconstruction by separable approximation. *IEEE Trans Sign Proc*, 57(7), 2479–2493, 2009.

[41] M. Duarte, S. Jafarpour, and R. Calderbank. Conditioning the Delsarte-Goethals frame for compressive imaging. *Preprint*, 2010.

[42] Brodatz Texture Database. Available at http://www.ux.uis.no/~tranden/brodatz.html.

[43] W. U. Bajwa, R. Calderbank, and S. Jafarpour. Revisiting model selection and recovery of sparse signals using one-step thresholding. To appear in *Proc. 48th Ann. Allerton Conf. Commun, Control, Comput*, 2010.

[44] L. Applebaum, S. Howard, S. Searle, and R. Calderbank. Chirp sensing codes: Deterministic compressed sensing measurements for fast recovery. *Appl Computa Harmonic Anal*, 26(2):283–290, 2009.

[45] W. Bajwa, R. Calderbank, and S. Jafarpour. Why Gabor Frames? Two fundamental measures of coherence and their role in model selection. *J Communi Networking*, 12(4):289–307, 2010.

[46] R. Calderbank, S. Howard, S. Jafarpour, and J. Kent. Sparse approximation and compressed sensing using the Reed-Muller Sieve. *Technical Report, TR-888-10*, Princeton University, 2010.

[47] I. Fodor. A survey of dimension reduction techniques. *LLNL Technical Report, UCRL-ID-148494*, 2002.

[48] R. G. Baraniuk and M. B. Wakin. Random projections of smooth manifolds. *Found Computa Math*, 9(1), 2002.

[49] X. Huo, X. S. Ni, and A. K. Smith. A survey of manifold-based learning methods. *Recent Adv Data Mining Enterprise Data*, 691–745, 2007.

[50] A. Andoni and P. Indyk. Near-optimal hashing algorithms for approximate nearest neighbor in high dimensions. *Commun ACM*, 51(1):117–122, 2008.

11 Data separation by sparse representations

Gitta Kutyniok

Modern data are often composed of two or more morphologically distinct constituents, and one typical goal is the extraction of those components. Recently, sparsity methodologies have been successfully utilized to solve this problem, both theoretically as well as empirically. The key idea is to choose a deliberately overcomplete representation made of several frames each one providing a sparse expansion of one of the components to be extracted. The morphological difference between the components is then encoded as incoherence conditions of those frames. The decomposition principle is to minimize the ℓ_1 norm of the frame coefficients. This chapter shall serve as an introduction to and a survey of this exciting area of research as well as a reference for the state of the art of this research field.

11.1 Introduction

Over the last few years, scientists have faced an ever growing deluge of data, which needs to be transmitted, analyzed, and stored. A close analysis reveals that most of these data might be classified as multimodal data, i.e., being composed of distinct subcomponents. Prominent examples are audio data, which might consist of a superposition of the sounds of different instruments, or imaging data from neurobiology, which is typically a composition of the soma of a neuron, its dendrites, and its spines. In both these exemplary situations, the data has to be separated into appropriate single components for further analysis. In the first case, separating the audio signal into the signals of the different instruments is a first step to enable the audio technician to obtain a musical score from a recording. In the second case, the neurobiologist might aim to analyze the structure of dendrites and spines separately for the study of Alzheimer-specific characteristics. Thus data separation is often a crucial step in the analysis of data.

As a scientist, three fundamental problems immediately come to one's mind:

(P1) What is a mathematically precise meaning of the vague term "distinct components"?

Compressed Sensing: Theory and Applications, ed. Yonina C. Eldar and Gitta Kutyniok. Published by Cambridge University Press. © Cambridge University Press 2012.

(P2) How do we separate data algorithmically?
(P3) When is separation possible at all?

To answer those questions, we need to first understand the key problem in data separation. In a very simplistic view, the essence of the problem is as follows: Given a composed signal x of the form $x = x_1 + x_2$, we aim to extract the unknown components x_1 and x_2 from it. Having one known data and two unknowns obviously makes this problem underdetermined. Thus, the novel paradigm of sparsity – appropriately utilized – seems a perfect fit for attacking data separation, and this chapter shall serve as both an introduction into this intriguing application of sparse representations as well as a reference for the state of the art of this research area.

11.1.1 Morphological Component Analysis

Intriguingly, when considering the history of Compressed Sensing, the first mathematically precise result on recovery of sparse vectors by ℓ_1 minimization is related to a data separation problem: The separation of sinusoids and spikes in [16, 11]. Thus it might be considered a milestone in the development of Compressed Sensing. In addition, it reveals a surprising connection with uncertainty principles.

The general idea allowing separation in [16, 11] was to choose two bases or frames Φ_1 and Φ_2 adapted to the two components to be separated in such a way that Φ_1 and Φ_2 provide a sparse representation for x_1 and x_2, respectively. Searching for the sparsest representation of the signal in the combined (highly overcomplete) dictionary $[\Phi_1 \mid \Phi_2]$ should then intuitively enforce separation provided that x_1 does not have a sparse representation in Φ_2 and that x_2 does not have a sparse representation in Φ_1. This general concept was later – in the context of image separation, but the term seems to be fitting in general – coined *Morphological Component Analysis* [36].

This viewpoint now measures the morphological difference between components in terms of the incoherence of suitable sparsifying bases or frames Φ_i, thereby giving one possible answer to (P1); see also the respective chapters in the book [33]. As already debated in the introduction (Chapter 1), one possibility for measuring incoherence is the *mutual coherence*. We will however see in the sequel that there exist even more appropriate coherence notions, which provide a much more refined measurement of incoherence specifically adapted to measuring morphological difference.

11.1.2 Separation algorithms

Going again back in time, we observe that long before [11], Coifman, Wickerhauser, and coworkers had already presented very inspiring empirical results on the separation of image components using the idea of Morphological Component Analysis, see [7]. After this, several techniques to actually compute the sparsest expansion in a composed dictionary $[\Phi_1 \mid \Phi_2]$ were introduced. In [31], Mallat and Zhang developed *Matching Pursuit* as one possible methodology. The study by Chen, Donoho, and Saunders in [6] then revealed that the ℓ_1 norm has a tendency to find sparse solutions when they exist, and coined this method *Basis Pursuit*.

As explained before, data separation by Morphological Component Analysis – when suitably applied – can be reduced to a sparse recovery problem. To solve this problem, there nowadays already exists a variety of utilizable algorithmic approaches; thereby providing a general answer to (P2). Such approaches include, for instance, a canon of greedy-type algorithms, and we refer to the introduction and Chapter 8 for further details. Most of the theoretical separation results however consider ℓ_1 minimization as the main separation technique, which is what we will also mainly focus on in this chapter.

11.1.3 Separation results

As already mentioned, the first mathematically precise result was derived in [11] and solved the problem of separation of sinusoids and spikes. After this "birth of sparse data separation," a deluge of very exciting results started. One direction of research are general results on sparse recovery and Compressed Sensing; here we would like to cite the excellent survey paper [4] and the introduction (Chapter 1).

Another direction continued the idea of sparse data separation initiated in [11]. In this realm, the most significant theoretical results might be considered firstly the series of papers [19, 10], in which the initial results from [11] are extended to general composed dictionaries, secondly the paper [23], which also extends results from [11] though with a different perspective, and thirdly the papers [3] and [14], which explore the clustering of the sparse coefficients and the morphological difference of the components encoded in it.

We also wish to mention the abundance of empirical work showing that utilizing the idea of sparse data separation often gives very compelling results in practice; as examples, we refer to the series of papers on applications to astronomical data [2, 36, 34], to general imaging data [32, 20, 35], and to audio data [22, 25].

Let us remark that also the classical problem of denoising can be regarded as a separation problem, since we aim to separate a signal from noise by utilizing the characteristics of the signal family and the noise. However, as opposed to the separation problems discussed in this chapter, denoising is not a "symmetric" separation task, since the characterization of the signal and the noise are very different.

11.1.4 Design of sparse dictionaries

For satisfactorily answering (P3), one must also raise the question of how to find suitable sparsifying bases or frames for given components. This search for "good" systems in the sense of sparse dictionaries can be attacked in two ways, either non-adaptively or adaptively.

The first path explores the structure of the component one would like to extract, for instance, it could be periodic such as sinusoids or anisotropic such as edges in images. This typically allows one to find a suitable system among the already very well explored representation systems such as the Fourier basis, wavelets, or shearlets, to name a few.

The advantage of this approach is the already explored structure of the system, which can hence be exploited for deriving theoretical results on the accuracy of separation, and the speed of associated transforms.

The second path uses a training set of data similar to the to-be-extracted component, and "learns" a system which best sparsifies this data set. Using this approach customarily referred to as *dictionary learning*, we obtain a system extremely well adapted to the data at hand; as the state of the art we would like to mention the K-SVD algorithm introduced by Aahron, Elad, and Bruckstein in [1]; see also [17] for a "Compressed Sensing" perspective to K-SVD. Another appealing dictionary training algorithm, which should be cited is the method of optimal directions (MOD) by Engan *et al.* [21]. The downside however is the lack of a mathematically exploitable structure, which makes a theoretical analysis of the accuracy of separation using such a system very hard.

11.1.5 Outline

In Section 11.2, we discuss the formal mathematical setting of the problem, present the nowadays already considered classical separation results, and then discuss more recent results exploiting the clustering of significant coefficients in the expansions of the components as a means to measure their morphological difference. We conclude this section by revealing a close link of data separation to uncertainty principles. Section 11.3 is then devoted to both theoretical results as well as applications for separation of 1-D signals, elaborating, in particular, on the separation of sinusoids and spikes. Finally, Section 11.4 focuses on diverse questions concerning separation of 2-D signals, i.e., images, such as the separation of point- and curvelike objects, again presenting both application aspects as well as theoretical results.

11.2 Separation estimates

As already mentioned in the introduction, data separation can be regarded within the framework of underdetermined problems. In this section, we make this link mathematically precise. Then we discuss general estimates on the separability of composed data, firstly without any knowledge of the geometric structure of sparsity patterns, and secondly, by taking known geometric information into account. A revelation of the close relation with uncertainty principles concludes the section.

In Sections 11.3 and 11.4, we will then see the presented general results and uncertainty principles in action, i.e., applied to real-world separation problems.

11.2.1 Relation with underdetermined problems

Let x be our signal of interest, which we for now consider as belonging to some Hilbert space $\mathcal{H}$, and assume that

$$x = x_1^0 + x_2^0.$$

Certainly, real data are typically composed of multiple components, hence not only the situation of two components, but three or more is of interest. We will however focus on the two-component situation to clarify the fundamental principles behind the success of separating those by sparsity methodologies. It should be mentioned though that, in fact, most of the presented theoretical results can be extended to the multiple component situation in a more or less straightforward manner.

To extract the two components from x, we need to assume that – although we are not given x_1^0 and x_2^0 – certain "characteristics" of those components are known to us. Such "characteristics" might be, for instance, the pointlike structure of stars and the curvelike structure of filaments in astronomical imaging. This knowledge now enables us to choose two representation systems, Φ_1 and Φ_2, say, which allow sparse expansions of x_1^0 and x_2^0, respectively. Such representation systems might be chosen from the collection of well-known systems such as wavelets. A different possibility is to choose adaptively the systems via dictionary learning procedures. This approach however requires training data sets for the two components x_1^0 and x_2^0 as discussed in Subsection 11.1.4.

Given now two such representation systems Φ_1 and Φ_2, we can write x as

$$x = x_1^0 + x_2^0 = \Phi_1 c_1^0 + \Phi_2 c_2^0 = [\,\Phi_1\,|\,\Phi_2\,] \begin{bmatrix} c_1^0 \\ c_2^0 \end{bmatrix}$$

with $\|c_1^0\|_0$ and $\|c_2^0\|_0$ "sufficiently small." Thus, the data separation problem has been reduced to solving the underdetermined linear system

$$x = [\,\Phi_1\,|\,\Phi_2\,] \begin{bmatrix} c_1 \\ c_2 \end{bmatrix} \qquad (11.1)$$

for $[c_1, c_2]^T$. Unique recovery of the original vector $[c_1^0, c_2^0]^T$ automatically extracts the correct two components x_1^0 and x_2^0 from x, since

$$x_1^0 = \Phi_1 c_1^0 \quad \text{and} \quad x_2^0 = \Phi_2 c_2^0.$$

Ideally, one might want to solve

$$\min_{c_1, c_2} \|c_1\|_0 + \|c_2\|_0 \quad \text{s.t.} \quad x = [\,\Phi_1\,|\,\Phi_2\,] \begin{bmatrix} c_1 \\ c_2 \end{bmatrix}, \qquad (11.2)$$

which however is an NP-hard problem. As already discussed in the introduction (Chapter 1), instead one aims to solve the ℓ_1 minimization problem

$$(\text{Sep}_s) \qquad \min_{c_1, c_2} \|c_1\|_1 + \|c_2\|_1 \quad \text{s.t.} \quad x = [\,\Phi_1\,|\,\Phi_2\,] \begin{bmatrix} c_1 \\ c_2 \end{bmatrix}. \qquad (11.3)$$

The lower case "s" in Sep_s indicates that the ℓ_1 norm is placed on the synthesis side. Other choices for separation are, for instance, greedy-type algorithms – and we refer the reader to the introduction (Chapter 1) and Chapter 8. In this chapter we will focus on ℓ_1

minimization as the separation technique, consistent with most known separation results from the literature.

Before discussing conditions on $[c_1^0, c_2^0]^T$ and $[\Phi_1 \mid \Phi_2]$, which guarantee unique solvability of (11.1), let us for a moment debate whether uniqueness is necessary at all. If Φ_1 and Φ_2 form bases, it is certainly essential to recover $[c_1^0, c_2^0]^T$ uniquely from (11.1). However, some well-known representation systems are in fact redundant and typically constitute Parseval frames such as curvelets or shearlets. Also, systems generated by dictionary learning are normally highly redundant. In this situation, for each possible separation

$$x = x_1 + x_2, \qquad (11.4)$$

there exist infinitely many coefficient sequences $[c_1, c_2]^T$ satisfying

$$x_1 = \Phi_1 c_1 \quad \text{and} \quad x_2 = \Phi_2 c_2. \qquad (11.5)$$

Since we are *only* interested in the correct separation and *not* in computing the sparsest expansion, we can circumvent presumably arising numerical instabilities when solving the minimization problem (11.3) by selecting a particular coefficient sequence for each separation. Assuming Φ_1 and Φ_2 are Parseval frames, we can exploit this structure and rewrite (11.5) as

$$x_1 = \Phi_1(\Phi_1^T x_1) \quad \text{and} \quad x_2 = \Phi_2(\Phi_2^T x_2).$$

Thus, for each separation (11.4), we choose a *specific* coefficient sequence when expanding the components in the Parseval frames; in fact, we choose the *analysis sequence*. This leads to the following different ℓ_1 minimization problem in which the ℓ_1 norm is placed on the *analysis* rather than the *synthesis* side:

$$(\text{Sep}_a) \qquad \min_{x_1, x_2} \|\Phi_1^T x_1\|_1 + \|\Phi_2^T x_2\|_1 \quad \text{s.t.} \quad x = x_1 + x_2. \qquad (11.6)$$

This new minimization problem can be also regarded as a mixed ℓ_1-ℓ_2 problem, since the analysis coefficient sequence is exactly the coefficient sequence which is minimal in the ℓ_2 norm. For more information, we refer to Chapter 2.

11.2.2 General separation estimates

Let us now discuss the main results of successful data separation, i.e., stating conditions on $[c_1^0, c_2^0]^T$ and $[\Phi_1 \mid \Phi_2]$ for extracting x_1^0 and x_2^0 from x. The strongest known general result was derived in 2003 by Donoho and Elad [10] and simultaneously by Gribonval and Nielsen [23] and used the notion of mutual coherence. Recall that, for a normalized frame $\Phi = (\varphi_i)_{i \in I}$, the *mutual coherence* of Φ is defined by

$$\mu(\Phi) = \max_{i,j \in I, i \neq j} |\langle \varphi_i, \varphi_j \rangle|. \qquad (11.7)$$

We remark that the result by Donoho and Elad was already stated in the introduction, however without a proof, which we will present in this section. For the convenience of the reader, we recall the relevant result.

THEOREM 11.1 ([10, 23]) *Let Φ_1 and Φ_2 be two frames for a Hilbert space $\mathcal{H}$, and let $x \in \mathcal{H}$. If $x = [\Phi_1|\Phi_2]c$ and*

$$\|c\|_0 < \frac{1}{2}\left(1 + \frac{1}{\mu([\Phi_1|\Phi_2])}\right),$$

then c is the unique solution of the ℓ_1 minimization problem (Sep_s) stated in (11.3) as well as the unique solution of the ℓ_0 minimization problem stated in (11.2).

Before presenting the proof, we require some prerequisites. Firstly, we need to introduce a slightly stronger version of the null space property than the one discussed in the introduction.

DEFINITION 11.1 *Let $\Phi = (\varphi_i)_{i \in I}$ be a frame for a Hilbert space $\mathcal{H}$, and let $\mathcal{N}(\Phi)$ denote the null space of Φ. Then Φ is said to have the* null space property of order k *if*

$$\|1_\Lambda d\|_1 < \frac{1}{2}\|d\|_1$$

for all $d \in \mathcal{N}(\Phi) \setminus \{0\}$ and for all sets $\Lambda \subseteq I$ with $|\Lambda| \leq k$.

This notion provides a very useful characterization of the existence of unique sparse solutions of the ℓ_1 minimization problem (Sep_s) stated in (11.3).

LEMMA 11.1 *Let $\Phi = (\varphi_i)_{i \in I}$ be a frame for a Hilbert space $\mathcal{H}$, and let $k \in \mathbb{N}$. Then the following conditions are equivalent.*

(i) For each $x \in \mathcal{H}$, if $x = \Phi c$ satisfies $\|c\|_0 \leq k$, then c is the unique solution of the ℓ_1 minimization problem (Sep_s) stated in (11.3), (with Φ instead of $[\Phi_1|\Phi_2]$).
(ii) Φ satisfies the null space property of order k.

Proof. First, assume that (i) holds. Let $d \in \mathcal{N}(\Phi) \setminus \{0\}$ and $\Lambda \subseteq I$ with $|\Lambda| \leq k$ be arbitrary. Then, by (i), the sparse vector $1_\Lambda d$ is the unique minimizer of $\|c\|_1$ subject to $\Phi c = \Phi(1_\Lambda d)$. Further, since $d \in \mathcal{N}(\Phi) \setminus \{0\}$,

$$\Phi(-1_{\Lambda^c} d) = \Phi(1_\Lambda d).$$

Hence

$$\|1_\Lambda d\|_1 < \|1_{\Lambda^c} d\|_1,$$

or, in other words,

$$\|1_\Lambda d\|_1 < \frac{1}{2}\|d\|_1,$$

which implies (ii), since d and Λ were chosen arbitrarily.

Secondly, assume that (ii) holds, and let c_1 be a vector satisfying $x = \Phi c_1$ with $\|c_1\|_0 \leq k$ and support denoted by Λ. Further, let c_2 be an arbitrary solution of $x = \Phi c$, and set

$$d = c_2 - c_1.$$

Then
$$\|c_2\|_1 - \|c_1\|_1 = \|1_{\Lambda^c} c_2\|_1 + \|1_\Lambda c_2\|_1 - \|1_\Lambda c_1\|_1 \geq \|1_{\Lambda^c} d\|_1 - \|1_\Lambda d\|_1.$$

This term is greater than zero for any $d \neq 0$ if
$$\|1_{\Lambda^c} d\|_1 > \|1_\Lambda d\|_1,$$
or
$$\frac{1}{2}\|d\|_1 > \|1_\Lambda d\|_1.$$

This is ensured by (ii). Hence $\|c_2\|_1 > \|c_1\|_1$, and thus c_1 is the unique solution of (Sep_s). This implies (i). $\square$

Using this result, we next prove that a solution satisfying $\|c\|_0 < \frac{1}{2}\left(1 + \frac{1}{\mu(\Phi)}\right)$ is the unique solution of the ℓ_1 minimization problem (Sep_s).

LEMMA 11.2 *Let $\Phi = (\varphi_i)_{i \in I}$ be a frame for a Hilbert space $\mathcal{H}$, and let $x \in \mathcal{H}$. If $x = \Phi c$ and*
$$\|c\|_0 < \frac{1}{2}\left(1 + \frac{1}{\mu(\Phi)}\right),$$
then c is the unique solution of the ℓ_1 minimization problem (Sep_s) stated in (11.3) (with Φ instead of $[\Phi_1|\Phi_2]$).

Proof. Let $d \in \mathcal{N}(\Phi) \setminus \{0\}$, hence, in particular,
$$\Phi d = 0;$$
thus also
$$\Phi^T \Phi d = 0. \tag{11.8}$$

Without loss of generality, we now assume that the vectors in Φ are normalized. Then, (11.8) implies that, for all $i \in I$,
$$d_i = -\sum_{j \neq i} \langle \varphi_i, \varphi_j \rangle d_j.$$

Using the definition of mutual coherence $\mu(\Phi)$ (cf. Introduction (Chapter 1) or (11.7)), we obtain
$$|d_i| \leq \sum_{j \neq i} |\langle \varphi_i, \varphi_j \rangle| \cdot |d_j| \leq \mu(\Phi)(\|d\|_1 - |d_i|),$$
and hence
$$|d_i| \leq \left(1 + \frac{1}{\mu(\Phi)}\right)^{-1} \|d\|_1.$$

Thus, by the hypothesis on $\|c\|_0$ and for any $\Lambda \subseteq I$ with $|\Lambda| = \|c\|_0$, we have
$$\|1_\Lambda d\|_1 \leq |\Lambda| \cdot \left(1 + \frac{1}{\mu(\Phi)}\right)^{-1} \|d\|_1 = \|c\|_0 \cdot \left(1 + \frac{1}{\mu(\Phi)}\right)^{-1} \|d\|_1 < \frac{1}{2}\|d\|_1.$$

This shows that Φ satisfies the null space property of order $\|c\|_0$, which, by Lemma 11.1, implies that c is the unique solution of (Sep_s). $\square$

We further prove that a solution satisfying $\|c\|_0 < \frac{1}{2}\left(1 + \frac{1}{\mu(\Phi)}\right)$ is also the unique solution of the ℓ_0 minimization problem.

LEMMA 11.3   *Let $\Phi = (\varphi_i)_{i \in I}$ be a frame for a Hilbert space $\mathcal{H}$, and let $x \in \mathcal{H}$. If $x = \Phi c$ and*

$$\|c\|_0 < \frac{1}{2}\left(1 + \frac{1}{\mu(\Phi)}\right),$$

then c is the unique solution of the ℓ_0 minimization problem stated in (11.2) (with Φ instead of $[\Phi_1|\Phi_2]$).

Proof. By Lemma 11.2, the hypotheses imply that c is the unique solution of the ℓ_1 minimization problem (Sep_s). Now, towards a contradiction, assume that there exists some $\tilde{c}$ satisfying $x = \Phi \tilde{c}$ with $\|\tilde{c}\|_0 \leq \|c\|_0$. Then $\tilde{c}$ must satisfy

$$\|\tilde{c}\|_0 < \frac{1}{2}\left(1 + \frac{1}{\mu(\Phi)}\right).$$

Again, by Lemma 11.2, $\tilde{c}$ is the unique solution of the ℓ_1 minimization problem (Sep_s), a contradiction. $\square$

These lemmata now immediately imply Theorem 11.1.

Proof of Theorem 11.1.   Theorem 11.1 follows from Lemmata 11.2 and 11.3. $\square$

Interestingly, in the situation of Φ_1 and Φ_2 being two orthonormal bases the bound can be slightly strengthened. For the proof of this result, we refer the reader to [19].

THEOREM 11.2   ([19]) *Let Φ_1 and Φ_2 be two orthonormal bases for a Hilbert space $\mathcal{H}$, and let $x \in \mathcal{H}$. If $x = [\Phi_1|\Phi_2]c$ and*

$$\|c\|_0 < \frac{\sqrt{2} - 0.5}{\mu([\Phi_1|\Phi_2])},$$

then c is the unique solution of the ℓ_1 minimization problem (Sep_s) stated in (11.3) as well as the unique solution of the ℓ_0 minimization problem stated in (11.2).

This shows that in the special situation of two orthonormal bases, the bound is nearly a factor of 2 stronger than in the general situation of Theorem 11.1.

11.2.3   Clustered sparsity as a novel viewpoint

In a concrete situation, we often have more information on the geometry of the to-be-separated components x_1^0 and x_2^0. This information is typically encoded in a particular clustering of the nonzero coefficients if a suitable basis or frame for the expansion of x_1^0 or x_2^0 is chosen. Think, for instance, of the tree clustering of wavelet coefficients of a point singularity. Thus, it seems conceivable that the morphological difference is encoded not only in the incoherence of the two chosen bases or frames adapted to x_1^0 and x_2^0, but in the interaction of the elements of those bases or frames associated with the clusters of significant coefficients. This should intuitively allow for weaker sufficient conditions for separation.

One possibility for a notion capturing this idea is the so-called *joint concentration* which was introduced in [14] with concepts going back to [16], and was in between revived in [11]. To provide some intuition for this notion, let Λ_1 and Λ_2 be subsets of indexing sets of two Parseval frames. Then the joint concentration measures the maximal fraction of the total ℓ_1 norm which can be concentrated on the index set $\Lambda_1 \cup \Lambda_2$ of the combined dictionary.

DEFINITION 11.2 *Let $\Phi_1 = (\varphi_{1i})_{i \in I}$ and $\Phi_2 = (\varphi_{2j})_{j \in J}$ be two Parseval frames for a Hilbert space $\mathcal{H}$. Further, let $\Lambda_1 \subseteq I$ and $\Lambda_2 \subseteq J$. Then the* joint concentration *$\kappa = \kappa(\Lambda_1, \Phi_1; \Lambda_2, \Phi_2)$ is defined by*

$$\kappa(\Lambda_1, \Phi_1; \Lambda_2, \Phi_2) = \sup_x \frac{\|1_{\Lambda_1} \Phi_1^T x\|_1 + \|1_{\Lambda_2} \Phi_2^T x\|_1}{\|\Phi_1^T x\|_1 + \|\Phi_2^T x\|_1}.$$

One might ask how the notion of joint concentration relates to the widely exploited, and for the previous result utilized mutual coherence. For this, we first briefly discuss some derivations of mutual coherence. A first variant better adapted to clustering of coefficients was the *Babel function* introduced in [10] as well as in [37] under the label *cumulative coherence function*, which, for a normalized frame $\Phi = (\varphi_i)_{i \in I}$ and some $m \in \{1, \ldots, |I|\}$, is defined by

$$\mu_B(m, \Phi) = \max_{\Lambda \subset I, |\Lambda| = m} \max_{j \notin \Lambda} \sum_{i \in I} |\langle \varphi_i, \varphi_j \rangle|.$$

This notion was later refined in [3] by considering the so-called *structured p-Babel function*, defined for some family $\mathcal{S}$ of subsets of I and some $1 \leq p < \infty$ by

$$\mu_{sB}(\mathcal{S}, \Phi) = \max_{\Lambda \in \mathcal{S}} \left(\max_{j \notin \Lambda} \sum_{i \in I} |\langle \varphi_i, \varphi_j \rangle|^p \right)^{1/p}.$$

Another variant, better adapted to data separation, is the *cluster coherence* introduced in [14], whose definition we now formally state. Notice that we do not assume that the vectors are normalized.

DEFINITION 11.3 *Let $\Phi_1 = (\varphi_{1i})_{i \in I}$ and $\Phi_2 = (\varphi_{2j})_{j \in J}$ be two Parseval frames for a Hilbert space $\mathcal{H}$, let $\Lambda_1 \subseteq I$, and let $\Lambda_2 \subseteq J$. Then the* cluster coherence *$\mu_c(\Lambda_1, \Phi_1; \Phi_2)$ of Φ_1 and Φ_2 with respect to Λ_1 is defined by*

$$\mu_c(\Lambda_1, \Phi_1; \Phi_2) = \max_{j \in J} \sum_{i \in \Lambda_1} |\langle \varphi_{1i}, \varphi_{2j} \rangle|,$$

and the cluster coherence $\mu_c(\Phi_1; \Lambda_2, \Phi_2)$ of Φ_1 and Φ_2 with respect to Λ_2 is defined by

$$\mu_c(\Phi_1; \Lambda_2, \Phi_2) = \max_{i \in I} \sum_{j \in \Lambda_2} |\langle \varphi_{1i}, \varphi_{2j} \rangle|.$$

The relation between joint concentration and cluster coherence is made precise in the following result from [14].

PROPOSITION 11.1 ([14]) *Let $\Phi_1 = (\varphi_{1i})_{i \in I}$ and $\Phi_2 = (\varphi_{2j})_{j \in J}$ be two Parseval frames for a Hilbert space $\mathcal{H}$, and let $\Lambda_1 \subseteq I$ and $\Lambda_2 \subseteq J$. Then*

$$\kappa(\Lambda_1, \Phi_1; \Lambda_2, \Phi_2) \leq \max\{\mu_c(\Lambda_1, \Phi_1; \Phi_2), \mu_c(\Phi_1; \Lambda_2, \Phi_2)\}.$$

Proof. Let $x \in \mathcal{H}$. We now choose coefficient sequences c_1 and c_2 such that

$$x = \Phi_1 c_1 = \Phi_2 c_2$$

and, for $i = 1, 2$,
$$\|c_i\|_1 \leq \|d_i\|_1 \quad \text{for all } d_i \text{ with } x = \Phi_i d_i. \tag{11.9}$$

This implies that

$$\|1_{\Lambda_1} \Phi_1^T x\|_1 + \|1_{\Lambda_2} \Phi_2^T x\|_1$$

$$= \|1_{\Lambda_1} \Phi_1^T \Phi_2 c_2\|_1 + \|1_{\Lambda_2} \Phi_2^T \Phi_1 c_1\|_1$$

$$\leq \sum_{i \in \Lambda_1} \left(\sum_{j \in J} |\langle \varphi_{1i}, \varphi_{2j} \rangle| |c_{2j}| \right) + \sum_{j \in \Lambda_2} \left(\sum_{i \in I} |\langle \varphi_{1i}, \varphi_{2j} \rangle| |c_{1i}| \right)$$

$$= \sum_{j \in J} \left(\sum_{i \in \Lambda_1} |\langle \varphi_{1i}, \varphi_{2j} \rangle| \right) |c_{2j}| + \sum_{i \in I} \left(\sum_{j \in \Lambda_2} |\langle \varphi_{1i}, \varphi_{2j} \rangle| \right) |c_{1i}|$$

$$\leq \mu_c(\Lambda_1, \Phi_1; \Phi_2) \|c_2\|_1 + \mu_c(\Phi_1; \Lambda_2, \Phi_2) \|c_1\|_1$$

$$\leq \max\{\mu_c(\Lambda_1, \Phi_1; \Phi_2), \mu_c(\Phi_1; \Lambda_2, \Phi_2)\} (\|c_1\|_1 + \|c_2\|_1).$$

Since Φ_1 and Φ_2 are Parseval frames, we have

$$x = \Phi_i(\Phi_i^T \Phi_i c_i) \quad \text{for } i = 1, 2.$$

Hence, by exploiting (11.9),

$$\|1_{\Lambda_1} \Phi_1^T x\|_1 + \|1_{\Lambda_2} \Phi_2^T x\|_1$$

$$\leq \max\{\mu_c(\Lambda_1, \Phi_1; \Phi_2), \mu_c(\Phi_1; \Lambda_2, \Phi_2)\} (\|\Phi_1^T \Phi_1 c_1\|_1 + \|\Phi_2^T \Phi_2 c_2\|_1)$$

$$= \max\{\mu_c(\Lambda_1, \Phi_1; \Phi_2), \mu_c(\Phi_1; \Lambda_2, \Phi_2)\} (\|\Phi_1^T x\|_1 + \|\Phi_2^T x\|_1). \qquad \square$$

Before stating the data separation estimate which uses joint concentration, we need to discuss the conditions on sparsity of the components in the two Parseval frames. Since for real data "true sparsity" is unrealistic, a weaker condition will be imposed. In the introduction (Chapter 1), compressibility was already introduced. For the next result, yet a different notion invoking the clustering of the significant coefficients will be required. This notion, first utilized in [9], is defined for our data separation problem as follows.

DEFINITION 11.4 *Let $\Phi_1 = (\varphi_{1i})_{i \in I}$ and $\Phi_2 = (\varphi_{2j})_{j \in J}$ be two Parseval frames for a Hilbert space $\mathcal{H}$, and let $\Lambda_1 \subseteq I$ and $\Lambda_2 \subseteq J$. Further, suppose that $x \in \mathcal{H}$ can be decomposed as $x = x_1^0 + x_2^0$. Then the components x_1^0 and x_2^0 are called δ-relatively sparse in Φ_1 and Φ_2 with respect to Λ_1 and Λ_2, if*

$$\|1_{\Lambda_1^c} \Phi_1^T x_1^0\|_1 + \|1_{\Lambda_2^c} \Phi_2^T x_2^0\|_1 \leq \delta.$$

We now have all ingredients to state the data separation result from [14], which – as compared to Theorem 11.1 – now invokes information about the clustering of coefficients.

THEOREM 11.3 ([14]) *Let $\Phi_1 = (\varphi_{1i})_{i \in I}$ and $\Phi_2 = (\varphi_{2j})_{j \in J}$ be two Parseval frames for a Hilbert space $\mathcal{H}$, and suppose that $x \in \mathcal{H}$ can be decomposed as $x = x_1^0 + x_2^0$. Further, let $\Lambda_1 \subseteq I$ and $\Lambda_2 \subseteq J$ be chosen such that x_1^0 and x_2^0 are δ-relatively sparse in Φ_1 and Φ_2 with respect to Λ_1 and Λ_2. Then the solution $(x_1^\star, x_2^\star)$ of the ℓ_1 minimization problem (Sep_a) stated in (11.6) satisfies*

$$\|x_1^\star - x_1^0\|_2 + \|x_2^\star - x_2^0\|_2 \leq \frac{2\delta}{1 - 2\kappa}.$$

Proof. First, using the fact that Φ_1 and Φ_2 are Parseval frames,

$$\|x_1^\star - x_1^0\|_2 + \|x_2^\star - x_2^0\|_2 = \|\Phi_1^T(x_1^\star - x_1^0)\|_2 + \|\Phi_2^T(x_2^\star - x_2^0)\|_2$$
$$\leq \|\Phi_1^T(x_1^\star - x_1^0)\|_1 + \|\Phi_2^T(x_2^\star - x_2^0)\|_1.$$

The decomposition $x_1^0 + x_2^0 = x = x_1^\star + x_2^\star$ implies

$$x_2^\star - x_2^0 = -(x_1^\star - x_1^0),$$

which allows us to conclude that

$$\|x_1^\star - x_1^0\|_2 + \|x_2^\star - x_2^0\|_2 \leq \|\Phi_1^T(x_1^\star - x_1^0)\|_1 + \|\Phi_2^T(x_1^\star - x_1^0)\|_1. \qquad (11.10)$$

By the definition of κ,

$$\|\Phi_1^T(x_1^\star - x_1^0)\|_1 + \|\Phi_2^T(x_1^\star - x_1^0)\|_1$$
$$= (\|1_{\Lambda_1}\Phi_1^T(x_1^\star - x_1^0)\|_1 + \|1_{\Lambda_2}\Phi_2^T(x_1^\star - x_1^0)\|_1) + \|1_{\Lambda_1^c}\Phi_1^T(x_1^\star - x_1^0)\|_1$$
$$+ \|1_{\Lambda_2^c}\Phi_2^T(x_2^\star - x_2^0)\|_1$$
$$\leq \kappa \cdot (\|\Phi_1^T(x_1^\star - x_1^0)\|_1 + \|\Phi_2^T(x_1^\star - x_1^0)\|_1) + \|1_{\Lambda_1^c}\Phi_1^T(x_1^\star - x_1^0)\|_1$$
$$+ \|1_{\Lambda_2^c}\Phi_2^T(x_2^\star - x_2^0)\|_1,$$

which yields

$$\|\Phi_1^T(x_1^\star - x_1^0)\|_1 + \|\Phi_2^T(x_1^\star - x_1^0)\|_1$$
$$\leq \frac{1}{1-\kappa}(\|1_{\Lambda_1^c}\Phi_1^T(x_1^\star - x_1^0)\|_1 + \|1_{\Lambda_2^c}\Phi_2^T(x_2^\star - x_2^0)\|_1)$$
$$\leq \frac{1}{1-\kappa}(\|1_{\Lambda_1^c}\Phi_1^T x_1^\star\|_1 + \|1_{\Lambda_1^c}\Phi_1^T x_1^0\|_1 + \|1_{\Lambda_2^c}\Phi_2^T x_2^\star\|_1 + \|1_{\Lambda_2^c}\Phi_2^T x_2^0\|_1).$$

Now using the relative sparsity of x_1^0 and x_2^0 in Φ_1 and Φ_2 with respect to Λ_1 and Λ_2, we obtain

$$\|\Phi_1^T(x_1^\star - x_1^0)\|_1 + \|\Phi_2^T(x_1^\star - x_1^0)\|_1 \leq \frac{1}{1-\kappa}\left(\|1_{\Lambda_1^c}\Phi_1^T x_1^\star\|_1 + \|1_{\Lambda_2^c}\Phi_2^T x_2^\star\|_1 + \delta\right). \tag{11.11}$$

By the minimality of $x_1^\star$ and $x_2^\star$ as solutions of (Sep$_a$) implying that

$$\sum_{i=1}^2 \left(\|1_{\Lambda_i^c}\Phi_i^T x_i^\star\|_1 + \|1_{\Lambda_i}\Phi_i^T x_i^\star\|_1\right) = \|\Phi_1^T x_1^\star\|_1 + \|\Phi_2^T x_2^\star\|_1$$
$$\leq \|\Phi_1^T x_1^0\|_1 + \|\Phi_2^T x_2^0\|_1,$$

we have

$$\|1_{\Lambda_1^c}\Phi_1^T x_1^\star\|_1 + \|1_{\Lambda_2^c}\Phi_2^T x_2^\star\|_1$$
$$\leq \|\Phi_1^T x_1^0\|_1 + \|\Phi_2^T x_2^0\|_1 - \|1_{\Lambda_1}\Phi_1^T x_1^\star\|_1 - \|1_{\Lambda_2}\Phi_2^T x_2^\star\|_1$$
$$\leq \|\Phi_1^T x_1^0\|_1 + \|\Phi_2^T x_2^0\|_1 + \|1_{\Lambda_1}\Phi_1^T(x_1^\star - x_1^0)\|_1 - \|1_{\Lambda_1}\Phi_1^T x_1^0\|_1$$
$$+ \|1_{\Lambda_2}\Phi_2^T(x_2^\star - x_2^0)\|_1 - \|1_{\Lambda_2}\Phi_2^T x_2^0\|_1.$$

Again exploiting relative sparsity leads to

$$\|1_{\Lambda_1^c}\Phi_1^T x_1^\star\|_1 + \|1_{\Lambda_2^c}\Phi_2^T x_2^\star\|_1 \leq \|1_{\Lambda_1}\Phi_1^T(x_1^\star - x_1^0)\|_1 + \|1_{\Lambda_2}\Phi_2^T(x_2^\star - x_2^0)\|_1 + \delta. \tag{11.12}$$

Combining (11.11) and (11.12) and again using joint concentration,

$$\|\Phi_1^T(x_1^\star - x_1^0)\|_1 + \|\Phi_2^T(x_1^\star - x_1^0)\|_1$$
$$\leq \frac{1}{1-\kappa}\left[\|1_{\Lambda_1}\Phi_1^T(x_1^\star - x_1^0)\|_1 + \|1_{\Lambda_2}\Phi_2^T(x_1^\star - x_1^0)\|_1 + 2\delta\right]$$
$$\leq \frac{1}{1-\kappa}\left[\kappa \cdot (\|\Phi_1^T(x_1^\star - x_1^0)\|_1 + \|\Phi_2^T(x_1^\star - x_1^0)\|_1) + 2\delta\right].$$

Thus, by (11.10), we finally obtain

$$\|x_1^\star - x_1^0\|_2 + \|x_2^\star - x_2^0\|_2 \leq \left(1 - \frac{\kappa}{1-\kappa}\right)^{-1} \cdot \frac{2\delta}{1-\kappa} = \frac{2\delta}{1-2\kappa}.$$

□

Using Proposition 11.1, this result can also be stated in terms of cluster coherence, which on the one hand provides an easier accessible estimate and allows a better comparison with results using mutual coherence, but on the other hand poses a slightly weaker estimate.

THEOREM 11.4 ([14]) *Let $\Phi_1 = (\varphi_{1i})_{i \in I}$ and $\Phi_2 = (\varphi_{2j})_{j \in J}$ be two Parseval frames for a Hilbert space $\mathcal{H}$, and suppose that $x \in \mathcal{H}$ can be decomposed as $x = x_1^0 + x_2^0$. Further, let $\Lambda_1 \subseteq I$ and $\Lambda_2 \subseteq J$ be chosen such that x_1^0 and x_2^0 are δ-relatively sparse in Φ_1 and Φ_2 with respect to Λ_1 and Λ_2. Then the solution $(x_1^\star, x_2^\star)$ of the minimization problem (Sep_a) stated in (11.6) satisfies*

$$\|x_1^\star - x_1^0\|_2 + \|x_2^\star - x_2^0\|_2 \leq \frac{2\delta}{1-2\mu_c},$$

with

$$\mu_c = \max\{\mu_c(\Lambda_1, \Phi_1; \Phi_2), \mu_c(\Phi_1; \Lambda_2, \Phi_2)\}.$$

To thoroughly understand this estimate, it is important to notice that both relative sparsity δ as well as cluster coherence μ_c depend heavily on the choice of the sets of significant coefficients Λ_1 and Λ_2. Choosing those sets too large allows for a very small δ, however μ_c might not be less than $1/2$ anymore, thereby making the estimate useless. Choosing those sets too small will force μ_c to become simultaneously small, in particular, smaller than $1/2$, with the downside that δ might be large.

It is also essential to realize that the sets Λ_1 and Λ_2 are a mere analysis tool; they do not appear in the minimization problem (Sep_a). This means that the algorithm does not care about this choice at all, however the estimate for accuracy of separation does.

Also note that this result can be easily generalized to general frames instead of Parseval frames, which then changes the separation estimate by invoking the lower frame bound. In addition, a version including noise was derived in [14].

11.2.4 Relation with uncertainty principles

Intriguingly, there exists a very close connection between uncertainty principles and data separation problems. Given a signal $x \in \mathcal{H}$ and two bases or frames Φ_1 and Φ_2, loosely speaking, an uncertainty principle states that x cannot be sparsely represented by Φ_1 and Φ_2 simultaneously; one of the expansions is always not sparse unless $x = 0$. For the relation to the "classical" uncertainty principle, we refer to Subsection 11.3.1.

The first result making this uncertainty viewpoint precise was proven in [19] with ideas already lurking in [16] and [11]. Again, it turns out that the mutual coherence is an appropriate measure for allowed sparsity, here serving as a lower bound for the simultaneously achievable sparsity of two expansions.

THEOREM 11.5 ([19]) *Let Φ_1 and Φ_2 be two orthonormal bases for a Hilbert space $\mathcal{H}$, and let $x \in \mathcal{H}$, $x \neq 0$. Then*

$$\|\Phi_1^T x\|_0 + \|\Phi_2^T x\|_0 \geq \frac{2}{\mu([\Phi_1|\Phi_2])}.$$

Proof. First, let $\Phi_1 = (\varphi_{1i})_{i \in I}$ and $\Phi_2 = (\varphi_{2j})_{j \in J}$. Further, let $\Lambda_1 \subseteq I$ and $\Lambda_2 \subset J$ denote the support of $\Phi_1^T x$ and $\Phi_2^T x$, respectively. Since $x = \Phi_1 \Phi_1^T x$, for each $j \in J$,

$$|(\Phi_2^T x)_j| = \left| \sum_{i \in \Lambda_1} (\Phi_1^T x)_i \langle \varphi_{1i}, \varphi_{2j} \rangle \right|. \tag{11.13}$$

Since Φ_1 and Φ_2 are orthonormal bases, we have

$$\|x\|_2 = \|\Phi_1^T x\|_2 = \|\Phi_2^T x\|_2. \tag{11.14}$$

Using in addition the Cauchy–Schwarz inequality, we can continue (11.13) by

$$|(\Phi_2^T x)_j|^2 \leq \|\Phi_1^T x\|_2^2 \cdot \left| \sum_{i \in \Lambda_1} |\langle \varphi_{1i}, \varphi_{2j} \rangle|^2 \right| \leq \|x\|_2^2 \cdot |\Lambda_1| \cdot \mu([\Phi_1|\Phi_2])^2.$$

This implies

$$\|\Phi_2^T x\|_2 = \left(\sum_{j \in \Lambda_2} |(\Phi_2^T x)_j|^2 \right)^{1/2} \leq \|x\|_2 \cdot \sqrt{|\Lambda_1| \cdot |\Lambda_2|} \cdot \mu([\Phi_1|\Phi_2]).$$

Since $|\Lambda_i| = \|\Phi_i^T x\|_0$, $i = 1, 2$, and again using (11.14), we obtain

$$\sqrt{\|\Phi_1^T x\|_0 \cdot \|\Phi_2^T x\|_0} \geq \frac{1}{\mu([\Phi_1|\Phi_2])}.$$

Using the geometric–algebraic relationship,

$$\frac{1}{2}(\|\Phi_1^T x\|_0 + \|\Phi_2^T x\|_0) \geq \sqrt{\|\Phi_1^T x\|_0 \cdot \|\Phi_2^T x\|_0} \geq \frac{1}{\mu([\Phi_1|\Phi_2])},$$

which proves the claim. □

This result can be easily connected to the problem of simultaneously sparse expansions. The following version was first explicitly stated in [4].

THEOREM 11.6 ([4]) *Let Φ_1 and Φ_2 be two orthonormal bases for a Hilbert space $\mathcal{H}$, and let $x \in \mathcal{H}$, $x \neq 0$. Then, for any two distinct coefficient sequences c_i satisfying $x = [\Phi_1|\Phi_2]c_i$, $i = 1, 2$, we have*

$$\|c_1\|_0 + \|c_2\|_0 \geq \frac{2}{\mu([\Phi_1|\Phi_2])}.$$

Proof. First, set $d = c_1 - c_2$ and partition d into $[d_{\Phi_1}, d_{\Phi_2}]^T$ such that

$$0 = [\Phi_1|\Phi_2]d = \Phi_1 d_{\Phi_1} + \Phi_2 d_{\Phi_2}.$$

Since Φ_1 and Φ_2 are bases and $d \neq 0$, the vector y defined by

$$y = \Phi_1 d_{\Phi_1} = -\Phi_2 d_{\Phi_2}$$

is nonzero. Applying Theorem 11.5, we obtain

$$\|d\|_0 = \|d_{\Phi_1}\|_0 + \|d_{\Phi_2}\|_0 \geq \frac{2}{\mu([\Phi_1|\Phi_2])}.$$

Since $d = c_1 - c_2$, we have

$$\|c_1\|_0 + \|c_2\|_0 \geq \|d\|_0 \geq \frac{2}{\mu([\Phi_1|\Phi_2])}.$$

□

We would also like to mention the very recent paper [39] by Tropp, in which he studies uncertainty principles for random sparse signals over an incoherent dictionary. He, in particular, shows that the coefficient sequence of each non-optimal expansion of a signal contains far more nonzero entries than the one of the sparsest expansion.

11.3 Signal separation

In this section, we study the special situation of signal separation, where we refer to 1-D signals as opposed to images, etc. For this, we start with the most prominent example of separating sinusoids from spikes, and then discuss further problem classes.

11.3.1 Separation of sinusoids and spikes

Sinusoidal and spike components are intuitively the morphologically most distinct features of a signal, since one is periodic and the other transient. Thus, it seems natural that the first results using sparsity and ℓ_1 minimization for data separation were proven for this situation. Certainly, real-world signals are never a pristine combination of sinusoids and spikes. However, thinking of audio data from a recording of musical instruments, these components are indeed an essential part of such signals.

The separation problem can be generally stated in the following way: Let the vector $x \in \mathbb{R}^n$ consist of n samples of a continuum domain signal at times $t \in \{0, \ldots, n-1\}$. We assume that x can be decomposed into

$$x = x_1^0 + x_2^0.$$

Here x_1^0 shall consist of n samples – at the same points in time as x – of a continuum domain signal of the form

$$\frac{1}{\sqrt{n}} \sum_{\omega=0}^{n-1} c_{1\omega}^0 e^{2\pi i \omega t / n}, \quad t \in \mathbb{R}.$$

Thus, by letting $\Phi_1 = (\varphi_{1\omega})_{0 \leq \omega \leq n-1}$ denote the Fourier basis, i.e.,

$$\varphi_{1\omega} = \left(\tfrac{1}{\sqrt{n}} e^{2\pi i \omega t / n} \right)_{0 \leq t \leq n-1},$$

the discrete signal x_1^0 can be written as

$$x_1^0 = \Phi_1 c_1^0 \quad \text{with } c_1^0 = (c_{1\omega}^0)_{0 \leq \omega \leq n-1}.$$

If x_1^0 is now the superposition of very few sinusoids, then the coefficient vector c_1^0 is sparse.

Further, consider a continuum domain signal which has a few spikes. Sampling this signal at n samples at times $t \in \{0, \ldots, n-1\}$ leads to a discrete signal $x_2^0 \in \mathbb{R}^n$ which has very few nonzero entries. In order to expand x_2^0 in terms of a suitable representation system, we let Φ_2 denote the Dirac basis, i.e., Φ_2 is simply the identity matrix, and write

$$x_2^0 = \Phi_2 c_2^0,$$

where c_2^0 is then a sparse coefficient vector.

The task now consists in extracting x_1^0 and x_2^0 from the known signal x, which is illustrated in Figure 11.1. It will be illuminating to detect the dependence on the number of sampling points of the bound for the sparsity of c_1^0 and c_2^0 which still allows for separation via ℓ_1 minimization.

The intuition that – from a morphological standpoint – this situation is extreme, can be seen by computing the mutual coherence between the Fourier basis Φ_1 and the Dirac basis Φ_2. For this, we obtain

$$\mu([\Phi_1 | \Phi_2]) = \frac{1}{\sqrt{n}}, \qquad (11.15)$$

and, in fact, $1/\sqrt{n}$ is the minimal possible value. This can be easily seen: If Φ_1 and Φ_2 are two general orthonormal bases of $\mathbb{R}^n$, then $\Phi_1^T \Phi_2$ is an orthonormal matrix. Hence

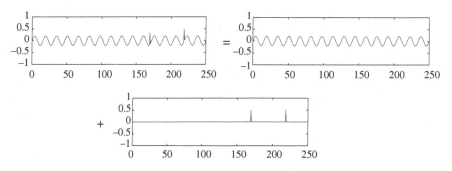

Figure 11.1 Separation of artificial audio data into sinusoids and spikes.

the sum of squares of its entries equals n, which implies that all entries can not be less than $1/\sqrt{n}$.

The following result from [19] makes this dependence precise. We wish to mention that the first answer to this question was derived in [11]. In this paper the slightly weaker bound of $(1+\sqrt{n})/2$ for $\|c_1^0\|_0 + \|c_2^0\|_0$ was proven by using the general result in Theorem 11.1 instead of the more specialized Theorem 11.2 exploited to derive the result from [19] stated below.

THEOREM 11.7 ([19]) *Let Φ_1 be the Fourier basis for $\mathbb{R}^n$ and let Φ_2 be the Dirac basis for $\mathbb{R}^n$. Further, let $x \in \mathbb{R}^n$ be the signal*

$$x = x_1^0 + x_2^0, \quad \text{where } x_1^0 = \Phi_1 c_1^0 \text{ and } x_2^0 = \Phi_2 c_2^0,$$

with coefficient vectors $c_i^0 \in \mathbb{R}^n$, $i = 1, 2$. If

$$\|c_1^0\|_0 + \|c_2^0\|_0 < (\sqrt{2} - 0.5)\sqrt{n},$$

then the ℓ_1 minimization problem (Sep_s) stated in (11.3) recovers c_1^0 and c_2^0 uniquely, and hence extracts x_1^0 and x_2^0 from x precisely.

Proof. Recall that we have (cf. (11.15))

$$\mu([\Phi_1|\Phi_2]) = \frac{1}{\sqrt{n}}.$$

Hence, by Theorem 11.2, the ℓ_1 minimization problem (Sep_s) recovers c_1^0 and c_2^0 uniquely, provided that

$$\|c_1^0\|_0 + \|c_2^0\|_0 < \frac{\sqrt{2} - 0.5}{\mu([\Phi_1|\Phi_2])} = (\sqrt{2} - 0.5)\sqrt{n}.$$

The theorem is proved. □

The classical uncertainty principle states that, roughly speaking, a function cannot both be localized in time as well as in frequency domain. A discrete version of this fundamental principle was – besides the by now well-known continuum domain Donoho–Stark uncertainty principle – derived in [16]. It showed that a discrete signal and its Fourier transform cannot both be highly localized in the sense of having "very few" nonzero entries. We will now show that this result – as it was done in [11] – can be interpreted as a corollary from data separation results.

THEOREM 11.8   ([16]) *Let $x \in \mathbb{R}^n$, $x \neq 0$, and denote its Fourier transform by $\hat{x}$. Then*

$$\|x\|_0 + \|\hat{x}\|_0 \geq 2\sqrt{n}.$$

Proof. For the proof, we intend to use Theorem 11.5. First, we note that by letting Φ_1 denote the Dirac basis, we trivially have

$$\|\Phi_1^T x\|_0 = \|x\|_0.$$

Secondly, letting Φ_2 denote the Fourier basis, we obtain

$$\hat{x} = \Phi_2^T x.$$

Now recalling that, by (11.15),

$$\mu([\Phi_1|\Phi_2]) = \frac{1}{\sqrt{n}},$$

we can conclude from Theorem 11.5 that

$$\|x\|_0 + \|\hat{x}\|_0 = \|\Phi_1^T x\|_0 + \|\Phi_2^T x\|_0 \geq \frac{2}{\mu([\Phi_1|\Phi_2])} = 2\sqrt{n}.$$

This finishes the proof.   □

As an excellent survey about sparsity of expansions of signals in the Fourier and Dirac bases, data separation, and related uncertainty principles as well as on very recent results using random signals, we refer to [38].

11.3.2   Further variations

Let us briefly mention the variety of modifications of the previously discussed setting, most of them empirical analyses, which were developed during the last few years.

The most common variation of the sinusoid and spike setting is the consideration of a more general periodic component, which is then considered to be sparse in a Gabor system, superimposed by a second component, which is considered to be sparse in a system sensitive to spike-like structures similar to wavelets. This is, for instance, the situation considered in [22]. An example for a different setting is the substitution of a Gabor system by a Wilson basis, analyzed in [3]. In this paper, as mentioned in Subsection 11.2.3, the clustering of coefficients already plays an essential role. It should also be mentioned that a specifically adapted norm, namely the mixed $\ell_{1,2}$ or $\ell_{2,1}$ norm, is used in [25] to take advantage of this clustering, and various numerical experiments show successful separation.

11.4 Image separation

This section is devoted to discussing results on image separation exploiting Morphological Component Analysis, first focussing on empirical studies and secondly on theoretical results.

11.4.1 Empirical results

In practice, the observed signal x is often contaminated by noise, i.e., $x = x_1^0 + x_2^0 + n$ containing the to-be-extracted components x_1^0 and x_2^0 and some noise n. This requires an adaption of the ℓ_1 minimization problem. As proposed in numerous publications, one typically considers a modified optimization problem – so-called *Basis Pursuit Denoising* – which can be obtained by relaxing the constraint in order to deal with noisy observed signals. The ℓ_1 minimization problem (Sep$_s$) stated in (11.3), which places the ℓ_1 norm on the *synthesis* side then takes the form:

$$\min_{c_1, c_2} \|c_1\|_1 + \|c_2\|_1 + \lambda \|x - \Phi_1 c_1 - \Phi_2 c_2\|_2^2$$

with appropriately chosen regularization parameter $\lambda > 0$. Similarly, we can consider the relaxed form of the ℓ_1 minimization problem (Sep$_a$) stated in (11.6), which places the ℓ_1 norm on the *analysis* side:

$$\min_{x_1, x_2} \|\Phi_1^T x_1\|_1 + \|\Phi_2^T x_2\|_1 + \lambda \|x - x_1 - x_2\|_2^2.$$

In these new forms, the additional content in the image – the noise – characterized by the property that it cannot be represented sparsely by either one of the two systems Φ_1 and Φ_2, will be allocated to the residual $x - \Phi_1 c_1 - \Phi_2 c_2$ or $x - x_1 - x_2$ depending on which of the two minimization problems stated above is chosen. Hence, performing this minimization, we not only separate the data, but also succeed in removing an additive noise component as a by-product.

There exists by now a variety of algorithms which numerically solve such minimization problems. One large class is, for instance, iterative shrinkage algorithms; and we refer to the beautiful new book [18] by Elad for an overview. It should be mentioned that it is also possible to perform these separation procedures locally, thus enabling parallel processing, and again we refer to [18] for further details.

Let us now delve into more concrete situations. One prominent class of empirical studies concerns the separation of point- and curvelike structures. This type of problem arises, for instance, in astronomical imaging, where astronomers would like to separate stars (pointlike structures) from filaments (curvelike structures). Another area in which the separation of points from curves is essential is neurobiological imaging. In particular, for Alzheimer research, neurobiologists analyze images of neurons, which – considered in 2-D – are a composition of the dendrites (curvelike structures) of the neuron and the attached spines (pointlike structures). For further analysis of the shape of these components, dendrites and spines need to be separated.

From a mathematical perspective, pointlike structures are generally speaking 0-D structures whereas curvelike structures are 1-D structures, which reveals their morphological difference. Thus it seems conceivable that separation using the idea of Morphological Component Analysis can be achieved, and the empirical results presented in the sequel as well as the theoretical results discussed in Subsection 11.4.2 give evidence to this claim.

To set up the minimization problem properly, the question arises which systems adapted to the point- and curvelike objects to use. For extracting pointlike structures, wavelets seem to be optimal, since they provide optimally sparse approximations of smooth functions with finitely many point singularities. As a sparsifying system for curvelike structures, two different possibilities have been explored so far. From a historical perspective, the first system to be utilized was *curvelets* [5], which provide optimally sparse approximations of smooth functions exhibiting curvilinear singularities. The composed dictionary of wavelets-curvelets is used in MCALab,[1] and implementation details are provided in the by now considered fundamental paper [35]. A few years later *shearlets* were developed, see [24] or the survey paper [27], which deal with curvilinear singularities in a similarly favorable way as curvelets (cf. [28]), but have, for instance, the advantage of providing a unified treatment of the continuum and digital realm and being associated with a fast transform. Separation using the resulting dictionary of wavelets-shearlets is implemented and publicly available in ShearLab.[2] For a close comparison between both approaches we refer to [29] – in this paper the separation algorithm using wavelets and shearlets is also detailed –, where a numerical comparison shows that ShearLab provides a faster as well as more precise separation.

For illustrative purposes, Figure 11.2 shows the separation of an artificial image composed of points, lines, and a circle as well as added noise into the pointlike structures (points) and the curvelike structures (lines and the circle), while removing the noise simultaneously. The only visible artifacts can be seen at the intersections of the curvelike structures, which is not surprising since it is even justifiable to label these intersections as "points." As an example using real data, we present in Figure 11.3 the separation of a neuron image into dendrites and spines again using ShearLab.

Another widely explored category of image separation is the separation of cartoons and texture. Here, the term cartoon typically refers to a piecewise smooth part in the image, and texture means a periodic structure. A mathematical model for a *cartoon* was first introduced in [8] as a C^2 function containing a C^2 discontinuity. In contrast to this, the term *texture* is a widely open expression, and people have debated for years over an appropriate model for the texture content of an image. A viewpoint from applied harmonic analysis characterizes texture as a structure which exhibits a sparse expansion in a Gabor system. As a side remark, the reader should be aware that periodizing a cartoon part of an image produces a texture component, thereby revealing the very fine line between cartoons and texture, illustrated in Figure 11.4.

---

[1] MCALab (Version 120) is available from http://jstarck.free.fr/jstarck/Home.html.
[2] ShearLab (Version 1.1) is available from www.shearlab.org.

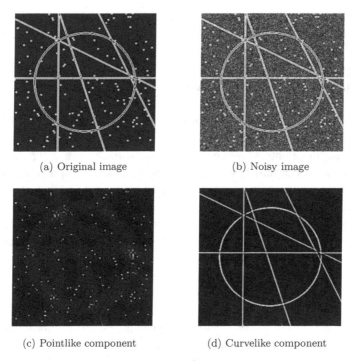

Figure 11.2 Separation of an artificial image composed of points, lines, and a circle into point- and curvelike components using ShearLab.

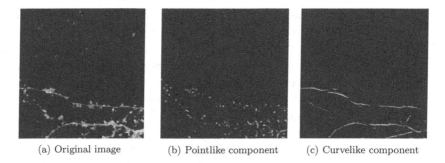

Figure 11.3 Separation of a neuron image into point- and curvelike components using ShearLab.

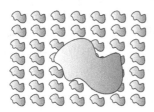

Figure 11.4 Periodic small cartoons versus one large cartoon.

(a) Barbara image   (b) Cartoon component   (c) Texture component

Figure 11.5 Separation of the Barbara image into cartoon and texture using MCALab.

As sparsifying systems, again curvelets or shearlets are suitable for the cartoon part, whereas discrete cosines or a Gabor system can be used for the texture part. MCALab uses for this separation task a dictionary composed of curvelets and discrete cosines, see [35]. For illustrative purposes, we display in Figure 11.5 the separation of the Barbara image into cartoon and texture components performed by MCALab. As can be seen, all periodic structure is captured in the texture part, leaving the remainder to the cartoon component.

11.4.2 Theoretical results

The first theoretical result explaining the successful empirical performance of Morphological Component Analysis was derived in [14] by considering the separation of point- and curvelike features in images coined the *Geometric Separation Problem*. The analysis in this paper has three interesting features. Firstly, it introduces the notion of cluster coherence (cf. Definition 11.3) as a measure for the geometric arrangements of the significant coefficients and hence the encoding of the morphological difference of the components. It also initiates the study of ℓ_1 minimization in frame settings, in particular those where singleton coherence within one frame may be high. Secondly, it provides the first analysis of a continuum model in contrast to the previously studied discrete models which obscure continuum elements of geometry. And thirdly, it explores microlocal analysis to understand heuristically why separation might be possible and to organize a rigorous analysis. This general approach applies in particular to two variants of geometric separation algorithms. One is based on tight frames of radial wavelets and curvelets and the other uses orthonormal wavelets and shearlets.

These results are today the only results providing a theoretical foundation to image separation using ideas from sparsity methodologies. The same situation – separating point- and curvelike objects – is also considered in [13] however using thresholding as a separation technique. Finally, we wish to mention that some initial theoretical results on the separation of cartoon and texture in images are contained in [15].

Let us now dive into the analysis of [14]. As a mathematical model for a composition of point- and curvelike structures, the following two components are considered: The function $\mathcal{P}$ on $\mathbb{R}^2$, which is smooth except for point singularities and defined by

$$\mathcal{P} = \sum_{i=1}^{P} |x - x_i|^{-3/2},$$

serves as a model for the pointlike objects, and the distribution $\mathcal{C}$ with singularity along a closed curve $\tau : [0,1] \to \mathbb{R}^2$ defined by

$$\mathcal{C} = \int \delta_{\tau(t)} dt,$$

models the curvelike objects. The general model for the considered situation is then the sum of both, i.e.,

$$f = \mathcal{P} + \mathcal{C}, \tag{11.16}$$

and the *Geometric Separation Problem* consists of recovering $\mathcal{P}$ and $\mathcal{C}$ from the observed signal f.

As discussed before, one possibility is to set up the minimization problem using an overcomplete system composed of wavelets and curvelets. For the analysis, radial wavelets are used due to the fact that they provide the same subbands as curvelets. To be more precise, let W be an appropriate window function. Then *radial wavelets* at scale j and spatial position $k = (k_1, k_2)$ are defined by the Fourier transforms

$$\hat{\psi}_\lambda(\xi) = 2^{-j} \cdot W(|\xi|/2^j) \cdot e^{i\langle k, \xi/2^j \rangle},$$

where $\lambda = (j, k)$ indexes scale and position. For the same window function W and a "bump function" V, *curvelets* at scale j, orientation ℓ, and spatial position $k = (k_1, k_2)$ are defined by the Fourier transforms

$$\hat{\gamma}_\eta(\xi) = 2^{-j\frac{3}{4}} \cdot W(|\xi|/2^j) V((\omega - \theta_{j,\ell}) 2^{j/2}) \cdot e^{i(R_{\theta_{j,\ell}} A_{2^{-j}} k)' \xi},$$

where $\theta_{j,\ell} = 2\pi\ell/2^{j/2}$, R_θ is planar rotation by $-\theta$ radians, A_a is anisotropic scaling with diagonal $(a, \sqrt{a})$, and we let $\eta = (j, \ell, k)$ index scale, orientation, and scale; see [5] for more details. The tiling of the frequency domain generated by these two systems is illustrated in Figure 11.6.

By using again the window W, we define the family of filters F_j by their transfer functions

$$\hat{F}_j(\xi) = W(|\xi|/2^j), \qquad \xi \in \mathbb{R}^2.$$

These filters provide a decomposition of any distribution g into pieces g_j with different scales, the piece g_j at subband j generated by filtering g using F_j:

$$g_j = F_j \star g.$$

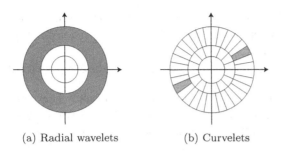

(a) Radial wavelets   (b) Curvelets

Figure 11.6 Tiling of the frequency domain by radial wavelets and curvelets.

A proper choice of W then enables reconstruction of g from these pieces using the formula

$$g = \sum_j F_j \star g_j.$$

Application of this filtering procedure to the model image f from (11.16) yields the decompositions

$$f_j = F_j \star f = F_j \star (\mathcal{P} + \mathcal{C}) = \mathcal{P}_j + \mathcal{C}_j,$$

where $(f_j)_j$ is known, and we aim to extract $(\mathcal{P}_j)_j$ and $(\mathcal{C}_j)_j$. We should mention at this point that, in fact, the pair $(\mathcal{P}, \mathcal{C})$ was chosen in such a way that $\mathcal{P}_j$ and $\mathcal{C}_j$ have the same energy for each j, thereby making the components comparable as we go to finer scales and the separation challenging at *each* scale.

Let now Φ_1 and Φ_2 be the tight frames of radial wavelets and curvelets, respectively. Then, for each scale j, we consider the ℓ_1 minimization problem (Sep_a) stated in (11.6), which now reads:

$$\min_{\mathcal{P}_j, \mathcal{C}_j} \|\Phi_1^T \mathcal{P}_j\|_1 + \|\Phi_2^T \mathcal{C}_j\|_1 \quad \text{s.t.} \quad f_j = \mathcal{P}_j + \mathcal{C}_j. \tag{11.17}$$

Notice that we use the "analysis version" of the minimization problem, since both radial wavelets as well as curvelets are overcomplete systems.

The theoretical result of the precision of separation of f_j via (11.17) proved in [14] can now be stated in the following way:

THEOREM 11.9 ([14]) *Let $\hat{\mathcal{P}}_j$ and $\hat{\mathcal{C}}_j$ be the solutions to the optimization problem (11.17) for each scale j. Then we have*

$$\frac{\|\mathcal{P}_j - \hat{\mathcal{P}}_j\|_2 + \|\mathcal{C}_j - \hat{\mathcal{C}}_j\|_2}{\|\mathcal{P}_j\|_2 + \|\mathcal{C}_j\|_2} \to 0, \quad j \to \infty.$$

This result shows that the components $\mathcal{P}_j$ and $\mathcal{C}_j$ are recovered with asymptotically arbitrarily high precision at very fine scales. The energy in the pointlike component is completely captured by the wavelet coefficients, and the curvelike component is completely contained in the curvelet coefficients. Thus, the theory evidences that the

Geometric Separation Problem can be satisfactorily solved by using a combined dictionary of wavelets and curvelets and an appropriate ℓ_1 minimization problem, as already the empirical results indicate.

We next provide a sketch of proof and refer to [14] for the complete proof.

Sketch of proof of Theorem 11.9. The main goal will be to apply Theorem 11.4 to each scale and monitor the sequence of bounds $\frac{2\delta}{1-2\mu_c}$ now dependent on scale. For this, let j be arbitrarily fixed, and apply Theorem 11.4 in the following way:

- x: Filtered signal f_j $(=\mathcal{P}_j + \mathcal{C}_j)$.
- Φ_1: Wavelets filtered with F_j.
- Φ_2: Curvelets filtered with F_j.
- Λ_1: Significant wavelet coefficients of $\mathcal{P}_j$.
- Λ_2: Significant curvelet coefficients of $\mathcal{C}_j$.
- δ_j: Degree of approximation by significant coefficients.
- $(\mu_c)_j$: Cluster coherence of wavelets-curvelets.

If
$$\frac{2\delta_j}{1-2(\mu_c)_j} = o(\|\mathcal{P}_j\|_2 + \|\mathcal{C}_j\|_2) \quad \text{as } j \to \infty \tag{11.18}$$

can be then shown, the theorem is proved.

One main problem to overcome is the highly delicate choice of Λ_1 and Λ_2. It would be ideal to define those sets in such a way that

$$\delta_j = o(\|\mathcal{P}_j\|_2 + \|\mathcal{C}_j\|_2) \quad \text{as } j \to \infty \tag{11.19}$$

and
$$(\mu_c)_j \to 0 \quad \text{as } j \to \infty \tag{11.20}$$

are true. This would then imply (11.18), hence finish the proof.

A microlocal analysis viewpoint now provides insight into how to suitably choose Λ_1 and Λ_2 by considering the wavefront sets of $\mathcal{P}$ and $\mathcal{C}$ in phase space $\mathbb{R}^2 \times [0, 2\pi)$, i.e.,

$$WF(\mathcal{P}) = \{x_i\}_{i=1}^P \times [0, 2\pi)$$

and
$$WF(\mathcal{C}) = \{(\tau(t), \theta(t)) : t \in [0, L(\tau)]\},$$

where $\tau(t)$ is a unit-speed parameterization of τ and $\theta(t)$ is the normal direction to τ at $\tau(t)$. Heuristically, the significant wavelet coefficients should be associated with wavelets whose index set is "close" to $WF(\mathcal{P})$ in phase space and, similarly, the significant curvelet coefficients should be associated with curvelets whose index set is "close" to $WF(\mathcal{C})$. Thus, using Hart Smith's phase space metric,

$$d_{HS}((b,\theta);(b',\theta')) = |\langle e_\theta, b-b'\rangle| + |\langle e_{\theta'}, b-b'\rangle| + |b-b'|^2 + |\theta-\theta'|^2,$$

where $e_\theta = (\cos(\theta), \sin(\theta))$, an "approximate" form of sets of significant wavelet coefficients is

$$\Lambda_{1,j} = \{\text{wavelet lattice}\} \cap \{(b,\theta) : d_{HS}((b,\theta); WF(\mathcal{P})) \leq \eta_j 2^{-j}\},$$

and an "approximate" form of sets of significant curvelet coefficients is

$$\Lambda_{2,j} = \{\text{curvelet lattice}\} \cap \{(b,\theta) : d_{HS}((b,\theta); WF(\mathcal{C})) \leq \eta_j 2^{-j}\}$$

with a suitable choice of the distance parameters $(\eta_j)_j$. In the proof of Theorem 11.9, the definition of $(\Lambda_{1,j})_j$ and $(\Lambda_{2,j})_j$ is much more delicate, but follows this intuition. Lengthy and technical estimates then lead to (11.19) and (11.20), which – as mentioned before – completes the proof. □

Since it was already mentioned in Subsection 11.4.1 that a combined dictionary of wavelets and shearlets might be preferable, the reader will wonder whether the just discussed theoretical results can be transferred to this setting. In fact, this is proven in [26], see also [12]. It should be mentioned that one further advantage of this setting is the fact that now a basis of wavelets can be utilized in contrast to the tight frame of radial wavelets explored before.

As a wavelet basis, we now choose *orthonormal Meyer wavelets*, and refer to [30] for the definition. For the definition of shearlets, for $j \geq 0$ and $k \in \mathbb{Z}$, let – the notion A_{2^j} was already introduced in the definition of curvelets – $\tilde{A}_{2^j}$ and S_k be defined by

$$\tilde{A}_{2^j} = \begin{pmatrix} 2^{j/2} & 0 \\ 0 & 2^j \end{pmatrix} \quad \text{and} \quad S_k = \begin{pmatrix} 1 & k \\ 0 & 1 \end{pmatrix}.$$

For $\phi, \psi, \tilde{\psi} \in L^2(\mathbb{R}^2)$, the *cone-adapted discrete shearlet system* is then the union of

$$\{\phi(\cdot - m) : m \in \mathbb{Z}^2\},$$

$$\{2^{\frac{3}{4}j}\psi(S_k A_{2^j} \cdot - m) : j \geq 0, -\lceil 2^{j/2}\rceil \leq k \leq \lceil 2^{j/2}\rceil, m \in \mathbb{Z}^2\},$$

and

$$\{2^{\frac{3}{4}j}\tilde{\psi}(S_k^T \tilde{A}_{2^j} \cdot - m) : j \geq 0, -\lceil 2^{j/2}\rceil \leq k \leq \lceil 2^{j/2}\rceil, m \in \mathbb{Z}^2\}.$$

The term "cone-adapted" originates from the fact that these systems tile the frequency domain in a cone-like fashion; see Figure 11.7(b).

As can be seen from Figure 11.7, the subbands associated with orthonormal Meyer wavelets and shearlets are the same. Hence a similar filtering into scaling subbands can be performed as for radial wavelets and curvelets.

Adapting the optimization problem (11.17) by using wavelets and shearlets instead of radial wavelets and curvelets generates purported point- and curvelike objects $\hat{W}_j$ and $\hat{S}_j$, say, for each scale j. Then the following result, which shows similarly successful separation as Theorem 11.9, was derived in [26] with the new concept of sparsity equivalence, here between shearlets and curvelets, introduced in the same paper as main ingredient.

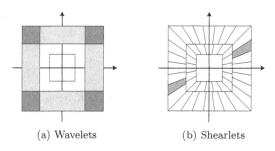

(a) Wavelets (b) Shearlets

Figure 11.7 Tiling of the frequency domain by orthonormal Meyer wavelets and shearlets.

THEOREM 11.10 ([26]) *We have*

$$\frac{\|\mathcal{P}_j - \hat{W}_j\|_2 + \|\mathcal{C}_j - \hat{S}_j\|_2}{\|\mathcal{P}_j\|_2 + \|\mathcal{C}_j\|_2} \to 0, \quad j \to \infty.$$

Acknowledgements

The author would like to thank Ronald Coifman, Michael Elad, and Remi Gribonval for various discussions on related topics, Wang-Q Lim for producing Figures 11.2, 11.3, and 11.5, and Roland Brandt and Fred Sündermann for Figure 11.3(a). Special thanks go to David Donoho for a great collaboration on topics in this area and enlightening debates, and to Michael Elad for very useful comments on an earlier version of this survey. The author is also grateful to the Department of Statistics at Stanford University and the Mathematics Department at Yale University for their hospitality and support during her visits. She acknowledges support by Deutsche Forschungsgemeinschaft (DFG) Heisenberg fellowship KU 1446/8, DFG Grant SPP-1324 KU 1446/13, and DFG Grant KU 1446/14.

References

[1] M. Aharon, M. Elad, and A. M. Bruckstein. The K-SVD: An algorithm for designing of overcomplete dictionaries for sparse representation. *IEEE Trans Signal Proc*, 54(11): 4311–4322, 2006.

[2] J. Bobin, J.-L. Starck, M. J. Fadili, Y. Moudden, and D. L. Donoho. Morphological component analysis: An adaptive thresholding strategy. *IEEE Trans Image Proc*, 16(11):2675–2681, 2007.

[3] L. Borup, R. Gribonval, and M. Nielsen. Beyond coherence: Recovering structured time-frequency representations. *Appl Comput Harmon Anal*, 24(1):120–128, 2008.

[4] A. M. Bruckstein, D. L. Donoho, and M. Elad. From sparse solutions of systems of equations to sparse modeling of signals and images. *SIAM Review*, 51(1):34–81, 2009.

[5] E. J. Candès and D. L. Donoho. Continuous curvelet transform: II. Discretization of frames. *Appl Comput Harmon Anal*, 19(2):198–222, 2005.

[6] S. S. Chen, D. L. Donoho, and M. A. Saunders. Atomic decomposition by basis pursuit. *SIAM J Sci Comput*, 20(1):33–61, 1998.

[7] R. R. Coifman and M. V. Wickerhauser. Wavelets and adapted waveform analysis. A toolkit for signal processing and numerical analysis. *Different Perspectives on Wavelets.* San Antonio, TX, 1993, 119–153, *Proc Symps Appl Math*, 47, Am Math Soc, Providence, RI, 1993.

[8] D. L. Donoho. Sparse components of images and optimal atomic decomposition. *Constr Approx*, 17(3):353–382, 2001.

[9] D. L. Donoho. Compressed sensing. *IEEE Trans Inform Theory*, 52(4):1289–1306, 2006.

[10] D. L. Donoho and M. Elad. Optimally sparse representation in general (nonorthogonal) dictionaries via l^1 minimization. *Proc Natl Acad Sci USA*, 100(5):2197–2202, 2003.

[11] D. L. Donoho and X. Huo. Uncertainty principles and ideal atomic decomposition. *IEEE Trans Inform Theory*, 47(7):2845–2862, 2001.

[12] D. L. Donoho and G. Kutyniok. Geometric separation using a wavelet-shearlet dictionary. *SampTA'09* (Marseilles, France, 2009), Proc., 2009.

[13] D. L. Donoho and G. Kutyniok. Geometric separation by single-pass alternating thresholding, preprint, 2010.

[14] D. L. Donoho and G. Kutyniok. Microlocal analysis of the geometric separation problem. *Comm Pure Appl Math*, to appear, 2010.

[15] D. L. Donoho and G. Kutyniok. Geometric separation of cartoons and texture via ℓ_1 minimization, preprint, 2011.

[16] D. L. Donoho and P. B. Stark. Uncertainty principles and signal recovery. *SIAM J Appl Math*, 49(3):906–931, 1989.

[17] J. M. Duarte-Carvajalino and G. Sapiro. Learning to sense sparse signals: Simultaneous sensing matrix and sparsifying dictionary optimization. *IEEE Trans Image Proc*, 18(7):1395–1408, 2009.

[18] M. Elad. *Sparse and Redundant Representations*, Springer, New York, 2010.

[19] M. Elad and A. M. Bruckstein. A generalized uncertainty principle and sparse representation in pairs of bases. *IEEE Trans Inform Theory*, 48(9):2558–2567, 2002.

[20] M. Elad, J.-L. Starck, P. Querre, and D. L. Donoho. Simultaneous cartoon and texture image inpainting using morphological component analysis (MCA). *Appl Comput Harmon Anal*, 19(3):340–358, 2005.

[21] K. Engan, S. O. Aase, and J. H. Hakon-Husoy. Method of optimal directions for frame design. *IEEE Int Conf Acoust, Speech, Sig Process*, 5:2443–2446, 1999.

[22] R. Gribonval and E. Bacry. Harmonic decomposition of audio signals with matching pursuit. *IEEE Trans Sig Proc*, 51(1):101–111, 2003.

[23] R. Gribonval and M. Nielsen. Sparse representations in unions of bases. *IEEE Trans Inform Theory*, 49(12):3320–3325, 2003.

[24] K. Guo, G. Kutyniok, and D. Labate. Sparse multidimensional representations using anisotropic dilation and shear operators. *Wavelets and Splines*, Athens, GA, 2005, Nashboro Press, Nashville, TN, 2006:189–201, 2006.

[25] M. Kowalski and B. Torrésani. Sparsity and persistence: Mixed norms provide simple signal models with dependent coefficients. *Sig, Image Video Proc*, to appear, 2010.

[26] G. Kutyniok. Sparsity equivalence of anisotropic decompositions, preprint, 2010.

[27] G. Kutyniok, J. Lemvig, and W.-Q. Lim. Compactly supported shearlets, *Approximation Theory XIII* (San Antonio, TX, 2010). *Proc Math*, 13:163–186, 2012.

[28] G. Kutyniok and W.-Q. Lim. Compactly supported shearlets are optimally sparse. *J Approx Theory*, 163:1564–1589, 2011.

[29] G. Kutyniok and W.-Q. Lim. Image separation using shearlets, *Curves and Surfaces*, Avignon, France, 2010, Lecture Notes in Computer Science, 6920, Springer, 2012.

[30] S. G. Mallat. *A Wavelet Tour of Signal Processing*, Academic Press, Inc., San Diego, CA, 1998.

[31] S. G. Mallat and Z. Zhang. Matching pursuits with time-frequency dictionaries. *IEEE Trans Sig Proc*, 41(12):3397–3415, 1993.

[32] F. G. Meyer, A. Averbuch, and R. R. Coifman. Multi-layered image representation: Application to image compression. *IEEE Trans Image Proc*, 11(9):1072–1080, 2002.

[33] J.-L. Starck, F. Murtagh, and J. M. Fadili. *Sparse Image and Signal Processing: Wavelets, Curvelets, Morphological Diversity*. Cambridge University Press, New York, NY, 2010.

[34] J.-L. Starck, M. Elad, and D. L. Donoho. Redundant multiscale transforms and their application for morphological component analysis. *Adv Imag Electr Phys*, 132:287–348, 2005.

[35] J.-L. Starck, M. Elad, and D. L. Donoho. Image decomposition via the combination of sparse representations and a variational approach. *IEEE Trans Image Proc*, 14(10):1570–1582, 2005.

[36] J.-L. Starck, Y. Moudden, J. Bobin, M. Elad, and D. L. Donoho. Morphological component analysis. *Wavelets XI*, San Diego, CA, SPIE Proc. 5914, SPIE, Bellingham, WA, 2005.

[37] J. A. Tropp. Greed is good: Algorithmic results for sparse approximation. *IEEE Trans Inform Theory*, 50(10):2231–2242, 2004.

[38] J. A. Tropp. On the linear independence of spikes and sines. *J Fourier Anal Appl*, 14(5-6):838–858, 2008.

[39] J. A. Tropp. The sparsity gap: Uncertainty principles proportional to dimension. *Proc 44th IEEE Conf Inform Sci Syst* (CISS), 1–6, Princeton, NJ, 2010.

12 Face recognition by sparse representation

Arvind Ganesh, Andrew Wagner, Zihan Zhou, Allen Y. Yang,
Yi Ma, and John Wright

In this chapter, we present a comprehensive framework for tackling the classical problem of face recognition, based on theory and algorithms from sparse representation. Despite intense interest in the past several decades, traditional pattern recognition theory still stops short of providing a satisfactory solution capable of recognizing human faces in the presence of real-world nuisances such as occlusion and variabilities in pose and illumination. Our new approach, called sparse representation-based classification (SRC), is motivated by a very natural notion of sparsity, namely, one should always try to explain a query image using a small number of training images from a single subject category. This sparse representation is sought via ℓ_1 minimization. We show how this core idea can be generalized and extended to account for various physical variabilities encountered in face recognition. The end result of our investigation is a full-fledged practical system aimed at security and access control applications. The system is capable of accurately recognizing subjects out of a database of several hundred subjects with state-of-the-art accuracy.

12.1 Introduction

Automatic face recognition is a classical problem in the computer vision community. The community's sustained interest in this problem is mainly due to two reasons. First, in face recognition, we encounter many of the common variabilities that plague vision systems in general: illumination, occlusion, pose, and misalignment. Inspired by the good performance of humans in recognizing familiar faces [38], we have reason to believe that effective automatic face recognition is possible, and that the quest to achieve this will tell us something about visual recognition in general. Second, face recognition has a wide spectrum of practical applications. Indeed, if we could construct an extremely reliable automatic face recognition system, it would have broad implications for identity verification, access control, security, and public safety. In addition to these classical applications, the recent proliferation of online images and videos has provided a host

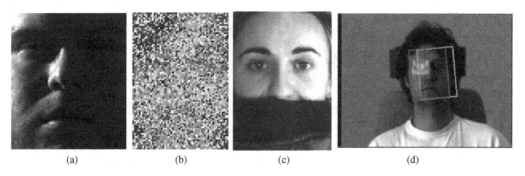

Figure 12.1 Examples of image nuisances in face recognition. (a) Illumination change. (b) Pixel corruption. (c) Facial disguise. (d) Occlusion and misalignment.

of new applications such as image search and photo tagging (e.g. Google's Picasa and Apple FaceTime).

Despite several decades of work in this area, high-quality automatic face recognition remains a challenging problem that defies satisfactory solutions. While there has been steady progress on scalable algorithms for face recognition in low-stake applications such as photo album organization,[1] there have been a sequence of well-documented failed trials of face recognition technology in mass surveillance/watch-list applications, where the performance requirements are very demanding.[2] These failures are mainly due to the challenging structure of face data: any real-world face recognition system must simultaneously deal with variables and nuisances such as illumination variation, corruption and occlusion, and reasonable amount of pose and image misalignment. Some examples of these image nuisances for face recognition are illustrated in Figure 12.1.

Traditional pattern recognition theory stops short of providing a satisfactory solution capable of simultaneously addressing all of these problems. In the past decades, numerous methods for handling a single mode of variability, such as pose or illumination, have been proposed and examined. But much less work has been devoted to simultaneously handling multiple modes of variation, according to a recent survey [52].[3] In other words, although a method might successfully deal with one type of variation, it quickly breaks down when moderate amounts of other variations are introduced to face images.

Recently, the theory of sparse representation and compressed sensing has shed some new light on this challenging problem. Indeed, there is a very natural notion of sparsity in the face recognition problem: one always tries to find only a single subject out of

---

[1] As documented, e.g., in the ongoing Labeled Faces in the Wild [26] challenge. We invite the interested reader to consult this work and the references therein.

[2] A typical performance metric that is considered acceptable for automatic mass surveillance may require both a recognition rate in the high 90s and a false positive rate lower than 0.01% over a database with thousands of subjects.

[3] The literature on face recognition is vast, and doing justice to all ideas and proposed algorithms would require a separate survey of comparable length to this chapter. In the course of this chapter, we will review a few works necessary to put ours in context. We refer the reader to [52] for a more comprehensive treatment of the history of the field.

a large database of subjects that best explains a given query image. In this chapter, we will discuss how tools from compressed sensing, especially ℓ_1 minimization and random projections, have inspired new algorithms for face recognition. In particular, the new computational framework can simultaneously address the most important types of variation in face recognition.

Nevertheless, face recognition diverges quite significantly from the common compressed sensing setup. On the mathematical side, the data matrices arising in face recognition often violate theoretical assumptions such as the restricted isometry property or even incoherence. Moreover, the physical structure of the problem (especially misalignment) will occasionally force us to solve the sparse representation problem subject to certain *nonlinear* constraints.

On the practical side, face recognition poses new nontrivial challenges in algorithm design and system implementation. First, face images are very high-dimensional data (e.g., a 1000×1000 gray-scale image has 10^6 pixels). Largely due to lack of memory and computational resource, dimensionality reduction techniques have largely been considered as a necessary step in the conventional face recognition methods. Notable holistic feature spaces include Eigenfaces [42], Fisherfaces [3], Laplacianfaces [25], and their variants [29, 10, 47, 36]. Nevertheless, it remains an open question: what is the optimal low-dimensional facial feature space that is capable of pairing with any well-designed classifier and leads to superior recognition performance?

Second, past face recognition algorithms often work well under laboratory conditions, but their performance would degrade drastically when tested in less-controlled environments – partially explaining some of the highly publicized failures of these systems. A common reason is that those face recognition systems were only tested on images taken under the same laboratory conditions (even with the same cameras) as the training images. Hence, their training sets do not represent well variations in illumination for face images taken under different indoor and outdoor environments, and under different lighting conditions. In some extreme cases, certain algorithms have attempted to reduce the illumination effect from only a single training image per subject [12, 53]. Despite these efforts, truly illumination-invariant features are in fact impossible to obtain from a few training images, let alone a single image [21, 4, 1]. Therefore, a natural question arises: How can we improve the image acquisition procedure to guarantee sufficient illuminations in the training images that can represent a large variety of real-world lighting conditions?

In this chapter, under the overarching theme of the book, we provide a systematic exposition of our investigation over the past few years into a new mathematical approach to face recognition, which we call *sparse representation-based classification* (SRC). We will start from a very simple, almost simplistic, problem formulation that is directly inspired by results in compressed sensing. We will see generalization of this approach naturally accounts for the various physical variabilities in the face recognition problem. In turn, we will see some of the new observations that face recognition can contribute to the mathematics of compressed sensing. The end result will be a full-fledged practical system aimed at applications in access control. The system is capable of accurately recognizing subjects out of a database of several hundred with high accuracy, despite large variations in illumination, moderate occlusion, and misalignment.

12.1.1 Organization of this chapter

In Section 12.2, starting with the simplest possible problem setting, we show how face recognition can be posed as a sparse representation problem. Section 12.3 discusses the possibility of solving this problem more efficiently by projecting the data into a randomly selected lower-dimensional feature space. In Sections 12.4 and 12.5, we then show how the SRC framework can be naturally extended to handle physical variabilities such as occlusion and misalignment, respectively. Sections 12.6 and 12.7 discuss practical aspects of building a face recognition system using the tools introduced here. Section 12.6 shows how to efficiently solve sparse representation problems arising in face recognition, while Section 12.7 discusses a practical system for acquiring training images of subjects under different illuminations. In Section 12.8, we combine these developments to give an end-to-end system for face recognition, aimed at access control tasks.

12.2 Problem formulation: sparse representation-based classification

In this section, we first illustrate the core idea of SRC in a slightly artificial scenario in which the training and test images are very well-aligned, but do contain significant variations in illumination. We will see how in this setting, face recognition can be naturally cast as a sparse representation problem, and solved via ℓ_1 minimization. In subsequent sections, we will see how this core formulation extends naturally to handle other variabilities in real-world face images, culminating in a complete system for recognition described in Section 12.7.

In face recognition, the system is given access to a set of labeled training images $\{\phi_i, l_i\}$ from the C subjects of interest. Here, $\phi_i \in \mathbb{R}^m$ is the vector representation of a digital image (say, by stacking the $W \times H$ columns of the single-channel image as a $m = W \times H$ dimensional vector), and $l_i \in \{1 \ldots C\}$ indicates which of the C subjects is pictured in the ith image. In the testing stage, a new query image $y \in \mathbb{R}^m$ is provided. The system's job is to determine which of the C subjects is pictured in this query image, or, if none of them is present, to reject the query sample as invalid.

Influential results due to [1, 21] suggest that if sufficiently many images of the same subject are available, these images will lie close to a low-dimensional linear subspace of the high-dimensional image space $\mathbb{R}^m$. The required dimension could be as low as nine for a convex, Lambertian object [1]. Hence, given sufficient diversity in training illuminations, the new test image y of subject i can be well represented as a linear combination of the training images of the same subject:

$$y \approx \sum_{\{j | l_j = i\}} \phi_j c_j \doteq \Phi_i c_i, \qquad (12.1)$$

where $\Phi_i \in \mathbb{R}^{m \times n_i}$ concatenates all of the images of subject i, and $c_i \in \mathbb{R}^{n_i}$ is the corresponding vector of coefficients. In Section 12.7, we will further describe how to select the training samples ϕ to ensure the approximation in (12.1) is accurate in practice.

In the testing stage, we are confronted with the problem that the class label i is unknown. Nevertheless, one can still form a linear representation similar to (12.1), now in terms of *all* of the training samples:

$$y = [\Phi_1, \Phi_2, \ldots, \Phi_C] c_0 = \Phi c_0 \in \mathbb{R}^m, \qquad (12.2)$$

where

$$c_0 = [\ldots, \mathbf{0}^T, c_i^T, \mathbf{0}^T, \ldots]^T \in \mathbb{R}^n. \qquad (12.3)$$

Obviously, if we can recover a vector c of coefficients concentrated on a single class, it will be very indicative of the identity of the subject.

The key idea of SRC is to cast face recognition as the quest for such a coefficient vector c_0. We notice that because the nonzero elements in c are concentrated on images of a single subject, c_0 is a highly *sparse* vector: on average only a fraction of $1/C$ of its entries are nonzero. Indeed, it is not difficult to argue that in general this vector is the sparsest solution to the system of equations $y = \Phi c_0$. While the search for sparse solutions to linear systems is a difficult problem in general, foundational results in the theory of sparse representation indicate that in many situations the sparsest solution can be exactly recovered by solving a tractable optimization problem, minimizing the ℓ_1-norm $\|c\|_1 \doteq \sum_i |c_i|$ of the coefficient vector (see [15, 8, 13, 6] for a sampling of the theory underlying this relaxation). This suggests seeking c_0 as the unique solution to the optimization problem

$$\min \|c\|_1 \quad \text{s.t.} \quad \|y - \Phi c\|_2 \leq \varepsilon. \qquad (12.4)$$

Here, $\varepsilon \in \mathbb{R}$ reflects the noise level in the observation.

Figure 12.2 shows an example of the coefficient vector c recovered by solving the problem (12.4). Notice that the nonzero entries indeed concentrate on the (correct) first subject class, indicating the identity of the test image. In this case, the identification is correct even though the input images are so low-resolution (12×10!) that the system of equations $y \approx \Phi c$ is underdetermined.

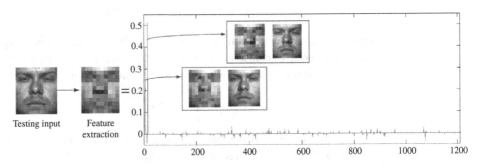

Figure 12.2 Sparse representation of a 12×10 down-sampled query image based on about 1200 training images of the same resolution. The query image belongs to Class 1 [46].

Once the sparse coefficients c have been recovered, tasks such as recognition and validation can be performed in a very natural manner. For example, one can simply define the concentration of a vector $c = [c_1^T, c_2^T, \ldots, c_C^T]^T \in \mathbb{R}^n$ on a subject i as

$$\alpha_i \doteq \|c_i\|_1 / \|c\|_1. \tag{12.5}$$

One can then assign to test image y the label i that maximizes α_i, or reject y as not belonging to any subject in the database if the maximum value of α_i is smaller than a predetermined threshold. For more details, as well as slightly more sophisticated classification schemes based on the sparse coefficients c, please see [46, 44].

In the remainder of this chapter, we will see how this idealized scheme can be made practical by showing how to recover the sparse coefficients c_0 even if the test image is subject to additional variations such as occlusion and misalignment. We further discuss how to reduce the complexity of the optimization problem (12.4) for large-scale problems.

12.3 Dimensionality reduction

One major obstacle to large-scale face recognition is the sheer scale of the training data: the dimension of the raw images could be in the millions, while the number of images increases in proportion to number of subjects. The size of the data is directly reflected in the computational complexity of the algorithm – the complexity of convex optimization could be cubic or worse. In pattern recognition, a classical technique for addressing the problem of high dimensionality is to project the data into a much lower dimensional feature space $\mathbb{R}^d$ ($d \ll m$), such that the projected data still retain the useful properties of the original images.

Such projections can be generated via principal component analysis [43], linear discriminant analysis [3], locality preserving projections [25], as well as less-conventional transformations such as downsampling or selecting local features (e.g., the eye or mouth regions). These well-studied projections can all be represented as linear maps $A \in \mathbb{R}^{d \times m}$. Applying such a linear projection gives a new observation

$$\tilde{y} \doteq Ay \approx A\Phi c = \Psi c \in \mathbb{R}^d. \tag{12.6}$$

Notice that if d is small, the solution to the system $A\Phi c = \tilde{y}$ may not be unique. Nevertheless, under generic circumstances, the desired *sparsest* solution c_0 to this system *is* unique, and can be sought via a lower complexity convex optimization

$$\min \|c\|_1 \quad \text{s.t.} \quad \|\Psi c - \tilde{y}\|_2 \leq \varepsilon. \tag{12.7}$$

Then the key question is to what extent the choice of transformation A affects our ability to recover c_0 and subsequently recognize the subject.

The many different projections referenced above reflect a long-term effort within the face recognition community to find the best possible set of data-adaptive projections.

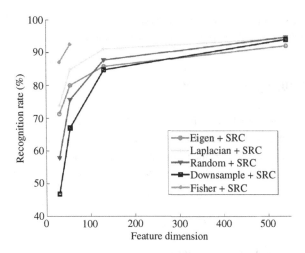

Figure 12.3 Recognition rates of SRC for various feature transformations and dimensions [46]. The training and query images are selected from the public AR database. © 2009 IEEE.

On the other hand, one of the key observations of compressed sensing is that *random projections* serve as a universal non-adaptive set of projections [9, 16, 17]. If a vector c is sparse in a known orthobasis, then ℓ_1 minimization will recover c from relatively small sets of random observations, with high probability. Although our matrix Φ is not an orthonormal basis (far from it, as we will see in the next section), it is still interesting to investigate random projections for dimensionality reduction, and to see to what extent the choice of features affects the performance of ℓ_1 minimization in this application.

Figure 12.3 shows a typical comparison of recognition rates across a variety of feature transformations A and feature space dimensions d. The data in Figure 12.3 are taken from the AR face database, which contains images of 100 subjects under a variety of conditions [33].[4] The horizontal axis plots the feature space dimension, which varies from 30 to 540. The results, which are consistent with other experiments on a wide range of databases, show that the choice of an optimal feature space is no longer critical. When ℓ_1 minimization is capable of recovering sparse signals in several hundred dimensional feature spaces, the performance of all tested transformations converges to a reasonably high percentage. More importantly, even random features contain enough information to recover the sparse representation and hence correctly classify most query images. Therefore, what is critical is that the dimension of the feature space is sufficiently large, and that the sparse representation is correctly computed.

It is important to note that reducing the dimension typically leads to decrease in the recognition rate; although that decrease is not large when d is sufficiently large. In large-scale applications where this tradeoff is inevitable, the implication of our investigation is that a variety of features can confidently be used in conjunction with ℓ_1 minimization. On the other hand, if highly accurate face recognition is desired, the original images

---

[4] For more detailed information on the experimental setting, please see [46].

themselves can be used as features, in a way that is robust to additional physical nuisances such as occlusion and geometric transformations of the image.

12.4 Recognition with corruption and occlusion

In many practical scenarios, the face of interest may be partially occluded, as shown in Figure 12.4. The image may also contain large errors due to self-shadowing, specularities, or corruption. Any of these image nuisances may cause the representation to deviate from the linear model $y \approx \Phi c$. In realistic scenarios, we are more likely confronted with an observation that can be modeled as

$$y = \Phi c + e, \tag{12.8}$$

where e is an unknown vector whose nonzero entries correspond to the corrupted pixels in the observation y, as shown in Figure 12.4.

The errors e can be large in magnitude, and hence cannot be ignored or treated with techniques designed for small noise, such as least squares. However, like the vector c, they often are *sparse*: occlusion and corruption generally affect only a fraction $\rho < 1$ of the image pixels. Hence, the problem of recognizing occluded faces can be cast as the search for a sparse representation c, up to a sparse error e. A natural robust extension to the SRC framework is to instead solve a combined ℓ_1 minimization problem

$$\min \ \|c\|_1 + \|e\|_1 \ \text{s.t.} \ y = \Phi c + e. \tag{12.9}$$

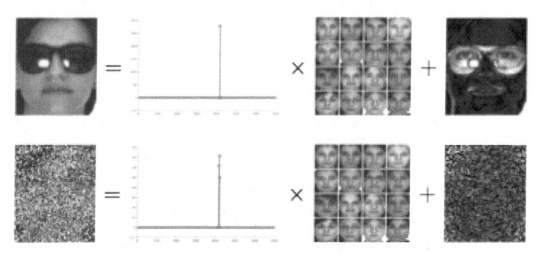

Figure 12.4 **Robust face recognition via sparse representation.** The method represents a test image (left), which is partially occluded (top) or corrupted (bottom), as a sparse linear combination of all the normal training images (middle) plus sparse errors (right) due to occlusion or corruption. The largest coefficients correspond to training images of the correct individual. Our algorithm determines the true identity (indicated with a box at second row and third column) from 700 training images of 100 individuals in the standard AR face database [46]. © 2009 IEEE.

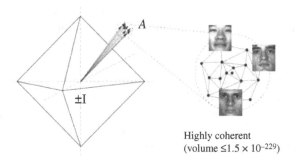

Figure 12.5 The cross-and-bouquet model for face recognition. The raw images of human faces expressed as columns of A are clustered with small variance [45]. © 2010 IEEE.

In [46], it was observed that this optimization performs quite well in correcting occlusion and corruption, for instance, for block occlusions covering up to 20% of the face and random corruptions affecting more than 70% of the image pixels.

Nevertheless, on closer inspection, the success of the combined ℓ_1 minimization in (12.9) is surprising. One can interpret (12.9) as an ℓ_1 minimization problem against a single combined dictionary $B \doteq [\Phi\ I] \in \mathbb{R}^{m \times (n+m)}$:

$$\min \ \|w\|_1 \ \text{s.t.} \ y = Bw, \qquad (12.10)$$

where $w = [c^T, e^T]^T$. Because the columns of Φ are all face images, and hence somewhat similar in the high-dimensional image space $\mathbb{R}^m$, the matrix B fairly dramatically violates the classical conditions for uniform sparse recovery, such as the incoherence criteria [15] or the restricted isometry property [8]. In contrast to the classical compressed sensing setup, the matrix B has quite inhomogeneous properties: the columns of Φ are coherent in the high-dimensional space, while the columns of I are as incoherent as possible. Figure 12.5 illustrates the geometry of this rather curious object, which was dubbed a "cross-and-bouquet" (CAB) in [45], due to the fact that the columns of the identity matrix span a cross polytope, whereas the columns of A are tightly clustered like a bouquet of flowers.

In sparse representation, the CAB model belongs to a special class of sparse representation problems where the dictionary Φ is a concatenation of sub-dictionaries. Examples include the merger of wavelet and heaviside dictionaries in [11] and the combination of texture and cartoon dictionaries in morphological component analysis [18]. However, in contrast to most existing examples, not only is our new dictionary B inhomogeneous, in fact the solution (c, e) to be recovered is also very inhomogeneous: the sparsity of c is limited by the number of images per subject, whereas we would like to handle as dense e as possible, to guarantee good error correction performance. Simulations (similar to the bottom row of Figure 12.4) have suggested that in fact the error e can be quite dense, provided its signs and support are random [46, 45]. In [45], it is shown that

As long as the bouquet is sufficiently tight in the high-dimensional image space $\mathbb{R}^m$, ℓ_1 minimization successfully recovers the sparse coefficients x from very dense ($\rho \nearrow 1$) randomly signed errors e.

For a more precise statement and proof of this result, we refer the reader to [45].

For our purposes here, it suffices to say that this result suggests that excellent error correction is possible in circumstances quite similar to the ones encountered in real-world face recognition. This is surprising for two reasons. First, as mentioned above, the "dictionary" in this problem dramatically violates the restricted isometry property. Second, the errors corrected can be quite dense, in contrast to typical results from compressed sensing in which the number of nonzero coefficients recovered (or errors corrected) is typically bounded by a small fraction of the dimension m [8, 17]. Interestingly, while the mathematical tools needed to analyze this problem are quite standard in this area, the results obtained are qualitatively different from classical results in compressed sensing. Thus, while classical results such as [8, 14] are inspiring for face recognition, the structure of the matrices encountered in this application gives it a mathematical flavor all its own.

12.5 Face alignment

The problem formulation in the previous sections allows us to simultaneously cope with illumination variation and moderate occlusion. However, a practical face recognition system needs to deal with one more important mode of variability: misalignment of the test image and training images. This may occur if the face is not perfectly localized in the image, or if the pose of the face is not perfectly frontal. Figure 12.6 shows how even small misalignment can cause appearance-based face recognition algorithms (such as the one described above) to break down. In this section, we will see how the framework of the previous sections naturally extends to cope with this difficulty.

To pose the problem, we assume the observation y is a warped image $y = y_0 \circ \tau^{-1}$ of the ground-truth signal y_0 under some 2-D transformation of domain τ.[5] As illustrated in Figure 12.6, when τ perturbs the detected face region away from the optimal position, directly solving a sparse representation of y against properly aligned training images often results in erroneous representation.

Nevertheless, if the true deformation τ can be efficiently found, then we can recover y_0 and it becomes possible to recover a relevant sparse representation c for y_0 with respect to the well-aligned training set. Based on the previous error correction model (12.8), the sparse representation model under face alignment is defined as

$$y \circ \tau = \Phi c + e. \qquad (12.11)$$

Naturally, one would like to use the sparsity as a strong cue for finding the correct deformation τ, solving the following optimization problem:

$$\min_{c,e,\tau} \|c\|_1 + \|e\|_1 \quad \text{s.t.} \quad y \circ \tau = \Phi c + e. \qquad (12.12)$$

---

[5] In our system, we typically use 2-D similarity transformations, $T = \mathbb{SE}(2) \times \mathbb{R}_+$, for misalignment incurred by face cropping, or 2-D projective transformations, $T = \mathbb{GL}(3)$, for pose variation.

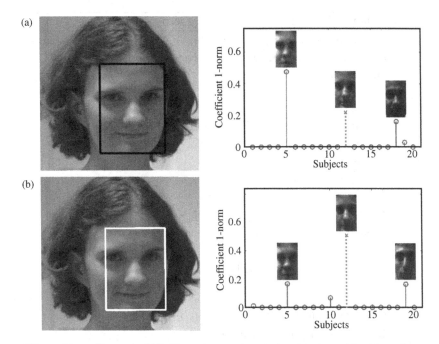

Figure 12.6 **Effect of face alignment [44].** The task is to identify the girl among 20 subjects, by computing the sparse representation of her input face with respect to the entire training set. The absolute sum of the coefficients associated with each subject is plotted on the right. We also show the faces reconstructed with each subject's training images weighted by the associated sparse coefficients. The dotted line corresponds to her true identity, Subject 12. (a) The input face is from Viola and Jones' face detector (the black box). The estimated representation failed to reveal the true identity as the coefficients from Subject 5 are more significant. (b) The input face is well-aligned (the white box) with the training images by our alignment algorithm, and a better representation is obtained. © 2009 IEEE.

Unfortunately, simultaneously estimating τ and (c,e) in (12.12) is a difficult nonlinear optimization problem. In particular, in the presence of multiple classes in the matrix Φ, many local minima arise, which correspond to aligning y to different subjects in the database.

To mitigate the above two issues, it is more practical to first consider aligning y individually to each subject k:

$$\tau_k^* = \arg \min_{c,e,\tau_k} \|e\|_1 \text{ s.t. } c \circ \tau_k = \Phi_k c + e. \tag{12.13}$$

Note that in (12.13), the sparsity of c is no longer penalized, since Φ_k only contains images of the same subject.

Second, if we have access to a good initial guess of the transformation (e.g., from the output of a face detector), the true transformation τ_k in (12.13) can be iteratively sought by solving a sequence of linearized approximations as follows:

$$\min_{c,e,\Delta\tau_k} \|e\|_1 \text{ s.t. } c \circ \tau_k^i + J_k^i \cdot \Delta\tau_k = \Phi_k c + e, \tag{12.14}$$

Algorithm 12.1 (Deformable SRC for face recognition [44])

1: **Input:** Frontal training images $\Phi_1, \Phi_2, \ldots, \Phi_C \in \mathbb{R}^{m \times n_i}$ for C subjects, a test image $y \in \mathbb{R}^m$, and a deformation group T considered.
2: **for** each subject k,
3: $\quad \tau_k^0 \leftarrow I$.
4: $\quad$ **do**
5: $\quad\quad \tilde{y}(\tau_k^i) \leftarrow \frac{y \circ \tau_k^i}{\|y \circ \tau_k^i\|_2}; \quad J_k^i \leftarrow \frac{\partial}{\partial \tau_k} \tilde{y}(\tau_k)\big|_{\tau_k^i};$
6: $\quad\quad \Delta \tau_k = \arg\min \|e\|_1$ s.t. $\tilde{y}(\tau_k^i) + J_k^i \Delta \tau_k = \Phi_k c + e$.
7: $\quad\quad \tau_k^{i+1} \leftarrow \tau_k^i + \Delta \tau_k;$
8: $\quad$ **while** $\|\tau_k^{i+1} - \tau_k^i\| \geq \varepsilon$.
9: **end**
10: Set $\Phi \leftarrow \left[\Phi_1 \circ \tau_1^{-1} \mid \Phi_2 \circ \tau_2^{-1} \mid \cdots \mid \Phi_C \circ \tau_C^{-1}\right]$.
11: Solve the ℓ_1 minimization problem:

$$\hat{c} = \arg\min_{c,e} \|c\|_1 + \|e\|_1 \text{ s.t. } y = \Phi c + e.$$

12: Compute residuals $r_k(b) = \|c - \Phi_k \delta_k(\hat{c})\|_2$ for $k = 1, \ldots, C$.
13: **Output:** identity$(y) = \arg\min_k r_k(c)$.

where τ_k^i is the current estimate of the transformation τ_k, $J_k^i = \nabla_{\tau_k}(y \circ \tau_k^i)$ is the Jacobian of $y \circ \tau_k^i$ with respect to τ_k, and $\Delta \tau_k$ is a step update to τ_k.[6]

From the optimization point of view, our scheme (12.14) can been seen as a generalized Gauss–Newton method for minimizing the composition of a non-smooth objective function (the ℓ_1-norm) with a differentiable mapping from transformation parameters to transformed images. It has been extensively studied in the literature and is known to converge quadratically in a neighborhood of any local optimum of the ℓ_1-norm [35, 27]. For our face alignment problem, we simply note that typically (12.14) takes 10 to 15 iterations to converge.

To further improve the performance of the algorithm, we can adopt a slightly modified version of (12.14), in which we replace the warped test image $y \circ \tau_k$ with the normalized one $\tilde{y}(\tau_k) = \frac{b \circ \tau_k}{\|y \circ \tau_k\|_2}$. This helps to prevent the algorithm from falling into a degenerate global minimum corresponding to zooming in on a dark region of the test image. In practice, our alignment algorithm can run in a multi-resolution fashion in order to reduce the computational cost and gain a larger region of convergence.

---

[6] In computer vision literature, the basic iterative scheme for registration between two *identical* images related by an image transformation of a few parameters has been long known as the Lucas–Kanade algorithm [31]. Extension of the Lucas–Kanade algorithm to address the illumination issue in the same spirit as ours has also been exploited. However, most traditional solutions formulated the objective function using the ℓ_2-norm as a least-squares problem. One exception prior to the theory of CS, to the best of our knowledge, was proposed in a robust face tracking algorithm by Hager and Belhumeur [24], where the authors used an iterative reweighted least-squares (IRLS) method to iteratively remove occluded image pixels while the transform parameters of the face region were sought.

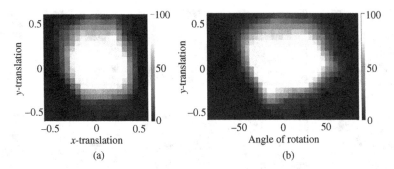

Figure 12.7 **Region of attraction [44].** Fraction of subjects for which the algorithm successfully aligns a manually perturbed test image. The amount of translation is expressed as a fraction of the distance between the outer eye corners, and the amount of in-plane rotation in degrees. (a) Simultaneous translation in x and y directions. More than 90% of the subjects were correctly aligned for any combination of x and y translations, each up to 0.2. (b) Simultaneous translation in y direction and in-plane rotation θ. More than 90% of the subjects were correctly aligned for any combination of y translation up to 0.2 and θ up to $25°$. © 2009 IEEE.

Once the best transformation τ_k is obtained for each subject k, we can apply its inverse to the training set Φ_k so that the entire training set is aligned to y. Then, a global sparse representation $\hat{c}$ of y with respect to the transformed training set can be sought by solving an optimization problem of the form (12.9). The final classification is done by computing the ℓ_2 distance between y and its approximation $\hat{y} = \Phi_k \delta_k(\hat{c})$ using only the training images from the kth class, and assigning y to the class that minimizes the distance.[7] The complete algorithm is summarized in Algorithm 12.1.

Finally, we present some experimental results that characterize the region of attraction of the proposed alignment procedure for both 2-D deformation and 3-D pose variation. We will leave the evaluation of the overall face recognition system to Section 12.8.

For 2-D deformation, we use a subset of images of 120 subjects from the CMU Multi-PIE database [23], since the ground-truth alignment is available. In this experiment, the training set consists of images under properly chosen lighting conditions, and the testing set contains one new illumination. We introduce artificial perturbation to each test image with a combination of translation and rotation, and use the proposed algorithm to align it to the training set of the same subject. For more details about the experiment setting, please refer to [44]. Figure 12.7 shows the percentage of successful registrations for all test images for each artificial perturbation. We can see that our algorithm performs very well with translation up to 20% of the eye distance (or 10 pixels) in both x and y directions, and up to $30°$ in-plane rotation. We have also tested our alignment algorithm with scale variation, and it can handle up to 15% change in scale.

For 3-D pose variation, we collect our own dataset using the acquisition system which will be introduced in Section 12.7. The training set includes frontal face images of each subject under 38 illuminations and the testing set contains images taken under densely

---

[7] $\delta_k(\hat{c})$ returns a vector of the same dimension as $\hat{c}$ that only retains the nonzero coefficients corresponding to subject k.

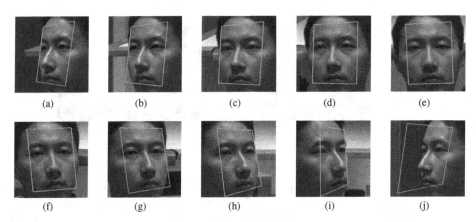

Figure 12.8 **Aligning different poses to frontal training images [44]. (a) to (i):** good alignment for poses from $-45°$ to $+45°$. **(j):** a case when the algorithm fails for an extreme pose ($> 45°$). © 2009 IEEE.

sampled poses. Viola and Jones' face detector is then used for face cropping in this experiment. Figure 12.8 shows some typical alignment results. The alignment algorithm works reasonably well with poses up to $\pm 45°$, which easily exceeds the pose requirement for real-world access-control applications.

12.6 Fast ℓ_1 minimization algorithms

In the previous sections, we have seen how the problem of recognizing faces despite physical variabilities such as illumination, misalignment, and occlusion fall naturally into the framework of sparse representation. Indeed, all of these factors can be addressed simultaneously by solving appropriate ℓ_1 minimization problems. However, for these observations to be useful in practice, we need scalable and efficient algorithms for ℓ_1 minimization.

Although ℓ_1 minimization can be recast as a linear program and solved to high accuracy using interior-point algorithms [28], these algorithms do not scale well with the problem size: each iteration typically requires cubic time. Fortunately, interest in compressed sensing has inspired a recent wave of more scalable, more efficient first-order methods, which can solve very large ℓ_1 minimization problems to medium accuracy (see, e.g., [40] for a general survey). As we have seen in Section 12.4, ℓ_1 minimization problems arising in face recognition may have dramatically different structures from problems arising in other applications of compressed sensing, and hence require customized solvers. In this section, we describe our algorithm of choice for solving these problems, which is essentially an augmented Lagrange multiplier method [5], but also uses an accelerated gradient algorithm [2] to solve a key subproblem. We draw extensively on the survey [48], which compares the performance of various solvers in the context of face recognition.

The key property of the ℓ_1 norm that enables fast first-order solvers is the existence of an efficient solution to the "proximal minimization":

$$S_\lambda[z] = \arg\min_c \lambda \|c\|_1 + \frac{1}{2}\|c-z\|_2^2, \tag{12.15}$$

where $c, z \in \mathbb{R}^n$, and $\lambda > 0$. It is easy to show that the above minimization is solved by *soft-thresholding*, which is defined for scalars as follows:

$$S_\lambda[c] = \begin{cases} c - \lambda, & \text{if } c > \lambda \\ c + \lambda, & \text{if } c < -\lambda \\ 0, & \text{if } |c| \leq \lambda \end{cases} \tag{12.16}$$

and extended to vectors and matrices by applying it element-wise. It is extremely simple to compute, and forms the backbone of most of the first-order methods proposed for ℓ_1 minimization. We will examine one such technique, namely, the method of *Augmented Lagrange Multipliers* (ALM), in this section. To keep the discussion simple, we focus our discussion on the SRC problem, although the ideas are directly applicable to the image alignment problem as well. The interested reader may refer to the Appendix of [48] for more details.

Lagrange multiplier methods are a popular tool in convex optimization. The basic idea is to eliminate equality constraints by adding an appropriate penalty term to the cost function that assigns a very high cost to infeasible points. The goal is then to efficiently solve the unconstrained problem. For our problem, we define the augmented Lagrangian function as follows:

$$L_\mu(c, e, \nu) \doteq \|c\|_1 + \|e\|_1 + \langle \nu, y - \Phi c - e \rangle + \frac{\mu}{2}\|y - \Phi c - e\|_2^2, \tag{12.17}$$

where $\mu > 0$, and ν is a vector of Lagrange multipliers. Note that the augmented Lagrangian function is convex in c and e. Suppose that $(c^\star, e^\star)$ is the optimal solution to the original problem. Then, it can be shown that for sufficiently large μ, there exists a $\nu^\star$ such that

$$(c^\star, e^\star) = \arg\min_{c,e} L_\mu(c, e, \nu^\star). \tag{12.18}$$

The above property indicates that minimizing the augmented Lagrangian function amounts to solving the original constrained optimization problem. However, this approach does not seem a viable one since $\nu^\star$ is not known a priori and the choice of μ is not evident from the problem. Augmented Lagrange Multiplier methods overcome these issues by simultaneously solving for $\nu^\star$ in an iterative fashion and monotonically increasing the value of μ every iteration so as to avoid converging to an infeasible point. The basic ALM iteration is given by [5]:

$$\begin{aligned} (c_{k+1}, e_{k+1}) &= \arg\min_{c,e} L_{\mu_k}(c, e, \nu_k), \\ \nu_{k+1} &= \nu_k + \mu_k(y - \Phi c_{k+1} - e_{k+1}), \end{aligned} \tag{12.19}$$

where $\{\mu_k\}$ is a monotonically increasing positive sequence. This iteration by itself does not give us an efficient algorithm since the first step of the iteration is an unconstrained convex program. However, for the ℓ_1 minimization problem, we will see that it can be solved very efficiently.

The first step to simplifying the above iteration is to adopt an alternating minimization strategy, i.e., to first minimize with respect to e and then minimize with respect to c. This approach, dubbed *alternating direction method of multipliers* in [22], was first used in [51] in the context of ℓ_1 minimization. Thus, the above iteration can be rewritten as:

$$\begin{aligned} e_{k+1} &= \arg\min_e L_{\mu_k}(c_k, e, \nu_k), \\ c_{k+1} &= \arg\min_c L_{\mu_k}(c, e_{k+1}, \nu_k), \\ \nu_{k+1} &= \nu_k + \mu_k(y - \Phi c_{k+1} - e_{k+1}). \end{aligned} \quad (12.20)$$

Using the property described in (12.15), it is not difficult to show that

$$e_{k+1} = \mathcal{S}_{\frac{1}{\mu_k}}\left[\frac{1}{\mu_k}\nu_k + y - \Phi c_k\right]. \quad (12.21)$$

Obtaining a similar closed-form expression for c_{k+1} is not possible, in general. So, we solve for it in an iterative procedure. We note that $L_{\mu_k}(c, e_{k+1}, \nu_k)$ can be split into two functions: $\|c\|_1 + \|e_{k+1}\|_1 + \langle \nu_k, y - \Phi c - e_{k+1}\rangle$ that is convex and continuous in x; and $\frac{\mu_k}{2}\|y - \Phi c - e_{k+1}\|_2^2$ that is convex, smooth, and has Lipschitz continuous gradient. This form of the $L_{\mu_k}(c, e_{k+1}, \nu_k)$ allows us to use a fast iterative thresholding algorithm, called FISTA [2], to solve for c_{k+1} in (12.20) efficiently. The basic idea in FISTA is to iteratively form quadratic approximations to the smooth part of the cost function and minimize the approximated cost function instead.

Using the above-mentioned techniques, the iteration described in (12.20) is summarized as Algorithm 12.2, where γ denotes the largest eigenvalue of $\Phi^T\Phi$. Although the algorithm is composed of two loops, in practice, we find that the innermost loop converges in a few iterations.

As mentioned earlier, several first-order methods have been proposed for ℓ_1 minimization recently. Theoretically, there is no clear winner among these algorithms in terms of the convergence rate. However, it has been observed empirically that ALM offers the best tradeoff in terms of speed and accuracy. An extensive survey of some of the other methods along with experimental comparison is presented in [48]. Compared to the classical interior-point methods, Algorithm 12.2 generally takes more iterations to converge to the optimal solution. However, the biggest advantage of ALM is that each iteration is composed of very elementary matrix-vector operations, as against matrix inversions or Gaussian eliminations used in the interior-point methods.

12.7 Building a complete face recognition system

In the previous sections, we have presented a framework for reformulating face recognition in terms of sparse representation, and have discussed fast ℓ_1 minimization algorithms

Algorithm 12.2 (Augmented Lagrange multiplier method for ℓ_1 minimization)

1: **Input:** $y \in \mathbb{R}^m$, $\Phi \in \mathbb{R}^{m \times n}$, $c_1 = 0$, $e_1 = y$, $\nu_1 = 0$.
2: **while** not converged ($k = 1, 2, \ldots$) **do**
3: $\quad e_{k+1} = \text{shrink}\left(y - \Phi c_k + \frac{1}{\mu_k}\nu_k, \frac{1}{\mu_k}\right)$;
4: $\quad t_1 \leftarrow 1, z_1 \leftarrow c_k, w_1 \leftarrow c_k$;
5: $\quad$ **while** not converged ($l = 1, 2, \ldots$) **do**
6: $\quad\quad w_{l+1} \leftarrow \text{shrink}\left(z_l + \frac{1}{\gamma}\Phi^T\left(y - \Phi v_l - e_{k+1} + \frac{1}{\mu_k}\nu_k\right), \frac{1}{\mu_k \gamma}\right)$;
7: $\quad\quad t_{l+1} \leftarrow \frac{1}{2}\left(1 + \sqrt{1 + 4t_l^2}\right)$;
8: $\quad\quad z_{l+1} \leftarrow w_{l+1} + \frac{t_l - 1}{t_{l+1}}(w_{l+1} - w_l)$;
9: $\quad$ **end while**
10: $\quad c_{k+1} \leftarrow w_l$;
11: $\quad \nu_{k+1} \leftarrow \nu_k + \mu_k(y - \Phi c_{k+1} - e_{k+1})$;
12: **end while**
13: **Output:** $c^* \leftarrow c_k, e^* \leftarrow e_k$.

to efficiently estimate sparse signals in high-dimensional spaces. In this section, we discuss some of the practical issues that arise in using these ideas to design prototype face recognition systems for access-control applications.

In particular, note that so far we have made the critical assumption that the test image, although taken under some unknown illumination, can be represented as a linear combination of a finite number of training illuminations. These assumptions naturally raise the following questions: *Under what conditions is the linear subspace model a reasonable assumption, and how should a face recognition system acquire sufficient training illumination samples to achieve high accuracy on a wide variety of practical, real-world illumination conditions?*

First, let us consider an approximation of the human face as a convex, Lambertian object under distinct illuminations with a fixed pose. Under those assumptions, the incident and reflected light are distributions on a sphere, and thus can be represented in a spherical harmonic basis [1]. The Lambertian reflectance kernel acts as a lowpass filter between the incident and reflected light, and as a result, the set of images of the object end up lying very close to a subspace corresponding to the low-frequency spherical harmonics. In fact, one can show that only nine (properly chosen) basis illuminations are sufficient to generate basis images that span all possible images of the object.

While modeling the harmonic basis is important for understanding the image formation process, various empirical studies have shown that even in the case when convex, Lambertian assumptions are violated, the algorithm can still get away with using a small number of frontal illuminations to linearly represent a wide range of new frontal illuminations, especially when they are all taken under the same laboratory conditions. This is the case for many public face databases, such as AR, ORL, PIE, and Multi-PIE. Unfortunately, in practice, we have observed that a training database consisting purely of frontal

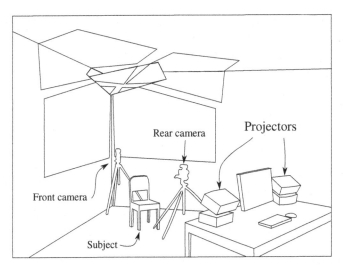

Figure 12.9 Illustration of the training acquisition system, which consists of four projectors and two cameras controlled by a computer. [44] © 2009 IEEE.

Figure 12.10 38 training images of a subject collected by the system. The first 24 images are sampled using the foreground lighting patterns, and the other 14 images using the background lighting patterns. [44] © 2009 IEEE.

illuminations is not sufficient to linearly represent images of a face taken under typical indoor and outdoor conditions. To ensure our algorithm works in practice, we need to more carefully acquire a set of training illuminations that are sufficient to linearly represent a wide variety of practical indoor and outdoor illuminations.

To this end, we have designed a system that can acquire frontal images of a subject while simultaneously illuminating the subject from all directions. A sketch of the system is shown in Figure 12.9. A more detailed explanation of this system is discussed in [44].

Based on the results of our experiments, the illumination patterns projected either directly on the subject's frontal face or indirectly on the wall correspond to a total of 38 training illumination images, as an example shown in Figure 12.10. We have observed that further acquiring finer illumination patterns does not significantly improve the image registration and recognition accuracy [44]. Therefore, we have used those illumination models for all our large-scale experiments.

12.8 Overall system evaluation

In this section, we present representative recognition results of our complete system on large-scale face databases. All the experiments are carried out using input directly obtained from the Viola and Jones' face detector, without any manual intervention throughout the process.

We use two different face databases to test our system. We first report the performance of our system on the largest public face database available that is suitable for testing our algorithm, the CMU Multi-PIE database [23]. This database contains images of 337 subjects across simultaneous variation in pose, expression, illumination, and facial appearance over time, and thus provides the most extensive test among all public databases. However, one shortcoming of the CMU Multi-PIE database for our purpose is that all the images are taken under controlled laboratory lighting conditions, restricting our choice of training and testing sets to these conditions, which may not cover all typical natural illuminations. Therefore, our goal of this experiment is to simply demonstrate the effectiveness of our fully automatic system with respect to such a large number of classes. We next test on a face database collected using our own acquisition system as described in Section 12.7. The goal of that experiment is then to show that with a sufficient set of training illuminations, our system is indeed capable of performing robust face recognition with loosely controlled test images taken under practical indoor and outdoor conditions.

For the CMU Multi-PIE database, we use all the 249 subjects present in Session 1 as the training set. The remaining 88 subjects are considered as "outliers" and are used to test our system's ability to reject invalid images. To further challenge our system, we include only 7 extreme frontal illumination conditions for each of the 249 subjects in the training, and use frontal images of all the 20 illuminations from Session 2–4 as testing, which were recorded at different times over a period of several months. Table 12.1 shows the result of our algorithm on each of the three testing sessions, as well as the results obtained using baseline linear-projection-based algorithms including Nearest Neighbor (NN), Nearest Subspace (NS) [30], and Linear Discriminant Analysis (LDA) [3]. Note that we initialize these baseline algorithms in two different ways, namely, from the output of the Viola and Jones' detector, indicated by a subscript "d," and with images which are aligned to the training with manually clicked outer eye-corners, indicated by a subscript "m." One can see in Table 12.1 that, despite careful manual registration, these baseline algorithms perform significantly worse than our system, which uses input directly from the face detector.

We further perform subject validation on the Multi-PIE database, using the measure of concentration of the sparse coefficients as introduced in Section 12.2, and compare this method to the classifiers based on thresholding the error residuals of NN, NS, and LDA. Figure 12.11 plots the receiver operating characteristic (ROC) curves, which are generated by sweeping the threshold through the entire range of possible values for each algorithm. We can see that our approach again significantly outperforms the other three algorithms.

Table 12.1. Recognition rates on CMU Multi-PIE database.

Rec. Rates	Session 2	Session 3	Session 4
LDA_d (LDA_m)	5.1 (49.4)%	5.9 (44.3)%	4.3 (47.9)%
NN_d (NN_m)	26.4 (67.3)%	24.7 (66.2)%	21.9 (62.8)%
NS_d (NS_m)	30.8 (77.6)%	29.4 (74.3)%	24.6 (73.4)%
Algorithm 12.1	**91.4%**	**90.3%**	**90.2%**

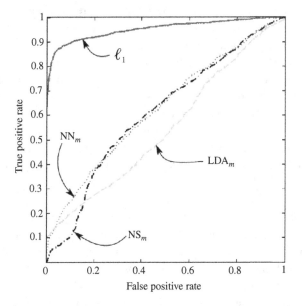

Figure 12.11 ROC curves for our algorithm (labeled as "ℓ_1"), compared with those for NN_m, NS_m, and LDA_m.

For experiments on our own database, we have collected the frontal view of 74 subjects without eyeglasses under 38 illuminations as shown in Section 12.7 and use them as the training set. For testing our algorithm, we have also taken 593 images of these subjects with a different camera under a variety of indoor and outdoor conditions. Based on the main variability in the test images, we further partitioned the testing set into five categories:

C1: 242 images of 47 subjects without eyeglasses, generally frontal view, under a variety of practical illuminations (indoor and outdoor) (Figure 12.12, row 1).
C2: 109 images of 23 subjects with eyeglasses (Figure 12.12, row 2).
C3: 19 images of 14 subjects with sunglasses (Figure 12.12, row 3).
C4: 100 images of 40 subjects with noticeable expressions, poses, mild blur, and sometimes occlusion (Figure 12.13, both rows).

Table 12.2. Recognition rates on our own database.

Test Categories	C1	C2	C3	C4	C5
Rec. Rates (%)	95.9	91.5	63.2	73.7	53.5

Figure 12.12 **Representative examples of categories 1-3**. One row for each category. [44] © 2009 IEEE.

Figure 12.13 **Representative examples of category 4**. Top row: successful examples. Bottom row: failures. [44] © 2009 IEEE.

C5: 123 images of 17 subjects with little control (out of focus, motion blur, significant pose, large occlusion, funny faces, extreme expressions) (Figure 12.14, both rows).

Table 12.2 reports the recognition rates of our system on each category. As one can see, our system achieves recognition rates above 90% for face images with general frontal views, under a variety of practical illuminations. Our algorithm is also robust to small amounts of pose, expression, and occlusion (i.e., eyeglasses).

Figure 12.14 **Representative examples of category 5**. Top row: successful examples. Bottom row: failures. [44] © 2009 IEEE.

12.9 Conclusion and discussion

Based on the theory of sparse representation, we have proposed a comprehensive framework/system to tackle the classical problem of face recognition in computer vision. The initial success of our solution relies on careful analysis of the special data structures in high-dimensional face images. Although our study has revealed new insights about face recognition, many new problems remain largely open. For instance, it is still not clear why the sparse representation based classification (SRC) is so discriminative for highly correlated face images. Indeed, since the matrix $\Phi = [\Phi_1, \Phi_2, \ldots, \Phi_C]$ has class structure, one simple alternative to SRC is to treat each class one at a time, solving a robust regression problem via the ℓ^1 norm, and then select the class with the lowest regression error. Similar to SRC, this alternative respects the physics of illumination and in-plane transformations, and leverages the ability of ℓ_1 minimization to correct sparse errors. However, we find that SRC has a consistent advantage in terms of classification percentage (about 5% on Multi-PIE [44]). One more sophisticated way to take advantage of class structure is by enforcing group sparsity on the coefficients c. While this may impair the system's ability to reject invalid subjects (as in Figure 12.11), it also has the potential to improve recognition performance [39, 32].

Together with other papers that appeared in the similar time frame, this work has inspired researchers to look into a broader range of recognition problems within the framework of sparse representation. Notable examples include image super-resolution [50], object recognition [34, 41], human activity recognition [49], speech recognition [20], 3-D motion segmentation [37, 19], and compressed learning [7]. While these promising works raise many intriguing questions, we believe the full potential of sparse representation for recognition problems remains to be better understood mathematically and carefully evaluated in practice.

References

[1] R. Basri and D. Jacobs. Lambertian reflectance and linear subspaces. *IEEE Trans Pattern Anal Machine Intell*, 25(2):218–233, 2003.

[2] A. Beck and M. Teboulle. A fast iterative shrinkage-thresholding algorithm for linear inverse problems. *SIAM J Imaging Sci*, 2(1):183–202, 2009.

[3] P. Belhumeur, J. Hespanda, and D. Kriegman. Eigenfaces vs. Fisherfaces: recognition using class specific linear projection. *IEEE Trans Pattern Anal Machine Intelli*, 19(7):711–720, 1997.

[4] P. Belhumeur and D. Kriegman. What is the set of images of an object under all possible illumination conditions? *Int J Comput Vision*, 28(3):245–260, 1998.

[5] D. Bertsekas. *Nonlinear Programming*. Athena Scientific, 2003.

[6] A. Bruckstein, D. Donoho, and M. Elad. From sparse solutions of systems of equations to sparse modeling of signals and images. *SIAM Rev*, 51(1):34–81, 2009.

[7] R. Calderbank, S. Jafarpour, and R. Schapire. Compressed learning: universal sparse dimensionality reduction and learning in the measurement domain. Preprint, 2009.

[8] E. Candès and T. Tao. Decoding by linear programming. *IEEE Trans Inform Theory*, 51(12), 2005.

[9] E. Candès and T. Tao. Near optimal signal recovery from random projections: universal encoding strategies? *IEEE Trans on Inform Theory*, 52(12):5406–5425, 2006.

[10] H. Chen, H. Chang, and T. Liu. Local discriminant embedding and its variants. *Proc IEEE Int Conf Comput Vision Pattern Recog*, 2005.

[11] S. Chen, D. Donoho, and M. Saunders. Atomic decomposition by basis pursuit. *SIAM Rev*, 43(1):129–159, 2001.

[12] T. Chen, W. Yin, X. Zhou, D. Comaniciu, and T. Huang. Total variation models for variable lighting face recognition. *IEEE Trans Pattern Anal Machine Intell*, 28(9):1519–1524, 2006.

[13] D. Donoho. Neighborly polytopes and sparse solution of underdetermined linear equations. Preprint, 2005.

[14] D. Donoho. For most large underdetermined systems of linear equations the minimal ℓ^1-norm near solution approximates the sparest solution. *Commun Pure Appli Math*, 59(6):797–829, 2006.

[15] D. Donoho and M. Elad. Optimally sparse representation in general (nonorthogonal) dictionaries via ℓ^1 minimization. *Proc Nati Acad Sci*, 100(5):2197–2202, 2003.

[16] D. Donoho and J. Tanner. Neighborliness of randomly projected simplices in high dimensions. *Proc Nati Acad Sci*, 102(27):9452–9457, 2005.

[17] D. Donoho and J. Tanner. Counting faces of randomly-projected polytopes when the projection radically lowers dimension. *J Am Math Soc*, 22(1):1–53, 2009.

[18] M. Elad, J. Starck, P. Querre, and D. Donoho. Simultaneous cartoon and texture image inpainting using morphological component analysis (MCA). *Appl Comput Harmonic Anal*, 19:340–358, 2005.

[19] E. Elhamifar and R. Vidal. Sparse subspace clustering. *Proc IEEE Int Conf Computer Vision Pattern Recog*, 2009.

[20] J. Gemmeke, H. Van Hamme, B. Cranen, and L. Boves. Compressive sensing for missing data imputation in noise robust speech recognition. *IEEE J Selected Topics Signal Proc*, 4(2):272–287, 2010.

[21] A. Georghiades, P. Belhumeur, and D. Kriegman. From few to many: illumination cone models for face recognition under variable lighting and pose. *IEEE Trans Pattern Anal Machine Intell*, 23(6):643–660, 2001.

[22] R. Glowinski and A. Marrocco. Sur l'approximation par éléments finis d'ordre un, et la résolution, par pénalisation-dualité d'une classe de problèmes de dirichlet nonlinéaires. *Rev Franc d'Automat Inform, Recherche Opérationnelle*, 9(2):41–76, 1975.

[23] R. Gross, I. Matthews, J. Cohn, T. Kanade, and S. Baker. Multi-PIE. *Proc IEEE Conf on Automatic Face Gesture Recog*, 2008.
[24] G. Hager and P. Belhumeur. Efficient region tracking with parametric models of geometry and illumination. *IEEE Trans Pattern Anal Machine Intell*, 20(10):1025–1039, 1998.
[25] X. He, S. Yan, Y. Hu, P. Niyogi, and H. Zhang. Face recognition using Laplacianfaces. *IEEE Trans Pattern Anal Machine Intell*, 27(3):328–340, 2005.
[26] G. Huang, M. Ramesh, T. Berg, and E. Learned-Miller. Labeled faces in the wild: a database for studying face recognition in unconstrained environments. Tech Rep 07-49, University of Massachusetts, Amherst, 2007.
[27] K. Jittorntrum and M. Osborne. Strong uniqueness and second order convergence in nonlinear discrete approximation. *Numer Math*, 34:439–455, 1980.
[28] N. Karmarkar. A new polynomial time algorithm for linear programming. *Combinatorica*, 4:373–395, 1984.
[29] T. Kim and J. Kittler. Locally linear discriminant analysis for multimodally distributed classes for face recognition with a single model image. *IEEE Trans Pattern Anal Machine Intell*, 27(3):318–327, 2005.
[30] K. Lee, J. Ho, and D. Kriegman. Acquiring linear subspaces for face recognition under variable lighting. *IEEE Trans Pattern Anal Machine Intell*, 27(5):684–698, 2005.
[31] B. Lucas and T. Kanade. An iterative image registration technique with an application to stereo vision. *Proc Int Joint Conf Artif Intell*, 3: 674–679, 1981.
[32] A. Majumdar and R. Ward. Improved group sparse classifier. *Pattern Recog Letters*, 31:1959–1964, 2010.
[33] A. Martinez and R. Benavente. The AR face database. Technical rep, CVC Technical Report No. 24, 1998.
[34] N. Naikal, A. Yang, and S. Sastry. Towards an efficient distributed object recognition system in wireless smart camera networks. *Proc Int Conf Inform Fusion*, 2010.
[35] M. Osborne and R. Womersley. Strong uniqueness in sequential linear programming. *J Aust Math Soc, Ser B*, 31:379–384, 1990.
[36] L. Qiao, S. Chen, and X. Tan. Sparsity preserving projections with applications to face recognition. *Pattern Recog*, 43(1):331–341, 2010.
[37] S. Rao, R. Tron, and R. Vidal. Motion segmentation in the presence of outlying, incomplete, or corrupted trajectories. *IEEE Trans Pattern Anal Machine Intell*, 32(10):1832–1845, 2010.
[38] P. Sinha, B. Balas, Y. Ostrovsky, and R. Russell. Face recognition by humans: nineteen results all computer vision researchers should know about. *Proc IEEE*, 94(11):1948–1962, 2006.
[39] P. Sprechmann, I. Ramirez, G. Sapiro, and Y. C. Eldar. C-HiLasso: a collaborative hierarchical sparse modeling framework. (To appear) *IEEE Trans Sig Proc*, 2011.
[40] J. Tropp and S. Wright. Computational methods for sparse solution of linear inverse problems. *Proc IEEE*, 98:948–958, 2010.
[41] G. Tsagkatakis and A. Savakis. A framework for object class recognition with no visual examples. In *Western New York Image Processing Workshop*, 2010.
[42] M. Turk and A. Pentland. Eigenfaces for recognition. *Proc IEEE Int Conf Comp Vision Pattern Recog*, 1991.
[43] M. Turk and A. Pentland. Eigenfaces for recognition. *J Cogn Neurosci*, 3(1):71–86, 1991.
[44] A. Wagner, J. Wright, A. Ganesh, Z. Zhou, and Y. Ma. Toward a practical automatic face recognition system: robust pose and illumination via sparse representation. *Proc IEEE Int Conf Comput Vision Pattern Recog*, 2009.

[45] J. Wright and Y. Ma. Dense error correction via ℓ^1-minimization. *IEEE Trans on Inform Theory*, 56(7):3540–3560, 2010.

[46] J. Wright, A. Yang, A. Ganesh, S. Sastry, and Y. Ma. Robust face recognition via sparse representation. *IEEE Trans Pattern Anal Machine Intelli*, 31(2):210–227, 2009.

[47] S. Yan, D. Xu, B. Zhang, H. Zhang, Q. Yang, and S. Lin. Graph embedding and extension: a general framework for dimensionality reduction. *IEEE Trans Pattern Anal Machine Intell*, 29:40–51, 2007.

[48] A. Yang, A. Ganesh, Z. Zhou, S. Sastry, and Y. Ma. Fast ℓ_1-minimization algorithms for robust face recognition. (Preprint) arXiv:1007.3753, 2011.

[49] A. Yang, R. Jafari, S. Sastry, and R. Bajcsy. Distributed recognition of human actions using wearable motion sensor networks. *J. Ambient Intelli Smart Environm*, 1(2):103–115, 2009.

[50] J. Yang, J. Wright, T. Huang, and Y. Ma. Image super-resolution as sparse representation of raw image patches. *Proc IEEE Int Conf Comput Vision Pattern Recog*, 2008.

[51] J. Yang and Y. Zhang. Alternating direction algorithms for ℓ_1-problems in compressive sensing. arXiv:0912.1185, 2009.

[52] W. Zhao, R. Chellappa, J. Phillips, and A. Rosenfeld. Face recognition: a literature survey. *ACM Comput Surv*, 35(4):399–458, 2003.

[53] S. Zhou, G. Aggarwal, R. Chellappa, and D. Jacobs. Appearance characterization of linear Lambertian objects, generalized photometric stereo, and illumination-invariant face recognition. *IEEE Trans Pattern Anal Machine Intell*, 29(2):230–245, 2007.

Index

ℓ_p-norm, 5
ℓ_0-norm, 5

active learning, 302
active vision, 302
adaptive vs. non-adaptive information, 270
Alltop sequence, 25
alternating direction method, 535
AMP, *see* aproximate message passing
analog discretization, 127
analog sensing, *see* Xampling
analog signals, 13
annihilating filter, 160
    equation, 160
    extended annihilating equation, 183
approximate isometries, 19, 214, 230
approximate message passing (AMP), 407–412
atom, 7
audio, 83
augmented Lagrange multipliers (ALM), 534

Babel function, 493
Back-DSP algorithm, 115
Bai–Yin's law, 229
basis, 6
basis mismatch, 129
Basis Pursuit, 486
Bayes' rule, 276–279
Bayesian experimental design, 275, 278
belief propagation, 404, 431
Bernoulli
    random matrices, 228
    random variables, 220
    random vectors, 224, 226
Bernstein-type inequality, 222
    non-commutative, 228, 236
best k-term approximation, 9
binary tree, 274
block-sparse, 10, 12, 375, 378

Cadzow denoising, 184
cartoon, 504, 506, 507
cluster coherence, 493, 494, 497, 507, 509
    definition, 493
clustered sparsity, 492
coefficients
    clustered, 495
cognitive radio, 104
coherence, 24
collaboration, 66
collaborative hierarchical Lasso, 80
collaborative hierarchical sparse model, 66
collaborative Lasso, 80
collaborative sparse model, 76
combinatorial group testing, 41
compressibility, 11
compressive distilled sensing (CDS), 298–301
Compressive Sampling Matching Pursuit, 367
    algorithm, 368
    for structured sparse models, 378
    recovery result, 369
concentration of measure, 230, 262
condition number, 211
continuous multiband signal, 376, 378
continuous to finite, 102
coordinate random vectors, 224, 226, 236
CoSaMP, *see* Compressive Sampling Matching Pursuit
covariance matrix, 224
    estimation, 240
covering argument, 232
covering numbers, 215
Cramér–Rao bound (CRB), 33, 177
    derivations, 202
    kernel comparison, 182
    periodic stream, 181
    simulations, 187
cross-and-bouquet model, 528
cross-polytope, 37, 313
cumulative coherence function, 493
curse of dimensionality, 271, 292, 293, 300
curvelets, 489, 504, 506–511

Dantzig selector, 39
data
    astronomical, 486
    audio, 484, 486, 500, 501

imaging, 484, 486, 503
multimodal, 484
neurobiological, 505
separation, 484
data streams, 41
decoupling, 246, 250
democratic measurements, 26
denoising, 270–273, 287, 289, 301, 486
Basis Pursuit, 503
dictionary, 7, 375
composed, 485, 493, 504
sparse, 486
dictionary learning, 487–489
dimensionality reduction, 525
Dirac basis, 500–502
discretization, *see* analog discretization
Distance-Preserving Property, 440
distilled sensing (DS), 286, 287, 289–293, 295, 297–301

effective rank, 241
entropy, 276–278, 280
expectation maximization, 71
exponential reproducing kernels, 166
CRB, 182
CRB derivation, 206
piecewise sinusoidal, 194
simulations, 188
external angle, 314

face recognition, 520
alignment, 529
occlusion, 527
system, 535
false discovery proportion (FDP), 288, 289, 291–293, 299, 301
finite rate of innovation (FRI), 13, 14, 149, 376
applications, 194
definition, 154
examples, 155
extensions, 194
history, 151
sampling scheme, 149
simulations, 185
Fourier basis, 486, 500–502
Fourier measurements, 259
frame, 6, 225, 244
dual frame, 7
equiangular tight frame, 25
Gabor frame, 25
Frobenius norm, 275, 297

Gamma function, 283
Gaussian
random matrices, 26, 228, 230
random vectors, 32–33, 224, 226
Gaussian mixture model, 66, 70
Gelfand width, 22
generative model, 278, 282
Geometric Separation Problem, 507, 509
Gershgorin circle theorem, 25
Gordon's theorem, 229
GP, *see* Gradient Pursuit
Gradient Pursuit, 352–353
algorithm, 353
recovery result, 359
Gram matrix, 24, 244, 250
graphical model, 394, 399, 428–434
Grassmann angle, 313, 320
Grassmann Manifold, 313
greedy algorithm, 39, 348
greedy pursuit, 349

Hadamard matrices, 260
hard edge of spectrum, 231
heavy-tailed
random matrices with independent columns, 249
random matrices with independent rows, 234
restricted isometries, 256
hierarchical Lasso, 79
hierarchical prior, 282, 284, 285
hierarchical sparse model, 66, 76
Hoeffding-type inequality, 220
hyperparameters, 279, 284, 285
hyperspectral images, 83

IHT, *see* Iterative Hard Thresholding
image restoration, 67
image separation, 485, 503, 504, 507
IMV, *see* infinite measurement vector
incoherence, 252
infinite measurement vector, 101
information gain, 276–278, 280, 285, 287
inner product, 5
instance optimality, 18, 36
internal angle, 313
inverse problem, 271–273
isotropic random vectors, 223, 224
iterative hard thresholding (IHT), 39, 362–363
algorithm, 363
recovery results, 363
iterative soft thresholding (IST), 409, 412, 413, 423

Johnson–Lindenstrauss lemma, 22
Johnson–Lindenstauss transform, 440
joint concentration, 493–495

Khinchine inequality, 221
non-commutative, 227
Kullback–Leibler (KL) divergence, 277

Lagrange multiplier, 273
Lagrangian, 273
Latala's theorem, 231
least absolute shrinkage and selection operator (LASSO), 28, 298, 299
low-rank matrices, 14, 43, 376, 378

manifold, 15
   Riemannian manifold, 15
Matching Pursuit, 350, 485
   algorithm, 350
maximum a posteriori (MAP) estimate, 273
MCALab, 504, 506
mean square error, 398, 400, 402, 412, 423, 425, 427
measurement budget, 271, 272, 275, 289–293, 296–300
message passing, 404–407
min-sum algorithm, 404, 405
minimax, 395, 399–404, 423
MMV, see multiple measurement vectors
model
   continuum, 507
   discrete, 507
model selection, 287, 298
model-based restricted isometry property, 379
modulated wideband converter, 108, 171
   hardware design, 112
Morphological Component Analysis, 485, 486, 503, 504, 506
MP, see Matching Pursuit
multiband model, 14
multiband sampling, 104
multiband spectra, 92
multichannel, 169
   history, 152
   infinite FRI, 172
   periodic FRI, 170
   semi-periodic FRI, 173
   simulations, 190
multiple measurement vector (MMV), 12, 42, 376, 385–389
   recovery result, 386
mutual coherence, 485, 489, 493, 497, 498, 500
   definition, 489
   lower bound, 500
mutual information, 278
MWC, see modulated wideband converter

net, 215
noise, 29–35, 176
   annihilating filter, see annihilating filter
   development, 151
   history, 152
   oversampling
      Cadzow, 184
      TLS, 184
   performance bounds
      bounded noise, 30–32
      continuous-time noise, 177
      Gaussian noise, 32–35
      sampling noise, 180
      sampling scheme, 177
noise sensitivity, 423
non-discovery proportion (NDP), 288, 289, 291–293, 299, 301
nonharmonic tones, 129
nonlinear approximation, 9
null space property (NSP), 17, 490
Nyquist-equivalent systems, 132
Nyquist-folding receiver, 141

OMP, see Orthogonal Matching Pursuit
onion peeling, 282
oracle estimator, 31
Order Recursive Matching Pursuit (ORMP), 357
Orthogonal Matching Pursuit (OMP), 39, 351
   algorithm, 351
   recovery result, 359
oversampling, see noise

parametric model, 15
pattern classification, 84
PCA, see principal component analysis
periodic mixing, 109
periodic nonuniform sampling, 106
phase transition, 38
PLA, see Projected Landweber Algorithm
PNS, see periodic nonuniform sampling
polynomial reproducing kernels, 168
   CRB, 182
   CRB derivation, 205
prefix code, 274
principal component analysis, 66, 72
probability distribution
   Gamma, 283
   Laplace, 282
   Student's t, 284
Projected Landweber Algorithm, 377
   algorithm, 377
   recovery result, 379–380
projection operator, 377
pursuit, see greedy pursuit

quantization, 21

radar sensing, 136
random demodulator, 128
random projections, 526
random subset, 246
rank aware order recursive matching, 388
refinement, 272, 287, 290, 291, 293, 294, 296–299

Index

Regularized Orthogonal Matching Pursuit, 356–357
relative sparsity, 495
replica method, 421, 428
restricted isometry property (RIP), 19–24, 252, 348, 380
ROMP, *see* Regularized Orthogonal Matching Pursuit
rotation invariance, 220
Rudelson's inequality, 227, 238, 257

sample covariance matrix, 240
sampling, 149
  equation, 149
  noise, 176
  noise-free, 159
sampling from matrices and frames, 242
Schatten norm, 227
second moment matrix, 223
selectors, 246
sensing energy, 276, 278, 281, 286
sensing matrix, 16
  random sensing matrix, 26
separation
  accuracy, 492
  algorithms, 485, 488
  analysis side, 489, 503
  image, 485, 503
  signal, 484, 499
  synthesis side, 488, 503
sequences of innovation, 122
sequential experiments, 276
sequential sampling/sensing, 270–272, 275, 278, 282, 285, 302
ShearLab, 504, 505
shearlets, 486, 489, 504, 506, 507, 510, 511
shift-invariant sampling, 98
signal recovery, 27
  ℓ_1-norm minimization, 27
  greedy algorithms, 39
signal separation, 499
sinc kernel, 159
  CRB, 182
  CRB derivation, 202
  simulations, 185
single measurement vector (SMV), 43
singular value decomposition, *see* SVD
singular values, 214
sinusoids, 485, 486, 499–502
sketching, 41
Slepian's inequality, 229
soft thresholding, 401–403, 407
soft-thresholding, 534
SoS kernel, 162
  CRB, 182
  CRB derivation, 204

  simulations, 185
source identification, 76
source separation, 76
SP, *see* Subspace Pursuit
spark, 16, 43
sparse recovery, 269, 270, 275, 296, 486
  analysis side, 489, 503
  synthesis side, 488, 503
sparse representation-based classification (SRC), 523
sparse shift-invariant, 99
sparsity, 8
  approximate sparsity, 11
  clustered, 492
  relative, 495
  sparse approximation error, 11
  structured sparsity, 12, 13
spectral norm, 215
  computing on a net, 216
spectrum slices, 105
spectrum-blind reconstruction, 106
spherical random vector, 225, 226
spikes, 485, 486, 499–502
stability, 20
Stagewise Orthogonal Matching Pursuit, 355–356
state evolution, 414–417
statistical physics, 421, 428
StOMP, *see* Stagewise Orthogonal Matching Pursuit
structured and collaborative sparse models, 66
structured models, 373
structured sparse models, 66, 70
structured sparsity, 274–275, 374, 375, 381
sub-exponential
  norm, 222
  random variables, 221
sub-gaussian
  norm, 219, 225
  random matrices with independent columns, 245
  random matrices with independent rows, 232
  random variables, 217, 219
  random vectors, 225
  restricted isometries, 26, 254
sub-matrices, 243, 260
Subspace Pursuit, 367
  algorithm, 369
  recovery result, 369
sum of sincs, 119
sum of sincs kernel, *see* SoS kernel
support recovery, *see* model selection
Support Vector Machines, 440
SVD, 184
  Cadzow, *see* Cadzow denoising
  TLS, *see* TLS
symmetrization, 237, 258

texture, 504, 506, 507
thresholding, 11
time-delay, 93
TLS, 184
total least-squares denoising, *see* TLS
tree-sparse, 375, 378

uncertainty principle, 485, 487, 498, 499, 502
   classical, 498, 501
   Donoho–Stark, 502
union of subspaces, 12, 94, 374–385
   sparse union of subspaces, 12, 13
universality, 26, 141
UoS, *see* union of subspaces

vector space
   normed vector space, 4

wavelets, 12, 486, 488, 502, 504, 507, 509, 510
   Meyer, 510, 511
   radial, 507–511
Welch bound, 24

Xampling, 13, 88, 94
   architecture, 96
   framework, 94
   X-ADC, 96
   X-DSP, 96

Printed in the United States
By Bookmasters